T0180029

The GENI Book

Rick McGeer • Mark Berman • Chip Elliott
Robert Ricci
Editors

The GENI Book

Editors
Rick McGeer
Chief Scientist, US Ignite
Washington, DC, USA

Chip Elliott
GENI Project Office
Raytheon BBN Technologies
Cambridge, MA, USA

Mark Berman
GENI Project Office
Raytheon BBN Technologies
Cambridge, MA, USA

Robert Ricci
School of Computing
University of Utah
Salt Lake City, UT, USA

ISBN 978-3-319-81596-1 ISBN 978-3-319-33769-2 (eBook)
DOI 10.1007/978-3-319-33769-2

© Springer International Publishing Switzerland 2016
Softcover reprint of the hardcover 1st edition 2016
This work is subject to copyright. All rights are reserved by the Publisher, whether the whole or part of the material is concerned, specifically the rights of translation, reprinting, reuse of illustrations, recitation, broadcasting, reproduction on microfilms or in any other physical way, and transmission or information storage and retrieval, electronic adaptation, computer software, or by similar or dissimilar methodology now known or hereafter developed.
The use of general descriptive names, registered names, trademarks, service marks, etc. in this publication does not imply, even in the absence of a specific statement, that such names are exempt from the relevant protective laws and regulations and therefore free for general use.
The publisher, the authors and the editors are safe to assume that the advice and information in this book are believed to be true and accurate at the date of publication. Neither the publisher nor the authors or the editors give a warranty, express or implied, with respect to the material contained herein or for any errors or omissions that may have been made.

Printed on acid-free paper

This Springer imprint is published by Springer Nature
The registered company is Springer International Publishing AG Switzerland

This book is dedicated to the families of the coeditors: Karen and Sean, Wu and Kashi, Emily, Samantha, and Libby, and Donielle and Michaela, whose unflagging support, perennial grace, and unending patience for workaholic husbands and fathers made both GENI and this book possible. We say thanks often—but we can't say it often enough. So, from each of us to each of you: thanks again. This book is for you.

Introduction

Background: Why GENI?

GENI represents the third wave, following the Grid and the Cloud, of the integration of the network into the computational infrastructure. The first wave, the Grid, focused on the application of distributed computing resources, typically supercomputer sites, towards the solution of a single problem. Essentially, it was an extension of batch processing to multiple sites, to more efficiently use large computing resources. It emerged in the late 1990s and was rapidly extended from scientific to business processing. The Cloud is of course quite familiar, and it refers to two dominant themes. The first is the per-hour rental of virtual machines or other computing resources; the second is the transfer of traditional desktop and enterprise applications to a server accessed over the network, with the Google office suite being perhaps the most prominent example. Of course, new applications are enabled by the Cloud that were unimaginable for the disconnected desktop. Media sharing is a prominent example of this class.

GENI differs from the Cloud and the Grid in that it is a platform for *distributed* applications. A distributed application differs from a Cloud application in that the network is central to the distributed application; it literally cannot exist without the network. While a Cloud application—such as, for example, Google Docs—logically runs on a single computer which happens to be accessed over the network, a GENI application or service can only run in a number of computing environments, geographically dispersed. The most prominent simple examples of this class of application are Content Distribution Networks, Distributed Storage Systems, multicast overlays, and wide-area collaborative exploration and creation systems, and collaborative gaming. The distinctive feature of these systems is that they require geographic distribution for one or a combination of a number of reasons. Perhaps the simplest of these reasons is resilience against local failure. In addition, some applications are inherently distributed, often because of the realities of geographically distant end users and data. Inherently distributed applications often require geographically distributed computing infrastructure to support high bandwidth or low latency to end users and data sources.

The central point about GENI is that the network becomes, not just a way for the user to access an application, but the central component of the application itself. This doesn't require just a different sort of computational platform; to be really effective, the application must have a different kind of network. GENI is the network that undergirds distributed applications, and it is a characteristic of the next generation of computational infrastructure.

Today's network is regarded as a network of simple pipes which carry bits between users and remote applications. The network for distributed applications is far richer and more complex; it consists of a large network of computing elements, and programs move seamlessly between these elements to provide service where required.

Though this sounds exotic, in fact it is simply a different assemblage and deployment of current Commercial-Off-The-Shelf (COTS) hardware and software. To a first approximation, what the developer sees is nothing more exotic than a collection of Linux VMs and containers, interconnected by a more-or-less standard network. However, she is able to allocate VMs and containers in specific places, not simply "somewhere in the Cloud," and she is able to configure the topology and priorities of the network between them. Simply put: she is able to design her own, application-specific, continent- and eventually world-wide network, deploy her application across it, and do so in a matter of moments.

This is an entirely new idea of computational infrastructure, though it is made from standard components. Up until now, the network and the computational service delivered over it were regarded as entirely separate components. The application writer had little control over the network topology, and could only influence packet delivery through the choice of transport protocol and some edge tweaking. Conversely, network engineers regarded the computational devices at the edge as foreign soil. The apotheosis of this attitude was found in the design of Content-Centric Networking. At an application level, CCN was easily achieved as an application-level protocol overlay on Content Distribution Networks. However, the CCN community spent an enormous amount of effort putting content information into the packet header, so that the network equipment could process it. It is a reasonable question on whether the performance penalty for doing content-based routing at the application level was sufficient to warrant the effort to do the application at lower levels of the protocol stack. However, the answer to this question is highly dependent on how tightly interwoven the network and application layers could be. If an application designer can control *where* the application points-of-presence are, and how application packets are routed from the user's host to the nearest application POP, the need to drive the application into the network stack is lessened.

As that example illustrates, there are two brutal realities of the computational infrastructure: network equipment can't be programmed, and computers can't forward packets quickly, and attempts to do either are deeply unnatural. This was ultimately why the ActiveNetworks program of the 1990s failed. This has led the networking community to ever-more complex control protocols to permit intelligent packet handling. But the only reason for intelligent packet handling is the relatively long distances packets must traverse between source and destination; a distributed cloud radically shortens that distance, and thus the demand for network equipment to perform functions better performed by a computer. In sum, the GENI infrastructure with distributed applications leads not only to more effective applications but also to a simpler network.

This overall design of a network, with ubiquitous standard computational components, is seen in many other places. Fifth-generation wireless networks ("5G") is an excellent example. The goal of 5G is gigabit bandwidth and millisecond latency to the wireless device. Of course there is no magic; the physics of wireless devices are well known and the coding schemes are close to the information limit. The only way to achieve the orders of magnitude in performance improvement anticipated in 5G is to radically change the network architecture, and this is exactly what the proposals in gestation at the various nations do. Specifically, all 5G wireless architecture proposals combine very small cells ("picocells") with a computational point-of-presence at the base station. This is the GENI architecture, again; in this case, the distributed applications are serving wireless devices.

The Network Function Virtualization movement in the telco industry is a similar example to the deployment of the GENI architecture. NFV was inspired at least in part by the deployment of carrier Content Distribution Networks, such as CoBlitz. The overarching architectural idea is to replace dedicated hardware with software running in virtual machines. This necessarily means deploying virtual machines over a distributed network infrastructure.

All of these similar architectural initiatives drive from a secular trend; the dramatic and continuing decline in the costs of computation against communication. The chart in Fig. 1, taken from Chap. 20, "The Ignite Distributed Collaborative Visualization System" shows the ratio of the price of a gigaflop of computation vs. a megabit/second of bandwidth. As can be seen from the figure, the ratio has declined from about 10 in 1998 to about 0.1 today, a decline of roughly two orders of magnitude. The most direct explanation for this trend is given in Chap. 20: point infrastructures such as computation follow a technology curve, whereas linear infrastructures follow an adoption curve, and the latter must always trail the former.

Paradoxically, as computation becomes more prolific and widespread, communication becomes much more of a dominant consideration in system design. This is because communication becomes the bottleneck in system performance. This is a secular trend throughout the computing industry, from chip design through, in our case, redesign of the Internet. In the case of chips, this has seen the rise over the past decade of multicore architectures and GPU-based vector computation, as increased parallelism becomes the performance driver rather than increasing clock rates. In single-server and data center systems, it has led to the redesign of the server around high-bandwidth memory systems and the data center around highly parallel massive data set searches and manipulations [1, 2], with an emphasis on Terasort rather than Linpack as a benchmark. This involved a radical change to both the memory hierarchy architecture and the design of very high-bandwidth,

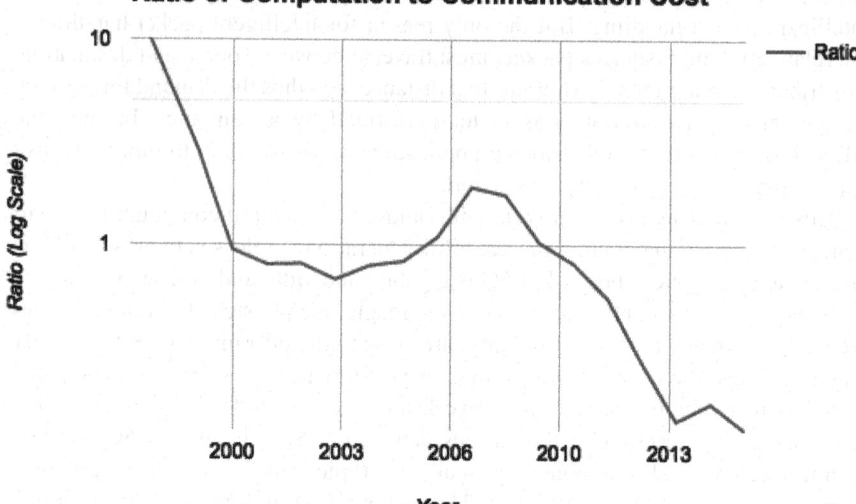

Fig. 1 Ratio of computation to communication cost

low-latency data center networks [3]. In the case of the wide-area network, it is the architecture described in this book: ubiquitous computational points-of-presence with a programmable network between them.

This redesign of the Internet architecture largely leaves the data plane untouched, and in fact radically simplifies the control plane. In fact, the obstacles to its adoption are largely cultural, social, and political rather than technical. We need to rethink our ideas about computation, communication, and data and information storage. Right now, any user of the Internet can tax the communication resources of almost any enterprise or institution; however, access to those institution's computing resources is tightly guarded. There are reasons other than cost, of course, but the dominant reason for this is because our computing systems grew up in an era of time-sharing, where computing was expensive and guarded. Access control was built into the systems from their inception; conversely, communication was unprotected and clumsy access controls retrofitted after the fact. From a cost perspective, this dominant theme of protecting computation but leaving communication open is exactly backwards.

As mentioned, there are other considerations, primarily data security, integrity of the computing environment, and fears of malicious use. But the Cloud has largely overcome those objections: an enormous number of enterprises entrust their data to third-party Cloud storage and do their computing on virtual machines running on the same hardware as an unknown and untrusted third party. A large number of enterprises, universities, and governments outsource their basic IT functions to third-party providers such as Google Apps for Enterprise. It would be an odd CIO indeed who worries about what a student might do with a VM, but will happily offload ERP functions to a cloud provider.

In sum, communication costs a lot more than computing, and we know a lot more about securing computing than we do about securing communication. It's time for GENI's Distributed Cloud.

How Did GENI Come To Be?

As usual, it started with the hackers. Come, be it admitted: academic computer scientists don't do new apps. We do exploit the properties of new technologies to come up with new infrastructures (see, for example, RAID [4] and NOW [5]). But by and large, computer scientists take services and applications hacked up in a hobbyist or commercial setting and build robust, scalable versions of the application or service.

So in the late 1990s people started to exploit the Internet, and a new breed of service known as "peer-to-peer" was born. It was initially popularized by the Napster file-sharing service, but its implications as a communications medium rapidly became apparent. Only a couple of years after Napster was founded, the first wide-area scalable indexing and storage system was devised [6]. A host of implementations followed, along with a large number of distributed applications and services: wide-area robust storage systems, content distribution networks, overlay multicast trees, etc.

This led to an immediate problem: how does one deploy such a system, at scale? In 2001, there was no platform available to deploy these new classes of systems. Rather, what was happening was that researchers were calling up their friends at other institutions, getting accounts on machines at their institutions—with heterogeneous configurations, different software installations, and so on—and then running an experiment. A system that took a few weeks to write might take months to deploy and test.

At an underground meeting at NSDI 2002, a group of researchers led by Larry Peterson of Princeton and David Culler of UC Berkeley devised a new infrastructure to serve as a community testbed. Each institution would agree to devote 2–3 × 86 servers to a community testbed, which would be centrally managed. To permit each researcher to create his own environment, nascent virtualization technology—Linux VServers—was employed to offer very lightweight virtual machines. David Tennenhouse, then head of Intel research, and Patrick Scaglia, who led the Internet and Computing Platforms division of HP Labs, agreed to form a consortium to fund the platform and grow it to several hundred sites worldwide. And the world's first Distributed Cloud, PlanetLab, was born.

PlanetLab grew rapidly, eventually reaching its current size of 1350 nodes at over 700 sites worldwide. More impressive was its immediate impact on the systems community; the vast majority of SOSP 2003 papers cited PlanetLab experiments just a year after the testbed was first built.

In early 2001, Jay Lepreau of the University of Utah and his staff and students devoted a cluster to network experimentation. The problem the Utah group was addressing was both similar and not to the problem addressed a year later by PlanetLab: the need to do short-run controlled experiments on new network protocols and services. Their Emulab platform became the world's first Cloud. It differed then and differs now from standard Clouds. Users are able to request hardware as a service, not simply virtual machines, and are able to finely control the emulated network between their nodes. As a result, it immediately became the premier experimental platform for controlled experiments on distributed systems and network protocols, and remains so today. It is described in detail in Chap. 2.

In September 2003, Dipankar (Ray) Raychaudhuri of Rutgers and his staff began the ORBIT program, a large-scale open-access wireless networking testbed for use by the research community working on next-generation protocols, middleware and applications. The ORBIT project continues to this day and has extensions for software-defined radio elements. Like Emulab, it is a shared testbed. Users log in to the ORBIT portal, and then construct an experiment, typically over ORBIT's 400-node (20 × 20) indoor radio grid facility. The testbed also includes an outdoor "field trial system" intended to support real-world evaluation for protocols validated on the emulator, and for application development involving mobile end users.

In 2005, UC Berkeley and the University of Southern California Information Sciences Institute collaborated to build a shared state-of-the-art scientific computing facility for security experimentation, the cyber DEfense Technology Experimental Research Laboratory (DeterLab). Based originally on the Emulab software stack, DeterLab has introduced a number of innovations to enhance scalability, reproducibility, and control of user experiments. Of course, since DeterLab is security focussed, some of its principal innovations are to ensure protection and isolation of experiments and protection of the world from running experiments. DeterLab offers security researchers the ability to "observe and interact with real malicious software, operating in realistic network environments at scales found in the real world." In other words, this is a facility where monsters are observed and experimented on; and so a primary concern, executed with enormous care and great success over more than a decade, is keeping the monsters safely penned while researchers discover how to neutralize them.

Shared experimental facilities such as ORBIT, Emulab, and DETER start from an economic and democratizing rationale—these facilities permit researchers from any institution to conduct experiments on best-in-world facilities, and it is far more efficient and effective for a funding organization to build a large shared facility rather than many small facilities. Not only does this permit researchers to run much larger-scale tests than would otherwise be possible, there are significant economies of scale. There have also been two major scientific benefits. The first is reproducibility. The availability of shared testbeds enables experimenters to report reproducible results which encourage subsequent validation: the test and the experimental facility are accessible by everyone. Moreover, for each of these facilities, simply running the facility and providing new scientific capabilities has been in and of itself a fecund source of research problems.

By 2006, the successes of these platforms were clear to the systems community and the National Science Foundation. Virtually every major experimental and research system built used one or more of these testbeds. In fact, use of at least two was the common case, because the platforms had complementary strengths. Emulab was an ideal system for short-run controlled experiments on new network systems and protocols in a laboratory setting. DETER, though similar to Emulab, had added crucial features to permit safe testing of security protocols, particularly under malware attack. PlanetLab was designed for long-running services and observations of services in the wide area.

However, Emulab, PlanetLab, and DETER had become victims of their own successes. By 2006 all three testbeds were under significant strain due to enormous demand. It wasn't uncommon for researchers to wait days or weeks to get free machines on Emulab or DETER, particularly as major conference deadlines approached. Because PlanetLab offered lightweight virtualization technology, its oversubscription did not appear as waiting times. But enough slices were active on the PlanetLab testbed at any time that load averages on PlanetLab machines could be over 20.

The systems Computer Science community then began to design a successor to these testbeds. The new system had to meet four major goals:

- Incorporate the controllability and flexibility of Emulab and DETER for short-run controlled experiments.
- Incorporate the geographic distribution of PlanetLab for long-running services and applications, particularly end-user-facing applications such as CDNs and multicast overlays.
- Incorporate the wireless aspects of ORBIT.
- Offer fine-grained control of the network and a principled and architectural approach to software control of the L2 and L3 networks.

Over the period 2006–2007 a group of 50 leading academic computer scientists in six working groups designed this system, producing a working prototype design for the National Science Foundation. In 2008 the NSF issued a call for a GENI Project Office (GPO) to manage the development of a prototype of the GENI system, which was won by BBN Technologies. In 2009, BBN led a community effort to develop this prototype, issuing contracts to universities and research organizations in the systems community to develop GENI.

Simultaneously with this was a happy Black Swan event—a revolutionary new technology, Software-Defined Networking. This concept, and its concrete realization, OpenFlow, grew from the Ethane project at Stanford University. Its most basic concept was that a software controller would load the routing tables of a network of L2 switches, permitting fine-grained software control of packet forwarding and QoS. This offered the key last piece that had been missing from the precursors of GENI: integration of the network into the computational infrastructure. OpenFlow immediately became a key component of the emerging GENI.

GENI's Community Development Approach

The entire community recognized GENI as a high-risk endeavor from the outset. At the time the GPO was initially stood up, it was by no means clear that GENI was technically feasible (or even well defined). Accordingly, the GPO chose a spiral-development approach to development, incrementally building, assessing, and redesigning GENI on a continuing basis, with a nominal spiral duration of one year, punctuated by three GENI Engineering Conferences (GECs) annually. Open to the interested public, the GECs provide impetus for community debate, information exchange, and development deadlines.

The GENI community embraced spiral development as a strategy to continuously confront the most pressing questions—technical and programmatic risks—of the day, with successive spirals addressing a sequence of vital questions. The interactions of dozens of development teams and an ongoing design and development effort driven by thrice-annual community meetings set the stage for rapid, if slightly raucous, progress. This community approach also gave rise to one of GENI's central execution strategies: whenever possible, pursue multiple implementations simultaneously.

Time period	Burning question	Key tactics
Spirals 1 and 2 (2008–2010)	"Is GENI technically feasible?"	Control frameworks and slicing
Spirals 2 and 3 (2009–2011)	"Can GENI be built at adequate scale with reasonable cost and effort?"	"GENI-enabling" equipment, federation, and meso-scale prototype
Spirals 3 and 4 (2010–2012)	"Will GENI be useful for research?"	Research-driven design, community outreach, and "GENI-enabling" tools
Spirals 5 and beyond (2012–)	"Will GENI transform the community?"	GENI racks and international federation

The very first spirals aimed to prove the technical feasibility of core GENI concepts. One such concept was a control framework that could manage multiple, heterogeneous suites of infrastructure. The second was an end-to-end "slice" construct that spanned such heterogeneous suites, interconnecting their diverse virtualization technologies. The GPO organized community projects into competing "clusters" (shown in Fig. 2). Projects then integrated within clusters to achieve four prototype GENI systems by the end of spiral 1. By the middle of Spiral 2, three of the major GENI systems (PlanetLab, ProtoGENI/Emulab, and OpenFlow) were capable of interoperation.

As the first technical hurdles were being overcome, the GENI community also confronted the central programmatic puzzle in GENI—how to afford construction and operation of a set of infrastructure that can support "at scale" research experimentation. The GENI meso-scale prototype presented an opportunity to test

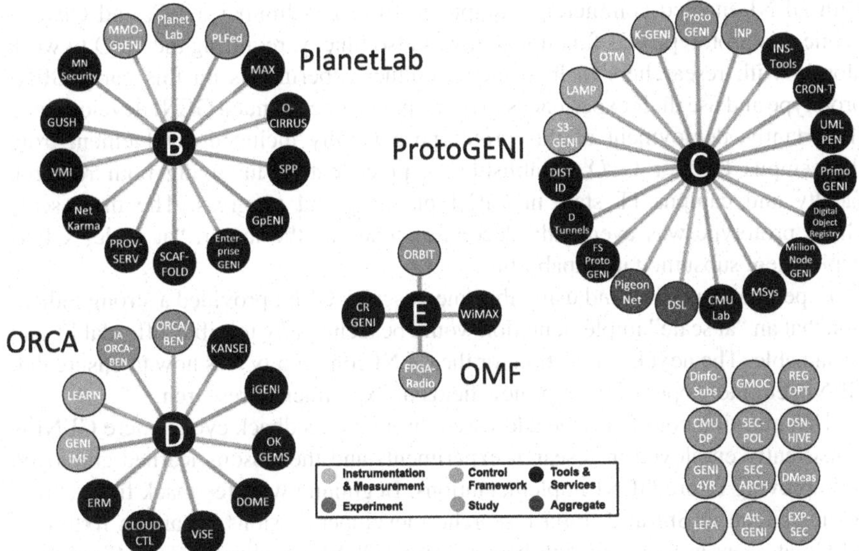

Fig. 2 Early GENI project clusters

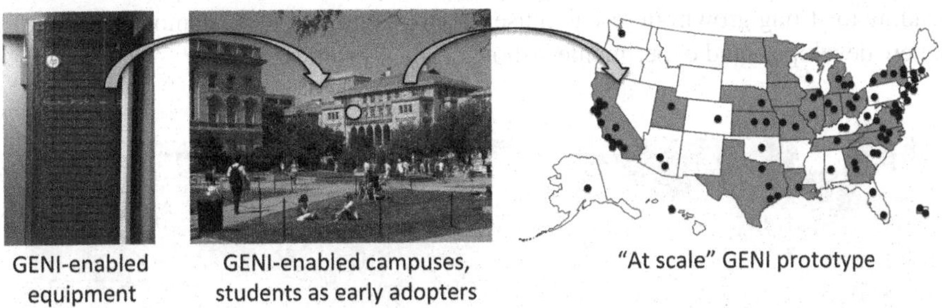

Fig. 3 The "GENI-enabled" campus strategy

the strategy of "GENI-enabling" campuses and research networks, as a way to overcome this challenge.

The strategy began by GENI-enabling existing testbeds, campuses, regional and backbone networks, cloud computing services, and commercial equipment. GENI could then incorporate these networks and services by federation, rather than constructing and operating a separate set of infrastructure for experimental research. Figure 3 depicts the plan: first GENI-enable commercial equipment, then use this equipment to create "GENI-enabled" campuses and the national backbones that can run GENI experiments on the same infrastructure as production networks. Finally, federate GENI-enabled campuses and networks to create "at scale" GENI.

The key hardware artifact of spirals 2 and 3 was a "meso-scale" version of this basic approach, spanning 14 campuses and 2 national backbones (Internet2 and NLR). The meso-scale prototype integrated PlanetLab, ProtoGENI, and OpenFlow, with GENI-enabled commercial equipment from HP, Juniper, NEC, and Quanta. While this prototype was functional, it was also finicky, requiring the GPO to work closely with researchers to help them conduct experiments on this early GENI prototype and use their experiences to refine plans for continued GENI development. Importantly, deployment of this prototype generally included involvement from the campus CIO or CTO, establishing a precedent of involving both research faculty and campus IT staff in GENI planning and progress. The meso-scale GENI prototype was eventually decommissioned as the larger, "at scale" GENI deployment subsumed its capabilities.

Experience building and using the "meso-scale" GENI provided a strong indication that an "at scale" implementation would be technically feasible, affordable, and sustainable. The next key question for the GENI community was how to ensure that GENI genuinely opens up major new fields of experimental research.

This question could only be addressed through a feedback cycle where GENI is consistently employed in research experiments and the lessons learned employed in improving future GENI implementations. Beginning with feedback from experiments begun in Spiral 2, joint researcher-developer sessions became a fixture of GECs, and research experiments began to drive GENI's evolving design. Significant outreach and support effort from the GPO, NSF, and the GENI development community encouraged GENI's rapid adoption by researchers in spirals 3 and 4, leading to strong growth in research use. Figure 4 shows a GEC demo night event, where developers and experimenters show off their progress.

Fig. 4 Demo night at GEC16, Salt Lake City, 2013

The "GENI-enabling" approach was also applied to popular research tools. Researchers are accustomed to working with specific tools, and their introduction to GENI was greatly eased by making GENI resources available through these familiar pathways. This approach began with the adoption and interoperation of precursor testbeds like PlanetLab, ProtoGENI/Emulab, and ORBIT and was extended to tools like the OMF control infrastructure and the Open Resource Control Architecture (ORCA) control framework.

As researchers began to experiment with the "meso-scale" GENI, they quickly became aware of its potential, as well as its limitations. Experimenters found great value in the key capabilities of the prototype, including slicing and deep programmability. They wanted a larger-scale deployment, with more programmable computation and network components throughout the GENI network. They needed additional automation to support dramatic growth in the number of simultaneous experiments.

The move to a larger GENI prototype began with basic GENI building blocks. Beginning in 2012, and continuing to the current GENI, campuses are GENI-enabled by deploying GENI racks, optional wireless base stations, and software-defined networks on campus. These resources are connected to a research backbone network and federated into the emerging nationwide GENI, where they are available to the entire GENI research community. The principles involved are consistent with the approach used in the "meso-scale" GENI, but the process is significantly simplified at each campus by the availability of GENI racks. A GENI rack includes computation (cluster of processors), storage, and an OpenFlow switch in a single deployable package, along with its associated control software. The rack provides the campus an entrée into the GENI federation. Additional campus resources, such as a science DMZ, may be federated as well, in keeping with the unique research needs of each campus.

As GENI grew within the USA, similar projects arose around the world. While each of these future Internet and distributed cloud (FIDC) testbeds has unique implementation and management aspects, there is strong motivation both to share ideas and software and to federate infrastructure, and GENI has been a leader in this area for several years. The globalization of FIDC concepts is an unfinished but highly promising chapter of the GENI story.

Organization of the Book

The book takes us through the GENI Project in its lifecycle in five parts. Part I describes the precursors of GENI that led to its development, with detailed histories of ORBIT, DETER, Emulab, and a discussion of the GENI idea from then NSF Assistant Director Peter Freeman. Part II describes the architecture of GENI as a set of control frameworks that interact and present the developer with a picture of a distributed cloud with a programmable network in between cloud nodes and describes how the specific precursors of GENI—PlanetLab, Emulab, and

ORBIT—were adapted into new complementary control frameworks within GENI. These chapters also discuss how new technologies, specifically emerging Cloud technologies and the new capabilities of software-defined networking, were adopted and integrated into the GENI framework and the specific control frameworks which made it up. Part III discusses the deployment of GENI as a nationwide infrastructure. Once the control frameworks were in place, GENI had to be made concrete and real. The control frameworks were integrated and deployed at 50 sites across the United States, in small, extensible clusters: "GENI Racks." These were interconnected by a programmable nationwide layer-2 network, the "Mesoscale Deployment." Once this was in place, GENI was ready to host applications and services. Part IV describes the applications of GENI to our society and profession, and the tools developed to use this infrastructure. GENI is not alone; it is one of several similar efforts worldwide. Part V discusses parallel and complementary efforts in Canada, Europe, and Asia, and the prospects for an international federation.

The story of GENI is far from done. We are now roughly where the NSFNet was in the late 1980s, with a few tens of sites connected by a nationwide backbone. As GENI transitions to the next phase of its life, which we believe will be an era of explosive growth, we recall the words of Vint Cerf as the ARPANET transitioned to become part of the Internet:

> It was the first, and being first, was best,
> but now we lay it down to ever rest.
> Now pause with me a moment, shed some tears.
> For auld lang syne, for love, for years and years
> of faithful service, duty done, I weep.
> Lay down thy packet, now, O friend, and sleep.
> -Vinton Cerf

Washington, DC — Rick McGeer
Cambridge, MA — Mark Berman
Cambridge, MA — Chip Elliott
Salt Lake City, UT — Robert Ricci

References

1. Ranganathan, P.: From microprocessors to nanostores: Rethinking data-centric systems. IEEE Comput. **44**(1) (2011)
2. Ousterhout, J. et al.: The case for RAMClouds: scalable high-performance storage entirely in DRAM. SIGOPS Operat. Syst. Rev., **43**, 4, 92–105 (2009)
3. Al-Fares, M., et al.: A scalable, commodity data center network architecture. Proc SIGCOMM. (2008)

4. Patterson, D. et al.: A case for redundant arrays of inexpensive disks (RAID). ACM Sigmod Record. (1988)
5. Anderson, T., et al.: A case for NOW (networks of workstations). IEEE Micro. (1995)
6. Ratnasamy, S., et al.: A scalable content-addressable network. Proc. SIGCOMM. (2001)

Acknowledgements

In his ballad "Calypso" about the research vessel of the undersea explorer Jacques Cousteau, John Denver penned these lyrics: "Aye Calypso, I sing to your spirit/The men who have served you so long and so well." The spirit of GENI is our colleagues: the hundreds of women and men who worked tirelessly over a decade to take this from an idea to the genesis of the next Internet. Before and above all else, we thank each and every one of them: it has been our great privilege and honor to know and work with each of them. This book is their ballad. Our research sponsors at the National Science Foundation have shown courage and vision and unflagging support over the years. We thank Suzi Iacono, Farnham Jahanian, Peter Freeman, Guru Parulkar, Gracie Narcho, Erwin Gianchandani, Bryan Lyles, Jim Kurose, and Jack Brassil for their leadership at NSF. We have been encouraged and supported by the White House Office of Science and Technology Policy, and thank particularly Tom Kalil. GENI would not exist without Larry Peterson, whose PlanetLab provided much of the technical inspiration for GENI and whose leadership was vital in starting the GENI project at NSF. We thank Larry and his early colleagues, particularly Tom Anderson, Scott Shenker, and Jon Turner.

We need to thank a friend no longer with us. The other large inspiration for GENI was Emulab, and that was the brainchild and passion of our close friend, and Rob Ricci's mentor, Jay Lepreau. Jay, sadly, passed away before this project was properly begun, but his spirit lives on in it, and we hope in us. Thanks, Jay, and we all still miss you terribly.

As we write these thanks, it occurs to us that many names which should be here are not—the list of those we should thank by name is so long that it would take many pages to enumerate them all, and we are certain to forget some. GENI is an enormous effort, and many people in many different roles played key parts in making it happen, and all were cheerful, graceful, and went far beyond the call of duty. This is inadequate, but the best we can do here: thanks. You know who you are, and so do we, and we will be forever grateful. Drinks are on us, the next time we meet.

GENI is supported under cooperative agreements from the US National Science Foundation. Any opinions, findings, conclusions, or recommendations expressed in this book are the authors' and do not necessarily reflect the views of the National Science Foundation.

Contents

Part I Precursors

The GENI Vision: Origins, Early History, Possible Futures 3
Peter A. Freeman

Precursors: Emulab 19
Robert Ricci and the Emulab Team

DETERLab and the DETER Project 35
John Wroclawski, Terry Benzel, Jim Blythe, Ted Faber,
Alefiya Hussain, Jelena Mirkovic, and Stephen Schwab

ORBIT: Wireless Experimentation 63
Dipankar Raychaudhuri, Ivan Seskar, and Max Ott

Part II Architecture and Implementation

GENI Architecture Foundation 101
Marshall Brinn

The Need for Flexible Community Research Infrastructure 117
Robert Ricci

A Retrospective on ORCA: Open Resource Control Architecture 127
Jeff Chase and Ilya Baldin

Programmable, Controllable Networks 149
Nicholas Bastin and Rick McGeer

4G Cellular Systems in GENI 179
Ivan Seskar, Dipankar Raychaudhuri, and Abhimanyu Gosain

Authorization and Access Control: ABAC 203
Ted Faber, Stephen Schwab, and John Wroclawski

The GENI Experiment Engine 235
Andy Bavier and Rick McGeer

Part III The GENI National Buildout

The GENI Mesoscale Network 259
Heidi Picher Dempsey

ExoGENI: A Multi-Domain Infrastructure-as-a-Service Testbed 279
Ilya Baldin, Jeff Chase, Yufeng Xin, Anirban Mandal, Paul Ruth, Claris Castillo, Victor Orlikowski, Chris Heermann, and Jonathan Mills

The InstaGENI Project 317
Rick McGeer and Robert Ricci

Part IV GENI Experiments and Applications

The Experimenter's View of GENI 349
Niky Riga, Sarah Edwards, and Vicraj Thomas

The GENI Desktop 381
James Griffioen, Zongming Fei, Hussamuddin Nasir, Charles Carpenter, Jeremy Reed, Xiongqi Wu, and Sergio Rivera P.

A Walk Through the GENI Experiment Cycle 407
Thierry Rakotoarivelo, Guillaume Jourjon, Olivier Mehani, Max Ott, and Michael Zink

GENI in the Classroom 433
Vicraj Thomas, Niky Riga, Sarah Edwards, Fraida Fund, and Thanasis Korakis

The Ignite Distributed Collaborative Scientific Visualization System 451
Matt Hemmings, Robert Krahn, David Lary, Rick McGeer, Glenn Ricart, and Marko Röder

US Ignite and Smarter Communities 479
Glenn Ricart and Rick McGeer

Part V GENI and the World

Europe's Mission in Next-Generation Networking with Special Emphasis on the German-Lab Project 513
Paul Müeller and Stefan Fischer

SAVI Testbed for Applications on Software-Defined Infrastructure 545
Alberto Leon-Garcia and Hadi Bannazadeh

Research and Development on Network Virtualization Technologies in Japan: VNode and FLARE Projects 563
Akihiro Nakao and Kazuhisa Yamada

Creating a Worldwide Network for the Global Environment for Network Innovations (GENI) and Related Experimental Environments 590
Joe Mambretti, Jim Chen, Fei Yeh, Jingguo Ge, Junling You, Tong Li, Cees de Laat, Paola Grosso, Te-Lung Liu, Mon-Yen Luo, Aki Nakao, Paul Müller, Ronald van der Pol, Martin Reed, Michael Stanton, and Chu-Sing Yang

Appendix: Additional Readings 633

Afterword: A Fire in the Dark 651

Contributors

Ilya Baldin Renaissance Computing Institute (RENCI)/UNC Chapel Hill, Chapel Hill, NC, USA

Hadi Bannazadeh Department of Electrical and Computer Engineering, University of Toronto, Toronto, ON, Canada

Nicholas Bastin Barnstormer Softworks Ltd. and University of Houston, Houston, TX, USA

Andy Bavier Princeton University and PlanetWorks, LLC, Princeton, NJ, USA

Terry Benzel USC Information Sciences Institute, Marina Del Rey, CA, USA

Jim Blythe USC Information Sciences Institute, Marina Del Rey, CA, USA

Marshall Brinn GENI Project Office, Raytheon BBN Technologies, Cambridge, MA, USA

Charles Carpenter Laboratory for Advanced Networking, University of Kentucky, Lexington, KY, USA

Claris Castillo Renaissance Computing Institute (RENCI)/UNC Chapel Hill, Chapel Hill, NC, USA

Jeff Chase Duke University, Durham, NC, USA

Jim Chen International Center for Advanced Internet Research, Northwestern University, Chicago, IL, USA

Heidi Picher Dempsey GENI Project Office, Raytheon BBN Technologies, Cambridge, MA, USA

Sarah Edwards GENI Project Office, Raytheon BBN Technologies, Cambridge, MA, USA

Ted Faber USC Information Sciences Institute, Marina Del Rey, Los Angeles, CA, USA

Zongming Fei Laboratory for Advanced Networking, University of Kentucky, Lexington, KY, USA

Stefan Fischer Institute of Telematics, University of Lübeck, Lübeck, Germany

Peter A. Freeman Georgia Institute of Technology, Atlanta, GA, USA

Fraida Fund NYU School of Engineering, New York City, NY, USA

Jingguo Ge China Science and Technology Network, Computer Network Information Center, Chinese Academy of Sciences, Beijing, China

Abhimanyu Gosain GENI Project Office, Raytheon BBN Technologies, Cambridge, MA, USA

James Griffioen Laboratory for Advanced Networking, University of Kentucky, Lexington, KY, USA

Paola Grosso University of Amsterdam, Amsterdam, The Netherlands

Chris Heermann Renaissance Computing Institute (RENCI)/UNC Chapel Hill, Chapel Hill, NC, USA

Matt Hemmings Computer Sciences Department, University of Victoria, Victoria, BC, Canada

Alefiya Hussain USC Information Sciences Institute, Marina Del Rey, CA, USA

Guillaume Jourjon NICTA, Australian Technology Park, Eveleigh, NSW, Australia

Thanasis Korakis NYU School of Engineering, New York City, NY, USA

Robert Krahn Y Combinator Research, San Francisco, CA, USA

Cees de Laat University of Amsterdam, Amsterdam, The Netherlands

David Lary Department of Physics, University of Texas at Dallas, Dallas, TX, USA

Alberto Leon-Garcia Department of Electrical and Computer Engineering, University of Toronto, Toronto, ON, Canada

Tong Li China Science and Technology Network, Computer Network Information Center, Chinese Academy of Sciences, Beijing, China

Te-Lung Liu National Center for High-Performance Computing, National Applied Laboratories, Hsinchu City, Taiwan

Mon-Yen Luo National Kaohsiung University of Applied Sciences, Kaohsiung, Taiwan

Joe Mambretti International Center for Advanced Internet Research, Northwestern University, Chicago, IL, USA

Anirban Mandal Renaissance Computing Institute (RENCI)/UNC Chapel Hill, Chapel Hill, NC, USA

Rick McGeer Chief Scientist, US Ignite, Washington, DC, USA

Olivier Mehani NICTA, Australian Technology Park, Eveleigh, NSW, Australia

Jonathan Mills NASA Center for Climate Simulation, Goddard Space Flight Center, Greenbelt, MD, USA

Jelena Mirkovic USC Information Sciences Institute, Marina Del Rey, CA, USA

Paul Müller Integrated Communication Systems Lab., Department of Computer Science, University of Kaiserslautern, Kaiserslautern, Germany

Akihiro Nakao The University of Tokyo, Tokyo, Japan

Hussamuddin Nasir Laboratory for Advanced Networking, University of Kentucky, Lexington, KY, USA

Victor Orlikowski Duke University, Durham, NC, USA

Max Ott NICTA, Sydney, Australia

Sergio Rivera P. Laboratory for Advanced Networking, University of Kentucky, Lexington, KY, USA

Ronald van der Pol SURFnet, Utrecht, The Netherlands

Thierry Rakotoarivelo NICTA, Australian Technology Park, Eveleigh, NSW, Australia

Dipankar Raychaudhuri WINLAB, Department of ECE, Rutgers University, 674 Rt. 1 South, North Brunswick, NJ 08902, USA

Jeremy Reed Laboratory for Advanced Networking, University of Kentucky, Lexington, KY, USA

Martin Reed University of Essex, Colchester, UK

Glenn Ricart US Ignite, Washington, DC, USA

Robert Ricci Flux Research Group, University of Utah, Salt Lake City, UT, USA

Niky Riga GENI Project Office, Raytheon BBN Technologies, Cambridge, MA, USA

Marko Röder Y Combinator Research, San Francisco, CA, USA

Paul Ruth Renaissance Computing Institute (RENCI)/UNC Chapel Hill, Chapel Hill, NC, USA

Stephen Schwab USC Information Sciences Institute, Marina Del Rey, Arlington, VA, USA

Ivan Seskar WINLAB, Department of ECE, Rutgers University, 674 Rt. 1 South, North Brunswick, NJ, USA

Michael Stanton Brazilian Research and Education Network—RNP, Rio de Janeiro, RJ, Brazil

Vicraj Thomas GENI Project Office, Raytheon BBN Technologies, Cambridge, MA, USA

John Wroclawski USC Information Sciences Institute, Marina Del Rey, CA, USA

Xiongqi Wu Laboratory for Advanced Networking, University of Kentucky, Lexington, KY, USA

Yufeng Xin Renaissance Computing Institute (RENCI)/UNC Chapel Hill, Chapel Hill, NC, USA

Kazuhisa Yamada NTT Network Innovation Lab, Musashino-shi, Japan

Chu-Sing Yang National Cheng-Kung University, Tainan City, Taiwan

Fei Yeh International Center for Advanced Internet Research, Northwestern University, Chicago, IL, USA

Junling You China Science and Technology Network, Computer Network Information Center, Chinese Academy of Sciences, Beijing, China

Michael Zink Department of Electrical and Computer Engineering, University of Massachusetts in Amherst, Amherst, MA, USA

Part I
Precursors

GENI, unlike Athena, did not spring fully grown and armed from the mind of the National Science Foundation. It was the fruit of a number of years of planning by the US and international distributed systems community, in careful consultation with the NSF. Moreover, and importantly, it was strongly influenced by a number of precursors, notably PlanetLab, Emulab, DeterLab, and ORBIT. These were (and still are) existing, highly successful shared distributed systems and networking testbeds, and it was their success that inspired GENI. It is fair to say that the original concept of GENI was "a distributed Emulab/DeterLab with: PlanetLab's ability to host long-running services and experiments in virtualized environments; control of the inter-site networking; and ORBIT's ability to incorporate wireless nodes". In the event, over the three years of planning and six years of construction GENI hewed to that vision remarkably closely.

Three of the four chapters in this section explore three of GENI's major precursors. The fourth chapter, the first in the volume, comes from former NSF CISE Assistant Director Peter Freeman, and explores the decision-making and planning process for GENI within NSF and the Distributed Systems Community.

Dr. Freeman's chapter focusses on the pre-history of GENI. By 2004, the inability of the distributed systems and networking communities to do research on the operational Internet was apparent, and the stunning success of PlanetLab was clear. Two of the founders of PlanetLab—Larry Peterson, then of Princeton, and Tom Anderson of the University of Washington—began a collaboration with leading networking researchers Scott Shenker of UC-Berkeley and Jonathan Turner of Washington University. This group worked with then new NSF Program Director Guru Parulkar, now at Stanford University, forming the "Gang of Four Plus One" to plan the continent-wide testbed of the next Internet that became GENI. Dr. Freeman walks us through the deliberations of this group, through the NSF-sponsored community planning group of 2007–2008, and the ultimate award of the GENI contract.

In chapter "Precursors: Emulab", Rob Ricci describes Emulab—in many ways, the world's first Cloud, and then and now the premier platform for controlled networking and distributed systems experimentation. Begun in 2001, Emulab is a hardware-as-a-service cluster at the University of Utah with elements devoted to controlled networking experimentation, primarily delay and traffic-shaping appliances. It eponymous software stack not only controls the Emulab cluster at Utah but a number of daughter clusters across the United States and beyond. The Emulab software stack forms the basis of the ProtoGENI control framework for GENI, and for a number of followon and related projects. The first of these was DeterLab, summarized in the next paragraph and described in chapter "DETERLab and the DETER Project". Others include the NSF CloudLab and Advanced Profile-Based Test Lab, both at Utah, and the NSF PRObE facility at Los Alamos.

A critical need for advanced cybersecurity testing was identified in the early 2000's, and in 2003 the University of California, Berkeley and USC's Information Sciences Institute collaborated to build a testbed at USC devoted to cybersecurity research, at scale and under realistic conditions. The DeterLab cluster at USC/ISI, described in chapter "DETERLab and the DETER Project", began with the Emulab software stack and is superficially similar to Emulab. However, it has been enhanced in multiple dimensions, with specific attention paid to scalability and security features. Both are absolute requirements for security experimentation; many attacks utilize load as a critical component, and cybersecurity testing necessarily involves the use of malware, worms, and viruses. Much as a biohazard facility requires special precautions, so to does a cybersecurity testing facility.

Another area problematic for experimentation is wireless testing, particularly at scale. The need for such testing motivated Dipankar (Ray) Raychaudhuri of Rutgers to construct ORBIT, a large-scale open-access wireless networking testbed consisting of a 'Wireless-Nodes-as-a-Service' 400-node indoor radio grid facility, It is described in chapter "ORBIT: Wireless Experimentation" and is the basis of the Wireless test facility of GENI.

The GENI Vision: Origins, Early History, Possible Futures

Peter A. Freeman

Abstract This paper presents the vision of GENI as first formulated at the National Science Foundation (NSF) in early 2004 and expanded during 2004–2007, identifies what forces shaped the basic idea during its formation, and comments on where it may go in the future. The paper describes motivations, concepts, and history—not technical details—that were in play between 2004 and 2007 as the GENI Project was being formulated and launched, and that continue today. Understanding the original vision and goals, basic ideas, and motivations of the GENI Project; the context in which it emerged; and the forces that shaped the Project will enable you to understand better the technical details and changes that occur in the future. I end with some comments about possible futures for GENI.

1 The Original Idea of GENI

1.1 The Objective

From the start, we thought of GENI as two, complementary lines of work (and support): Research on future architectures for the Internet and a robust infrastructure for experimenting with new and innovative infrastructural and application ideas.

© Peter A. Freeman, 2015. The name 'GENI' wasn't created until over a year after the effort started at NSF—we initially called it "CIRI—Clean-slate Internet Re-Invention Initiative." A second version was "GEENI—Global Experimental Environment for Networking Investigations." The name was shortened to GENI in mid-2005 and today it is mostly an unexpanded acronym standing for Global Environment for Networking Innovations.

P.A. Freeman (✉)
Georgia Institute of Technology, Atlanta, GA 30332, USA
e-mail: freeman@cc.gatech.edu

© Springer International Publishing Switzerland 2016
R. McGeer et al. (eds.), *The GENI Book*, DOI 10.1007/978-3-319-33769-2_1

GENI comprises two components: the GENI Research Program[1] *and the experimental GENI Facility. It is intended to catalyze a broad community effort that will engage other agencies, other countries, and corporate entities*[2].

For the most part, I believe these two objectives have been held to by those in leadership positions at NSF, in the GENI Project Office (GPO), and in the networking community. For a variety of reasons, building a continental-scale network on which novel ideas could be implemented and experimented with has seemed to dominate. At the same time, ideas generated in building the GENI Facility, in the research supported by the Future Internet Architecture (FIA) program started in 2010, and in other research programs may ultimately develop into exactly the kind of innovative network architectures we envisioned in the first place. The focus of the Project so far is natural because it is almost always more compelling to build a "tangible" architecture than to experiment with abstract ideas, and because most people working on a project must focus on one small aspect of the overall effort.

1.2 Expansion of the Objective

A very important expansion of the original objective took place in 2005 as we understood more fully the potential for the GENI Facility and the absence of much, if any, of a body of scientific knowledge to support new network and applications engineering design. Quoting from the same SIGCOMM announcement as above:

To have significant impact, innovative research and design ideas must be implemented, deployed, and tested in realistic environments involving significant numbers of users and hosts. The Initiative includes the deployment of a state-of-the-art, global experimental GENI Facility that will permit exploration and evaluation under realistic conditions.

What we were calling for, in effect, was the development of a scientifically based body of design knowledge.

We elaborated on this idea internally in the context of our application for NSF Major Research Equipment and Facilities Construction (MREFC) funding,[3] but regrettably did not make this as specific and public as we could have. Comments in this paper may help rectify that oversight.

[1]The GENI Research Program was not launched until 2010, in the form of the Future Internet Architecture (FIA) solicitation from NSF. http://www.nets-fia.net/.

[2]*The GENI Initiative*, NSF announcement distributed at SIGCOMM 2005 on August 25, 2009.

[3]This is funding distinct from the research funds controlled by individual NSF directorates, and is used for projects whose initial cost would be a substantial proportion of a directorate's annual budget.

1.3 Origins of the GENI Idea

No one person can claim credit for "inventing" GENI. Initially it was formulated from ideas that were emerging in the networking research community ("the community") through a long series of workshops. Those of us at NSF had the privilege and responsibility of harvesting ideas from the entire community and turning them into a specific project. Community involvement has continued to generate ideas as the GENI Facility has evolved, and is one of the strengths of the Project.

GENI didn't suddenly appear as a full-blown, well thought out project on a specific date, but it seems appropriate to mark the start of the direct GENI effort as being April 2004. That was when Guru Parulkar, a Program Director in CISE, made a presentation to Deborah Crawford, Deputy AD/CISE, and me. His presentation reviewed some of the emerging problems with the Internet (e.g. security, quality of service, capacity, connection between digital and physical worlds) and the difficulty or impossibility of creating the digital world so many people were, and still are, envisioning. Parulkar then went on to review the thinking and work to date of the community and especially the "Gang of Four"[4] on ways to deal with what they and others viewed as a looming "brick wall" in front of continued Internet functional expansion. The presentation of the basic ideas for a significant experimental facility also included significant thinking and contribution of Parulkar. He used the term "clean-slate" in describing what needed to be done.

The vision that Parulkar presented to us incorporated three fundamental ideas:

- Slicing (having multiple networks simultaneously using the same routers, with the ability to switch between the networks dynamically);
- Virtualization (of a network);
- Programmability (of routers).

None of these ideas were new to computer science, of course, and in some ways were in use in networking, but the combination was new. The power of a network based on these ideas and incorporating the latest optical and other technology led us to envision an entirely new form of "Internet." That caused us to sometimes reply to those asking what our objective was that we were "Reinventing the Internet."

While ultimately that was exactly what we were aiming at in the long run, it led to some misperceptions that we had to correct. One misperception was that we wanted to replace wholesale the current Internet with GENI (not true).

A second, somewhat partially accurate perception, was that we were critiquing the current technical details of the Internet. It is true that based on what the community was saying, we tended toward the view that current structures could

[4]The "Gang of Four" consisted of Tom Anderson (Univ. of Washington), Larry Peterson (Princeton), Scott Shenker (Berkeley), and Jon Turner (Washington University). Parulkar had been a professor at Washington University and successful entrepreneur. At the time he was a program director at NSF and we often referred internally to the "Gang of Four Plus One" since he contributed his own ideas as well.

not be adequately modified to deal with the emerging imperatives (see Motivations below); but never did we believe that they had not been extremely successful in adapting to demands on the Internet (and individual networks) that were never envisioned when those structures were created. Any disagreement was solely over the ability of current structures to continue to evolve to meet new demands adequately.

A third misperception was that NSF would build the new network (not true). NSF is not an operational agency in the sense of building and maintaining infrastructure or equipment for public use or to provide a public service. When NSF does build something like a network, it is solely for the support of scientific research and education (NSF's mission)—as was the case with NSFNET. When it becomes clear that the object has a broader, possibly commercial, public utility, it is then "spun out" to control/ownership by others. This is exactly what happened in 1995 when NSFNET, which had merged with ARPANET for research use, was turned over to public control and use, becoming what today we call "the Internet."

A second set of points arising from the community helped shape the GENI Project including:

- The belief that continued improvements to individual components and protocols of the Internet were not going to suffice to meet the societal demands being placed on it[5];
- The belief that the ideas brought together by Parulkar offered a path to an important new technical basis for the Internet;
- The fact that early networks and versions of today's Internet had been developed in environments in which experimentation occurred almost daily on operational networks (sometimes disastrously!), and that the ability was lost when the Internet and operational networks in general were commercialized;
- The fact that there was little theory and few models that would permit experimentation with new network architectures *in silico*.

A third set of points arising from my long experience as a computer scientist and Crawford's experience as a computer engineer and NSF official also helped shape the Project:

- Belief in the importance of experimentation in developing complex, computer based systems of all types;
- Broad understanding of the importance of networking to all aspects of society;
- Deep understanding of how projects and major programs are approved, funded, and then carried out in NSF and in the U.S. Government in general;
- Knowledge of the trends and current status of funding for research in computer science, especially the loss at that time of DARPA funding for basic CS research;
- Knowing the importance of making sure that a major project addressed both internal NSF objectives and external societal objectives.

[5]A decade later we still don't have robust security on the Internet, for example.

As with any technically and organizationally complex project involving many players, there were undoubtedly other factors not listed here. These, however, are the ones that stand out to me 10 years later.

1.4 Motivations

Major projects don't get approved just because they may be technically or scientifically interesting. There are always more interesting things to do than there are available funds. So, what were some of our major motivations in driving this project forward?

Among the driving forces in society that were becoming very pronounced (especially to those of us in the U.S. Government who had to answer frequently and personally to concerns from Congress) were the needs to:

- Build in security and robustness;
- Bridge the gap between the physical and virtual worlds (mobile, wireless, and sensor networks);
- Control and manage other critical infrastructures;
- Provide ease of operation and usability;
- The need to enable new classes of societal-level services and applications.

Even without these societal driving forces, all of which involved challenging technical problems that had been identified by the community, it was clear to all of us in the technology R&D community that several issues needed addressing, including:

- Inherent architectural limitations of current networking and Internet architectures and structures to enable future development;
- Ossification of the current Internet architecture;
- The push of "unrelated" technology developments (e.g. optical switching, mobile devices);
- The pull of enticing and important new applications (e.g. telemedicine).

Today, to some extent at least, all of these driving forces are known by everyone in the community. They are also discussed elsewhere this book.

These ideas arose from societal needs that were increasing (security, more innovations to drive economic development) due to exponentially increasing usage of networks by non-experts and technical assessments and projections from the networking research community. These were the people (along with others) that were largely responsible for the development of the Internet prior to 1995 (and to some extent since).

The ideas were brought together, expanded, and modified by people at NSF. These individuals were all knowledgeable in one or more aspects of the networking domain, were excited about the project and its possibilities, and devoted their full energy over a sustained period to bring about the promise of a successful execution of the basic objectives. The leadership of CISE, and to some extent NSF, was personally involved and leading the project for the first several years and, thus, were able to direct substantial resources to it.

The community was also very excited about the project, although by no means unanimously,[6] and a substantial number of senior researchers became engaged in the project in various ways, often contributing substantially of their time and resources to help push it forward.

1.5 Overview of GENI's Early Days (Years)

A detailed recounting of the history of the intellectual origins of GENI is interesting and important, but beyond the scope of this paper. Figure 1 provides an overview of the history.

Beginning in 1995 when NSFNET was turned over to broad public control and funding, there were a number of workshops and meetings, which viewed retrospectively, constituted a review of the community's work in creating the Internet. Around 1999, the community's attention started to turn toward the future and what needed to be done to insure that their creation, although no longer under their control, could continue to grow and meet the demands being placed on it.

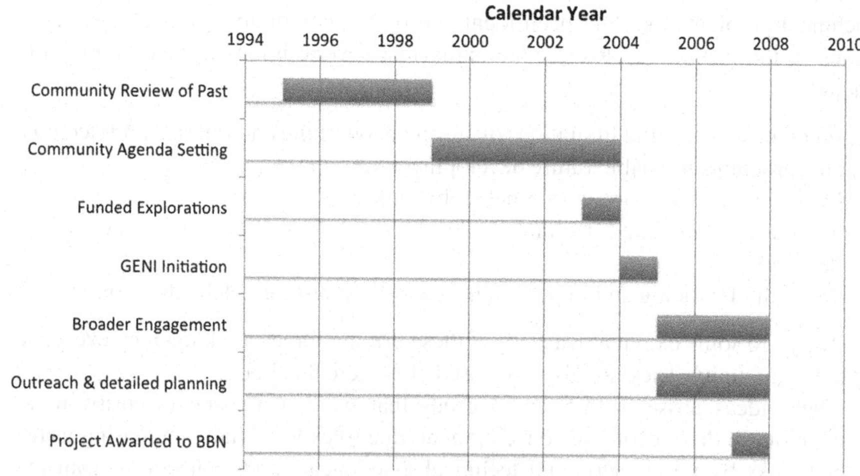

Fig. 1 Approximate Stages in the Origin of GENI up to 2008

[6]Notably, at a meeting in mid-2005, a group of very senior networking pioneers reviewed our plans as of that date and, while some expressed strong support ("Just do it!" said one), some expressed deep reservations about attempting the Project at all ("The Internet has evolved well so far without this!").

This agenda setting continued up to early 2004. Some new networking research programs at NSF that were amplified around 2002[7] were for the first time in almost a decade supporting some significant experimental developments—some of which led directly to the ideas collected and shaped by Parulkar.

Community involvement in setting the research agenda has continued throughout and was extremely helpful in formulating the GENI Project. In the early development of the GENI ideas, inputs from the community were focused by NSF. The initiation of GENI as described above consumed most of 2004. By 2005 a number of efforts to engage more of the community began to pay off and by the end of the year, a substantial number of people were actively engaged in GENI planning and initial projects; additional people, including most of the senior people identified with networking, had been explicitly asked to comment on the Project and to offer their inputs.

After publicly announcing the Project at SIGCOMM-2005, we began a concerted effort to discuss the Project with industry, broader reaches of academia, Congress, and other parts of the U.S. Government; this continued in 2006. This was complemented by additional planning focused on technical details and obtaining major funding.

In early 2006 we issued a solicitation for proposals to fund a consortium to engage the broad computing research community in generating ideas for major initiatives in areas other than networking; we saw the GENI Project as the first such initiative. The solicitation included the following:

> *One of the first responsibilities of the CCC will be guiding the design of the Global Environment for Networking Innovations (GENI). ... The GENI facility is expected to increase the quality and quantity of experimental research outcomes supported by CISE, and to accelerate the transition of these outcomes into products and services to enhance economic competitiveness and secure the Nation's future.*[8]

The Computing Research Association (CRA) received the award in mid-2006 to create CCC and by the end of 2006, they had an interim leadership team of computing community volunteers in place; early in 2007 they formed a committee of networking community volunteers to produce a GENI Science Plan. This was our way of explicitly turning intellectual control over to representatives of the computing research community.

Later in 2006, we issued a solicitation to create a GENI Project Office to guide and perform the engineering work necessary to create a GENI Facility.

As with the related CCC solicitation, the objectives were clearly stated, including:

[7]Guided by Larry Landweber, Senior Advisor to the AD. Larry was also instrumental in helping motivate and involve a broad swath of the networking community here and abroad in the GENI Project.

[8]*Computing Community Consortium (CCC): Defining the Large-Scale Infrastructure Needs of the Computing Research Community*, NSF Solicitation 06-551, http://1.usa.gov/1IOdAQq.

… It is anticipated that the GPO will then have full responsibility for overseeing the construction of the facility, ensuring that GENI is delivered on time and within budget. Upon successful GENI construction and commissioning, the GPO may subsequently operate the facility in service to the computing research community.[9]

The Project Office solicitation attracted considerable interest among major organizations and the final stage of reviews of the proposals was scheduled for early 2007. An award was made to BBN Technologies in mid-2007 and by the end of 2007 the GPO was in operation and continues to the present (late 2015).

To recap, our intent was to let the computing research community broadly, and especially the networking community, drive the research priorities and policies for the GENI Research Program and to have an experienced, professional organization oversee and perform critical parts of the construction of the GENI Facility. By the middle of 2007 both of these mechanisms were in place and beginning operation.

While the initial committee created by the CCC to guide GENI made an honest effort to fulfill the charge from NSF, those efforts were not continued in a way that would ultimately fulfill the vision and plans we had laid. The GPO, on the other hand, has performed their role extremely well and by engaging a broad swath of the community filled in the vacuum left by the CCC in this area. (The performance of the CCC in other areas, notably robotics, has been more in tune with their original charge.)

Other factors, both before and after 2007, have impacted the evolution of the GENI Project as it stands today. That is the subject of the next section of this paper.

2 What Has Shaped the GENI Project?

Several forces have shaped the Project so far, and in one form or another will continue to do so in the future:

- Engagement of the technical community;
- Leadership from the community and from funders;
- Exogenous technical developments;
- Availability of funding;
- Organizational dynamics;
- Societal and practical imperatives.

This is not a sociological paper,[10] but a few comments may be helpful to you in understanding GENI, its objectives, and some of the forces at play in its evolution.

---

[9] *Global Environment for Networking Innovations (GENI): Establishing the GENI Project Office (GPO) (GENI/GPO).* NSF Solicitation 06-601, http://1.usa.gov/1TJQKdi.

[10] There have been studies of GENI, however, for example: Kirsch, Laurie J. and Slaughter, Sandra A., "Managing the unmanageable: How IS research can contribute to the scholarship of cyber projects." Journal of the Association for Information Systems, Vol. 14 (2013), No. 4, p. 198–214.

The engagement of a broad swath of the networking community, in one way or another, is clearly the most important single factor in the origination, evolution, and ultimate success (or failure) of GENI. Without the ideas that originated in the work and analyses of the community, there would be no GENI. NSF has always been a community-driven agency for its basic directions and programs, modulated by the guidance of NSF program and management personnel almost all of whom have expertise in one or more areas of research and education. That is what gives NSF its long-term success in supporting fundamental scientific investigations and development of the advanced tools and facilities to carry out those investigations. GENI is an excellent example of this.

The lesson is to listen to the best minds in the field, while developing new and unique ideas and approaches.

The ultimate responsibility for a line of investigation or the construction of an advanced facility belongs to the community, not NSF, because it is NSF's policy to rely on a scientific community to carry out investigations and build needed facilities. If a community does not come together and provide the leadership necessary to carry a project forward, then the chances a project will fail increase dramatically.

If this second factor had not been present, GENI would never have happened. The fact that a small cohort of the most respected and active networking researchers (the "Gang of Four" referred to above) were engaged and willing to come together to help blend their ideas with those of others was fortuitous. This has not always been the case in other instances.

The community must make sure that informed and engaged members from the field take time to serve the Nation and their community by serving at NSF.

Equally important was the willingness of a large number of networking researchers (many not engaged in the Project then or now) to meet with us to critique the Project—sometimes quite severely—and add their ideas to the mix we were trying to harvest and shape into a coherent project. These ranged from the late Paul Baran to some not even yet born in 1959 when he started his seminal work on survivable communication networks!

With all due humility, the other part of the equation that was critical was having leadership at NSF willing and able to see the importance of enabling more networking research, to see the potential value of the technical ideas that were emerging, to be willing to lead and take risks, to listen carefully to and use any and all critiques, and to broaden the objectives where needed. Again, this was most fortuitous and one doesn't always have a good alignment of program management and senior leadership at NSF (or any agency) as some subsequent personnel changes have shown during the evolution of GENI.

The need for community involvement extends from the occasional reviewer or informal consultant to the leadership of NSF overall. It is the responsibility of a community's senior leadership to insure that appropriate people serve at NSF.

Good leadership makes a difference. If a scientific community wants its ideas listened to then it must take the responsibility of providing appropriate leaders to NSF and other agencies.[11]

Every reader can understand the impact of rapid technological changes external to a project on its course over time. For example, the progress of optical technology was not necessarily foreseen by the technical projects that influenced the start of GENI. Similarly, the exponential increase in the capabilities and usage of mobile devices also was not immediately factored in to our plans.

The availability of funding for a project is always important, but in my opinion not a complete "showstopper." If the ideas had been brought together as they were, then they likely could have eventually found funding from some source to carry them forward. Again, though, luck played a big role in permitting us to devote significant funds for GENI and the experimental programs that preceded it because it came at a time that our budget was growing and we were already devoting a larger fraction of those funds to networking and security research. While the availability of funding has varied in the 10 years since the GENI Project started, CISE nonetheless has spent a very large amount from its operating budget on the two components of the Project (Research and Facility Construction).

Undertaking an "audacious" project of the scale of GENI takes a large amount of funding and time.

A close corollary is that it is always better to have multiple sources of funding, something the leadership has not been able to do in any significant ways until very recently. This is changing in the form of donated resources and parallel efforts coordinated to what is happening in GENI—specifically, the participation of many campuses using their own funding.

Organizational dynamics is yet another important, but not critical, factor in the origination and evolution of GENI. The primary example of this was our decision to seek funding from the special equipment fund in the NSF Budget (the MREFC account that pays for research ships, telescopes, and other very expensive research instruments). Ultimately, our attempt failed, primarily due to organizational dynamics internal to NSF. Had we not made this attempt—which had been encouraged by top management—progress that has taken ten years might have come more quickly.

Sometimes forces beyond the technical merit or the scope of authority of a project's leadership dictate its progress in ways that cannot be prevented.

We have already discussed some of the societal imperatives that influenced us (e.g. security, the need to enhance innovation broadly). Perhaps the most pervasive, but fortunately not a showstopper, in the past decade has been the financial crisis starting in 2007–2008 that has slashed budgets, made industry more cautious, and generally reduced people's desire to take risks on the future.

[11]We attempted to engage DARPA, DoE Office of Science, and other agencies in a broad GENI effort. Unfortunately, we were not successful in that effort.

Ultimately we are all part of a larger society that in the aggregate may have more important, immediate concerns than supporting a future oriented project.

3 What Does the Future Hold?

As Yogi Berra (or, was it Confucius?) said "Predictions are very hard—especially about the future!"

We hope that the original vision of developing a useful "instrument" (the GENI Facility) for experimentation will succeed and enable the development of a body of verifiable knowledge on which future networks and applications can be built. However, if you read the list of factors above that have shaped the GENI Project and that will continue to do so, then you know that the future of something as large and complex as this project is far too hard to predict with any accuracy.

We can invoke some standard platitudes about technological (and scientific) developments, including:

- The time to achieve impact of a development is often much longer than initially predicted, but at the same time may eventually be much greater than ever imagined;
- The impact of a development often turns out to be nothing like what the originator imagined;
- Size of a project $\neq$ eventual impact;
- External factors often alter a project;
- Vision and leadership must be maintained for the life of a project.

The first three of these "truths" and others like them have been shown true in development after development. They need no explanation since examples are abundant, but they are worth keeping in mind for everything from your specific work to an entire project. The last two points (and others) were discussed above, but stand out to me when considering GENI and its evolution.

So, is there anything else to say about the future of the GENI Project? Yes!

I believe there are several potential pitfalls or ways in which the project could be driven into failure by those of you in the community in whose hands its future lies. Specifically,

- Insufficient funding;
- Lack of community engagement;
- Absence of focused and effective leadership;
- Insufficient, positive engagement by industry;
- Absence of broad and useful experimentation.

3.1 Funding

Insufficient funding is, in my opinion, a short-term issue, yet potentially deadly. The current efforts of the GPO are preparing a useful infrastructure installed in a number of leading universities and research sites and connected by donated fiber from Internet2. It is critical that this be completed and operation supported in the mid-term. If the infrastructure is useful then funding mechanisms will be created outside the framework of total government support and the critical period will have passed.

This is exactly analogous to what happened when NSF turned over NSFNET to public use and support in 1995. While financial support and technical viability will be tenuous in the early years, as it was with the Internet, if GENI is useful then its future in some form will be assured.

3.2 Community Involvement

Without the continued engagement of the networking research community, the original vision will surely not be fulfilled. So far, the GENI Project, with the leadership of GPO, has been very successful in engaging large swaths of the networking research community from the most experienced and senior to just beginning graduate students. As the GPO finishes its current job and fades from the scene, some active mechanism outside of NSF is needed to insure community engagement.

The CCC was set up by NSF as a mechanism that would do this, initially for GENI, for a number of research areas in computing. As noted above, this has not happened for GENI in a sustained manner. If the CCC does not step in to provide focused leadership in the coming months and years, the networking community will have to do it in an ad hoc manner. This is not necessarily a bad outcome. In fact, it is the modality used to operate other large, NSF-supported research facilities where a whole new organization is set up to operate the facility.

There is another, more fundamental need for community involvement in the coming years in the, as yet unfulfilled, quest to develop a scientific basis for the design of future networks and applications. This is not a quest that everyone in the community is interested in pursuing, nor capable of doing. Building an immediately useful system or subsystem is a very engaging activity that can have great impact in the short term. That activity should never be looked down upon, but at the same time it should not be all that we as a community, most especially those of us working in academia and industrial research labs, undertake. If we don't communicate in verifiable ways our theories, experimental results, and developed wisdom to those that come after us then our students and their students and those that devote their efforts to building operational networks and applications are doomed to make the same mistakes we have made. They deserve to make their own, new mistakes, not just repeat ours!

Some in the networking community over the past 50+ years have done fundamental, essential, and deep thinking about the nature of networking and the behavior of real networks and the applications that utilize them. Some of it has been recorded in ways that are accessible to future generations of designers, but I fear that the press of immediately applying that knowledge means that a good bit of it is being lost. This work needs to be augmented by experiments that utilize the GENI Facility, documented in archival form, and utilized to develop further a scientific basis for network and application design in a virtuous cycle of theoretical development, experimentation, and further theory refinement—the cycle that has been so productive over the years for researchers in other scientific fields, such as theoretical and experimental physicists.

3.3 Leadership

This is ultimately a community responsibility, not just in the performance and focusing of technical work as discussed above, but also in encouraging and enabling community members to serve in leadership positions in government and industry that have the authority to support and advance network research.

It is also a responsibility of community members to help educate leaders in many positions, not just those on the front lines of networking research. Members of Congress, senior officials in many government agencies, CEO's, engineering managers, and the general public all need to understand the value of networking research and, for the long term, the value of developing a basis of solid, scientific knowledge upon which future networks and applications with the desired levels of security, robustness, and so on can be built.

The parallels to other branches of technology and science are abundant. You need only think about two that everyone in the modern world depends on—medical care based on deep and fundamental scientific investigations and the construction of large buildings. We all understand how scientific research has and is continuing to revolutionize the medical care available to people, even in the most remote corners of the world (much of it actually employing a lot of advanced engineering). If you've ever thought about how skyscrapers are built and how large buildings are sometimes situated on what otherwise appears to be less than firm ground and how in the advanced world we rarely hear of buildings collapsing of their own accord, then you know there must be a considerable base of scientifically developed knowledge on which the necessary foundation engineering is built.

If the networking community does not provide the leadership to develop a more solid scientific basis, then our "buildings" will continue to fail, just as buildings constructed without the benefit of modern, scientifically based knowledge sometimes do. The essential ingredient of leadership, of course, is that it be focused on important and obtainable goals that are pursued over whatever time period is required. Part of the job of enlightened leadership is to make sure there is a balance between the near term imperatives and the longer term, essential goals.

3.4 Industry Engagement

Industry engagement to date has been modest, at best. That's understandable, but moving forward it is essential that industry becomes positively engaged. Again, the current situation is very reminiscent of the transition period between the purely research oriented days of the ARPANET and NSFNET and the late 1990s when the commercial value of the Internet became apparent.

Yet, the situation from here forward is different from the early days of the Internet. While some of the developments that may come out of GENI work may spark strong public demand and thus commercial interest, building a scientific base of knowledge will not happen in a time span short enough to make the effort of obvious value to short-term focused industry. I call this "death by a thousand small successes." How to avoid this outcome is an open question.

There is a darker side to industry involvement, of course. There are multi-billion dollar companies built on the current networking technology. The "innovators dilemma" comes into play here because there are very powerful forces that don't want change since it may threaten their current business models. I don't need to name names or point out examples—the news is filled every day with arguments about net neutrality, control of telecommunications pipes, and so on.

This is another area where research community involvement and leadership will be essential. If we don't educate industry leaders about the value of a more rigorous network and application engineering, and at the same time work with industry to obtain shorter-term economic benefits from our work, then we have no one but ourselves to blame if future generations do not have the scientific basis that we know is essential to future success.

3.5 Useful Experimentation

The final potential failure mode should be clear to you now. The vision that sparked GENI was not just of a continental-scale research facility that could be used to develop some new networking technology and applications. The vision included experimentation aimed at developing a scientific base of knowledge for future design efforts so that our entire field will be raised to a new level of capability.

The community, leadership, funding, and commercial interests must stay aligned and focused on the near-term and the long-term goals. It is an open question if that will happen.

4 Conclusion

The GENI Project came from the ideas, analyses, visions, and efforts of many. It was shaped into what it is today and continues to evolve with the ideas and efforts of many more. There have already been some technical successes and the beginnings of meaningful experimentation. At the moment, the path ahead, while unclear, seems devoid of serious roadblocks; but they could appear at any moment.

The real win and lasting contribution will be to deliver on the initial objective of new architectures and a scientific basis of knowledge for continued and enhanced progress in networking of value to our entire society.

It's up to you!

Acknowledgements I have already acknowledged that without the networking community there would be no GENI. My colleagues at NSF, principally Guru Parulkar, Deborah Crawford, and Larry Landweber, deserve a very large amount of credit for bringing GENI into existence. In addition, their comments on this paper and an earlier, longer version have been very helpful. The editors of this volume have also contributed useful suggestions on the paper.

Peter A. Freeman I left NSF in early 2007 as leadership for the Project transitioned away from direct NSF control and to the community. I am now Emeritus Dean and Professor at the Georgia Institute of Technology.

Precursors: Emulab

Robert Ricci and the Emulab Team

Abstract One of the precursors of the GENI project is Emulab, a testbed effort that has been ongoing at the University of Utah since 1999. Emulab is both the name of a *testbed control system*, and the name of a *particular facility* built using that system. The Emulab facility is housed at the University of Utah, but is available to researchers worldwide—thousands of users have run hundreds of thousands of experiments over the lifetime of the testbed. The Emulab software is open-source, and has been used to bring up dozens of experimental facilities at institutions around the world. Some of these, like the Utah facility, are open to the public for the purposes of research and educations; others are run by individual institutions for their own use, which may include product R&D, classified work, etc.

One of the precursors of the GENI project is Emulab, a testbed effort that has been ongoing at the University of Utah since 1999. Emulab is both the name of a *testbed control system* [29], and the name of a *particular facility* [10] built using that system. The Emulab facility is housed at the University of Utah, but is available to researchers worldwide [12]—thousands of users have run hundreds of thousands of experiments over the lifetime of the testbed. The Emulab software is open-source, and has been used to bring up dozens of experimental facilities at institutions around the world [11]. Some of these, like the Utah facility, are open to the public for the

---

The Emulab project was founded by Jay Lepreau, who led it from 1999 until his death due to cancer in 2008. As of 2016, the Emulab team includes: Keith Downie, Jonathon Duerig, Dmitry Duplyakin, Eric Eide, David Johnson, Mike Hibler, Dan Reading, Leigh Stoller, Kirk Webb, and Gary Wong. Over the last 16 years, dozens of people have worked on Emulab, including Christopher Alfeld, David G Andersen, David Anderson, Kevin Atkinson, Grant Ayers, Chad Barb, Srikanth Chikkulapelly, Steve Clawson, Austin Clements, Cody Cutler, Russ Fish, Daniel Montrallo Flickinger, Daniel Gebhardt, Shashi Guruprasad, Fabien Hermenier, Ryan Jackson, Abhijeet Joglekar, Xing Lin, Nikhil Mishrikoti, Ian Murdock, Yathindra Naik, Mac Newbold, Tarun Prabhu, Raghuveer Pullakandam, Prashanth Radhakrishnan, Srikanth Raju, Pramod Sanaga, Timothy Stack, Matt Strum, Weibin Sun, Kevin Tew, Brian White, and Kristin Wright.

R. Ricci (✉)
Flux Research Group, School of Computing, University of Utah, Salt Lake City, UT, USA
e-mail: ricci@cs.utah.edu

© Springer International Publishing Switzerland 2016

R. McGeer et al. (eds.), *The GENI Book*, DOI 10.1007/978-3-319-33769-2_2

Fig. 1 Early Emulab builders. *Left to right*: Mac Newbold, Logan Axson, Christopher Alfeld, Kristin Wright, Jay Lepreau, Mike Hibler

purposes of research and educations; others are run by individual institutions for their own use, which may include product R&D, classified work, etc.

The Emulab system was originally developed as the "Utah Network Testbed" by the Flux Research Group under the direction of Jay Lepreau. The facility's initial purpose was to solve a problem that the group itself had: research in the area of computer systems (including networks) requires a great deal of hands-on experimentation, and performing those experiments necessarily means managing the equipment on which they will be run. It was clear, however, that the needs of the Flux group with respect to infrastructure were by no means unique, and the group decided to open the testbed to the wider research community in early 2000. Similarly, after several years of operating one facility, it became clear that others would like to run their own, similar infrastructure, and the software was generalized to run at other sites, with the first two being at the University of Kentucky [16] and Georgia Tech. Since then, the Emulab software has become the basis for a number of testbeds with different focuses, including: NSF's CloudLab [23] (cloud computing), GENI [2, 3] (federation), PhantomNet [28] (mobile networking), PRObE [17] (large scale systems) and Apt [24] (adaptability and repeatability); DARPA's National Cyber Range (security); and DHS's DETERLAB [8] (security). The Emulab facility and codebase are key parts of the nationwide GENI infrastructure and several international federations in Europe, Brazil, Japan, and South Korea.

Fig. 2 Wiring in one of the Utah Emulab clusters

As its name suggests, Emulab was originally designed with emulation as its primary purpose: by this, we mean running "real" code on real hosts, interacting with a network that is "real" in the sense that it is constructed of real hardware such as switches and NICs, but which may be artificially manipulated in order to create effects such as latency, limited delay, etc. in a controllable manner. While emulation remains one of Emulab's key use cases, it has grown far beyond this original focus, and supports other environments such as simulation, live wide-area network experimentation, wireless networks, and more.

Thought its lifetime, the development of Emulab has been supported primarily by the National Science Foundation, with additional support from DARPA. Additional vendor contributions have come from Cisco, Intel, Compaq, Microsoft Research, Novel, Nortel, and HP. As a result of this support, users have never paid to use the Emulab facility at Utah or for using the source code to build other facilities. Emulab has also benefited greatly from the support of the University of Utah, in the form of machine room space, power, and cooling, as well as support from the IT organization for its unusual network needs and usage patterns. Emulab has never "stood still," with each successive grant being used to add new, unique features such as wireless experimentation, support for virtualization, and federation.

1 Running Experiments on Emulab

To run an experiment on Emulab, a user describes the network he or she would like to run an experiment on. This specification is written in a dialect of the language that is used for the *ns-2* simulator [26], and is thus typically called an "NS file." An example of such a file can be seen in Fig. 3. An NS file includes specifications of the nodes, links, and events that define the experiment. The user submits this NS file to Emulab, which attempts to find a free set of resources that match what is requested. For the most part, users do not ask for *specific* physical machines, and repeated runs of the same experiment might map onto a different set of physical hosts.

For nodes, the specifications of interest include the type of hardware to run on, the operating system to run, and possibly additional software packages to install. Emulab offers some flexibility in the specification of hardware: many users simply ask for a "pc," which gives Emulab the flexibility to assign any one of the numerous hardware types it might control. Or, users can be more specific, referring to a specific type, if it has features that they require, or if they are trying to run on the same type repeatedly for consistency. Operating systems and software are often specified in terms of *disk images*, which Emulab automatically loads onto the hosts using a custom scalable multicast protocol. Emulab itself provides standard disk images for several Linux distributions, FreeBSD, and Windows, and users may also make their own (usually by customizing one of the facility images.)

In terms of links, the user may specify either point-to-point links or multipoint LANs, creating a *topology*. One of Emulab's key features is that these links can

```
# Import definitions of Emulab-specific commands; Emulab also provides a
# stubbed-out version of this file for use inside of the ns-2 simulator
source tb_compat.tcl

# The Simulator object, from ns-2, encapsulates the representation of the
# topology
set ns [new Simulator]

# Scripts can include variables that can be modified to alter the topology or
# behavior of an experiment
set num_pcs 16

set lan_string ""

# Full OTcl language features, such as loops and string manipulation, are
# available
for {set i 1} {$i <= $num_pcs} {incr i} {

    # Create a new node, requesting a specific operating system
    # and type of hardware
    set node($i) [$ns node]
    tb-set-node-os $node($i) FBSD-STD
    tb-set-hardware $node($i) pc3000
    append lan_string "$pc(${i})_"

}

# Put all nodes into a 1Gbps LAN with no traffic shaping
set lan0 [$ns make-lan "$lan_string" 1000Mb 0ms DropTail]

# Indicate the the topology definition is complete (in ns-2, this would start
# the simulation)
$ns run
```

Fig. 3 An example of an "NS file" used to describe a topology in Emulab

have traffic shaping parameters attached to them: bandwidth, base latency (typically modeling propagation delay), base packet loss (before congestive losses), and queuing discipline. These enable a cluster that resides entirely within a single datacenter to emulate networks that cross large distances. Typically, experimenters set these to values that are representative of real networks, but one of the strengths of Emulab is that they can be set to *any* value (within the limits of testbed hardware), enabling sensitivity analyses, "what if" experiments, etc. Links in the request are implemented in Emulab using VLANs, and if any traffic shaping parameters are specified, a node running the Dummynet [4] emulator is transparently inserted into the middle of the link to realize them.

Most experimenters use Emulab interactively; that is once their nodes are ready, they log in via `ssh` and use the command line to launch their experiments. However, they may also specify a set of *events* in the NS file. These events can run at specified to run at certain times (relative to the successful reservation of resources), or be grouped into *timelines* that can be launched on-demand at any point during the experiment. Events can run programs, modify traffic shaping parameters, take nodes or links up or down, and more. Experiments that are specified in this way can be run as *batch* experiments, in which Emulab waits for sufficient resources to become available, instantiates the experiments, and lets the events run to completion.

Once the user has submitted an NS file, Emulab *swaps in* (a term borrowed from virtual memory) that experiment on physical hardware. This process typically takes a few minutes; time varies depending on the number of nodes in the experiment, whether custom disk images need to be loaded, etc. Emulab *maps* the requested topology onto the resources available at the time: this means finding nodes that match the user's specification, and finding paths across the physical network that meet the requirements of the requested links. In general, users do not expect to get the same nodes for different runs of the same experiment, but can request specific *types* of nodes to get some assurance that successive experiments are run on machines that are effectively identical for most purposes. Experiments typically last hours to days; Emulab encourages users to hold resources for only as long as they are actively using them. When a user is done with a particular allocation of resources, he or she can either *swap out* the experiment or *terminate* it. Swapping out preserves the experiment definition from the NS file (but not node, storage, or network state) so that it can easily be swapped back in later.

2 The Emulab Control Infrastructure

The Emulab control infrastructure (shown in Fig. 4) consists of three kinds of hosts: one `boss` node, one `ops` node, and zero or more `sub-boss` nodes. (In recent installations, `boss` and `ops` are typically run in two VMs on one physical host.) The collection of hardware controlled in a typical Emulab installation includes *nodes*, one or more *switches*, and devices to provide management control over the nodes (power controllers, serial console concentrators, etc.) Emulab controls

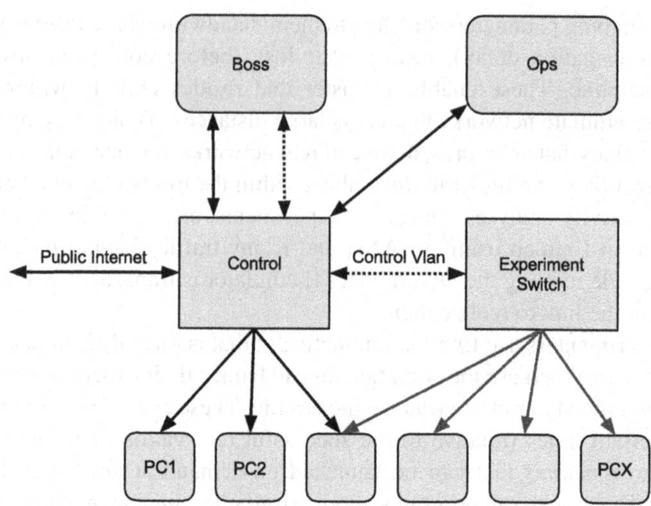

Fig. 4 Emulab's control infrastructure

nodes by controlling their boot process and through disk imaging [7, 14, 20]; it control switches by configuring them through SNMP, NETCONF, and other network configuration protocols.

The `boss` node hosts most of Emulab's critical control infrastructure: the database, webserver, DNS server, various boot and imaging servers, etc. Many infrastructure control interfaces, such as power controllers, switches' SNMP interfaces, etc. are protected by VLANs, so that they can only be controlled from `boss`. Due to its sensitive nature, users are not given shell access to `boss`. The primary interactions that `boss` has with nodes and switches occur during experiments swap in and swap out, as `boss` manages the boot and image loading processes of the nodes, and it is during the swap process that `boss` configures switches. While the resource needs for a `boss` server are not extreme, it can become the bottleneck for instantiating large experiments. For large installations (more than two or three hundred nodes), some work can be offloaded to `sub-bosses`, which can handle boot and imaging services for a subset of the cluster. `sub-bosses` provide only read-only services (such as answering DHCP queries and distributing disk images); this design allows us to avoid complicated state synchronizations between the various `boss` and `sub-boss` nodes.

The main functions of `ops` are to give users a place to get a shell independent of particular experiments and to act as a central fileserver. In some Emulab installations that do not have public IP addresses, `ops` acts as a sort of a bastion host, where users must log in before they can reach the (private) control network interfaces of the nodes in their experiments. As discussed later, we discourage users from over-reliance of the network filesystem hosted on `ops`, but it does serve useful functions with respect to user home directories, etc.

3 Distinguishing Features of Emulab

Emulab is far from the only system for managing collections of servers. In recent years, cloud management systems such as OpenStack [27] and Eucalyptus [18] have become popular ways to manage virtualized resources. More contemporaneous with Emulab's initial conception, various cluster management toolkits such as ROCKS [25], xCat [9], and Grid [13] systems have a long history as well. In the research infrastructure space, management systems such as PlanetLab [6], the ORBIT Management Framework [19], and ORCA [5] are also used to run testbeds.

What makes Emulab unique among these systems is the emphasis it places on three things: scientific fidelity, bare-metal resource provisioning, and the network as a first-class concern. This has led Emulab to make a number of different design decisions, in terms of the makeup of its physical infrastructure, the architecture of its control software, and the way that users interact with its resources.

3.1 Focus on Scientific Fidelity

One of Emulab's design goals from the beginning has been a focus on scientific fidelity. This means that, to the extent possible, experiments run on Emulab should not contain artifacts that are the result of other concurrent use of the facility. This is quite different from the goal of satisfying service-level agreements (SLAs), as clouds aim to do, and leads to a number of properties that distinguish it from other datacenter or infrastructure-as-a-service offerings.

The first consequence of this philosophy is an emphasis on allocation and provisioning of resources at a "bare-metal" level. Most experiments are run on nodes that are completely dedicated to a single experiment at a time. While this is more challenging to provision that a virtualized environment or one with shared hosts, it means that processes belonging to different users do not compete for CPU time, memory, I/O bandwidth, etc. It also gives users direct access to hardware and full control over the operating system, features that are critical to researchers in operating systems and networks.

Second, all Emulab nodes are connected to two networks: one "control network" and one "experiment network" [29]. The control network is shared by all experiments at once, and it is connected to the public Internet. It is over this network that users log into nodes, that remote filesystems are mounted, and users are encouraged to run traffic that coordinates their experiment, collects logs, etc. Emulab makes to attempt to provide isolated performance on the control network. The experiment network is isolated: it is configured to match the topology requested by the experimenter, and experiments do not see each others' traffic on this network. Experiments may run whatever protocols they wish, at whatever speed they would like, on this network. While cost constraints prevent this network from having full bisection bandwidth, one of the key features of Emulab's resource mapper is that it uses information about the topology submitted by the user to find an embedding that minimizes use of bottleneck links in the physical switching topology [21]. Thus, experimenters can have confidence that they are competing only minimally with other users on this network.

Third, the focus on fidelity affects Emulab's strategy towards storage. Many clusters make heavy use of shared, network-mounted filesystems. While this is convenient for users, it is less desirable from a fidelity standpoint: any experiments whose behavior is affected by the performance of filesystem I/O become vulnerable to interference from other users. Thus, while Emulab does offer some shared filesystems, they are relatively small and low-performance, and users are strongly encouraged to rely on the local disk of each node for their experiments. Each physical node has one or more local disks, and like the nodes themselves, they are not shared with any other simultaneous experiments. This does make persisting large datasets across experiments more difficult, but Emulab offers features for easy disk image creation, which helps alleviate this problem for many users.

Finally, it is critically important that Emulab provide users with the resources having the performance that the user requested: if an experiment is un-knowingly run under the wrong network conditions, it can produce an incorrect result. These kinds of events have occurred in Emulab in the past, due to mis-wired networks, bugs in the provisioning software, etc. Thus, Emulab incorporates a piece of software called "linktest" [1] which is run on new experiments. Linktest has the job of ensuring that the topology that Emulab has instantiated is the same as that requested by the experimenter, and that any link shaping parameters (delay, bandwidth, etc.) are within set tolerances of what was requested. Linktest takes a completely separate code path from the standard Emulab provisioning software to guard against bugs in the Emulab software—it even goes so far as to use a completely separate parser for NS files. The result is a high assurance that the user has received the network that they requested.

3.2 *Focus on Multi-Tenant, Bare-Metal Allocation*

Emulab has always focused primarily on the provisioning of bare-metal, rather than virtualized, resources. It has, over time, added the ability to provision virtual machines and container-based operating systems [15], but its core strength continues to lie in bare metal. This is for three reasons: first, as discussed above, this provides an environment with higher fidelity and lower probability of artifacts due to shared use. Second, Emulab's original use case (and one of it major uses still) is for development of low-level code such as operating system kernels, hypervisors, software switches, and other code that requires direct hardware access. Third, Emulab development began before the current generation of virtualization technologies was mature.

This means that the provisioning systems in Emulab are focused on controlling the booting of physical machines. Emulab retains control by booting all nodes off of the network (and not allowing users to change this setting.) A simple, small bootloader checks to see what it should do next: this might mean continuing to boot from the disk, loading a kernel from the network for disk imaging, etc. Emulab "cleans" machines between users by loading a default disk image every time a node is released from an experiment. It does *not* attempt a secure erase of the entire disk, as this would be extremely time consuming and is overkill for most research use. One interesting aspect of this process is that, for security reasons, Emulab needs to be sure that its own disk loader program has been run and has loaded the correct image, rather than malware impersonating the disk loader, leaving backdoors, etc. behind. To this end, we have developed the Trusted Disk Loading System (TDLS) that uses the Trusted Platform Module (TPM) to attest to the secure booting of the disk loader [7].

One consequence of the focus on bare metal is on the efficiency of allocation. Put simply, virtual allocation can be more efficient, because it shares hosts between multiple simultaneous tenants and can over-subscribe resources to take advantage of users who do not fully utilize their allocations. Virtual allocations can also be finer-grained. However, these properties would violate Emulab's need for artifact-free experimentation. Combined with the fact that Emulab does not charge for use (providing no incentive for users to minimize their resource consumption), this means that Emulab must take measures to ensure that resources are used efficiently. The major strategy that Emulab employs in this regard is idle monitoring: the standard Emulab OS images (and thus, most user images that derive from them) include daemons that monitor for CPU and network activity, remote logins, etc. If an experiment goes idle, the user is notified, and has some set period (typically a few hours) to return to using it before resources are automatically reclaimed. Of course, it is possible for users to trick the idle detection system, for example by running programs that consume CPU cycles but do no useful work, but we have found that most users are good citizens and that this puts sufficient pressure on most of them to behave responsibly.

3.3 The Network as a First-Class Entity

Because Emulab began life with an emphasis on network experimentation, it has always viewed the network as a first-class entity; whereas many cluster management systems attempt to abstract over the network, or view network setup as a side effect of provisioning compute resources, Emulab gives the network topology equal importance. One consequence of this emphasis is that hosts, storage, and networks are specified *together* in Emulab NS files. This means that Emulab is less suited to the "elastic" provisioning style of clouds, but excels at uses cases in which one needs to capture a description of the whole environment.

Emulab's topology specification language is designed to handle both large broadcast domains (eg. LANs) and point-to-point links, and is thus amenable to complex topologies. In contrast to, say, cloud providers, who have no information about the intended communication patterns of their tenants, Emulab knows exactly how much bandwidth it must provision, and where. Its network mapper, called `assign` [21], maps the links and LANs of the submitted topology on to available hardware, ensuring that every link in the submitted topology is mapped to sufficient capacity on the network.

Emulab uses simple VLANs to segregate traffic on the experiment network, since they map directly to the abstraction provided to users and are available on all managed Ethernet switches. VLANs ensure that traffic from one experiment is not visible in others, enabling experimenters to use whatever addresses, protocols, etc. they wish within their experiments. Emulab does have some capabilities for exposing switch programming primitives, such as OpenFlow, but it views these as features offered to users rather than as mechanisms for provisioning. The Emulab facility at Utah also includes a set of switches that are allocatable to users in the same way that PCs are: for the duration of the experiment, the switch is reserved for the exclusive use of the experimenter and is under their full control. These switches are connected to a set of layer-1 switches, which perform simple forwarding without any Ethernet protocol processing, making them ideal for experiments that examine or modify the Ethernet layer.

4 The Evolution of Emulab into ProtoGENI

One of the key features that was historically lacking from the Emulab codebase was support for federation. Each of the "classic" Emulab clusters is an island unto itself; if an experimenter wants to use more than one, he or she must apply for accounts on each, and no support is provided for moving files, disk images, etc. between clusters.

This changed significantly with Emulab's participation in the GENI project. For this project, Emulab evolved to become ProtoGENI [22]. The name stems from the fact that it was initially viewed prototype of GENI; this is now anachronistic, as Emulab now contains one of the most complete implementations of the GENI concepts and APIs.

The Emulab software and the ProtoGENI software are one and the same; in essence, ProtoGENI is an alternative interface to Emulab. It is a new set of APIs, specified by GENI, and the ecosystem of tools that exists on top of those APIs. Underneath, when these APIs are invoked, they are mapped onto existing internal Emulab features and concepts. For example, the GENI notion of "slices" [3] map on to Emulab "experiments." Emulab does have its own set of APIs that pre-date the GENI project; however, they were never heavily used by users, and they did not include any notion of federated access.

ProtoGENI includes both a set of "native" ProtoGENI APIs and the "official" GENI APIs. The "native" APIs predate the standardization of the GENI APIs, and were among the many influences on the GENI standards. Thus, the two sets of APIs are similar, and offer similar feature sets, but are not completely identical.

ProtoGENI does choose not to expose some features from Emulab that are problematic for federation. For example, Emulab includes a shared filesystem that is available on all nodes. This works fine in a cluster, where all nodes have relatively fast and reliable access to the fileserver, but it becomes a major liability when it must be exported across the wide area, so ProtoGENI does not expose this feature. Similarly, Emulab includes a publish-subscribe event system that works smoothly within one cluster, but is hard to federate across the wide area; this, however is a feature we may attempt to re-introduce in the future.

The addition of the GENI/ProtoGENI features has opened a new era for Emulab; it used to be that each Emulab installation operated completely independently, and it is now possible to create portals that offer access to many of them at the same time. Also, while it used to be the case that almost all features in Emulab had to be offered by the Emulab software itself, the opening up of a richer set of APIs means that a greater number of tools, developed for GENI, can now be used on it.

5 Lessons from Emulab

Emulab has always taken the approach of making hardware and new software features available to users as soon as possible, even when those features are not yet complete, and/or there are still features planned that will add more "polish" or user-friendliness. This strategy has been highly successful; for most new features, there exist a set of "power users" who are the early adopters, and by their use, provide feedback on how future development should be directed. We have sometimes found that features that we thought were incomplete, in fact, already offer users everything they need, or have found that certain features were not as valuable as we had thought. The GENI project has adopted a similar approach with its "spiral" development.

It has been critical to develop a community of users around Emulab. Overall, the burden of user support for the facility is relatively light given the size of its userbase and the complexity of the features that it makes available. We attribute this in part to the fact that users within an institution or collaborative group tend to talk with each other, sharing stories of what has worked, not worked, workarounds for problems encountered, etc. The flipside of this lesson is that Emulab has put such an emphasis on user privacy that it has perhaps erred too much on the side of making it difficult for users to publicly share with each other. Emulab adopted a public user mailing list fairly late in its development, and sharing artifacts such as NS files across projects can be problematic as they often depend on private resources such as fileserver space. This is an issue we are working to address with ongoing development activities [24].

Emulab's decision to use a full programming language (NS scripts, written in OTcl) as its topology specification language has been both a major asset and limiting factor. It is extremely useful to be able to use the constructs of a full programming language, such as loops for building regular network topologies, and conditionals for constructing scripts that are flexible and parameterizable. On the other hand, it means that these descriptions are not, themselves, directly machine-manipulable;

Emulab essentially "compiles" NS files that are submitted to it to an internal format, which users do not have direct access to. This means that generating NS files programmatically is possible (though awkward, as code generation often is), but that writing tools such as GUIs to manipulate them after they have been written is nearly impossible. Emulab does have a topology GUI, written in Java, but that GUI is only able to understand a very small subset of the language, and cannot be used to view or manipulate NS files that have loops, conditionals, etc.

Going forward, a way to keep the "best of both worlds" in terms of a rich, user-friendly topology specification is to have a well-defined declarative representation free of complex programming language constructs, but which (unlike Emulab's internal representation) is exposed to users. This representation can be used as a "compiler target" for one (or more) higher-level languages that offer the flexibility and programmer-friendliness of ns-2. This is essentially the design choice that GENI has made: the low-level representation is the RSpec, which can be generated with `geni-lib`, which is a library for Python that (among other things) helps construct RSpecs.

It was also originally hoped that by using a language that came from a network simulator, we would make it easy for users to move back and forth between simulation and emulation. In practice, this has not materialized. There are many potential reasons for this: perhaps the communities that value these methodologies are non-overlapping, perhaps the topology description itself is not enough to ease the transition, or perhaps there is simply little call for such transitions. Regardless, in retrospect, picking ns-2 as a base language in Emulab was not a poor choice, but direct, native integration of emulation and simulation is not a design feature that Emulab is likely to pursue in the future.

We believe that, in large part, Emulab's success can be attributed to the fact that the Flux Research Group built a facility designed around its own needs, and those needs were reasonably representative of a larger research community.

6 The Future of Emulab

One of the main themes of Emulab's continued development is as a *platform* for the management of testbeds, clusters, clouds, and other facilities. The Emulab team cannot possibly, by itself, develop infrastructure that provides the features needed for the entire computer science research community; what it can do, however, is develop a base layer of hardware and software on top of which others can build infrastructure that meets their own communities' needs. Because it provisions at a bare-metal level, and because it takes a holistic view of the network, storage, and other aspects of the facility, it is in a good position to have other types of infrastructure deployed on top, using (in GENI terms) "slices." Additionally, because of the federation features it now has thanks to GENI, it is relatively easy to deploy such infrastructure across cluster and organizational boundaries.

The direction of Emulab's development is perhaps best typified by two follow-on testbeds built by the Flux Research Group: Apt [24] and CloudLab [23].

Apt, the Adaptable Profile-Driven Testbed, is a facility that is customizable with "profiles." A profile represents an encapsulation of the environment necessary to run a particular experiment, or a class of experiments. Each of these profiles can thus be thought of as a mini-testbed, running on top of Apt's hardware infrastructure. Profiles can be created by experts in particular domains, and shared with others. For example, an expert in database systems could install a standard set of database software, workload generators, reference datasets, etc., and share it with her or his community; likewise, someone from the high-performance computing community could do the same for domain science MPI-based code. Indeed, the Apt cluster is shared between computer scientist and the University of Utah's Center for High Performance Computing, with nodes being moved back and forth between the two sides on demand. Profiles can be shared either as semi-permanent facilities that many users access simultaneously, or their definitions can be shared, via hyperlink, such that every user gets their own instantiation, as happens with Emulab experiments. This is a powerful way to share research code and data; for example, the authors of a paper can create a profile that encapsulates everything needed to repeat the experiments in their paper, and provide this link in the paper itself or on their website.

CloudLab is a facility built for researchers who need their "own cloud" in order to conduct their work. While public and private clouds are extremely useful for many types of research, they have major limitations when it comes to transparency and control of software and resources at the bottom of the stack. There are certain elements that, by their nature, are considered part of the infrastructure, and cannot be changed (or even, typically, observed) by users; these include the hypervisor, the network, and the storage systems. For researchers who want to study and improve those parts of the stack, simply using "someone else's" cloud is not enough. Using Emulab for provisioning and federation, CloudLab is a facility where researchers can have exactly this sort of access. Like a real cloud, CloudLab is physically distributed, with sites at the University of Utah, the University of Wisconsin Madison, and Clemson University. Each of these clusters is an autonomous Emulab instance, and they are federated with each other and with GENI. So far, the infrastructure is a great success, with over 800 users running experiments in its first year of operation.

References

1. Anderson, D.S., Stoller, L., Hibler, M., Stack, T., Lepreau, J.: Automatic online validation of network configuration in the Emulab network testbed. In: Proceedings of the Third IEEE International Conference on Autonomic Computing (ICAC 2006) (2006)
2. Bastin, N., Bavier, A., Blaine, J., Chen, J., Krishnan, N., Mambretti, J., McGeer, R., Ricci, R., Watts, N.: The InstaGENI initiative: an architecture for distributed systems and advanced programmable networks. Comput. Netw. **61**, 24–38 (2014)

3. Berman, M., Chase, J.S., Landweber, L., Nakao, A., Ott, M., Raychaudhuri, D., Ricci, R., Seskar, I.: GENI: a federated testbed for innovative network experiments. Comput. Netw. **61**, 5–23 (2014)
4. Carbone, M., Rizzo, L.: Dummynet revisited. ACM SIGCOMM Comput. Commun. Rev. **40**(2), 12–20 (2010)
5. Chase, J., Grit, L., Irwin, D., Marupadi, V., Shivam, P., Yumerefendi, A.: Beyond virtual data centers: toward an open resource control architecture. In: International Conference on the Virtual Computing Initiative (ICVCI) (2009)
6. Chun, B., Culler, D., Roscoe, T., Bavier, A., Peterson, L., Wawrzoniak, M., Bowman, M.: PlanetLab: an overlay testbed for broad-coverage services. ACM SIGCOMM Comput. Commun. Rev. **33**(3), 3–12 (2003)
7. Cutler, C., Hibler, M., Eide, E., Ricci, R.: Trusted disk loading in the Emulab network testbed. In: Proceedings of the Third Workshop on Cyber Security Experimentation and Test (CSET) (2010)
8. DeterLab: Cyber-security experimentation and testing facility (web site). Information Sciences Institute, University of Southern California. http://www.deterlab.net (2016). Accessed Jan 2016
9. Extreme Cluster/Cloud Administration Toolkit. http://www.xcat.org (2016). Accessed Jan 2016
10. Emulab.net: Network emulation testbed web site. Flux Research Group, School of Computing, University of Utah. http://www.emulab.net (2016). Accessed Jan 2016
11. Emulab.net: Other Emulab testbeds. Flux Research Group, School of Computing, University of Utah. https://wiki.emulab.net/Emulab/wiki/OtherEmulabs (2016). Accessed Jan 2016
12. Emulab.net: Projects that have actively used emulab.net. Flux Research Group, School of Computing, University of Utah. http://www.emulab.net/projectlist.php3 (2016). Accessed Jan 2016
13. Foster, I., Kesselman, C.: The Grid: Blueprint for a New Computing Infrastructure. Morgan Kaufmann Publishers, San Francisco (1999)
14. Hibler, M., Stoller, L., Lepreau, J., Ricci, R., Barb, C.: Fast, scalable disk imaging with frisbee. In: Proceedings of the USENIX Annual Technical Conference. USENIX (2003)
15. Hibler, M., Ricci, R., Stoller, L., Duerig, J., Guruprasad, S., Stack, T., Webb, K., Lepreau, J.: Large-scale virtualization in the Emulab network testbed. In: Proceedings of the USENIX Annual Technical Conference (2008)
16. Laverell, W.D., Fei, Z., Griffioen, J.N.: Isn't it time you had an Emulab? In: Proceedings of the 39th SIGCSE Technical Symposium on Computer Science Education (2008)
17. NMC Probe (Web site). http://www.nmc-probe.org (2016). Accessed Jan 2016
18. Nurmi, D., Wolski, R., Grzegorczyk, C., Obertelli, G., Soman, S., Youseff, L., Zagorodnov, D.: The eucalyptus open-source cloud-computing system. In: IEEE/ACM International Symposium on Cluster Computing at the Grid (2009)
19. Ott, M., Seskar, I., Siracusa, R., Singh, M.: ORBIT testbed software architecture: supporting experiments as a service. In: Proceeding of IEEE Tridentcom (2005)
20. Ricci, R., Duerig, J.: Securing the Frisbee multicast disk loader. In: Proceedings of the First Workshop on Cyber Security and Test (CSET) (2008)
21. Ricci, R., Alfeld, C., Lepreau, J.: A solver for the network testbed mapping problem. ACM SIGCOMM Comput. Commun. Rev. **33**(2), 65–81 (2003)
22. Ricci, R., Duerig, J., Stoller, L., Wong, G., Chikkulapelly, S., Seok, W.: Designing a federated testbed as a distributed system. In: Proceedings of the 8th International ICST Conference on Testbeds and Research Infrastructures for the Development of Networks and Communities (Tridentcom) (2012)
23. Ricci, R., Eide, E., The CloudLab Team.: Introducing CloudLab: scientific infrastructure for advancing cloud architectures and applications. USENIX ;login: **39**(6), 36–38 (2014)
24. Ricci, R., Wong, G., Stoller, L., Webb, K., Duerig, J., Downie, K., Hibler, M.: Apt: A platform for repeatable research in computer science. ACM SIGOPS Oper. Syst. Rev. **49**(1), 62–69 (2015)
25. Rocks Cluster Distribution. http://www.rocksclusters.org (2016). Accessed Jan 2016

26. The NS-2 User Manual. http://www.isi.edu/nsnam/ns/ (2016). Accessed Jan 2016
27. The OpenStack Website. http://www.openstack.org (2016). Accessed Jan 2016
28. The PhantomNet Testbed. http://www.phantomnet.org (2016). Accessed Jan 2016
29. White, B., Lepreau, J., Stoller, L., Ricci, R., Guruprasad, S., Newbold, M., Hibler, M., Barb, C., Joglekar, A.: An integrated experimental environment for distributed systems and networks. In: Proceedings of the USENIX Symposium on Operating System Design and Implementation (OSDI). USENIX (2002)

DETERLab and the DETER Project

John Wroclawski, Terry Benzel, Jim Blythe, Ted Faber, Alefiya Hussain, Jelena Mirkovic, and Stephen Schwab

1 Introduction

This chapter describes the DETER Project and its centerpiece facility DETERLab. DETERLab is a large-scale, shared, and open modeling, emulation, and experimentation facility for networked systems, developed and operated as a national resource for cyber-security experimentation. The Project itself has three major components:

- A research and development program focused on the creation and deployment of advanced technologies and methodologies for experimental research in cyber-security, with particular focus on the security of large-scale, networked, cyber and cyber-physical systems;
- Development and operation of the DETERLab facility as a resource for cyber-security researchers and educators, and as a technology transfer and deployment vehicle for new experimental research technologies and methodologies as they emerge;
- A program of community evangelization and outreach, intended to coalesce and strengthen experimental research communities in these areas.

Hosted primarily at the University of Southern California's Information Sciences Institute, the core DETER Project is a collaboration of researchers at USC/ISI and University of California at Berkeley, with participation and use since its inception by over 8000 researchers, developers, educators, and students spanning some 275 universities and corporations across 40+ countries. The Project's primary sponsor throughout its existence has been the US Department of Homeland Security, with

J. Wroclawski (✉) • T. Benzel • J. Blythe • T. Faber • A. Hussain • J. Mirkovic
USC Information Sciences Institute, Marina Del Rey, Los Angeles, CA 90292, USA
e-mail: jtw@isi.edu

S. Schwab
USC Information Sciences Institute, Marina Del Rey, Arlington, VA 90292, USA

© Springer International Publishing Switzerland 2016

R. McGeer et al. (eds.), *The GENI Book*, DOI 10.1007/978-3-319-33769-2_3

substantial additional support from the US National Science Foundation and the US Defense Advanced Projects Research Agency, and further support from industrial and international sponsors.

2 Project History

The DETER Project was conceived and initiated in reaction to the first widely reported large-scale DDoS attack on the Internet in 2000 [1]. This event brought a realization of the urgent need for high quality, forward-looking, experimentally grounded cybersecurity research in the domain of interconnected and networked systems, to counter increasingly sophisticated threats and concerns. In response, a community-wide study [2] proposed the creation of a DDoS-focused experimental networking testbed isolated from the Internet. This study, together with insights drawn from the University of Utah's pioneering Emulab testbed effort, a workshop sponsored by the US National Science Foundation [3], and general consideration of desirable properties for a cyber security testbed, provided the initial inputs to the design of the DETER testbed facility. After a brief period of development, the DETERLab testbed became operational in March 2004, and the first DETER Community Workshop was held in November 2004. Since that date, the Project's program of testbed research and development, facility operation, and community outreach engagement has continued to evolve for over 15 years, to the present writing in late 2015.

2.1 Project Evolution

Since its inception, DETERLab's mission has been to provide a general purpose, capable, and usable environment in which to carry out well framed, meaningful cybersecurity experiments, including "risky" experiments that could not be safely conducted on the open Internet. This goal is not static. Over time, user needs for increased functionality, scale, complexity, diversity, and repeatability within the experimental infrastructure have driven a continuing evolution of DETER project objectives. This evolution can be described in terms of three distinct, but overlapping, phases.

Phase 1 focused on designing and building the DETER testbed facility to establish an immediate operating capability. This design included definition of the initial testbed architecture, together with corresponding hardware and software infrastructure. Concurrently, operating procedures and processes to achieve the required safety and level of service were created.

Phase 2 saw maturation of the testbed through use and expansion, growth of the supported research communities, and a greatly increased breadth of activity. DETER project researchers and community collaborators increasingly turned their focus towards the creation of new technologies and methodologies in support of experimental cybersecurity research, focusing on such areas as experiment automation, benchmarking, and malware containment.

As each of these technologies emerged, it was tested and deployed in the DETER testbed facility. The result was an evolution from the DETER testbed to DETERLab, a shared virtual lab composed of the underlying testbed resources, technology for using and managing the resources as test fixtures, and a growing variety of tools and services for experiment support.

Phase 3 continued work to establish DETERLab as a premier environment for advanced cybersecurity experimentation, primarily in the context of complex networked and cyber-physical systems. In this phase the project team has focused on increasing the sophistication and capabilities of the DETERLab instrument, extending DETERLab's capabilities to support experimental research in cyber-physical, as well as purely cyber, systems, and providing strong support to the community through development of end-to-end usage scenarios reference materials.

In parallel with DETERLab's advance in the research community, the project engaged a rapidly growing community of DETER users focused on education. To support these users the project team developed specialized DETERLab capabilities targeting the educational environment, and initiated a community-wide project to create DETERLab-based instructional curriculum modules for use in classroom settings.

Even as DETERLab continues to evolve, the project team is presently engaged in looking towards new directions for cyber experimentation. Building on the project's 15 year history of research, development and operational experience, members of the DETER project team have focused on a new and broader vision for Cyber Experimentation of the Future [4]. We are working across several early stage projects and community projects to develop a strategic plan for attaining this vision.

3 Objectives

The DETER Project's fundamental objective is to develop and make available a powerful, easy to use, open platform to support experimental cybersecurity research, with particular focus on experiments that involve large, complex, networked or distributed cyber and cyber-physical systems.

The substance of any such platform is that it allow the experimenter to create, manipulate, and observe an experimental system they wish to study, under sufficiently controlled conditions and with sufficient richness and accuracy to support the desired experiment. In essence, the platform becomes a controllable, configurable, approachable modeling environment for the system being studied.

In light of DETER's intended domain of application, a number of more specific objectives were identified as part of the initial testbed design. These included:

Fidelity: Fidelity is often viewed as identical to "realism"—that is, researchers wishing for experimental environments that achieve fidelity to real existing networks and to the Internet in particular. However, the true objective is more sophisticated, because what is actually needed for a viable experimental platform is the ability to achieve sufficient modeling accuracy for the specific phenomena being studied in a particular experiment, within an environment that may or may not correspond to any existing real network.

In particular, for researchers desiring to understand the evolution and future potential of a technology, the most useful of experiments are often those that involve the creation of hypothetical environments that may exist at some point, but do not, or even can not, exist in reality at the time of the experiment. For example, DETERLab has recently been used to study properties of the Tor [5] anonymity network across a range of assumptions about Internet topology, number and location of Tor nodes, and usage patterns that differ widely from what is actually observed in today's Internet and the existing Tor network.

Validity: An experiment is valid if it is executing as the experimenter intended and in conformance with the experiment description.[1] There are many possible causes for validity loss, e.g., infrastructure resource overload or misallocation, software bugs, host system crashes, missing software, or configuration errors. Because ensuring validity is a major problem when running large, complex experiments, the DETER project identified early in its work the need for explicit validity management mechanisms.

A particularly important class of validity management ensures that limitations of the testbed itself do not accidentally distort an experiment. For example, a researcher's experiment studying distributed denial of service (DDoS) might become highly misleading if hidden, unintentional bandwidth or computational resource limits within the experiment were created by limitation of the testbed's physical resources, rather than through limits defined by the experiment description. Validity management mechanisms can identify and call out violations of these required experimental conditions, alerting the researcher to potential failures of experiment validity.

Scale: The experimental facility must be able to support experiments at sufficient scale to be representative—that is, to capture complex scale-related effects, to demonstrate appropriate scaling properties of a growing system, and the like. Scale may be considered as an aspect of fidelity, but was broken out in the DETER project's requirements analysis because of its unique and concrete effects on testbed design and implementation. Section 4 of this chapter discusses a number of technologies that contribute to DETERLab's ability to create, manipulate, and execute experiments of significant scale. Leveraging these technologies independently and in combination, today's DETERLab's facility is able to support a range of experimental scenarios with as many as 100,000+ modeled nodes overall, while at the same time emulating key elements within a large-scale experiment at high fidelity on physical hardware.

---

[1]Note that *validity* and *fidelity* are separate, though related, properties.

Safety: The DETER project's work explicitly targets experimentation in cyber-security, including the study of risky malware, worst case system behaviors, and the like. Early planning for the DETER testbed therefore centered on analysis of the security risks and corresponding mitigations [6]. Risks considered included intrusion into the experiments of other users or the testbed infrastructure itself, extrusion (escape) of malware into the Internet, and accidental or deliberate DDoS attacks on the Internet.

Repeatability: Fundamental to scientific study is the ability to reproduce and build on the results of others. A simplistic view of this objective would lead to a goal of deterministic execution—that is, that an experiment, once executed, could be "freeze-dried" and then run in an identical environment at a later time, to produce identical results. While this goal was adopted as an original objective of the DETERLab facility, it quickly became apparent that, in the study of any physical system of significant complexity that evolves over time, the concept of maintaining an identical environment to obtain identical results proves insufficient. Consequently, the DETER project has recently focused on more sophisticated framings of the problem, which focus on the maintenance of defined repeatability invariants through the lifetime of a series of experiments.

Flexibility: For maximum flexibility in experiment creation, researchers should ideally be able to modify any algorithm or behavior of any modeled device. The DETER testbed's initial solution was to focus on complete programmability of the facility at OSI layer 3 and up, using general-purpose computing devices to emulate both network devices and end-systems. As technology has advanced, DETERLab has preserved this basic principle but moved to incorporate additional classes of programmable devices into the infrastructure, as well as broadening the use of general-purpose computing devices. As examples, today's DETERLab incorporates high-performance NetFPGA devices [7] for researchers requiring hardware-level emulation of key system elements, while also moving to integrate a performance-optimized implementation of Open vSwitch [8] for those requiring a flexible, software-based implementation of standard or experimental network switch functions.

User-Centric Perspective: Finally, the DETER project has over time increasingly emphasized the premise that, to meet project goals, the capabilities provided to DETERLab users can not be limited to traditional "testbed" functions, but must also stretch to include a new class of high-level tools that facilitate, enable, and guide successful experimental research. A key observation is that experimental research methodologies are themselves an area of active research and rapid advance. As the DETERLab facility's capabilities grow increasingly capable, and as new research methodologies emerge, presenting these advances in an approachable manner and guiding users towards valid, meaningful experimentation have become central concerns of the project.

4 DETERLab Technologies

In support of the objectives outlined above, the DETER Project has, in developing DETERLab, adopted or created a number of technologies that, brought together, implement a modern, integrated experimental facility for cybersecurity research. In this section, we review a core subset of these DETERLab technologies. More in-depth review of DETERLab's technical goals and constituent technologies, including some not discussed here, may be found in [9–12]. For readers interested in the evolution of DETER over time, Ref. 13 presents a discussion of the testbed's architecture, technologies, and intended uses as they were imagined in 2007. References 14–21 each focus on a more specific technology or capability developed within the DETER project framework.

4.1 Core Technologies

To create the initial version of the DETER testbed, analysis of the goals described in Sect. 3 led to the choice of a cluster testbed architecture, following the example of the Emulab cluster testbed then-recently introduced by Jay Lepreau at the University of Utah [22].

Emulab's design centers around a pool of experiment nodes implemented as x86 servers, linked through multiple Ethernet interfaces via a high-performance "backplane", a bank of Ethernet VLAN switches. The testbed control software loads ("swaps in") an experiment by allocating testbed resources (nodes and links), loading the specified node software, and controlling the VLAN switch to "plumb" the allocated nodes together into the specified network topology.

Although extensively modified, this core architecture remains at the heart of DETERLab today. DETERLab continues to utilize a configuration of some 500+ server-class x86 machines as its core computational engine, interconnects these nodes with high performance SDN-capable 10 Gb/port Ethernet switches, and relies on an extended version of Emulab's central resource allocation function "assign()" to map the facility's physical compute and communication resources to individual experiments on demand.

The initial DETERLab testbed differed in one fundamental way from Emulab. While Emulab users can directly log into each allocated node from the Internet and Emulab experimental nodes can send arbitrary data into the Internet, DETERLab experimental nodes are blocked from direct external IP connectivity. In keeping with DETER's isolation and containment requirements as a cybersecurity testbed, DETER's configuration required that users perform a double SSH login to reach their experiment nodes. More recently, DETERLab has added a Controlled Internet Access capability, which creates narrowly sculpted holes in the containment firewall, allowing specific experimental nodes for approved users to communicate with explicitly defined Internet hosts.

Following the initial standup, DETERLab's hardware and software base has been significantly modified and extended. Utah and DETER have gone separate ways with respect to major system elements, and the source code repositories have diverged. Notwithstanding this evolution, many basic concepts remain aligned, with DETERLab remaining clearly recognizable as an evolutionary branch of the original Emulab concept.

4.2 *Containers*[2] *for Scale and Fidelity*

In the original Emulab and early versions of DETER, a fundamental concept was the use of an individual emulation node (i.e., an x86 server) within the testbed to emulate or model each emulated entity (e.g. a host, router, switch, etc.) within the researcher's experiment. In this approach, the modeling fidelity of the emulated entity is potentially quite high, to the point of the emulation node running the actual code of the entity being emulated. Conversely, a limitation of the approach is that the maximum entity count in the experimenter's model is essentially bounded by the number of hardware nodes in the testbed.

To meet DETER Project goals, this structural limitation is unacceptable. DETER focuses on supporting large-scale experimentation, because experimentation at this level is essential for exploring and understanding Internet-scale phenomena. Many key research areas, such as botnet dynamics and evolution or user anonymity and monitoring, can only be studied fruitfully in the very large. No approach that depends on a one-to-one mapping between emulated entities and emulation hardware can come close to meeting this requirement.

At the same time, however, many larger-scale experiments do not require high fidelity in every part of the experimental apparatus. Instead, what is required is that each of the elements included in an experiment provide an *appropriate* level of modeling fidelity for that element of that experiment. Importantly, otherwise identical experimental configurations may differ significantly in the fidelity required of each element, depending on the purpose of the experiment.

This point is illustrated by Fig. 1. The figure shows two representations of an identical experimental network topology, deployed for two different purposes. In the figure, the size of each network element represents the degree of modeling fidelity required for that element if the experiment is to be meaningful. The first experiment, designed to study BGP security, is concerned with detailed behaviors among router nodes in the network. The second, focused on worm propagation, shares the exact

[2]In retrospect, DETER's selection of the name "containers" was an unfortunate choice, because of the industry's subsequent adoption of the "container" name to describe a class of lightweight virtualization and software packaging technologies. DETER containers and industry containers are not the same thing and do not serve identical goals, although there is some significant overlap of ideas between the two.

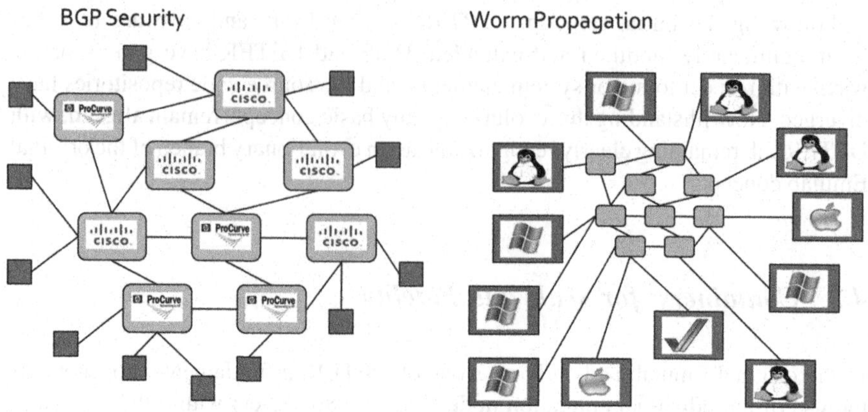

Fig. 1 Two perspectives on a common experiment topology

same network and end-system layout, but requires detailed modeling of edge node behavior, but only highly abstract modeling of intermediate router behavior.

The recognition that most large-scale experiments fall into this "multi-resolution" category is crucial to creating a laboratory facility that can support extremely large, complex, yet accurate and meaningful experiments. In the remainder of this section we describe DETERLab's approach to supporting this capability.

To begin, we adopt the metaphor that the DETERLab facility and its cluster of computers should be seen as providing a single pool of undifferentiated computing power, rather than as a collection of distinct and logically separate nodes, each designated to emulate a particular element within an experiment. Our objective then becomes one of fluidly allocating just enough "computrons" from the pool to different elements within an experiment to implement each part of the experiment at the needed level of fidelity. While limitations such as I/O granularity will, in practice, always preclude treating the computing power provided by a cluster as completely fluid, the metaphor accurately describes our idealized objective. The technical challenge then becomes reaching as close to this objective as possible within the limits of a practical system.

Ultimately, we would like this process to be largely automatic. That is, we would like the experimental infrastructure to *understand* what level of fidelity was required for each element of an experiment, based on extracting the researcher's intent for the experiment from the experiment description. Then, the facility would use this understanding to guide its use of model selection, virtualization, resource allocation, and other scaling techniques in realizing the experiment.

It is important to recognize that, because the researcher's goal is often to create a specific experiment to study an unusual phenomenon, achieving the above objective in the general case involves detailed understanding of the researcher's intent, as well as potentially unconventional application of the various available techniques. For example, a DETERLab experiment may aim to model some normally secondary or tertiary phenomenon at high fidelity to study how that effect can lead to security or robustness failures, while a mainstream deployment of the same technologies would assume that this secondary effect was entirely ignorable.

Supporting our top-level objective requires advance in two major areas. The first and most challenging is extracting sufficient understanding of fidelity requirements from experiment descriptions and other expressions of researcher intent. This problem is compounded by the fact that the experiment designer herself may not fully understand the required dimensions and metrics of fidelity, particularly when the purpose of the experiment is to study unexpected phenomena. However, model validity checking and related concepts offer some directions forward in this area, and as of this writing are an active area of work for the DETER Project.

A second, more straightforward, area of required advance focuses on development of the actual technologies required to create and deploy experiment elements operating at many different points within the fidelity/computation tradeoff space, once the fidelity goals for each element are understood. The ability to move widely within this tradeoff space is essential to DETERLab's support of extremely large scale, yet still useful and valid, experiments. DETERLab's *container* system provides foundational capability in this regard.

Phase 1 of this system focuses on flexible, fine-grain allocation of computing power to different elements within an experiment, while also providing a range of choices for other dimensions of modeling fidelity such as isolation, memory and I/O resource availability. The system provides control over allocation of computational resources in two dimensions: first, in the selection of *container type* used to implement each element within an experiment, and second, by controlling the allocation and packing of containers onto physical compute nodes. Each container type operates at a different point in the fidelity/efficiency/overhead space. However, all types share a common instantiation, configuration, and control interface, facilitating the construction and management of experiments that include multiple types.

DETERLab currently offers the following container types, each with different fidelity characteristics in multiple dimensions:

- Type 1: Dedicated physical machine. Very high fidelity model of a physical machine.
- Type 2: QEMU [23] based virtual machine. Lower fidelity model than Type 1, but full hardware modeling. Multiple instances per physical machine.
- Type 3: OpenVZ [24] Linux container-based virtualization. Software abstraction, minimal hardware modeling. Shared kernel structures. Roughly an order of magnitude more instances per machine than Type 2 QEMU-based virtualization.
- Type 4: ViewOS [25] process. Lower fidelity software-level abstraction, but more instances per machine.
- Type 5: Co-routine threads. Very low fidelity in multiple dimensions: minimal isolation, timing control, etc. Extremely efficient, thousands of element models per physical CPU.

In addition to specifying the container type that each element in an experiment will use, the system allows the experiment designer to control the packing of containers onto physical machines very flexibly. For example, a researcher can specify that each experiment element representing a router in their experiment should be implemented as an OpenVZ container and packed to a level that provides 1 "standard" CPU unit per container, while each element representing an intrusion detection system should be implemented as QEMU virtual machine to provide hardware-level emulation, and provided with 10 standard CPU units per VM. Though these knobs for controlling the fidelity vs. scalability tradeoff are fairly coarse, experience has shown them to be effective and approachable by a significant body of DETERLab users.

Phase 2 of DETERLab's container system development is focused on providing the experiment designer with similar fidelity/scalability tradeoffs in the space of networking and communication resources. Modern network interfaces provide two key capabilities not present when DETER was first developed. These are the ability to support very low level, fine-grained provisioning of network capacity, and support for I/O virtualization strategies such as SR-IOV. Building on these capabilities we are able to provide the experiment designer with access to network interface models at several levels of fidelity, ranging from direct physical hardware to highly abstract software interfaces, and then, as above, flexibly map the model provided to the experimenter into a physical realization as needed.

By creating the ability to move flexibly and at fine grain in the tradeoff spaces of computing and communications fidelity vs. scalability, DETER researchers are provided with a mechanism to create experiments at very large scale, yet with key elements modeled with very high fidelity. Today, DETERLab's facility is able to support many experimental scenarios with order 100,000+ modeled nodes overall, at the same time modeling some nodes with perfect fidelity on physical hardware.

4.3 Federation

As early as in NSF's foundational testbed report [3] the potential for interconnection or *federation* of testbeds operated by different entities was recognized. Depending on circumstances, there are three major benefits to a federation: (1) the ability to provide the researcher with access to substantially more resources, or to some remote hardware or software resource that would otherwise be unavailable; (2) the ability to share resources among participating organizations for efficiency, and (3) the ability to encourage collaboration and community building across researchers and organizations.

DETERLab achieves these benefits through implementation of a powerful federation framework, the *DETERLab Federation Architecture*, or DFA [15, 16]. The DFA allows researchers to create "DETER-like" experimental environments that incorporate resources from many providers, including but not limited to DETERLab itself. Policies for the use of each resource are set by the facility or researcher that makes it available. This collaborative control is what distinguishes federation from other forms of sharing, and is key to its success. Owners of federated resources do

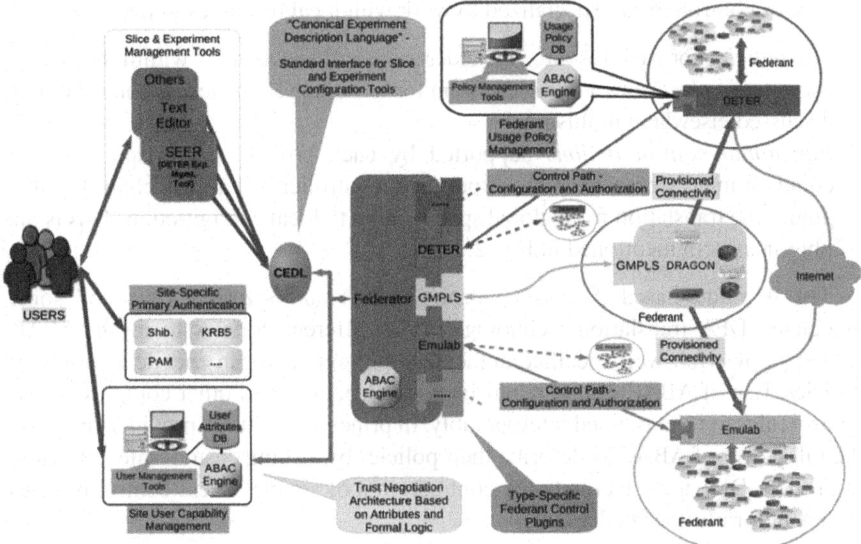

Fig. 2 The DETER Federation Architecture (DFA)

not relinquish control over their use when sharing them with the community, which often simplifies sharing of the resources at all.

Federated experiments are "DETER-like" in that resources and facilities outside DETERLab may impose different constraints or have different capabilities than those of DETERLab itself. A federated experiment embodies an environment as similar to a DETERLab experiment as the federated environment will support, subject to the researcher's and resource providers' objectives and preferences.

The overall architecture of the DETER Federation Architecture is shown in Fig. 2. Users may design experiments using a variety of low- or high-level tools, but in each case the ultimate result is an *experiment description*, expressed in a standard representation, that describes all aspects of the desired experimental environment. At present, this standard representation is an extensively augmented version of the NS file format originally adopted by Emulab to define its experiment configurations.

This representation is passed to the *federator*, which acquires resources from different providers to be used by an experiment and builds a cohesive experimental environment from these resources. The federator is composed of two parts. The first is a logically centralized *experiment controller* running on behalf of the user, which is responsible for partitioning the experiment across resource providers, gathering resources from each provider, and creating the overall configuration. The second is a set of distributed agents called *access controllers*, affiliated with individual resource providers, which are responsible for allocating local resources at each provider in accord with local rules and configuring these resources as needed to form the shared environment.

Access controllers are specialized to work with local resources in two ways:

- *Use policies* for each resource provider are expressed as rules within the ABAC access control system, a flexible system for reasoning about access control policy discussed elsewhere in this book.
- *Functional control actions* supported by each provider are mapped into a common interface between the experiment controller and access controller that guides the translation from global specification to local configuration. This is the plug-in system mentioned in Fig. 2.

The Attribute-based Access Control (ABAC) authorization logic and some associated DFA translation technology allow different policies to mesh. ABAC [26] is a powerful, well-specified authentication logic that can encode a variety of policies. Use of ABAC allows sophisticated delegations and other complex access control rules to be expressed interoperably. In principle, resource providers may use the full power of ABAC to describe their policies by writing custom rulesets, while in practice DFA tools facilitate the configuration of simple access control policies sufficient for most testbeds.

In addition to specifying a standard workflow and vocabulary for constructing federated experiments, the plug-in interface allows access controllers to advertise their capabilities to the experiment controller and researchers. An access controller can describe which DETERLab-like features it supports—e.g., isolation, shared file systems. In addition, the plug-in interface enables access controllers to advertise and accept configuration for services unknown to DETERLab. This lets researchers tinker with new features quickly.

The DETER Federation Architecture has proven powerful and adaptable, allowing DETERLab users to construct federated experiments spanning diverse systems, including as examples the ProtoGENI GENI control framework, the OSCARS secure circuit configuration system developed for ESnet and now used by Internet2 and over 50 research networks globally, NICT (Japan)'s StarBed network emulation testbed, and several cyber-physical systems laboratories. A particularly unique and useful access controller is the *desktop federation controller*, which enables an individual desktop computer to create or join a federated experiment. This lightweight controller gives researchers full, direct access to entire large, multi-facility federated experiments from a local desktop, using locally running software such as experiment design and construction tools, visualization and data analysis facilities, and presentation software for demonstrations.

4.4 Experiment Orchestration

The technologies described above relate to the initial creation of an environment in which a DETERLab experiment is executed: basic allocation of testbed resources to the experimenter, mechanisms that support orders-of-magnitude experiment scaleup over that obtainable with more traditional approaches, and use of federation to support sharing and cross-coupling of testbeds.

Once these issues are addressed, attention turns to the dynamic aspects of configuring and running the experiment. Experimenters face a variety of needs concerning configuring the experimental apparatus, feeding the experiment input data and events, observing and controlling the experiment's evolution over time, and collecting experimental data during and after experiment execution.

In early testbeds, these tasks were typically supported through ad-hoc approaches, often relying on remote login, shell scripts, and similar tools. Experience demonstrated, however, that as experiment scale and complexity grows, these methods quickly become infeasible. For DETERLab, the tipping point was reached with the development of the container subsystem, which shifted the scale of feasible experiments by three orders of magnitude, from a few tens or hundreds of elements to tens or hundreds of thousands. A new approach was required.

In addressing this need, the DETER Project took the view that the challenge of managing and monitoring large-scale, complex experiments is a substantial research problem in its own right. This is particularly true when the requirement is recognized to develop approaches suitable for researchers of widely varying background and creating experiments for widely different purposes. Rather than attempting to create tools that solved the entire problem, the project adopted the architectural approach of creating modularity. Work focused first on a common, low-level experiment management substrate suitable for large, complex experiments, with the intent that this substrate could then serve as an implementation vehicle for higher-level tools, developed both by the DETER project itself and by the larger research community.

This lower-level substrate is now available in DETERLab as the MAGI Experiment Orchestration System. MAGI[3] provides the infrastructure needed to support workflow tools that instrument and control an experiment. In concept, MAGI plays a role in the DETER ecosystem similar to that played by orchestrators such as Ansible, Chef, Puppet, and Fabric [27–30] in the Cloud Computing world. MAGI differs from these orchestrators primarily in its focus on scale and its specialization to experiment requirements and semantics, rather than general-purpose cloud computing. An example of this specialization may be found in MAGI's design objective of orchestrating experiments of up to hundreds of thousands of elements, through the utilization of efficient group communication and explicit implementation of sparse partial ordering semantics.

The MAGI framework has three main components: a *messaging substrate* that provides scalable and reliable mechanisms to communicate across federated and containerized experiment nodes; *agent modules* to enact different behaviors in the experiment; and an *orchestrator* that executes a workflow and provides deterministic

[3]MAGI is an abbreviation for "Montage Agent Infrastructure."

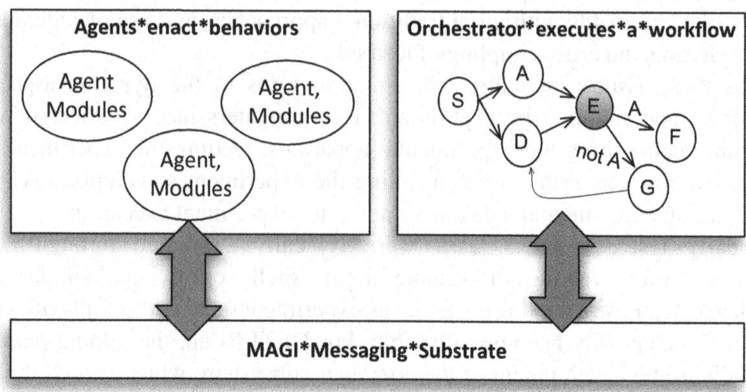

Fig. 3 MAGI System Architecture

control for the experiment. Figure 3 shows the components of MAGI and how they provide an execution environment for large scale experiments. We discuss each component briefly below.

MAGI's **messaging substrate** provides an overlay control network to deploy, monitor, and control the entities in an experiment. This network provides seamless access to all types of nodes in a DETERLab experiment, across federated testbeds and within containerized environments. MAGI is based on group communication semantics. Agents can join and leave groups dynamically and any member of a group can send messages to other members of the group. From the implementation perspective, a MAGI daemon is located on each physical computing node to manage the overlay transport and route messages to agents. Different implementations of the daemon support different container and federation environments. Consequently, the messaging substrate is logically viewed as a cloud providing a well defined, unified group communication semantic across all experiment nodes irrespective of each node's implementation strategy.

Messages are sent to and received from MAGI agents. Messages sent to agents are known as events while messages received from agents are triggers. Event messages direct agents to change their current state. For example, a message could request a web traffic client agent to change the distribution of the requested objects. Trigger messages report the current state of the agent. For example, a trigger could report the set of neighbors to an agent's peers. This bi-directional messaging enables incorporating feedback from the experiment for control and is used extensively for orchestration.

A MAGI **agent** enacts a specific behavior in an experiment. It can be an *actuator*, for example a web server agent responding to requests from client agents and generating traffic on the experiment links. An agent can also be a *sensor*, e.g., for measuring and reporting the number of packets seen on a particular link in an experiment. Agents interact with other agents and MAGI tools through the messaging substrate, using event and trigger messages. Every agent in the MAGI framework support a suite of standard methods, such as configure, start, and stop. In addition, agents may support agent-specific functions. MAGI supports loading and unloading agents on experiment nodes dynamically.

DETERLab provides a library of agents that implement basic functions such as traffic generation and monitoring. The library includes agents to generate web, ftp, ssh, VoIP, and IRC traffic and malicious DoS attack traffic.

In addition to these off-the-shelf agents, the MAGI library provides code fragments, helper classes and other tools to support extension of standard agents and development of experiment-specific agents. At the time of writing, MAGI supports Python, C and Java based agents, realized as threads or processes depending on environment and available resources. Each agent module has an interface definition (IDL) file that describes the inputs, output, and methods supported by the agent. The IDL can be parsed by supporting tools to ensure the agents are correctly configured and controlled.

The MAGI **orchestrator** uses agents and the messaging substrate to implement the "control flow" of an experiment, providing deterministic control over the various experiment. The experiment procedure is expressed as a stream of events and triggers, capturing the partial orderings required to provide correct experiment semantics. Events activate specific behaviors in the agents, while triggers provide event-based and time-based synchronization points for control flow. Complex control flows may be specified by creating trigger message expressions using boolean operators. Error recovery methods based on timeouts and recovery actions are also provided.

An experiment's control flow is expressed in a low-level standard format called Agent Activation Language (AAL). AAL is a YAML-based descriptive language with three main directives. The *group* directive defines the required groups on the messaging substrate in the experiment nodes. This directive is used to map sets of one or more experiment nodes to a group name. The *agent* directive defines a required functional behavior on a set of experiment nodes. An agent definition has a name key to identify the agent, a group key to define where it is deployed, a path key to indicate the location of the agent implementation, and an arguments key that can be passed to the agent during deployment. The *event stream* directive provides the main body of AAL. It is a list of events and triggers that are parsed and executed by the orchestrator tool. A procedure typically contains multiple event streams. Event streams execute concurrently and are synchronized with each other using triggers.

In addition to the orchestrator, MAGI provides two further base-level tools that allow deploying, configuring, and visualizing the experiment. The bootstrap tool is used to deploy, install, and configure the software libraries and the messaging substrate on experiment nodes. It is typically run once, when resources are first allocated to an experiment. A graph tool is used to collect basic measurements from the sensor agents and display the current status on the experiment for real time and offline analysis.

4.5 *Multi Party Experiments*

In the world of security research it is fundamental that factions with different powers, interests and knowledge compete for control of equipment or information. Attackers try to gain access to computers, botnets try to gather machines and organize, governments try to identify forbidden communications, and cryptanalysts try to break codes.

Central to many meaningful experiments in such domains is that the experimental platform supports scenarios where different parties can interact, with each limited to only the scope of information or environment they would see in the real world. At the same time, a benefit of the laboratory experiment environment is that a "full" complete view can also be made available, to facilitate experiment construction, control, and assessment by researchers. DETERLab implements this capability, calling such an environment a multi-party experiment.

DETERLab's multi-party experiment facility erects and enforces constraints on what the various parties to the experiment can see and do. Each party that enters the environment does so from a different perspective and with different constraints on what they can see and manipulate. We call each such perspective a "view" or "worldview". Further, DETERLab permits these views to overlap, for example, allowing an evaluator to view the experiment as a whole at the same time that other groups are competing inside the experiment.

Figure 4 shows a visualization of a multi-party experiment in progress in DETERLab. This experiment implements a multi-party security game. In this game there are two researchers: The Attacker who designs an attack worm, launches the worm that attacks the network, and monitors the attack progress; and the Defender, who runs a defense service, monitors the server load, and deploys additional servers to absorb the load. The playing field as seen by the Attacker is shown in the lower left view, while that seen by the Defender is shown in the lower right view. Note that these views represent the *only* information directly available to these two parties—that is, network routing protocols, measurement tools, etc, operated by either party have direct access to only a portion of the full scenario. As in real life, any knowledge the parties desire about the rest of the world must be learned indirectly.

The third view depicted in the figure is that of the Proving Ground, or overall playing field for the experiment. This is the unified scenario made up of the Attacker's environment, the Defender's environment, and other network structure not directly visible to either. The topology of the full playing field is unknown to the two researchers, and the environment managed by the Proving Ground can change dynamically. As can be seen, DETERLab's multi-party experiment capability allows researchers to construct scenarios that closely model the decentralized, limited-information-flow, collaborative-competitive world of actual networked systems.

Composition of worldviews for a multi-party experiment occurs at several levels. Parties may know different parts of an experiment layout, or see the same

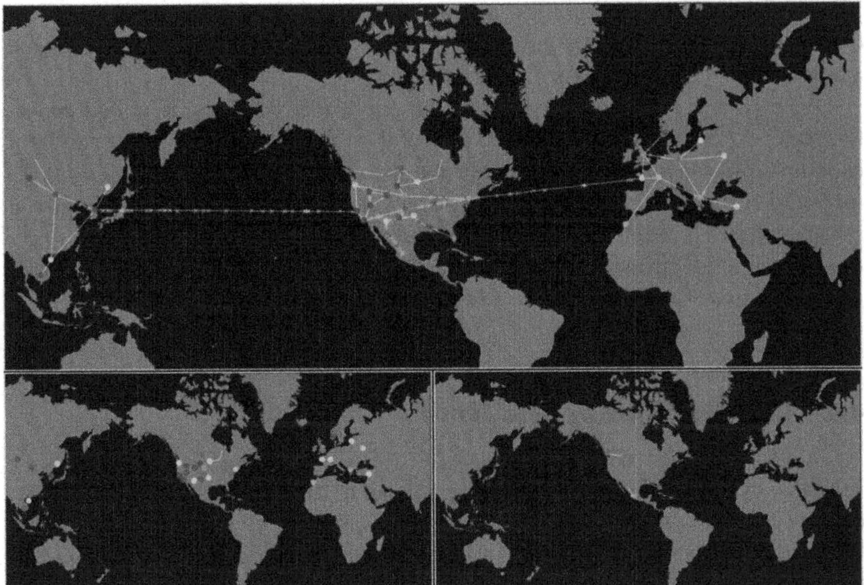

Fig. 4 A DETERLab Multi Party Experiment

experiment layout differently. For example, two parties may be able to access the same nodes on a subnet, but identify them with different DNS names or IP addresses. Different parties may have direct access to parts of the overall experiment topology, but see other parts of the topology only through whatever measurement or analysis tools they can deploy. Similarly, parties may share access to the same experiment elements, but have different rights to view and manipulate them.

Providing the full richness of a multi-party experiment depends on characterizing the different viewpoints, interpreting them as access control rules, and realizing the experiment in accordance with these rules. It is interesting to note the parallel between this objective and that of federation, described earlier. In each case, policies for access to different experimental resources may be defined separately, and then brought together in a coherent whole by the federation system.

Consequently, DETERLab implements multi-party experiments as a special form of federation. To construct a multi-party experiment, the federation system constructs an experiment that uses resources allocated to each of the parties, configured appropriately, and interconnects them. Here, the different groups federating are not disparate facilities that contribute resources, but different groups within DETERLab that control sub-experiments.

This layout is captured and created through the federation system's use of ABAC at the sub-experiment level. Each sub-experiment within the overall scenario is itself, recursively, an experiment that has its access policy captured by ABAC. The federation system creates the sub-experiments based on those policy specifications, just as it does between facilities, and then integrates them together to form the overall experiment. In addition, the process may be applied recursively to create a hierarchical multi-party experiment.

DETERLab's federation system can be applied directly, as described above, to create experiments implementing a useful subset of all possible multi-party experiment semantics. However, certain non-hierarchical scenarios cannot be implemented directly in this fashion. If two parties see the same element differently, a richer approach is required. At the time of writing, work continues within the project to implement this richer multi-party semantic, while the federation-based approach is in active use within DETERLab's research communities.

4.6 Modeling Human Behavior

To date, cybersecurity and networking research testbeds have focused almost entirely on supporting experiments that study hardware or software-level system behaviors. Yet, an important aspect of system-level security is that the processes and tools put in place to defend these systems will be used by individual humans, who may not be well informed about security risks and vulnerabilities, who are typically not expert in the use of the tools, and who have other jobs to perform while using them. Consequently, many successful attacks to date have relied on some aspect of human behavior, for example phishing attacks that trick users into installing malware, or social engineering attacks that convince users to reveal passwords. Human behavior can also lead to significant deviations and inaccuracies in simple, technically focused predictions about the value of defenses and success rate of attacks.

For these reasons we seek to provide DETERLab experimenters with the ability to incorporate models of human behavior, as well as software and hardware behavior, into their experiments. To do this, the DETER project has developed an agent-based system that models many aspects of human behavior relevant to security and networked system experiments, and incorporated this system into DETERLab. We describe the objectives and design of this system below.

Although in many cases human behavior is hard to predict, some aspects are driven by relatively predictable factors and may in turn form the basis of useful predictions about behavior. These factors include limited knowledge about security, reliance on common misconceptions and flawed analogies when considering security questions, and human processing biases, such as the effects of limited attention. In addition, users may have particular difficulty following some parts of a security protocol because of an interaction with their work protocols and routines, leading to distraction or "cognitive overload". These interactions may be hard to predict when thinking only about security tools but become much clearer when additional modeling to represent the human's more primary, non-security concerns is incorporated.

DETERLab's agent-based human behavior modeling uses these regularities to predict when security systems are likely to be used inappropriately, or even deliberately disabled or avoided. Within a DETERLab experiment, this allows more accurate prediction of the effects of a tool that relies on human use or defends against a human-centered attack. In general, the results can also help in designing more effective approaches given the reality of human behavior. Below, we outline our *dash* agent architecture, developed to implement this capability, and briefly describe two projects that aim to build models of human behavior from observations in a number of real scenarios and that use *dash* to implement and validate these models.

4.6.1 The Dash Agent Platform

In modeling, we take the view that end users are goal-driven and, given sufficient time, can make rational decisions about security based on their knowledge about vulnerabilities and risks as well as their perceptions of its impact on the task at hand. We then develop a suite of software agents that are capable of emulating relevant human behaviors under these constraints when interacting with software, and make these agents available to DETERLab experimenters.

The main elements of this agent system are:

- a flexible plan executor that chooses tasks to perform based on the modeled human's goals, as well as re-assigning tasks if goals change priority or current plans are failing,
- a cognitively plausible representation for incorrect or incomplete views of the world, particularly about security,
- a fall-back mechanism to choose actions quickly when the agent perceives time pressure or other kinds of stress that prevent rational decision-making.

The *dash* agent platform provides tools to meet these requirements with the following components. First, a reactive planning system [31] chooses actions based on the agent's goal utilities and costs and re-plans when the world changes significantly. Second, an implementation of mental models [32] is used to capture alternative models that agents may have for the world. This approach is well suited to analogies like physical security or healthcare that have been shown to be influential in studies of user models [33]. Finally, the approach uses a dual-process architecture, in which two components compete to determine the agent's next action: a deliberative component that uses planning and mental models and an instinctive component that matches actions to situation elements [34]. *Dash* has been demonstrated to capture several repeatable and significant aspects of human behavior about security in a single framework [35–37].

While *dash* provides a plausible model of human security behaviors as reported by a number of researchers, the DETER project team is also working on more detailed investigations and modeling of behaviors in a number of scenarios, with the aim of extending and further validating the model. In one collaborative project with researchers at Dartmouth and the University of Pennsylvania we are studying end user circumvention of security, particularly in healthcare domains [38]. In these extreme cases, users disable or avoid security measures that have been put in place, either deliberately or unwittingly.

We have collected a large number of examples of circumvention from interviews, observation and the literature and categorized them in terms of mismatches between the security designer's view of the world and that of the user. This broadly fits the *dash* model in which the circumvention is a rational decision based on the user's view of the world. While results from an abstract study will be available in *dash* to use in a number of domains, we also plan to build more realistic scenarios in which we can gather observational data from individuals performing regular tasks that may conflict with security goals. In this work the DETER project team is itself using DETERLab as an environment in which to test user behavior safely in potentially risky situations.

Summarizing, human behavior cannot be ignored when designing or testing security systems. We have isolated a small set of repeatable behaviors and shown that we can capture them in the *dash* software agent platform. We continue experimental work with collaborators that validates the *dash* architecture, provides direction for extending it and also provides real examples and behavioral data that can be made available to other researchers wishing to include behavioral models in their research. *Dash*, along with the models and behavioral data that come from our current work, is now available within DETERLab for the larger community to use within their experiments.

5 A DETERLab Use Case

In this section we present a recent use case for the DETERLab facility, as the host testbed for a collaborative, multi-institution effort to prototype, evaluate, and demonstrate the capabilities of future large-scale distributed cyber-physical (CPS) systems. This work was carried out in the context of the *SmartAmerica Challenge.* The SmartAmerica Challenge, hosted by the 2013–2014 US Presidential Innovation Fellows program [39] and US National Institute of Standards and Technology (NIST), aimed to exhibit the potential benefits of these systems to society at large, in such areas as public safety, energy delivery and sustainability, healthcare, mobility, and overall quality of life.

A key goal of the Challenge was to catalyze advances in these domains by providing a showcase venue for researchers and innovators to demonstrate prototypes of future large-scale cyber-physical systems with the potential to achieve these benefits, leveraging collaborations and federations of existing CPS technologies, research projects, and test beds. By showcasing the potential of these systems in a highly visible, spotlight venue, the Challenge aimed to further accelerate the already-growing interest in CPS and "smart cities" across a broad mix of policy, standards, and business communities throughout the country.

DETERLab hosted a SmartAmerica team, "SmartEnergy", focused on electric power distribution and the smart grid. The team included academic institutions (USC/ISI, North Carolina State University, Iowa State University, UNC), a DOE-supported national research lab (NREL), and industry participants (Scitor Corporation, NAI). The team developed and presented two demonstrations during the Challenge. The first showed how a wide area, distributed, approach to grid monitoring yields national-scale resilience to large scale disruptions caused by malicious actors or unpredictable acts of nature, while the second explored the effects of distributed denial-of-service (DDOS) attacks on a smart power grid and some defenses against these effects. As an example use case for DETERLab, we discuss the first of these in more detail below.

Working in collaboration with North Carolina State's FREEDM Systems Center, the DETER project team showed how transitioning from traditional centralized wide-area monitoring mechanisms used in current state-of-art power systems to a more distributed architecture creates enhanced resiliency and robustness against cyber and physical disruptions.

We demonstrated a completely distributed communication and computational architecture for wide-area oscillation monitoring, created through the federation of three independent testbeds: a PMU-based[4] power grid simulation and modeling testbed at North Carolina State, an ExoGENI [40] testbed at UNC Chapel Hill, and the DETER cyber-security testbed at USC. In this architecture, estimators located at the control centers of various utility companies, emulated in the DETER and ExoGENI testbeds, run local optimization algorithms using local PMU data supplied from NC State simulators. The estimators communicate with other estimators across the federated testbeds to reach a global solution.

This system, once constructed, was subjected to a series of faults and attacks on the communication links. The team developed MAGI agents for estimators and used the DETERLab library of agents to create traffic, faults, and attacks. The MAGI orchestration framework was used to conduct several hundred experiment runs to demonstrate how the estimators can coordinate with each other to keep the grid running even if several of these links go down due to the attack.

For the SmartAmerica Challenge, the team used these capabilities to carry out and demonstrate a real-time simulation of a five-area model of the US west coast electrical grid.

Figure 5 is a snapshot of the visualization used to present this distributed estimator experiment. Dots on the map represent power grid substations. The experiment conducted within the simulation focused on a fundamental concern of wide-area power grid control, damping oscillations between groups of generators within the grid. Left unchecked, oscillatory power swings between groups of generation facilities can lead to catastrophic results such as blackouts. To conduct

[4]A phasor measurement unit (PMU) is a device that measures the power at different points in an electrical grid using a common time source for synchronization. PMUs are recognized as one of the most important measuring devices in future power systems [41].

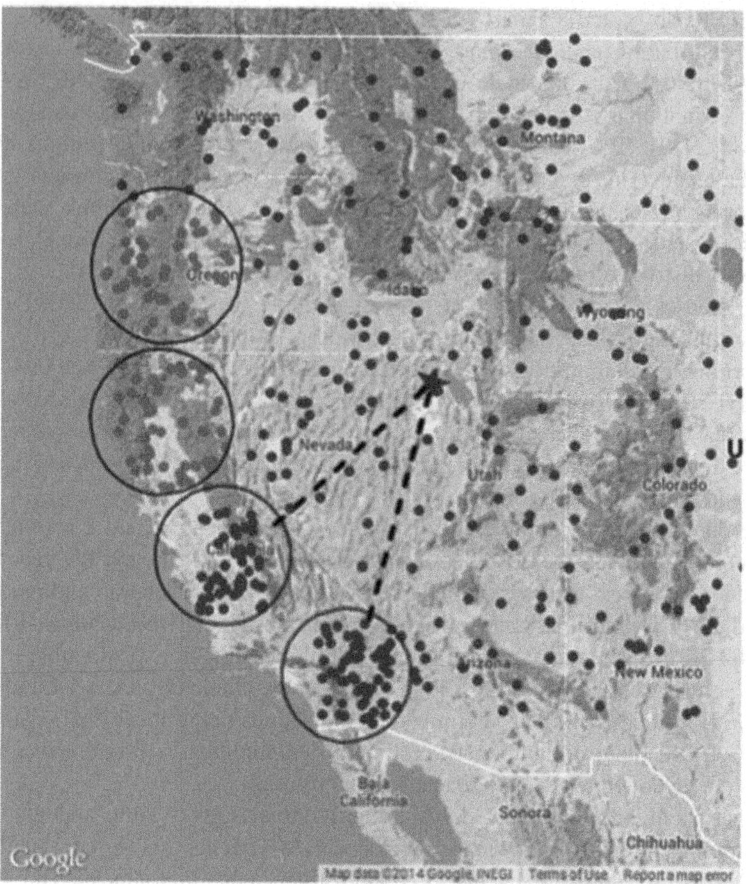

Fig. 5 Visualization of distributed estimator experiment

this sort of control, oscillations must first be detected, then the frequency and mode of oscillation determined and finally a control action that damps the oscillation must be computed.

The current state of the art is a *centralized* process. Of concern to power engineers, this centralization makes the process extremely vulnerable to DDoS attacks or network partitioning events. If the servers running detection and control algorithms are cut off from the rest of the grid, the stability of the system may be easily compromised.

As an alternative, the system demonstrated at Smart America distributed the detection and control logic across multiple servers that effectively partitioned the grid into individual control areas. The physical system under simulation was subjected to destabilizing oscillatory behavior while the cyber system supporting the control infrastructure was subjected to attack. In Fig. 5, the red dots indicate two control areas whose monitoring and control servers have been lost. Our demonstration showed that even in this case the other areas are able to coordinate, and then to compute a global damping solution that prevents the destabilizing oscillations from compromising the overall stability of the grid.

6 DETERLab in Education

Although DETERLab remains fundamentally focused as a research testbed, it has found an increasingly visible second role as a widely available shared resource for project-oriented, hands-on cybersecurity education. DETERLab-based exercises allow students to explore advanced concepts related to current cybersecurity research areas, as well as teaching practical skills such as configuring secure systems, performing forensics after an intrusion, and configuring security devices, all in a highly accessible and engaging environment. Experience has shown that DETERLab offers a number of potential advantages over the use of local university labs for such security exercises.

Among the most valuable of these are benefits related to sharing and reuse of educational materials. DETERLab's exercises are easily reusable across multiple sites and institutions, since they are all developed within a common, widely accessible environment. Because exercise configurations can be archived and reused by others, educators find it simple to share exercises with others and use DETER-Lab's exercises for repeat offerings courses. DETERLab's substantial resources are freely available to the educational community, offering significant benefits to community members unable to support local laboratories on their own. To realize these advantages, DETERLab has developed and now provides educators with a public portal to access shared exercises and to contribute new material to the larger educational community.

DETERLab also offers powerful functional and convenience advantages to educators. DETERLab offers substantial hardware resources and is able to support exercises and activities of significant, realistic size and complexity. The facility was designed explicitly to contain risk from experiments involving live malware and other risky behaviors, allowing these topics to be included in educational material. DETERLab's automated support for housekeeping functions such as environment configuration, OS load, and application installation simplifies exercise development and allows the educator to focus on essentials. Finally, DETERLab's remote access capabilities allow students to carry out lab work at their own schedule, from their dorms, labs, or homes.

To advance educational use of DETERLab, the DETER project has carried out activities in two distinct dimensions: development of education-specific technical capabilities and development of publicly available, shared curriculum materials. Initially, the project focused on extending the technical capabilities of the facility to meet unique educational needs. Mechanisms added to the testbed in this phase included administrative permission delegation and access control models suited to classroom use, the ability to batch-create and batch-delete student accounts based on data from external sources, and new allocation and scheduling mechanisms that manage and share resources across educational activities at different granularities (individual student, team, or class) without compromising the facility's research users.

Development of public teaching materials began in 2009, funded by a grant from the National Science Foundation. These materials include a collection of student exercises [42] that leverage use of DETERLab, together with teacher materials that aid adoption of these exercises in classes and enable teachers to troubleshoot and help their students with assignments. These materials are made available to all interested parties from DETERLab's Education Web page [43]. More recently, the project has initiated development of classroom and Web materials to be used in cybersecurity lectures. Also initiated in this timeframe is the development of a new category of lightweight class competitions—class capture-the-flag exercises or CCTFs [44]. Supported by Intel and the National Science Foundation, these materials are again publicly available on DETERLab's Education Web page.

The Project's focus on education, initiated in 2009 and continuing to the present writing, has led to a great increase in the size and diversity of DETERLab's educational users. Figure 6 shows the number of active classes per semester utilizing DETERLab since 2006. From a norm of less than five in 2009, usage has grown to a typical number of roughly 50 classes per semester at present. Over 11 years, DETERLab has been used by nearly 150 distinct courses (in some cases repeated over several semesters), by nearly 100 institutions in 20 countries. Overall, more than 4000 students have benefited from DETERLab in class settings during this time.

7 Looking to the Future

As described above and throughout this book, the 15 years from 2000 to 2015 have been a period of tremendous development in the reach and capability of infrastructures for experimentation in the cybersecurity, CPS, distributed systems and networked systems realms. Significant advances, both in experimental platform *technology* and in experimental platform *availability*, have changed the face of experimentation as a core element of research and education in these areas.

At the same time, much of this progress has focused on expanding the capabilities and availability of experimental infrastructures per se, with somewhat less attention paid to the broader ecosystems in which such infrastructure lives. As these core infrastructures reach a point of wide availability and continue to advance in functionality, it becomes a moment to pause and reflect on future directions.

Under the auspices of a study activity funded by the US National Science Foundation, a community effort facilitated by SRI International and USC/ISI engaged in such a reflection through 2014 and early 2015, producing as its key output a report on Cybersecurity Experimentation of the Future [4]. This report presents a

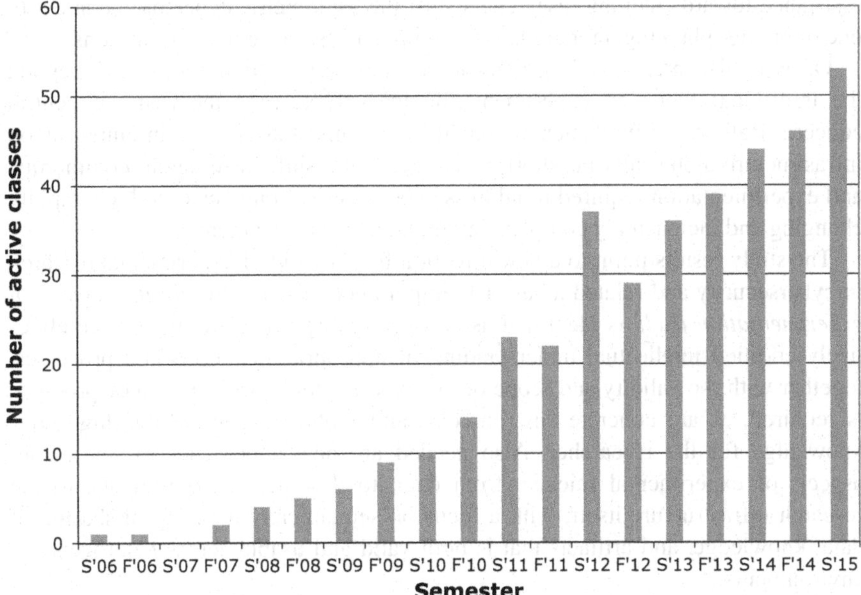

Fig. 6 Growth of DETERLab's use in education over time

strategic plan intended to catalyze generational advance in the field of experimental cybersecurity research, together with an enabling roadmap describing discrete steps towards realizing this goal. Although the charge to the group centered on experimentation in the cybersecurity arena, the breadth of the group's background and assessment, as well as the intersection of concerns across the cybersecurity, networking, distributed systems, and cyber-physical systems communities suggests the relevance of the group's conclusions to experimentation across each of these areas.

The overarching finding of this community activity is that transformational progress in three distinct, yet synergistic, areas is required to achieve the objective of generational advance:

1. Fundamental and broad intellectual advances in the field of experimental *methodologies and techniques*, with particular focus on methodologies targeting complex systems and human-computer interactions.
2. New approaches to rapid and effective *sharing of data and knowledge*, and information synthesis that accelerates multi-discipline and cross-organizational knowledge generation and community building.
3. Continued research, development, and deployment of *advanced, accessible experimentation infrastructures* and infrastructure capabilities.

Taken together, these areas, as embodied in the roadmap, paint a vision for a new generation of experimental cybersecurity research—one that offers powerful assistance towards helping researchers shift the asymmetric cyberspace context to one of greater planning, preparedness, and higher assurance fielded solutions.

Of particular interest is the conclusion of the study community to reach beyond the traditional focus on experimentation infrastructure as the issue of central concern. Rather, the fundamental conclusion of this study is that an emphasis on infrastructure alone falls far short of achieving the shift in research, community, and experimentation required to address cybersecurity concerns in today's rapidly changing and increasingly complex, interconnected environment.

The study results point to a new direction for the field of experimental research in cybersecurity and related areas. The importance of research *into the science of experimentation itself* is identified as an overarching need. Stronger, more effectively codified intellectual understanding of appropriate experimental processes, together with the validity and scope of the experimental conclusions they produce, is required, as are concrete environments and tools that capture this high-level knowledge for the researcher. Also needed are new approaches to sharing all aspects of experimental science, from data, to designs, to experiments, to the research infrastructure itself, with a focus on semantically meaningful sharing of data, knowledge, and artifacts that is both valid and usable across heterogeneous environments.

The conclusion is that strong, coupled, and synergistic advances across each of the areas outlined above—fundamental methodological development, fostering and leveraging communities of researchers, and in the capabilities of the infrastructure supporting that research—has the potential to transform experimental research in cybersecurity and related domains, and represents a generational grand challenge for the field in the coming years.

8 Conclusion

In this chapter we described the DETER Project, a research, development and experimental facility operation effort spanning over 10 years at the time of this writing. We discussed the historical progression of the project from initial facility development to advanced scientific instrument for cyber security experimentation, and discussed a sampling of key DETERLab technologies and their use. Finally, we briefly discussed new directions, looking beyond today's DETER project and towards cyber experimentation of the future.

References

1. Kessler, G.C.: Defenses against distributed denial of service attacks. Available at http://www.garykessler.net/library/ddos.html. Also included in Bosworth, S., Kabay, M.E., Whyne, E. (eds.) Computer Security Handbook. John Wiley & Sons, March 2014

2. Hardaker, W., Kindred, D., Ostrenga, R., Sterne, D., Thomas, R.: Justification and requirements for a national DDoS defense technology evaluation facility. NAL Report #02-052, Network Associates Laboratories, Rockville, MD, July 2002
3. NSF workshop on network research testbeds. Workshop Report, October 2002. http://gaia.cs.umass.edu/testbed_workshop
4. Balenson, D., Tinnel, L., Benzel, T.: Cybersecurity experimentation of the future (CEF): catalyzing a new generation of experimental cybersecurity research. Available at http://cyberexperimentation.org
5. Dingledine, R., Mathewson, N., Syverson, P.: Tor: the second-generation onion router. In: Proceedings of the 13th USENIX Security Symposium, August 2004
6. Ostrenga, R., Schwab, S., Braden, R.: A Plan For Malware Containment In The DETER testbed. In: Proceedings of the DETER Community Workshop on Cyber Security Experimentation and Test, August 2007
7. Lockwood, J.W., McKeown, N., Watson, G., Gibb, G., Hartke, P., Naous, J., Raghuraman, R., Luo, J.: NetFPGA—an open platform for gigabit-rate network switching and routing, MSE 2007, San Diego, June 2007. Further information available at http://netfpga.org
8. Pfaff, B., Pettit, J., Koponen, T., Jackson, E.J., Zhou, A., Rajahalme, J., Gross, J., Wang, A., Stringer, J., Shelar, P., Amidon, K., Casado, M.: The design and implementation of open vSwitch. In: Proceedings of the 12th USENIX Symposium on Networked Systems Design and Implementation (NSDI 2015), Oakland, CA, 4–6 May 2015
9. Benzel, T., Braden, B., Faber, T., Mirkovic, J., Schwab, S., Sollins, K., Wroclawski, J.: Current developments in DETER cybersecurity testbed technology. In: Proceedings of the Cyber Security Applications & Technology Conference for Homeland Security (CATCH 2009), March 2009
10. Benzel, T.: The science of cyber-security experimentation: the DETER project. In: Proceedings of the Annual Computer Security Applications Conference (ACSAC) '11, Orlando, FL, December 2011
11. Benzel, T., Wroclawski, J.: The DETER project: towards structural advances in experimental cybersecurity research and evaluation. J. Inform. Process. **20**(4), 824–834 (2012)
12. Mirkovic, J.: Benzel, T.V., Faber, T., Braden, R., Wroclawski, J.T., Schwab, S. The DETER project: advancing the science of cyber security experimentation and test. In: Proceedings of the IEEE HST '10 Conference, Waltham, MA, November 2010
13. Benzel, T., Braden, R., Kim, D., Joseph, A., Neuman, C., Ostrenga, R., Schwab, S., Sklower, K.: Design, deployment, and use of the DETER testbed. In: Proceedings of the DETER Community Workshop on Cyber Security Experimentation and Test, August 2007
14. Faber, T., Ryan, M.: Building apparatus for multi-resolution networking experiments using containers. ISI Technical Report ISI-TR-683 (2011)
15. Faber, T., Wroclawski, J., Lahey, K.: A DETER federation architecture. In: Proceedings of the DETER Community Workshop on Cyber Security Experimentation and Test, August 2007
16. Faber, T., Wroclawski, J.: A federated experiment environment for Emulab-based testbeds. In: Proceedings of Tridentcom (2009)
17. Mirkovic, J., Sollins, K., Wroclawski, J.: Managing the health of security experiments. In: Proceedings of the Cyber security Experimentation and Test (CSET) Workshop, July 2008
18. Schwab, S., Wilson, B., Ko, C., Hussain, A.: SEER: a security experimentation environment for DETER. In: Proceedings of the DETER Community Workshop on Cyber Security Experimentation and Test, August 2007
19. Viswanathan, A., Hussein, A., Mirkovic, J., Schwab, S., Wroclawski, J.: A semantic framework for data analysis in networked systems. In: Proceedings of the 8th USENIX Symposium on Networked Systems Design and Implementation, NSDI, April 2011
20. Wroclawski, J., Mirkovic, J., Faber, T., Schwab, S.: A two-constraint approach to risky cyber security experiment management. Invited paper at the Sarnoff Symposium, April 2008
21. Lahey, K., Braden, R., Sklower, K.: Experiment isolation in a secure cluster testbed. In: Proceedings of the Cyber security Experimentation and Test (CSET) Workshop, July 2008

22. White, B., Lepreau, J., Stoller, L., Ricci, R., Guruprasad, S., Newbold, M., Hibler, M., Barb, C., Joglekar, A.: An integrated experimental environment for distributed systems and networks. In: Proceedings of the 5th Symposium on Operating Systems Design & Implementation, pp. 255–270, December 2002
23. Bellard, F.: QEMU, a fast and portable dynamic translator. In: Proceedings of the USENIX 2005 Annual Technical Conference, April 2005, pp. 41–46
24. OpenVZ Containers Website, http://openvz.org
25. Gardenghi, L., Goldweber, M., Davoli, R.: View-OS: a new unifying approach against the global view assumption. Lecture Notes in Computer Science, vol. 5101/2008, Computational Science—ICCS 2008. Further information available at http://virtualsquare.org
26. Faber, T., Schwab, S., Wroclawski, J.: Authorization and access control: ABAC. In: The GENI Book, Springer International Publishing Switzerland, 2016, doi:10.1007/978-3-319-33769-2_10
27. Ansible Documentation. http://docs.ansible.com/, version of January 2016.
28. Chef Documentation. https://learn.chef.io/, version of January 2016
29. Fabric Documentation. http://www.fabfile.org, version of January 2016
30. Pupper Documentation, https://puppetlabs.com/, version of January 2016
31. Bratman, M.: Intention, plans, and practical reason (1987)
32. Johnson-Laird, P.: Mental models (1983)
33. Wash, R.: Folk models of home computer security. In: Proceedings of the Sixth Symposium on Usable Privacy and Security (SOUPS) (2010)
34. Stanovich, K.E.: Who is Rational? Studies of Individual Differences in Reasoning. Psychology Press, Hove (1999)
35. Blythe, J., Camp, J.L.: Implementing mental models. In: Proceedings of IEEE Symposium Security and Privacy Workshops (SPW), pp. 86–90 (2012)
36. Blythe, J.: A dual-process cognitive model for testing resilient control systems. In: Proceedings of Resilient Control Systems (ISRCS), 2012 5th International Symposium, 2012
37. Kothari, V., Blythe, J., Smith, S., Koppel, R.: Agent-based modeling of user circumvention of security. In: Proceedings of the 1st International Workshop on Agents and CyberSecurity (2014)
38. Blythe, J., Koppel, R., Smith, S.W.: Circumvention of security: good users do bad things. IEEE Security & Privacy **11**(5), 80–83 (2013)
39. Presidential Innovation Fellows Program, https://www.whitehouse.gov/innovationfellows, version of October 2015
40. Baldin, I., Chase, J., Xin, Y., Mandal, A., Ruth, P., Castillo, C., Orlikowski, V., Heermann, C., Mills, J.: ExoGENI: a multi-domain infrastructure-as-a-service testbed. In: GENI: Prototype of the Next Internet. Springer (2016)
41. Nuqui, R.F.: State estimation and voltage security monitoring using synchronized phasor measurement. Ph.D. Dissertation, Virginia Polytechnic Institute, Blacksburg, VA, July 2001. "Simulations and field experiences suggest that PMUs can revolutionize the way power systems are monitored and controlled" (via Wikipedia)
42. Mirkovic, J., Benzel, T.: Teaching cybersecurity with DETERLab. IEEE Security and Privacy Magazine, January/February 2012, vol. 10, no. 1, pp. 73–76 (invited paper)
43. DETERLab Education Web page, http://education.deterlab.net, version of October 2015
44. Mirkovic, J., Peterson, P.A.H.: Class capture-the-flag exercises. In: Proceedings of the USENIX Summit on Gaming, Games and Gamification in Security Education (2014)

ORBIT: Wireless Experimentation

Dipankar Raychaudhuri, Ivan Seskar, and Max Ott

Abstract This chapter presents an overview of the ORBIT testbed for wireless experimentation. ORBIT is an NSF supported community testbed for wireless networking which provides a variety of programmable resources for at-scale reproducible experimentation as well as real-world outdoor trials. The centerpiece of the ORBIT testbed is the 400-node "radio grid" deployed at the Rutgers Tech Centre facility in North Brunswick, NJ. The radio grid enables researchers to conduct reproducible experiments with large numbers of wireless nodes over a wide range of radio technologies, densities and network topologies. The ORBIT system architecture is outlined and technical details are given for the radio grid's key hardware and software components including the radio node platforms, software defined radios, RF measurement system, switching and computing backend and the ORBIT management framework (OMF). Additional ORBIT resources including special purpose sandboxes and the outdoor WiMax campus deployment are also described. The experimental interface and scripting tools for running an experiment on ORBIT are outlined, and examples of a few representative experiments which have been run on the ORBIT testbed are summarized. The chapter concludes with a view of ORBIT's evolution and future upgrade path along with an explanation of how it links to the overall GENI project.

1 Introduction

Experimental verification of emerging network protocols and software at scale is a critical need for the computer science and engineering research communities. Basic research on networking has historically relied on simulation tools (such as ns2 [1] and Opnet [2]) but there is a growing consensus on the need to supplement simulation results with "real-world" experiments that capture the complexities of user traffic, router implementations and physical-layer channel properties. This is

D. Raychaudhuri (✉) • I. Seskar
WINLAB, Department of ECE, Rutgers University, 671 Rt. 1 South, North Brunswick, NJ 08902, USA
e-mail: ray@winlab.rutgers.edu

M. Ott
NICTA, Sydney, Australia

© Springer International Publishing Switzerland 2016
R. McGeer et al. (eds.), *The GENI Book*, DOI 10.1007/978-3-319-33769-2_4

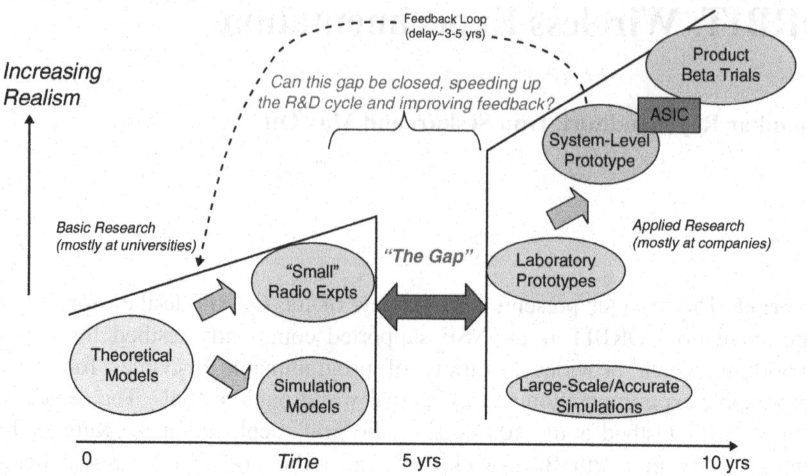

Fig. 1 Typical wireless R&D cycle with "gap" between academic research and product trials

particularly true for wireless networks in which protocol performance depends strongly on time-varying radio channel properties due to signal fading, interference and mobility. While many research groups do attempt to build prototype systems for experimental evaluation of protocols, even a small network requires considerable effort and cost, without providing a sufficient level of scale and reproducibility necessary to support the scientific process. Figure 1 below looks at the typical R&D life cycle for wireless systems going from theory to small-scale experiments and simulations to laboratory prototypes and system level prototypes for field trial. This is the methodology that has been followed for development of previous generations of wireless technology such as 3G cellular, but there is a significant gap [3] between small experiments that can be carried out by academic research groups and larger technology trials mostly limited to large companies. This gap leads to longer product cycles, now about 7–10 years for wireless, and it would be desirable to improve research methodologies to close the gap at least to an extent and thereby speed up the innovation cycle.

The above considerations have motivated development of various shared network research testbeds [4–14] as multiuser experimental facilities intended to enable at-scale and realistic experimentation by early stage/academic researchers. These testbeds were conceived as community infrastructure which would reduce the barrier of entry for networking researchers and make it possible for them to carry their results to a more comprehensive level of validation and evaluation than currently possible. The first generation of testbeds for network architecture and protocol evaluation include PlanetLab [12] for wired overlay networks, Emulab [13] for wired/wireless protocol emulation, and the ORBIT testbed [13] for at-scale wireless experimentation which is the focus of this chapter.

Fig. 2 The ORBIT radio grid deployed at Rutgers Tech Center Facility

The ORBIT open access testbed for next-generation wireless networking at Rutgers University was originally developed under the National Science Foundation's NRT (Network Research Testbeds) program during the period 2003–2007 to address the challenge of supporting realistic and reproducible wireless networking experiments at scale [15]. A notable result of the NRT project was the successful construction and community release of the 400-node ORBIT radio grid facility (Fig. 2) that enables remote users to conduct reproducible experiments with large numbers of programmable wireless nodes set up to emulate specifiable real-world network topologies and mobility scenarios [16–18]. The ORBIT radio grid was first made available to research users on an informal basis in October 2005, and since then, has rapidly become an NSF sponsored community resource for evaluation of emerging wireless network architectures and protocols. There are currently over 1000 registered users worldwide who have conducted a cumulative total of over 60,000 × 1–2 h experiments on the testbed to date—the user base includes both expert users who work with network drivers, as well as researchers moving up from ns2, ns3 or Opnet based simulation to ORBIT emulation. The ORBIT architecture (both in terms of hardware and software aspects) has also served as a reference model for development of wireless capabilities in the GENI national-scale networking research infrastructure project [19].

In the sections that follow, we describe the design requirements for ORBIT, the testbed's overall architecture and key hardware/software components in detail. A few examples of research experiments which have been run on the ORBIT testbed are given to illustrate its capabilities. The chapter concludes with a perspective on future evolution of the testbed.

2 Design Requirements

The development of a general-purpose open-access wireless multi-user experimental facility poses significant technical challenges related to recreating the wireless networking environment in a reproducible manner. The main design problem is that of capturing physical world characteristics such as wireless device density, location and mobility and the associated radio frequency channel as a function of time. In addition, wireless systems tend to exhibit complex interactions between the physical, medium access control and network layers, so that strict layering approaches often used to simplify wired network prototypes cannot be applied here.

Some of the basic characteristics of radio channels that need to be incorporated into a viable wireless network testbed include:

- Physical world realism in terms of density of devices and radio propagation
- Ability to incorporate user location characteristics and mobility patterns
- Radio physical layer bit-rates and error-rates reflecting actual signal-to-noise and interference phenomena at the receiver
- Realistic medium access control (MAC) layer reflecting interfering nodes, carrier sensing thresholds, traffic prioritization, and so on
- Programmability of radio PHY and MAC layer to the extent feasible

A flexible wireless network testbed must be capable of realistically incorporating the above characteristics, while permitting experimentation with a range of radio physical and MAC layers, network topologies and protocol options. For the testbed to be useful, it should be scalable and cover a sufficiently broad range of wireless network research problems that might be anticipated over the next 5–10 years.

Some examples of systems or protocol designs that help to understand the overall design space under consideration are:

- Heterogeneous wireless networks, including high-speed cellular (2.5G, 3G or 4G), wireless local-area networks (802.11a, b, g, n etc.), and wireless personal area networks (Bluetooth, 802.15.3, etc.). The testbed should support evolution in radio technology from 802.11n to 802.11ac or ad, and should accommodate new technologies such as LTE and 5G as they emerge. The framework should permit the user to set up a complete system with a mix of radio technologies and experiment with protocols for mobility control, resource management, inter-network handoff, quality-of-service (QoS), service security, etc.

- Future mobile Internet service scenarios based on cellular and WiFi radio access in conjunction with new protocols for authentication, mobility management delay-tolerant, routing, multi-homing, content delivery, etc. The testbed should permit the end-user to define various usage scenarios and network topologies, and then experiment with new protocols for discovery, routing, mobility management, caching, security, etc.
- Mobile ad hoc networks (MANET), typically based on 802.11 × WLAN radios, extended to support multi-hop ad hoc routing protocols such as AODV [20] and DSR [21]. The testbed needs to support a moderately large number of ad hoc wireless nodes capable of running alternative MANET protocols under consideration. Accurate modeling of node mobility is a key requirement for this scenario.
- Wireless personal area networks (WPAN) in which multiple radio technologies such as 802.11× and 802.15.3 are used to create a high-performance network for use in the home or office, as a body-area network or as a desktop network for connecting multiple computing devices. The testbed should thus support multiple radio technologies and permit the user to specify typical usage scenarios and to experiment with protocol alternatives.
- Internet-of-Things (IoT) applications in which large numbers of low-power sensor/actuator devices are interconnected for applications such as environmental monitoring, industrial control/logistics or security. In addition, such networks may require evaluation of alternative networking and software models.
- Dynamic spectrum access or cognitive radio networking scenarios in which software-defined radios (SDR) are used to form adaptive systems capable of sharing radio spectrum via sensing, interference avoidance, networked collaboration and other techniques. Research on this emerging class of radio systems involves scale (in terms of number of nodes), programmable radio functionality and accurate spectrum measurement.

In addition to the above, an open-access community testbed has several general service and user level requirements including:

- Open access over the Internet with a user portal which provides experimenter access and tools, documentation needed to use the testbed effectively
- Flexibility/programmability of experiments across a range of radio technologies, network topologies, layer 2, 3 and 4 protocols and usage scenarios
- Shared use across a large community of researchers, implying the need for resource scheduling and/or virtualization of hardware and computing resources
- High-level experiment scripting language and component libraries for ease of use by users
- Experiment execution tools for code downloading, run time controls and status monitoring
- Tools for measurement and automated collection of data from experiments

The ORBIT testbed was designed to meet or approach each of the design goals outlined above. The testbed is centered around the concept of the "radio grid" which

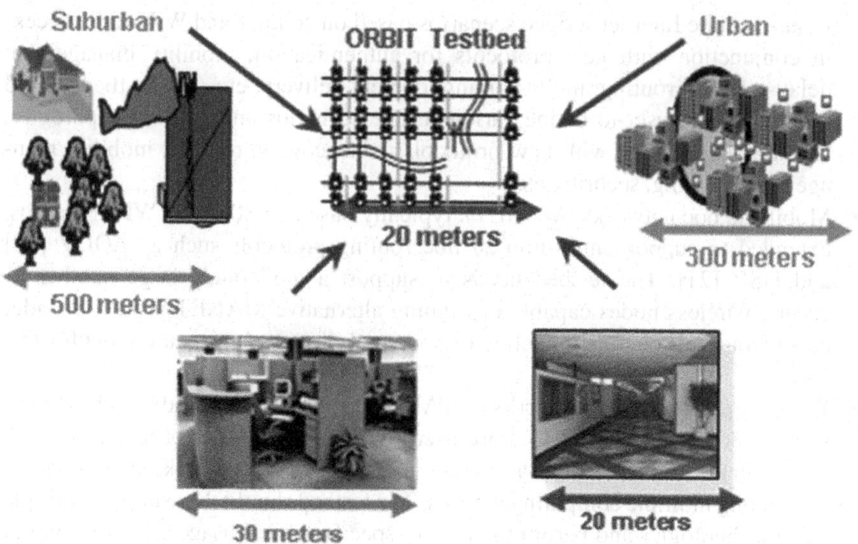

Fig. 3 Topology mapping concept in ORBIT radio grid

enables reproducible and at-scale experimentation of with a set of wireless nodes physically located in a two-dimensional space. The key idea of the radio grid is the mapping of a real-world wireless network topology to a set of nodes on the grid with noise injection used to emulate increasing distance (i.e. "stretching space"). Once the right set of nodes and noise injection settings are selected, it is possible to run experiments on the radio grid and obtain reproducible results across multiple runs. As shown in Fig. 3, this allows the same 20×20 radio grid to be used for emulating various topologies including indoor office, suburban and urban outdoors via proper selection of nodes and noise parameters. Note that in contrast to outdoor wireless experiments, the resulting emulation is reproducible in the sense that the same parameter settings should result in similar system performance over multiple runs.

In addition to the radio grid, the testbed has two other kinds of facilities to support the full life cycle of an experiment. For early stage experimenters, the testbed provides a number of "sandboxes" intended for initial development and code debugging thus reducing the load on the main radio grid. For late stage experimenters, the testbed includes an "outdoor ORBIT" deployment that consists of open/programmable access points, base stations and client devices intended to support real-world end users with mobile devices and applications. The outdoor testbed (Fig. 4) makes it possible to smoothly migrate network code from emulation to real-world evaluation all within the same software framework provided by ORBIT. Further details about the testbed design and key components are given in the sections that follow.

Fig. 4 Outdoor ORBIT deployment on Rutgers Busch Campus

3 ORBIT Testbed Technical Details

3.1 ORBIT System

The ORBIT testbed consists of a set of programmable wireless networking resources including sandboxes for small-scale experimentation and debugging, the ORBIT radio grid emulator for large-scale reproducible experimentation, and the ORBIT outdoor network for increased realism and real-world end-users. All testbed resources share a common software framework in order to enable researchers to migrate code from one resource to another as the scale or realism of the experiment increases.

The ORBIT large-scale radio grid emulator [14–18] consists of an array of $\sim 20 \times 20$ open-access programmable nodes each with multiple 802.11a,b,g or other (Bluetooth [22], Zigbee [23], GNU [24]) radio cards. The radio nodes are connected to an array of backend servers over a switched Ethernet network, with separate physical interfaces for data, management and control. Interference sources and spectrum monitoring equipment are also integrated into the radio grid as shown in Fig. 5. Users of the grid emulator system log into an *experiment management server* which executes experiments involving network topologies and protocol software specified using an ns2 like scripting language. A *radio mapping algorithm* [25, 26] which uses controllable noise sources spaced across the grid to emulate the effect of physical distance is used to map real-world wireless network scenarios to specific nodes in the grid.

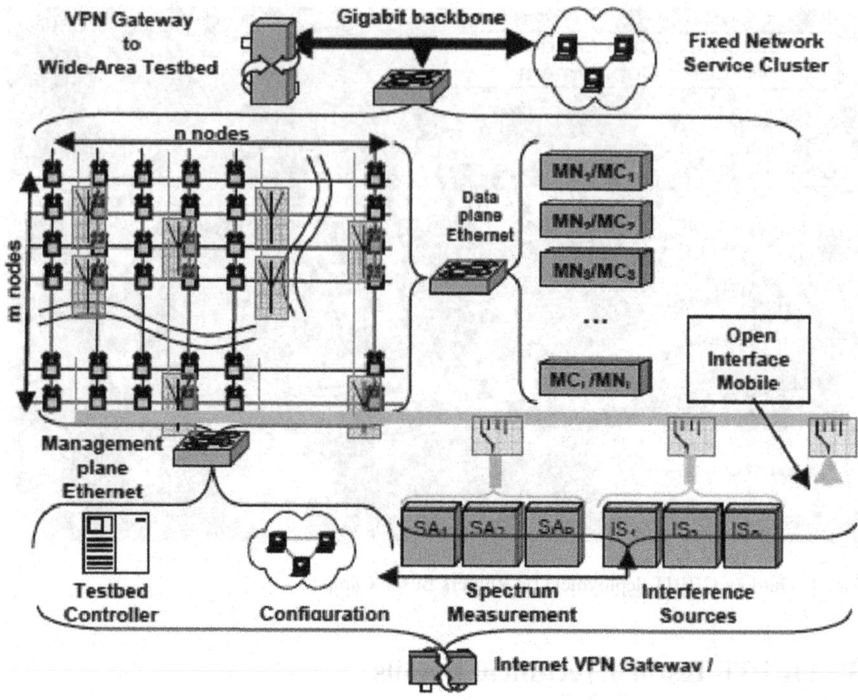

Fig. 5 ORBIT radio grid architecture

The radio grid outlined above was implemented in several phases starting in September 2003. First, a 64-node prototype grid was set up in an available laboratory environment with the objective of validating the design and testing its experimental capabilities with a few internal users. After validating feasibility with the small-scale prototype, the 400-node ORBIT radio grid was installed in a custom-built facility (shown earlier in Fig. 2) with ~5000 sq-ft of RF shielded space and ceiling pre-wiring to support power and Ethernet connectivity to the 20×20 grid. The ORBIT nodes in this setup are suspended from the ceiling with grid spacing of 1 m; each node is equipped with two 802.11a,b,g radio cards and optionally Bluetooth, Zigbee and GNU radios. The facility also includes a separate control room and a separate switching and server rack area for the testbed's backend equipment. The floor area is kept clear for use by robotic nodes or sensor deployments needed for certain mobility and Internet of Things experiments respectively.

The 400-node radio grid testbed was first released for experimental use in October 2005. Since then, the ORBIT testbed has gone through two equipment upgrade cycles in order to meet emerging experimental requirements and stay on the "Moore's Law" curve for computing and switching platforms. The first set of upgrades (carried out between 2008–2010) focused on adding USRP [27] and

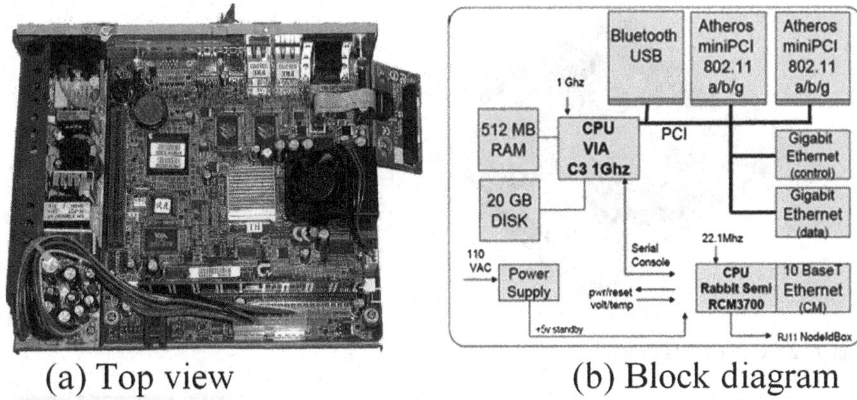

Fig. 6 First Gen ORBIT radio node. (**a**) Top view. (**b**) Block diagram

USRP2 software-defined radios (SDR) to the grid, adding programmable OpenFlow switches to the testbed backend, and introducing outdoor WiMax capability. The second upgrade (2010–2014) was aimed at replacing the original ORBIT radio nodes with current i7 platforms with CPU speeds sufficient for wideband SDR support while also introducing a small number of second generation SDR units.

In the following sections, we provide an overview of the main hardware and software components that constitute the ORBIT testbed.

3.2 ORBIT Hardware

ORBIT radio node platform: ORBIT radio nodes serve as the primary computing platform for user experiments. The Gen 1 node (Fig. 6) was custom designed around a 1 GHz VIA C3 processor with 512 MB of RAM and 20 GB of hard disk. Each platform also includes two wired 1000 BaseT Ethernet interfaces for experimental data and control. In addition, each node contains a CM (chassis manager) module with Ethernet connectivity to be used for remote monitoring and rebooting of the node independent of experiments running on the main processor. A batch of 500 commercial grade ORBIT nodes with this design were manufactured with the help of an industry partner and deployed on the 400-node radio grid. These so called "ORBIT yellow nodes" were also used as vehicular nodes and for external ORBIT deployments in the field or at partner sites.

During 2010–2014, an infrastructure upgrade project resulted in a new second generation node (see Fig. 7) based on off-the-shelf mini-ITX form factor motherboards with a combination of interfaces (mini-PCI and mini-PCI Express and USB 3.0) that are used to plug in standard wireless devices (Wi-Fi, WiMAX, Bluetooth etc.). In addition, platforms are equipped with at least one full high-speed bus interface enabling expansion with high performance radio devices. Each

Fig. 7 2nd Gen ORBIT Node

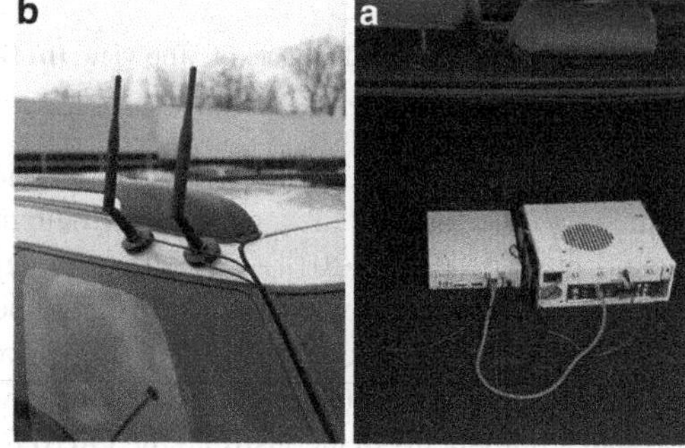

Fig. 8 Outdoor vehicular deployments with second gen ORBIT node. (**a**) New ORBIT node in vehicle trunk. (**b**) External antenna installation

of the nodes is equipped with at least one 802.11 a/b/g/n device and one combo 802.11/802.16 device (Intel WiMAX/Wi-Fi Link 5350 device).

A vehicular use version of the second generation ORBIT radio node has also been produced using LV-67B and F processor boards. Figure 8 shows a typical vehicular node deployment for use with the outdoor network. This version of the radio node includes both WiFi and WiMax interfaces as needed to support heterogeneous wireless experiments using outdoor WiFi AP's and the open WiMax base station.

Chassis Manager: The ORBIT Chassis Manager (CM) is a simple, reliable, platform-independent subsystem for managing and autonomously monitoring the status of each node in the ORBIT network testbed. As shown in Fig. 9, each ORBIT grid node consists of one Radio Node with two radio interfaces, two Ethernet interfaces for experiment control and data, and one Chassis Manager (CM) with a separate Ethernet network interface. The Radio nodes are positioned about 1 m apart in a rectangular grid. Each CM is tightly coupled with its Radio Node host. CM subsystems are also used with non-grid support nodes. The CMC is the control and monitoring manager for all CM elements of ORBIT. An "Experiment Controller" (EC), also referred to as the "node handler", is the ORBIT system component that configures the grid of Radio Nodes (through the CMC service) for each experiment. The non-Grid elements of the ORBIT lab are not normally in the management domain of the EC.

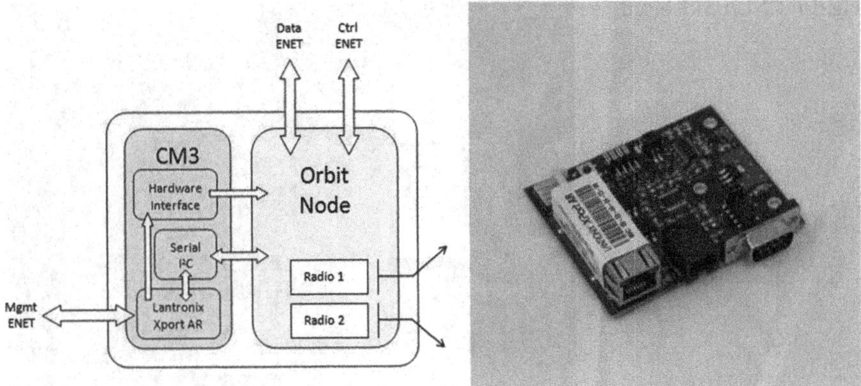

Fig. 9 ORBIT Node Chassis Manager (CM) and photo of the CM board

Each Chassis Manager is used to monitor the operating status of one node. It can determine out-of-limit voltage and temperature alarm conditions, and can regain control of the Radio Node when the system must return to a known state. Managing a system in this manner reduces the human resources needed to monitor hundreds of nodes. The CM subsystem also aids debugging by providing telnet to the system console of the Radio Node, as well as telnet access to a CM diagnostic console.

802.11a,b,g,n Radios: ORBIT uses Atheros and Intel 802.11a,b,g wireless cards for many short-range radio experiments. The drivers for these cards are compatible with rest of the ORBIT software framework and are regularly upgraded with new open source releases from the Linux community. Additionally, ORBIT facilitates asynchronous *get* and *set* operations of PHY and MAC parameters on a per-packet basis in order to provide experimenters with control over key parameters.

RF instrumentation: The ORBIT grid includes equipment for measurement of radio signal levels and supports injection of various types of artificial RF interference (white noise, colored noise, microwave oven like noise etc.) inside the grid. The interference generator is based on the RF Vector Signal Generator while the spectrum measurements are done using Vector Signal Analyzers. The noise injection framework is integrated into the management framework and runs as a service on ORBIT which has also been used as a means of topology creation for evaluating performance of multi-hop ad hoc networks [28].

GNU USRP2 Radios: The GNU Universal Software Radio Peripheral (USRP/USRP2) software radio board has been interfaced with the ORBIT node via the USB 2.0 interface (see Fig. 10). The USRP provides a set of RF daughter boards to perform analog RF up and down conversion, 1 million gates of FPGA, typically used to convert to and from complex baseband, 4 high speed A/Ds (64 MS/s 12-bit), 4 high speed D/As (128 MS/s 14-bit) and a USB 2.0 controller chip. This GNU/URSP setup is available on ~50 ORBIT radio grid nodes and a sandbox, and this capability is being used extensively by the ORBIT experimenter community.

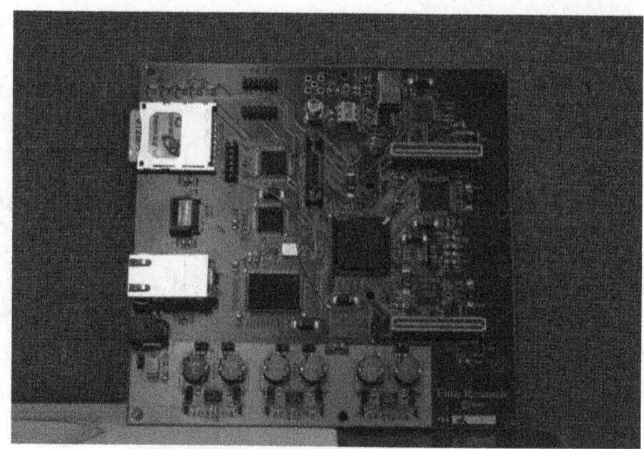

Fig. 10 GNU/USRP board

Fig. 11 2nd Generation SDR Radio Board

Second Gen Wideband Software Radios: A number of second generation SDR's (see Fig. 11) have also been deployed on the grid to support wideband experiments beyond the capability of USRP radios. The SDR hardware being used was designed, integrated and validated at U Colorado and WINLAB with support from the GENI program [19]. The SDR module consists of two boards—an off-the shelf Avnet FPGA system card, and wideband SDR radio front ends custom-made by Radio Systems Technology. The system card supports variety of Xilinx Virtex 5 components (LX50T, SX50T,SX95T, LX110T and LX155T) two programmable LVDS clock generators, EXP expansion slot, 64 MB DDR2 SDRAM, 256 MB DDR2 SODIMM RAM, 16 MB Flash, RS-232 serial port, Cypress USB 2.0 Controller, Two GbE PHYs and multiple GTP Interfaces.

WiMAX Radios: 4G cellular capability was added to the ORBIT testbed starting in 2009 using WiMAX (802.16e) base stations from NEC Corp. and mobile WiMAX client devices. The NEC WiMAX base-station hardware (photo in Fig. 12) is a 5U rack based system which consists of multiple Channel Cards (CHC) and a Network Interface Card. The shelf can be populated with up to three channel cards, each supporting one sector for a maximum of three sectors. The BS operates in

Fig. 12 WiMAX equipment deployed for ORBIT outdoor testbed

the 2.5 GHz or the 3.5 GHz bands and can be tuned to use either 5, 7 or 10 MHz channels. At the MAC frame level, 5 ms frames are supported as per the 802.16e standard. The TDD standard for multiplexing is supported where the sub-channels for the Downlink (DL) and Uplink (UL) can be partitioned in multiple time-frequency configurations. The base-station supports standard adaptive modulation schemes based on QPSK, 16QAM and 64QAM. The interface card provides one Ethernet Interface (10/100/1000) which is used to connect to a high performance controller PC. The base station has been tested for radio coverage and performance in realistic urban environments—typical coverage radius is ~3–5 km, and peak service bit-rates achievable range from 15–30 Mbps depending on operating mode and terrain.

The 802.16e base station allocates time-frequency resources on the OFDMA link with a number of service classes as specified in the standard—these include unsolicited grant service (UGS), expedited real time polling service (ertPS), real-time polling service (rtPS), non-real time polling (nrtPS) and best effort (BE). The radio module as currently implemented includes scheduler support for the above service classes in strict priority order, with round-robin, or weighted round-robin being used to serve multiple queues within each service class. The GENI slice scheduling module implemented in the external PC controller is responsible for mapping "*Rspec*" requirements (such as bandwidth or delay) to the available 802.16e common packet layer services through the open API. Slices which do not require bandwidth guarantees can be allocated to the nrtPS class, while slices with specific bandwidth requirements can be allocated to the UGS category.

Figure 13 shows the software architecture of the WiMax base station which has been designed to support virtualization of radio resources. The controller provides support for multiple slices assigned to the GENI WiMAX node. Each slice runs within its own virtual machine [29] (using software such as UML—User Mode Linux). Each VM is capable of providing multiple virtual interfaces, so that programs loaded on a slice that runs within a virtual machine can emulate its own

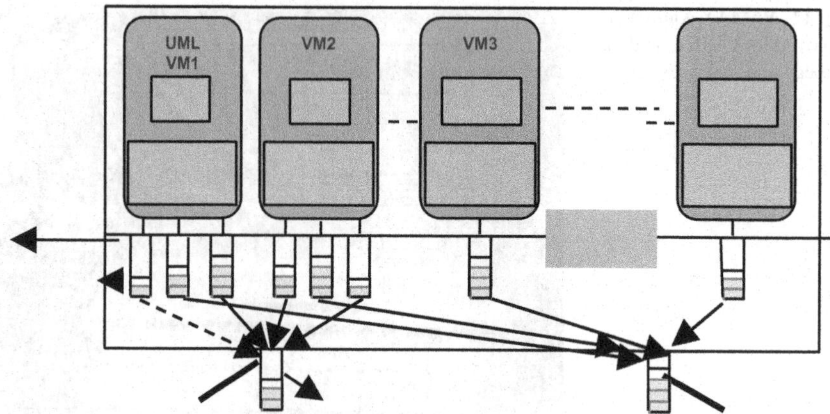

Fig. 13 GENI Base Station Node Controller (GBSN) architecture

router and perform IP routing. Virtual interfaces are mapped to physical interfaces based on the next hop for a virtual interface. The controller receives IP packets from the base station on the R6+ interface. When a packet is received, it is forwarded to the appropriate slice for further processing. The outgoing IP packets from a slice are placed on queues specific to a virtual interface. Outgoing packets on virtual interfaces mapped to the layer 2 interface of the WiMAX base station are tagged so that they can be assigned traffic class and bandwidth parameters (BE, ertPS, rtPS etc.) [30] as determined by the flow CID (connection ID).

The L2 base station control software on the external controller provides APIs to both control Layer 2 parameters and also to receive L2 specific information from the base station. An experimenter's program within a slice (obtained through the OMF control interface) can use these APIs to modify L2 parameters as well as receive L2 specific data both at load time and also at run time.

Open LTE Base Stations: In view of the importance of LTE for cellular access, an ongoing upgrade to ORBIT aims to migrate current WiMax equipment in ORBIT to LTE and also provide experimenters with the capability of implementing "soft LTE" on software defined radios in the grid. In keeping with the programmability and virtualization requirements of ORBIT and GENI testbeds, open LTE base stations under development are also based on the open VSwitch model used for WiMax. The basic idea is to replace all the LTE GW, MME, Handoff and other functionality with software modules in the Aggregate Manager with southbound interfaces to the base station hardware and northbound interfaces to the access network. The plan calls for software-based LTE implementations, either open source (e.g. bellard.org [31]) or commercial (e.g. Amarisoft [32]) on either USRP2 platform or second generation SDR platforms described earlier.

Backend Servers and Network Equipment: Backend servers in ORBIT are used for a number of important functions. They provide the web-based experimenter interface used to access ORBIT, support multiple user accounts, serve as repositories

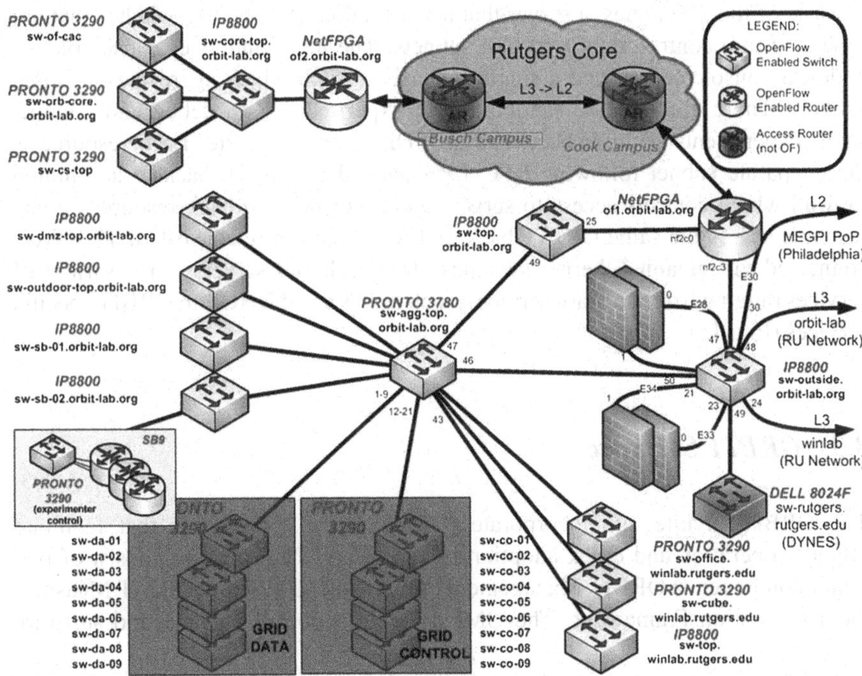

Fig. 14 ORBIT switching infrastructure with OpenFlow upgrade

for code and data, and run the experiment controller software necessary for experiment imaging, execution and measurements. Backend servers can also be used by experimenters to accelerate protocol stack computations carried out on the ORBIT radio node. Also, servers are used to support experiment virtualization and federation with wired network testbeds such as GENI, Emulab, and PlanetLab as required.

ORBIT uses an SDN switching backplane based on OpenFlow switches including the NEC IP8800 and Pronto. The programmable OpenFlow feature helps to improve manageability of the backend infrastructure and can be used to implement priorities between control and data traffic switching, etc. OpenFlow is also a good platform for adding virtualization to the radio grid testbed since each virtual network can be dynamically configured with its own VLAN service, etc. Figure 14 shows the overall connectivity of the ORBIT testbed to the Rutgers core network and on to the Internet2/GENI national backbones. ORBIT currently has a 1 Gbps dedicated connection to Internet2 via the Magpi/Internet2 PoP in Philadelphia and an increase to 10 Gbps is planned.

ORBIT Radio Grid Resources: ORBIT testbed resources currently consist of 436 nodes, 26 Servers, and 48 OpenFlow (SDN) Ethernet switches. Nodes, servers, and switches accessible to the experimenters are grouped into ORBIT resources which are referred to as "grid", "sb1" through "sb8" and "outdoor". The main grid consists of the 400 nodes, a server that acts as a console, and 30 switches that are separated into control, data, and CM subnets. The nine sandboxes consist of two nodes, a console server, and a single managed switch which aggregates all three subnets. There are separate sandboxes for experimenters to debug and conduct small experiments with WiMax/cellular, SDR, OpenFlow, etc. Each resource is on a separate subnet following RFC 1981 and all are routed back to a common firewall which controls access to servers and other infrastructure resources. Each resource shares the same ORBIT back-end which currently consists of 17 servers connected via gigabit Ethernet switches. The back-end servers run a variety of services ranging from standard services, such as DNS and DHCP, to ORBIT specific services (Fig. 15).

3.3 ORBIT Software

The ORBIT architecture incorporates key software components that facilitate efficient operation and quick adoption by end users. The software consist of two major components: ORBIT Management Framework (OMF) and ORBIT Measurement Collection Framework. The OMF is a collection of services and software

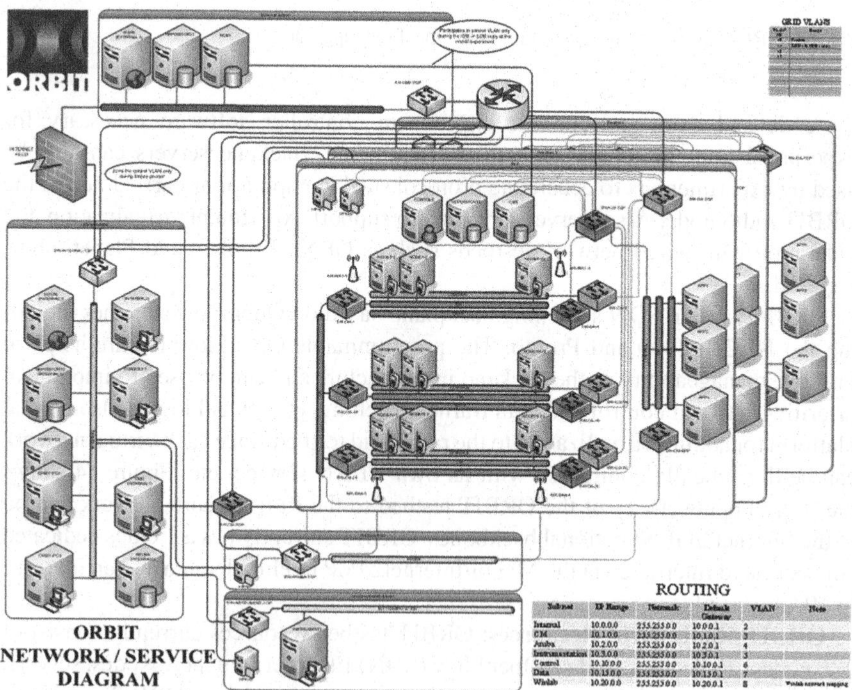

Fig. 15 ORBIT resources overview

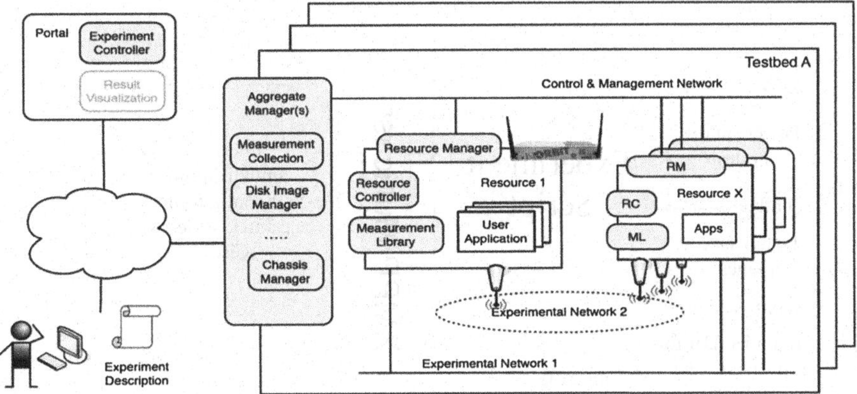

Fig. 16 ORBIT software overview

components that is used for testbed management and experiment coordination. Major OMF components are: (1) collection of Aggregate Managers (AMs) that are in charge of various system level components (like Chassis Manager Controller (CMC) AM, Disk Image Manager, etc.); (2) ORBIT User portal (Control Panel) that is facilitating user interaction with various AMs and is also displaying the state of the testbed; (3) Experiment Controller (EC) that is in charge of orchestrating and executing experiments; and (4) Resource Controller (RC) that is running on each node—see Fig. **16**.

The *Chassis Manager Controller (CMC)* controls and monitors all CMs in the ORBIT grid and hence facilitates remote controllability of the ORBIT testbed. The *ORBIT User Interface* handles secure user access and scheduling of experiments on the 400-node main grid and the smaller sandboxes. Users schedule experiment time on testbed resources, using a web interface access through www.orbit-lab.org. Users have complete control over testbed resources during their assigned time slot. This is achieved by means of firewalls and sets of security rules that govern all traffic to and from each testbed. The *Experiment Management Service* software (Fig. 17) is an optional feature provided to ease usability by certain classes of experimenters, particularly those who are trying to larger-scale experiments or are migrating up from ns2/ns3 simulation to more realistic emulation studies. This part of the software was designed to provide higher level abstractions to experimenters in order to make it easier to use the testbed. The core of this service is the *Experiment Controller (EC)* which orchestrates the described experiment by issuing appropriate commands to the specified nodes, and keeping track of their execution. It communicates with the Resource Controller (RC) *t* software component that resides on each node. The experiment is specified in the form of a *Ruby* script with testbed specific extensions (OEDL—ORBIT Experiment Description Language). During the execution, EC interprets the OEDL statements in the script and disseminates relevant execution instructions over logical transport channel (xmpp/multicast) to the specified nodes. The EC-RC pair (a.k.a. *NodeHandler-NodeAgent)* thus enables

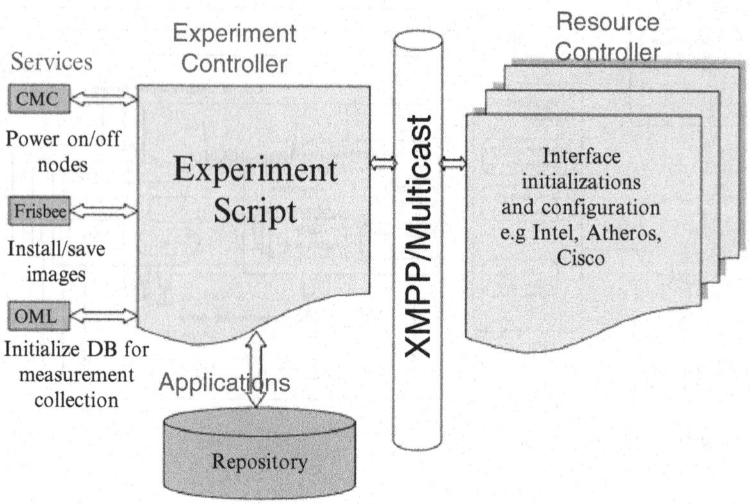

Fig. 17 Experiment management service

automation of experiments, and configuration of wireless and application parameters. An *experiment server* maintains the *Ruby* scripts of experiment descriptions. ORBIT's *Frisbee server* [33] loads the hard disk images onto all the specified nodes using a scalable multicast protocol.

ORBIT's *Measurement Collection Framework* (Fig. 18) is a special feature designed to facilitate collection of experimental results. Data collection efficiency is important for experiments involving large numbers of nodes in which traffic from measurement traces can overload the control network and management servers. The core of this framework is the ORBIT Measurement Library (OML). Over the years, OML went through multiple iterations in order to better support scaling. In the original implementation, the collection service used a 2-tier database with a fast Berkeley DB as front-end, and standard mysql as backend. The latest implementation supports two database backends: SQLite and PostgreSQL. The *ORBIT Measurement Library* (OML) [18] provides API's that may be incorporated in the user's experiment software to automate result collection. The data collection operates in a separate wired subnet that does not interfere with the actual experiment. All OML measurements are stored in a database at runtime for subsequent post-processing by the experimenter, while freeing the testbed facility for the next experiment. Users can also specify the granularity of measurements (sample or time based) and some preprocessing of results (sum, max, min, averages etc). In addition to direct database access on the repository server, experimenters can use the Result AM and it's REST API for remote access to experiment results.

In addition to RC, ORBIT node software consists of vendor provided device drivers some of which are also accessible through a high-level API called *Libmac* (Fig. 19) [34] which provides application-level access to PHY/MAC parameters.

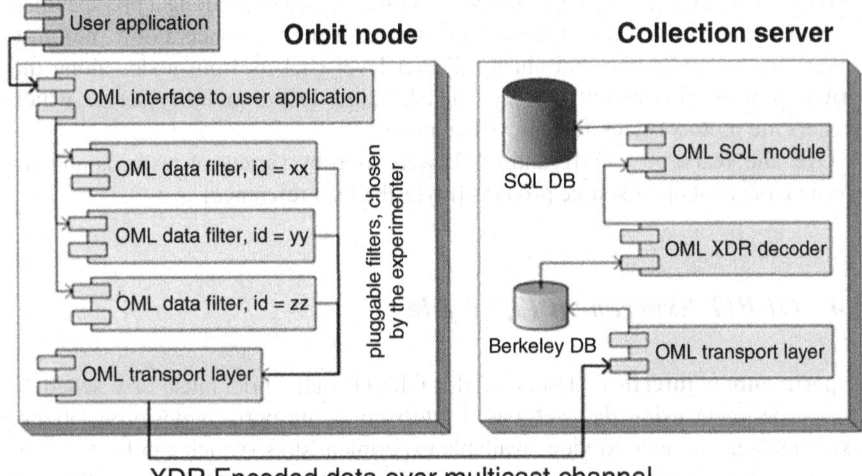

Fig. 18 ORBIT measurement framework

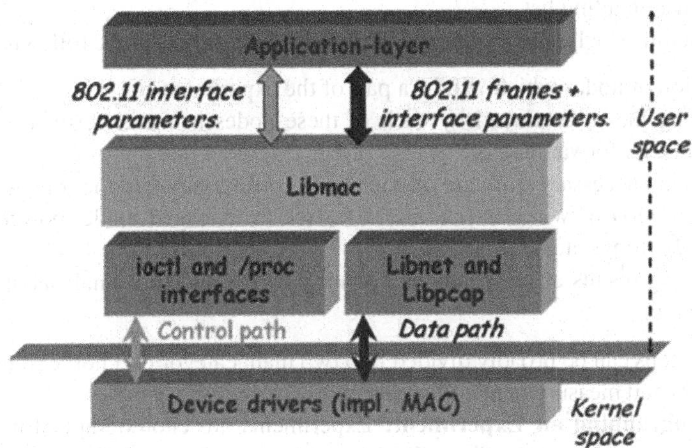

Fig. 19 Node Software architecture and Libmac

Using this library, applications may inject and capture MAC layer frames, manipulate wireless interface parameters at both aggregate and per-frame levels, and communicate wireless interface parameters over the air on a per-frame basis.

In addition, the *ORBIT Traffic Generator* (OTG) is a tool for generating configurable traffic used to load the network and measure performance. Its default operation includes the OML library to collect common cross-layer parameters on a per-packet basis during experiment run-time.

Additional experimental support services in ORBIT include a *topology configuration service* which sets noise injection parameters on the grid to create specified topologies [25, 26] Another available feature is *mobility emulation* [35] based on a software switching approach that moves a "virtual" radio node along a pre-specified trajectory. The virtual radio node is implemented using the concept of a "mobility server" in the grid's backend cluster. Driver level packets from nodes along the virtual path are forwarded over switched Ethernet to the mobility server which anchors the protocol stack for that virtual node.

Over the years, both OMF and OML were used by variety of testbeds and are now independent open-source projects [mytestbed.net reference].

3.4 ORBIT Experiment Life-Cycle

Experimenters interface: Users of the ORBIT radio grid must first schedule their experiment using the web-based interface at the portal www.orbit-lab.org. Experimenters are able to view available experiment slots (which can be 1–2 h in duration) on a web page and make a reservation for their desired slot (Fig. **20**). The current allocation policy is first-come-first-served (FCFS), and during busy periods, some users will negotiate priority requests or slot swaps on an informal basis using the orbit-user mailing list.

Once a time slot has been obtained, an experiment consists of the following steps:

1. Selection of nodes which will be a part of the experiment
2. Selecting the roles played by each of these nodes in the experiment (sender, receiver, AP, forwarder etc)
3. Deploying necessary software on each node corresponding to the role they play
4. Configuration of wireless interfaces (ad-hoc or managed mode, power levels, channel settings etc)
5. Collecting results at run-time and collating them (statistical analysis or simple time plots)

These steps can be broadly divided into two main categories: choreographing an experiment and measurement collection.

Choreographing an Experiment: Experiments are choreographed using the experiment management service (*EC* software) outlined earlier. The user specifies the experiment using a Ruby script which defines nodes, their respective roles and the specific measurements to be collected. The script also includes parameters and traffic settings, which can be set for the whole experiment, or can be dynamically varied during the course of an experiment. It is noted that ORBIT also allows expert users to bypass the *OMF* and write their own console software for experimental control and node setup.

A sample experimental script is shown in Fig. 21 on the left. In this experiment, node 1–2 sends UDP datagrams of 1024 bytes at the rate of 300 Kbps to the receiver 1–4. The wireless settings use 802.11b with the receiver acting as an AP (this is done using the "Master" mode on the card) and the sender is the client (using the setting "Managed" on the card. Note how the actual interfaces are abstracted (w0) to hide the hardware specific interface nomenclature (e.g. Atheros based cards show up as *athX* whereas Intel and Cisco cards show up as *ethX*).

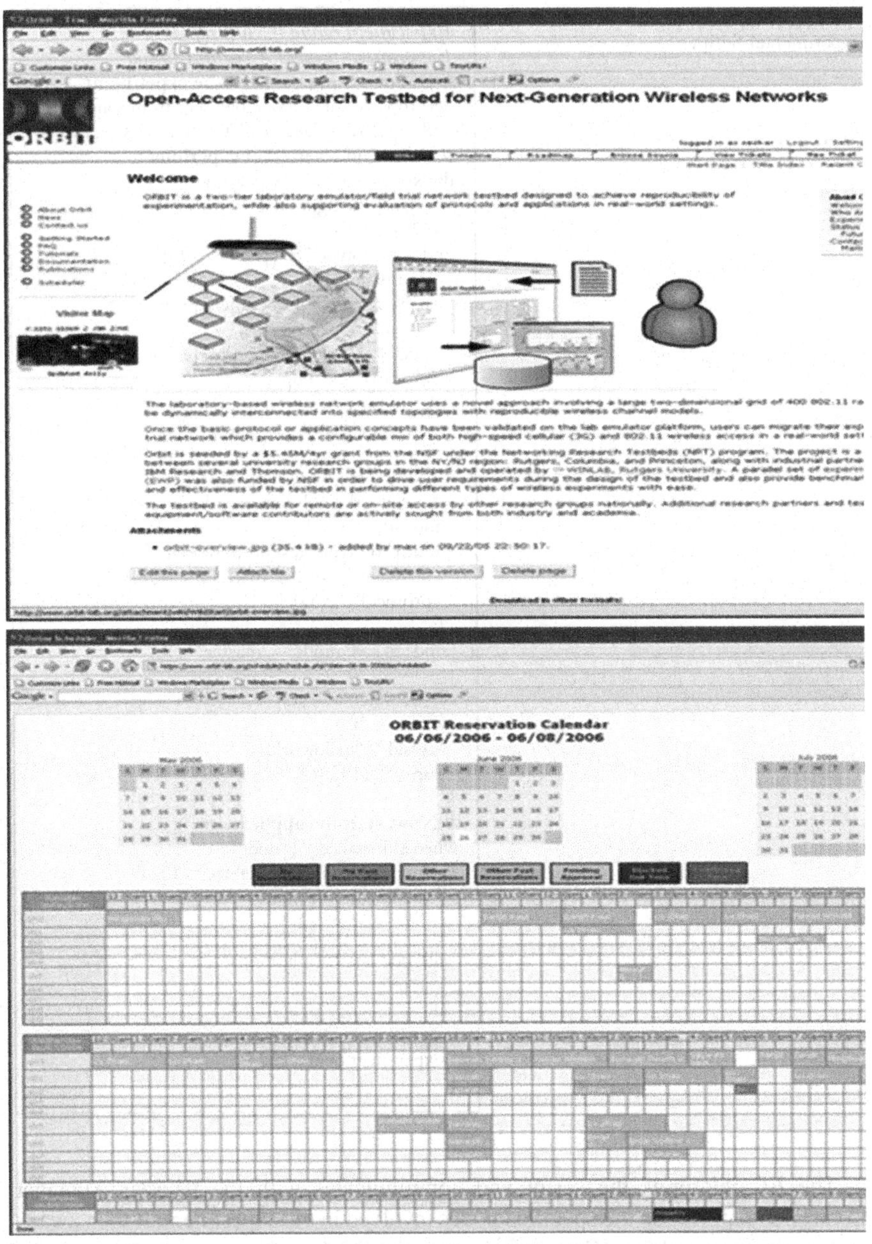

Fig. 20 Experiment scheduling interface on ORBIT web portal

Fig. 21 Sample Experimental Script

```
Experiment.name = "tutorial-1"
Experiment.project = "orbit:tutorial"
# Define settings used in the experiment
defProperty('rate', 300, Kbps sent from sender')
defProperty('packetSize', 1024, 'Packet size, bytes)
# Define nodes used in experiment
defNodes('sender', [1,2]) {|node|
# assume the right image to be on disk
  node.image = nil
  node.prototype("test:proto:sender", {
   'destinationHost' => '192.168.1.4',
   'packetSize' => Experiment.property
               ("packetSize"),
   'rate' => Experiment.property("rate"),
   'protocol' => 'udp'
  })
  node.net.w0.mode = "managed"
}
defNodes('receiver', [1,4]) {|node|
  # assume the right image to be on disk
  node.image = nil
  node.prototype("test:proto:receiver" , {
   'hostname' => '192.168.1.4',
   'protocol' => 'udp'
  })
  node.net.w0.mode = "master"
}
allNodes.net.w0 {|w|
  w.type = 'b'
  w.essid = "helloworld"
  w.ip = "%192.168.%x.%y"
}
# Now, start the application
whenAllInstalled() {|node|
  Experiment.props.packetSize = 1024
  Experiment.props.rate = 300

  allNodes.startApplications
  wait 60
  allNodes.stopApplications
  wait 10
  Experiment.done
}
```

The *EC* interprets the script and communicates with the Chassis Manager Controller to power on the specified nodes involved in the experiment. It then awaits nodes to boot up and the *RC* to report back to the *EC*. In software terminology, this is like a barrier implementation that waits until all nodes have reported back to the *EC*.

Once the *RC* have reported back, the *EC* then requests the *RC* to apply initial configuration settings for the wireless interfaces. Note that the *RC* deduce which wireless card is installed on the nodes, load the appropriate driver module and issue commands to configure the same. The testbed currently supports Atheros-based and Intel-based 802.11a/b/g cards. Thus, the *RC* provides a simple abstraction of a wireless interface to the experimenter. After the interfaces have been configured, the *EC* directs each node to launch the application based on its specified roles.

Experimental Measurements: During the reserved slot, users are allowed access to the ORBIT console (through SSH) and can run their experiments using scripts similar to the example outlined above from the command-line. As explained earlier, the measurement capability is implemented via the OML software [18]. In order to use the measurement framework, the user only needs to invoke simple library calls.

By enabling a type-safe transport layer, OML also supports reporting of standard data types such as *int*, *float*, *double*, *string*, etc. The measurements can be either time based or sample based. In addition, the OML framework allows run-time filters to be applied to either of these measurement techniques to report minimum, maximum, average or sum of time-based or sample-based measurements. A separate run-time and post-experiment database allows users to view results during experiment run-time as well as to archive them for future retrievals and offline analysis as shown in Fig. 22. The reader is referred to [36] for examples of specific ORBIT experiments and their results.

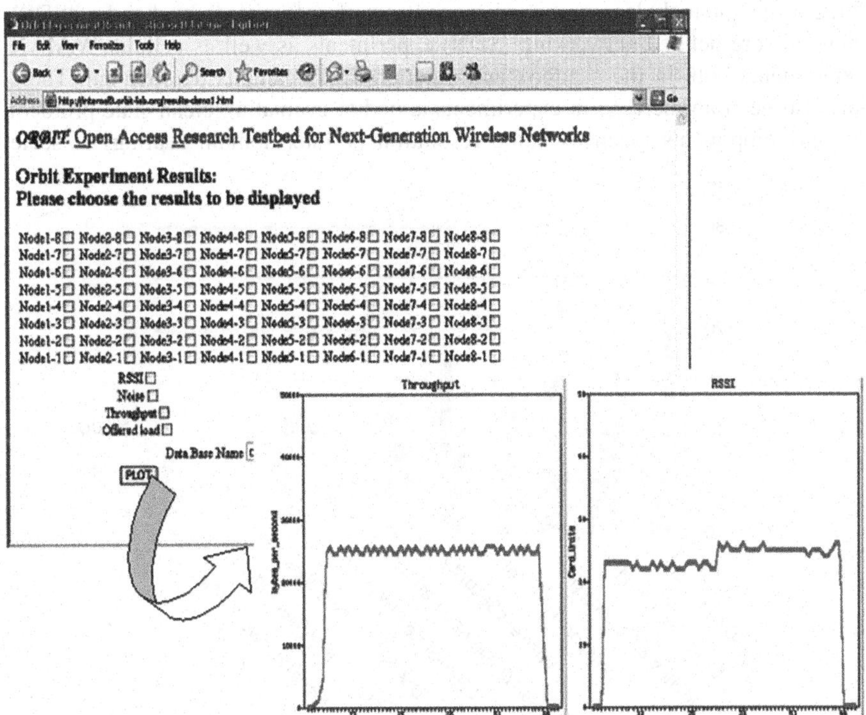

Fig. 22 Runtime measurement interface

4 Experimental Research Enabled by ORBIT

ORBIT serves as a community testbed for a broad cross-section of researchers working on next-generation wireless networks and the future Internet. The range of research topics currently supported by ORBIT testbed experiments include:

- Next-generation wireless networks (dynamic spectrum, cognitive radio, mesh, cross-layer)
- Future internet architectures (hybrid networks, mobility protocols, delay-tolerant networks, content services, transport-layer protocols)
- Security in wireless systems (secure routing, security with physical layer reinforcements, denial-of-service in wireless networks)
- New wireless and mobile services (peer-to-peer, content caching, automotive safety, location-aware applications, mobile social networks)

The actual mix of experiments running on the ORBIT facility is user-driven, and over time may be expected to reflect changing priorities due to emerging architectural themes or technologies of interest to the research community. Figure 23 provides a summary of the experiment mix running on the testbed by category in ~2007, 2012 and 2015 (projected). From the figure, it can be seen that the 2012 snapshot includes a growing percentage of software-defined radio (SDR) and software-defined networking (SDN) experiments as well as cellular/WiMAX experiments both in the sandbox and outdoor GENI campus deployment. There has also been an increase in experiments aimed at evaluating clean-slate protocol design components (such as name resolution or inter-domain routing) at scale

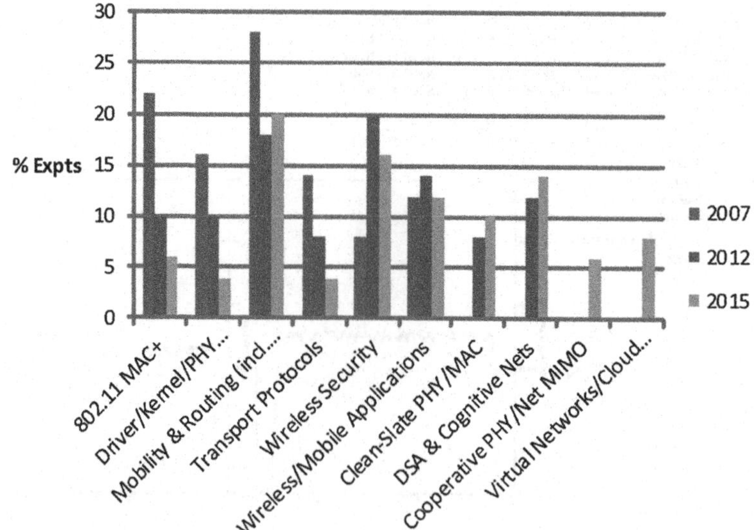

Fig. 23 Summary of Experiments Run on ORBIT

with ∼100's to ∼1000's of routers. Wireless network security, dynamic spectrum access and radio resource management for dense WiFi networks also continue to be important classes of experiments generated by the wireless research community. Looking at the estimates for 2015, we anticipate an increase in several categories including dynamic spectrum/SDR, cooperative PHY/network MIMO, clean-slate Internet architecture, virtual networks and cloud mobile applications.

While it is not possible to provide an exhaustive list of experiments, a few representative samples of experiments run on the ORBIT testbed are given in the subsections below:

4.1 Radio Channel Signature Based Encryption

Secure communication over wireless networks is an important research area. Because of the broadcast nature of the wireless channel, it is possible for an adversary to hear communication between two entities. Therefore, strong encryption is necessary to encrypt all data transmitted wirelessly. Encryption requires a secret key exchange between the two parties interested in communicating. However, an adversary might be able to "hear" the key exchange process. In [37], the authors propose a novel scheme for extracting the encryption key from the time-varying, stochastic mapping between the transmitted and received signals, without an explicit key establishment process. This mapping is both, location-specific and reciprocal, i.e., the mapping is the same whether Alice or Bob is the transmitter. Further, this time-varying mapping de-correlates over distances of the order of half a wavelength which is 6.25 cm for a 2.4 GHz transmission. Thus the reciprocity and fast de-correlation properties of fading channels, allow us to generate a common, secret cryptographic key at Alice and Bob such that Eve gets no information about the key.

Extensive measurements were performed on the ORBIT testbed using available software-defined radio (SDR) nodes located at three points on the grid (i.e. corresponding to Alice, Bob and Eve). Nodes constantly send probe messages to detect the time varying nature of their channels. Received signal strength indicators (RSSI) reported by the receivers (see Fig. 24) were used to derive estimates of the scalar channel response. The observed RSS values confirm the de-correlation and reciprocity properties. The level crossing and excursion profiles of the RSSs are then analyzed to extract a sequence of bits that form the encryption key. Experiments confirmed that the level crossing technique can provide a reliable method for extracting secret bits at about ∼1 secret bit per second in an indoor environment [37]. The ORBIT radio grid setup enables PHY-assisted security experiments of the kind presented above.

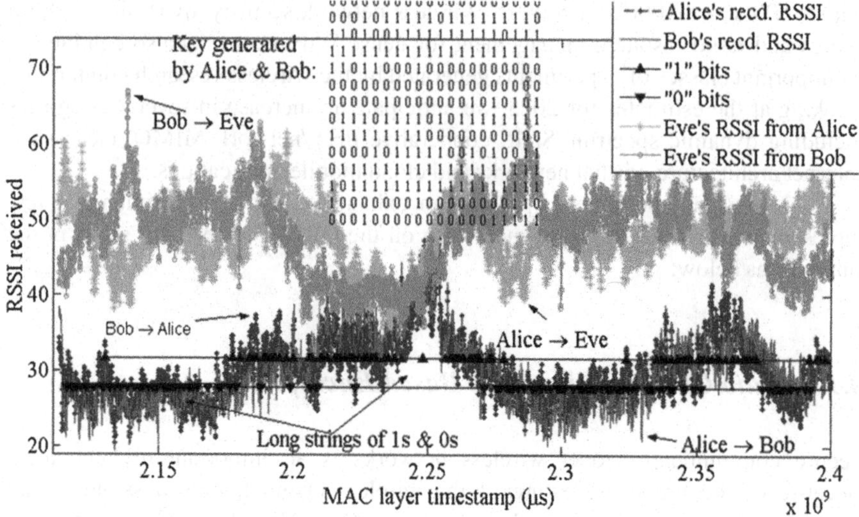

Fig. 24 Received signal strength indicator (RSSI) traces from channel based key extraction experiment

4.2 Dynamic Spectrum Coordination in Dense Multi-Radio Environments

Dynamic spectrum coordination between heterogeneous wireless systems is an important open research problem in view of recently opened "TV white space" band [38] and other proposed shared-use unlicensed bands [39]. One approach to inter-system coordination is the use of a common spectrum coordination channel (CSCC) [40] which provides a control mechanism for neighboring radios to identify each other and coordinate spectrum usage. The use of a CSCC protocol enables radios to use a variety of spectrum coordination algorithms that are designed to control interference and maintain system throughput.

A prototype implementation of the CSCC concept was built on ORBIT, using multi-radio nodes with Bluetooth and Wi-Fi radios. The goal was to verify the use of CSCC control messaging to support dynamic spectrum coordination policies, and to estimate the overheads due to control. The experimental setup on ORBIT used an 802.11b radio for exchange of CSCC control, and a second Wi-Fi or Bluetooth radio for data transfer. In the first set of experiments, radio nodes in a moderately dense environment (~10 nodes per 100 sq-m) engage in pairwise UDP streaming sessions or TCP file transfers using either Bluetooth or Wi-Fi in the same unlicensed 2.4 GHz band. The experimental setup (i.e. ORBIT node with dual radios, and sample topology) is shown in Fig. 25.

A number of alternative spectrum sharing policies were evaluated, showing significant system capacity gains when some form of dynamic coordination is

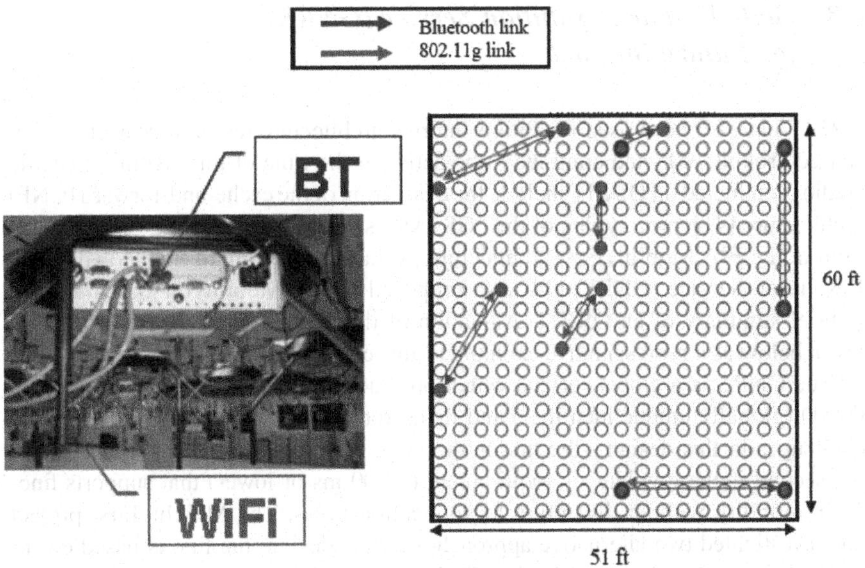

Fig. 25 Experimental Setup in ORBIT for CSCC Experiments with Bluetooth and Wi-Fi Nodes

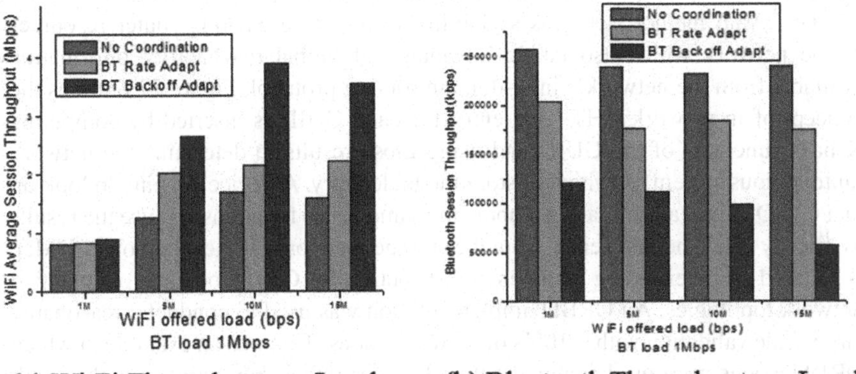

Fig. 26 Sample Experimental Results for CSCC Experiment

implemented. Coordination algorithms considered include "BT backoff" in which Bluetooth defers in time whenever a Wi-Fi transmission is detected, as well as "BT rate adapt" in which permissible Bluetooth rate is computed from neighboring Wi-Fi node information received over the control channel. From the sample results given in Fig. 26, it is observed that significant improvements to Wi-Fi throughput are achieved with CSCC coordination, and the gain can be as large as $\sim 2\times$ over the case with no coordination.

4.3 Global Name Resolution Service (GNRS) for Future Internet

ORBIT has also been used for future Internet architecture research projects which started around 2005 and continue to be active at the time of this writing. Specific studies conducted on ORBIT include the evaluation of the cache-and-forward (CNF) architecture [41], prototyping of the "GSTAR" storage-aware routing protocol [42] used in the FIA MobilityFirst architecture [43] and validation of the Global Name Resolution Service (GNRS) [44] also proposed for the MobilityFirst architecture. A brief summary of an ORBIT evaluation of the MobilityFirst GNRS at scale is given below as a representative example of this class of experiments.

The GNRS is a global service which provides the dynamic binding between a GUID (globally unique identifier) and its current network locators. There are two challenges in this design—the first is the large scale with billions of objects, and the second is the low latency requirement (~100 ms or lower) that supports fine-grain mobility without disruption to application flows. The MobilityFirst project has investigated two alternative approaches to this design: the first is based on in-network router distributed hash table (DMap [44]), and the second uses a distributed overlay service with locality-aware replication for each name and partitioning across names so as to optimize latency while respecting capacity constraints [45].

The DMap method achieves scalability by utilizing existing router resources in the network while also taking advantage of global reachability information available from the network's inter-domain routing protocol. Figure 27 outlines the concept of in-network DHT—the entry for each GUID is inserted by computing K hash functions of the GUID and using those results to determine the network (autonomous system) at which to store the table entry. A device wishing to look up that GUID entry can similarly compute the same K has functions and use the results to directly fetch the table entry from the storage locations. The evaluation of DMap is focused on determining the latency distribution for GUID lookups for specified network topologies. An ORBIT implementation was used to conduct a reasonably large-scale validation with ~100's of network nodes. This was an example in which ORBIT nodes were used as wired network nodes rather than wireless nodes, with the SDN interconnection fabric used to set up different topologies. The Jellyfish [46] model was used to map realistic Internet topologies on to the ORBIT grid - each grid node represents an AS, and the number of nodes for different layers and links between different layers is proportional to Internet AS-level topology. Internet datasets from DIMES [47] and CAIDA [48] were used to generate realistic network topologies. Figure 28 shows some experimental results obtained from such ORBIT experiments, indicating 90 % latencies in the range of 200 ms, somewhat higher than predicted by simulation. Further experiments aimed at validating the GNRS over a global scale network are currently being conducted using GENI.

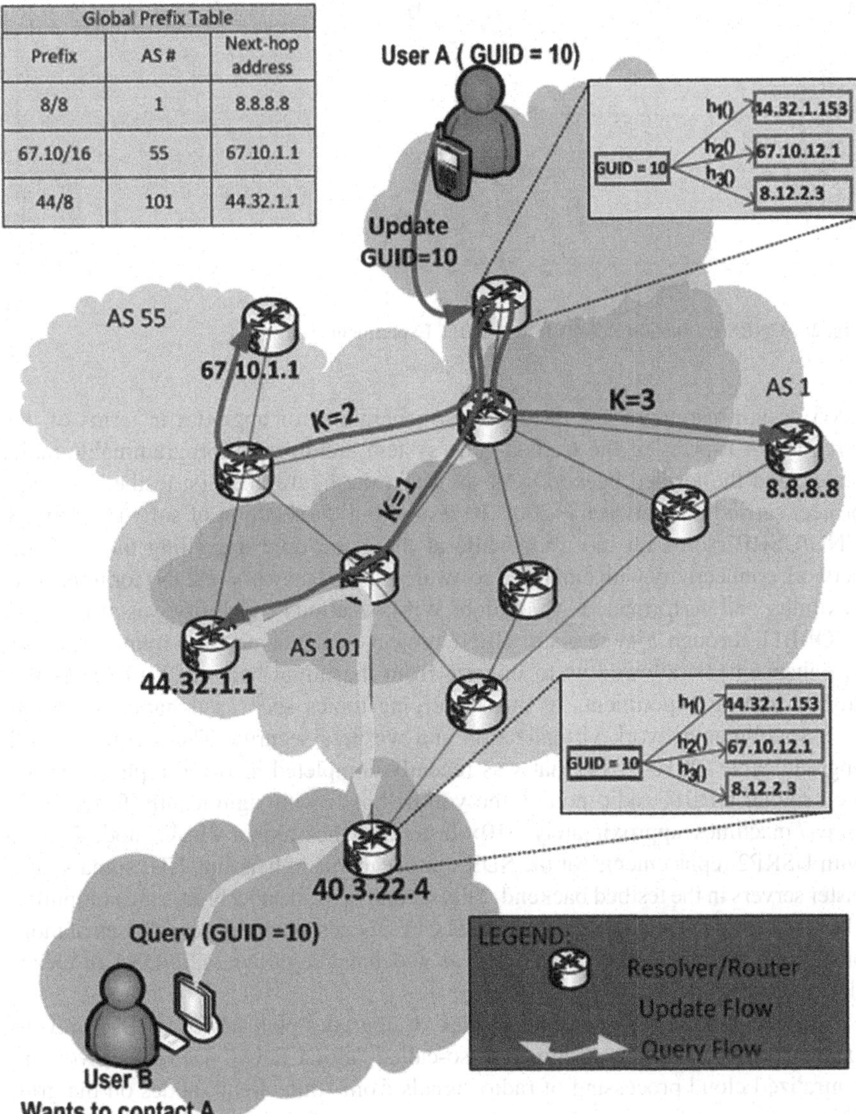

Fig. 27 DMap GNRS example with K=3 DHT

5 ORBIT Evolution and Future Upgrades

The ORBIT project team has followed a phased long-term strategy for continually upgrading the testbed equipment from both Moore's Law obsolescence and experimental requirements perspectives. Figure 29 shows how the original circa

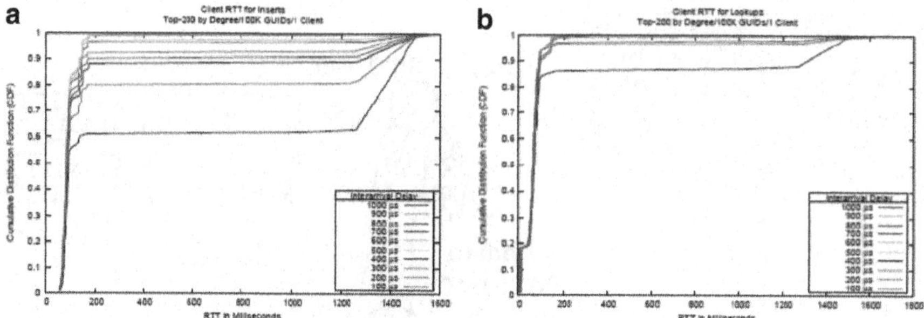

Fig. 28 GNRS Evaluation Results from ORBIT Experiments

2005 design has now gone through two generations of upgrades in terms of the three major aspects of the testbed—the system architecture, programmable radio nodes, and the testbed backend. As shown in the figure, the first testbed upgrade project carried out between ~2007-10 resulted in the addition of software-defined GNU/USRP radios on the grid, while at the same time upgrading the backend network connectivity with emerging software-defined networks (SDN) for improved flexibility and performance. An outdoor WiMax/cellular capability was also added to ORBIT through a synergistic GENI project in the same time-frame. This set of enhancements allowed us to migrate from the initial base of WiFi MAC and ad-hoc routing experiments to then emerging topics such as dynamic spectrum access, cellular network virtualization and wireless security. The second testbed upgrade cycle (2010–2014) that was recently completed involved replacement of the original ORBIT radio node ("the yellow box") with significantly faster dual-core i7 machines (approximately ~10× faster than the original VIA C3 nodes) along with USRP2 replacements for the SDR units (capable of handling LTE signals) and faster servers in the testbed backend. This second upgrade now brings the computing platforms used in the radio grid to current CPU speeds, enabling new classes of more processing intensive experiments such as wideband cognitive radio/DSA or clean-slate radio PHY/MAC.

As shown in the figure, the next phase of upgrades planned for ORBIT aims to extend the system-level capability to so-called "cloud RAN" scenarios involving centralized cloud processing of radio signals from "thin client" nodes on the grid. This scenario is of particular interest for emerging cellular network architectures in view of large potential gains obtained from multi-node cooperation at the physical layer [49]. The planned upgrade includes the use of an array of computing blades with FPGA acceleration as required to handle the large computational workload associated with cloud RAN scenarios. The cloud processing upgrade planned for the ORBIT backend is also expected to be useful for a broad class of emerging mobile cloud experiments involving mobile devices connected to cloud processing clusters co-located with edge networks. The next phase of ORBIT will also involve addition of emerging radio technologies, initially LTE and subsequently others such as 802.11ad and future 5G standards.

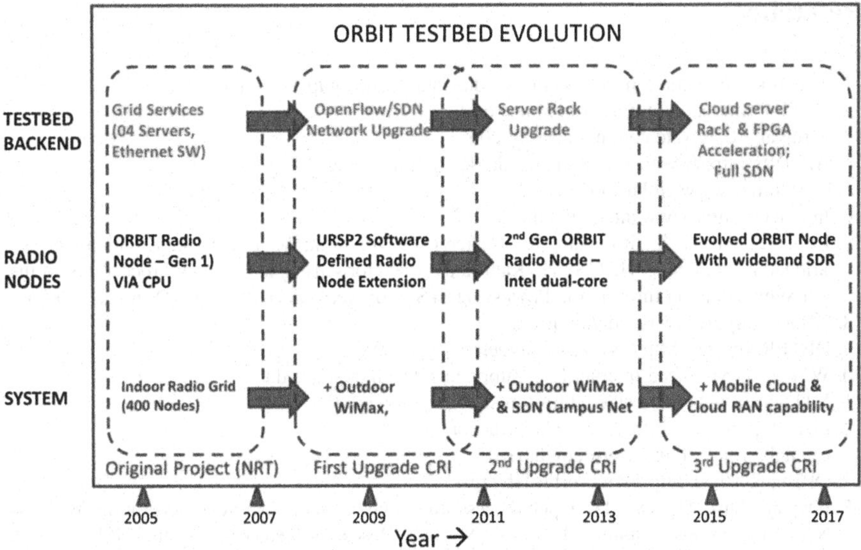

Fig. 29 Phased equipment upgrade strategy for ORBIT testbed

6 Links to GENI Project

ORBIT was one of the important testbed precursors which provided guidance for the GENI project particularly in terms of technologies and software for programmable wireless networks. Early in the GENI planning process [50], it was widely recognized that wireless/mobile scenarios are of growing importance to the Internet and this needs to be reflected in the experimental infrastructure to be built. In the ten years since GENI was first being planned, wireless has grown in importance with over 1 billion smartphones in use worldwide and mobile data traffic now exceeding that from fixed PC's. These trends imply the need for an experimental GENI infrastructure with extensive wireless networking capabilities at the edge, similar in spirit to the goals of the ORBIT testbed. ORBIT developers were active in the early definition of the GENI system, and proposed several components such as the open WiMAX/LTE campus network, software-defined radios for dynamic spectrum studies, vehicular networking support and others. The ORBIT radio node, WiMAX base station and OMF controller/software have been used quite extensively in the GENI network and the control protocols and experimental interfaces have been harmonized to work with the overall GENI framework. OMF continues to be used for control and management of wireless networks in GENI, and the OML (ORBIT measurement library) has been generalized to both wired and wireless networks in GENI. ORBIT resources have been fully federated into GENI and the radio grid, outdoor campus network and sandboxes are currently available to users through the GENI portal.

References

1. Network Simulator 2. http://nsnam.isi.edu/nsnam/index.php/Main_Page
2. OPNET simulator. http://www.opnet.com/
3. Parulkar, G.: private communication, 2005
4. ABONE. http://www1.cs.columbia.edu/dcc/nestor/abone/
5. Grid: http://www.globus.org
6. Internet2: http://www.internet2.edu
7. Ertin, E., Arora, A., Ramnath, R., Nesterenko, M., Naik, V., Bapat, S., Kulathumani, V., Sridharan, M., Zhang, H., Cao, H., Kansei: a testbed for sensing at scale. In: Proceedings of the 4th Symposium on Information Processing in Sensor Networks (IPSN/SPOTS track) (2006).
8. XBone. http://www.isi.edu/xbone/
9. DETER Testbed. http://www.isi.edu/deter
10. Wisconsin Advanced Internet Laboratory. http://wail.cs.wisc.edu/
11. MIT sensor network testbed. http://mistlab.csail.mit.edu/
12. PlanetLab project. http://www.planet-lab.org/
13. Emulab project. http://www.emulab.net/
14. ORBIT Testbed. http://www.orbit-lab.org
15. Raychaudhuri, D.: ORBIT: Open-Access Research Testbed for Next-Generation Wireless Networks, proposal submitted to NSF Network Research Testbeds Program, NSF award # ANI-0335244, 2003-07, May 2003.
16. Raychaudhuri, D., Seskar, I., Ott, M., Ganu, S., Ramachandran, K., Kremo, H., Siracusa, R., Liu, H., Singh, M.: Overview of the ORBIT radio grid testbed for evaluation of next-generation wireless network protocols. In: Proceedings of the IEEE Wireless Communications and Networking Conference (WCNC) (2005).
17. Ott, M., Seskar, I., Siracusa, R., Singh, M.: ORBIT testbed software architecture: supporting experiments as a service. In: Proceedings of IEEE Tridentcom 2005, Trento, Italy, February 2005.
18. Singh, M., Ott, M., Seskar, I., Kamat, P.: ORBIT measurements framework and library (OML): motivations, design, implementation, and features. In: Proceedings of IEEE Tridentcom 2005, Trento, Italy, February 2005.
19. Peterson, L.: GENI: Global environment for network investigations. ACM SIGCOMM '05, August 2005.
20. AODV—Ad hoc On-Demand Distance Vector Routing. http://moment.cs.ucsb.edu/AODV/
21. DSR—Dynamic Source Routing Protocol. http://www.cs.cmu.edu/~dmaltz/dsr.html
22. Bluetooth special interest group. https://www.bluetooth.org/
23. Zigbee alliance. http://www.zigbee.org/
24. GNU Radio Project. http://www.gnu.org/software/gnuradio
25. Lei, J., Yates, R., Greenstein, L., Liu, H.: Wireless link SNR mapping onto an indoor testbed. In: Proceedings of IEEE Tridentcom 2005, Trento, Italy, February 2005.
26. Lei, J., Yates, R., Greenstein, L., Liu, H.: Mapping link SNRs of wireless mesh networks onto an indoor testbed. In: Proceedings of IEEE Tridentcom 2006, Barcelona, Spain, March 1–3, 2006
27. Universal Software Radio Peripheral (USRP). http://www.ettus.com/downloads/usrp_1.pdf
28. Kaul, S., Gruteser, M., Seskar, I.: Creating wireless multi-hop topologies on space-constrained indoor testbeds through noise injection. In: IEEE Tridentcom, March 2006.
29. Raychaudhuri, D.: Proof-of-concept Prototyping of Methods for Wireless Virtualization and Wired-Wireless Testbed Integration. supplement to NSF Award ANI-0335244, June 2006.
30. IEEE 802.16 Working Group. IEEE Standard for Local and Metropolitan Area Networks, Part 16: Air Interface for Fixed Broadband Wireless Access Systems. *IEEE Std* 802 (2004).
31. LTE Base Station Software. http://bellard.org/lte/
32. Off-The-Shelf 4G Network. http://www.amarisoft.com/

33. Hibler, M., Stoller, L., Lepreau, J., Ricci, R., Barb, C.: Fast scalable disk imaging with Frisbee. In: Proceedings of the 2003 USENIX Annual Technical Conference, June 2003.
34. LibMac—A user-level C library. https://www.orbit-lab.org/browser/libmac/trunk
35. Ramachandran, K., Kaul, S., Mathur, S., Gruteser, M., Seskar, I.: Towards large-scale mobility emulation through spatial switching on a wireless grid. In: E-WIND Workshop (held with ACM SIGCOMM) (2005).
36. Ganu, S., Seskar, I., Ott, M., Raychaudhuri, D., Paul, S.: Architecture and framework for supporting open-access multi-user wireless experimentation. In: Proceedings of International Conference on Communication System Software and Middleware (COMSWARE 2006), Delhi, India, January 2006.
37. Li, Z., Xu, W., Miller, R., Trappe, W.: Securing wireless systems via lower layer enforcements. In: Proceedings of the 2006 ACM Workshop on Wireless Security (WiSe) (2006).
38. In the Matter of Unlicensed Operation in the TV Broadcast Bands: Third Memorandum Opinion and Order, April 2012. http://hraunfoss.fcc.gov/edocs_public/attachmatch/FCC-08-260A1.pdf
39. Extending LTE Advanced to Unlicensed Spectrum-White Paper, Qualcomm Inc., December 2013. https://www.qualcomm.com/media/documents/files/white-paper-extending-lte-advanced-to-unlicensed-spectrum.pdf
40. Jing, X., Raychaudhuri, D.: Spectrum Co-existence of IEEE 802.11b and 802.16a Networks using the CSCC Etiquette Protocol. In: Proceedings of IEEE DySPAN'05, Baltimore, MD, November 8–11, 2005.
41. Paul, S., Yates, R., Raychaudhuri, D., Kurose, J.: The cache-and-forward network architecture for efficient mobile content delivery services in the future internet. In: Proceedings of IEEE Innovations in NGN: Future Network and Services (2008).
42. Nelson, S., Bhanage, G., Raychaudhuri, D.: GSTAR: Generalized storage-aware routing for mobilityfirst in the future mobile internet. In: Proceedings of ACM MobiArch 2011.
43. MobilityFirst Future Internet Architecture Project. http://mobilityfirst.winlab.rutgers.edu/
44. Vu, T., Baid, A., Zhang, Y., Nguyen, T., Fukuyama, J., Martin, R., Raychaudhuri, D.: DMap: a shared hosting scheme for dynamic identifier to locator mappings in the global internet. In: Proceedings of the 32nd International Conference on Distributed Computing Systems (ICDCS 2012).
45. Sharma, A., Tie, X., Uppal, H., Venkataramani, A., Westbrook, D., Yadav, A.: A global name service for a highly mobile internetwork. In: Proceedings of ACM SIGCOMM (2014)
46. Siganos, G., Tauro, S., Faloutsos, M.: Jellyfish: a conceptual model for the AS Internet topology. J. Netw. Commun. **8**(3), 339–350 (2006)
47. Shavitt, Y., Shir, E.: DIMES—Letting the Internet Measure Itself. http://www.netdimes.org/
48. CAIDA: The Cooperative Association for Internet Data Analysis. http://www.caida.org/
49. Venkatesan, S., Lozano, A., Valenzuela, R.: Network MIMO: overcoming interference in indoor wireless systems. In: Proceedings of IEEE Asilomar Conference on Signals, Systems and Computers (2007)
50. Raychaudhuri D., Gerla M.: New Architectures and Disruptive Technologies for the Future Internet: The Wireless, Mobile and Sensor Network Perspective, *Report of NSF Wireless Mobile Planning Group (WMPG) Workshop*, August 2005

Part II
Architecture and Implementation

Once the initial planning phase had completed and the GENI Project Office had been established by the National Science Foundation, the hard work of building GENI began. As mentioned in the introduction, the GENI Project Office used a Spiral strategy to build GENI—building rough prototypes of a number of ideas, learning, discarding ideas and approaches that were not proving fruitful, and deepening the commitments in those areas and approaches that showed promised.

By the end of Spiral Two the architecture had reached a close approximation to its ultimate form; it was of course refined as it was deployed in Spiral Three and beyond, but the shape was clear then. This section details this final architecture, and its various components.

The original architectural principles of "Slices"—virtual networks of virtual execution environments, or "Slivers"—were clear from PlanetLab, Emulab, and DeterLab. In the latter two the terminology was slightly different, but the concepts were the same. However, a number of issues remained to be resolved, including:

- PlanetLab, Emulab, ORBIT, and DeterLab all functioned under the aegis of a single controlling and administrative agency. But GENI would be far more complex: each individual component would be under the control of its hosting campus; the control software would be provided and administered by remote academic organizations; and a third organization would monitor and provide day-to-day support of the deployment. How could the tasks and authority be decentralized, but coordinated enough that users could easily form slices across many different administrative and control domains?
- GENI experimenters would be connecting together heterogenous collections of resources of disparate underlying types, each "slivered" with potentially different and uncoordinated virtualization techniques (e.g., a virtual machine is fundamentally unlike a portion of OpenFlow flowspace). How could the slice concept be extended and implemented to maintain appropriate levels of isolation and performance when stitching together these peculiar combinations of resources?

- Administratively, slices in PlanetLab and experiments in Emulab and DeterLab all had fairly similar objectives and functioned easily under the same AUP. However, GENI expected to encompass a much wider range of experiments and services than any of its precursors had, and from many sites which operated in different legal and social environments; how could an appropriate decentralized authorization and access policy be crafted?
- GENI envisioned embedding multiple control regimes, appropriate to different experiments and services; how could such nested control regimes be accommodated?
- GENI envisioned deeply programmable networking in the wild, something attempted by none of its precursors; this was aided by the serendipitous invention of OpenFlow early in GENI's life. How could programmable, controllable networks be accommodated in GENI?

In this section, the final architecture of GENI and its details will be described. The eventual form of the architecture consisted of a network of Aggregate Managers (loosely, Cloud Control systems) which offered resources to Slice Managers. Here, these concepts and their concrete realizations are described.

In chapter "GENI Architecture Foundation", GENI Chief Architect Marshall Brinn describes the overall architecture of GENI and the roles each of these actors play. Of these, the workhorses are the Aggregate Managers, who roughly fill the role of a Cloud manager such as OpenStack. The two most prominent aggregate managers in GENI are described in chapters. "The Need for Flexible Mid-scale Computing Infrastructure" and "A Retrospective on ORCA: Open Resource Control Architecture".

In chapter "The Need for Flexible Mid-scale Computing Infrastructure", Rob Ricci of the University of Utah makes the case for flexible management of infrastructure. In GENI, this role is performed by ProtoGENI, the Aggregate Manager based on Emulab which is the Aggregate Manager for the InstaGENI racks discussed in the deployment section. Flexibility is the hallmark of ProtoGENI, which offers hardware-as-a-service, VMs-as-a-service, and even other aggregates as a service.

In chapter "A Retrospective on ORCA: Open Resource Control Architecture", Jeff Chase and Ilia Baldine describe ORCA, the aggregate manager underlying the ExoGENI racks. In the GENI deployment, ORCA installations focus on flexible and efficient allocation of virtual machines.

An entirely novel feature of GENI is the deployment of wide-area software-defined networks. GENI has been a major driver of OpenFlow deployments, from the initial campus trials to the wide-area deployments with the GENI Racks. In chapter "Programmable, Controllable Networks", Nick Bastin and Rick McGeer make the case for Software-Defined networking (really, application-controlled routing) as an essential element of the next Internet and discuss the experiences from the GENI deployments.

A novel feature of GENI is the seamless incorporation of wired and wireless testbeds, unifying the ORBIT testbeds with the fixed-link wired backbones. In chapter “4G Cellular Systems in GENI”, Ivan Seskar and his colleagues discuss this integration.

GENI requires a decentralized, robust system of authorization and access control that doesn’t rely on a single centralized authority. In chapter “Authorization and Access Control: ABAC”, Ted Faber, Steve Schwab, and John Wroclawski of USC/ISI describe such a system, Attribute-Based Access Control (ABAC) which offers unforgeable decentralized authorization keys.

Embedding PlanetLab’s long-running lightweight services in ProtoGENI’s flexible infrastructure was an early GENI design decision: this fit well with ProtoGENI’s flexibility and PlanetLab’s ability to live light on the land. In the event, it was not until the mesoscale deployment that this could come to fruition, but it proved worth the wait. Late adoption and implementation meant that PlanetLab-on-GENI could take full advantage of modern Cloud technologies, and introduce something wholly new: an embedded infrastructure that could be instantiated across infrastructures. In chapter “The GENI Experiment Engine”, Andy Bavier and Rick McGeer introduce the GENI Experiment Engine, a modern instantiation of the PlanetLab ideas utilizing modern containerization, deployment, and orchestration technologies, and one that is designed to be instantiated on one or more underlying VMaaS or HaaS platforms.

GENI Architecture Foundation

Marshall Brinn

1 Introduction

> The purpose of the GENI Architecture is to facilitate trusted exchange of resources.

This description has the benefits of being concise and, perhaps, complete. But it raises some questions, which this chapter hopes to address:

- What objects and principals are involved in this *exchange of resources*?
- What makes an exchange *trusted* and why is this important?
- What does it mean to *facilitate* exchange?

1.1 Facilitating Trusted Exchange of Resources

When we speak of resources in a GENI context, we are referring to infrastructure or services that provide compute, storage and transport capabilities. These may be physical devices such as a PC, a hardware network switch or a disk, or virtual such as a VM, a software switch or a cloud-based storage service.

GENI seeks to mediate the desires of two communities with respect to such resources:

M. Brinn (✉)
GENI Project Office, Raytheon BBN Technologies, 10 Moulton St. Cambridge, MA 02138, USA
e-mail: mbrinn@bbn.com

© Springer International Publishing Switzerland 2016
R. McGeer et al. (eds.), *The GENI Book*, DOI 10.1007/978-3-319-33769-2_5

- *Resource providers*: People, organizations or institutions who own or manage available resources
- *Resource consumers*: Experimenters, educators, engineers who wish to use resources to perform research, classwork or development

Both providers and consumers have motivation to exchange, that is, to enter into agreements by which consumers may use the resources of providers. Consumers wish to have access to resources for their own purposes, while providers may be paid or otherwise incentivized to provide these resources [1].

Several barriers stand in the way of an efficient market for the exchange of resources. First, providers and consumers may not know each other or know how to find one another. Second, beyond incentives, the provider wants assurances that the consumer will use the resources in a responsible, safe manner, i.e. not bringing risk of damage to the underlying physical resources or the resources let to other consumers. Finally, the consumer wants assurances that the resources are reliable, secure and performant to advertised specifications.

Overcoming these barriers requires a trust on the part of the provider and consumers. At small scale (a small number of providers and consumers), this can be managed by relationships between individuals. However, as the number of providers and consumers grows, this approach becomes unscalable: there will be $M * N$ relationships to be developed to support exchange between M providers and N consumers.

A trusted third party is thus required, one who is trusted by both provider and consumer and can vouch for the integrity of the other side. Such a third party needs to maintain only $M + N$ such relationships. Finally, this third party can provide services to make the discovery, allocation, configuration, and return of resources possible in an efficient and uniform manner. Providing these services is what we mean by *facilitating* the exchange of resources (Fig. 1).

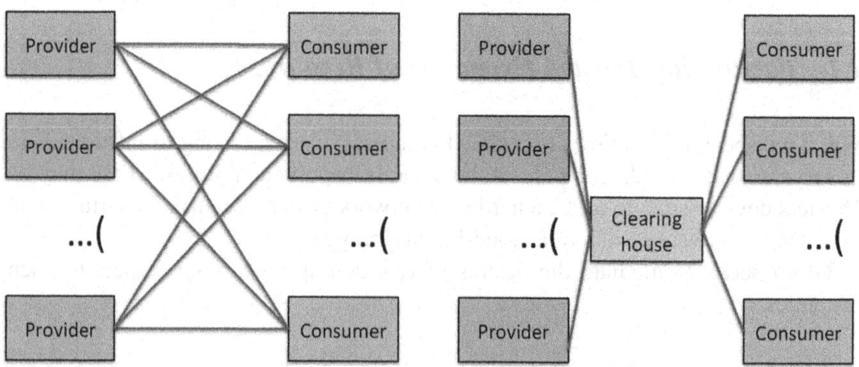

Fig. 1 A trusted third party such as a Clearinghouse manages the scalability of trust relationships required among resource providers and consumers

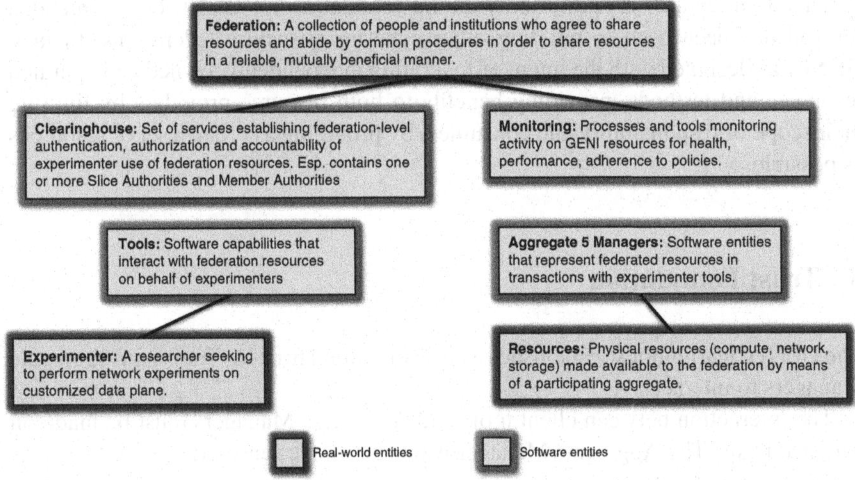

Fig. 2 The GENI Federation consists of people and software that represent their interests

2 GENI Federation

The GENI Federation is the trusted third party required to enable reliable resource exchanges among large numbers of resource providers and consumers.

The exchange of resources, and the trust that enables such exchanges, are human activities. There are negotiations and agreements of terms and conditions, assurances, consequences and accountability. People and organizations enter into federations in order to facilitate these exchanges according to these terms.

In GENI [2], we have three sets of human parties that make up the federation, the resource providers, the resource users and the federation organization itself. These in turn, are represented in the federation architecture by software entities that represent their interests and act on their behalf to effect their desires.

A *federation architecture* is a set of software services that codify and enforce these agreements. These services are the Clearinghouse, the Aggregate Manager and Client Tools. Each of these will be described in greater detail later in the chapter but a brief introduction of terms is appropriate here:

- The ***Clearinghouse*** (CH) is a set of services representing the federation, establishing statements of trust, mutually recognized identity, policy-based authorization, and accountability.
- ***Aggregate Managers*** are services that represent the interests of different resource providers, providing access to resources to trusted requesting users.
- ***Client tools*** represent consumers who seek to access resources from federated services (Fig. 2).

This architecture serves not only as the foundation for GENI, but is intended for and has been used in building other cyberinfrastructure federations. In fact, GENI was designed with the intent of federating independently owned and operated resources and testbeds, providing benefits to both resource providers by limiting their scope of responsibility and consumers by providing as broad a set of resources as possible.

3 Trust Foundation

The GENI Clearinghouse establishes a basis of mutual trust for Tools and Aggregate Managers to interact.

The interaction between client tools and Aggregate Managers must be made on a trusted basis. The Aggregate Manager wants a reliable sense of:

- *Authentication*: Who is making this request?
- *Authorization*: In what context should I allow access to my resources?
- *Accountability*: Who is responsible if something bad happens on my resources when they are used?

There are many possible software mechanisms to establish the trust to answer these questions and support the trust required to complete these transactions. GENI uses public key infrastructure (PKI), and specifically X.509 certificates and keys as the basis of authentication. Furthermore, SSL/TLS is used as the basis of trusted service interaction [3–5].

An SSL-based interaction requires that the initiator (the client tool) encrypt its transactions with its private key and pass its X.509 public cert (including its public key) along with its message. The receiver (the service provider) then checks that the certificate is in a trusted chain of certificates.

X.509 certificates and keys are provided by Clearinghouse services. The certificates are signed by the Clearinghouse authority's private key, and assert some identifying data (an email address, a URN, a UUID) about the bearer of the certificate and associated private key.

The act of including the Clearinghouse authority's root certificate in the trust bundle of the service is, therefore, a reflection of the human act of federation. An Aggregate Manager includes the set of Clearinghouses it trusts, and anyone bearing a certificate/key from any such Clearinghouse may speak to that Aggregate Manager.

> An Aggregate Manager may belong to many Clearinghouses, and thus accept requests from users from different federations.

To be specific, while GENI is a Federation with its own Clearinghouse Authority services, so are Emulab [6], Fed4Fire [7] and several other comparable projects. Many Aggregates can and do chose to be members of one or many of these Federations.

In GENI, the act of providing a certificate to a user is not an automatic process but a human process, representing some degree of vetting and, implicitly, vouching that the user is who the user claims to be. GENI is entrusted by the National Science Foundation to support network and computer science research and education. To that end, GENI accepts requests from users with accounts at an academic institution participating in the InCommon federation. Additionally, GENI accepts requests from users to use GENI federated resources and one job of the GENI Federation staff is to check that the requesters are who they claim to be, that they are at a recognized academic institution and that they are in an appropriate research program or lab.

The question of authorization has elements that are both distributed and centralized. The Clearinghouse provides signed statements called *Credentials* that may be used by Aggregate Managers in their authorization decisions. These credentials are statements from the Clearinghouse indicating a set of roles or rights the Clearinghouse asserts the user has in a particular context. The Aggregate Manager will typically use these credentials as part of its authorization decision. That said, each Aggregate Manager is independent and autonomous, and can make whatever authorization decisions it chooses.

GENI supports two particular types of credentials:

- *SFA Credential*: These are credentials granting privilege or roles to a given user in a given context (typically a slice). This is an example of role-based access control (RBAC) and conforms to the SFA format [8].
- *ABAC Credential*: These are credentials asserting attributes about a given user (e.g. "X has admin privileges on slice Y" or, generally, "X is a member of the set Z"). This supports attribute-based access control by mixing ABAC assertion and policy statements and conforms to the standard ABAC format [9].

Accountability will be discussed below as we look at Federation Monitoring services.

4 GENI Concepts

Before we provide more detail on GENI Services, definitions of key GENI concepts are in order:

An ***Aggregate*** (or Aggregate Manager or AM) is a service representing a set of resources and the interests and policies of the provider of those resources. It provides services for allocating, deleting, configuring, renewing, and deleting resources using the GENI Aggregate Manager (AM) API, as will be discussed below.

An ***Authority*** is a Clearinghouse Service which generates statements (typically signed XML documents) that are trusted by Aggregates of the federation. Example authorities are the Slice Authority and Member Authority, as discussed below.

A ***Member*** is a registered GENI user, with an assigned certificate, UUID, URN at a given Member Authority.

A ***Sliver*** is a resource or set of resources provided by an Aggregate to requester through invocations of the AM API. These slivers may be a whole physical resource (e.g. a bare metal machine), or a virtualized piece of a physical resource (e.g. a virtual machine) or a combination of these.

A ***Slice*** is a collection of slivers in which resource requests are performed. The Slice is both an *accounting* mechanism for knowing what slivers where allocated to a given user at a given time, but also an *isolation* mechanism. That is, it is assumed that resources in the same slice have visibility to one another while slivers in different slices are logically (and, ideally, physically) isolated from one another.

A ***Project*** is a grouping of Slices for a particular purpose. A Project is managed by a lead, typically a professor or head of a laboratory or equivalent, and all slices in a project are associated with some common research or educational goal.

5 GENI Services

This section details the specific services and interfaces that make up the different services provided in the context of a GENI federation.

5.1 Federation API

The Federation API [10] is an interface for defining federation services, commonly negotiated by representatives of GENI, Emulab and Fed4Fire in 2013. This section provides a high-level view of these services.[1]

All Federation API calls are XMLRPC/SSL and thus require invocation with the invoking member's private key and passing their certificate.

The Service Registry (SR) is a service providing a dictionary of services available in a given Federation Clearinghouse. The SR provides the following API:

[1] See http://groups.geni.net/geni/wiki/CommonFederationAPIv2 for details.

Method	Arguments	Description
get_version	None	Return version of SR API including extensions and supported object model
lookup_member_authorities	Options: Query specifying which MA's to match	Return list of MA's with matching and filter criteria specified in options
lookup_slice_authorities	Options: Query specifying which SA's to match	Return list of SA's with matching and filter criteria specified in options
lookup_aggregates	Options: Query specifying which AM's to match	Return list of aggregates with matching and filter criteria specified in options
get_trust_roots		Return list of trust roots trusted by authorities and aggregates of the federation associated with this Clearinghouse

The Member Authority (MA) is a service providing information for creating, modifying and looking up information about registered users at a given Federation CH. If not otherwise indicated, 'member_urn' is a unique identifier of a particular member, as is 'member_uid' (a unique identifier of a different non-human-readable form), 'key_id' is a unique identifier for an SSH key pair, and 'options' is the set of fields describing a given member or SSH key pair. The MA provides the following API:

Method	Arguments	Description
get_version	None	Return version of MA API including extensions and supported object model
create_member	Options: Member details (email, name, etc.)	Create new Member with given details
update_member_info	Member_urn, Options:	Update information associated with Member
lookup_member_info	Options:	Lookup information about member. Note: the members and specific details about the member will tend to be tightly controlled by authorization policy
get_credentials	None	Get credentials representing MA's sense of roles and rights of members independent of slice context
create_key	Options	Create or upload SSH key pair for given member

(continued)

Method	Arguments	Description
delete_key	Key_id	Delete given SSH key pair from member
update_key	Key_id, Options:	Modify given SSH key pair info for given member
lookup_keys	Options:	Lookup SSH key pair information matching given query criteria
create_certificate	Options: CSR provided or not	Create an X.509 cert (and optionally private key if no CSR provided)
add_member_privilege	Member_uid, privilege	Add privilege to given member (e.g. LEAD, ADMIN)
revoke_member_privilege	Member_uid, privilege	Remove privilege from given member
add_member_attribute	Member_urn, attribute_name, attribute_value	Add attribute to given user
remove_member_attribute	Member_urn, attribute_name	Remove attribute from given user

The Slice Authority (SA) is a service providing information for creating, modifying, and looking up information about slices and projects. If not otherwise indicated, the 'slice_urn' argument is the unique identifier for a given slice (likewise for 'sliver_urn' and 'project_urn'), 'credentials' represents the credentials provided from the MA indicating the good-standing and role of a given user, and 'options' provides details on the given slice, slice membership, sliver info, project, project_membership being created or updated. The SA provides the following API:

Method	Arguments	Description
get_version	None	Return version of SA API including extensions and supported object model
create_slice	Credentials, Options	Create a new slice in a given project
update_slice	Slice urn, credentials, Options	Update a given slice with new details specified by options
get_credentials	Slice_urn, credentials	Get credentials representing SA's sense of roles and rights of members in slice context
modify_slice_membership	Slice_urn, credentials, options	Modify (add, change, delete) membership/role for members in a slice
lookup_slice_members	Slice_urn	Lookup members/roles of a given slice

(continued)

Method	Arguments	Description
lookup_slices_for_member	Member_urn	Lookup slices to which a given member belongs
create_sliver_info	Options	Create a record of sliver info (sliver, slice, aggregate, time, expiration)
delete_sliver_info	Options	Delete a record of sliver info
update_sliver_info	Sliver_urn, options	Update sliver info record
create_project	Options, credentials	Create a new project (typically a privileged operation for members with "Project_Lead" privilege reflected in their credential)
modify_project_membership	Project_urn, options	Modify (add, delete, change) membership/role for members in an project
lookup_project_members	Project_urn	Lookup members/roles for a given project
lookup_projects_for_member	Member_urn	Lookup projects to which a given member belongs
lookup_project_attributes	Project_urn, options	Lookup attributes associated with given project
add_project_attribute	Project_urn, options	Add a given attribute to a given project
remove_project_attribute	Project_urn, options	Remove an attribute from a given project

5.2 *Aggregate Manager (AM) API*

The Aggregate Manager API [11] provides the interface by which client tools (representing users) can request, configure, renew and delete resources from an aggregate resource provider.

This section provides a high-level view of this API.[2] All calls take a list of credentials as an argument. These are typically a UserCredential provided by a previous call to the Clearinghouse Member Authority or a SliceCredential provided by a previous call to the Clearinghouse Slice Authority. Additionally, a dictionary of options is provided to most calls. These arguments are not included in the 'Arguments' table below.

AM API calls are made to a particular Aggregate Manager. They are XMLR-PC/SSL calls and thus must be invoked by a user with a private key and a certificate generated by a CA trusted by the given AM.

[2]More details on Versions 2 and 3 of the GENI AM API can be found at http://groups.geni.net/geni/wiki/GAPI_AM_API_V2 and http://groups.geni.net/geni/wiki/GAPI_AM_API_V3.

Method (V2)	Method (V3)	Arguments	Description
GetVersion	GetVersion		
ListResources	ListResources		This is the 'no slice' version of ListResources, providing an 'advertisement RSpec'
CreateSliver	Allocation, Provision	Slice_urn: The slice into which to add sliversRSpec: The request RSpec indicating which resources to allocateUsers: Information (SSH login info, e.g.) about users for created compute resources	Create a given resources specified in the request RSpec in the context of the given slice. Set up any compute accounts with specified user accounts.In V3, we separate this into an 'allocate' (prepare) and 'provision' (initialize and configure) calls
	Perform Operational Action	Slice_urn(v2)/urns(v3): List of sliver URNs (or slice URN) for which to perform action on slivers.Action: Type of action to perform (e.g. reboot, create_image)	Perform a sliver-specific action on a running sliver instance (V3 only)
SliverStatus	Status	slice_urn (V2)/urns (v3)	Retrieve current operational status for a given set of slivers or all slivers in a given slice at an AM
ListResources	Describe	slice_urn (V2)/urns (v3)	The slice-specific version of ListResources: provide a manifest RSpec with all resources associated with a given slice
RenewSliver	Renew	slice_urn (V2)/urns (v3)expiration_time	Renew lease on resources to given expiration time if possible
DeleteSliver	Delete	slice_urn (V2)/urns (v3)expiration_time	Delete resources associated with given slice (or specific sliver urn's, in V3)
Shutdown	Shutdown	Slice_urn	Shutdown given slice and all associated slivers at this AM

We should note here that many features associated with allocated GENI topologies, e.g. Deep Programmability, Sliver isolation, cross-site stitching, are features provided by individual aggregates, or orchestrated by tools between aggregates. They are not, per se, features of the GENI Federation architecture and are not detailed in this chapter.

5.2.1 GENI Resource Specifications (RSpecs)

The GENI AM API requires three particular kinds of specifications to describe availability, requirements or descriptions of resources at an Aggregate Manager. RSpecs are XML documents that adhere to the http://www.geni.net/resources/rspec schema, plus recognized extensions.

These are described below.[3] Examples are provided without XML preface and namespace detail for brevity and clarity.

Advertisement RSpec. A description of the available resources at an Aggregate. These are provided by a ListResources (no-slice) call. These may include bare-metal machines, virtual machines, IP address space, VLAN space, disk images and other compute or network specific resource features. The following is a simple advertisement RSpec, showing the availability of two (fake) compute nodes:

```
<rspec  type="advertisement">
  <node component manager id="urn:publicid:geni:gpo:gcf+authority+am"
        component name="a70786ad-8c9c-4fb0-a1d1-b33a8e74d0bd"
        component id="urn:publicid:IDN+geni:gpo:gcf+fakevm+a70786ad-8c9c-4fb0-
a1d1-b33a8e74d0bd" exclusive="false">
    <available now="true"/>
  </node>
  <node component manager id="urn:publicid:geni:gpo:gcf+authority+am"
        component name="370bd049-e3eb-4964-a381-6f2ec038918a"
        component id="urn:publicid:IDN+geni:gpo:gcf+fakevm+370bd049-e3eb-4964-
a381-6f2ec038918a" exclusive="false">
    <available now="true"/>
  </node>
</rspec>
```

Request RSpec. A set of requirements for resources to be created at a given AM. These are required to the CreateSliver (or Allocate in V3) call. This may include (compute) nodes of given types and images, links and interfaces and other extension-specific details. The following is a simple advertisement RSpec, requesting a single compute node:

```
<rspec type="request">
  <node client id="expl-host1" exclusive="false">
    <sliver type name="m1.small" >
    <disk image description="" name="ubuntu-12.04" os="Linux" version="12"/>
    </sliver type>
  </node>
</rspec>
```

Manifest RSpec. A description of all resources allocated at a given AM for a given Slice. This is returned from a CreateSliver (or Provision in V3) call or a ListResources (with slice_urn, or Describe in V3) call. This will include the content from the Request RSpec plus provision-specific details such as IP addresses, allocated VLANs and other extension-specific details. The following is a simple manifest RSpec showing the allocation of a single compute node:

[3] More details on GENI RSpecs can be found at http://www.protogeni.net/ProtoGeni/wiki/RSpec.

```
<rspec type="manifest">
    <node client id="exp1-host1"
          component id="urn:publicid:IDN+geni:gpo:gcf+fakevm+ce92868e-5eea-4ca2-
817c-700bd19670b2"
          component manager id="urn:publicid:geni:gpo:gcf+authority+am"
          sliver id="urn:publicid:IDN+geni:gpo:gcf+sliver+fakevmce92868e-5eea-
4ca2-817c-700bd19670b2-ch-mb-gpolab-bbn-com--COUNTFOO"/>
</rspec>
```

5.3 Monitoring Services

The GENI Federation seeks to provide assurances to resource providers that their resources aren't being ill-used, intentionally or otherwise. It also seeks to provide assurances to resource consumers that the resources they've been provided by Aggregate Managers will be reliable.

Towards these ends, GENI provides a set of services that engender trust on the part of users towards resources provided by GENI, and accountability of users for actions taken on these resources.

GENI requires that all Federation services, including Aggregate Managers, provide data to a central monitoring service. These data consist of health status and load information (on CPU, storage, network interfaces) as well as current topology information (which slivers are connected to which across aggregates).

In this way, the monitoring service can serve as the basis of an *alerting* mechanism (indicating resources that are experiencing abnormal loads or patterns). It can further be the source of *forensics support*, by which the person associated with a misbehaving sliver or slivers can be identified. It can support a *slice shutdown* service by which a given resource can be shutdown, nullifying its ongoing impact on other virtual or physical resources.

From there, Federation or Aggregate-local policy can take corrective action, ranging from a warning, to deleting the resources to revoking the credentials of the user or even those of the supervisor.

6 Tools

Of course, humans cannot directly invoke calls to the Federation of AM API's: it requires software tools to invoke these calls on their behalf. Given that these calls use SSL as an authentication and encryption mechanism, callers must have access to their private key.

Good 'key hygiene' (keeping private materials private) requires that private keys stay on local machines and not transit the network. GENI therefore supports two kinds of tools that manage the problem of keeping private keys private in two different ways.

Desktop Tools. In this case, the tool is run locally on the user's desktop. The tool is pointed to the user's certificate and private key which reside locally. The cert is sent as part of the SSL handshake but the key is only used to sign/encrypt the message and never leaves the desktop. Examples of such tools include *omni* (a python command-line tool for invoking arbitrary Federation and AM API calls) and *jFed* (a Java application run on the user's machine that provides graphical interfaces to create and view allocated topologies).

Hosted Tools. In this case, a tool runs remotely on web server. There are several ways to enable such a configuration:

- *Providing the private key*. While not recommended, a tool can require that the user upload (or cut-and-paste) their private key to the tool and then the tool may 'speak as' the user.
- *Using the tool's key*. The tool may be configured with its own certificate and private key and 'speak as' the tool itself. This has the advantage of not requiring divulging the user's private key. But it loses any sense of accountability.
- *Using a speaks-for credential*. GENI supports using a credential (an ABAC statement, in fact) signed by the user that a given tool may speak for the user, possibly limited to certain contexts. If the tool speaks with its own cert and key and such a credential is present in the list of credentials given to the API call, the Federation and AM services will authorize and account the call to the user, not the tool. This approach preserves accountability and private key security and is thus both a novel and preferred approach for supporting hosted tools (Fig. 3).

6.1 The GENI Portal

The GENI Portal is an example of a hosted tool. The Portal supports a Shibboleth configuration that is federated with several Identity providers including those from the InCommon Federation, CAFé in Brazil and an IdP provided by the GENI Program Office. Users authenticate to the Portal through one of these Shibboleth IdPs to establish a single-sign-on (SSO) session [12].

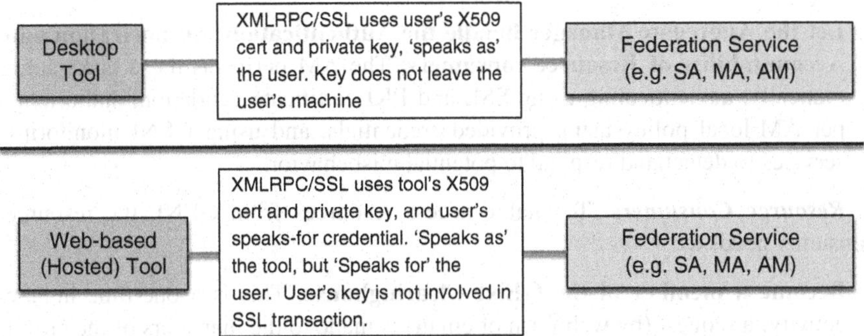

Fig. 3 Two kinds of tools for invoking Federation or Aggregate Manager APIs: desktop and hosted tools

The Portal can then operate on behalf of the user in one of two ways:

- Users that have created a speaks-for credential (i.e. a statement that the portal can speak for the user) store these credentials in the portal, and the portal retrieves these credentials in the context of a Shibboleth authenticated session. It can then use the speaks-for credential along with the Portal's own cert and key to invoke Federation and AM API calls on the user's behalf based on user interface actions.
- Users that have not created a speaks-for credential may allow the Portal to create a certificate and private key for that user which the Portal can extract from its own private database and use to 'speak as' the user. This mode is popular (and only recommended) for short-term engagements such as tutorials and users first getting acquainted to GENI.

7 Summary

We have described the pieces required to establish trusted resource exchange in GENI. In this section, we will put them together into a single scenario. We will note that some pieces are human activities required to establish trust; others are automated processes to preserve trust. Additionally, some of these steps are one-time "set-up" steps, and others are ongoing or repeated steps for each allocation request.

Resource Providers. To enable resources to be exchanged in a trusted manner, the resource provider needs to:

1. **Stand up a GENI Aggregate Manager service**. This is a one-time human activity: installing and configuring the GENI AM service and registering it with the GENI Service Registry.
2. **Trust the GENI Clearinghouse**. This is a one-time human activity: installing the GENI Certificate Authority (CA) root in the AM's trust root bundle. In indicates that the AM will trust users invoking AM calls using a certificate issued by this CA.
3. **Let the Aggregate Manager handle the Authentication, Authorization and Accountability of Resource consumers.** The AM performs these tasks automatically, authenticating using SSL and PKI certificate validation, authorizing per AM-local policy using provided credentials, and using GENI monitoring services to detect and respond to potential misbehavior.

Resource Consumers. To enable access resources from GENI, the resource consumer needs to:

1. **Become a member of the GENI Clearinghouse**. This is a one-time human activity: a request (by web form or email) is made to the managers of the GENI

Clearinghouse who vet that the member satisfies the policy requirements for GENI membership. If so, an account is created. In addition, an account at an academic institution participating in the InCommon Federation is sufficient to be given membership in GENI.

2. **Acquire GENI-generated identity credentials**. This is a one-time human activity: receiving the X.509 certificate and private key is an out-of-band process, often facilitated by tools such as the GENI Portal, that differs among GENI installations.
3. **Gain membership to a GENI Project**. This is a human process. If a GENI member wants to be a Project Lead (capable of creating a new project), a request is made and vetted by the managers of the GENI Clearinghouse. Otherwise, the member must find an existing project lead (typically a professor or lab manager) to add this member to an existing project.
4. **Gain membership to a GENI Slice**. Once a GENI member is a member of a project, that member may create a slice using GENI tools and Clearinghouse Slice Authority operations.
5. **Use a GENI Tool to get a Slice Credential from GENI Clearinghouse**. The Slice Authority service will provide a requesting authenticated tool with Slice Credentials to slice members, which indicate membership and role in a slice.
6. **Use a GENI Tool to request resources from a GENI Federated Aggregate Manager**. Using the AM API (using the private key and certificate plus Slice and perhaps other credentials), request resources of the Aggregate Manager. The request will succeed or fail based on the AM's available resources and its local policies for authorization and quotas.
7. **Use the resources**. This step is rests above the GENI Architecture. The resources are the consumer's to configure and use freely. That said, GENI is involved in at least two ways. First, the configuration of resources places SSH public keys and accounts on compute nodes allowing login to these resources. Second, GENI monitoring checks the resource consumption and traffic generation patterns to ensure against misbehaving software operating on the allocated slivers.
8. **Cleanup**. Use the AM API to renew slivers as needed, and then delete them at the Aggregate when done.

The following diagram summarizes these steps and the transactions with GENI services required to establish trusted resource exchange between resource providers and resource consumers. The providers and consumers do not need to know or trust one another: they merely need to mutually trust the GENI Clearinghouse (Fig. 4).

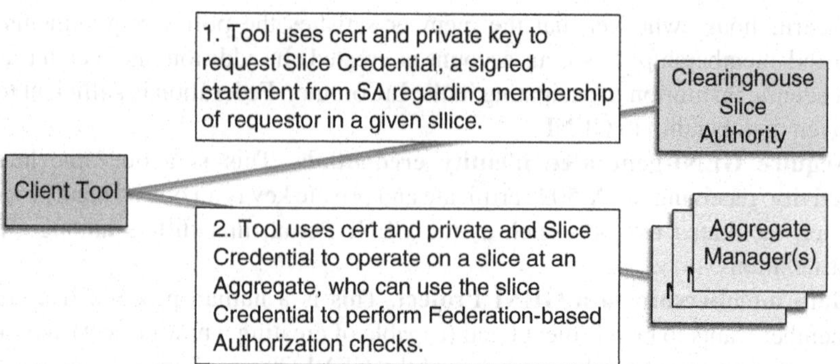

Fig. 4 Requesting resources from an Aggregate Manager requires passing a Slice Credential from the Clearinghouse Slice Authority from which the AM may make policy-based authorization decisions

References

1. Brinn, M., Bastin, N., Bavier, A., Berman, M., Chase, J., Ricci, R.: Trust as the foundation of resource exchange in GENI, TRIDENTCOM (2015)
2. Berman, M., Chase, J.S., Landweber, L., Nakao, A., Ott, M., Raychaudhuri, D., Ricci, R., Seskar, I.: GENI: a federated testbed for innovative network experiments. Comput. Netw. **61**, 5–23 (2014)
3. Freier, A., Karlton, P., Kocher, P.: RFC 6101: the Secure Sockets Layer (SSL) protocol, version 3.0 [Online]. http://tools.ietf.org/html/rfc6101 (2011)
4. Cooper, D., Santesson, S., Farrell, S., Boeyen, S., Housley, R., Polk, W.: RFC 5280: Internet X.509 public key infrastructure certificate and Certificate Revocation List (CRL) profile. http://tools.ietf.org/html/rfc5280 (2008)
5. International Telecommunication Union. ITU-T Recommendation X.509: Information Technology—Open Systems Interconnection—The Directory: public-key and attribute certificate frameworks. https://www.itu.int/rec/T-REC-X.509-201210-I/en (2012)
6. White, B., Lepreau, J., Stoller, L., Ricci, R., Guruprasad, S., Newbold, M., Hibler, M., Barb, C., Joglekar, A.: An integrated experimental environment for distributed systems and networks. SIGOPS Oper. Syst. Rev. **36**(SI), 255–270 (2002)
7. Tim, W., Vermeulen, B., Vandenberghe, W., Demeester, P., Taylor, S., Baron, L., Smirnov, M., et al.: Federation of internet experimentation facilities: architecture and implementation. In: European Conference on Networks and Communications, Proceedings, pp. 1–5 (2014)
8. Peterson, L., Sevinc, S., Lepreau, J., Ricci, R., Wroclawski, J., Faber, T., Schwab, S., Baker, S.: Slice-based facility architecture. http://svn.planet-lab.org/attachment/wiki/WikiStart/sfa.pdf (2009)
9. ABAC Development Team. ABAC. http://abac.deterlab.net/
10. Brinn, M., Duerig, J., Helsinger, A., Mitchell, T., Ricci, R., Rother, T., Stoller, L., Van de Meerssche, W., Vermeulen, B., Wong. G.: Common Federation API, version 2. http://groups.geni.net/geni/wiki/CommonFederationAPIv2 (2013)
11. GENI Aggregate Manager API, version 3 [Online]. http://groups.geni.net/geni/wiki/GAPI_AM_API_V3 (2012)
12. Morgan, R.L., Cantor, S., Carmody, S., Hoehn, W., Klingenstein, K.: Federated security: the shibboleth approach. EDUCAUSE Q. **27**(4), 8–22 (2004)

The Need for Flexible Community Research Infrastructure

Robert Ricci

Abstract Many areas of computing research have strong empirical components, and thus require testbeds, test networks, compute facilities, clouds, and other infrastructure for running experiments. The most successful facilities of these types are those built by the communities that need them: domain experts are in the best position to ensure that infrastructure they design meet the needs of their communities. The observation that we make in this chapter is that the hardware, and in many cases, software, infrastructure needs that underlie many of these facilities are remarkably similar. This points out the opportunity to build infrastructure that supports a wide range of computing research domains in an easy to use, cost effective, and low-risk manner. This chapter describes our vision for the future of computing research infrastructure.

1 Introduction

Many areas of computing research require, or can greatly benefit from, infrastructure for running experiments. For example, empirical evaluation is required for credibility in big data, networking, storage, cloud computing, operating systems, data mining, image analysis, databases and many more areas. Infrastructure is of course also critical to computation-heavy domain sciences. Our experience with Emulab [20], which supports primarily networking research, is that there is never enough hardware infrastructure to support all potential users, no matter how much is added [10]. We therefore see a clear and pressing need for flexible infrastructure—not only to support research communities that are the historic users of such infrastructure, but to reach out to other research communities as well.

Computing research moves quickly. Evolving infrastructure takes more time. It is therefore imperative to avoid making long-term commitments to particular software platforms in the planning and construction of large-scale infrastructure. For example, if we were to consider long-term research infrastructure in 2010, we would have likely missed supporting big data computing; a few years before that, and we

R. Ricci (✉)
The Flux Research Group, School of Computing, University of Utah,
Salt Lake City, UT, USA
e-mail: ricci@cs.utah.edu

© Springer International Publishing Switzerland 2016

R. McGeer et al. (eds.), *The GENI Book*, DOI 10.1007/978-3-319-33769-2_6

would have missed cloud computing; a few years earlier, virtualization. Because of the high capital investment required for infrastructure, it should be designed to last as long as possible. What is needed, then, is not simply facilities that target certain domain areas: instead, what is needed is a *meta-infrastructure* where researchers can go to build their own testbeds. (For the purposes of this chapter, we use the word "infrastructure" to refer to the underlying meta-infrastructure, and "testbed" or "facility" as shorthand for the variety of layers that can be built on top.)

Meta-infrastructure would offer researchers an environment that gives them top-to-bottom control of the software and hardware, and just as importantly, instrumentation at those layers. Without these properties, which are missing from commercial clouds and "off-the-shelf" clusters and datacenters, many researchers are limited to making incremental improvements within the bounds of existing commercial technologies rather than pursuing transformative research. To truly enable scientific advancement, researchers need facilities that offer them the possibility of access to the base layers of compute, network, and storage resources.

There are many existence proofs that infrastructure can be shared by simultaneous users [4, 8, 9, 16, 20]. What is critical to understand is the level of sharing which is acceptable, because sharing inevitably leads to artifacts caused by other users of the infrastructure. In some research areas, it is the output of a simulation that is of interest, not the running time of the experiment. These users can easily tolerate resource sharing. In others, aggregate run times (e.g. total download times, order-of-magnitude experiments in algorithms, etc.) are important, but fine-grained details are not. These users can generally tolerate artifacts that arise from infrastructure sharing, though they do need to be alerted if the artifacts are large enough to potentially affect their experiments—otherwise, scientific conclusions may be incorrect. Others depend on fine-grained measurements, where there is virtually no tolerance of interference from other experiments, and some researchers require the maximum computing power available, ruling out virtual machines and other middleware for sharing that imposes performance penalties. The more isolation an experiment requires, the further down the hardware/software stack the isolation must be implemented. At the limit, it should be possible to provide "bare metal" access to the resources in the infrastructure. Since the meta-infrastructure's goal is to provide for a diversity of testbeds on top, it follows that the meta-infrastructure should support bare metal. It is possible to build remove less isolated (and therefore cheaper) testbeds on top of bare metal, but it is impossible to build a strongly isolated facility on top infrastructure with weaker isolation. Adoption of technologies like virtual machines for the "base" layer of meta-infrastructure would necessarily compromise scientific fidelity for many users. At the same time, because virtualization *is* so prevalent and useful, meta-infrastructure should be capable of assisting the testbeds it hosts in provisioning and managing VMs.

There are also existence proofs of using taking general-purpose infrastructure and running testbeds and services on it that specialize it to particular uses. For example, in the GENI world, the GENI Experiment Engine [2] offers an alternative environment to GENI experimenters; it runs on top of InstaGENI racks without needing physical infrastructure of its own. Several different user tools provide alternative interfaces to the GENI infrastructure, including the GENI portal [18], the GENI Desktop [19], and LabWiki [12]. In the commercial space, there are numerous services that rent resources from clouds and other datacenters, and re-sell them to end users.

We have one existence proof of the potential for federation (autonomous infrastructure cooperating to act as a cohesive whole) in the form of GENI [14, 15]. Other countries have adopted similar strategies, including the Fed4FIRE [6] project in Europe and the FIBRE [7] project in Brazil. Two of the main motivations for federation are that (1) it allows many parties to contribute resources, spreading hardware costs, and (2) it is a natural model for operation of infrastructure that is spread across the country or world.

In the remainder of this chapter, we elaborate on our positions regarding meta-infrastructure, risk and cost reduction, and strategies for supporting a diverse user base.

2 Meta-Infrastructure

There are two main components to building infrastructure for research in computing, the *hardware* platform and the *software* environment that runs on the hardware. Note that "software environment" refers not to users' own software that they are evaluating, but to software that is "part of the testbed." This could include virtualization software, user interfaces such as web interfaces or APIs, standard or custom libraries such as MPI, or tools for orchestrating experiments such as scientific workflow systems.

We start with three observations:

1. The hardware needed by many research communities is quite similar: some combination of compute, networking, and storage. The hardware can generally be purchased from a commercial vendor (with careful input by the community as to its specification).[1]
2. Far more diversity is found in the software environment, and it is much harder to build a one-size-fits-all solution. Sometimes, the software can be acquired from a vendor, but more commonly, researchers have specialized needs that vendors do not have incentive to meet, so that software must be custom-built or modified from the original.
3. The most useful infrastructure projects are those whose software (and occasionally hardware) environment is built by the communities that need them—domain experts are in the best position to put together an environment that matches

[1]There is still clearly a need for smaller infrastructure for research communities that do not follow this pattern. In some cases, this need can be met by augmenting part of the infrastructure with specialized hardware (such as GPUs or other accelerators), but in others (such as wireless and mobile networking), building separate testbeds makes more sense.

their own and their colleagues' needs. In our own personal experience, we built Emulab because it was something we needed; it has turned out that it is what thousands of other researchers and educators need as well.

This points to a clear strategy for community infrastructure: build *meta-infrastructure that has a common hardware base, and empowers individual research communities to build their own infrastructure on top.* Meta-infrastructure would be operated using a "base" control framework, which would take care of common tasks such as provisioning resources, managing security, and ensuring the fidelity of the infrastructure. We expect that only a small number of users would interact directly with this base layer. Instead, experts in domains such as big data, HPC, cloud computing, security, networking, operating systems, simulation, etc. would build software environments on top which would cater to their own communities.

We are motivated to move in this direction by a simple conclusion which took us years to reach. As Emulab became more mature, and its userbase grew beyond network researchers, we began to notice a trend. We realized that more and more users who needed specialized features were not asking us to build these features into Emulab itself (or doing so themselves.) They were increasingly building these features *on top of* Emulab's existing features. Though Emulab had some virtual machine capabilities, some users wanted something slightly different, and so would allocate bare metal and build their own VMs. Though Emulab offered some basic features for running security experiments, they would build their own, highly sophisticated ones [17], on top. They would build what amounted to domain-specific testbeds by providing libraries and syntactic sugar for the experiment specification language. They would install cloud platforms such as Eucalyptus [13] on top of Emulab. At first we resisted this trend, believing that we should endeavor to support everyone directly with our platform. As this pattern became more common, however, we realized that it was a trend to embrace: trying to satisfy everyone would not scale in terms our effort, and would make Emulab itself too bulky. And ultimately, *we* were not the right ones to build these features, as we were not experts in those domains and would be likely to get things wrong. We have come to believe that the best path to follow is one of *empowering the experts in research domains to easily build and share software infrastructure for their own domains.*

Thus, we propose a distinct role for domain experts in community infrastructure. Meta-infrastructure should provide a layer of software between the underlying infrastructure and the ultimate users that enables experts to adapt the infrastructure to support experiments in a particular scientific domain, thereby making the infrastructure more valuable to its users. This is a significant departure from the way that testbeds are built today—it adds a new layer between the infrastructure and the testbed(s) that run atop it. The advantage is that it simultaneously makes infrastructure more useful and significantly decreases the associated cost and risk, as discussed in the next section.

Figure 1 shows a diagram of how meta-infrastructure might be designed. At the bottom is a meta control system that controls the base layer of allocation, assigning different quantities of resources to the testbeds shown on the right. These testbed

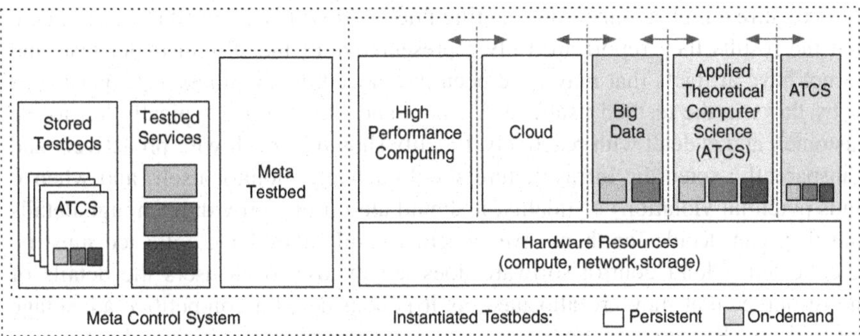

Fig. 1 Architecture of meta-infrastructure

are of two types. *Persistent* testbeds are long-lived, multi-user facilities such as clouds and HPC clusters. *On-demand* testbeds are relatively short-lived, perhaps lasting less than a day and instantiated to run experiments by one individual or group. Both type of testbeds use can optionally use *testbed services* (shown on the left) provided on top of the control system to perform common functions such as provisioning of physical machines, account management, and security management. The meta control system also exposes a meta-testbed for those who wish to use the meta-infrastructure directly, and offers a service for storing testbed descriptions and software when not in use.

In order for this scheme to work, the base control layer must be flexible enough to move resources between different testbeds as their needs vary. The more efficient this re-provisioning process, the more efficient the use of resources. Re-provisioning on the order of minutes is already possible [11, 20]. This provisioning should be done at the lowest layer of the software/hardware stack possible, allowing for maximum flexibility for the testbeds running on top.

It is tempting to consider off-the-shelf cloud control software in this role; however, there are important differences in needs. First, it naturally (though not necessarily) places users of the meta-infrastructure inside of virtual machines; while this should certainly be one option, many research topics either cannot be investigated at all (including research on building cloud infrastructure and virtualization) or suffer serious performance penalties (such as many types of HPC jobs) when run in this kind of environment. Second, it binds us up-front to particular technology decisions (i.e. VM technology and cloud software) and will be hard to change over time, running the risk that we make a poor choice in this fast-evolving area, and get stuck with it. This may be acceptable for small, purpose-built testbeds, but is not appropriate for large facilities that need the flexibility to serve a diverse population of researchers.

Finally, and most importantly, cloud infrastructure simply does not meet the standards of a scientific instrument because there are subtle, but critically important, differences in the needs of typical cloud users and researchers. While cloud providers provide service level agreements (SLAs) to their customers, research users require a higher standard: scientific fidelity. Researchers need to be confident that the results they report are truly representative of the system under test, and do not have artifacts that may have been introduced by simultaneous experiments or by flaws in the testbed itself. SLA compliance detection is typically left to the customer, and is dealt with reactively. Fidelity should be dealt with proactively and transparently: scientific infrastructure should actively monitor itself, and when it finds potential violations of fidelity, it should alert users, providing enough details that they can decide for themselves whether results they have collected must be re-collected. Cloud control software does not expose to its users the details of the resources that they are allocated, as it is considered a competitive advantage not to do so. Researchers, however, need transparency, as it is required for a deep understanding of the performance of their systems.

3 Risk and Cost Reduction

Building infrastructure for computing research involves significant cost, and with that cost comes risks: the infrastructure may not turn out to be what researchers truly need, it may fail to attract users, and it may not age well as technologies improve and research directions change.

Some kinds of research infrastructure, such as computational clusters, are well established, and their usage models, applications, value, and risks are fairly well understood. This is *not* true of infrastructure for emerging research areas, nor is it true for research ares whose infrastructure needs to evolve rapidly. In these areas, it can take quite some time for the value (or lack thereof) of infrastructure to become apparent and for it to see high levels of use. Building a large userbase for novel infrastructure is fundamentally a bottom-up process of community building: to the majority of researchers, the most convincing argument for adopting a new facility is seeing other researchers who are successful at using it. While it is certainly possible to "advertise" new infrastructure, results are louder than words, and it is ultimately initial success that breeds later success. One way for a community of users to begin to build is highly visible publications from the infrastructure's early users, which can take years of planning, development, submission, rejection, revision, and acceptance. Another is for researchers who find the facility valuable as PhD students to graduate and move to other institutions, spreading knowledge of and expertise with the facility in their new faculty or post-doc positions. This process has a time horizon that can be 5 years or more.

This leaves us with a "chicken and egg" problem: we clearly cannot afford the risk of long-term investments in infrastructure before its value is demonstrated, yet at the same time, it can take many years for the infrastructure to prove its worth. We believe the best way to break this deadlock is to build meta-infrastructure. This keeps the initial investment in a new facility low: instead of investing large amounts in capital equipment to build a brand-new facility, we can invest in "human capital" for the development of facilities targeted at specific research areas. (Additional equipment investment will be periodically required to keep the infrastructure current.) Facility development projects can begin with a small allocation of resources in already-established infrastructure; those that prove to be valuable can be given more resources. Those that fail to gain traction can eventually have their resources re-allocated to more successful testbeds, or to other new testbeds just starting out. We can afford to try out more novel, higher-risk facility development projects because we need not commit to them long term, or run the risk that we cut them off too soon, before their value is proved or disproved.

Because the meta control system need not directly support particular research domains, it can evolve largely independent of research trends, and will rarely be on the critical path to supporting new types of testbeds. Testbeds and other facilities built on top of the meta-infrastructure can evolve rapidly or slowly along with their research communities, and many strategies can be tried in parallel. For example, if a new cloud computing framework appears, an instance of it can be brought up in parallel to the older infrastructure, and as usage demonstrates their relative value, one can be kept up and the other shut down, or both can continue to run to serve different constituencies.

4 Maximizing Research

It goes without saying that the goal of a research infrastructure project is to maximize the amount of useful research that is produced. We believe that the best way to do this, particularly early in the lifecycle of an infrastructure project, is to aim to maximize the diversity of research that it supports. (Later in the lifecycle of the infrastructure, when demand from users far exceeds the available resources, processes for determining user limits and priorities become more necessary.) Because much valuable research comes from the "long tail," and it can be quite difficult to predict the success of projects and experiments, giving access to many and diverse users is an excellent strategy for maximizing the research output of the facility.

The first fundamental guiding principle in determining access policy should be that *the level of administrative process that a user encounters should be proportional to the level of use*. Concretely, what this means is that a user just trying out the infrastructure for the first time, or who has very modest resource needs should not have to go through a laborious application or lengthy approval process—our experience is that what infrastructure operators see as lightweight processes are viewed by those who are not yet certain of the value of the infrastructure as large hurdles, discouraging use. As time goes on and light users of the facility see it as a critical part of their research and become heavier users, it is appropriate to involve them in a more heavyweight approval process. At this point, the user is more invested in the infrastructure and unlikely to be deterred.

A second way to maximize usage is by flexible resource usage policies in which usage (i.e. quantity and duration of resource allocation) is dealt with reactively, rather than proactively. It is very difficult, and potentially harmful, to set a priori limits to resource usage. One reason for this is the wide range of users that would be supported by meta-infrastructure; what constitutes "light use" for a programming language researcher is quite different from light use from a big data researcher. Even within programming languages, needs vary greatly—some language researchers require only a single node and no network, while others work on languages for parallel cluster resources that require large numbers of cores distributed across many physical PCs in order to evaluate. The result is that a priori policies cannot be anything but arbitrary, and researchers getting started on the facility may be unable to make any progress under a regime of proactive enforcement. We believe that a better policy is to have reactive resource use policies to deal with outliers; i.e. heavy users may be asked to go through the heavyweight approval process to continue their level of resource usage.

One technique we successfully use in Emulab is *idle detection*; we use various metrics to detect allocated resources that are not being used. When a user is found to be under-utilizing their allocated resources, they are automatically reclaimed (after a warning). Users who are chronic under-users are contacted by the operations staff. Since most experiments are conducted by students or post-doctoral researchers, we have found that a short note to a supervisor goes a long way to curbing problems of this sort, since faculty do not want to be seen as bad citizens. This mechanism has been extremely effective in making sure that resources in Emulab are well-utilized.

A third way that meta-infrastructure can maximize usage is by being adaptable to differing research domains. Researchers are more likely to adopt infrastructure with a user interface that is appropriate to their domain. This may take the form of interfaces for software that already exists, or it may take the form of interfaces specially-built for experimenter facilities.

Understandably, researchers would rather concern themselves with their research than with infrastructure for experiments, and if the latter proves to take much time or effort, they will prefer to put their effort into the former—even if this means making sacrifices in the scope, scale, or evaluation of their research. This is a human "problem" with a straightforward technical solution: if we build infrastructure that empowers users in a domain to build their own testbeds, and then share those testbeds with others, the effort necessary to run experiments in that domain goes down drastically. Conversely, the diversity of users, and thus the research output of the infrastructure, will go up, and the infrastructure will support a broader range of computing research. For example, a user with experience with HPC schedulers is more likely to user a facility that exports that familiar interface than one that has its own interface, and a researcher not familiar with using testbeds at all would prefer to use an interface only exposing the details relevant to her or his own domain.

Finally, the software developed for the meta-infrastructure control system, and as many of the higher-level facilities as possible, should be released as open source. Our personal experience is that we built the Emulab software for use with our own facility, but that it quickly became clear that it suited others' needs as well. We released it as open source, and there are now approximately 50 installations of the software worldwide [5]; some support research, some support education, and some support internal R&D activities in industry.

5 Conclusion

In summary, we believe that there is a demonstrated need for flexible research infrastructure, and that it can and should reach out to a very broad community of computing researchers. Our experience suggests that it should be approached from the perspective of meta-infrastructure. Many testbed facilities can share the same underlying infrastructure with varying levels of resource isolation and scientific fidelity. This also creates a role for domain experts, who are in the best position to design testbed facilities for their own domains. Doing so amortizes cost and reduces risk. At the same time, it will help to maximize the number and diversity of users, maximizing research output.

In closing, it is also worth noting that research infrastructure is an interesting object of study in its own right: studies of its use can be used to inform the design of future infrastructure, both research and commercial. We have published a study analyzing half a million network topologies submitted to Emulab [10] in order to improve the physical design of infrastructure. We have also studied disk image usage in an effort to produce improved disk imaging for cloud infrastructures [1].

Acknowledgements The ideas in this chapter are heavily informed by discussions throughout the years with dozens of other Emulab designers and implementors, members of the GENI developer community, and other testbed designers. They are also informed by discussions with users of those facilities.

References

1. Atkinson, K., Wong, G., Ricci, R.: Operational experiences with disk imaging in a multi-tenant datacenter. In: Proceedings of the Eleventh USENIX Symposium on Networked Systems Design and Implementation (NSDI) (2014)
2. Bavier, A., Chen, J., Mambretti, J., McGeer, R., McGeer, S., Nelson, J., O'Connell, P., Ricart, G., Tredger, S., Coady, Y.: The GENI experiment engine. In: 26th International Teletraffic Congress (2014)
3. Brinn, M., Bastin, N., Bavier, A., Berman, M., Chase, J., Ricci, R.: Trust as the foundation of resource exchange in GENI. In: Proceedings of the 10th International Conference on Testbeds and Research Infrastructures for the Development of Networks and Communities (Tridentcom) (2015)
4. Chun, B., Culler, D., Roscoe, T., Bavier, A., Peterson, L., Wawrzoniak, M., Bowman, M.: PlanetLab: an overlay testbed for broad-coverage services. ACM SIGCOMM Comput. Commun. Rev. **33**(3), 3–12 (2003)
5. Emulab.net: Other Emulab testbeds. Flux Research Group, School of Computing, University of Utah. https://wiki.emulab.net/Emulab/wiki/OtherEmulabs (2016). Accessed Jan 2016

6. Fed4FIRE web site. http://www.fed4fire.eu (2016). Accessed Jan 2016
7. FIBRE: Future Internet testbeds experimentation between Brazil and Europe. http://www.fibre-ict.eu (2016). Accessed Jan 2016
8. FutureGrid: A distributed testbed, exploring possibilities with clouds, grids, and high performance computing (web site). http://portal.futuregrid.org (2016). Accessed Jan 2016
9. GENI Project Office, BBN Technologies. GENI: Exploring networks of the future (web site). http://www.geni.net (2016). Accessed Jan 2016
10. Hermenier, F., Ricci, R.: How to build a better testbed: lessons from a decade of network experiments on emulab. In: Proceedings of the 8th International ICST Conference on Testbeds and Research Infrastructures for the Development of Networks and Communities (Tridentcom) (2012). Awarded Best Paper
11. Hibler, M., Ricci, R., Stoller, L., Duerig, J., Guruprasad, S., Stack, T., Webb, K., Lepreau, J.: Large-scale virtualization in the Emulab network testbed. In: Proceedings of the USENIX Annual Technical Conference (2008)
12. LabWiki web page. http://labwiki.mytestbed.net/
13. Nurmi, D., Wolski, R., Grzegorczyk, C., Obertelli, G., Soman, S., Youseff, L., Zagorodnov, D.: The eucalyptus open-source cloud-computing system. In: IEEE/ACM International Symposium on Cluster Computing at the Grid (2009)
14. Ricci, R., Duerig, J., Stoller, L., Wong, G., Chikkulapelly, S., Seok, W.: Designing a federated testbed as a distributed system. In: Proceedings of the 8th International ICST Conference on Testbeds and Research Infrastructures for the Development of Networks and Communities (Tridentcom) (2012)
15. Ricci, R., Wong, G., Stoller, L., Duerig, J.: An architecture for international federation of network testbeds. IEICE Trans. **E96-B**(1), 2–9 (2013). Invited paper
16. Ricci, R., Wong, G., Stoller, L., Webb, K., Duerig, J., Downie, K., Hibler, M.: Apt: a platform for repeatable research in computer science. ACM SIGOPS Oper. Syst. Rev. **49**(1), 100–107 (2015)
17. Schwab, S., Wilson, B., Ko, C., Hussain, A.: SEER: a security experimentation environment for deter. In: DETER Community Workshop on Cyber Security Experimentation and Test (2007)
18. The GENI Portal. http://portal.geni.net/
19. The GENI Desktop. https://genidesktop.netlab.uky.edu/ (2016). Accessed Jan 2016
20. White, B., Lepreau, J., Stoller, L., Ricci, R., Guruprasad, S., Newbold, M., Hibler, M., Barb, C., Joglekar, A.: An integrated experimental environment for distributed systems and networks. In: Proceedings of the USENIX Symposium on Operating System Design and Implementation (OSDI). USENIX (2002)

A Retrospective on ORCA: Open Resource Control Architecture

Jeff Chase and Ilya Baldin

Abstract ORCA is an extensible platform for building infrastructure servers based on a foundational leasing abstraction. These servers include Aggregate Managers for diverse resource providers and stateful controllers for dynamic slices. ORCA also defines a brokering architecture and control framework to link these servers together into a federated multi-domain deployment. This chapter reviews the architectural principles of ORCA and outlines how they enabled and influenced the design of the ExoGENI Racks deployment, which is built on the ORCA platform. It also sets ORCA in context with the GENI architecture as it has evolved.

1 Introduction

The Open Resource Control Architecture (ORCA) is a development platform and control framework for federated infrastructure services. ORCA has been used to build elements of GENI, most notably the ExoGENI deployment [4]. In ExoGENI the ORCA software mediates between GENI user tools and the various infrastructure services (IaaS) that run a collection of OpenFlow-enabled cloud sites with dynamic layer-2 (L2) circuit connectivity.

ORCA is based on the SHARP resource peering architecture [12], which was conceived in 2002 for federation in PlanetLab [11] and related virtual infrastructure services [5, 8] as they emerged. ORCA incorporates the Shirako resource leasing toolkit [14, 18, 20] and its plug-in extension APIs. It incorporates the research of three Duke PhD students [13, 16, 23], which was driven by a vision similar to GENI: a network of federated resource providers enabling users and experimenters to build custom *slices* that combine diverse resources for computing, networking, and storage.

J. Chase (✉)
Duke University, Durham, NC, USA
e-mail: chase@cs.duke.edu

I. Baldin
Renaissance Computing Institute (RENCI)/UNC Chapel Hill, Chapel Hill, NC, USA

© Springer International Publishing Switzerland 2016

R. McGeer et al. (eds.), *The GENI Book*, DOI 10.1007/978-3-319-33769-2_7

When construction of GENI began in 2008, ORCA was selected as a candidate control framework along with three established network testbeds: PlanetLab, Emulab, and ORBIT. ORCA was the only candidate framework that had been conceived and designed from the ground up as a platform for secure federation, rather than to support a centrally operated testbed. At that time ORCA was research software with no production deployment, and so it was more speculative than the other control framework candidates. We had used it for early experiments with elastic cloud computing [10, 17–20, 24], but we were just beginning to apply it to network resources [2, 9].

The GENI project was therefore an opportunity to test whether we had gotten ORCA's architecture and abstractions right, by using it to build and deploy a multi-domain networked IaaS system. The crucial test was to show that ORCA could support advanced network services, beginning with RENCI's multi-layer network testbed (the Breakable Experimental Network—BEN) in the Research Triangle region.

During the GENI development phase (2008–2012), participants in the GENI *Cluster D* group, led by RENCI, built software to control various infrastructures and link them into a federated system using ORCA. For example, the Kansei group built KanseiGenie [21], an ORCA-enabled wireless testbed. Through this phase ORCA served as a common framework to organize these efforts and link them together. This was possible because ORCA was designed as an *orchestration platform* for diverse resource managers at the back end and customizable access methods at the front end, rather than as a standalone testbed itself.

ORCA was based on the premise that much of the code for controlling resources in a system like GENI would be independent of the specific resources, control policies, and access methods. The first step was to write the common code once as a generic *toolkit*, keeping it free of assumptions about the specific resources and policies. The second step was to plug in software adapters to connect the toolkit to separately developed IaaS resource managers, which were advancing rapidly outside of the GENI effort.

The RENCI-led team used ORCA's plug-in extension APIs to implement various software elements later used in ExoGENI. They include: a control system and circuit API for the BEN network, modules to link ORCA with off-the-shelf cloud managers, storage provisioning using commercial network storage appliances, VLAN-sliced access control for OpenFlow-enabled network dataplanes, adapters for various third-party services, and a front-end control interface that tracked the GENI API standards as they emerged. Section 2.5 summarizes some ORCA plugins used in ExoGENI; the accompanying chapter on ExoGENI [1] discusses these examples in detail.

A key outcome of the BEN experience was a methodology for describing network resources and configurations declaratively using a powerful logic-based *semantic resource description language* (NDL-OWL) [2, 3], originally for use by the BEN plug-in modules (Sect. 2.4). The language represents attributes and relationships (e.g., attachment points, connectivity, protocol compatibility, layering) among virtual infrastructure elements and network substrate resources. Descriptions in the language (models) drive policies and algorithms to co-schedule compute and network resources and interconnect them according to their properties and dependencies. NDL-OWL is now used for all resources in ExoGENI. The NDL-OWL support was a substantial effort in itself, but we were able to add logic-based resource descriptions to ORCA and use them to build ExoGENI without modifying the ORCA core.

This approach enabled us to demonstrate key objectives of GENI—automated embedding and end-to-end assembly (*stitching*) of L2 virtual network topologies spanning multiple providers—by early 2010 [3, 22], well before GENI had defined protocols to enable these functions. ORCA already defined a protocol and federation structure similar to what was ultimately adopted in GENI; we essentially used that structure to link together third-party back-end resource managers as they appeared, and control them through their existing APIs. This philosophy carried through to the ExoGENI effort when it was funded in 2012: the *exo* prefix refers in part to the idea of incorporating resources and software from outside of GENI and exposing their power through GENI APIs.

The remainder of this chapter outlines the ORCA system in more detail, illustrating with examples from ExoGENI. Section 2 gives an overview of ORCA's abstractions and extension mechanisms, and the role of logic-based semantic resource descriptions. Section 3 summarizes ORCA's architecture for federating and orchestrating providers based on *broker* and *controller* services. Section 4 sets ORCA in context with the GENI architecture as it has evolved.

2 Overview of the ORCA Platform

ORCA and GENI embody the key concepts of slices, slivers, and aggregates derived from their common heritage in PlanetLab. An *aggregate* is a resource provider: to a client, it appears as a hosting site or domain that can allocate and provision resources such as machines, networks, and storage volumes. An Aggregate Manager (AM) is a service that implements the aggregate's resource provider API. A *sliver* is any virtual resource instance that is provisioned from a single AM and is named and managed independently of other slivers. Slivers have a lifecycle and operational states, which a requester may query or transition (e.g., **shutdown**, **restart**).

Client tools call the AMs to allocate and control slivers across multiple aggregates, and link them to form end-to-end environments (slices) for experiments or applications. A *slice* is a logical container for a set of slivers that are used for some common purpose. Each sliver is a member of exactly one slice, which exists at the time the sliver is created and never changes through the life of the sliver. The slice abstraction serves as a basis for organizing user activity: loosely, a slice is a unit of activity that can be enabled, disabled, authorized, accounted, and/or contained.

In ORCA we refer to a client of the AM interface as a Slice Manager (SM). Each request from an SM to an AM operates on one or more slivers of exactly one slice. The role of a slice's SM is to plan and issue AM requests to build the slice to order with suitable end-to-end connectivity according to the needs of an experiment or application. ORCA is a toolkit for building SMs and AMs by extending a generic core, and linking them together in a federation using SM *controller* extensions and common services for resource brokering, user identity, and authorization.

This section summarizes the ORCA abstractions and principles, focusing on how to use ORCA to implement diverse AMs. Figure 1 illustrates an ORCA aggregate and the elements involved in issuing a sliver lease to a client SM. ORCA bases all resource management on the abstraction of *resource leases* (Sect. 2.1). The core leasing engine is generic: ORCA factors support for specific resources and policies into replaceable *extension modules* ("plugins") that plug into the core (Sect. 2.2). The extensions produce and/or consume declarative *resource descriptions* that represent information needed to manage the resources (Sect. 2.4). The common leasing abstractions and plug-in APIs facilitate implementation of AMs for diverse resources (Sect. 2.5).

2.1 Resource Leases

ORCA resource leases are explicit representations of resource commitments granted or obtained through time. The lease abstraction is a powerful basis for negotiating and arbitrating control over shared networked resources. GENI ultimately adopted an equivalent leasing model in the version 3.0 API in 2012.

A lease is a promise from an aggregate to provide one or more slivers for a slice over an interval of time (the *term*). Each sliver is in the scope of exactly one lease. A'lease is materialized as a machine-readable document specifying the slice, sliver(s), and term. Each lease references a description (Sect. 2.4) of the slivers and

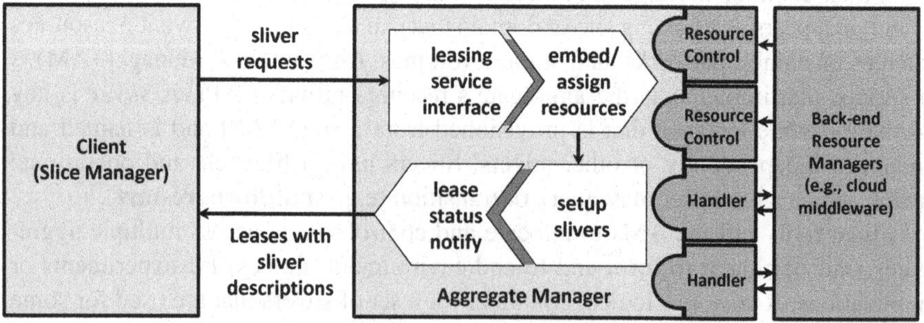

Fig. 1 Structure of an ORCA aggregate. An Aggregate Manager (AM) issues sliver leases (Sects. 2.1 and 2.3) to clients called Slice Managers (SMs). The AMs are built from a generic leasing server core in the ORCA platform (*light shade*). The core invokes plug-in extension modules (*dark shade*) for resource-specific functions of each aggregate (Sect. 2.2). These extensions may invoke standard APIs of off-the-shelf IaaS resource managers, e.g., OpenStack. Their resource control functions are driven by logical descriptions of the managed resources (Sect. 2.4). ExoGENI uses this structure for diverse aggregates, including network providers (Sect. 2.5)

the nature of the access that is promised. Leases are authenticated: ORCA leases are signed by the issuing AM. Lease contracts may be renewed (*extended*) or *vacated* prior to expiration, by mutual agreement of the SM and AM. If an SM abandons a sliver, e.g., due to a failure, then the AM frees the resources when the lease expires.

ORCA is based on SHARP [12], which introduced a two-step leasing API in which the client first obtains an approval to allocate the resources (a *ticket*), and then redeems the ticket to claim the resources and provision (instantiate) the slivers. A ticket is a weaker promise than a lease: it specifies the promised resources abstractly. The AM assigns (binds) concrete resources to fill the ticket only when it is redeemed. In ORCA (as in SHARP) the tickets may be issued by *brokers* outside of the aggregates (Sect. 3.2).

By *separating allocation and provisioning* in this way, the leasing API enables a client to obtain promises for resources at multiple AMs cheaply, and then move to the redeem step only if it succeeds in collecting a resource bundle (a set of tickets) matching its needs. The two-step API is a building block for grouped leases and *atomic co-allocation*—the ability to request a set of slivers such that either the entire request succeeds or it aborts and no sliver is provisioned. The AM may commit resources cheaply in advance, and then consider current conditions in determining how to provision the resources if and when they are needed.

From the perspective of the AMs, leases provide a means to control the terms under which their resources are used. The resource promises may take a number of forms expressible in the logic, ranging along a continuum of assurances ranging from a hard physical (e.g., bare metal) reservation to a weak promise of best-effort service over the term. By placing a time bound on the commitments, leases enable AMs to make other promises for the same resources in the future (*advance reservations*).

From the perspective of the SMs, leases make all resource allotments explicit and visible, enabling them to reason about their assurances and the expected performance of the slice. Since the SM may lease slivers independently of one another, it can modify the slice by adding new slivers and/or releasing old slivers, enabling *elastic* slices that grow and shrink according to need and/or resource availability. Various research uses of the ORCA software experimented with elastic slice controllers (Sect. 3.3), building on our early work in adaptive resource management for hosting centers [7, 8]. ExoGENI supports elastic slices by using the native ORCA APIs internally, due to limitations of the early GENI APIs, which did not support elastic slices.

2.2 Extension Modules

ORCA is based on a generic reusable leasing engine with dependencies factored into stackable plug-in extension modules (the "plugins") [18]. The core engine tracks lease objects through time in calendar structures that are accessible to the extensions. For example, an AM combines the leasing engine with two types of extensions that are specific to the resources in the aggregate:

- **`ResourceControl`.** The AM core upcalls a `ResourceControl` policy module periodically with batches of recently received sliver requests. Its purpose is to `assign` the requests onto the available resources. The batching interval is a configurable parameter. It may defer or reject each request, or approve it with an optional binding to a resource set selected from a pool of matching resources. The module may query the attributes of the requester, the state of the resources and calendar, other pending requests, and the history of the request stream.
- **`Handler`.** The AM core upcalls a `Handler` module to `setup` a sliver after approval, or `teardown` a sliver after a lease is closed (expired, cancelled, or vacated). Resource handlers perform any configuration actions needed to implement slivers on the back-end infrastructure. The handler API includes a `probe` method to poll the current status of a sliver, and a `modify` method to adjust its properties.

An ORCA AM may serve multiple types of slivers by combining multiple instances of these modules, which are indexed and selectable by sliver type. Each upcall runs on its own thread and is permitted to block, e.g., for configuration actions in the handler, which may take seconds or minutes to complete. The extensions post their results asynchronously through lease objects that are shared with the leasing core.

2.3 Leasing Engine

The ORCA lease abstraction defines the behavior of a resource lease as a set of interacting state machines on the servers that are aware of it. The lease state machines have well-defined states and state transition rules specific to each type of server. Figure 2 illustrates typical states, transitions, and actions.

The core engine within each server serializes state machine transitions and commits them to stable storage. After a transition commits, it may trigger asynchronous actions including notifications to other servers, upcalls to extension modules, and various other maintenance activities.

Lease state transitions and their actions are driven by the passage of time (e.g., sliver `setup` at the start of the term and `teardown` at the end of the term), changes in status of the underlying resources (e.g., failure), decisions by policy extension modules, and various API calls.

Cross-server interactions in the leasing system are asynchronous and resilient. After a failure or server restart, the core recovers lease objects and restarts pending actions. The extensions may store data blobs and property lists on the lease objects. The core upcalls each extension with the recovered lease objects before restarting any actions. The servers and extensions are responsible for suppressing any effects from duplicate actions that completed before a failure but were restarted or reissued on recovery.

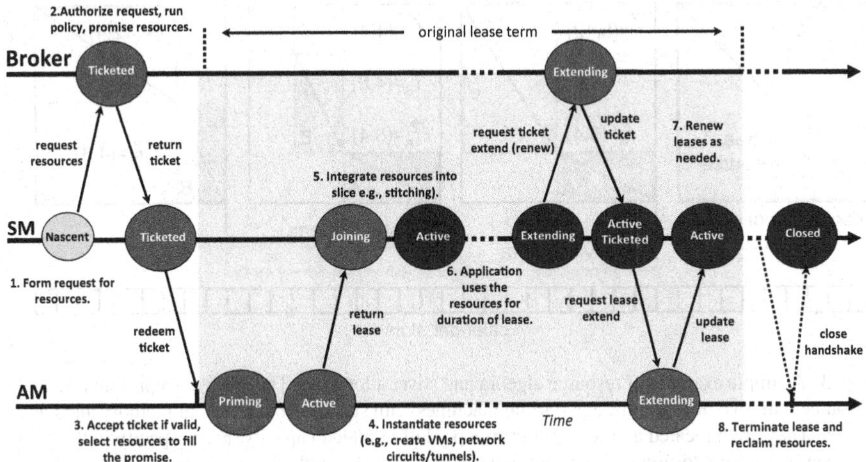

Fig. 2 Lease states and transitions. Interacting state machines representing a single ORCA lease at three servers: the SM that requested the resource, the broker that issued the ticket, and the provider (AM) that sets up the slivers, issues the lease, and tears down the resource when the lease expires. Each state machine passes through various intermediate states, triggering policy actions, local provisioning actions, and application launch. This figure is adapted from [18]

2.4 Resource Descriptions

The ORCA platform makes it possible to build new AMs quickly by implementing the `Handler` and `ResourceControl` modules. Since the leasing core and protocols are generic, there must be some means to represent resource-specific information needed by these modules. This is achieved with a data-centric API in which simple API requests and responses (`ticket/redeem/renew/close`) have attached *descriptions* that carry this content. The descriptions contain statements in a declarative language that describe the resources and their attributes and relationships.

The description language must be sufficiently powerful to describe the *resource service* that the aggregate provides: what kinds of slivers, sizes and other options, constraints on the capacity of its resource pools, and interconnection capabilities for slivers from those pools. It must also be able to describe resources at multiple levels of abstraction and detail. In particular, clients describe their *sliver requests* abstractly, while the aggregate's descriptions of the *provisioned slivers* are more concrete and detailed. GENI refers to these cases as *advertisement*, *request*, and *manifest* respectively.

In ORCA the descriptions are processed only by the resource-specific parts of the code, i.e., by the extension modules. The core ignores the descriptions and is agnostic to their language. An ORCA resource description is a set of arbitrary strings, each indexed by a key: it is a property list. ORCA defines standard labels for distinct property lists exchanged in the protocols, corresponding to the advertisement, request, and manifest cases.

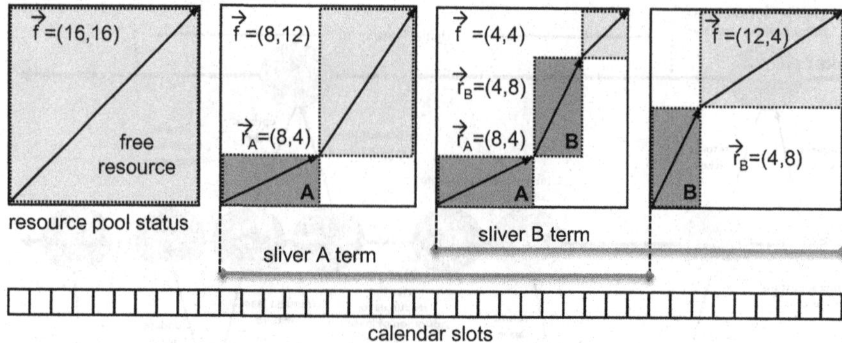

Fig. 3 A simple example of resource algebra and sliver allocation. These slivers represent a virtual resource with two dimensions, e.g., virtual machines with specified quantities of memory and CPU power. They are allocated and released from a pool of bounded capacity. Free capacity in the pool is given by vector addition and subtraction as slivers are allocated and released

To support meaningful resource controls, the description language must enable a *resource algebra* of operators to split and merge sliver sets and resource pools. Given descriptions of a resource pool (an advertisement) and a set of slivers, a resource algebra can determine if it is feasible to draw the sliver set from the pool, and if so to generate a new description of the resources remaining in the pool. Another operator determines the effect of releasing slivers back to a pool.

The original SHARP/Shirako lease manager [18] used in ORCA described pools as quantities of interchangeable slivers of a given type. A later version added resource controls using an algebra for multi-dimensional slivers expressed as vectors [13], e.g., virtual machines with rated CPU power, memory size and storage capacity, and IOPS. Figure 3 depicts a simple example of resource algebra with vectors.

In the GENI project we addressed the challenge of how to represent complex network topologies and sliver sets that form virtual networks over those topologies. For this purpose we adopted a logic language for *semantic resource descriptions*. Logical descriptions expose useful resource properties and relationships for inference using a generic reasoning engine according to declarative rules. In addition to their expressive power, logical descriptions have the benefit that it is semantically sound to split and combine them, because they are sets of independent statements in which all objects have names that are globally unique and stable. For example, a logical slice description is simply a concatenation of individual sliver descriptions, each of which can be processed independently of the others. Statements may reference objects outside of the description, e.g., to represent relationships among objects, including graph topologies.

To this end, the RENCI team augmented the Network Description Language [15] with a description logic ontology in the OWL semantic web language. We called the resulting language NDL-OWL [3]. We used NDL-OWL to describe infrastructures orchestrated with ORCA: BEN and other networks and their attached edge resources, including virtual machine services. For example, NDL-OWL enables us to enforce semantic constraints on resource requests, express path selection and topology embedding as SPARQL queries over NDL-OWL semantic resource descriptions, manage capacity constraints across sequences of actions that allocate and release resources (an "algebra"), check compatibility of candidate interconnections (e.g., for end-to-end VLAN tag stitching, Sect. 3.4), and generate sequences of handler actions to instantiate slivers automatically from the descriptions. These capabilities are implemented in extension modules with no changes to the ORCA core.

2.5 Building Aggregates with ORCA

We used ORCA and NDL-OWL to build a collection of Aggregate Managers (AMs) for back-end resource managers from other parties. In ExoGENI, these include two off-the-shelf cloud resource managers: OpenStack for Linux/KVM virtual machines and xCAT for bare-metal provisioning. These systems expose local IaaS APIs to allocate and instantiate resources. Each AM runs one or more `Handler` modules that invoke these back-end control APIs. The AM structure makes their resources available through the leasing APIs, and provides additional functions to authorize user requests and connect a slice's slivers on the local aggregate with other resources in the slice.

We augmented the cloud aggregates with additional back-end software to function as sites in a *networked IaaS federation* under a separate NSF SDCI project beginning in 2010. The added software includes a caching proxy for VM images retrieved by URL, linkages to an off-the-shelf OpenFlow access control proxy (FlowVisor) to enable slices to control their virtual networks, and custom OpenStack extensions for dynamic attachment of VM instances to external L2 circuits. We also added handlers to invoke storage provisioning APIs of third-party storage appliances. These elements are independent of the cloud manager: the AM handlers orchestrate their operation.

We also implemented AMs for network management to provide *dynamic circuit service* for a network of cloud sites under ORCA control. Most notably, the control software for BEN and its L2 circuit service were implemented natively as ORCA extensions in 2009; the AM handlers issue direct commands to the vendor-defined APIs on the BEN network elements. ExoGENI also includes other circuit AMs that proxy third-party L2 circuit services from national-footprint backbone providers, including the OSCARS services offered by ESNet and Internet2. For these systems the AM handler calls the circuit API under its own identity; the circuit provider does not know the identities of the GENI users on their networks. In effect, the provider implicitly delegates to the AM the responsibility to authorize user access, maintain user contacts and billing if applicable, and provide a kill switch. This approach was easy to implement without changing the circuit providers: the AM interacts with the provider using its standard client APIs.

In ExoGENI, these various additions permit users to obtain resources at multiple OpenStack sites, without the need to register identities and/or images at each site. The network AMs and OpenStack cloud site AMs together enable ExoGENI slices to connect their resources at multiple federated cloud sites into a slice-wide virtual L2 network topology, use OpenFlow SDN to control the topology, and interconnect their OpenStack VMs with other resources outside of OpenStack. The broker and controller elements of ORCA (Sect. 3) can select the sites to satisfy a request automatically, without the user having to know the sites.

Finally, for ExoGENI we implemented an AM to control *exchange points*—RENCI-owned switches installed at peering centers where multiple transit providers come together (e.g., Starlight). These switches implement VLAN tag translation: the exchange AM uses this capability to stitch circuits from different providers into a logical end-to-end circuit, expanding the connectivity options for circuits on an ExoGENI slice data plane. The handler for the exchange AM issues direct commands to the vendor API on the switches, similarly to the BEN control software.

3 Orchestration and Cross-Aggregate Resource Control

Section 2 described how we can build diverse aggregates with ORCA by plugging resource-specific and policy-specific extension modules into a common leasing core, and accessing their resources via common leasing protocols. This approach can apply to any aggregate whose resources are logically describable. It helps to deliver on a key goal of GENI: support for diverse virtual resources from multiple providers (aggregates).

But GENI's vision of a provider federation goes beyond that: it is also necessary to coordinate functions across aggregates in the federation. ORCA defines two kinds of coordinating servers that are not present in the GENI architecture: brokers and controllers. The *brokers* (Sect. 3.2) issue tickets for slivers on aggregates: they facilitate resource discovery and cross-aggregate resource control. The *controllers* (Sect. 3.3) run as extension modules within the SMs to manage slices: each slice is bound to exactly one controller, which receives notifications for all events in the slice and issues sliver operations on the slice. In general, ORCA controllers run on behalf of users to manage their slices: they have no special privilege.

These servers play an important role in ExoGENI. In particular, the ExoGENI controllers manage topology embedding and slice assembly (Sect. 3.4). They also implement the user-facing GENI API and proxy requests from GENI users for ExoGENI resources (Sect. 3.5). Proxying requires them to check authorization for GENI users, so the ExoGENI AMs are configured to trust these special controllers to perform this function.

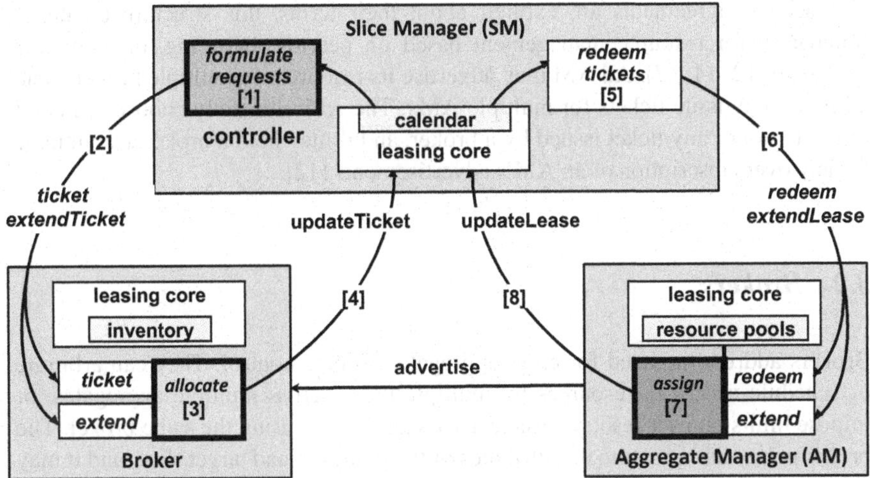

Fig. 4 Interacting servers in the ORCA resource control plane, and their policy control modules. The AMs advertise their resource pools to one or more brokers. Brokers accept user requests and run policy for ticket allocation. The SM controllers select and sequence resource requests for the slice to the brokers and AMs. The AMs provision slivers according to the tickets and other directives passed with each request

3.1 ORCA Resource Control Plane

Brokers and controllers are built using the same leasing platform as the AMs. The ORCA toolkit design recognized that these servers have key structural elements in common: a dataset of slices and slivers; timer-driven actions on lease objects organized in a calendar; similar threading and communication models; and lease state machines with a common structure for plug-in extension modules.

Figure 4 illustrates the three types of interacting servers (actors) in the ORCA framework and their policy modules. The brokers run policy extensions—similar to the AM `ResourceControl` modules—to allocate sliver requests against an inventory of advertisements, and to issue tickets for those slivers. Controllers run as extension modules within generic Slice Manager (SM) servers. The protocol messages carry resource descriptions produced and consumed by the policy modules. All messages are signed to ensure a verifiable delegation path.

An ORCA deployment may combine many instances of each kind of server. ORCA was conceived as a *resource control plane* in which multiple instances of these servers interact in a deployment that evolves over time. The ORCA toolkit combines a platform to build these servers and a control framework to link them together. These linkages (e.g., delegations of control over advertised resources from AMs to brokers) are driven by configuration files or administrative commands, or programmatically from an extension module. Any delegation of resource control is represented by a ticket or lease contract with an attached resource description.

Since all agreements are explicit about their terms, this structure creates a foundation for resource management based on peering, bartering, or economic exchange [12, 14, 17]. An AM may advertise its resources to multiple brokers, and a broker may issue tickets for multiple AMs. The AM ultimately chooses whether or not to honor any ticket issued by a broker, and it may hold a broker accountable for any oversubscription of an AM's advertisements [12].

3.2 *Brokers*

Brokers address the need for cross-aggregate resource control. They can arbitrate or schedule access to resources by multiple users across multiple aggregates, or impose limits on user resource usage across aggregates using the same broker. The broker policy has access to the attributes of the requester and target slice, and it may maintain a history of requests to track resource usage by each entity through time.

Brokers also offer basic support for co-allocation across multiple aggregates, including advance reservations. This support is a key motivation for the two-step ticket/lease protocol in SHARP and ORCA. In particular, an ORCA broker may receive resource delegations from multiple aggregates and issue co-allocated tickets against them. Because all allocation state is kept local to the broker, the co-allocation protocol commits locally at the broker in a single step.

The key elements that enable brokers are the separation of allocation and provisioning (Sect. 2.1) and the "resource algebra" of declarative resource descriptions (Sect. 2.4). Given the algebra, the processing for ticket allocation can migrate outside of the AMs to generic brokers operating from the AM's logical description. The AM must trust the broker to perform this function on its delegated resource pools, but it can always validate the decisions of its brokers because the tickets they issue are redeemed back to the AM for the provisioning step [12].

The logical resource descriptions enable advertisements at multiple levels of abstraction. In practice, most AMs advertise *service descriptions* rather than their internal substrate structure. For example, ExoGENI network AMs hide their topology details: they advertise only their edge interconnection points with other domains and the bandwidth available at each point. Abstract advertisements offer more flexibility and autonomy to the AMs, who may rearrange their infrastructures or adjust sliver bindings to respond to demands and local conditions at the time of service.

Each AM chooses how to advertise its resources in order to balance the risks of ticket rejection or underutilization of its resources. For example, an unfortunate side effect of abstract advertisements is that brokers may issue ticket sets that are not feasible on the actual infrastructure, particularly during periods of high utilization. Ticket rejection by the AM is undesirable and disruptive, but it is unavoidable in the general case, e.g., during outages. An AM may hold unadvertised capacity in reserve to mask problems, but this choice leaves some of its resources unused.

Alternatively, an AM may advertise redundantly to multiple brokers. This choice reduces the risk of wasting resources due to broker failure at the cost of a higher risk of overcommitment and ticket rejection.

To summarize, the flexible delegation model enables a continuum of deployment choices that balance the local autonomy of resource providers with the need for global coordination through the brokering system. Resource contracts and logical descriptions enable AMs to delegate varying degrees of control over their resources to brokers that implement policies of a community. At one end of the continuum, each AM retains all control for itself, effectively acting as its own broker, and consumers negotiate resource access with each provider separately. At the other end of the continuum a set of AMs federate using common policies implemented in a central brokering service. These are deployment choices, not architectural choices. In ExoGENI all AMs advertise to a central broker, but each cloud site also serves some resources locally.

3.3 Controllers

The slice controllers in ORCA match the Software-Defined Networking and Infrastructure (SDN/SDI) paradigm that is popular today. Like SDN controllers they control the structure of a virtual network (a slice) spanning a set of low-level infrastructure elements. They issue commands to define and evolve the slice, and receive and respond to asynchronous notifications of events affecting the slice. Like other ORCA servers, the controller/SM is stateful: it maintains a database recording the status of each slice, including active tickets and leases and any pending requests. It is the only control element with a global view of the slice.

One simple function of the controller is to automate sliver renewal ("meter feeding") as leases expire, to avoid burdening a user. The controller may allow leases to lapse or formulate and inject new lease requests according to a policy. In the early work, the controller and SM were conceived as the locus of automated adaptation policy for elastic slices and elastic services running within those slices [10, 17–20, 24]. (This is why the SM was called *Service Manager* in the early papers.) For example, a 2006 paper [20] describes the deployment of elastic grid instances over a network of virtual machine providers, orchestrated by a "grid resource oversight controller" (GROC). The grid instances grow and shrink according to their observed load.

To assist the controller in orchestrating complex slices, the ORCA leasing engine can enforce a specified sequencing of lease setup actions issued from the SM. The controller registers dependencies by constructing a DAG across the lease objects, and the core issues the lease actions in a partial order according to the DAG. This structure was developed for controllers that orchestrate complex hosted services [18], such as the GROC, but it has also proved useful to automate stitching of network connectivity within slices that span aggregates linked by L2 circuit networks [3], as described in Sect. 3.4 below.

In particular, the leasing engine redeems and instantiates each lease before any of its successors in the DAG. Suppose an object has been ticketed but no lease has been received for a redeem predecessor: then the engine transitions the ticketed object into a **blocked** state, and does not fire the redeem action until the predecessor lease arrives, indicating that its setup is complete. The core upcalls the controller before transitioning out of the **blocked** state. This upcall API allows the controller to manipulate properties on the lease object before the action fires. For example, the controller might propagate attributes of the predecessor (such as an IP address or VLAN tag returned in the lease) to the successor, for use as an argument to a configuration action.

3.4 Automated Stitching and Topology Mapping

ExoGENI illustrates how controllers and their dependency DAGs are useful to plan and orchestrate complex slices. In particular, the controllers automate end-to-end circuit stitching by building a dependency DAG based on semantic resource descriptions.

The controller first obtains the description for each edge connection point between domains traversed by links in a slice. The descriptions specify the properties of each connection point. In particular, they describe whether each domain can produce and/or consume *label* values (e.g., VLAN tags) that name an attachment of a virtual link to an interface at the connection point. A domain with translation capability can act as either a label producer or consumer as needed.

Stitching a slice involves making decisions about which domains will produce and which will consume labels, a process that is constrained by the topology of the slice and the capabilities of the domains. Based on this information, the controller generates a *stitching workflow DAG* that encodes the flow of labels and dependencies among the slivers that it requests from these domains. A producer must produce before a consumer can consume. The controller traverses the DAG, instantiating resources and propagating labels to their successors as the labels become available.

Figure 5 illustrates with a hypothetical scenario. The NLR circuit service is a producer: its circuits are compatible for stitching only to adjacent domains with consumer or tag translation capability. The resulting DAG instantiates the NLR circuit first, and obtains the produced VLAN tag from the sliver manifest returned by the domain's AM. Once the tag is known, the controller propagates it by firing an action on the successor slivers at the attachment point, passing the tag as a parameter. Each domain signs any labels that it produces, so that downstream AMs can verify their authenticity. A common broker or other federation authority may function as a trust anchor. In extreme cases in which VLAN tag negotiation is required, e.g. among adjacent "producer" domains, it is possible to configure the broker policy module to allocate a tag from a common pool of values.

The ExoGENI controllers also handle inter-domain topology mapping (embedding) for complex slices [22]. A controller breaks a requested topology down into

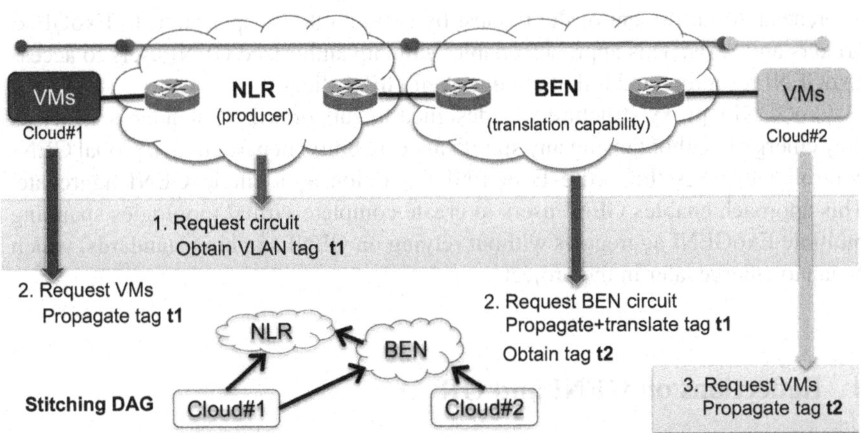

Fig. 5 Dependency DAG and stitching workflow for an end-to-end L2 circuit scenario. NLR/Sherpa chooses the VLAN tag at both ends of its circuits, BEN has tag translation capability, and the edge cloud sites can accept tags chosen by the adjacent provider. The connection point descriptions yield a stitching workflow DAG. The controller traverses the DAG to assemble the circuit with a partial order of steps

a set of paths, and implements a shortest-feasible-path algorithm to plan each path, considering compatibility of adjacent providers in the candidate paths as described above. To plan topologies, the controller uses a `query` interface on the brokers to obtain and cache the complete advertisements of the candidate network domains. It then performs path computation against these logical models in order to plan its sliver requests. If a path traverses multiple adjacent producer domains, it may be necessary to bridge them by routing the path through an exchange point that can translate the tags. After the controller determines the inter-domain path, the network domain AMs select their own intra-domain paths internally at sliver instantiation time.

Topology embedding is expensive, so it is convenient to perform it in the SM controllers. The SMs and controllers are easy to scale because they act on behalf of some set of independent slices: as the number of slices grows it is easy to add more SMs and distribute the control responsibility for the slices across the SMs.

3.5 *GENI Proxy Controller*

ExoGENI SMs run special GENI controller plugins that offer standard GENI APIs to GENI users. The GENI controllers run a converter to translate the GENI request specification (RSpec) into NDL-OWL. The converter also checks the request for compliance with a set of semantic constraints, which are specified declaratively in NDL-OWL. If a request is valid, the SM and its controller module act as a proxy to orchestrate fulfillment of the request by issuing ORCA operations to ExoGENI brokers and AMs. This approach enables suitably authorized GENI users to access ExoGENI resources and link them into their GENI slices.

ExoGENI's proxy structure was designed to support GENI standards easily as they emerged, without losing any significant capability. In particular, a global GENI controller exposes the entire ExoGENI federation as a single GENI aggregate. This approach enables GENI users to create complete virtual topologies spanning multiple ExoGENI aggregates without relying on GENI stitching standards, which began to emerge later in the project.

4 Reflections on GENI and ORCA

This section offers some thoughts and opinions on the GENI-ORCA experience. We believe that the ORCA architecture has held up well through the GENI process. We built and deployed ExoGENI as a set of extension modules with few changes to the ORCA core. Although the ORCA software itself is not used outside of ExoGENI, the GENI standards have ultimately adopted similar solutions in all areas of overlap.

In particular, the latest GENI API standard (3.0) is similar to the ORCA protocol, with per-sliver leases, separate `allocate` and `provision` steps, dynamic stitching, abstract aggregates with no exposed components, elastic slices with adjustable sliver sets, and a decoupled authorization system. Beyond these commonalities, GENI omits orchestration features from ORCA that could help meet goals of GENI that are still incomplete. It also adopts policies for user identity and authorization, which are outside the scope of the ORCA architecture but are compatible with it.

The remaining differences lie in the data representations carried with the protocol—the languages for resource descriptions and for the credentials that support a request. In particular, GENI uses a resource description language (RSpec) that is not logic-based. RSpec may prove to be more programmer-friendly than NDL-OWL, but it is decidedly less powerful, and it rests on weaker foundations.

These differences are primarily interoperability issues rather than architectural issues or restrictions of the protocol itself. The version 3.0 GENI API is open to alternative credential formats including (potentially) broker-issued tickets, by mutual agreement of the client and server. In principle the protocol is open to alternative resource description languages as well.

4.1 Platforms vs. Products + Protocols

In retrospect, ORCA's toolkit orientation set us apart from the GENI project's initial focus on standardizing protocols to enable existing network testbeds to interoperate. Many of our colleagues in the project understood ORCA as another testbed provider, rather than as a platform to federate and orchestrate diverse providers. They focused on the infrastructure that ORCA supported, which at that time was limited to Xen virtual machine services [18]. There was less interest in the toolkit itself, in part because ORCA used a different language (Java) and tooling than the other GENI clusters. Our focus on the toolkit—and the architectural factoring of GENI-relevant functions and APIs that it embodied—clashed with the primacy of the protocol standards, which were seen as the key to interoperability.

Even so, the ORCA toolkit allowed the GENI *Cluster D* team to accelerate development by using generic ORCA servers to "wrap" existing back-end systems and call them through their existing APIs. The result looked much like the structure ultimately adopted in GENI (Sect. 4.2), but using the ORCA protocols rather than the GENI standards. The ORCA experience suggests that the lengthy GENI development phase could have been shortened by focusing on wrappers and adapters in the early spirals, rather than on the protocols.

Moreover, if the wrappers are standardized, then it is possible to change the protocols later by upgrading the wrappers. We found that it is easier to stabilize the plug-in APIs for the toolkit than the protocols themselves. For example, ORCA uses an RPC system (Axis SOAP) that has never served us well and is slated for replacement. The GENI standards use XMLRPC, which is now seen as defunct.

Although it is always difficult to standardize protocols, interoperability in a system like ORCA or GENI is less about protocols than about data: machine-readable descriptions of the principals and resources. In both systems the protocols are relatively simple, but the messages carry declarative resource descriptions and principal credentials, which may be quite complex. Our standards for these languages will determine the power and flexibility of the systems that we build (Sect. 4.4).

4.2 Federation

The key differences in control framework architectures relate to their approaches to federating the aggregates. In general, the aggregates themselves are IaaS or PaaS services similar to those being pursued by the larger research community and in industry. The problem for GENI is to connect them.

The GENI framework takes a simple approach to federation: it provides a common hierarchical name space (URNs) and central authorities to authorize and monitor user activity. GENI leaves orchestration to users and their tools and it does not address cross-aggregate resource control (Sect. 4.3). Even so, the GENI community invested substantial time to understand the design alternatives for federation, reconcile terminologies, and specify a solution. Various architectures were proposed for GENI to factor identity management and authorization functions out of the standalone testbeds and into federation authorities, but a workable convergence did not emerge until 2012.

The GENI solution—so far—embodies a design principle also used in ORCA. The AMs do not interact directly; instead, they merely delegate certain functions for identity management, authorization, and resource management to common coordinator servers. The coordinators issue signed statements certifying that clients and their requests comply with federation policy. The AM checks these statements before accepting a request.

For example, the GENI Clearinghouse authorities approve users, authorize slices, and issue credentials binding attributes to users and slices. User tools pass their credentials to the AMs. These mechanisms provide the common means for each AM to verify user identity and access rights for the GENI user community. The coordinators in ORCA/ExoGENI include brokers and the GENI controllers, which are trusted by the AMs to check the GENI credentials of requests entering the ORCA/ExoGENI enclave.

As originally conceived, ORCA AMs delegate these user authorization functions to the brokers: if a community broker issues a ticket for a request, the AM accepts the user bearing the ticket. It is the responsibility of the broker to authorize each user request in its own way before granting a ticket. More precisely, the ORCA architecture left the model for user identity and authorization unspecified, and it is fully compatible with GENI's policy choices. However, these choices should remain easily replaceable in any given deployment (Sect. 4.4).

4.3 Orchestration

GENI has not specified any coordinator functions beyond checking user credentials. In particular, GENI has not adopted brokers or any form of third-party ticketing to enable cross-aggregate resource management.

Although the sponsor (NSF) has voiced a desire to control user resource usage across multiple aggregates, GENI has defined no alternative mechanism for this purpose. Importantly, AMs and controllers are not sufficiently powerful to meet this need without some structure equivalent to brokers. The local policy of any AM may schedule or limit allocation of its own resources, but it has no knowledge or control over allocations on other aggregates. Similarly, any limit that an unprivileged controller imposes on resource usage by a slice is voluntary, because the controller acts as an agent of the slice and its owners.

Slice controllers are also not part of the GENI architecture. GENI was conceived as a set of protocols and service interfaces: the client software to invoke these interfaces was viewed as out of scope. Instead, the idea was that a standard AM API would encourage an ecosystem of user tools to grow organically from the community. To the extent that computation is needed to orchestrate cross-aggregate requests for a given slice—such as topology mapping—those functions were conceived as central services provided to the tools through new service APIs. We believe that it is more flexible and scalable to provide these functions within the tools. ExoGENI shows that it is possible to do so given sufficiently powerful resource descriptions and a platform for building the tools.

Over time it became clear that the GENI client tools must be stateful to provide advanced slice control functions. In particular, tools must maintain state to implement timer-driven sliver renewal, multi-step atomic co-allocation, "tree" stitching across aggregates, and elastic slices that adapt to changing demands. Later in the project, the GENI Project Office developed an extensible tool called `omni` and a Web portal to proxy requests from stateless tools into GENI. These clients have steadily incorporated more state and functionality. It seems likely that they will continue to evolve in the direction of ORCA's stateful slice controllers.

The ORCA view is that a federated infrastructure control plane is a "tripod": all three server types—aggregate/AM, controller/SM, and broker—are needed for a complete system. The factoring of roles across these servers is fundamental. The AMs represent the resource providers, and are the only servers with full visibility and control of their infrastructures. The SMs represent the resource consumers, and are the only servers with full visibility and control over their slices. The brokers and other authorities (e.g., the GENI clearinghouse) mediate interactions between the SMs and AMs: they are the only servers that can represent policy spanning multiple aggregates.

4.4 Description Languages

Our experience with ORCA and GENI deepened our view that the key problems in federated infrastructure—once the architecture is put right—are largely problems of description. This understanding is a significant outcome of the GENI experience.

GENI differs from other infrastructure services primarily in its emphasis on diverse infrastructure, rich interconnection, and deep programmability. It follows that the central challenges for GENI are in describing "interesting" resources and in processing those descriptions to manage them.

The early development phase of GENI was marked by an epic debate on `dev@geni.net` about whether a common framework for diverse resource providers is even possible. It is perhaps still an open question, but if the answer is yes, then the path to get there involves automated processing of rich resource descriptions. To incorporate a new resource service into an existing system, we must first describe the service and its resources in a way that enables generic software to reason about the space of possible configurations and combinations.

For example, the ORCA experience shows that it is easy to incorporate current cloud systems and third-party transit network providers as GENI aggregates through an adaptation layer if we can describe their resources logically. Powerful logical descriptions also enable the various coordinator functions (Sect. 4.3) in ORCA/ExoGENI. One lesson of this experience is that AM advertisements do not in general describe the infrastructure *substrate*, as the GENI community has understood them, but instead describe infrastructure *services*, which are even more challenging to represent and process. For example, AMs may proxy or "resell" resources from other providers whose substrate they do not control, or they may offer various mutually exclusive options for virtual sliver sets within the constraints of a given substrate resource pool.

Our ongoing research focuses on declarative representations for resources and trust, and their role in automating management of resources and trust. For example, in recent work we have shown how to specify trust structure and policy alternatives for a GENI deployment concisely and precisely in SAFE declarative trust logic [6]. Statements in the logic are embedded in credentials; a generic compliance checker validates credentials according to policy rules, which are also expressed in the logic.

With this approach, the GENI architecture can be implemented as a set of autonomous services (e.g., the AMs) linked by a declarative trust structure that is represented in about 150 lines of scripted SAFE logic. The various coordinator roles and trust relationships are captured in declarative policy rules rather than as procedures or assumptions that are "baked in" to the software. We believe that this approach balances low implementation cost with flexibility for deployments to accommodate diverse policies of their members, evolve their structures and policies over time, and federate with other deployments.

Acknowledgements This document is based upon work supported by the US National Science Foundation through the GENI Initiative and under NSF grants including OCI-1032873, CNS-0910653, and CNS-1330659, and by the State of North Carolina through RENCI.

References

1. Baldin, I., Castillo, C., Chase, J., Orlikowski, V., Xin, Y., Heermann, C., Mandal, A., Ruth, P., Mills, J.: Exogeni: a multi-domain infrastructure-as-a-service testbed. In: GENI: Prototype of the Next Internet. Springer, New York (2016)
2. Baldine, I., Xin, Y., Evans, D., Heerman, C., Chase, J., Marupadi, V., Yumerefendi, A.: The missing link: putting the network in networked cloud computing. In: International Conference on the Virtual Computing Initiative (2009)
3. Baldine, I., Xin, Y., Mandal, A., Heerman, C., Chase, J., Marupadi, V., Yumerefendi, A., Irwin, D.: Autonomic cloud network orchestration: A GENI perspective. In: GLOBECOM Workshops: 2nd IEEE International Workshop on Management of Emerging Networks and Services (MENS 2010) (2010)
4. Baldine, I., Xin, Y., Mandal, A., Ruth, P., Yumerefendi, A., Chase, J.: ExoGENI: a multi-domain infrastructure-as-a-service testbed. In: TridentCom: International Conference on Testbeds and Research Infrastructures for the Development of Networks and Communities (2012)
5. Braynard, R., Kostić, D., Rodriguez, A., Chase, J., Vahdat, A.: Opus: an overlay peer utility service. In: Proceedings of the 5th International Conference on Open Architectures and Network Programming (OPENARCH) (2002)
6. Chase, J., Thummala, V.: A guided tour of SAFE GENI. Technical Report CS-2014-002, Department of Computer Science, Duke University (2014)
7. Chase, J.S., Anderson, D.C., Thakar, P.N., Vahdat, A.M., Doyle, R.P.: Managing energy and server resources in hosting centers. In: Proceedings of the 18th ACM Symposium on Operating System Principles (SOSP), pp. 103–116 (2001)

8. Chase, J.S., Irwin, D.E., Grit, L.E., Moore, J.D., Sprenkle, S.E.: Dynamic virtual clusters in a grid site manager. In: Proceedings of the Twelfth International Symposium on High Performance Distributed Computing (HPDC) (2003)
9. Chase, J., Grit, L., Irwin, D., Marupadi, V., Shivam, P., Yumerefendi, A.: Beyond virtual data centers: toward an open resource control architecture. In: Selected Papers from the International Conference on the Virtual Computing Initiative (ACM Digital Library) (2007)
10. Chase, J., Constandache, I., Demberel, A., Grit, L., Marupadi, V., Sayler, M., Yumerefendi, A.: Controlling dynamic guests in a virtual computing utility. In: International Conference on the Virtual Computing Initiative (2008)
11. Chun, B., Culler, D., Roscoe, T., Bavier, A., Peterson, L., Wawrzoniak, M., Bowman, M.: Planetlab: an overlay testbed for broad-coverage services. SIGCOMM Comput. Commun. Rev. **33**(3), 3–12 (2003)
12. Fu, Y., Chase, J., Chun, B., Schwab, S., Vahdat, A.: SHARP: an architecture for secure resource peering. In: Proceedings of the 19th ACM Symposium on Operating System Principles (2003)
13. Grit, L.E.: Extensible resource management for networked virtual computing. Ph.D. thesis, Duke University Department of Computer Science (2007)
14. Grit, L., Irwin, D., Yumerefendi, A., Chase, J.: Virtual machine hosting for networked clusters: building the foundations for "Autonomic" orchestration. In: Proceedings of the First International Workshop on Virtualization Technology in Distributed Computing (VTDC) (2006)
15. Ham, J., Dijkstra, F., Grosso, P., Pol, R., Toonk, A., Laat, C.: A distributed topology information system for optical networks based on the semantic web. J. Opt. Switch. Netw. **5**(2–3), 85–93 (2008)
16. Irwin, D.: An operating system architecture for networked server infrastructure. Ph.D. thesis, Duke University Department of Computer Science (2007)
17. Irwin, D., Chase, J., Grit, L., Yumerefendi, A.: Self-recharging virtual currency. In: Proceedings of the Third Workshop on Economics of Peer-to-Peer Systems (P2P-ECON) (2005)
18. Irwin, D., Chase, J.S., Grit, L., Yumerefendi, A., Becker, D., Yocum, K.G.: Sharing networked resources with Brokered leases. In: Proceedings of the USENIX Technical Conference (2006)
19. Lim, H., Babu, S., Chase, J.: Automated control for elastic storage. In: IEEE International Conference on Autonomic Computing (ICAC) (2010)
20. Ramakrishnan, L., Grit, L., Iamnitchi, A., Irwin, D., Yumerefendi, A., Chase, J.: Toward a doctrine of containment: grid hosting with adaptive resource control. In: Proceedings of the Supercomputing (SC06) (2006)
21. Sridharan, M., Zeng, W., Leal, W., Ju, X., Ramanath, R., Zhang, H., Arora, A.: From Kansei to KanseiGenie: architecture of federated, programmable wireless sensor fabrics. In: Proceedings of the ICST Conference on Testbeds and Research Infrastructures for the Development of Networks and Communities (TridentCom) (2010)
22. Xin, Y., Baldine, I., Mandal, A., Heerman, C., Chase, J., Yumerefendi, A.: Embedding virtual topologies in networked clouds. In: 6th International Conference on Future Internet Technologies (CFI 2011) (2011)
23. Yumerefendi, A.R.: System support for strong accountability. Ph.D. thesis, Duke University Department of Computer Science (2009)
24. Yumerefendi, A., Shivam, P., Irwin, D., Gunda, P., Grit, L., Demberel, A., Chase, J., Babu, S.: Toward an autonomic computing testbed. In: Workshop on Hot Topics in Autonomic Computing (HotAC) (2007)

Programmable, Controllable Networks

Nicholas Bastin and Rick McGeer

Abstract We describe OpenFlow, a first step on the road to networks which are fully integrated into the IT infrastructure ecosystem. We review the history of OpenFlow, its precursors, its design and initial implementations. We discuss its use within the GENI project and the applications and services developers have built on the OpenFlow platform. Finally, we review the implementation issues with OpenFlow, and consider extensions and the next generation of Software-Defined Networking.

1 Integrating the Network into IT Infrastructure

Though the network has been a key component of computational infrastructure, it has remained curiously isolated from the rest of information technology, residing in both an administrative and technological silo. Network functions are performed by a collection of specialized devices—routers, switches, firewalls, etc.—which are configured entirely independently of the hardware and the software of the computers connected to the network. In most organizations, the network is administered by a separate organization from the IT staff, and the sets of skills of the people in the two organizations are almost completely distinct. Network administrators are not typically expert in computer hardware and software, and computer system administrators as well as most application developers regard the network as a black box which delivers packets with some semblance of reliability. There are few platforms for integrated monitoring of both network and information technology, and almost no APIs which a developer can use to influence network forwarding or satisfaction of dynamic application requirements.

This is an artifact of decisions made in the very early days of networking, when the only network services were bulk data transfer and low-bandwidth connectivity to remote services (e.g., telnet). At this time, it was argued persuasively that the

N. Bastin (✉)
Barnstormer Softworks Ltd. and University of Houston, Houston, TX, USA
e-mail: nbastin@uh.edu

R. McGeer
Chief Scientist, US Ignite, 1150, 18th St NW, Suite 900, Washington, DC 20036, USA
e-mail: rick.mcgeer@us-ignite.org

© Springer International Publishing Switzerland 2016
R. McGeer et al. (eds.), *The GENI Book*, DOI 10.1007/978-3-319-33769-2_8

network should be an opaque, application-insensitive, featureless pipe which simply carried bits from source to destination, and would offer no guarantees of loss, latency, bandwidth, or indeed any properties at all. This “end-to-end argument” [39] held that since applications were so variegated, there was no single network policy or property that could satisfy them all, and hence these properties should be guaranteed and enforced at the endpoints.

Even from its earliest days, this overly simplistic view of the network was problematic. While early networks were largely devoted to file transfer of one sort or another, there were always multiple services and protocols in the network, and these services required differentiated handling. Even in the most nascent networks, there were significant distinctions between monitoring and file-transfer services, and as bandwidth expanded, more services became available. Moreover, the network has grown increasingly discontinuous over time, giving rise to a family of proxies and middle-boxes. Initially thought of as workarounds, these have become increasingly viewed as central to the Internet architecture [19] as the network itself has become more complex. What was conceived as a bulk data transfer service for a small community of technologists has become the all-inclusive communications medium for a civilization. Unsurprisingly, the transition was not seamless. That it happened at all is a testament to the robust, scalable design of Cerf and Kahn and the ingenuity of the legions of networking scientists and engineers who followed.

More problematic from an administrative perspective, the network has become highly dynamic. When the basic elements of TCP/IP were designed, all computing resources were fixed. Personal computing was largely confined to research laboratories, portable computing was unheard of, and there were relatively few sites. Hence the network was largely static and computers largely spent their lives on a single local area network. Similarly, the numbers and addresses of local area networks were largely fixed. Indeed, the network was so simple that IP addresses could be managed single-handedly by Jon Postel [43], and to a large extent were. Today, however, not only are there billions of devices, mobility and portability have become the rule and not the exception, breaking the original de facto model of long-duration addressing.

Further, it was anticipated that new applications and services would be added to the network only rarely, and each new application would require global agreement among all devices on the subnetwork that supported the application. Hence, the Internet Engineering Task Force, whose job it is to collaboratively design new network services in such a way that they can be globally implemented. We believe that it is safe to aver that this organization is not universally regarded as the exemplar of agility. In fact, the noted computer and network scientist Scott Shenker has bluntly described this design process as “crazy”.[1] From the perspective of the Internet's original assumptions, it isn't—and this loose organization has largely

[1] See: https://www.youtube.com/watch?v=eXsCQdshMr4.

served the community well for far longer than one might have imagined—but in the ensuing generation the world has changed beyond the imagination of anyone on the original design team.

Moreover, the new applications and services introduced in this manner could interact in unpredictable ways, and since each new feature was enabled for all traffic if it was enabled for any, many features were by default disabled for many sites. IP Multicast was, for many years, a notorious example, but the technological roadside is littered with considerable amounts of lesser-known detritus as well.

The Internet was designed with little regard for security, which, from the perspective of 1980, was entirely reasonable. Security was the business of endpoint systems—friendly, easy-to-deal with objects running Unix-based operating systems. It was inconceivable in 15 short years that the Internet would be scaled in the billions of devices, with the vast majority running operating systems that did not rival Unix or Unix-derived operating systems in transparency and security. There were two central issues. First, the vast majority of computers attached to the network were not nearly as secure as the network designers imagined that they would be. Second, network control protocols were themselves subject to attack of various forms. These ranged from simple resource-exhaustion attacks (e.g., distributed denial-of-service) and exploitation of explicit vulnerabilities in Internet control protocols (e.g., Border Gateway Protocol routing attacks, or the Kaminsky Attack on the Domain Name Service), to those that compromised otherwise secure protocols by exploiting the increasing technical naivete of the end user.

In sum, the Internet has now drifted far from its original design assumptions, and it has been kept going through an ongoing series of patches. The size and complexity of this effort can be seen by counting the number of Internet standards. There are over 7400 IETF standards ("Request for Comments"), including 33 for the Lightweight Directory Access Protocol (LDAP) alone—of which at least 25 are still active. Moreover, administering each site became a complex undertaking, largely because many specialized services required special-purpose subnets which crossed broadcast-domain boundaries, or Virtual Local Access Networks (VLANs). The cross-domain nature of VLANs added substantial complexity to administering a site, Network configurations for a site could amount to several thousand lines of configuration, and simply verifying the properties of a configuration became a cottage industry. Physically moving a computer or network device could require several hundred lines of configuration changes to a network.

There have been a variety of attempts to introduce application-specific programmability into the network, following John Ousterhout's dictum that "wonderful things happen when an infrastructure becomes programmable".[2] A good survey can be found in [29]. This started with DARPA Active Networks program in the 1990s, which attempted to allow individual users—or groups of users—to inject customized programs into the nodes of the network. the goal was to enable a massive increase in the complexity and customization of the computation that is performed

[2]Remark in a presentation given ca 1995, confirmed in a recent personal communication.

within the network. In effect, the idea was to turn each router or switch into a customizable middle-box. There were a variety of concerns with this idea, notably security [49]. The Active Networks community then turned to overlay networks, which treated the underlying network infrastructure as a black box and built the application-specific network entirely on computers at the endpoints. This offered the properties desired for the application-specific network but sacrificed performance in a number of dimensions. To a first approximation, the primary quality-of-service properties of a network, notably latency and bandwidth, can only be controlled with application-specific control of the routers and switches on a network. A Distributed Hash Table such as Tapestry [51] or Chord [44] can guarantee message delivery in a small number of overlay hops, but the latency of each hop is dependent on the underlying network, and so too the overall performance of the application.

The need for tight application control of the network may seem revolutionary, but it is a natural consequence of the evolution of applications and services in the era of the network. As has been observed elsewhere in this book, applications are increasingly becoming distributed systems where components of the application are separated by tens to hundreds of milliseconds. For distributed systems, the network often becomes the dominant element in system performance, and hence the application designer must exert positive control on the network.

US Ignite Chief Technology Officer Glenn Ricart points out that packets sent from his home in Salt Lake City to his colleagues at the University of Utah, just a few miles away, are routed through Kansas City, a round-trip distance of over 2000 miles. A mile in fiber is 5 ms, so this adds a minimum of 10 ms in transit time. For bulk data transfer, this is rarely a problem; for a distributed system application, it can be deadly.

Control of paths in the network for a distributed system is similar to control of data layout on disk for a storage-centric application such as a database. This is vital to a storage designer, and in fact most of the effort in database schema design goes to ensuring that the data layout is tailored to the expected activity of the application. For example, transactions-oriented systems are optimized to rapidly write all of the data in a transaction, so the ideal is to put all of the data in a single table, stored in row-major order. Analytics applications seek over all the rows of the database, so they are either written in a star topology—one column per table—or in a column-oriented database, where the database is stored in column-major order. Systems with many simultaneous reads and writes often use two file systems—one optimized for write and one for read, and reformatting software which moves data from the write-oriented system to the read-oriented system.

The level of control that a storage system designer takes for granted is perhaps best exemplified by the story told about the great database pioneer Jim Gray. Gray collaborated with the noted astronomer Alex Szalay on the SkyServer, an ambitious project to make the Sloan Digital Sky Survey widely available [46]. In a tribute [45] Szalay recounted Gray's first visit to the SkyServer data center. "Jim then came to Baltimore to look over our computer room and within 30 s declared, with a grin, we had the wrong database layout. My colleagues and I were stunned. Jim explained later that he listened to the sounds the machines were making as they operated; the disks rattled too much, telling him there was too much random disk access."

It goes without saying that a storage system designer would find it completely unacceptable if there were a Storage Administrator, who determined the layout of data on disk by application-agnostic configuration. But a distributed systems designer faces precisely that situation today, because the network is designed for efficient network administration, operation, and policy enforcement, not application performance. Ricart's packets take a detour through Kansas City because of peering agreements and the desire of each Autonomous System to have relatively few border points.

Internet routing today is similar to Federal Express' original method of package delivery. FedEx originally routed all packages through Memphis, even if the source and destination addresses were neighbors in San Francisco. Senders and receivers never noticed the difference; the added latency in package delivery was well within the bounds of the package-delivery application. Similarly, neighbors who are email correspondents never notice that their packets detour across a continent merely to cross the street. But when the application changed in the real world, so did the tolerance for detours. Telephone users could easily tell the difference between a satellite call and terrestrial routing—the latency of a satellite call was unacceptable for voice communication. The Internet is similarly moving from bulk data delivery to interactive applications—we are moving from an era of package delivery to an era of video calls.

If the Internet were being designed today, it would be designed far differently. In the ideal case, it would be designed around the paradigm of private, application-specific, highly-dynamic, secure networks on a common substrate. These application-specific networks would be controlled from a "single pane of glass", and the networks and switches along the path would be reconfigured to handle the traffic from this application. Designing such a network from a clean slate is not an insuperable task. However, we don't have a clean slate; the long and torturous introduction of IPv6 is a cautionary tale on just how difficult it is to introduce changes to an existing network architecture. Even when the need for such change is clear—and with considerable consensus on the base requirements—the opportunity presents itself so infrequently that such attempts are mired with new features that massively expand the scope, resulting in implementations and deployments bogged down in technical complexity and social inertia.

The key lesson of previous attempts is that ubiquity and ease of introduction are paramount concerns. Network innovations that are likely to succeed should require minimal changes to existing network equipment; a firmware upgrade is the reasonable ceiling on acceptable changes. Further, the change should address an acknowledged shortcoming of the network in meeting today's bulk-delivery service, not tomorrow's distributed-systems needs. Network operators need to see a cure for current pains.

Enter OpenFlow. The origins of what is now called OpenFlow trace their roots back to a 2007 Sigcomm paper [4] introducing a system called *Ethane*, which proposed a central controller for managing the "admittance and routing of flows" across an enterprise network. A switch is composed of two distinct entities: the *data plane* (or *forwarding plane*), which is a high-bandwidth backplane that forwards packets as directed by routing tables; and the *control plane*, an embedded computer which maintains and updates the routing tables through a combination of distributed path-finding algorithms and configuration. The central insight of the Ethane paper was that the control plane could be moved largely off-switch, updating the routing tables directly from configuration. The central manager which implemented the control planes for all the switches offered the operator both much greater control of the network and a simpler programming interface.

The ideas behind Ethane were not particularly novel—even in the late 1990s central management was available for LANs, if all of the equipment on the LAN was purchased from a single vendor. Since such a homogeneous environment is not common, adoption rates for this management technology were fairly low.

The novel feature in Ethane was bringing fine-grained flow management to multi-vendor network systems, both wireless and wireline. This meant that Ethane could potentially be used in "brownfield" deployments—that is, that existing network equipment could support its installation. This was potentially revolutionary. At the time of Ethane the only implementations of centralized layer 2 network control were single vendor, and production deployments were largely restricted to wireless networks where enterprises could enjoy the ease of "greenfield" deployment of a completely new infrastructure. On these networks the value of centralized control had been demonstrated. Ethane's potential was to extend this to the mission-critical multi-vendor wireline networks that formed the backbone of campus IT deployments. This vision is still potent, though a number of practical realities, detailed later in this chapter, still leave this promise largely in the future.

The heart of Ethane and its multi-vendor capabilities came from a simple switch abstraction. The Ethane switch was a device which matched on a subset of the header bits (including wildcard entries), and when a packet matched performed one or more of four actions: drop, rewrite the header bits, output to a specific port, or output to the controller. The actions were common to all switches; and by careful choice of header matching bits a wide variety of vendor switches could be supported.

2 The OpenFlow Protocol

Ethane had three components: an implementation of the matching rules on the switch, a network security and flow management application resident on a host, and a middleware element, resident on the host, which functioned as the network API for the application. In the original Ethane implementation, the switch portion had been implemented on a NetFPGA system, and the middleware and application components were in a single block of code.

By December 2007, HP Labs and Stanford had collaborated to implement the switch side of Ethane on a commercial switch, the HP 5406. This was a significant experiment, because it demonstrated that the switch side of Ethane was practical in existing networks with a firmware upgrade, and it offered a road map for other vendors to support this protocol. For this reason, the three components of Ethane were broken out into standalone artifacts. The switch side of Ethane became the OpenFlow Switch Specification; the middleware component became the Network Controller, and the first implementation was the middleware in Ethane, which was christened NOX. Ethane itself became simply the first of many applications on the NOX/OpenFlow stack.

The factorization and the outline of the OpenFlow protocol was described in a seminal *Computer Communications Review* article in March of 2008 [26] and the NOX controller was described in the same venue the following month [11]. The initial target was experimentation on campus networks, an explicit attempt to define an open, programmable switch platform with sufficient bandwidth and port density to be used in a campus wiring closet, with isolation so that operational campus flows could be carried on the same switch fabric with experimental flows without interference.

This was significant development occurred in late 2008, when the OpenFlow Switch Specification Version 0.8.9 [47] was published. Based strongly on Ethane, OpenFlow specified a common API for programming abstract flow tables that could be used across devices regardless of their vendor. The specification codified only the interface between network elements (switches) and the controller—the policy language and higher level functions present in the Ethane paper were left to those writing controllers leveraging this new API. Critical to the continuing development of OpenFlow was the fact that at least one vendor (HP) had devoted resources to developing an OpenFlow agent for existing hardware, allowing for testing on port-dense commodity network hardware and not just the specialized FPGA-based devices that had been used for Ethane.

2.1 Brief Summary of OpenFlow and the OpenFlow Protocol

The goal of OpenFlow was to expose an abstraction of a switch's forwarding plane which would be valid across most commercial switches and which would permit application control of forwarding on a per-flow basis. The abstraction identified was a routing table, which consisted of a list of pairs *(specification, action)*. A *specification* is a ternary integer, which specifies the values of selected bits of the header; a '2' in a field indicates a wildcard. The *action* field specified one of: **drop** the packet; **output the packet on port** *j*, for specified *j*; **send the packet to the controller** for further processing; **rewrite the header bits to** *value*.

A comparison of an OpenFlow vs. a Classic switch can be found in Fig. 1. It shows this is a factoring of a classic switch; the control plane is brought off-switch to a centralized controller. The switch control plane is conceptually replaced by a secure communication channel to an off-switch controller, though in practice a rudimentary control plane remains in place.

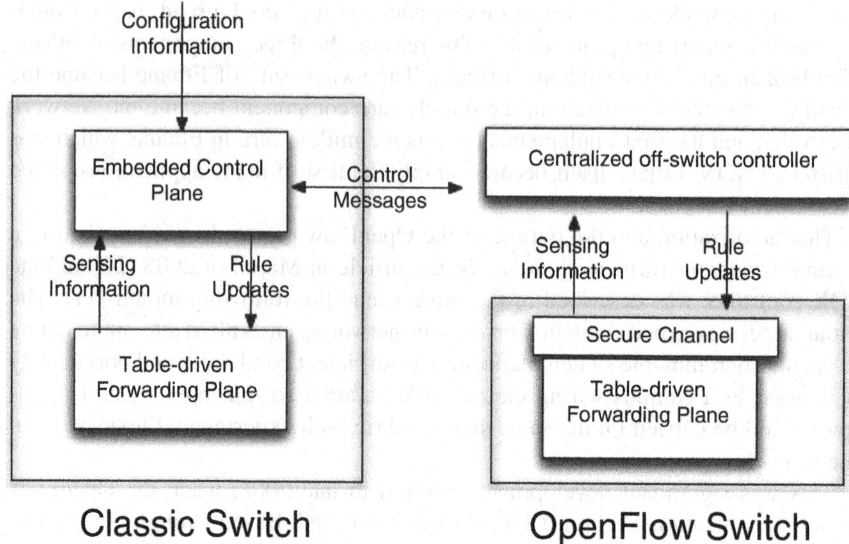

Fig. 1 Classic vs. OpenFlow switch

Optionally, more than one rule could match a given packet, for (for example) multicast, or rewrite-and-forward.

The rule tables were communicated to the switch securely via a secure, encrypted, authenticated protocol, with packets signed by the controller.

This factoring of the control plane does far more than make the switch more transparent; it radically simplifies and makes more controllable the network. A conceptual picture of a classic vs OpenFlow network is shown in Fig. 2. As can be seen from the diagram, the classic network's autonomic control and per-device configuration is replaced by a centralized controller.

A full OpenFlow specification can be found in [47]. Its attraction to both industrial operators and academic researchers had less to do with the specifics of OpenFlow (aside from the obvious criteria: it could be implemented and desired network functions easily implemented) than for the promises that seemed inherent in it. These included the explicit exposure of the network control plane, which previously had been partially autonomic and partially configuration; a standardized, vendor-independent interface to the network control plane; logical centralization of the control plane across the entire network, rather than dealing with a separate control plane on each switch, and, perhaps most important, the *potential of virtualization of the network.* This last promised is still not fully realized; the difficulties are described later in this chapter.

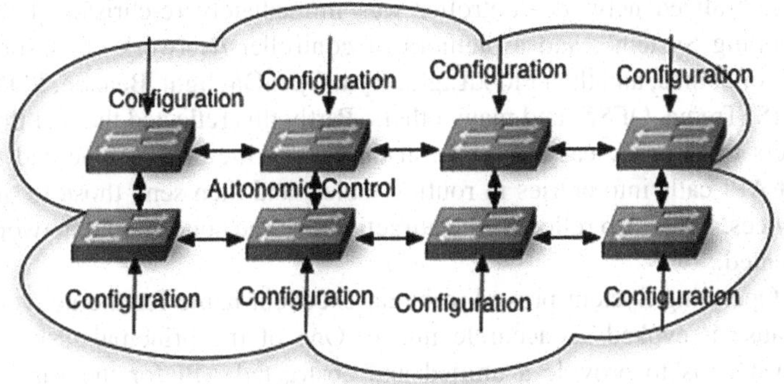

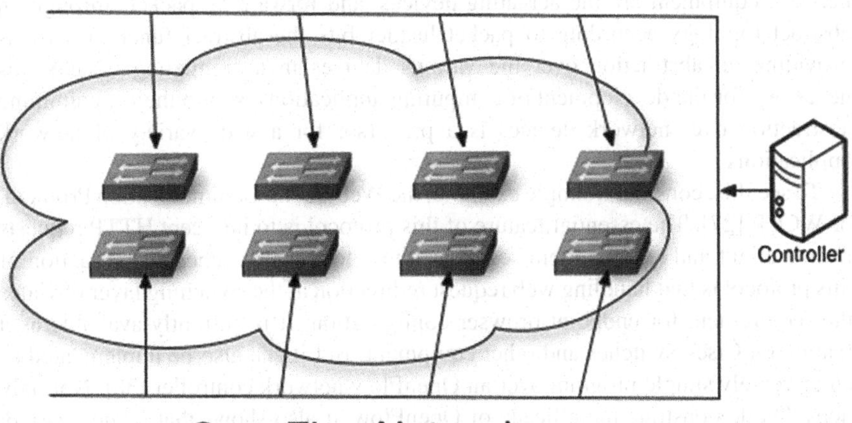

Fig. 2 Classic vs. OpenFlow network

2.2 *Promises of OpenFlow*

The operator community, with an immediate problem of cost explosion to solve, found vendor agnosticism and exposure of the control plane particularly attractive. The hope was that specifying a standard interface to the data, or forwarding, plane, would standardize and thus commoditize network equipment in the way that the Lintel platform standardized computing servers in the 1990s and early 2000s. Further, making the control plane explicit meant that new network features and services could be implemented in software by the operators themselves, rather than as a very-high-value-added feature on each piece of equipment.

The research and campus IT communities focused on the potential virtualization of the network and the idea of a single, logically-centralized network controller.

The logically-centralized network controller was immediately re-christened the "Network Operating System", and a plethora of controllers/Network OS's soon emerged: NOX, of course, and then FloodLight, Ryu, OpenDaylight, Beacon, POX, Jaxon, Mul, IRIS, Trema, OESS, and many others. Partly this reflected the fact that trivial controllers were pretty easy to write: at the end of the day, all one had to do was convert API calls into entries in routing tables, and then send those to the appropriate devices, but it also reflected the attraction that the concept of a network controller presented.

A Network Operating System proved to be an excellent name for the network controller, because it evoked an accurate image. One of the principal tasks of an operating system is to provide a unified and abstracted API for the various logical functions of a computer (storing to disk, displaying on screen, sending to network) and then issuing the actual commands to the appropriate devices to accomplish the programmer's task. In this analogy, the individual switches and network equipment are the actuating devices, and forwarding packets through an abstract topology according to packet header bits the abstract function. Just as providing an abstraction over the various devices in a computing system was necessary for the development of computing applications, so too the programming abstraction over network devices is a precursor for a wide variety of network applications.

To see this, consider a simple example, the Web Cache Communication Protocol, or WCCP [27]. The essential feature of this protocol is to intercept HTTP requests from a client and redirect them seamlessly to a local web cache. The attraction of this protocol is that handling web request redirection at the switching layer obviates the requirement for endpoint browser configuration. It is currently available as a feature on Cisco switches and other equipment. But it can also be implemented as an extremely simple program over an OpenFlow network controller [38]. Not only does this demonstrate the efficacy of OpenFlow, it also shows that a wide variety of new network applications can be implemented with a small amount of endpoint support and simple network applications. For example, Content-Centric Networking [15] can be easily implemented using a similar combination of endpoint support and controller-based redirection at the switching layer.

This culminated in the release of the Open Network Operating System [30], a high-availability network operating system for service providers.

Virtualization of the network is a potential consequence of the directed handling of flows by a controller. To see this, note that each flow and many sets of flows in the network are uniquely identified by logical expressions on header bits, precisely the matching criteria for OpenFlow. All virtualization requires is that the rules generated for each application refer only to the set of flows involved in the application. This is a restriction that can be relatively easily enforced by any network controller. The specific technical requirement is that the product of two logic functions, each expressed in sum-of-products form, be empty; this is a test that can be performed in time proportional to the product of the sizes of the logic functions.

Network virtualization was central to the design and motivation of OpenFlow and, more generally, Software Defined Networking. Recall that the fundamental motivation was to permit experimental and operational network traffic to run over the same wires and equipment, without interference. In other words, experimental and operational traffic would run on isolated virtual networks over a common physical network.

As has been seen in operating systems, virtualization is an exceptionally powerful primitive, and so it has been in networks. Not only can experimental and operational traffic be isolated, so too can flows from different applications, using the same mechanism. This is particularly important in applications where strong guarantees of isolation are required, such as in the Payment Card Industry Data Security Standard (PCI DSS) [32], which requires an isolated network from point-of-sale terminal to bank. An OpenFlow implementation of PCI DSS was demonstrated by Stanford in 2010, and formed an important use case in campus deployments [6].

The virtual, isolated networks enabled by a Network Operating System under SDN are referred to by the term "slices", originally coined in the context of PlanetLab and adopted by GENI. It was soon recognized that the ability to manipulate slices offered tremendous possibilities in a large number of contexts for network applications and services. For example, virtual machine migration in a slice became straightforward; all that was required was updating a few flow table entries in the slice's space in the various networking tables.

An example of a slice network appears in Fig. 3. As can be seen, there are three separate applications, or slices, in this network, and each has a different topology on the same underlying (mesh) topology. Moreover, each slice has different admission

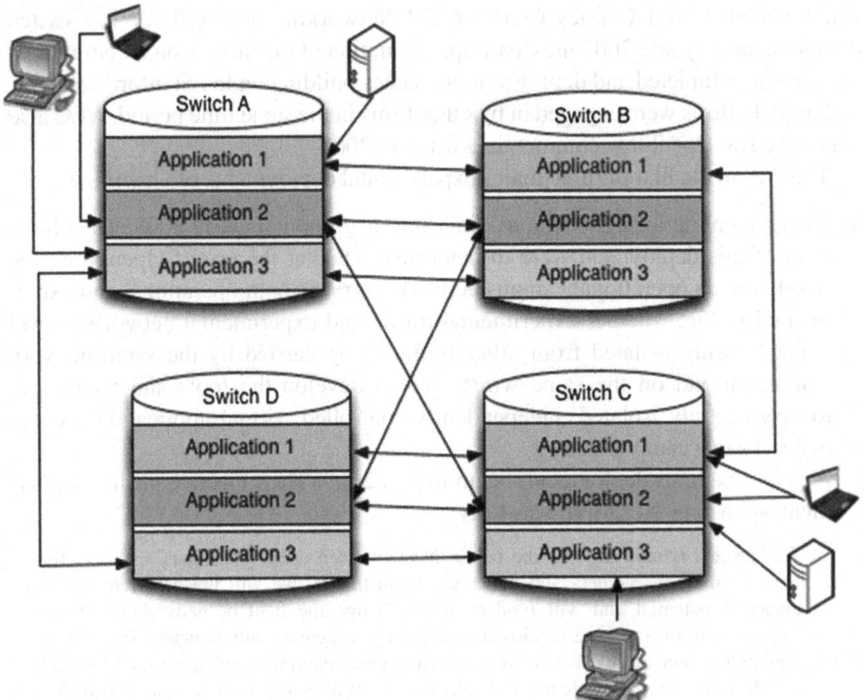

Fig. 3 A network sliced between applications

control, as shown by the connectivity of the three represented computers to Switch A and Switch C.

It should be noted that even though OpenFlow and similar technologies enabled a new paradigm of "Software Defined Networking", in fact there was no more (in fact, there was perhaps less) software in a "Software Defined Network" than in a classic network. However, in an SDN the software is open, and largely developer-written, rather than closed and vendor written. In a real sense, it can perhaps be better defined as an Application-Defined Network, not a Transport-Defined Network.

3 Initial Implementations and Campus Experiments

In the fall of 2007, the success of the Ethane program indicated that trials of OpenFlow should begin in a production campus environment. This required support for OpenFlow on commercial switches. In the fall of 2007, Nick McKeown of Stanford approached one of us (McGeer, then at HP Labs) about porting OpenFlow to an HP switch and donating a number of those switches to Stanford for a trial. The timing was fortuitous: a team of researchers at HP Labs, including McGeer, had just embarked on a project to build a next-generation network platform. The team quickly made OpenFlow a focus of the project, and two HP Labs researchers, Jean Tourrilhes and Praveen Yalagandula, implemented OpenFlow on an HP 5406 switch. Gary Campbell, Chief Technologist of HP Enterprise, Greg Astfalk, HP Chief Scientist, and Charles Clark of HP Networking arranged for the switch donation., and by late 2007 the first implementation of OpenFlow on a commercial switch was completed and deployed in the Gates building on the Stanford campus.

Parallel efforts were engaged in by other firms in the same time period. NEC also announced an OpenFlow commercial switch in 2008.

This began the first of three major experimental deployments of OpenFlow.

1. Deployment on a single campus: Stanford, beginning in early 2008. The objectives of this deployment were to determine whether the use of OpenFlow was feasible in an operational campus network, carrying both operational and experimental traffic; whether experimental traffic and experimental network control could be truly isolated from other traffic being carried by the same network equipment and on the same wires; and to develop the tools and techniques to operate truly isolated, independently-controlled, virtual application-specific networks in a campus setting .
2. An eight-campus deployment, beginning in 2009. The objective of this deployment, sponsored by GENI, was lofty:

> This project is motivated by the belief that if we can open up campus networks for innovation by researchers and network administrators, we will unleash tremendous untapped potential and will fundamentally change the field of networking. If we can move from a culture of closed, proprietary, expensive infrastructure (with very long innovation cycles) to an open infrastructure enabling rapid innovations by all stakeholders, we will create a market place for ideas allowing the best ideas to win [6].

In other words, the primary goal here was to determine the value of the OpenFlow-enabled applications and services in a campus setting, and to further expand the technological developments of the initial one-campus deployment.

3. Deployment across the wide-area backbone through the GENI Mesocale deployment [7]. The objective of this deployment was to open OpenFlow and SDN up to experimenters across the wide area, and to determine what tools and technologies were required to knit OpenFlow domains together. This deployment augmented the campus deployment substantially. This deployment extended GENI's hardware OpenFlow deployments to over 50 campuses and approximately 10 regional networks, by means of the OpenFlow switch in each GENI rack [1, 25]. This deployment, a collaborative effort of GENI, Internet-2, and the National Lambda Rail, formed the largest and most ambitious OpenFlow deployment undertaken to date.

Given access to OpenFlow-capable (albeit beta quality) firmwares for readily available hardware platforms, in 2009 the Clean Slate Program at Stanford University pushed to deploy OpenFlow on the GENI WAN (by placing user-programmable OpenFlow switches at Internet2 and National Lambda Rail (NLR) POPs in the WAN paths between campuses).

Starting with a small scale WAN deployment available to experimenters on the GENI test bed would provide an environment for development and testing of OpenFlow applications and provide a strong proof-of-concept. The GENI deployment was remarkable, in that it represented a multi-site trial for a new foundation for networking within 2 years of its invention. This created significant interest in a number of communities—within 2 years the Open Network Foundation (ONF) had been established with 17 charter members from industry (quickly growing to more than double that number within months), pledging to develop and adopt SDN technologies in their products. The deployment and user experience reality in GENI, further, gave real world experiences that would ultimately inform both industry actors and future test bed SDN deployment considerations.

One measure of the interest in the OpenFlow platform was the number of research-firmware shipments from HP Labs' Open Networking Group. Groups wishing to run OpenFlow on HP hardware in this period used a standard HP 5406 switch, and then get the HP Labs implementation of OpenFlow for the 5406. Open Networking Group co-lead Sujata Banerjee eventually shipped well over 50 copies of her group's warranty-less "as is" firmware, to both commercial and academic users. A large number of the initial GENI campus trials used the 5406.

Among the earliest learnings from the initial Stanford deployment was that virtualization was also required in the controller—the controller had to enforce slicing the network between competing applications, both in the addressing and performance dimensions. To some extent this had already been known; in fact, the Virtual LAN, or VLAN, permitted re-use of IP addresses. However, the Stanford deployment showed that competing applications had to be rate-limited, both in data plane and controller traffic.

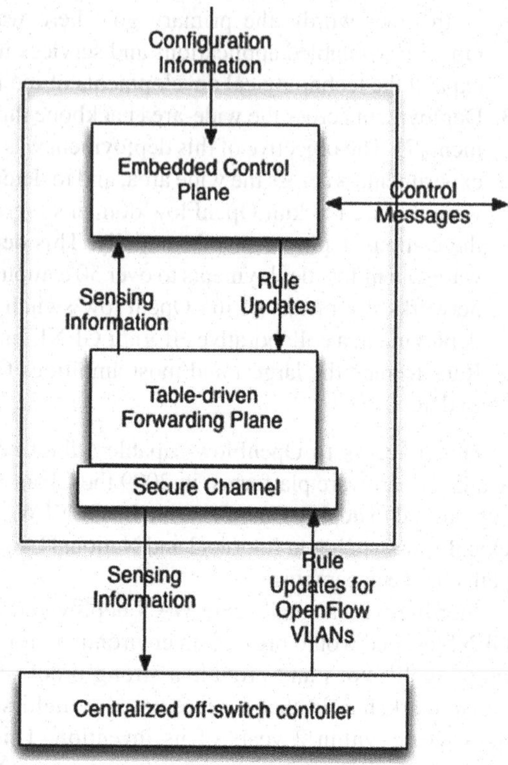

Fig. 4 A hybrid switch

4 Using OpenFlow in a Multi-Tenant Network

As a multi-user test bed, any OpenFlow deployment on GENI would need to facilitate shared usage of the devices and connectivity. Plain time-sharing would have been a straightforward option, albeit inefficient and costly—if only a single experimenter could use the OpenFlow resources at a time he would be consuming nationwide transit circuits that were scarce resources. This would have the effect of limiting OpenFlow to a small group of patient and dedicated experimenters. In order to provide access to OpenFlow resources to a wider audience it would be necessary to support concurrent usage of the test bed by as many users as possible.

Fortunately, this need had already been satisfied at a base level in two ways—both out of a desire to increase possible deployment and research using existing networks, but through measurably different approaches.

4.1 *Hybrid Switching*

The first hardware vendor to implement the 0.8.9 specification—HP—did so in a way that optionally allowed for only *partial* OpenFlow control of the switch. This was accomplished by enabling OpenFlow on a per-VLAN basis—allowing a single OpenFlow instance to control a non-VLAN-aware logical switch, which was itself an abstraction of a single VLAN in the switch hardware. All ports configured for the OpenFlow VLAN would be advertised to the controller, and any packets received on those ports in the OpenFlow VLAN would be handled according to the OpenFlow specification, while all other traffic on these ports would continue to be handled with existing protocol behavior. There were some caveats to this approach—certain protocols like the Spanning Tree Protocol (STP) and the Link Level Discovery Protocol (LLDP)which operated outside of a single VLAN would not be handled by the OpenFlow instance, and packets would not appear to be VLAN tagged to the OpenFlow instance—but it was a powerful way to safely add rudimentary OpenFlow support to an existing network on a trial basis while still maintaining all existing forwarding logic. Further, as mentioned above, one of the major goals of OpenFlow was to subsume in the open control plane protocols such as LLDP and STP.

The OpenFlow specification did not initially afford this possibility, but the compelling nature of such a simple deployment path into existing operational networks meant that the 1.0 specification evolved to include language referring to the possibility of vendors supporting such an abstraction. Later the community, and the standard-setting ONF, would focus considerable effort on the best way to codify support for and the behavior of such devices, with the 1.1 and later specifications including increasingly more specific language for handling of "OpenFlow-Hybrid" switches.

A conceptual diagram of a Hybrid OpenFlow switch is shown in Fig. 4. Traffic on non-OpenFlow VLANs is handled in a Classic Switch fashion, through a combination of classic configuration and autonomic control. Traffic for OpenFlow VLANs is governed by OpenFlow configuration messages.

As HP continued to support this mode of operation it was possible to create multiple OpenFlow instances on a single device. When configured to use different experimenter controllers this allowed multiple concurrent OpenFlow experiments to use the same device, isolated on the data plane by VLAN tag. While there were practical limits on the number of instances that could be handled simultaneously by the switch CPU and forwarding tables, as well as the number of unique VLANs that might be available across the entire GENI WAN, this was a powerful functionality that would allow for at least a moderate level of resource sharing.

4.2 FlowVisor

Concurrent with the initial availability of OpenFlow switch firmwares was an effort to design and implement a network virtualization layer for such devices, called FlowVisor [40–42]. The original OpenFlow architecture envisaged a single controller which would partition a network among various applications, each of which had an isolated virtual network. The purpose of FlowVisor was to create and enforce truly isolated virtual networks, and offer each of these to a separate controller. FlowVisor replaced a single controller with a hierarchy of controllers, where each controller presented a controller interface on its southbound, or network-facing, side, and a network interface on its northbound, or application-facing, side. In the words of the FlowVisor paper, FlowVisor acted as a transparent proxy sitting between multiple controllers and the network. The heart of FlowVisor was a simple insight: the unit of isolation in the network was the header bits allocated to the northbound controllers on each switch, the available bandwidth and inter-switch latency on the links in the presented virtual network, the controller/network bandwidth for control-path updates, on both the northbound and southbound sides, switch CPU for slow-path traffic, and TCAM entries for fast-path traffic. FlowVisor maintains, for each network slice, the allowable values of header bits (the "flowspace") assigned to the slice on each network switch, and ensures that the slice's rule-table entries lie within that flowspace. It also enforces static limits on flow-table entries per slice and per switch. For bandwidth limitation it uses priority VLAN tag bits per slice. For dynamic resources (switch CPU, controller/switch bandwidth), it does root-cause analysis of the events which cost the dynamic resource, and monitors and controls them.

An example of a FlowVisor deployment is shown in Fig. 5. In this figure, each controller sees an OpenFlow network with a restricted flowspace; each switch sees a single OpenFlow controller, FlowVisor. FlowVisor seamlessly intercepts control traffic between the controllers and the network, ensuring mutual isolation. Each controller sees a simplified view of the network, as shown in Fig. 6; FlowVisor is entirely transparent.

OpenFlow and FlowVisor thus correspond to two separate layers of abstraction, as shown in Fig. 7. OpenFlow abstracts the details of the underlying switches, presenting a uniform commodity switch with fully-controllable routing; FlowVisor abstracts the other controllers, to present a virtual single-tenant network of OpenFlow switches to a controller.

While FlowVisor had a rich feature set and research road map—slicing/virtualization of every key metric in the network, such as device CPU, hardware table space, topology, bandwidth, etc.—the most important feature for a test bed was the singular ability for multiple upstream controllers to manage the same device. In this way, FlowVisor functioned as a mux/demuxing proxy for OpenFlow control connections, allowing multiple concurrent users of the same physical resource.

While hybrid firmwares could often be capable of running multiple concurrent OpenFlow instances and thus supporting multiple controllers (provided one could

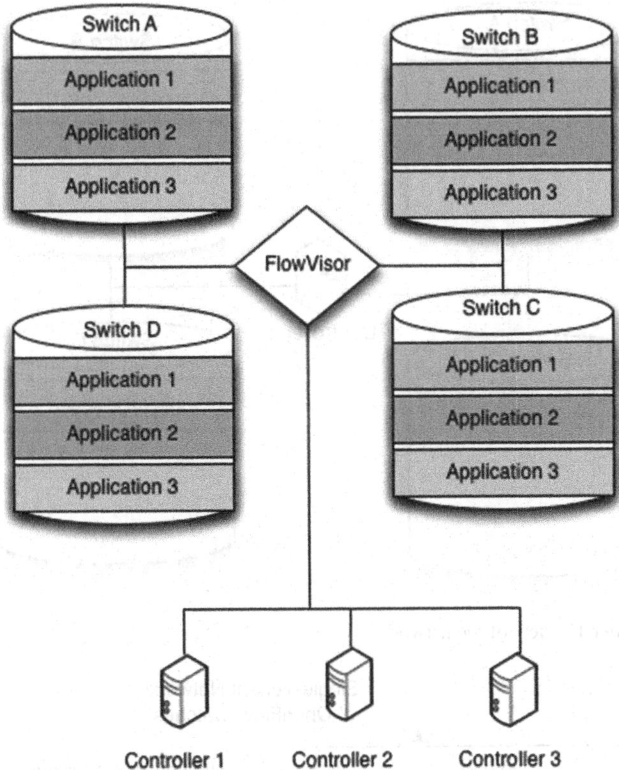

Fig. 5 FlowVisor in the network

operate within the limited mechanisms for defining data plane discrimination between "slices"), FlowVisor offered this functionality to all OpenFlow devices. FlowVisor users also benefited from the more aggressive software development release cycle of an off-device proxy, rather than the long development and test cycle of hardware firmwares, allowing for more rapid innovation and experimentation with interfaces and mechanisms for crafting isolation boundaries between users.

4.3 *Software Datapaths*

While not seriously considered at the outset for GENI deployment, Open vSwitch provided an OpenFlow-based control channel in 2009 and could be deployed on commodity x86 hardware. Earlier packages had a long history of providing routing functionality on *BSD and Linux—Zebra/Quagga, Click, etc.—and while fully functional soft-switching was not as commonly deployed at the time, it would obviously become critical in cloud infrastructures. As we will discuss in later sections, software virtual switches ultimately became an important part of test bed usage, for a wide variety of reasons.

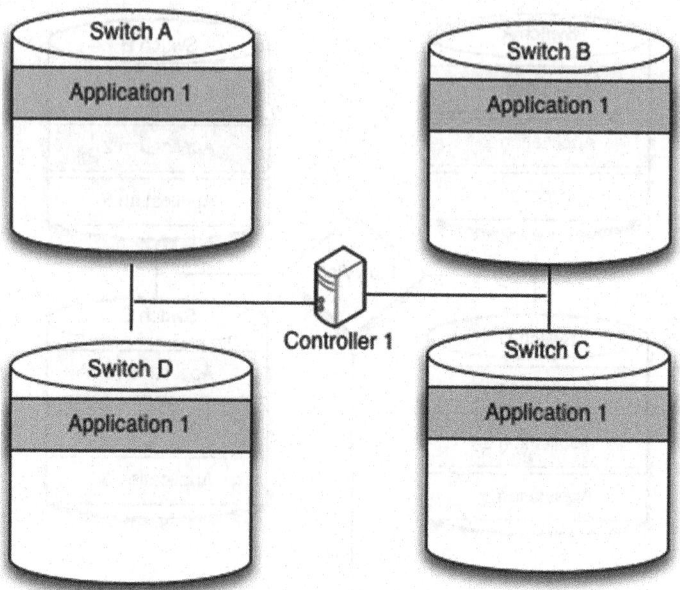

Fig. 6 Controller 1's view of the network

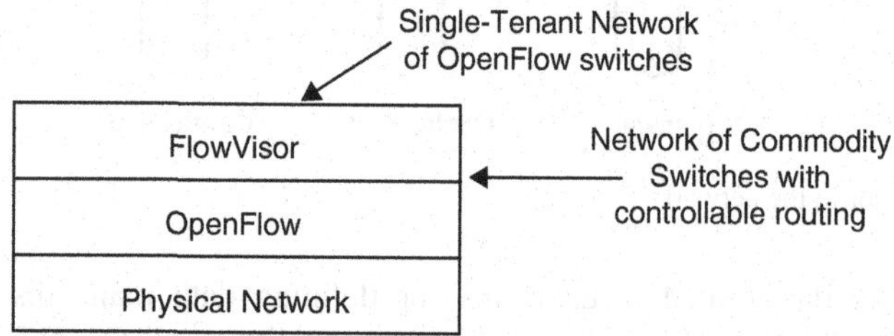

Fig. 7 Abstraction layers in a FlowVisor network

5 Integration with GENI

Through agreements with Internet2 and National Lambda Rail (NLR), the GENI Project Office (GPO) was able to place OpenFlow switches at ten core network locations across the continental United States—five with NLR and five with Internet2, with a cross-connect between the networks at a shared facility in Atlanta—to build out a small scale OpenFlow-enabled WAN. (As a result of the Internet2 shutdown of the ION circuit service and the earlier operational shutdown of NLR the last remnants of this network were disabled in May of 2015, although topology information is still available for historical purpose [10].)

This provided dedicated network hardware for the WAN test bed, although physical connectivity was still provided via common infrastructure used by each provider for many customers. Given that reality, and the provisioning available at the time, GENI was afforded a limited number of VLANs across the WAN test bed topology (as VLANs at each connection point were allocated out of a pool shared by all users of I2/NLR). Granting each GENI user even a single VLAN would severely limit the number of concurrent experiments, so FlowVisor was used to further slice individual VLANs into distinct L3 address space for each user. This allowed for an essentially unlimited number of L3 allocations on the same topology (functionally limited only by the table space in each device, and not the available address space).

At each campus edge, local resource islands (particularly GENI Racks) had far more VLAN space available, and as such experimenters could be allocated their own hybrid OpenFlow instance from the hardware directly (where such functionality was available based on vendor hardware in place), instead of being proxied through FlowVisor. This combined with users provisioning VM resources for deploying their own software datapaths to make a complex but functional end-to-end SDN networking environment.

6 Experimenter Experience

Ultimately the user experience for SDN experiments on GENI varied by the resource used and the way it was virtualized/sliced, leading to a litany of support issues as well as research constraints. These experiences should inform any SDN research environment being created today, as well as an understanding of some of the subtler implications of incremental brownfield deployment of SDN solutions in production networks.

6.1 Fundamental Infrastructure Issues

A serious but initially-unforeseen problem was that the underlying WAN L2 paths were not "clean", even across circuit services such as those provided by I2 and NLR, but particularly at their borders. Transit devices had underlying assumptions about how forwarding should work (MAC Learning), and would produce legacy control frames within user slices (LLDP, STP, etc). Some of these problems could be cleaned up by finding device owners and changing device configurations, but this was not always possible—particularly for MAC learning. This had the consequence of making certain classes of novel SDN research impossible to perform on the GENI SDN substrate, as intervening devices could be confused and drop packets or broadcast them to undesirable destinations.

Virtualized compute resources at the topological edge also typically involved paths that didn't look like what the experimenter imagined. Even in cases where the experimenter was not using FlowVisor or WAN circuits—they were using a local private OpenFlow instance in a VLAN or set of VLANs—the virtualization stacks in use would insert a logical L2 switch between the experimenter's VM and the physical NIC, which would mean that while experimenters believed they had a logical topology of a number of VMs connected to a single OpenFlow device, in reality they had a single OpenFlow device connected to a set of non-OpenFlow devices, and needed to engineer their traffic patterns accordingly. Technologies such as SR-IOV could be used to mitigate these problems, but were (and are still today) available on only a limited number of test bed virtualization hosts.

On top of legacy protocol issues in experiment paths, network devices are not substantially more powerful than their intended workload. Unlike today's larger and larger compute servers which are at this point designed to be sharded and split between many shifting workloads (having generally more memory, CPU, disk, etc. than are necessary for any one application), this is not true of network hardware, and trying to share these resources makes this problem readily apparent.

For hybrid switch implementations at the edge this could manifest in overloaded management CPUs. OpenFlow requires a constant always-on control connection (preferably using TLS) for each instance, which is outside the original design requirements of existing hardware, where the management endpoint was used to serve configuration and monitoring needs (SNMP, NetFlow, telnet, etc.). Each OpenFlow instance running on the device not only required separate storage and maintenance of table state, but also a response time and reliability beyond that of legacy management tasks (which were often *also* still being used to monitor and provision the underlying infrastructure). This could result in dropped control connections and flow table thrash/flapping as controllers reconnected and tried to re-synchronize their state.

Similarly, switch table size was often designed by the vendor to fit the original market for the device, and certainly not intended to be split between experiments which were designed around the notion that each logical device in a topology was a blank slate with an isolated set of resources. This was compounded by the fact that early OpenFlow implementations tended to leverage only the Access Control List (ACL) table of a switch for inserting rules—OpenFlow before the 1.1 specification could not expose multiple tables to the controller, so the path of least resistance for firmware developers was to expose the single most functional table, which was also typically the smallest. Even in multi-table implementations certain devices were not intended to be used with the tens of thousands of MAC addresses created by synthetic experiment workloads. This meant that at relatively low throughput (hundreds of megabits or less), two or three concurrent experiments could overwhelm at least one resource vector on a hardware device, leading to undefined behavior.

6.2 *Virtualization/Slicing Issues*

On top of infrastructure issues that are somewhat unavoidable and can only be remedied over the long time frames of firmware deployment and hardware refresh, the virtualization and slicing software adversely impacted end-user experience as well. While FlowVisor was well-maintained during the first few years of deployment, certain original design decisions were in retrospect problematic, and other solutions clashed with security and isolation concerns.

As it was a proxy for OpenFlow only, FlowVisor did not allow the end-user direct access to the hardware device, which led to a litany of issues. As OpenFlow was the only channel FlowVisor had to the upstream controller, any proxy errors were communicated using OpenFlow error messages, making it difficult to determine whether an error message came from the hardware device or from the proxy. Ultimately this often meant that an experimenter would be required to contact an administrator to look through both FlowVisor and device logs in order to isolate any problems, placing both substantial burden on the experimenter and the local administrators. Similarly, as FlowVisor did not proxy other management or monitoring mechanisms, experimenters could not augment their controller with data from SNMP, NetFlow, etc., meaning that measurement-driven research was difficult or impossible to accomplish on the test bed in some cases.

6.3 Lessons for the Future

The history of SDN infrastructure on GENI is a path populated with obstacles, but this history provides many valuable lessons about crafting a quality environment for SDN research.

Despite demands from experimenters, it is hard to imagine a useful shared test bed being constructed from commodity hardware platforms. Devices are being produced today with more powerful CPUs, solving some problems with the high demands OpenFlow can place on a switch, but table size will continue to be an extremely limiting factor for sharing resources across multiple projects. Researchers requiring direct hardware access will need to use environments that allow exclusive hardware usage, rather than a larger shared test bed. Even in that case, using commodity hardware will limit the forward-looking function of research, and keeping abreast of just the commodity state-of-the-art would require a large amount of funding to continue to refresh hardware.

Leveraging FPGA and NPU hardware is one powerful way to allow a research platform to move beyond any current commodity deployment, and is certainly worth exploring based on the needs and goals of a future test bed. There are limitations—both FPGA and NPU platforms have a high barrier to entry on both cost and difficulty to work with, from difficult programming models to practical legal non-disclosure issues—but flexible hardware opens up research avenues for experiments with significant performance requirements. Limited deployments of both technologies on GENI (NetFPGA [5, 21, 28], Dell-SDP [13]) have received only small amounts of use, given the development and topology constraints and inability to foster community support due to NDA requirements, but persistent experimenters have produced some quality results.

Ultimately the most widely used resource in GENI for SDN has been software switches—no FlowVisor-sliced resources have been used by experimenters since mid-2013, and while local hybrid instances get some use they are of limited value as they still only provide OpenFlow 1.0 capability and limited table sizes. Current virtual switch deployments by experimenters on compute resources still suffer from many of the same underlying infrastructure issues—MAC learning bridges in paths, LLDP/STP packet production—but judicious use of tunnels or label shims can mitigate this problem for experiments that require it. A relatively new service on GENI—VTS—facilitates the orchestration of clean label-isolated Software-Defined Infrastructures (SDI) on the existing test bed leveraging both hardware and software resources, which has seen over 1000 slices created in the last 6 months of 2015 alone, pointing the way to a compelling solution for many researchers.

Future expansion of SDN research goals on GENI and similar test beds will have to strongly consider the supported use cases and funding available, and provision accordingly. Software-Defined Infrastructure capabilities will expand the ability to do SDN research, regardless of whether the underlying infrastructure is itself SDN-enabled. Ultimately hardware-infrastructure-as-test-bed-resource is a solution that should be reserved for only the most well funded and performance-sensitive experiments.

7 New Opportunities with OpenFlow and SDN

The experience of using OpenFlow and SDN in campus deployments and with GENI in the wide area not only taught experiences of deployment, but also illustrated a number of use cases and advantages for OpenFlow and SDN far beyond the original experimentation vision.

Persistent Addresses Across Broadcast Domains Many end-host services and applications are intolerant of connection disruption. Good examples are ssh, voip, video streaming, and games. For this reason, both servers and clients in these applications tend to be tied to fixed network locations. In various special circumstances (e.g., cellular networks) network operators can manipulate routing tables to maintain persistent addresses, but in general this has not been possible. Managing the migration of addresses across broadcast domains was one of the first demonstrations of OpenFlow, at SIGCOMM 2008; it has since been used in campus deployments to offer seamless migration between wired and wireless networks and migration of virtual machines in data centers.

Partial Deployment of SDN in Enterprise Networks It has been observed [20] that many of the benefits of SDN deployments can be achieved in an enterprise with partial SDN deployments, where traffic crosses only one SDN-enabled switch. This enables partial and incremental upgrades of networks from classic to SDN networking, reducing the barriers to adoption.

Verification of Network Configuration The forwarding plane of a switch is stateless; the actual forwarding of packets by a switch is done entirely on the basis of its forwarding tables. Mathematically, a switching network is therefore equivalent to a combinational logic network, and its verification properties are therefore in $\mathcal{NP}$ [23]. A practical implementation of network verification of OpenFlow networks was given in [17], which demonstrated its practicability on campus networks, and a comprehensive approach in [18]. A survey of verification of software-defined networks is given in [50].

Safe and Secure Network Updates The open transparency of the network control plane and the relatively low latency of network updates offered a prospect of reliable network updates (that is, updates where routing invariants were maintained throughout the update process). Reitblatt et al. [36] offered the first reliable algorithm, with guaranteed correctness (all packets arrived in order at all destinations) at the expense of TCAM space; however, this algorithm was guaranteed correct independent of update arrival times of the various switches in the network. McGeer [22] offered a different procedure which guaranteed correctness under all schedules, using the same criterion of correctness as [36]. McGeer [24] and Katta et al. [16] took a different approach, giving *schedules* for correct updates; McGeer [24] used a slightly different correctness criterion, namely the maintenance of network verification invariants throughout the update process.

Open Architecture for Middleboxes A middlebox, to a first approximation, is nothing more than a computer running some service or other (email sniffing, web proxy, etc) fronted by a switch which acts as a filter, bringing down the incoming packet stream to something a computer can handle. Given a programmable switch in the network, an easy, programmable architecture for middleboxes. Combined with the insights of [19], this offers a fundamentally new paradigm for Internet architecture: a collection of stateful proxies coupled with stateless switches throughout the network, where the proxies act to provide in-network computational services on a local domain. The InstaGENI racks used as the backbone of the GENI mesoscale deployment are the prototypes of this universal, programmable middlebox which, in turn, will be the backbone of the next Internet. A diagram of classic vs the new universal middlebox is shown in Fig. 8.

Power- and Server-Aware Routing Routing is currently done independently of conditions at the end-host. If the network is a black box to end-hosts, the end-hosts—their power consumption, load, memory usage, etc.—are a black box to the network. Heller et al. [14] proposed a method of incorporating end-host usage, power, and load information into routing, offering clients the best server among many for a services, incorporating both end-host and routing information.

More Insightful and Finer-Gained Network Monitoring One advantage of OpenFlow was that both network routing became completely transparent and the state of the switches was transparent to the network controller. A unified host-based network monitoring platform was therefore available to provide finer-grained but less intrusive network measurements. OpenNetMon [48] was one of many network monitoring procedures made possible by OpenFlow.

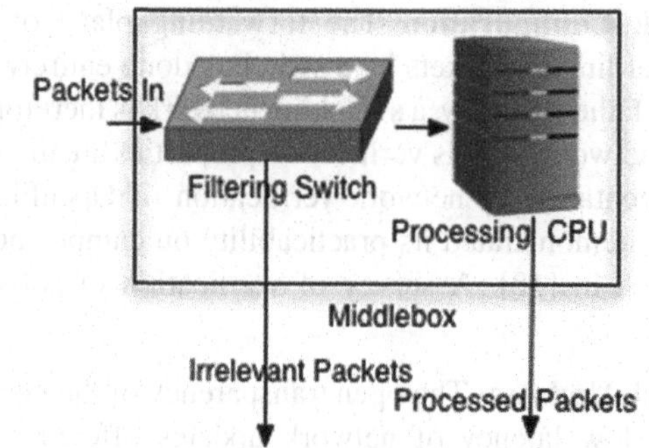

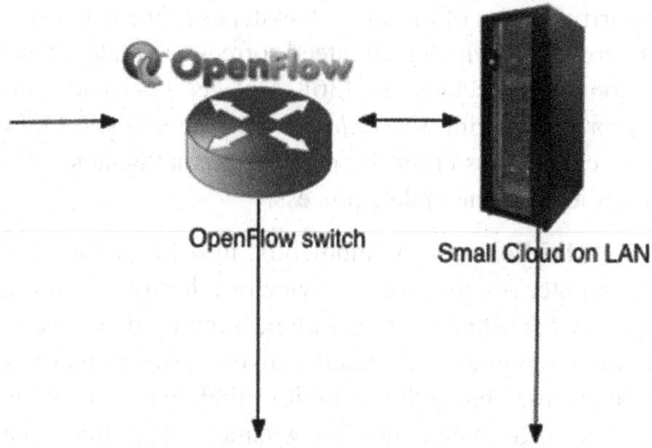

Fig. 8 Classic and universal middlebox

Better Security, Including Admission Control on a Per-Flow Basis This was the original OpenFlow application with Ethane. Though not unanticipated, it will have perhaps the most profound immediate impact on networking and distributed systems. To date, admission control has been done at the endpoint or on an indiscriminate basis by a firewall. For example, many enterprise systems block external access to most internal systems, and then stand up Virtual Private Networks (VPNs) to grant access on a selective basis to authenticated external users. The VPN is a misnomer; it is neither private, nor virtualized, nor a network. Rather, it is an encrypted tunnel over the public Internet, a glorified https connection. A network with OpenFlow switches everywhere offers the prospect of truly private, virtualized networks, dynamically established and dynamically modified, instantiated over the wide area. In effect, the enterprise intranet can be extended seamlessly over the wide area. This is merely the coarsest, and most immediate application of this capability: in the future, any user will be able to construct his own instance-specific network for any application, as easily as he today sets up a conference call.

Enabler for Network Function Virtualization *Network Function Virtualization* (NFV) is a technology of significant recent interest in the telecommunications industry. In the words of the industry white paper on the subject [8]: "Network Functions Virtualization aims to transform the way that network operators architect networks by evolving standard IT virtualization technology to consolidate many network equipment types onto industry standard high volume servers, switches and storage, which could be located in Datacentres, Network Nodes and in the end user premises". In other words, dump the special-purpose equipment and switches that currently run telco functions and replace them with something a lot cheaper, more transparent, flexible, and scalable: x86 servers running Linux VMs. This is made concrete in the current Open Networking Lab/AT&T collaboration "Central Office Re-architected as a Data Center" [31], which uses as its base technology the OpenStack-based XOS [33, 34] and ON.LAB's Open Network Operating System (ONOS) as the SDN controller [2]. In this, CORD combines sophisticated cloud technology with SDN, as both GENI envisioned and as envisioned in [8]: "(NFV) approaches relying on the separation of the control and data forwarding planes as proposed by SDN can enhance performance, simplify compatibility with existing deployments, and facilitate operation and maintenance procedures".

Telco motivation to pursue NFV is largely rooted in dramatic reductions in internal CAPEX and OPEX: bluntly, NFV is viewed primarily as a way to make current telco operations much cheaper and more efficient. However, in time this may become the physical realization of the distributed, open cloud envisioned by GENI: a Central Office as a Datacenter is not readily distinguishable from the Digital Town Square described in [37], and opening this up to third-party developers will make over-the-top service providers into telco customers.

The recent introduction of OpenFlow has inspired a plethora of papers, and the tide does not seem to be receding. The networking and distributed systems community has only begun to explore the possibilities inherent in a programmable, transparent, verifiable network.

8 SDN: The Next Generation

The promises and opportunities of Software-Defined Networking are too compelling for this technology to fail: widespread adoption of SDN is inevitable. As always, it is happening much more slowly than enthusiasts hope. Greg Papadopoulos of New Enterprise Associates points out that new technologies follow a well-known curve: first, there is the peak of unrealistic expectations, followed by the "valley of disappointment" where the technology is dismissed as a flash in the pan, finally followed by a period of slow but inexorable growth which eventually outstrips all expectations.

As we write this, we're in the Valley of Disappointment for SDN. It's been an enormous success in greenfield deployments such as data centers (and even non-greenfield data centers in advanced firms such as Google). It has shown tremendous promise in the carrier space, and (given the usual slow rate of technology introduction by carriers) it is more or less on track there. But the original motivation for OpenFlow was to rationalize campus enterprise networks, and there the adoption rate has been very slow. Moreover, adoption across the wide area, required for the application-specific private networks that we envision, is currently non-existent. We forecast the next Internet will be characterized by the inexorable growth of the campus, and then inter-campus, SDN.

This ultimate phase of inexorable growth happens when the tools and technologies conceived during the unrealistic expectations phase have matured to the level where they can be deployed and reliably used. This phase is now upon us. There are three major thrusts of technology development which are now nearing maturity:

- **A "Northbound" Open Networking API and network-specification language, or languages**. Just as the relational model of storage required SQL before databases could be used reliably in application programs, so too does SDN required one (or more) network specification and programming languages to bring this technology to the application developer and network administrator. A number of recent candidates have emerged, including FreNetIc [9, 35] more recently P4 [3] among many others. One thing that is certain is that the ultimate successful language will be state-free in order to preserve the validation and verification properties of an open data plane.
- **Better match of the switch ASICs and processing to SDNs**. Switches generally don't present the simple abstraction of the data plane anticipated by SDN. In particular, naive OpenFlow implementations are profligate with the most expensive resource in a switch, the TCAM, and fail to use other, cheaper resources such as prefix matching, and don't accurately mirror the switch packet-processing pipeline. In order for SDN to succeed, the switch processing pipeline must be made more transparent and effective use of limited memories must be used. Intel's DPDK is a very promising start, and more general purpose silicon will emerge as greater parallelism on silicon becomes available.
- **Software-defined Exchanges**. A number of researchers and industrial operators have observed that OpenFlow can simplify the operation of Internet Exchange points [12] and hence can simplify and enrich the Border Gateway Protocol. It is certainly the case that such a "Software-Defined Exchange" will make the routing and operation of the current Internet far more flexible and efficient. However, this only scratches the surface of the possibilities inherent in this.

The vision of the Internet put forward in this chapter, and more generally in this book; the vision that is at the heart of GENI, and that inspired PlanetLab, is of erasing the boundaries between network and application and eliminating the distance between service and user. Calit2 Director Larry Smarr speaks of using high-bandwidth networks to create a world where "distance is eliminated".[3] Smarr's vision was of a world where the *perception* of distance was eliminated between user and service. By moving programs to universal middleboxes throughout the network, we can *truly* eliminate the distance between user and service, by siting services close to users, wherever they happen to be.

To make this vision a reality, a Software-Defined Exchange will have to be richer than that envisioned by Gupta et al. [12]. In this new exchange, users will send specifications of slices with management and orchestration information across the exchange point, to autonomously instantiate a network of services in a remote Autonomous System. The GENI RSPEC, with extensions for orchestration and management, is a first attempt at the information that must be put through such an exchange, and the GENI AM API is an early prototype of the implementation of such an exchange.

As with any new technology, the reaction to the introduction of OpenFlow and SDN was highly optimistic, and the prospect of immediate solutions to longstanding problems seemed imminent. And as always, this early optimism met the realities of brownfield deployment, existing hardware mismatched to the technology, existing IT policies and skill sets, etc. Revolutions in IT do not happen overnight.

But they do happen. The promise of SDN is real, and the tools and technologies to deploy SDN successfully in enterprise and service provider arenas are under continuous development. Hardware is becoming more compatible with SDN technologies. The trial deployments of OpenFlow, particularly under GENI, have been critical in developing the experience necessary for enterprise and service provider deployments of the near future of network technology.

Acknowledgements We have been fortunate to enjoy the support of a number of brilliant colleagues and a vibrant ecosystem throughout the course of this project. Sujata Banerjee was instrumental in ensuring that many of the commercial and academic researchers who wanted to experiment with OpenFlow could do so on commercial switches. Charles Clark of HP Networking was the first person to suggest hybrid switching and using OpenFlow on specific VLANs. Nick McKeown and Guru Parulkar of Stanford were unfailingly supportive throughout this process. The legions of GENI users and experimenters provided invaluable feedback. The initiative of the campuses first involved in the initial OpenFlow trials was critical, and Nick Feamster, Jennifer Rexford, Russ Clark, and Ron Hutchins. were notably helpful.

References

1. Baldin, I., Chase, J., Xin, Y., Mandal, A., Ruth, P., Castillo, C., Orlikowski, V., Heermann, C., Mills, J.: Exogeni: a multi-domain infrastructure-as-a-service testbed. In: GENI: Prototype of the Next Internet. Springer, New York (2016)

[3]See, for example, http://lsmarr.calit2.net/multimedia?vid=VqAjLalPEmQ.

2. Berde, P., Gerola, M., Hart, J., Higuchi, Y., Kobayashi, M., Koide, T., Lantz, B., O'Connor, B., Radoslavov, P., Snow, W., et al.: ONOS: towards an open, distributed SDN OS. In: Proceedings of the Third Workshop on Hot Topics in Software Defined Networking, pp. 1–6. ACM, New York (2014)
3. Bosshart, P., Daly, D., Gibb, G., Izzard, M., McKeown, N., Rexford, J., Schlesinger, C., Talayco, D., Vahdat, A., Varghese, G., Walker, D.: P4: programming protocol-independent packet processors. SIGCOMM Comput. Commun. Rev. **44**(3), 87–95 (2014)
4. Casado, M., Freedman, M.J., Pettit, J., Luo, J., McKeown, N., Shenker, S.: Ethane: taking control of the enterprise. In: Proceedings of ACM SIGCOMM (2007)
5. Covington, G.A. Naous, J., Erickson, D., Mckeown, N.: Implementing an openflow switch on the NetFPGA platform. In: Proceedings of ANCS (2008)
6. Davy, M., Parulkar, G., van Reijendam, J., Schmiedt, D., Clark, R., Tengi, C., Seskar, I., Christian, P., Cote, I., China, G.: A case for expanding openflow/SDN deployments on university campuses. http://archive.openflow.org/wp/wp-content/uploads/2011/07/GENI-Workshop-Whitepaper.pdf, (2011)
7. Dempsey, H.: The GENI mesoscale network. In: GENI: Prototype of the Next Internet. Springer, New York (2016)
8. ETSI. Network functions virtualisation: an introduction, benefits, enablers, challenges & call for action. In: SDN and OpenFlow World Congress (2012)
9. Foster, N., Guha, A., Reitblatt, M., Story, A., Freedman, M., Katta, N., Monsanto, C., Reich, J., Rexford, J., Schlesinger, C., Walker, D., Harrison, R.: Languages for software-defined networks. IEEE Commun. Mag. **51**(2), 128–134 (2013)
10. G.M.-O. Center: Geni Openflow Map. http://gmoc.grnoc.iu.edu/uploads/a5/b4/a5b452ec193c769a309d5adcbe801ecd/OF-INT-BB-14-Dec-2012.png (2012)
11. Gude, N., Koponen, T., Pettit, J., Pfaff, B., Casado, M., McKeown, N., Shenker, S.: Nox: towards an operating system for networks. ACM SIGCOMM CCR **38**(3), 105–110 (2008)
12. Gupta, A., Vanbever, L., Shahbaz, M., Donovan, S.P., Schlinker, B., Feamster, N., Rexford, J., Shenker, S., Clark, R., Katz-Bassett, E.: SDX: a software defined internet exchange. In: Proceedings of the 2014 ACM Conference on SIGCOMM, pp. 551–562. ACM, New York (2014)
13. Gurkan, D., Dane, L., Bastin, N.: Split data plane switches on GENI. http://groups.geni.net/geni/raw-attachment/wiki/GEC20Agenda/EveningDemoSession/1959_GEC20SDPonGENI.pdf (2012)
14. Heller, B., Seetharaman, S., Mahadevan, P., Yiakoumis, Y., Sharma, P., Banerjee, S., Mckeown, N.: Elastictree: saving energy in data center networks. In: Proceedings of IN NSDI (2010)
15. Jacobson, V., Mosko, M., Smetters, D., Garcia-Luna-Aceves, J.: Content-centric networking. Whitepaper, Palo Alto Research Center, pp. 2–4 (2007)
16. Katta, N.P., Rexford, J., Walker, D.: Incremental consistent updates. In: Proceedings of the Second ACM SIGCOMM Workshop on Hot Topics in Software Defined Networking, pp. 49–54. ACM, New York (2013)
17. Kazemian, P., Varghese, G., McKeown, N.: Header space analysis: Static checking for networks. In: Proceedings of the 9th USENIX Conference on Networked Systems Design and Implementation, NSDI'12, pp. 9–9. USENIX Association, Berkeley, CA (2012)
18. Khurshid, A., Zhou, W., Caesar, M., Godfrey, P.: Veriflow: verifying network-wide invariants in real time. ACM SIGCOMM Comput. Commun. Rev. **42**(4), 467–472 (2012)
19. Knutsson, B., Peterson, L.: Transparent proxy signalling. J. Commun. Netw. **3**(2), 164–174 (2001)
20. Levin, D., Canini, M., Schmid, S., Schaffert, F., Feldmann, A., et al.: Panopticon: reaping the benefits of incremental SDN deployment in enterprise networks. In: Proceedings of USENIX ATC (2014)
21. Lockwood, J.W., Mckeown, N., Watson, G., Gibb, G., Hartke, P., Naous, J., Raghuraman, R., Luo, J.: Netfpga - an open platform for gigabit-rate network switching and routing. In: Proceedings of MSE '07, pp. 3–4 (2007)

22. McGeer, R.: A safe, efficient update protocol for openflow networks. In: Proceedings of the First Workshop on Hot Topics in Software Defined Networks, HotSDN '11, pp. 61–66. ACM, New York (2011)
23. McGeer, R.: Verification of switching network properties using satisfiability. In: ICC Workshop on Software-Defined Networks (2012)
24. McGeer, R.: A correct, zero-overhead protocol for network updates. In: Proceedings of the Second ACM SIGCOMM Workshop on Hot Topics in Software Defined Networking, pp. 161–162. ACM, New York (2013)
25. McGeer, R., Ricci, R.: The instaGENI project. In: GENI: Prototype of the Next Internet. Springer, New York (2016)
26. McKeown, N., Anderson, T., Balakrishnan, H., Parulkar, G., Peterson, L., Rexford, J., Shenker, S., Turner, J.: Openflow: enabling innovation in campus networks. ACM SIGCOMM CCR **38**(2), 69–74 (2008)
27. McLaggan, D.: Web cache communication protocol v2, revision 1. http://tools.ietf.org/html/draft-mclaggan-wccp-v2rev1-00 (2012)
28. Naous, J., Bolouki, S.: Netfpga: reusable router architecture for experimental research. In: Proceedings of the ACM Workshop on Programmable Routers for Extensible Services of Tomorrow PRESTO '08, pp. 1–7. ACM, New York (2008)
29. Nunes, B., Mendonca, M., Nguyen, X.-N., Obraczka, K., Turletti, T., et al.: A survey of software-defined networking: past, present, and future of programmable networks. IEEE Commun. Surv. Tutorials **16**(3), 1617–1634 (2014)
30. ON.LAB. Introducing ONOS - a SDN network operating system for service providers. http://onosproject.org/wp-content/uploads/2014/11/Whitepaper-ONOS-final.pdf (2014)
31. ON.LAB. Central office re-architected as a datacenter (cord). http://onrc.stanford.edu/protected%20files/PDF/ONRC-CORD-Larry.pdf (2015)
32. P.C.I.S.S. Council. Payment card industry data security standard requirements and security assessment procedures version 2.0. https://www.pcisecuritystandards.org/documents/pci_dss_v2.pdf (2010)
33. Peterson, L.L.: Opencloud: a showcase for cloud applications, SDN and NFV. http://ftp.tiaonline.org/Technical%20Committee/CCSC/2014.03.27/CCSC-20140327-05%20-%20Larry%20Peterson%20-%20OpenCloud%20A%20Showcase%20for%20Cloud%20Application,%20SDN%20and%20NFV.pdf (2014)
34. Peterson, L., Baker, S., De Leenheer, M., Bavier, A., Bhatia, S., Nelson, J., Wawrzoniak, M., Hartman, J.: XOS: an extensible cloud operating system. In: Proceedings of BigSystem (2015)
35. Reich, J., Monsanto, C., Foster, N., Rexford, J., Walker, D.: Modular SDN programming with Pyretic. USENIX ;login **38**(5), 128–134 (2013)
36. Reitblatt, M., Foster, N., Rexford, J., Schlesinger, C., Walker, D.: Abstractions for network update. In: Proceedings of the ACM SIGCOMM 2012 Conference on Applications, Technologies, Architectures, and Protocols for Computer Communication, pp. 323–334. ACM, New York (2012)
37. Ricart, G., McGeer, R.: US ignite and smarter GENI cities. In: GENI: Prototype of the Next Internet. Springer, New York (2016)
38. Sakurauchi, Y., McGeer, R., Takada, H.: Openweb: seamless proxy interconnection at the switching layer. Int. J. Netw. Comput. **1**(2), 157–177 (2011)
39. Saltzer, J.H., Reed, D.P., Clark, D.D.: End-to-end arguments in system design. ACM Trans. Comput. Syst. (TOCS) **2**(4), 277–288 (1984)
40. Sherwood, R., Gibb, G., Yap, K.-K., Appenzeller, G., Casado, M., McKeown, N., Parulkar, G.: Flowvisor: a network virtualization layer. Technical report, OPENFLOW-TR-2009-1, Open Network Foundation (2009)
41. Sherwood, R., Gibb, G., Kobayashi, M.: Carving research slices out of your production networks with openflow. ACM SIGCOMM CCR **40**(1), 129–130 (2010)
42. Sherwood, R., Gibb, G., Yap, K.-K., Appenzeller, G., Casado, M., McKeown, N., Parulkar, G.: Can the production network be the testbed? In: Operating Systems Design and Implementation (OSDI) (2010)

43. Society, I.: A ten-year tribute to Jon postel. http://www.internetsociety.org/what-we-do/grants-and-awards/awards/postel-service-award/ten-year-tribute-jon-postel, (2008)
44. Stoica, I., Morris, R., Karger, D., Kaashoek, M.F., Balakrishnan, H.: Chord: a scalable peer-to-peer lookup service for internet applications. In: Proceedings of SIGCOMM'01, pp. 149–160 (2001)
45. Szalay, A.S.: Jim gray, astronomer. Commun. ACM **51**(11), 58–65 (2008)
46. Szalay, A.S., Gray, J., Thakar, A., Kunszt, P.Z., Malik, T., Raddick, J., Stoughton, C., van den Berg, J.: The SDSS skyserver: public access to the sloan digital sky server data. In: Proceedings of the 2002 ACM SIGMOD International Conference on Management of Data, Madison, WI, 3–6 June 2002, pp. 570–581 (2002)
47. The Openflow Switch Specification. http://OpenFlowSwitch.org (2009)
48. Van Adrichem, N.L., Doerr, C., Kuipers, F., et al.: Opennetmon: network monitoring in openflow software-defined networks. In: 2014 IEEE Network Operations and Management Symposium (NOMS), pp. 1–8. IEEE, New York (2014)
49. Wetherall, D.: Active network vision and reality: lessons from a capsule-based system. In: Symposium on Operating Systems Principles (1999)
50. Zhang, S., Malik, S., McGeer, R.: Verification of computer switching networks: an overview. In: Proceedings of the 10th International Conference on Automated Technology for Verification and Analysis, ATVA'12, pp. 1–16. Springer, Berlin/Heidelberg (2012)
51. Zhao, B.Y., Kubiatowicz, J., Joseph, A.D.: Tapestry: an infrastructure for fault-tolerant wide-area location and routing. Technical report, UC-Berkeley (2001)

4G Cellular Systems in GENI

Ivan Seskar, Dipankar Raychaudhuri, and Abhimanyu Gosain

1 Introduction

Open, programmable networks are an important enabler for the future Internet because of their ability to support flexible experimentation and to evolve functionality as new network architectures are deployed on a trial basis. The NSF supported GENI initiative is an ongoing effort to build a national scale open programmable network using a combination of open switching, routing and wireless technologies. The main features of open networking devices used in such testbeds are: (a) an open API which provides access to link-layer technology parameters; (b) downloadable programmability of protocols used at the network layer; (c) virtualization of network resources such as routers and base stations in order to enable multiple simultaneous experiments; and (d) observability of key performance measures such as throughput and packet loss. At the start of the GENI project, it became clear that wireless edge networks and mobile devices are critically important to the future Internet, indicating the need for open programmable wireless access technologies that can be deployed to supplement the virtualized routers and server racks described in other chapters. As a first step, wireless access based on open/programmable WiFi access points has been provisioned into various campus deployments associated with GENI (see for example, the ORBIT testbed described in Chapter 4). Although WiFi is an important mode of access, an increasing proportion of Internet traffic originates

I. Seskar (✉) • D. Raychaudhuri
WINLAB, Department of ECE, Rutgers University, 671 Rt. 1 South, North Brunswick, NJ 08902, USA
e-mail: seskar@winlab.rutgers.edu

A. Gosain
GENI Project Office, Raytheon BBN Technologies, 10 Moulton street, Cambridge, MA 02138, USA
e-mail: agosain@bbn.co

© Springer International Publishing Switzerland 2016

R. McGeer et al. (eds.), *The GENI Book*, DOI 10.1007/978-3-319-33769-2_9

from cellular devices such as smartphones, motivating consideration of open cellular systems using the latest available technologies such as 4G WiMax and LTE.

The main goal of 4G wireless deployment in GENI was to address the two key issues: (1) providing campus-wide GENI wireless coverage for opt-in users; and (2) offering programmable wireless networking capabilities which reflect the growing importance of mobility service scenarios in the Internet. The WiMAX base station kit was designed to directly address both of these needs by providing wireless/mobile access together with the ability to attract opt-in users over a relatively large coverage area 25–50 sq-km, sufficient to cover a significant portion of many university campuses. With the availability of 802.16e PC cards [1] and mobile handsets [2] as commodity products, the setup makes it possible to support large numbers of mobile or fixed end-users on a campus as needed for certain classes of experiments. Specific examples of experimental research that are supported by a GENI WiMAX device include: mobile network routing, hybrid P2P and wide-area access, vehicular networking, transport-layer protocols for wireless, cross-layer optimization of transport and link scheduling, location-aware applications, content delivery networks and wireless network security. Of course, these deployments can also serve as the "last mile" for any wired network protocol experiment that would benefit from opt-in users who do not have access to a GENI enabled Ethernet connection. 802.16e base station products were viewed as a good starting point for the addition of 4G wireless into GENI.

In 1960s, the FCC designated the so called Instructional Television Fixed Service (ITFS) consisting of a band of 20 microwave channels totaling 120 MHz bandwidth in the 2.5–2.7 GHz spectrum for local credit granting educational institutions. The primary purpose of this allocation was for educational institutions to deliver live or pre-recorded video instruction to multiple sites within school districts and to higher education branch campuses. Originally, the authorization was for one-way, line of sight analog TV operation and each institution was required to carry at least 40 h of programming per week. Over the years, two FCC rulings had huge impact on the actual use of this spectrum: in the late 70s FCC allowed commercial use of the spectrum (i.e. leasing by commercial entities from the institutions that had excess capacity) while in the late 90s, the rules were changed to allow for two-way and cellular like operation opening the door for wireless data delivery. In the later ruling, FCC also reduced the 40 h requirement to "5 % of channel capacity"; the service was also renamed to Educational Broadband Service (EBS) [3]. Finally, in 2003 FCC was petitioned and subsequently allowed use of ITFS/EBS spectrum for wireless broadband service.

On the technology side, the project was initially focused on developing and deploying wide-area wireless experimentation services with WiMAX technology. The fact that the equipment was readily available at the time, is inherently IP based, and that it operates in EBS spectrum (i.e. that quite a few institutions already owned spectrum) was a reason for initial choice of WiMAX as a wide-area wireless technology for GENI deployment. In the 6 years since the projects started, LTE technology, owing to its prevalence in commercial world and more recently availability of equipment in the EBS spectrum (most notably the fact that commercial handsets started supporting 2.6 GHz TDD variant of LTE), was chosen as a technology for the next round of deployment. It is noted here that the project had to deal with a fundamental technical challenge of modifying available 4G cellular technologies to separate the basic radio access functionality from the 3GPP protocol stack in order to allow for flexible experimentation with new network protocols. The concept of an "open base station" which can be plugged in to an arbitrary (virtual, programmable) network infrastructure is still an evolving one, though the basic idea has gained considerable momentum on the wired network side with the emergence of software-defined network (SDN) standards. Just as wired network devices such as Ethernet switches and routers can be made programmable with SDN, our concept for programmable wireless in GENI is to modify existing 4G base stations to expose an "open API" and then migrate the functionality to an external controller which can be programmed and virtualized.

On the legal side, the project had to deal with the spectrum licensing issues. Given the shortage of spectrum, two major telecom companies (Sprint and Clearwire) began offering WiMAX on EBS frequencies leased from schools and other license holders. The caveat was that FCC rules required such leases to reserve 5 % of system capacity for license holders' educational mission, but were otherwise free to use the system as they saw fit. Initially each campus was either applying for a separate experimental FCC license for WiMAX operation or using existing EBS allocation (in cases where spectrum was not leased to carriers). One of the administrative issues is that experimental licenses have requirements for coordinating with the primary owner. The more permanent solution was found in Q2 2013 when, through Rutgers University master agreement with Clearwire, GENI WiMAX base stations at 13 campuses across the United States were allowed to operate as a unified experimental system that is closely coordinated with Sprint (which acquired Clearwire in second half of 2013).

Another major achievement of the project was that, for advanced experimentation, even in the case of NEC base station, that was designed for use with a pre-standard gateway, the project was able to unbundle the basic layer-2 functionality of the device and make it accessible through an external control "open API". This enabled development of GENI control software on an external (PC based) controller, substantially meeting layer 2,3 programmability and virtualization requirements in a manner that is very similar to the software approach used for wired GENI routers.

2 Deployment

GENI WiMAX kit design and development was carried out in Spirals I while larger scale deployments were carried out in the Spiral III and Spiral VI. The current deployment sites are shown in Fig. 1. The original GENI WiMAX design was based on the first generation NEC base stations while the second round of deployments was based on the second generation WiMAX base stations from Airspan. At the end of Spiral IV, the project started development of the LTE kit for the next round of 4G deployment in GENI.

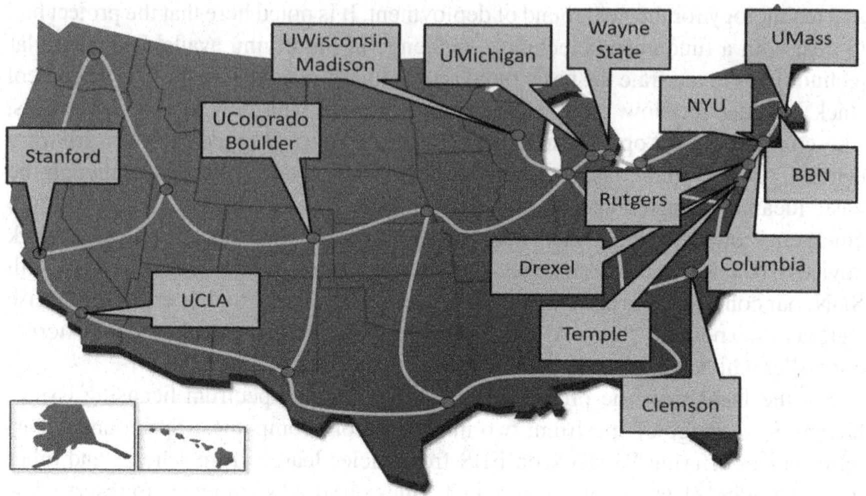

Fig. 1 GENI 4G deployment sites

Also, the service (software) development is closely coordinated with the EU FP7 FIRE project "FIRE LTE testbeds for open Experimentation" (FLEX) that is developing a range of LTE experimentation facilities in Europe [4] and using the same testbed management framework.

2.1 Spiral III Deployment Sites

Spiral III deployments were carried out at seven University campuses and at the GENI Project office in Cambridge, MA. Each site received one NEC Profile A base station, one 12 dbi Omni directional antenna, one high performance HP rack server and 10 AWB US210 USB dongles. All outdoor deployments provided 360° campus coverage in a 2–3 km radius, depending on urban or suburban terrain surrounding the campus. The sites also acquired five USB WiMAX modem adapters that had an open source Linux driver to allow for finer grain control on the client side and use with opt-in laptops and other compatible devices.

2.2 Spiral IV Deployment Sites

In Spiral IV, 13 Airspan Profile C base stations were deployed at five existing sites and five new campuses. Two of the deployments were multi-cell/

multi-sector operations: Wayne State University and Clemson University supporting both vehicular nodes for mobility experiments and opt-in end users for service-level evaluations. Each of these sites deployed multiple base stations: Wayne State University in downtown Detroit area covering two major highways around the campus and Clemson University covering sections of the sub-urban area in Greenville, South Carolina and their campus downtown. In addition to the BS and 2×4 17 dbi MIMO antenna, each site received a full GENI WiMAX kit consisting of one console (ORBIT) node, three client (ORBIT) nodes with WiMAX modems and multiple Samsung Galaxy S2 handsets.

3 Typical Deployment Architecture

Figure 2 shows a schematic of a typical GENI 4G deployment. As shown, one or more base stations are typically connected directly to a base station controller over L2 or L3 network. That same controller is in turn connected to a GENI access network with layer 2 switched connectivity using Ethernet or optical fiber technology.[1] The figure also indicates three distinct interfaces associated with a deployment: (a) the RF interface (WiMAX or LTE) between clients and the base station, (b) the controller interface (R6 for WiMAX and S1/X1 for LTE) and (c) the GENI facing network interface.

3.1 NEC WiMAX Base Station

The NEC Release 1 (PassoWings) first generation WiMAX base-station hardware that was cornerstone of the initial GENI 4G deployment, is shown in Fig. 3.

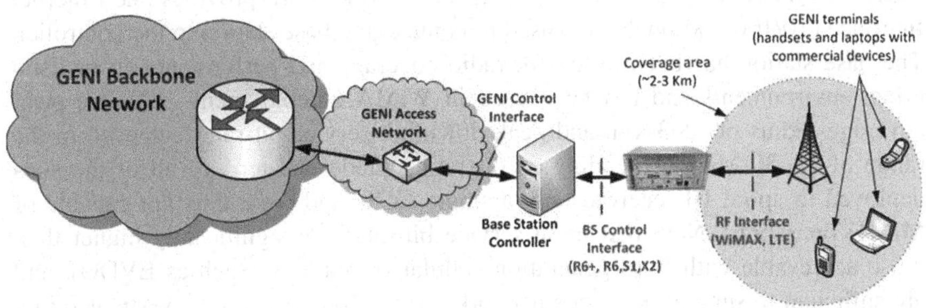

Fig. 2 4G deployment architecture

[1]Remote deployments that lack direct GENI backbone connectivity can be connected to the core over the L2TP tunnel through one of the other sites or any GENI rack machines.

Fig. 3 NEC PasoWing basestation

It consists of the Indoor (IDU) and Outdoor (ODU) units. The IDU is a 5U rack based system with two types of cards: Channel Cards (CHC) and a Network Interface Card. The shelf can be populated with up to three CHCs, each supporting one sector for a maximum of three sectors. Each channel card (sector card) can be connected through a fiber with a pair of ODU units (bottom white box in Fig. 3) that are typically mounted close to the antenna. The BS operates in the 2.5 GHz or the 3.5 GHz bands and can be tuned to use either 5, 7 or 10 MHz channels. At the MAC frame level, 5 ms frames are supported as per the 802.16e standard. The TDD standard for multiplexing is supported where the sub-channels for the Downlink (DL) and Uplink (UL) can be partitioned in multiple time-frequency configurations. The base-station supports standard adaptive modulation schemes based on QPSK, 16QAM and 64QAM. The interface card provides one Ethernet Interface (10/100/1000) which is used to connect the base station to the controller. The base station has been tested for radio coverage and performance in realistic urban environments and has been used in WiMAX deployments with a typical coverage radius of 2–3 km, and peak downlink service bit-rates achieved in the range of 15–30 Mbps depending on operating mode and terrain (all of the sites deployed in spiral III received only a single ODU and were thus not capable of MIMO operation). Note that these service bit-rates are significantly higher than those achievable with third generation cellular technology (such as EVDO), and are sufficient to support advanced network service concepts to be investigated in a typical wireless GENI experiment.

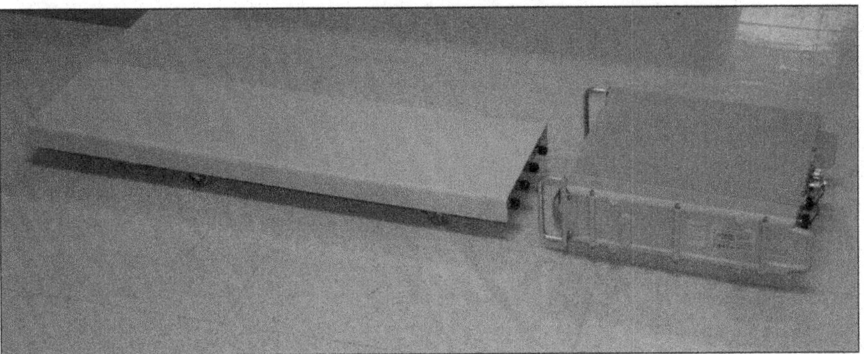

Fig. 4 Airspan base station

3.2 *Airspan WiMAX Base Station (Fig. 4)*

The Air4G-W24 (a.k.a. MacroMAXe) is a highly-integrated second-generation WiMAX base station manufactured by Airspan that was chosen for Spiral IV deployment. As opposed to a single RF front-end that covers the entire 2.5–2.7 GHz band as in case of the NEC base station, Air4G-W24 is offered in three different RF variants: 2510 Lo (low band covering 2496–2570 MHz), 2510 Mid (mid-band with 2560–2630 MHz) and 2510 Hi (hi-band covering 2620–2690 MHz) with maximum transmit power of 43 dBm (2 × 40 dBm) and EIRP of 61 dBm. As shown in Fig. 2, in addition to the actual base station, the setup includes 2.3–2.7 GHz 90° Quad X-Polar panel antenna with −4° (downward) tilt and effective gain of 17.0 dBi. Air4G-W24 also supports mode for fixed/nomadic applications which do not require support for handovers (i.e. no need for asn-gw deployment). It is a IEEE802.16e-2005 Wave 2 compliant device that supports the two main MIMO downlink configurations:

- Matrix A: space-time coding (STC) in which the base station transmits each data symbol twice with slightly different coding which is a form of diversity that increases range/error rate but does not affect channel bit-rate.
- Matrix B: vertical encoding (2 × 2 MIMO) where the date is split among the two antennas which theoretically doubles the bit-rate.

On the physical layer, the base station feature set includes both 512 and 1024 OFDMA, configurable downlink/uplink split, QPSK, 16QAM and 64 QAM on both downlink and uplink, as well as, fractional frequency reuse, open and closed loop power control (power adjustment range of 20 dB with 1 dB steps) and fast feedback (ACKCH for H-ARQ and CQICH for MIMO). It supports all five standard service flow types (BE, NRT, ERT, RT and UGS) with up to 32 service flows per single mobile station and up to 4096 service flows and up to 256 mobile stations per 10 MHz channel. The Air4G-W24 base station has power consumption of 370 W under the full load.

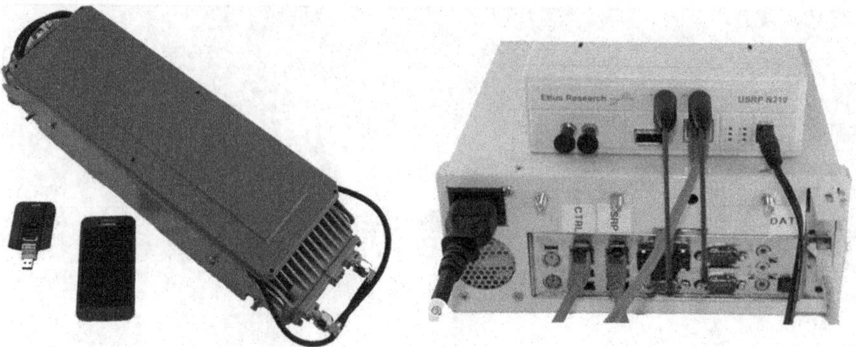

Fig. 5 LTE base-stations

3.3 LTE Base Stations

In recent years, LTE has emerged as a dominant wide-are wireless system. The majority of LTE deployments in the US are using paired spectrum (i.e. Frequency Division Duplex – FDD) and, given the scarcity, experimental licenses and permission for operation from primary licenses holder are much harder to obtain. Fortunately, recent push by equipment vendors to support an unpaired version of LTE (i.e. Time Division Duplexing—TDD) that is using mid-band EBS RF spectrum, opened a smooth path for transition from WiMAX to LTE. The two base stations that are used for GENI LTE support (Fig. 5) are: Airspan AirSynergy LTE and Amarisfot LTE 100.

AirSynergy LTE is a production grade base station with dual 30dBm (2×1 W) transmitters with support for full range of channel bandwidth: 1.4, 3, 5, 10, 15 and 20 MHz. The base station antenna is a multi-element cross polarized (dual slant) design which can be used in directional or omni modes of operation with average gain of 2 or 8 dBi respectively. AirSynergy supports QPSK, 16QAM and 64QAM modulations on both downlink and uplink with all modulation and coding schemes (MCS) defined in 3GPP TS 36.211.

The Amarisoft LTE 100 is an example of new breed of pure Software Defined Radio (SDR) eNodeB implementations. It is a user-space software solution running under Linux OS on a commodity PC that uses RF head-end (Ettus Research/NI USRP N2x0) front-end for RF conversion. It is a LTE release 9 compliant that supports both FDD and TDD configurations and full range of bandwidths (1.4, 3, 5, 10, 15 and 20 MHz). It also implements the MAC, RLC, PDCP and RRC protocol layers and intra eNodeB, "S1" or "X2" handovers. Due to flexible nature of used SDR front-end (USRP with SBX daughter-card that covers 400 MHz–4 GHz) and the configurability of software based solution, this platform can be, in addition to all of the standard LTE frequencies, tuned to arbitrary frequency within the range of RF

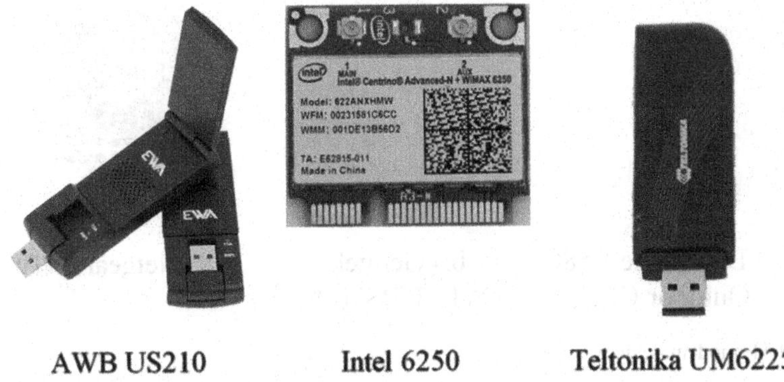

Fig. 6 WiMAX modem devices

front-end which makes it ideal for experimentation with new frequency allocations. It also exposes large number of control parameters through configuration files which are made available to the experimenters via the LTE Aggregate Manager.

3.4 4G Client Devices

Over the years, GENI WiMAX deployed all three classes of client devices: (a) external and internal modems, (b) mobile phones and (c) WiFi gateways. In Fig. 6, the three dominant modem devices are shown.

The AWB US210 is a IEEE802.16e-2005 WiMAX Wave 2 compatible device that was shipped with the WiMAX kit as a reference device. The shipped variant (US210-2.5), operates in 2.496–2.696 GHz range and supports 5, 7, 8.75, 10 MHz channel bandwidths. The device has one transmit and two receive antennas and supports mobile connectivity for speeds of up to 30 km speed with (combined) peak rates of up to 33 Mbps.

Intel® Centrino® Advanced-N + WiMAX 6250 [5] is a PCIe form factor half MiniCard. It is a combination device that includes both WiFi and WiMax radios. On the WiFi side, it supports 2 × 2 Tx/Rx streams with maximum speed of 300 Mbps and supports both 2.4 and 5 GHz operation. It also has an open source Linux driver that enables flexible experimentation on the client side.

Teltonika UM6225 is an IEEE802.16e-2005 WiMAX Wave 2 compatible USB modem with somewhat unique design since it has an embedded CPU. Among other features, it supports 2 transmit streams closed-loop diversity, both Matrix A and B MIMO types and HARQ category 7 with max downlink rate of 40 Mbps. The device supports two operational modes: NAT-ed Ethernet device with non-routable address or as bridged Ethernet device.

The new generations of TDD LTE client devices have also been tested with the Airspan Airsynergy Base stations. As shown in Fig. 7, all devices have drivers for

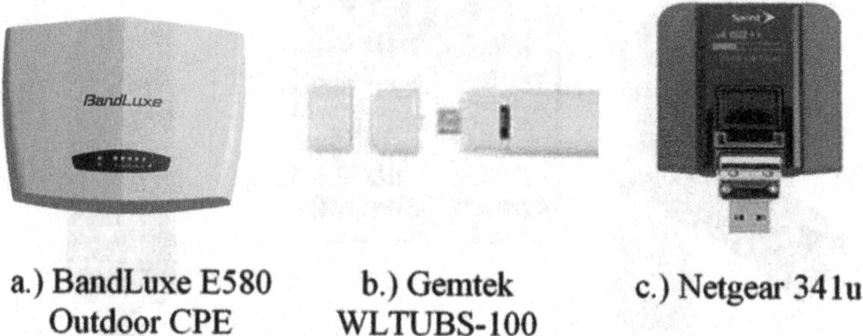

a.) BandLuxe E580 Outdoor CPE b.) Gemtek WLTUBS-100 c.) Netgear 341u

Fig. 7 LTE modem devices

the Linux OS providing flexibility for experimenters. (a) Bandrich BandLuxe E580 outdoor CPE supports both Router and bridge mode and has an embedded 10–13 dBi directional MIMO antenna set. It supports upto 20 MHz bandwidth with data rates upto 10 Mbps DL and 50 Mbps UL. (b) Gemtek WLTUBS-100 is a TDD LTE USB dongle using the Sequans baseband processor chip. It complies with the 3GPP Release 8 Category 3, providing data rates up to 100 megabits per second (Mbps) in downlink and 50 megabits per second (Mbps) in uplink. (c) The Netgear 341u is a USB based LTE modem. It emulates a USB router, and supports NAT and other standard features. It supports LTE Bands 25, 26, and 41 (TDD).

The ubiquity of mobile phones and their popularity as a development tool for experimenters is well known. As such, the GENI WiMAX project provided sites with a number of HTC Evo 4G and Samsung SII WiMAX handheld devices shown in Fig. 8. These Android [6] operating system based handsets are ideal for research use since, in addition to open-source nature of the OS and the abundance of open-source applications, have alternative (open-source) ROMs allowing for even greater customization. GENI wireless sites received these handsets with custom applications like: WiMAX frequency switching application by Rutgers University, spectrum sensing app from University of Wisconsin, range and throughout measurement application by Clemson University pre-installed. These handsets also seamlessly connect to both the GENI network as well as SciWiNet [7] (MVNO running on top of Sprint 3G/4G network).

WiMAX + Wifi gateways are also deployed at various GENI WiMAX sites to provide high speed 4G connectivity to legacy Wifi devices and increase the coverage range for non-WiMAX radios. Greenpacket DX Indoor/Outdoor Modem is a Wave 2 compliant device used in GENI. On the WiMAX backhaul, the device operates in the 2.3, 2.5–2.7 and 3.5 GHz spectrum and on the WLAN side is 802.11b/g/n compliant. On the WiMAX backhaul, it has a maximum Tx power of 25 dBm and an Omni directional antenna with a gain of 5 dBi.

	HTC Evo 4G (WiMAX)	Samsung Galaxy SII (WiMAX)

CPU	Qualcomm Scorpion @ 1GHz	Qualcomm QSC6085 @ 1.4GHz (dual core)
Storage	512MB LPDDR1/ 1GB ROM	1GB RAM/ 16 GB ROM
Battery	Removable 1500 mAh	1650 mAh
Connectivity	Dual-band CDMA/EVDO Rev. A (800 1900 MHz) □ WiMAX 802.16e □ 802.11b/g/n, Bluetooth 2.1 + EDR	□ GSM + UMTS (800 1900 2100 Mhz) □ WiMAX 802.16e □ 802.11 b/g/n, Bluetooth 4.0 + HS + FM
WiMAX Features	□ 2.3-2.4, 2.5-2.7, 3.3-3.8 GHz □ 2 X 23 dBm transmit power □ DL MIMO: MRC, Matrix A + MRC, Matrix B □ UL MIMO: Matrix A □ Tx Diversity □ Fast feedback □ Fast scanning	□ 2.5-2.7 GHz □ Supports all Mobile WiMAX Wave2 Profiles □ 2x2 MIMO □ Matrix A & Matrix B □ Space Time Coding (STC) □ Low power
WiMAX Performance	Combined UL/DL: > 40 Mbps	Max DL: 40 Mbps Max UL: 15 Mbps

Fig. 8 WiMAX handheld devices

4 GENI Wireless Site Management Framework

The GENI wireless site management framework is an extension of Orbit Management Framework (OMF). OMF was originally developed for the ORBIT wireless testbed at Rutgers University. GENI project has extended this framework to operate with its network and resource technologies.

As shown in Fig. 9 each site deployment consists of a console machine, three bare metal client nodes and a number of WiMAX handsets. Two of these client nodes have a fixed power supply. The third client has a 12 V power supply so that it can function as a mobile node in a vehicular environment. The console machine functions as a portal/gateway to allow experimenters to login to the client resources and manages identity, security and resource allocation functions. It is running Linux OS with a set of standard services (DHCP,DNS, LDAP, etc.) and the OMF packages needed for operating the setup as a testbed.

This controller manages the base station(s) as well as performs layer 2 and layer 3 processing of client packets. The core of the base station controller software revolves

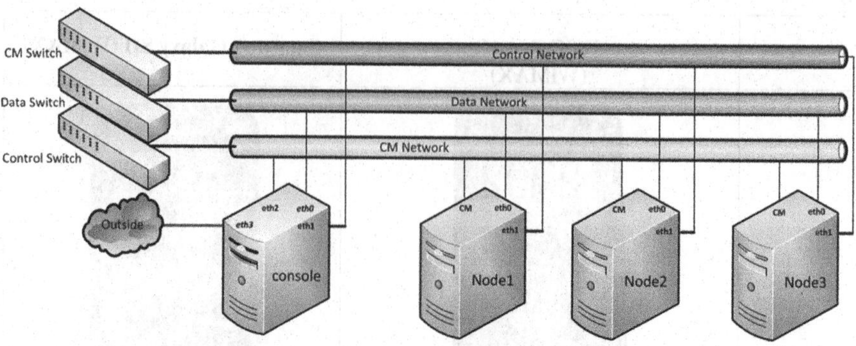

Fig. 9 Spiral IV site kit

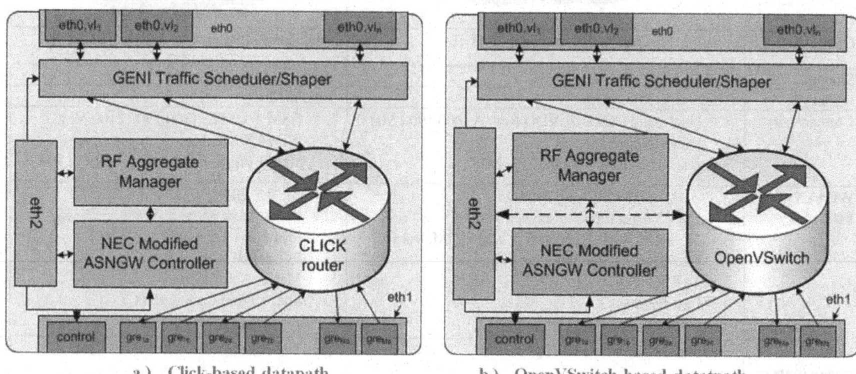

Fig. 10 GENI WiMAX controller

around three components: (a) Access Service Network Gateway (asn-gw), (b) GENI (WiMAX, LTE) RF Aggregate Manager (a.k.a. wimaxrf, lterf) and (c) one or more datapath routers.

802.16 standards introduce three choices for access network implementation (the "Profiles"): A, B and C. The main difference between these is the place at which certain functionalities are implemented [8]: in Profile A, the handoff control and radio resource control (RRC) are implemented in the asn-gw; in Profile B, most of the functions are implemented in the base station (even if access network implementation is distributed among multiple base stations); in Profile C, both handoff and RRC are implemented in the base station; NEC base stations are Profile A while Airspan base stations are Profile C. The main role of asn-gw in GENI deployments is to provide connectivity (and in general case mobility) management across one or more base-stations. The asn-gw exchanges control and management information with the base-station over the R6 bearer logical interface (802.16 standard interface for communication between WiMAX base station and the controller) through control port on the south-side interface in Fig. 10.

4.1 RF Aggregate Manager

The RF Aggregate Manager is the experimenter-facing component that is used to configure and manage the GENI wireless controller. It is a standard OMF Aggregate Manager with REST-like web based interface that allows experimenters to get and set all exposed parameters of the base station and asn-gw. It performs three sets of functions: (1) connection manager functionality (the default implementation is Simple Authorization Manager that is only using client MAC address for authentication), (2) base station and asn-gw parameter getters/setters and (3) datapath configuration and management functions. This service also exposes hooks for interfacing with the GENI slices.

One of the objectives of the GENI wireless controller development was to enable experimentation with the handoff algorithms. To support this, the whole range of additional handoff related control functions was exposed through the RF AM REST interface for both Profile A and Profile C base stations enabling a range of (system independent) handoff solutions (e.g. University of Wisconsin and Clemson University handoff implementations as described in Sect. 5).

4.2 Datapath Management

The 802.16e base station allocates time-frequency resources on the OFDMA link with a number of service classes as specified in the standard—these include unsolicited grant service (UGS), expedited real time polling service (ertPS), real-time polling service (rtPS), non-real time polling (nrtPS) and best effort (BE). The radio module as currently implemented includes scheduler support for the above service classes in strict priority order, with round-robin, or deficit round-robin being used to serve multiple queues within each service class. These packet queuing and service scheduling features provide adequate granularity for virtualization of radio resources used by each slice in GENI [9]. It is noted here that OFDMA in 802.16e with its dynamic allocation of time-frequency bursts provides resource management capabilities qualitatively similar to that of a wired router with multiple traffic classes and priority based queuing. The GENI aggregate manager is responsible for mapping "*Rspec*" requirements (such as bandwidth or delay) to the available 802.16e common packet layer services. Slices which do not require bandwidth guarantees are allocated to the BE class, while slices with specific bandwidth requirements (for quantitatively oriented experiments, for example) are allocated to the other categories. In GENI wireless controller, each slice contains a datapath "router" implementation that is either Click [10] or OpenVSwitch [11] based (RF Aggregate Manager allows for other datapath implementations). These datapath implementations are in charge of handling all of the data packets for a set of clients that belong to the slice. When a data packet is received by the controller, it is classified based on the client MAC address and routed to appropriate datapath where

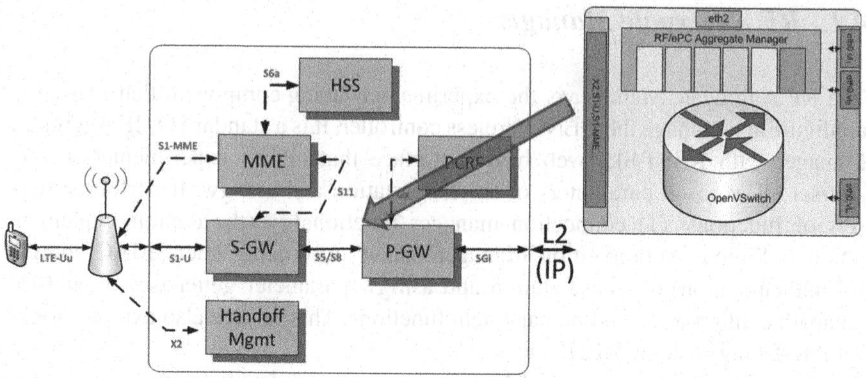

Fig. 11 LTE controller

it is processed by either Click or an OpenVSwitch software router and forwarded to the appropriate slice (VLAN) for further processing. Outgoing packets on virtual interfaces mapped to the layer 2 interface of the WiMAX base station can also be tagged so that they can be assigned traffic class and bandwidth parameters (BE, ertPS, rtPS etc.) as determined by the flow connection identifier (CID).

In keeping with the programmability and virtualization requirements of GENI testbeds, an open LTE base station controller is under development using the same model that was used for WiMax as shown in Fig. 11. The basic idea is to replace all the LTE GW (gateway), MME (Mobility Management Entity), Handoff and other functionality with software modules in the Aggregate Manager with southbound interfaces to the base station hardware and northbound interfaces to the access network.

4.3 *Virtualization*

As part of the development of an open API WiMax base station, we addressed the problem of virtualization of wireless resources. Isolation of resources between multiple virtual slices is complicated by the fact that the capacity of a radio channel varies with signal strength at the mobile clients being served. In addition, because wireless devices are mobile, the network topology and interference regions change with time implying the need for virtualization methods which are dynamic and can respond in real time to changes in available resources (Fig. 12).

Our implementation for WiMax uses load adaptive traffic shaping in an Open-VSwitch framework to maintain fairness between virtual networks when the channel approaches saturation. An example result with two virtual networks on WiMAX is shown in Fig. 13. The figure shows how the bit-rate of slice 1 varies with time due to signal fluctuations, and the corresponding bit-rate of slice 2 is also affected severely

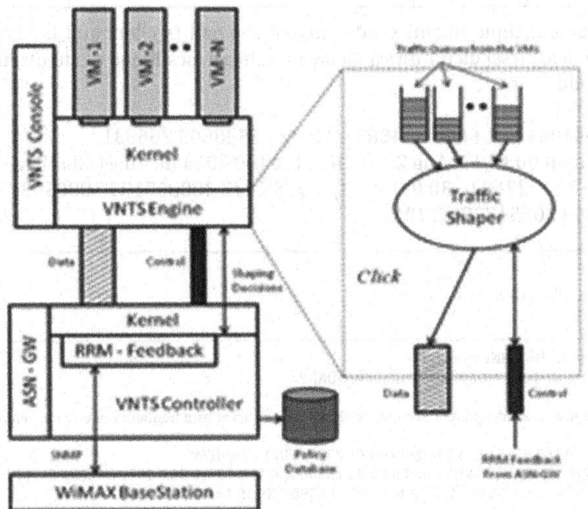

Fig. 12 Architecture of virtual WiMax base station in GENI

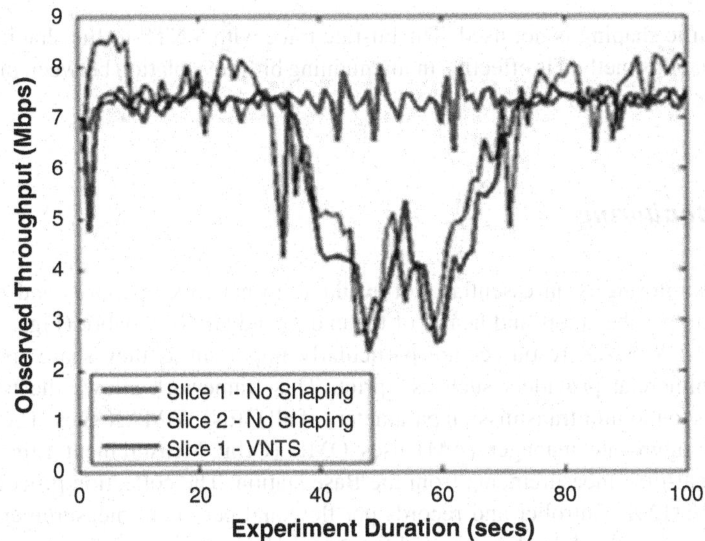

Fig. 13 Bit-rate traces for VNs in experimental WiMAX network

```
<FIELDS>oml_tuple_id oml_sender_id oml_seq oml_ts_client oml_ts_server
bsid ma mac ulrssi ulcinr dlrssi dlcinr mcsulmod mcsdlmod dlsdu ulsdu
dlpdu lpdu
</FIELDS>
<ROW>440547 21 6674 5548804.206924 5548805.266831
44:51:db:00:00:01 Fri Aug 22 16:48:04 -0400 2014 00:1d:e1:3b:4f:9a -
107.75 -7.5 -32768.0 -30.0 QPSK 1/2 QPSK 1/2 380857179 388059377
24053448 30551329</ROW>
```


Fig. 14 Local measurements


```
<DATABASE ExperimentID="GENI-wimaxrf">
<QUERY>select * from wimaxrf_basestation limit 10</QUERY>
<RESULT>
<FIELDS>oml_tuple_id oml_sender_id oml_seq oml_ts_client oml_ts_server bsid frequency power noclient ulsdu ulpdu dlsdu
dlpdu</FIELDS>
<ROW>1 1 1 61.860506 62.811513 44:51:db:00:00:03 2590.0 38.15 2</ROW>
<ROW>2 1 2 62.427304 63.376368 44:51:db:00:00:01 2590000.0 100.0 2 0.0 96.0 1992.0 90880.0</ROW>
<ROW>3 1 3 121.909544 122.859186 44:51:db:00:00:03 2590.0 38.15 1</ROW>
...
```


Fig. 15 Global measurements

when traffic shaping is not used. The bit-rate trace with VNTS traffic shaping [12] shows that this method is effective in maintaining bit-rate isolation between multiple virtual network slices.

4.4 Monitoring

GENI monitoring is an essential tool available to network operators and experimenters to get the status and health of the end to end GENI network (Figs. 14 and 15). GENI WiMAX resources are particularly important as they share spectrum with commercial providers such as Sprint. This framework allows the network providers to monitor transmission parameters of all GENI WiMAX sites. The GENI WiMAX aggregate manager (AM) uses OML (Orbit Measurement Library) to collect real-time measurements from the Base station. The collection process runs on the WiMAX controller and records per flow and per client measurements and stores them in a local database. In addition to wireless client related measurements, the AM uses OML for both Layer 2 and 3 base station aggregate measurements. The service extends the GENI Monitoring framework and stores the measurements in a local datastore, which is queried periodically by a global collector and all active measurements are stored in a global database. The measurements collected include: WiMAX base station transmit power, center frequency, Downlink and Uplink modulation scheme and number of clients associated with each BS.

4.5 *Portal Integration and Account Federation*

GENI WiMAX experimenters have access to all 13 WiMAX sites for experimentation using their account at the GENI portal and clearinghouse [13]. This allows an experimenter to reserve either a single campus testbed or even multiple campus testbeds connected via the Internet2 [14] research backbone for an experiment. In a deviation from common GENI philosophy, the GENI WiMAX resources are time sliced and only one project can access the resources at a time. This is primarily done to avoid any conflict in modifying the WiMAX base station parameters by multiple experiments. The resource reservation is performed by using web-based scheduler at each site independently. Account federation between the portal and individual sites is performed in two steps: The GENI clearinghouse periodically communicates with the ORBIT site account management service over HTTPS to sync account and project information. The ORBIT login service is responsible for syncing account events with the remaining 12 sites and resolves any discrepancies that may arise.

4.6 *Integration with GENI Rack*

GENI wireless resources shown in Sect. 4 are representative of single site deployment. The rollout of GENI racks (both ExoGENI and InstaGENI) at these sites has allowed them to include the high performance computing and Internet2 connectivity into their experimenter offerings as shown in Fig. 16. The basestation controller supports arbitrary number of datapaths each of which corresponds to a separate slice and is identified with a VLAN. In all sites that have GENI racks, these VLANs are carried through a single Ethernet trunk, through the campus or even external network, to the local rack's dataplane switch where they can be distributed to various computing or networking components. This also enables the aggregation function for the local rack for wireless client traffic and provides support for local and multi-site complex topologies to experimenters. At least two VLANs are mandatory at each campus deployment: "wireless-local" and "wireless-multipoint". The wireless-local VLAN carries traffic that terminates on the local campus and is the default VLAN that RF Aggregate Manager(s) use for its clients. As the name implies, the wireless-multipoint VLAN carries traffic that is passed onto permanent Internat2 Advanced Layer 2 Service (AL2S) [15] slice dedicated for wireless experimentation that connects all of the wireless enabled campuses. This facilitates experiments that span the entire GENI wireless footprint, as well as, enables management functions that are typically used by wireless carriers (e.g. multi-site coordination, over-the-air device provisioning, profile updates, etc.). Given that the rack switches are SDN capable, and the fact that both wireless datapath is also based on SDN components (and the abundance of computing resources at each site) resulting collection of resource introduces additional flexibility in the wireless control plane and is enabling significantly more complex experimentation as shown in Fig. 17 (whether itspans multiple wireless deployments of heterogeneous wireless technologies on a single camps or multiple campuses that are connected over one or more computing/networking slices).

5 Experimentation

Given that GENI WiMAX deployment was one of the first cases of research community getting their hands on a wide-area cellular system, the first series of experiments and research results were focused mostly on service development and performance evaluation. Initial group of papers introduced the design of the open BS architecture [12, 16] while the second set of reports was on coverage for the initial deployment [17]. In [18], a series of experiments was performed on two GENI WiMAX campuses (NYU Poly and UMass) under various wireless signal conditions and network traffic patterns and the performance of several popular wireless Internet applications was characterized while in [19] the authors evaluate the performance of a novel video delivery service in typical mobile environment. The complexity of these experiments continued to rise and client side mobility management and handoff techniques were developed at Clemson University and University of Wisconsin. The implementation on the client side device was based on a Floodlight OpenFlow [7] controller, which managed client data flows over multiple radio interfaces. This handoff technique was demonstrated in mobility scenarios across the Clemson University campus and later developed into a tutorial and offered at multiple GENI Engineering conferences.

One of the first production uses of a WiMAX deployment was in Brooklyn by the ParkNet project [20]. As shown in Fig. 18, a fleet of eight vehicles drove the streets around MetroTech Center collecting and transmitting live data about availability of parking spaces along the streets. Each vehicle was equipped with a mobile node with external side-sweeping sensor pack (external attachment on the side of the car in Fig. 16) and an Intel 6250 client device. The node ran an OML-ised ParkNet application that collected measurements from ultrasonic sensor, video camera and an on-board GPS receiver, pre-processing and aggregating the data and streaming the results over NYU Poly WiMAX connection to the remote collection server at Rutgers campus in North Brunswick. The interesting feature of this deployment was the use of disconnected OML mode which supported opportunistic transmission of collected measurements as vehicles entered the WiMAX coverage area (i.e. OML proxy that buffers the data while node is not in the coverage area and delivers it as soon as connection is re-established). This deployment scenario was also demonstrated live during the plenary session of GEC9.

The research use of wireless infrastructure was quickly followed by the educational use with a number of tutorials and classroom courses [21–24]. NYU-Poly and WINLAB, Rutgers have conducted a series of WiMAX tutorials at GENI Engineering conferences to familiarize the community with 4G wireless experimentation environment including experiment orchestration and scripting language,

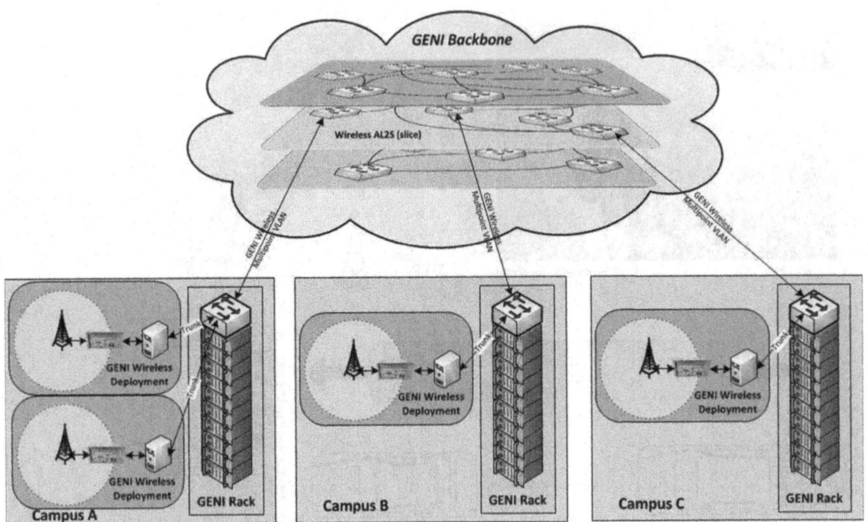

Fig. 16 GENI 4G resources integrated with GENI rack(s)

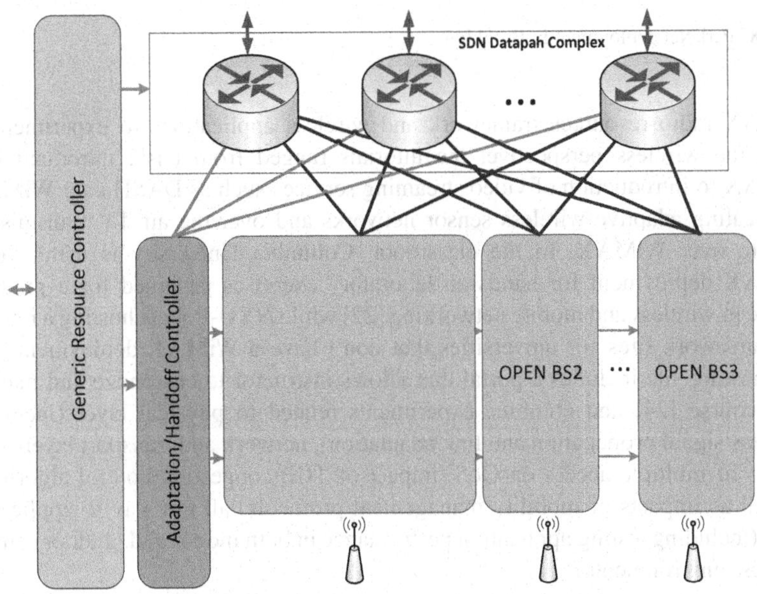

Fig. 17 GENI software defined wireless edge

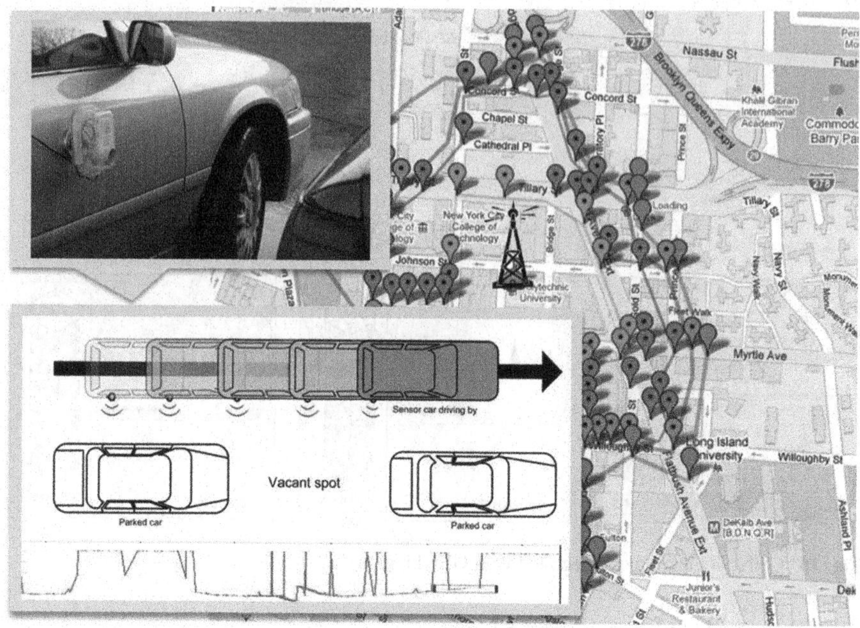

Fig. 18 ParkNet deployment in Brooklyn

WiMAX radio resources framework and wireless applications to experimenters. From the wireless perspective, the tutorials ranged from basic introduction to WiMAX to introduction of video streaming services such as DASH over WiMAX, self-healing adaptive wireless sensor networks and over the air TV transmission capture over WiMAX. In the classroom, Columbia University is using GENI WiMAX deployment for hands-on laboratory exercises designed for a graduate course in wireless and mobile networking [22] while NYU-Poly is hosting a number of coursework sites for universities that don't have a WiMAX deployment [23]. The hosting site includes a portal that allows instructor to customize and manage each course [24] and combine experiments related to physical layer (including wireless signal propagation and link adaptation), network and transport layers (like effects of multiple access on QoS, impact of TCP congestion control algorithms as well as impacts of mobility management protocols) all the way to application layer (including testing application performance in both indoor and outdoor campus wireless environments).

More recently, a new long-term use case of GENI deployments is starting to appear. One such deployment consists of provisioned Xen VMs (1 GB mem and a 2.9 GHz core) on InstaGENI racks across seven sites (U.Utah, U.Wisconsin, UIUC, NYSERNET, GPO-BBN, NYU/NYU-Poly, Rutgers-WINLAB) (Fig. 19). Each of the nodes is running a MF (MobilityFirst) software router and naming service. The core interconnection between machines is set up using a multi-point VLAN provided by Internet2's advanced layer-2 services (AL2S) and the edge network at

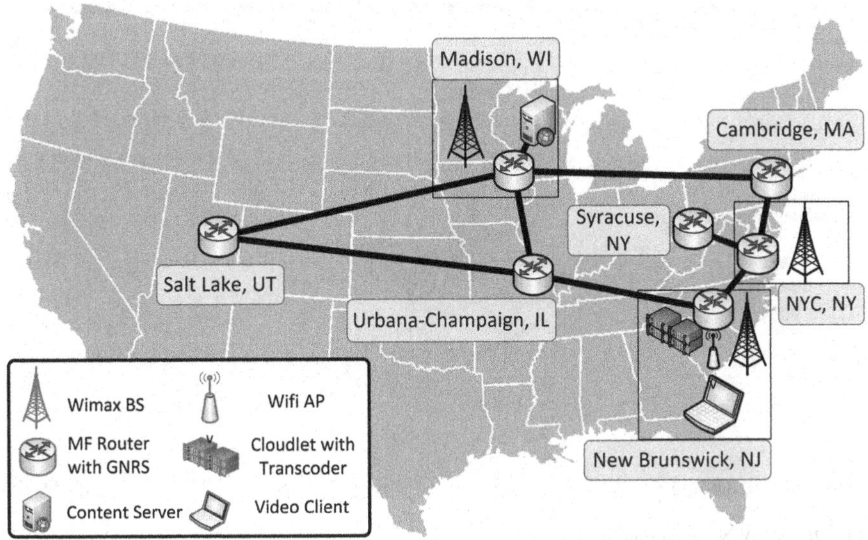

Fig. 19 MobilityFirst Deployment

three of these sites (Rutgers, NYU and Wisconsin) is made up of clients connected over the WiMAX and WiFi networks. The wireless edge access is provided by a combination of MF edge routers deployed on the WiMAX kit nodes (that are also acting as WiFi APs) and MF edge routers deployed on the campus GENI rack machines that are connected to the WiMAX controllers (at Wisconsin, Rutgers and NYU Polytechnic).

6 Extending GENI Cellular Coverage Using SciWiNet

GENI Wireless (WiMAX) deployments are restricted to 13 campuses across the nation. As the popularity of this infrastructure has grown the scale and complexity of experiments has also increased. GENI needed support for real world mobility scenarios and experiments that require a large coverage footprint. A NSF exploratory project to evaluate the efficacy of an MVNO (Mobile Virtual Network Operator) model to provide a wireless testbed for the academic research community "Science Wireless Network for the Research Community" (SciWiNet) [7] was funded. The SciWiNet project currently offers cellular data services to researchers using Sprint's 3G, WiMAX, and LTE networks providing country-wide service availability. A coverage map of the Clemson University area is shown in Fig. 20. The project also has a number of devices (modems and phones) that are loaned to research project/institutions on demand and it also honors a BYOD (bring your own device) system to alleviate stress on its resources. Given the synergy with

Fig. 20 SciWiNet coverage portal

WiMAX campus deployments, most of the early users were members of the GENI community. Vehicular experiments conducted at Wayne State University to capture high resolution video for mapping a city used SciWiNet USB dongles to offload data to a public server, when the cars were not in range of their campus WiMAX base stations.

SciWiNet has also developed an Android app that enables on demand switching between the two network providers so that registered devices can connect to either GENI WiMAX, Sprint WiMAX, Sprint 3G, or WiFi networks. This tool was demonstrated at GEC 20 by showcasing seamless handoffs between heterogeneous GENI and Sprint networks based on signal strength measurements.

References

1. Raychaudhuri, D., et al.: Overview of the ORBIT radio grid testbed for evaluation of next-generation wireless network protocols. In: Wireless Communications and Networking Conference. www.orbit-lab.org (2005)
2. Paul, S., Yates, R., Raychaudhuri, D., Kurose, J.: The cache-and-forward network architecture for efficient mobile content delivery services in the future internet. To Appear in ITU-T Next Generation Networks (NGN) Conference, Geneva (May 2008)
3. Federal Communications Commission. http://wireless.fcc.gov/services/index.htm?job=service_home&id=ebs_brs
4. FIRE LTE testbeds for open experimentation (FLEX). http://www.flex-project.eu/.
5. Intel 6250. http://ark.intel.com/products/59468/Intel-Centrino-Advanced-N--WiMAX-6250-Dual-Band
6. Android Operating System. www.android.com
7. SciWiNet: http://www.sciwinet.org

8. WiMAX Forum Network Architecture: Stage 2: Architecture Tenets, Reference Model and Reference Points [Part 2]. WMF-T32-003-R010v05 (March 2009)
9. Wireless virtualization in GENI. GDD-06-17. www.geni.net/GDD/GDD-06-17.pdf (2006)
10. Kohler, E., Morris, R., Chen, B., Jannotti, J., Frans Kaashoek, M.: The Click modular router. ACM Trans. Comput. Syst. **18**(3), 263–297 (2000)
11. Pettit, J., Gross, J., Pfaff, B., Casado, M., Crosby, S.: Virtual switching in an era of advanced edges. In: 2nd Workshop on Data Center—Converged and Virtual Ethernet Switching (DC-CAVES), ITC 22, 6 September 2010
12. Bhanage, G., Daya, R., Seskar, I., Raychaudhuri, D.: VNTS: a virtual network traffic shaper for air time fairness in 802:16e slices. In: Proceedings of IEEE ICC—Wireless and Mobile Networking Symposium, South Africa (May 2010)
13. GENI Portal. https://portal.geni.net
14. The Internet2 Community. http://www.internet2.edu/
15. Internet2 AL2S. http://noc.net.internet2.edu/i2network/advanced-layer-2-service.html
16. Bhanage, G., Seskar, I., Mahindra, R., Raychaudhuri, D.: Virtual base station: architecture for an open shared WiMAX framework. In: Proceedings of the ACM SIGCOMM VISA Workshop, New Delhi (August 2010)
17. http://groups.geni.net/geni/wiki/GEC10WiMaxCampusDeployment
18. Fund, F., Wang, C., Korakis, T., Zink, M., Panwar, S.: GENI WiMAX performance: evaluation and comparison of two campus testbeds. In: Research and Educational Experiment Workshop (GREE), 2013 Second GENI, pp. 73, 80, 20–22 (March 2013)
19. Fund, F., Wang, C., Liu, Y., Korakis, T., Zink, M., Panwar, S.S.: Performance of DASH and WebRTC video services for mobile users. In: Proceedings of the 2013 20th International Packet Video Workshop, San Jose, California, USA, pp. 1–8 (2013)
20. Mathur, S., Jin, T., Kasturirangan, N., Chandrasekaran, J., Xue, W., Gruteser, M., Trappe, W.: ParkNet: drive-by sensing of road-side parking statistics. In Proceedings of the 8th International Conference on Mobile Systems, Applications, and Services (MobiSys '10) (2010)
21. http://groups.geni.net/geni/wiki/GEC19Agenda/WimaxTutorial
22. http://groups.geni.net/geni/wiki/GEC14Agenda/WiMAXTutorial
23. http://groups.geni.net/geni/wiki/GEC18Agenda/LabWikiAndOEDL
24. WiMAX GEC Tutorials Repository. https://github.com/mytestbed/gec_demos_tutorial

Authorization and Access Control: ABAC

Ted Faber, Stephen Schwab, and John Wroclawski

1 Introduction

GENI's goal of wide-scale collaboration on infrastructure owned by independent and diverse stakeholders stresses current access control systems to the breaking point. Challenges not well addressed by current systems include, at minimum, support for distributed identity and policy management, correctness and auditability, and approachability. The Attribute Based Access Control (ABAC) system [1, 2] is an attribute-based authorization system that combines attributes using a simple reasoning system to provide authorization that (1) expresses delegation and other authorization models efficiently and scalably; (2) provides auditing information that includes both the decision and reasoning; and (3) supports multiple authentication frameworks as entry points into the attribute space. The GENI project has taken this powerful theoretical system and matured it into a form ready for practical use.

ABAC facilitates authorization decisions by providing rules under which actors in the system, called principals, prove that they have certain attributes necessary for accessing resources. Which attributes are required for a given resource is a matter of policy, to be defined and encoded with statements that are meaningful to stakeholders yet precise enough for automatic determination of authorization. ABAC represents delegation of various forms in scalable and separable ways that can be reasoned about formally. This section introduces key security concepts and challenges underlying large-scale decentralized systems, such as GENI, and illustrates the ideas behind ABAC.

T. Faber • J. Wroclawski
USC Information Sciences Institute, Los Angeles, CA, USA

S. Schwab (✉)
USC Information Sciences Institute, Arlington, VA, USA
e-mail: schwab@isi.edu

© Springer International Publishing Switzerland 2016

R. McGeer et al. (eds.), *The GENI Book*, DOI 10.1007/978-3-319-33769-2_10

First, we introduce *principals*, *attributes*, and *rules* of reasoning and delegation that support authorization decisions. Using ABAC, principals can represent an individual or larger organization. An attribute is a property of a principal, created by the assertion of other principals. Each principal may define their own attribute names, and issue statements assigning those attributes as needed to express security policies. Assertions are represented as a digitally signed statement, called a credential. Principals can use a range of systems to authenticate themselves, as well as using public-key cryptography to securely exchange credentials.

A principal's requests will be the subject of authorization decisions based on attributes asserted about it by other principals. Two classes of *delegation* rules, introduced in the example below and more formally described later in the chapter, enable very concise policies to assign attributes to large numbers of principles distributed across many organizations. Service providers combine attributes with rules of delegation and inference to check security policy. Enforcement decisions are ultimately made based on whether a requestor has presented the necessary attributes, assigned either directly or indirectly via delegation rules, to *authorize* a request or action.

We next illustrate, with examples drawn from GENI, how fine-grained access control may be enforced via policies. For example, an individual GENI researcher, Ted Faber, may request to use resources controlled by policies established by authorities such as the NSF, the University of Southern California (USC), or the GENI Project Office (GPO). USC (a principal) may say (assert) that Ted Faber (a principal) is a local GENI user (attribute), and enforce a policy at USC that grants access to a local GENI Rack by checking for this attribute.

Delegation may govern attributes to set policy. For example, the GPO may assert that all USC GENI users are also "*GPO prototypers*." This rule delegates authority to USC to add to the set of GPO prototypers. In this case the delegated attribute (GPO prototypers) is given to principals who also possess the delegating attribute (USC GENI user). Resource providers can, for example, grant access to slices at many GENI Racks, by enforcing a policy that checks for GPO prototypers.

Finally, a principal may delegate at one level of indirection. The GPO may assert that any NSF PI (any principal that the NSF says is a PI by signing a credential) can designate another principal as a GENI user and that user (principal) will furthermore be treated as a GPO prototyper. The NSF can affect the set of GPO prototypers by adding or removing assertions that a particular principal is a PI.

Note the flexible and fine-grained nature of decision-making afforded by using attributes, in contrast to the limited set of choices possible in less expressive identification and access control schemes. For example, to grant access to a server or resource in existing systems, each user might be given a login account, or have their individual ssh public key added to a list of authorized keys enabling login to a privileged account. In the worst case, a user may require root access. While this situation may violate best practices, early versions of GENI required users to upload their ssh private key to tools on third-party servers, granting those tools access to their GENI resources to perform tasks such as automating data collection during an experiment. Using ABAC, resource owners set policy via the set of attributes required rather than the set of user identities granted carte blanche.

The advantage of fine-grained policies is amplified by the indirection allowed via delegation rules. Adopting this approach eliminates an inherent source of insecurity: the need to grant overly general privilege (e.g. root access via sudo) because there is no precise way to name and enable a user to perform a specific action or access a specific resource. Moreover, a central authority is not essential to track all users and their associated privileges. Rather, different principals (users, system administrators, resource owners, etc.) may make local decisions and the collective policy defined by all these stakeholders will determine privileges.

In our GENI example, the delegated attribute (GPO prototyper) is delegated to principals who possess a set of attributes (e.g. *P* GENI user for many different principals *P*). That set is defined in terms of an authorizer attribute (NSF PI). Any principal with the *authorizer* attribute can assign the *delegated* attribute by assigning their local version of the delegating attribute (*P* GENI user where *P* has the NSF PI attribute). This links the authorizer attribute to the delegating attributes, and is a potent form of ABAC delegation called a linked attribute.

The distinction between high-level policies expressed in ABAC rules and attributes, and the low-level implementation of credentials, signing formats, validation schemes and inference algorithms provides another benefit. The authentication-and-authorization (AA) logic is entirely separated from the implementation of the facility or service itself, such as a GENI aggregate manager that uses ABAC as its authorization engine. Furthermore, a record of the decisions made and the attributes and rules that factored into each decision are always available for post-mortem audit, compliance verification, or pro-active forecasting of the impact of a policy change.

Until an authorization decision needs to be made, all of the relevant credentials can be kept locally and brought together just-in-time. Principals can also pass them around so they are pre-positioned when needed, or upload them to retrieval services for ease of accessibility. For example, when the NSF designates a PI, it may send them the signed attribute credential and also forward a bulk set of certificate updates to a central organization, e.g. a GENI clearinghouse.

Reliance on signed credentials carrying attributes instead of identities also reduces the need for each facility to maintain and securely manage a separate authentication database. Rather than supporting a large number of users directly and increasing the security risks due to account break-ins through password theft, ABAC enables a much smaller and more manageable set of policy credentials to be signed, stored and shared as needed. More importantly, potentially vulnerable passwords are replaced with credentials, offering a path to improved security overall.

This chapter describes work on Attribute-Based Access Control (ABAC) that addresses the challenge of creation, management, and implementation of rich, 'audience- appropriate' authorization and access control policy management mechanisms suitable for GENI. We argue that such next-generation policy mechanisms offer a powerful tool for securing GENI as well as future national-scale cyber infrastructures such as those identified and supported by NSF and other U.S.

Government agencies in a manner that provides effective security while fostering wide-spread use and catalyzes collaboration.

The ABAC system, currently in the process of being integrated in GENI, supports authorization policy expression and enforcement mechanisms that provide:

- Formally grounded policy definition and interpretation. ABAC is based upon rigorous underlying theory and logical formalisms and semantics. Logical underpinnings are embedded deeply within the system, while users are only exposed to authorization concepts appropriate to their role and domain of expertise;
- Capability to define common vocabulary across communities and organizations. Common, well understood vocabulary may be rapidly adopted for entities, resources, and privileges within common use cases, while preserving the extensibility required to support diverse specialized policies for specialized sub-communities;
- Auditability of requests, authorizations, and policy changes. ABAC decisions result in tangible proofs of authorization derived from distributed policies, or explicit indications of what policies or insufficient privileges resulted in a request being denied;
- Library implementations suitable for incorporation into a range of GENI and future cyber infrastructures. ABAC software provides a compact library implementation and language bindings for several of the standard programming languages used throughout the Networks, Grid, Cloud and Cyber infrastructure communities.

Together, these capabilities provide a strong foundation for the implementation of strong, secure authorization and access control capabilities within large-scale, federated cyber infrastructures, while simultaneously facilitating the key objectives of flexible collaboration and local control. This chapter describes development and prototyping efforts pursued under the GENI Trial Integration Environment in DETER (TIED) project in re-implementing and delivering the ABAC authorization system [1] as a mature technology providing an expressive, practical, distributed authorization system based on formal logic. The remainder of this chapter is structured as follows. The next section elaborates on GENI's authorization requirements and how ABAC satisfies those needs, followed by a section that presents an introduction to the logical formalism that underpins ABAC and a detailed example of using ABAC to encode GENI's "speaks-for" authorization policy. This background material is followed by sections discussing system design issues related to the incorporation of a distributed authorization framework into GENI; a description of the ABAC architecture; and a brief overview of the current ABAC implementation. The chapter concludes with a brief synopsis of future directions and work.

2 GENI Authorization Requirements

GENI's primary goal is to create a distributed laboratory for large scale networking experimentation, composed as a federation of resources owned and managed by many independent organizations. Because resources are provided and managed locally, GENI spreads out the cost of administration and maintenance. Resources are supplied to researchers on-demand in accord with each resource owners' policy.

This means that an effective authorization system for GENI must be able to express resource utilization policies that are created by distributed actors, gather and apply distributed policies when considering requests, and to produce auditable information about the reasons underlying the authorization decisions to assure stakeholders that all parties are respecting their agreements.

2.1 GENI Authorization Needs

The benefits of the GENI distributed, collaborative structure can only be realized if the system as a whole can easily incorporate new researchers and new equipment and make decisions about who can use resources. Different contributors may have different requirements on who can use what resources. Though convention and agreements can minimize these differences, the GENI cyber infrastructure as a whole will be more powerful if it can potentially include interesting resources that require more restrictive access policies than those of the most generic nodes and networks present in the resource pool.

To support GENI's growth and incorporation of national and international resources, the system must be able to accept contributed resources and services, and admit users from many different institutions of diverse types with minimal a priori negotiations. These institutions will have different policies for how their resources are used and means of identifying their users.

The authorization system is responsible for finding the rights of an identified user, finding the policy that governs access to an independently managed resource and determining if that user can carry out an action on it. GENI's size and service model requires an expressiveness that local systems often do not. In particular, the resources it allocates may be used by a set of researchers with different rights. Those rights may be delegated from leaders of a project to other members and simplicity requires that these delegations be managed locally and respected globally.

In addition to delegating rights between researchers, GENI supports long-running services that operate on users' behalf. Users must be able to delegate tools the right to act on their behalf, e.g. to *speak for* them. This *speaks-for* right is a particularly interesting requirement for the GENI authorization system, and is discussed further in Sect. 3.5.

3 Attribute Based Access Control and ABAC

Speaking generically, an attribute based access control or authorization system is any system that makes decisions about authorization based on some set of explicit attributes associated with the entity seeking authorization. One widely known form of attribute based access control is role based access control, in which the decision to allow access to a resource (for use, for configuration, or for some other purpose) is based on the *role* the requestor seeking access is playing—for example, "Sheila is granted access to administer the LDAP directory because she holds the 'DIRADM' attribute", which states that she is acting in the Directory Administrator role. Attribute based access control systems typically contrast with *identity based* access control systems, in which the fundamental information exchanged between requester and requestee, and the consequent action taken by the requestee, is based on *who* the requester is, rather than the attributes the requester may hold.[1]

Somewhat confusingly, the acronym ABAC,[2] standing for Attribute Based Access Control, is also the name given to one specific, and particularly elegant, attribute based access control system by its original developers. ABAC was developed at the theoretical level in the early 2000s at Stanford and NAI Labs[3] to address distributed authorization using simple predicate logic [1]. ABAC allows service requesters and providers to attach attributes to principals in the system, define rules for deriving one attribute from others, and express those attributes and rules in a common logical framework. ABAC then employs formal logics and proofs to implement authorization decisions based on these rules, and provides interested parties with auditable evidence of the rationale behind these decisions in the form of completed proofs.

ABAC's logic is designed around the concept of principals assigning attributes to other principals, directly or through delegation rules. Attributes are scoped by the principal assigning the attribute, meaning that two attributes with the same name, but assigned by different principals, are different attributes. Delegation rules are also issued by principals and define how an attribute scoped by the issuing principal can be derived from other attributes.

Both service requesters and service providers may be ABAC principals. Services are bound to attributes (usually attributes scoped by the service provider), so that when a principal requests a service the principal providing the service does so only if the requesting principal has the appropriate attribute. Consequently, the authorization decision consists, ultimately, of proving through formal logic that the

[1]Strictly speaking, identity based systems are a subset of attribute based systems, because "identity" can be viewed as an attribute.

[2]In this chapter, use of the capitalized ABAC acronym always refers to the specific ABAC system, rather than attribute based access control systems generally.

[3]Later renamed McAfee Research. Subsequently, this research lab was acquired by SPARTA, Inc. and operated as the Security Research Division of SPARTA.

requesting principal has the attributes required to obtain the service. We discuss ABAC's authorization logics in significantly more detail below.

To simplify integrating many practical systems' notions of principals, ABAC imposes only three constraints on principal semantics. First, a principal must be able to prove its identity. Second, when two principals refer to a third by identity, they always refer to the same unique principal. Third, a principal must be able to issue assertions about attributes that are unambiguously bound to it. A public key cryptosystem such as RSA [3, 4], where the identity of a principal is its public key, meets these constraints. A principal can perform a challenge/response authentication to prove it holds the private key; a public key always refers to the same principal; and principals can issue cryptographically signed attribute assertions.[4]

Because principals are so simple and do not require significant coordination to generate, they can be created easily and without appeal to a central authority. This is a critical requirement for a system that must scale to national or global size. Many elements can be principals in such a system, and even if we only consider humans, decentralized assignment and management is key to a scalable distributed approach.

ABAC implements a fully distributed system *authorization policy*. The authorization policy of any system utilizing ABAC is distributed because a principal's direct assignment of attributes and derivation rules are managed by the issuing principal. The system's policy as a whole is the union of those distributed policy fragments (attributes and rules) but for any given decision, only the relevant rules must be consulted. Changes to a given principal's policy are only relevant to those dealing with that principal, e.g. making a resource or service request from that specific GENI aggregate manager.

Importantly, the structure of ABAC attributes—specifically, that each attribute is scoped to a principal, and that each rule assigns one attribute—allows the ABAC logic designers to ensure that a principal is assured that it can obtain all of the data relevant to an authorization decision if each principal's store of assertions can be located [2]. This insures that the correct decision can always be reached, even with the distributed management of policy.

Finally, ABAC's well-defined, logic-based framework implements a common, system-wide semantics within which authorization decisions can be clearly and unambiguously expressed and evaluated. This use of formal logic ensures that decisions are clear and transparent, and allows for both extremely simple and extremely sophisticated authorization policies to be implemented, as required by each specific use case. To increase this flexibility, ABAC defines a *family* of logics that form a hierarchy of increasingly more complex predicate logics, each reducible to datalog [5]. Because of their close relationship to role-based authorization principles, these logics are referred to as *Role-based Trust management*, or RT, logics. RT logics are discussed further below.

[4]A misbehaving principal can undermine these properties, e.g., by sharing a private key. ABAC assumes good behavior of principals.

3.1 ABAC and GENI

ABAC meets GENI's needs because it is flexible with respect to identity representation, designed to resolve distributed policy, structures its logic to make policy discovery feasible, and as mentioned provides a proof structure that supports both unambiguous decision-making and auditing. Though GENI does not currently require this capability, ABAC also allows authorization to make use of restricted information—for example security clearances or sensitive attributes—in ways that tightly limit direct and indirect disclosure.

Conceptually, ABAC's authorization logic can be applied to principals identified and authenticated by a number of different systems, independently or simultaneously. One of our contributions to GENI is providing bindings from GENI's identity system, based on X.509 certificates, to an ABAC logic system. The GENI infrastructure allows researchers to bind to X.509 identity certificates from other identity services, including Shibboleth and the InCommon [6] attribute framework.

In GENI, most of the authorization decisions can be encoded in the simplest of ABAC's RT logics, RT0. Section 4 of this chapter discusses *libabac*, a concrete implementation of the ABAC system. This software distribution supports core ABAC functionality and a robust, efficient RT0 prover, that been integrated into the GENI software base, and is in wide use today for this purpose.

3.2 ABAC Logics

ABAC presents a family of logics designed to be simple to reason about while capturing useful authorization abstractions. All the logics are based on attaching principal-scoped attributes to other principals. The logics primarily differ in the extent to which attributes can be parameterized and the rules used to delegate attributes.

Here and in subsequent sections, we give an overview of these RT authorization logics. The reader interested in a more detailed discussion of these logics is referred to Refs. 1, 2. Readers primarily focused on system implementation issues may wish to review these sections quickly before moving to Sect. 4.

ABAC defines a family of five RT logics. These include:

- **RT0**: a basic delegation logic that attaches un-parameterized attributes to principals. The basic delegation rules are direct assignment, simple delegation and linked delegation. This logic is described in detail in Sect. 3.3.
- **RT1**: RT0 extended with typed parameters attached to the attributes. Attribute parameters can be used to scope the delegation rules and further control how attributes are assigned. Described in Sect. 3.4.
- **RT2**: RT2 adds the ability to attach attributes to non-principals and reason about them. This allows one to reason about RT1 parameters using RT0 delegation rules. RT2 is described in Sect. 3.4.

- **RTT**: RT2 with the addition of a delegation rule that express consensus among some number of principals. This logic is not further discussed below.
- **RTD**: RTD with the addition of a delegation rule to delegate attributes to principals only within a specific context. This logic is not further discussed below.

Each of these logics can be expressed as datalog rules. Datalog is a negation-constrained, safe prolog subset that is efficient to implement [5].

All the logics scope their attributes by principal and share the property that a principal making a query can always ensure it can discover all delegation rules needed to reason about a request [2].

At present most of GENI's authorization needs can be met using RT0, though some forms of authorization are more elegantly and compactly represented in RT2. The GENI community has primarily focused on implementing and using these simpler logics.

The remainder of this section describes RT0, RT1, and RT2 in enough detail to give the reader a feel for their expressive power and notation. We also comment on how we use RT0 to meet GENI's needs when RT2 might be more elegant. Finally we present an extended example that shows how to use ABAC logic to express a complex GENI authorization feature, the "speaks-for" right.

3.3 RT0 Logic

ABAC's RT0 logic allows one to attach an attribute to a principal, define a direct delegation rule and define a rule linking the possession of an attribute to the ability to delegate attributes. This section introduces the notation and semantics.

In ABAC's logic an attribute is a string attached to a principal by another principal. Using a GENI example, if an aggregate manager identified as *AM* wishes to attach the *ListResources* attribute to a user identified as *U*, we say that *AM* has attached *AM.ListResources* to *U*. Only *AM* can assign attributes from the *AM* space. Furthermore *AM1.ListResources* and *AM2.ListResources* are distinct.

There are three ways to attach an attribute to a principal:

1. Direct assignment.
   Meaning: *U* has attribute *AM.ListResources.*
   Notation: $AM.ListResources \leftarrow U$
2. Delegation.
   Meaning: All principals with attribute *AM2.ListResources* have *AM1.List Resources.* Notation: $AM1.ListResources \leftarrow AM2.ListResources$
3. Linked Delegation.
   Meaning: Any principal *P* with the *AM2.Linked attribute* can assign the *AM1.ListResources* attribute by assigning the *P.ListResources* attribute.
   Notation: $AM1.ListResources \leftarrow (AM2.LinkedResources).ListResources$

Direct assignment is straightforward. A principal binds an attribute to another principal. If we take *AM.ListResources* to indicate the ability to invoke the *ListResources* operation on *AM*, the example in case 1 above is interpreted to assert that *AM* has explicitly granted that ability to user *U*.

In the second example, *AM1* has expressed a rule delegating the ability to assign principals the *AM1.ListResources* attribute to *AM2*. In turn, *AM2* exercises that delegation by assigning its *AM2.ListResources* attribute. Consequently, any principal that knows both

`AM1.ListResources←AM2.ListResources`

and

`AM2.ListResources←U`

can conclude that *U* has *AM1.ListResources*.

In some cases, *AM1* and *AM2* will want to closely coordinate such a delegation. In others, however, *AM2* may be entirely oblivious to the delegation. If *AM2* is a well-known certifier, or *AM1* and *AM2* have a pre-existing general relationship where they agree on the semantics of *ListResources*, there is no need or reason to discuss each specific delegation. In any case, ABAC does not require any coordination to make the delegation.

The last example above adds a second indirection. This delegates *AM1.List Resources* to a number of other principals that have an attribute assigned by *AM2*, rather than to *AM2* itself. In this case a principal must know that

`AM1.ListResources← (AM2.Linked).ListResources`

and

`AM2.Linked←P and P.ListResources←U`

to conclude that *U* has *AM1.ListResources*.

Linked delegation is best viewed as allowing a principal to appoint agents. An agent is another principal that can assign an attribute on the first principal's behalf. The example above illustrated one principal (*AM1*) directly delegating that authority to the agents of another (*AM2*). A ruleset of the form

```
AM2.ListResources← (AM2.Linked).ListResources
AM1.ListResources←AM2.ListResources
```

lets *AM2* express its creation of agents and *AM1* delegate to this second principal. The first rule is controlled by *AM2*, because it controls the *AM2.ListResources* attribute, while the second rule is controlled by *AM1*.

The requirements for a delegation—the right hand side of the arrows above—can include conjunctions. For example in a scenario involving a Clearing House (*CH*) and Slice Authority (*SA*),

`AM.CreateSlice←CH.CreateSlice ∩ SA.CreateSlice`

asserts that *AM* will assign the *AM.CreateSlice* attribute to a principal that has demonstrated it has both *CH.CreateSlice* and *SA.CreateSlice*. The intersection symbol ∩ underscores that the conjunction in the attribute interpretation is a set intersection in the set inclusion sense.

In practice, each of these declarations—the assignment of an attribute or the creation of a delegation rule, is expressed in an ABAC *credential*. The simplest credential is a signed statement of the rule or assignment in RT0 logic, signed by the principal that controls the attribute being assigned or delegated. That is, a credential is signed by the principal whose identity is attached to the attribute on the left side of the arrow. In general, a credential may express one or several RT0 rules. Such credentials carry the assertions that form the basis of proofs in the ABAC system, and are consumable by any entity that can verify the signatures.

3.4 *RT1 and RT2*

RT1 adds typed parameters to attributes and the ability to reason about them. Rather than reasoning about the *AM.ListResources* attribute which might allow a principal to list any kind of resource, an RT1 rule can further scope that attribute by binding it to a named subset of resources, e.g., a particular GENI slice: *AM.ListResources(Slice1)*.

The parameters can be integers, floating-point numbers, dates, times and enumerations. The enumerations can be closed enumerations ('read', 'write', 'execute') or open-ended—for example, any principal name or file name.

ABAC can reason using parameters in three ways:

Case 1: Using literal parameters:

AM1.ListResources("Slice1") ← *AM2.ListResources("Slice1")*

Here any principal that has attribute *AM2.ListResources*, parameterized by the literal string "Slice1" also has *AM1.ListResources* parameterized by "Slice1". A principal that can list the resources of "Slice1" from principal *AM2* can also do so from principal *AM1*.

Case 2: Named parameters, implicitly constrained:

AM1.ListResources(?Slice) ← *AM2.ListResources(?Slice)*

Prefixing the parameter name with a ? marks it as a variable; the requirement for a match is that the parameter must have the same value on both sides of the assignment.[5]

For example, of a principal has attribute *AM2.ListResources("Slice1")* this rule implies that the principal also has attribute *AM1.ListResources("Slice1")*, just as in case 1. It also means that a principal with attribute *AM2.ListResources ("Some other slice")* also has attribute *AM1.ListResources("Some other slice")*.

[5]The typing is implicit. AM1.ListResources and AM2.ListResources must have direct assignments made so the system can determine the type and types must be consistent. This is a place where the theoretical nature of the ABAC papers is abundantly clear. In our implementation of RT2 we added syntax to declare types of parameters and perform explicit type checking.

To insure that the rules are tractable, any parameter name on the left hand side must also appear somewhere on the right hand side. A rule such as *AM1.ListResources(?Slice)*←*AM2.KnownUser* is illegal.

Case 3: Named parameters, explicitly constrained (the constraint set follows the :):

AM1.ListResources(?Slice)←*AM2.ListResources(?Slice:["Slice1", "Slice2", "Slice3"])*

This means that any principal that possesses attribute *AM2.ListResources("Slice1")* also has attribute *AM1.ListResources("Slice1")*, and likewise for attributes parameterized by "*Slice2*" or "*Slice3*". A principal that is granted attribute *AM2.ListResources("Slice4")* is not granted any new *AM1.ListResources(?Slice)* attribute as a result.

The constraint sets must be finite, though the ABAC papers describe several syntaxes to enumerate those sets.[6]

RT1 lets policy writers naturally express authorization to certain objects for certain operations scoped by principal. It is easy to manage one principal granting another the rights to read certain objects even through complicated delegation.

RT2 relaxes the limitation that the set used to constrain parameters must be a static parts of the delegation rule. RT2 attaches principal-scoped attributes to parameter values and allows the sets of parameter values defined by those attributes to constrain parameter variables. These sets of parameter values are called *o-sets* (object sets) in the ABAC descriptions.

To see how they work, consider this rule:

AM1.ListResources(?Slice)←*AM2.ListResources(?Slice:AM2.ValidSlice)*

This rule says that a principal that has *AM2.ListResources()* for any slice name that has the attribute *AM2.ValidSlice* also has *AM1.ListResources()* for that slice. This is basically the same rule as example 3 above, except that the set of valid slice names is dynamic.

As with principal attributes, slice name attributes are assigned by ABAC RT0 logic rules.[7] The *AM2* principal can directly declare a slice name to have *AM2.ValidSlice* by issuing the rule:

```
AM2.ValidSlice← "Slice5"
```

The *AM2* principal may also delegate the ability to designate valid slice names to principal *AM3*:

```
AM2.ValidSlice←AM3.ValidSlice
```

Or delegate that right to a dynamically defined set of principals:

```
AM2.ValidSlice← (AM2.SliceNamer).ValidSlice
```

[6]For example, there is syntax for referring to the principal being evaluated when looking at a parameterized linking role. This is useful, but well beyond the scope of this document. The interested reader is referred to [1] Sections 3.1 and 3.3.

[7]In the TIED ABAC RT2 library, we use distinct notation for o-set rules and attribute rules, to ensure that each is represented by a unique type.

Moving from RT0 to RT1 requires a reasoning engine of sufficient expressive power to operate on constrained parameters. Moving from RT1 to RT2 does not require any expansion of the reasoning engine's abilities, simply the application of those features to sets of parameter values as well as sets of principals.

The rules for encoding current GENI access are generally only scoped to a given slice or sliver name. In an RT1 implementation, we would express this as *AM.DeleteSliver(uuid)*. But even without an RT1 implementation, we can express this scoping within the name of an RT0 attribute, such as *AM.DeleteSliver_uuid* .

Because the rules for GENI access never require arithmetic or other operations, but only matching, and only principals who issue scoped attributes need to re-interpret them, we can express a these rules in RT0 and use a simpler reasoning engine. In the case study below, we adopt RT1 as the policy language for clarity. In practice, our implementation of "speaks-for" is based on RT0, leveraging our understanding of the GENI system to ensure soundness when these rules are, as described above, expressed in RT0 rather than RT1 format. Many GENI documents refer to this convention as RT1 Lite.

3.5 Case Study: GENI Authorization and Speaks-for

To motivate the adoption of ABAC as the primary authorization system for the GENI AM API, the authors and their team demonstrated how the existing GENI authorization model can be expressed in ABAC logic. A first step in this direction was to implement current GENI policy—including a new "speaks-for" feature—using ABAC logic [7].

This section delves into considerable detail showing how to express GENI authorization checks and "speaks-for" in RT1. The intent is to provide enough detail to convey the expressive power of the logic and illustrate the usefulness of ABAC's abstractions.

The GENI authorization model centers on specific named privileges for accessing GENI objects such as slices and resources. Resource providers grant these privileges to GENI principals (users) by issuing signed certificates that encode the privileges. These privileges are ad hoc and tied to the service definitions, and a given GENI credential could assign multiple such privileges. The specific list of privileges corresponds to operations or groups of operations supported by the GENI APIs. The GENI authorization model also allows further delegation of privileges to other principals, provided the GENI credential carrying the privilege grants that right.

The "speaks-for" privilege is an additional privilege intended to be used as follows. A user wishes to use a tool to access GENI services, but, due to security concerns, does not want to upload their identity certificate and private key to that tool, which may be a web service.

Instead, the user issues a GENI credential granting the "speaks-for" privilege to the tool (which is itself an ABAC principal). The tool includes that credential in its requests. Consequently, the GENI services will, for authorization purposes, treat these requests from the tool as though they came directly from the user.

This differs from delegation in three ways:

1. Semantically, a tool operating under "speaks-for" authority is exercising the user's authority under close supervision. The user is taking the action through the tool and the user is responsible for the actions. In contrast, a delegated privilege is exercised independently by the recipient of that delegation. The user who has been delegated authority is responsible for its use, not the delegator.
2. All credentials may be used in conjunction with "speaks-for" authority. In contrast, credential issuers need not issue delegatable privileges.
3. A tool requires far fewer speaks-for privileges when compared to delegated privileges. (For example, a tool need not have all of a user's slice credentials to look up the status of all the user's slices.)

3.5.1 Semantics of GENI Privilege Credentials

This section describes the content and use of GENI privilege credentials [18] as used to implement speaks-for privileges using ABAC in this case study.

A GENI privilege credential encodes a set of statements of the form "The issuer of this credential (a principal) gives the owner of the credential (a principal) these privileges (strings) with respect to the target (a principal)." The privilege strings are defined with respect to the GENI APIs.

For each of the privileges, the additional optional right to delegate that privilege to others is indicated with a Boolean value. For example, a slice authority (issuer) can grant the resolve privilege to a GENI user (owner) on a given slice (target). The issuer is always the principal that signed the credential. The target and owner are given explicitly as X.509 certificates.

Under the GENI authorization semantics, a credential chain is used to encode delegation. If the credential is delegated, the original credential granting the delegatable privilege, called the base credential, is included verbatim in a new credential signed by the owner of the base credential. The owner of the base credential is the issuer of the new, delegating, credential. This new credential assigns privileges to a new owner. The new credential is valid if the base one is, if the delegated rights are marked delegatable (e.g. true), and the expiration time of the new credential does not extend beyond the expiration time of the base credential. If multiple delegations are performed, then the credential chain grows to reflect these delegations.

3.5.2 GENI Policy in RT1

Next we describe RT1 rules that express the GENI authorization policy. The policy and credential formats are entwined, and we cannot speak of one independently without the other.

Taking a different approach from the description of GENI privilege credentials above, we first describe how to encode an ABAC policy that supports "speaks-for" and subsequently extend the policy to add delegation. Speaks-for requires simpler rules to encode.

GENI Privileges with Speaks-for in RT1

For a given service request, the aggregate manager (AM) service provider knows the principal making the request, the target of the request, and which privilege is required to execute it. The service provider initializes a prover with its policy encoded as RT1 rules, augmented with any additional RT1 rules conveyed with the request. Finally, the initialized prover is queried with the question required for authorization: "does the principal making this request have the proper privilege?"

Let's examine how to encode this question in ABAC, starting with "the proper privilege." The RT1 attribute *AM.privilege(Target)* means that AM believes principals in that set have a specific privilege with respect to Target. For example, the privilege to issue *resolve* on a slice S would be the RT1 attribute *AM.resolve(S)*.

When a principal *P* requests an operation that requires resolve rights on slice *S*, the provider *AM* asks the prover if *P* is a member of *AM.resolve(S)*—or in other words, if *P* has the attribute *AM.resolve(S)*.

Service providers do not issue credentials conferring *AM.privilege()* directly to users. Instead, these privileges are inferred based on RT1 rules in the local policy. While RT1 can express complex delegation, our use case presents a series of simple delegations.

There exists a collection of issuers that each service provider trusts, and hence believes RT1 statements signed by these issuers. For each Issuer that *AM* trusts, it includes a policy rule of the form:

`AM.privilege(Target)` ← `Issuer.privilege(Target)`

There is one rule of this form for each privilege covered by AM's policy. E.g., for the privileges *resolve* and *info*, there exist corresponding rules

`AM.resolve(Target)` ← `Issuer.resolve(Target)`

`AM.info(Target)` ← `Issuer.info(Target)`

Issuers, such as Slice Authorities and Clearinghouses, issue credentials to users. Here we describe how RT1 expresses a credential as multiple RT1 statements. Because GENI policy allows all privileges to be transferred to another entity using "speaks-for", the ABAC translation of a credential issuing a privilege to a user P is expressed as:

1. *Issuer.privilege(Target)* ← *Issuer.speaks_for(P)*
2. *Issuer.speaks_for(P)* ← *P*
3. *Issuer.speaks_for(P)* ← *Issuer.TrustedTool* ∩ *P.speaks_for(P)*

These RT1 statements mean:

1. The Issuer says that anyone that speaks for P can exercise the privilege as P. Note that this means that the speaker-for, P, must also have the right under these rules. If we want only the actual principal (and whoever that principal delegates to) to have the right (1) can be modified to read *Issuer.privilege(Target)* ← *P*.
2. The Issuer says P speaks for itself.
3. The Issuer says that any entity that both the Issuer believes is a trusted tool and that P says speaks for P, speaks for P.

When an Issuer signs and delivers a GENI credential assigning privilege with respect to Target, it is making those three statements in ABAC. The first line is repeated for each privilege in the credential; the last two are added to the prover's state only once per credential.

When a user P issues a speaks-for credential for a tool T, that credential is translated into RT1 as:
`P.speaks_for(P)` ← `T`

To recap concretely in GENI terms: if T makes a request including credentials carrying all the following assertions, the corresponding rules express the policy at AM:

GENI Statement	ABAC RT1 Rules
AM trusts *Issuer* about resolve on *Target*	*AM.resolve(Target)* ← *Issuer.resolve(Target)*
Issuer has delivered to *P* a GENI privilege credential assigning resolve on *Target*	*Issuer.resolve(Target)* ← *Issuer.speaks_for(P)Issuer.speaks_for(P)* ← *PIssuer.speaks_for(P)* ← *P.speaks_for(P)*
P has issued a "speaks-for" credential to tool *T*	*P.speaks_for(P)* ← *T*
The *Issuer* trusts tool *T*	*Issuer.TrustedTool* ← *T*

When *AM* decides if *T* can proceed (e.g., if *T* is in *AM.resolve(Target)*), the proof chain supporting this inference will be:
`AM.resolve(Target)` ← `Issuer.resolve(Target)` ← `Issuer.speaks_for(P)` ← `P.speaks_for(P)` ← `T`

4 Implementing ABAC—The *libabac* System

This section describes *libabac*, an implementation of ABAC suitable for use in large, decentralized, heterogeneous distributed system designs. *Libabac* was developed by the authors and their team for use within the DETER Cybersecurity Testbed [8] and later GENI, and is used by both of these systems today.

At time of writing, the core *libabac* distribution includes:

- Libabac itself, a linkable C/C++ library
- Perl and Python bindings to libabac
- A standalone java implementation
- *creddy*, a command line credential management tool

Two additional general-purpose tools that build on *libabac* are also available:

- *crudge* is a visual editor for ABAC policies and proofs
- *credential printer* is an XMLRPC service to convert credentials from standard system formats to a readable text representation

Libabac is both a system design architecture and a concrete implementation. At the time of writing, the concrete implementation supports a subset of the complete design architecture. This subset is sufficient to fully implement the distributed authorization policies used in both the GENI and DETER projects. However, it is expected that additional architectural elements will be incorporated into the implementation over time. Sections 4.1 and 4.2 of this chapter present system design considerations and the *libabac* system architecture, respectively, while Sect. 4.4 discusses the current implementation as incorporated into the GENI system at the time of writing.

4.1 System Design Issues

This section discusses some concrete design requirements and approaches arising from the core ABAC concept, and from issues touched on in Sects. 1 and 2. We discuss the requirements on principals, on information representation, on protocol integration and on the negotiation and collection of information.

4.1.1 Principal Requirements

In an authorization system, *principals* are the key entities on behalf of which the system carries out its function. In an attribute-based authorization system, a principal can do two things: it can ask for the system to take an action on its behalf, and it can assert attributes about itself or another principal.

For a distributed authorization system to scale, the creation of principals must be decentralized; no single entity can vet all principals or issue all identities. Consequently, what is required is a format for principal identifiers that can be issued, assigned, and managed in a decentralized fashion, and can meet three semantic requirements:

1. A principal can prove its identity to another entity.
2. If two parties reference a principal by identity, they are referencing the same principal.
3. A principal can assert attributes about other principals.

A natural implementation of these requirements is to create a public/private key pair, and make the principal's identity the public key. The principal keeps the private key secret and can prove its identity by responding to cryptographic challenges and make assertions by signing them. As long as the key space is large enough, collision probabilities can be made effectively zero, insuring a unique identity. [8]

Such a principal identifier contains minimal information in and of itself. Importantly, there is no information about the role that the principal plays, its relationship with other principals, or even a short human-readable name to print. Instead, such information is created and maintained by system applications. In particular, authorization attributes are part of the authorization framework.

Binding a Principal to a Request

Defining a principal representation sets the stage for binding principles to attributes that are used to make an authorization decision, but such a decision is also predicated on knowing which principal is requesting the action. A request must be bound to a principal and then any required attributes proven about the principal before the requestee will allow the action.

This binding is, strictly speaking, outside the responsibility of the authorization system. Formally, the system needs to know only what attribute must be proven about what principal—the binding of attributes and principals to a protocol request itself is application-specific. However, the presence of principal identifiers meeting the three requirements enumerated above requirements simplifies this task. In particular:

An interactive, channel-based request—that is, a request arriving over a specific communication channel established to support the request—can include a challenge from the requestee system to the principal that establishes the principal's identity using Property 1. Requests made across this channel can then be attributed to this principal. Network protocols based on TLS connections typically make use of this approach.

For non-interactive requests, some form of explicit binding of identity to request is required. Many practical implementations of principals—including the public/private key implementation mentioned above—allow for non-interactive binding of a principal to a request message. For example, such a message could be digitally signed and timestamped.

[8] Alternatively, a public key *fingerprint* can be utilized as the identity, if the benefits of the smaller identifier outweigh the increased collision probability.

4.1.2 Representing Attributes and Rules

Any distributed authorization system must represent the various attributes and rules that are communicated between system elements in a format that is comprehensible to each element and can be transmitted between elements effectively. Such a representation must communicate the semantics of the rule and also validate its source. In other words, when presented with a concrete representation of this logical object, a principal must be able to determine both what it means and that it is valid. We call this object a *credential.*

To support validation, the asserting principal must be bound tightly to each credential, typically through use of a digital signature. Consequently, reformatting a credential or converting between credential representations to obtain compatibility is challenging, because reformatting the credential necessarily invalidates its signature. If the conversation action is required at a point where the asserting principal is not available to resign the new version, the converted version cannot be signed correctly.

The implication for practical distributed authorization systems is that a single, or at worst a very small number, of *canonical credential formats* must be defined, which are understood by all participants, and in which credentials can be exchanged between principals and services.

While architecturally sound, this approach is unfortunately problematic from an implementation perspective, because a number of existing credential formats are already in some use today, without necessarily being either semantically sufficient or widespread enough to be adoptable as a canonical format for our purposes. At the same time, it is problematic to assume that existing credentials can be converted to some new canonical format without losing key signature information.

As examples, certain modern authentication and authorization systems, such as X.509 [9, 10] and Shibboleth/SAML [11] include the ideas of attributes and represent them naturally. In these cases, the "local" attributes render naturally into desired ABAC system attributes from a semantic perspective, but are represented in two distinct, concrete formats. Consequently, neither existing format is ideal as a canonical representation, due to the conversion issue.

Many authorization systems are based simply on identity, or alternatively do not export internal attributes used for access control, such as group membership or access control lists. For example, Kerberos acts primarily as a trusted introducer between services and clients. In a Kerberos [12] environment each service makes independent access control decisions based on trusted identity of the requesting Kerberos principal, together with internally stored authorization policy information and/or information acquired elsewhere (e.g., system group files). To integrate such information into a distributed attribute-based authorization system, we need to provide an additional translation from local information to credentials.

One way to address this discrepancy is for identity management domain to provide attribute servers. An attribute server explicitly provides credentials based on local identity. For example a Kerberized attribute service would provide, on request by a client validated by identity, a signed set of credentials in a format the distributed authorization system understands. Similar services could be provided by enterprise systems based on password control or keyed by a PGP key. In some cases, the credential server may even generate credentials and an identity for the requesting principal. In practice, the GENI Clearinghouse is such a server.

4.1.3 Negotiation

Semantically expressive attribute-based authorization in a fully distributed environment may, in general, require a multi-party, multi-phase negotiation to complete an authorization decision. This is in contrast with the simple table lookup that characterizes local authorization. In addition, principals may need to gather credentials from others to complete the proof chains necessary to make decisions.

For an authorization system designed with this perspective, negotiation is part of the process of requesting service. As the authorization function is typically embedded within a some larger overall system, it is this larger system that must implement the protocol support for negotiated authorization function if this function is to be supported directly. Often legacy systems using pre-existing network protocols will need to be integrated as well, which demands an alternative implementation tactic. We briefly discuss those cases.

Protocol Support for Multi-Phase Negotiation

A protocol designed to support multi-phase authorization directly must include messages and function codes that indicate that a negotiation needs more information. To implement this function the application protocol must support, or be adaptable to support, a workflow as follows:

1. A service requester collects local credentials and includes them in the request.
2. The request recipient (“Server”) passes request credentials and local credentials to a reasoning engine. The server library may also attempt to gather more information from its environment to support the reasoning process.
3. If necessary, the reasoning engine returns a failure code indicating that more information is necessary and an encoding of the proof so far (as credentials).
4. The server returns to the requestor a code indicating more work is required and the proof so far.
5. The requester calls its local routines that attempt to extend the proof.
6. If the local routines have extended the proof, the requester can make the request again with the additional information.
7. When neither party can extend the proof, the request fails. Alternatively, when the proof is completed, it succeeds.

Legacy Applications and Pre-Proving

If an application utilizes existing network protocols that are not amenable to the multi-phase strategy set out above, another choice is to implement a separate pre-approval service using the same principal that offers the service. The requester asks for the same service it would request directly and interacts with the pre-approval service using a protocol designed to carry out negotiation. Once the protocol completes, either the requester and pre-approval service know that the request will not be allowed, or the requester holds the set of credentials necessary to be authorized for the action.

If the pre-approval service is conducted by the same entity providing the service, or the pre-approval service and the actual service can communicate securely, the server can be seeded with the complete proof and its result, and the requester need only provide its identity. The server binds the request to the identity and is able to "complete" the proof and authorize service for the requester.

4.1.4 Negotiating with Sensitive Data

A distributed authorization system may wish to support the notion that principals hold credentials that they may not be willing to share with all principals.

Control of these credentials is essentially a recursive application of the authorization system. A principal that holds a sensitive credential might only reveal it to a second principal if that principal can prove it has a particular required attribute.

If one principal sees that a negotiation can only be extended by revealing a sensitive credential, it inserts a sub-proof into the ongoing negotiation with the other endpoint as the target principal and the required attribute for the sensitive credential as the goal. If that sub-proof is completed, the sensitive credential can be introduced and the main proof continued.

These controls place several requirements on the negotiation. First, it introduces the requirement that all servers be principals. If controlled data is not involved, there is no reason from the negotiation semantics perspective that the server needs to be a principal.[9] The authorization decision is just a matter of collecting the proper public credentials that prove the requester has the required attribute. Because all this information is public, there is no need for the requester to restrict which entities receive its credentials. IN the case of controlled credentials, however, the requirement that all participants in a negotiation potentially be able to prove their identity mandates that all participants be principals.

Similarly, support for controlled credentials forces negotiations to be pairwise. If negotiations were even three-way, a server *Server* could construct a proof that contained information from *Client1* that *Client2* did not have permission to see, or

[9]There may be other reasons to make the server a principal; mutual authentication is rarely a mistake.

vice versa. This proof is useful to the server, but neither of the parties has permission to view it. Such a scenario undermines the transparency and logging value of the system.

It is an open problem to design a system to correctly control information in this kind of scenario. Consequently, the *libabac* system currently restricts negotiations to two parties.

4.2 *libabac* Software System Architecture

This section describes key aspects of the *libabac* software system architecture. It presents the basic software modules, processes, and interfaces used to implement ABAC-style authorization within *libabac*, and discusses topics related to integration of *libabac* into larger software packages. The architecture takes into account the full complexity of the ABAC authorization function, including multi-step negotiations, use of credentials gathered from third parties and negotiating with sensitive credentials.

At the time of writing this chapter, only a portion of the conceptual architecture has been implemented in the publicly available *libabac* code base. Section 4.4 provides some additional detail about the implemented *libabac* functionality.

Figure 1 shows the core components of the *libabac* implementation architecture. The application communicates with a *prover* that reasons about credentials, a *credential access controller* that controls access to sensitive credentials, and an asynchronous discovery subsystem that finds publically available credentials. These modules are linked directly with the application. Separately, the implementation provides a *credential discovery daemon*, implemented as an independent service that gathers credentials at the request of an application.

The subsystems that can be linked directly with the application (the components inside the dashed rectangle) are referred to as the *endpoint engine*. We describe key aspects of the endpoint engine's operation below.

4.2.1 Basic Operation

Applications interface with the endpoint engine primarily through the core module. The application passes relevant credentials and proof targets into the core. The core coordinates the actions of the various *libabac* subsystems to carry out as much of the proof as possible, and return either a result or a partial proof as discussed in Section "Negotiation".

From a user's perspective, the endpoint engine is primarily focused on manipulating *contexts* in which proofs are carried out. Contexts can be created, copied, deleted, and have credentials loaded into them. Each context encapsulates the knowledge directly accessible to the prover.

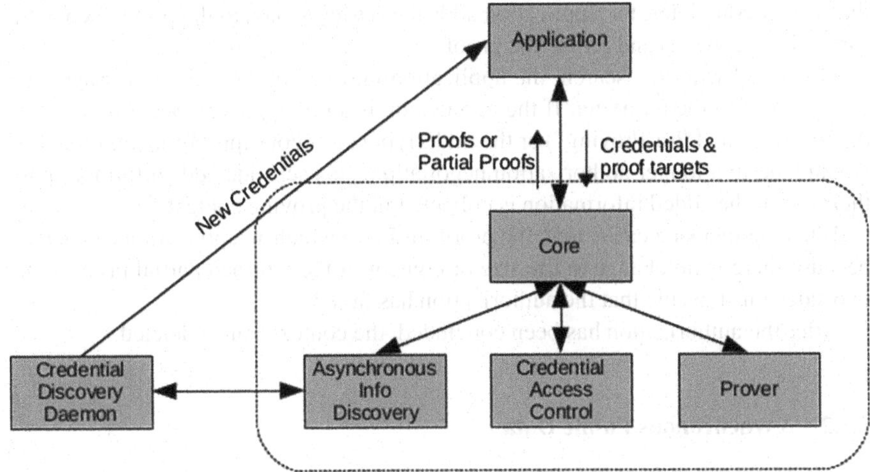

Fig. 1 *libabac* architecture components

Credentials are opaque to the user. The core provides interfaces for adding credentials to a context as required and returns them as a proof, but the user is not required to understand credential encoding. The core interface validates the credentials. An interface is provided for a user to create new credentials from an encoding-independent data structure.

When an application starts up, it creates a base context containing the credentials that define its local delegation rules as well as any other information it has stored locally or has public access to. When a request comes in, the base context is cloned into a *proving context* for the particular authorization being requested. The application binds a principal to the request as described in Section "Binding a Principal to a Request", and looks up the relevant authorization attribute for the requested action from its configuration. It then adds any credentials in the request to the proving context and makes a call into the library for the proof. The call includes the context, the target principal and the target attribute.

If the proof succeeds, the list of credentials constituting the complete proof is returned, with an indication that the proof is complete and successful. The proof is logged, the action authorized, and the proof returned to the requester for their logging purposes (assuming that the application interface allows that).

If the proof fails, a list of credentials that the prover believes are relevant to extending the proof are returned, along with an indication that the proof has failed. At this point the response is up to the application, but the architecture admits several options.

Particularly, the application may conclude that the request should not be authorized and return an appropriate error. Alternatively, the application may request an asynchronous search of public data, driven by the partial proof information. When that search completes, the application adds the retrieved data to the proving context (or the base context) and retries the proof.

In parallel with this search, the application may return the failure message and partial proof to the requester. If the application is a pre-approver (Section "Legacy Applications and Pre-Proving") or the underlying protocol supports negotiation this would lead to a longer authorization negotiation. As each side adds information to their proof, the added information is collected in the proving context.

When neither side can extend the proof further—which an application can detect because there is no change in the size or content of the returned partial proof—the two sides must decide that the authorization has failed.

After the authorization has been concluded, the context can be deleted.

4.2.2 Asynchronous Public Data

The interface to asynchronous public data is one of the most interesting and least solidified aspects of the current *libabac* architecture. Accessing publicly available data can require long latency and may require understanding and supporting multiple externally defined query protocols. The application may want to hide this latency or place it in parallel with endpoint negotiations. The current *libabac* design encapsulates each of these alternatives.

As Fig. 1 shows, a separate process gathers credentials. It encapsulates the knowledge of outside query protocols and carries out searches asynchronously. A standard interface is defined and provided between the endpoint engine's asynchronous information discovery module and the system search daemon.

To fully meet system goals, the interface between daemon and discovery system must be generic and extensible. The discovery system must be able to guide the search, by suggesting attested attributes to search for and likely storage sites, as well as an indication of how exhaustively to search. Requests to the daemon also include an identifier so the application can route the return information to the right context. The daemon returns a list of results to the application, including the query identifier for demultiplexing them.

An important implementation consideration is the method by which these returned results are delivered to the application. In order to make *libabac* useful in as many system implementation environments as possible, it is desirable to support both traditional single threaded applications as well as more complex threaded and event-driven programming models. The *libabac* design must not limit the application designer's choices unnecessarily.

Libabac's solution is to make the interface to the asynchronous info discovery module as simple as possible. In the Unix/Linux implementation, the application makes a synchronous call into the library with hints to the search daemon, and implementations return a communications socket on which the list of returned results will be delivered. The socket itself binds requests to responses. A single-threaded application can simply block waiting for the response, while threaded and event driven programming systems should be able to easily integrate checking a socket into their event loops and thread schedules. As the architecture matures, adapters will be included to link the response sockets into additional commonly used programming environments.

4.2.3 Controlling Sensitive Credentials

Figure 1 also shows the *Credential Access Control* block for controlling access to sensitive credentials. This block encapsulates a data vault that provides a storage facility for controlled credentials. Credentials are added to the store through the same software interface as they are added to a context, but a required attribute is also associated with each. In order to release the credential, the principal must hold the attribute.

Generally, authorization negotiations continue as previously described until the endpoint can make no further progress using its unconstrained credentials. The application can then clone the proving context and try adding each of the controlled credentials to see if the proof progresses—the prover will tell if a credential is useful, as it will be included in the output partial proof. If a controlled credential would be useful, the application will instead add the target attribute and the other principal to the list of proof targets and return that request to the other side. When the sub-goal is met—again, the prover will tell the application this—the controlled credential can be added to the mix.

In order to fully support the use of controlled credentials, the library needs to support the idea of pseudo-credentials. A pseudo-credential is a credential that the application does not have, but that would trigger a sub-proof request if the application did. The absence of a challenge could be taken as the absence of a protected credential.

Pseudo-credentials must be handled like credentials by the prover, except that a proof involving a pseudo-credential is invalid. Applications add pseudo-credentials to the controlled store, and treat them like credentials to trigger the extra proofs that hide the leakage.

4.2.4 Representing Partial Proofs

A partial proof is the representation of the current proof goals and the current state of their satisfaction. A partial proof represents the state of an authorization negotiation in progress, as described in Section "Negotiation". The representation of a partial proof includes:

- The negotiation partners (one of which should be this entity)
- The current attribute graph (represented as credentials)
- The proof "goals"—a set of pairs indicating which attributes are to be proven about which principals
- A list of target attribute/principal pairs that will terminate the proof process

While the attribute graph shows the inferences that have been made, they are only meaningful with respect to a given goal. Any ABAC engine will connect the credentials into the same graph, but unless they agree on the proof goals, they cannot agree that it is complete, or on meaningful ways to advance the proof. There may be more than one proof goal because access control may introduce subordinate proofs, as we describe below.

The list of terminating credentials is required to indicate that some attribute proofs may be discarded if the negotiators are able to prove the real attributes of interest after discarding a subordinate objective. Only the prover cares about this list.

Although not currently part of the system design, it will likely be useful to include control interfaces that tell the prover to try to prove any one of the requested attributes or to prove all of them that it can. The prover may be able to prove a set of credentials as a set rather than serially.

4.3 Integration

The endpoint engine described above implements one side of the negotiation. An authorization decision is generally made as a result of a request for some service, and it is in that context that an endpoint engine must be embedded. Generally, we would like the endpoint engine to minimally constrain the design and operation of the larger application or service that incorporates it. Figure 2 shows two *libabac* endpoint engines working on behalf of a client and server.

The applications, client and server, primarily communicate with one another, passing requests and partial proofs between one another as described in Sect. 4.1.

If the application is using the pre-proving strategy of Section "Legacy Applications and Pre-Proving", the legacy client and server calls out to an application set up as in Fig. 2. That application then passes the success or failure of the negotiation back to the server in a way that the legacy system understands.

4.4 A libabac Implementation

At the time of writing, a publicly available, open-source software package implementing key elements of the *libabac* conceptual system architecture has been developed by the authors and their team [13]. This implementation offers support for multiple operating systems and programming languages, and is in production use by the DETER Cybersecurity Testbed [8, 14], as well as by the GENI Program Office's GENI toolchain. While presently providing only a subset of the full *libabac* architecture, the implementation has proven robust, flexible, and useful across a number of complex distributed authorization scenarios.

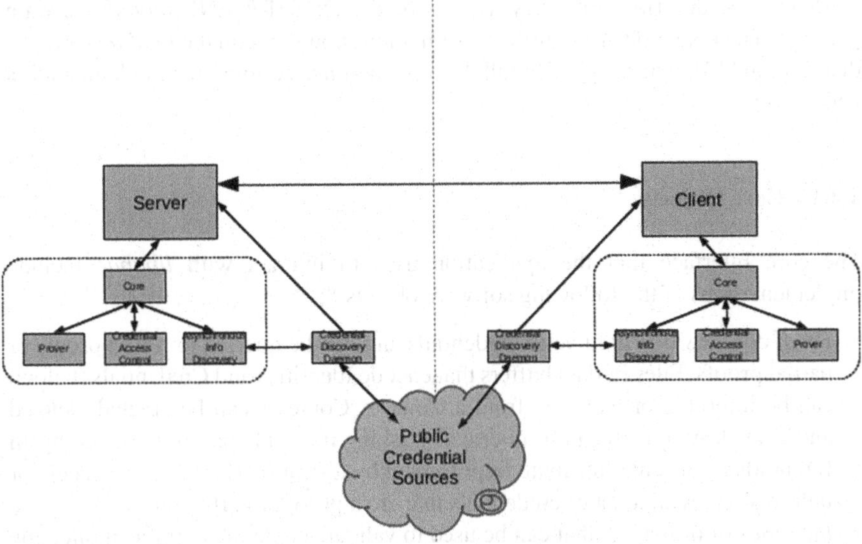

Fig. 2 Communicating applications

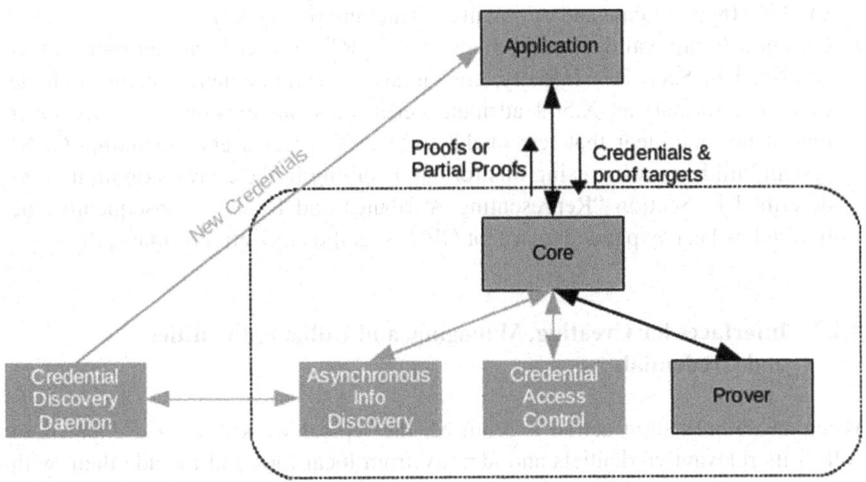

Fig. 3 *libabac* implementation components

Figure 3 shows the subset of the *libabac* conceptual system architecture available in the implementation available at time of writing. This includes an implementation of the core functions module and two independent provers—one for RT0 and one for RT2. The core and RT0 prover are well tested and in wide use within the DETER and GENI systems. The RT2 prover is presently considered experimental, with further development expected in the future.

Below, we describe some key aspects of the DETER/GENI *libabac* software package. The reader desiring further information should obtain the *libabac* software distribution [13], which includes full documentation describing the implementation and its use.

4.4.1 Core Objects

The core interface that the application uses to interact with *libabac* include implementations of the following software objects[10]:

- *Proof contexts* that hold valid credentials and can be queried to return proofs or partial proofs. Files or data buffers that encode Identities and Credentials (below) can be imported or exported from a Context. Contexts can be created, deleted and cloned. A context can be queried by asking if a principal (represented by an ID) holds a particular attribute (represented by a string). The result is success or failure along with a list of credentials that are a proof or partial proof.
- *Identities* of principals that can be used to validate credentials. If the application has appropriate information—a private key—the identity object can be used to create new credentials. These are created from a file or data buffer containing a valid X.509 certificate and optionally a matching private key.
- *Credentials* are valid attested rules (either RT0 or RT2, as appropriate) as described in Sect. 3.3. Initially, credentials were implemented using a single canonical format: as X.509 attribute certificates, an existing standard. Over time it became clear that few of the software's target users—including GENI system builders—were using this format, resulting in the conversion limitations described in Section "Representing Attributes and Rules". Consequently, the library has been expanded to accept GENI specific credential formats [19].

4.4.2 Interfaces for Creating, Managing, and Utilizing Identities and Credentials

When an authorization action is required, the typical workflow for a client is to collect its relevant credentials and identity from local files and include them with a request to a server. A server typically has a context initialized with the service's policies. When a request is passed to it, the server clones the context, imports the request's additional information, and queries the more complete context for the appropriate attribute.

The *libabac* software package provides system interfaces to support this workflow and variants. These interfaces are provided by the "core" module within the software.

[10]These are objects in the software engineering sense, containing and providing both executable methods and data.

Interfaces are also provided to create identities and credentials on the fly. Users can utilize the following interfaces to work with identities and credentials. In addition, *libabac* is packaged with simple pre-built command line and GUI tools that can perform these tasks. While intended primarily as examples, these tools are functional in their own right.

- The *Identity* object described above includes interfaces to create new identities as well. Once a new ID is created, there are interfaces to return the data needed to store the identity and its keys in files.[11]
- An *Attribute* object represents an RT0 or RT2 statement that will be turned into a credential. It features a head, representing the right hand side of the rule (the attribute being assigned) and one or more tails that represent the conditions used to assign the attribute. The head and tail are Role objects. The Attribute object can have Roles attached to it and then be baked into a Credential. Once baked, the credential data can be returned and added to a Context.
- A *Role* object represents a term in an RT0 or RT2 statement. It can be a principal (valid only on the right hand side of a type 1 rule), an attested role (valid as either a left hand side, or on the right hand side in a type 2 rule), or a linked role (valid as a right hand side of a type 3 rule). The rule types are described in Sect. 3.3.

Attributes with multiple Role objects attached to the right hand side are intersection rules.

A user who wishes to create a new identity and a credential based on it takes the following steps:

1. Create a new identity object without loading any data. The core interface will create a new ID and private key. The user may choose to save this data locally.
2. Create a new Attribute object with the ID as a basis, and the Attribute to be assigned as a Role.
3. Create Role objects for the element(s) on the right hand side and attach them to the Attribute.
4. Call the Attribute's bake interface to create a credential.
5. Get credential data from the Attribute object and either add it to a Context and carry out proofs, or save it to local storage for later use.

4.5 *libabac* Adoption

The *libabac* implementation discussed here is presently in use by both the GENI system and the DETER Cybersecurity Testbed. This section outlines the use of ABAC authorization and *libabac* in these systems. In both cases, ABAC concepts and the *libabac* implementation are integral to the construction, operation, and evo-

[11] In some sense this is extraneous code, as any X.509 toolkit can create an identity certificate and key files, but we have found the unified interface to be helpful.

lution of a large, heterogeneous distributed system with increasingly decentralized resource ownership and policy control.

4.5.1 Use in GENI

ABAC and *libabac* were selected by the GENI effort in 2013 to replace GENI's original ad-hoc authorization approach. This decision followed a lengthy evaluation and consideration process by the GENI Architecture Group, in which the group studied the power of the logics and state of the implementation before deciding to integrate *libabac* into the code base.

Currently GENI uses ABAC in a number of places.

- A tool to create GENI speaks-for credentials in the format libabac can process (Section "GENI Privileges with Speaks-for in RT1")
- The GENI Clearinghouse [15] evaluates these credentials when authorizing requests from tools
- The GENI Clearinghouse generates GENI credentials [17] that encode ABAC credentials directly.
- The Clearinghouse uses ABAC directly to authorize clearinghouse operations
- The GPO designers are adding ABAC authorization to Aggregate Managers, the components that directly allocate resources
- The GPO has built standalone tools for generating credentials

These systems use a mix of *libabac* software and tools written to GENI specifications that the authors and their team worked with the GPO and other stakeholders to produce. Additional tools that generate or manipulate GENI credentials, and have their own infrastructure for generating signed XML [16] have adapted that infrastructure to produce *libabac*-compatible credentials.

4.5.2 Use in DETER

The *libabac* developers are members of the DETER Cybersecurity Testbed project and have actively integrated ABAC ideas and *libabac* implementations into this facility. *Libabac* is currently fundamental to two core DETER software systems:

- The DETER Federation System (DFS) is a key element of the DETER facility. The Federation System allows DETER to create experiments that span a wide range of cyber- and cyber-physical testbed environments, each with its own use policies. DFS uses ABAC for all authorization decisions between federants. All coordination of these operations in DETER is supported directly by *libabac*.
- DETER developers are defining and implementing a new unified System Programming Interface for the DETERLab testbed and other testbed clusters that run DETER software. In this system, all testbed policies are encoded in ABAC and decisions are made using libabac.

5 Conclusions and Future Directions

GENI developers have explored the problem of distributed authorization in a national-scale distributed system and made several significant contributions to that area as part of addressing GENI's challenges. These include:

- Identifying ABAC as a viable logic to support distributed authorization.
- Definition of a software architecture to support the full expression of such a system.
- Detailed analysis of GENI's authorization needs and how they are met by both the logic and the architecture.
- A robust implementation of core ABAC functions and RT0 logic that is in current use in both GENI and DETER [8] and that will form the basis for future development in both systems.
- Prototyping an RT2 prover and studying the limitations of that prototype.

Of crucial importance, mainline system managers and operators, rather than only a small group of experts, must be able to generate policies and credentials that meet the needs of their organization and system resources. ABAC, as implemented by *libabac*, provides a firm basis for this activity, by implementing a well-defined logic that supports distributed decision making and auditing. What is needed next is a collection of tools that make this technical capability available to a much wider range of potential users.

Designing such policy and credential generating tools is the authors' key near-term future objective. These tools must be significantly more intuitive to policy designers than are existing tools that manipulate ABAC logic, which poses a challenge to policy designers and user interface designers alike. Though we have been successful in creating prototypes that capture ABAC logics directly and simple grouping and attribute assignment, we recognize that tools will need to intuitively represent more complex constructs to be useful.

Overall, GENI designers succeeded in showing that a sophisticated distributed logic can be practically applied to a national-scale system, laid out the architectural structure for future development and identified steps that will make ABAC more widely applicable.

References

1. Li, N., Mitchell, J.C., Winsborough, W.H.: Design of a role-based trust management system. In: Proceedings of the 2002 IEEE Symposium on Security and Privacy (May 2002)
2. Li, N., Winsborough, W.H., Mitchell, J.C.: Distributed credential chain discovery in trust management (extended abstract). In: Proceedings of the Eighth ACM Conference on Computer and Communications Security (CCS-8), pp. 156–165 (November 2001)
3. Callas, J., Donnerhacke, L., Finney, H., Shaw, D., Thayer, R.: Open PGP Message Format. RFC 4880 (November 2007)

4. Rivest, R., Shamir, A., Adleman, L.: A method for obtaining digital signatures and public-key cryptosystems. Commun. ACM **21**(2), 120–126 (1978)
5. Huang, S.S., Green, T.J., and Loo, B.T.: Datalog and emerging applications: an interactive tutorial. In: Proceedings of the 2011 ACM SIGMOD International Conference on Management of Data (SIGMOD '11), pp. 1213–1216. New York, NY, USA (June 2011)
6. Internet 2, InCommon: InCommon Basics and Participating in InCommon. http://www.incommon.org/docs/guides/InCommon_Resources.pdf. Retrieved Aug 2014
7. TIED Team: GENI-Compatible ABAC Credentials. http://groups.geni.net/geni/wiki/TIEDCredentials. Retrieved Aug 2014
8. ProtoGENI Team: Privileges in the Reference Implementation. http://www.protogeni.net/ProtoGeni/wiki/ReferenceImplementationPrivileges. Retrieved Aug 2014
9. Benzel, T.: The science of cyber-security experimentation: the DETER project. In: Proceedings of the Annual Computer Security Applications Conference (ACSAC) '11, Orlando, FL (December 2011)
10. Cooper, D., Santesson, S., Farrell, S., Boeyen, S., Polk, W.: Internet X.509 Public Key Infrastructure Certificate and Certificate RevocationList (CRL) Profile. RFC 5280 (May 2008)
11. Yee, P.: Updates to the Internet X.509 Public Key Infrastructure Certificate and Certificate Revocation List (CRL) Profile. RFC 6818 (January 2013)
12. Shibboleth Consortium: Shibboleth 3—A New Identity Platform. https://shibboleth.net/consortium/documents.html. Retrieved Aug 2014
13. Kohl, J., Neuman, C.: The Kerberos Network Authentication Service (V5). Internet RFC 1510 (September 1993)
14. TIED Team Libabac Software Distribution. http://abac.deterlab.net. Retrieved Aug 2014
15. The DETER Team: The DETER Federation Architecture. http://fedd.deterlab.net/wiki/FeddAbout. Retrieved Aug 2014
16. TIED Team: GENI ABAC Credentials. http://groups.geni.net/geni/wiki/TIEDABACCredential. Retrieved Aug 2014
17. GENI Program Office: Clearinghouse. http://groups.geni.net/geni/wiki/GeniClearinghouse. Retrieved Aug 2014
18. GENI Program Office: GENI Credentials. http://groups.geni.net/geni/wiki/GeniApiCredentials. Retrieved Aug 2014
19. Bartel, M., Boyer, J., Fox, B., LaMacchia, B., Simon, E.: XML Signature and Processing, 2nd edn. W3C Recommendation. http://www.w3.org/TR/xmldsig-core/ (June 2008)

The GENI Experiment Engine

Andy Bavier and Rick McGeer

Abstract The GENI Experiment Engine (GEE) is a lightweight, easy-to-use Platform-as-a-Service on GENI inspired by PlanetLab. The GEE offers one-click creation of *slicelets* (sets of lightweight containers), single-pane-of-glass orchestration and configuration of slice execution, an integrated intra-slice messaging system, and will soon offer a wide-area file system, and an integrated reverse proxy mechanism. A key design goal of the GEE was simplicity: it should be possible for a new user to get up-and-running with GEE in less than 5 min. The GEE is constructed as an overlay on GENI resources and is available to all GENI users.

1 Introduction and Motivation

GENI is a distributed, highly-flexible Infrastructure-as-a-Service (IaaS) Cloud with deeply-programmable networking. This platform offers great power and flexibility to its users, experimenters, and application developers. However, GENI's general configuration mechanisms provide more power than necessary for some classes of users. One class is users studying how distributed applications can leverage programmable networks; examples of such applications are content distribution networks, distributed hash tables, wide-area stores, network observation platforms, distributed DNS, distributed messaging services, multicast overlays, and wide-area programming environments. Experience with PlanetLab [4, 25] indicates that many such applications can be built using fully-connected networks of lightweight operating system containers, rather than the more expensive virtual machines, dedicated physical servers, and stitched network topologies that GENI allocates. GENI could support more users in this class by making it easy for them to allocate lightweight resources for running their experiments. Another class contains novice users that can be intimidated by the complexity of GENI, for example its authentication and authorization mechanisms that involve multiple keys and

A. Bavier (✉)
Princeton University and PlanetWorks, LLC, Princeton, NJ, USA
e-mail: acb@cs.princeton.edu

R. McGeer
Chief Scientist, US Ignite, 1150, 18th St NW, Suite 900, Washington, DC 20036, USA
e-mail: rick.mcgeer@us-ignite.org

© Springer International Publishing Switzerland 2016

R. McGeer et al. (eds.), *The GENI Book*, DOI 10.1007/978-3-319-33769-2_11

certificates to demonstrate agreement from multiple authorities. GENI could attract more novice users by providing a quick and simple on-ramp for GENI. These considerations point to the need to have, within GENI, a lightweight, easy-to-use infrastructure for potentially long-duration use of inexpensive resources.

Lower-level as-a-service platforms easily support higher-level as-a-service platforms. IaaS platforms support overlay Platforms-as-a-Service (PaaS), and this is exploited both in the academic and commercial sectors. The overlay platforms are always specializations of the underlying infrastructure: one can limit the capabilities and flexibility in an overlay, for ease of use and to encourage the use of specific types of resources; it's difficult to enhance capabilities not present in the underlying platform. GENI was specifically designed to permit the construction of overlay Clouds built within GENI itself; after all, Cloud research is a major driving use case for GENI.

This made our strategy clear: to construct an easy-to-use PaaS Cloud within GENI that offered long-duration slices of containers, distributed throughout the GENI infrastructure. It is straightforward to construct lightweight, easy-to-use, PaaS infrastructures on top of highly-customizable, IaaS infrastructures as overlays; one simply instantiates a slice with appropriate configuration choices within the underlying infrastructure and then hands out nested, lightweight slicelets within the underlying slice. This offers multiple advantages: the underlying infrastructure can provide most of the services required (keeping nodes up, maintaining network connections, authenticating users) while the nested service focuses on providing the specific functionality it was designed to offer.

PlanetLab is a successful example of this type of infrastructure. PlanetLab is a Cloud that offers long-duration slices of distributed containers, with a large user base, a decade of 24/7 operation, and a mature toolchain. However the PlanetLab code base itself is dated. PlanetLab was one of the first Clouds in the world, and came of age before the underlying virtualization and distribution technologies were fully mature. As a result, PlanetLab adapted nascent and developing technologies for its early implementation, such as VServers as a containerization technology. Since that time, technology has evolved; containerization technology has become a standard part of Linux distributions, various specialized container tools have become available (e.g., Docker [15]), and a significant number of Cloud management platforms have emerged. These platforms and technologies subsume many of the functions provided by the PlanetLab platform.

Given our familiarity with the PlanetLab toolchain, our implementation strategy for GEE had three parts. First, bootstrap the system using the classic PlanetLab code base running on GENI, and construct the GEE as an overlay on PlanetLab-on-GENI. Second, refactor the PlanetLab code base by replacing its components with open-source Cloud technologies and GENI services as they became mature. In other words, we designed the platform to *evolve*, while maintaining architectural consistency, from being based on PlanetLab to running a modern software stack. Third, sharpen and focus the platform by extending the range of services offered on the GEE, to make it easy for users to create and deploy slices, and to enrich the set of tools available to running experiments and applications. These new tools, where possible, would be deployed in slices, much as we deployed all services that didn't absolutely have to be part of the PlanetLab distribution in slices on classic PlanetLab. We also hoped to influence the future direction of the PlanetLab platform by demonstrating a similar system running on a modern code base.

The remainder of this chapter is organized as follows. In Sect. 2 we describe the user-level view of the services offered by the GENI Experiment Engine. In Sect. 3 we outline the architecture of the GENI Experiment Engine; Sect. 4 describes the current implementation of the GEE and how it has evolved over time. Section 5 relates experiences deploying an actual application on the GEE. In Sect. 6 we discuss related work and similar infrastructures. In Sect. 7 we share some final thoughts and consider where these infrastructures will go next.

2 A User's View of the GEE

The GENI Experiment Engine (GEE) is designed as a restricted, easy-to-use programming platform on GENI. Each user allocates a *slicelet* of lightweight containers connected to the public Internet as well as interconnected by a private L2 network. The user can choose the disk image that runs in his container from a curated set; advanced users can provide their own image. A rich set of services are available from within the slicelet for building and deploying experiments. Our fundamental mantra is that it should be easier to configure, deploy, and run an experiment than it is to design and write it. In the extreme, this translates into the "five-minute rule": one should be able to compile, deploy, and run a "Hello, World" experiment in 5 min.

One inspiration for the GENI Experiment Engine is the Google App Engine [29]. This is a high-level platform-as-a-service (PaaS) API that offers access to automated scalability through the Google infrastructure. The developer writes a standard program, typically a web page to act as a front end to the application, and a small configuration file. An SDK permits prototyping the application on the developer's personal computer, and then a simple command-line tool can be invoked to upload the program.

The GEE is intended for several purposes:

- To permit high-in-the-stack distributed systems experimenters to use GENI without having to allocate virtual machines, configure virtual networks, write Resource Specifications (RSPECs), configure VMs, etc.
- To provide single-pane-of-glass control of an experiment from the user's desktop.
- To provide a GENI-wide filesystem-like storage infrastructure for GEE Experimenters.
- To provide a messaging infrastructure for GEE experiments.
- To provide shared, application-level HTTP server access to the public Internet.

We derived these features from an analysis of a number of applications and demonstrations of GENI that we had built over the years, notably TransCloud [5] at GEC-10 and TransGeo at GEC-16. After both of these demonstrations, one of the most common questions that we were asked was "Can my experiment use the infrastructure you built for that system?" People wanted access to more than the various GENI aggregates we used; they meant the specific set of application-level services that we built to undergird our demonstrations, and the deployment engines that we used to deploy our application across GENI.

From these questions, the idea for the GEE was born. Specifically, we asked ourselves which chunks of our demonstrations could be re-used by other experimenters, and what tools and services would we have found useful in building and deploying these demonstrations; then we asked what barriers to deployment we encountered and how we could remove those.

The easiest way to get a feel for an architecture like GEE is to consider its usage. To use the GEE, a user logs in to the GEE portal using his GENI credentials. The GEE portal stores no user information or credentials; instead, OpenID [26] is used to call back to the GENI portal, and the user's returned email is used as the userid for the purposes of the GEE. The user is then presented with his dashboard, where he can choose his slicelet's disk image from a list and, with the click of a single button, allocate a GEE slicelet. When this process is completed (typically within 30 s), a download link to a small zip file appears on his dashboard for download. The zip file contains these items:

- An SSH keypair (public and private key) for the slicelet.
- An SSH config file, enabling simple login to the slicelet from the command line, e.g., *ssh -F ssh-config ig-princeton*.
- A Fabric [16] file whose environment imports the SSH config file and lists all the nodes in the slicelet.
- An Ansible [2] config file.
- An Ansible inventory file listing all the nodes in the slicelet.
- Ansible playbooks for starting the GEE Message Service in the slicelet, and for generating a Python file linking symbolic names to containers' private IP addresses.
- A README file.

Only the first item (the private key) is required to access the slice. The rest are convenience items to make it easy for the user to access and configure the containers in his slicelet, and deploy and run his application. Once the user has the private key he is immediately able to ssh into nodes in his slicelet and configure them using standard Linux tools. The user can leverage any SSH-based tool of his choosing to populate or control his slice. The GEE provides starter configurations for Fabric and Ansible; using either of these tools makes upload and execution easy and quick (roughly, as easy as uploading a Python program to the Google App Engine).

Scalable control of large Cloud instances is a requirement, and a number of tools have emerged in recent years to enable that. The fundamental goal is single-pane-of-glass control of multiple running nodes. We support two, in the sense of providing pre-packaged configuration and host scripts: Fabric and Ansible.

Fabric is a slicelet-control tool with imperative semantics. It is a Python wrapper around SSH commands that automates the execution of both remote and local commands. We have pre-loaded the Fabric file with a number of commands to both introduce the user to Fabric and to provide out-of-the-box functionality for his slicelet. For instance, typing: "fab nmap" runs a script on each host that reports the reachable IP addresses on the private network. If additional software is required to run an experiment (beyond that pre-installed on a typical Ubuntu distribution) then the commands to install it can be added to the "fab install" hook. Commands to run the experiment are added to the "fab run" hook.

The other tool that the GEE leverages for slicelet control is Ansible. Ansible has declarative semantics: users describe the desired configuration of the nodes in their slicelet in a markup language, and Ansible executes the necessary commands to build the desired configuration. Ansible is an IT automation and orchestration tool that leverages built-in modules to perform system tasks on remote systems. Ansible

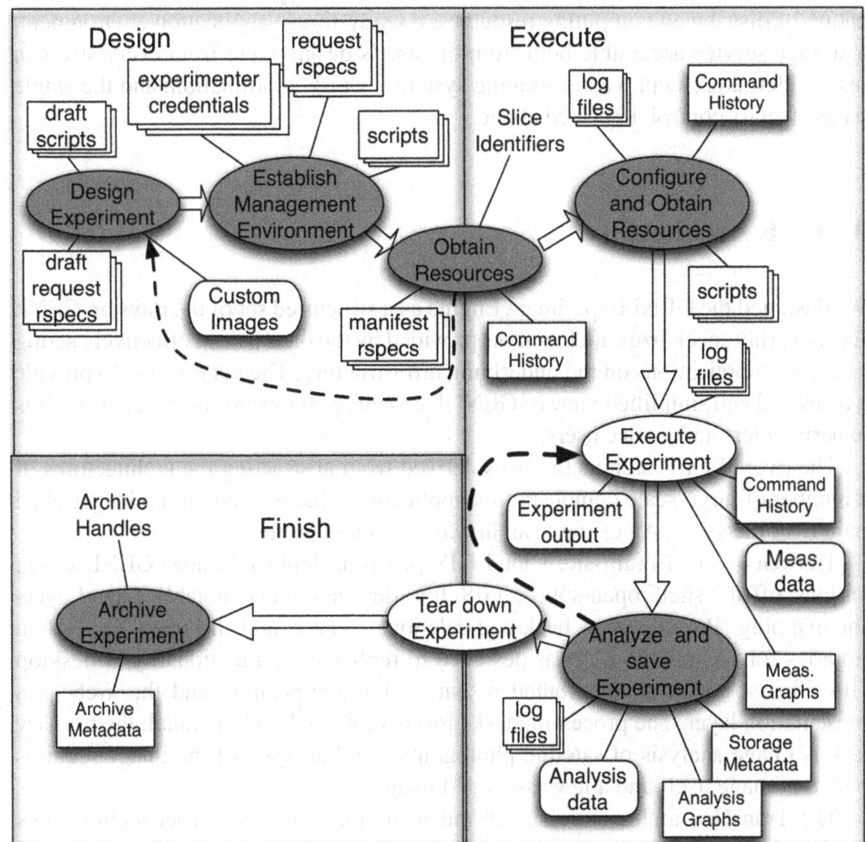

Fig. 1 GENI workflow

modules make it straightforward to install packages, copy files, and run scripts on all the nodes in the GEE. Ansible tasks can be executed from the command-line (e.g., "ansible -i ansible-hosts -m ping") or from within an Ansible playbook.

To see the simplified workflow from a user's perspective, consider the GENI Experiment Workflow shown in Fig. 1. In this figure, we have annotated the workflow items. The dark bubbles represent configuration and deployment activities, and it is easy to see that these steps dominate GENI experiment workflows. The light bubbles are experimental activities. The fundamental goal of the GEE is to automate all but the light bubbles.

GENI is a distributed system, and so a network will be required for most experiments. The processes in the right-hand side box throw off a great deal of data, which must be conveniently stored, so a global storage system is a requirement. Finally, distributed systems require some form of coordination through distributed messaging, so a global communications bus is a requirement. And, of course, ubiquitous execution environments are inherent in any slice.

These user requirements defined the GEE tightly. The platform required fast allocation of distributed computing resources; a ready-to-use application environment; a storage service accessible both from the user's desktop and from every sliver in his slicelet; a pre-deployed messaging system for task coordination; and the single pane-of-glass control described above.

3 GEE Architecture

We designed the GENI Experiment Engine as a structured set of overlays on GENI. Each overlay specializes the functions provided by the ones below, effectively acting as a set of applications on the underlying infrastructure. The net effect is to provide a restricted and simplified view of GENI that is targeted towards distributed systems experimenters and novice users.

The overall GEE architecture was derived from abstracting the architectures of a number of large-scale demonstration applications that we and our colleagues had built over the years. An example architecture is show in Fig. 2.

TransGeo was a multi-site Cloud GIS platform deployed under GENI. Using various off-the-shelf open-source GIS libraries and tools, notably OpenLayers for mapping, PostGIS as a back-end relational server, and the openGIS Python libraries for computation, it was designed to replicate the functionality of desktop GIS clients, but using distributed resources for computation and the Web as a presentation layer. The processing tasks for this job are highly parallel; the primary task is image analysis of satellite photographs, and analysis of the image requires only the image itself and a few associated images.

The TransGeo architecture shared much in common with earlier architectures we had built, and thus it was a good candidate to abstract the interfaces and determine what could be generalized for a PaaS platform for GENI. We examined

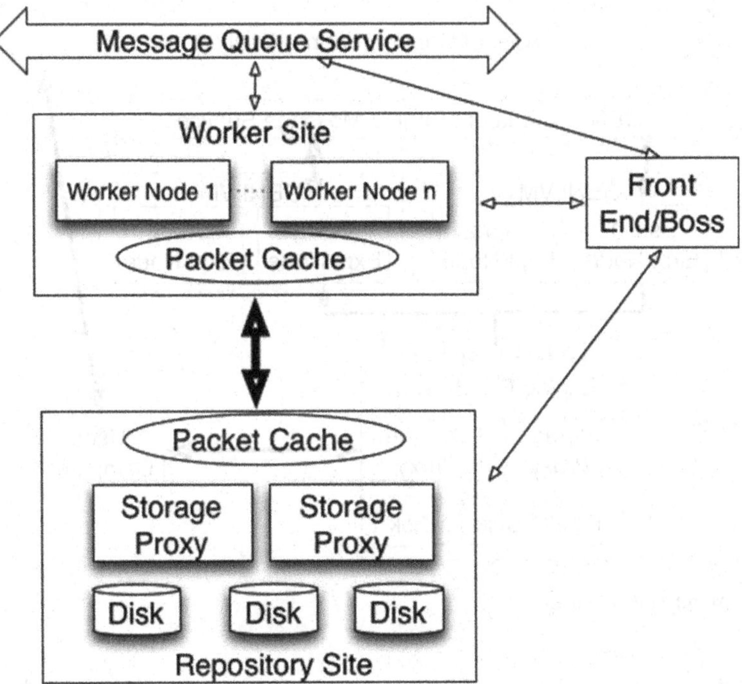

Fig. 2 TransGeo architecture

the TransGeo experiment architecture to determine which elements were common across multiple similar infrastructures, and what was specific to TransGeo. In this architectural diagram, the Message Queue Service and set of Repository Sites represented facilities that TransGeo required, but were not essential elements developed by the TransGeo team; they represented generic distributed facilities which could be used by any similar experiment. The Front End/Boss Node was a head node designed to show results and control the experiment; though it was done by us, it was not an element deployed in the TransGeo slice—it could have been placed on any convenient site on the commodity Internet. The only code specific to TransGeo was the Worker Node code, a few thousand lines of Python programs and several tens of thousands of lines of generic Python third-party libraries. The resulting abstracted picture is shown in Fig. 3.

The GEE is our generalized version of the TransGeo architecture. It is configured as a set of four services: a Compute Service, which allocates and configures GEE slicelets; a Storage Service, which offers a filesystem interface onto a distributed store; a Message Service, which offers a simple mechanism for slivers within a slicelet to send messages to each other; and a Reverse Proxy, which offers outbound HTTP access to slivers within a slicelet. Slice control is done through off-the-shelf orchestration engines, Fabric and Ansible. These services all rely on a persistent, GENI-wide layer-2 network, the GEE Network.

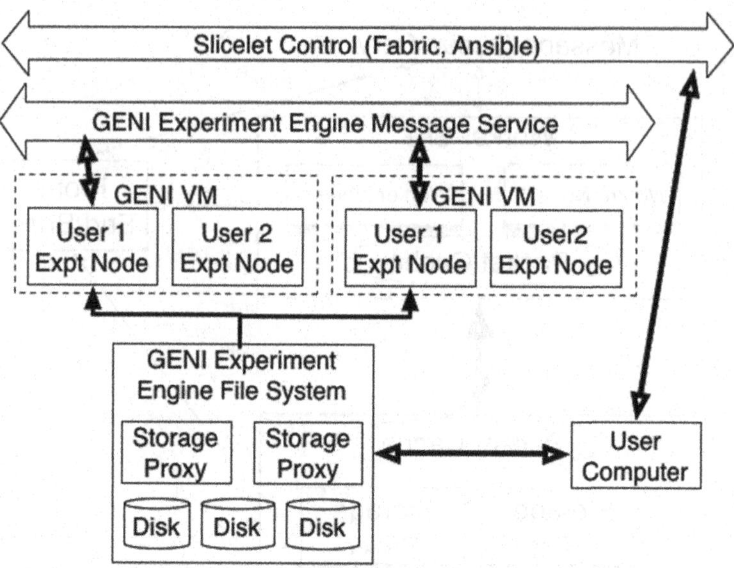

Fig. 3 Architecture of GEE

Of the four services, the GEE Compute Service is offered through the GEE Portal: it hands out slicelets of lightweight containers, with some additional software pre-installed. The GEE Storage Service and the GEE Message Service are offered through loadable Python libraries. The GEE Filesystem service is just Syndicate [22, 23]: once the library is loaded the developer can simply issue standard filesystem calls. The library itself then makes REpresentational State Transfer (REST) calls to a network of storage proxies to store and retrieve data. The Message Service is simply a server which can be loaded into the slicelet, and a client library; a user activates the server on whichever nodes in the slicelet she prefers through a simple fabric command. The Reverse Proxy Service runs in a slicelet, and controls HTTP ports on the routable interfaces of the GEE nodes. A slicelet registers to use the reverse proxy service through the GEE Portal. After that, HTTP requests to that slicelet's sliver are routed by the reverse proxy to the sliver's HTTP server.

In the GENI Experiment Engine we have adopted the PlanetLab philosophy that, wherever possible, services should be provided within slicelets running on the GEE. This is one example of a key architectural idea underlying the GEE: reducing dependency on both the underlying implementation and our own legacy code base. This is a solution to the Innovator's Dilemma [14]: we need to continuously modernize and update our existing implementation while minimizing disruption of user services. For this reason, wherever possible we factor services away from our implementation code base, both pushing services down into the underlying infrastructure and up into slices. Further, we minimize the contact surface between our legacy code base and overlying and underlying services. How this principle was applied will become clear in the next section.

4 GEE Implementation

The implementation of the GENI Experiment Engine evolved over time in accordance with the architecture described in Sect. 3. We bootstrapped the GEE using PlanetLab-on-GENI, an instantiation of the PlanetLab code base running on bare-metal machines obtained from GENI. This early version of the GEE provided the GEE Compute Service using PlanetLab slivers, the GEE Network Service using a GENI virtual topology, and had minimal support for the other services required by our architecture. Over time we reimplemented the Compute and Network Services, and added support for new services, while continually maintaining a facility that could be used by the GENI community. To do this we leveraged GENI slices: the production GEE ran in one GENI slice and we would develop the next version of GEE in another slice. Once the development version was ready, we cut the GEE Portal over to use it and helped existing GEE users to migrate to the new infrastructure. In this way we were able to develop the GEE with only infrequent disruption of user experiments.

The base GEE infrastructure went through three major revisions:

1. Compute Service as an overlay on PlanetLab-on-GENI; Network Service at L2, provided by GENI topology formed from stitched links.
2. Compute Service as an overlay created by the Fabric, Ansible, Docker (FAD) Architecture (more on this below) on GENI VMs; Network Service at L3, provided by a GENI topology formed from EGRE tunnels.
3. Compute Service as an overlay created by FAD on GENI VMs with more resources (e.g. CPU cores, memory, disk space); Network Service at L2 that we built ourselves using Open vSwitch running inside the VMs.

The rest of the section describes the current state of the various services that compose the GEE (as of this writing) as well as discusses the issues that caused us to significantly evolve the implementation from one revision to the next.

4.1 The GEE Portal

The GEE Portal (at http://www.gee-project.org) is the primary means by which users access and manipulate their resources on GEE. The user is redirected to the GENI Portal to log in. After logging in at the GENI Portal he is returned to his dashboard, which initially has a single button: an invitation to “Get a Slicelet”. The user also is presented with a dropdown list of curated images that he can load into the slicelet to be allocated (e.g, various flavors of Ubuntu, CentOS, and Fedora);

Fig. 4 GEE portal user dashboard

power users can supply an arbitrary Docker image as well. If the user clicks on the button (or already has a slicelet), he is directed to a dashboard with three options: free his slicelet, renew his slicelet, or download the file containing the slicelet helper files. A screenshot of the user dashboard is shown in Fig. 4.

Authentication and user access were questions that we considered carefully. One fundamental design goal was to offer the use of GEE to any user with GENI access, without maintaining a separate database of authentication information. Indeed, the goal was to retain no authentication information for a user of any sort. This was chosen for reasons of user convenience, maintainability, and user security. Users, once they have registered with GENI, should not need to add themselves to a separate database. Further, delegating authentication promotes maintainability, and not keeping user authentication information afforded attackers one fewer place to obtain SSH keys and passwords.

To authenticate users on the GEE portal, GEE uses an OpenID callback to the GENI portal, obtaining the minimum information needed to create and maintain user slices—the user's email address, which was the only indexing information used in the GEE portal database. OpenID authentication was present in all revisions of the GEE. In v1 we were able to leverage the PlanetLab database to store the minimal information that GEE maintains for each user. In v2 and later we did not have this option so we used a MongoDB [21] database.

The GEE Portal creates a public/private SSH keypair for each slicelet; the user must download a tarball containing the keypair prior to accessing the slicelet containers. The GEE creates this use-once, or "burner" key for two primary reasons: speed and security. Speed is obvious: interaction with the GEE Portal is streamlined because the user does not need to upload his own public key. Also, v1 of the GEE used the PlanetLab code base deployed on GENI bare-metal machines; using a burner key allowed us to pre-allocate a pool of slices and pre-propagate the keys to avoid the 15-min delay in creating PlanetLab slices. Security is nearly as obvious: if a user's slicelet is compromised, or the use-once key is discovered, all that is compromised is the user's slicelet. The GEE Portal retains no credential from the user of any sort, and therefore cannot be a vector for compromise of any user information or credentials. Similarly, compromise of a user's personal SSH key won't result in an attacker gaining access to a GEE slicelet. Use-once keys are the infrastructure equivalent of hotel room cardkeys; they are allocated when the slicelet is instantiated, used only to access the slicelet, and are destroyed when the slicelet is de-allocated. As a result, they come with many fewer security concerns than do standard keys, just as a hotel is completely unconcerned with travelers departing with room cardkeys in their pockets.

The GEE Portal's interfaces with both the GENI Portal and, in v1, with PlanetLab-on-GENI were deliberately minimized: in the case of the GEE Portal, it is roughly 15 lines of node.js code, primarily configuring a Passport module for OpenID. In the case of PlanetLab-on-GENI, it was six scripts written to the PLC API. These interfaces were kept simple so that the GEE can be extended over other infrastructures easily, and to enable changes to the underlying services to be transparent to GEE users.

Another goal we had for the GEE was to make it easy for anyone to bring up their own GEE infrastructure on any Cloud provider. Our approach was to structure the GEE Portal as two Docker containers: one running the portal code (written primarily in node.js) and the other containing the MongoDB database. Docker streamlines the process of bring up a new GEE Portal to launching two containers. This structure also simplifies management of a production GEE Portal by providing snapshot/rollback capabilities, and making it trivial to deploy new versions of the web server code in production.

4.2 *The GEE Compute Service*

The GEE Compute Service is a simple compute overlay on the GENI infrastructure. For each slicelet, a Docker container is launched inside a VM obtained from the Aggregate Manager at a particular GENI site. Currently the GEE is active at 20 GENI sites. Figure 5 shows the distribution of GEE v3 across the United States.

The "five-minute rule" has dominated our design consideration. The GEE drew inspiration from PlanetLab, and v1 leveraged the PlanetLab code base. However one problem with PlanetLab is that it can take up to 15 min to initialize a PlanetLab slice across the infrastructure. Our requirement for the GEE was that it take less than a minute to initialize the slicelet, install the SSH keys required for access, and pre-install the software required for bootstrapping the experiment. At a minimum, we pre-install: Python; pip; the GEE Filesystem Python library; a package manager (e.g., apt or yum as appropriate); and a use-once public key. A number of other services, such as the GEE Message Service, can be activated with a simple Fabric command. The effect of all of this is that the slice is usable as soon as the use-once slice private key is downloaded; the user won't have to wait for slice configuration or key propagation.

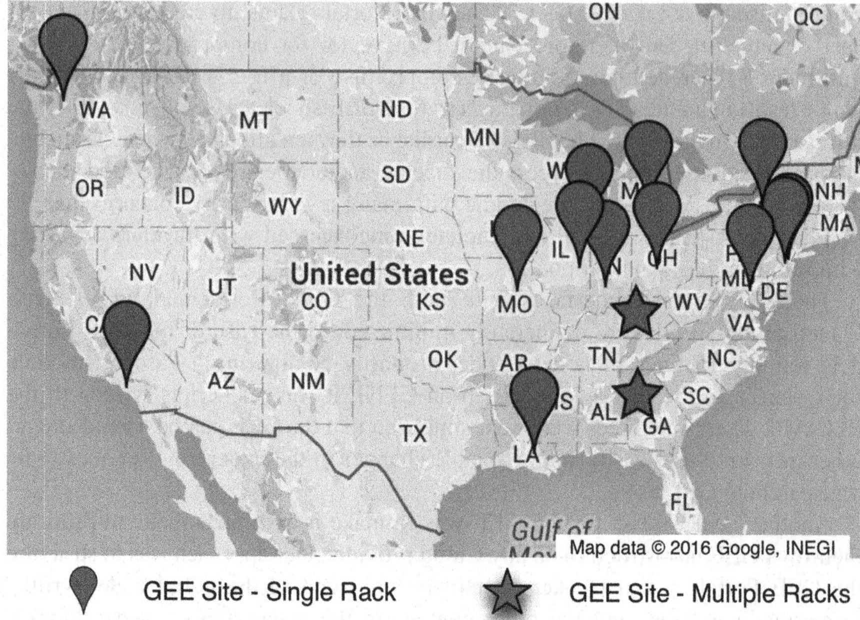

Fig. 5 GEE current deployment

We maintained the PlanetLab convention of using the slice name as the user name on the slivers of the slice (e.g., *slice25*). Universities haven't agreed on a common naming scheme for GENI slivers, and the experimenter shouldn't need to know the servers on which his GEE slicelet is instantiated. Thus, we implemented a DNS service which standardizes GEE node names and abstracts away all but the location information for a GEE slicelet's slivers. One can log in to *slice25* on the GEE node running in Georgia Tech's InstaGENI rack as follows:

```
$ ssh -i ./id_rsa slice25@gatech.gee-project.net
```

This illustrates our overall naming scheme: gee-project.org is used for our control nodes: www.gee-project.org is the address of the GEE Portal. The DNS domain gee-project.net is used for slivers. However, the authors are lazy and even the above is rather too much typing to login to a GEE slicelet. The GEE portal provides a *ssh-config* file along with the public key that provides short nicknames for each node in the slicelet, as well as specifying the user name and private key. Assuming that the helper files downloaded from the GEE Portal are in the current directory, the user can log into the Georgia Tech node as:

```
$ ssh -F ssh-config ig-gatech
```

Version 1 of the GEE was deployed as an overlay on PlanetLab-on-GENI. Likewise, PlanetLab-on-GENI was an overlay service running over ProtoGENI [28]

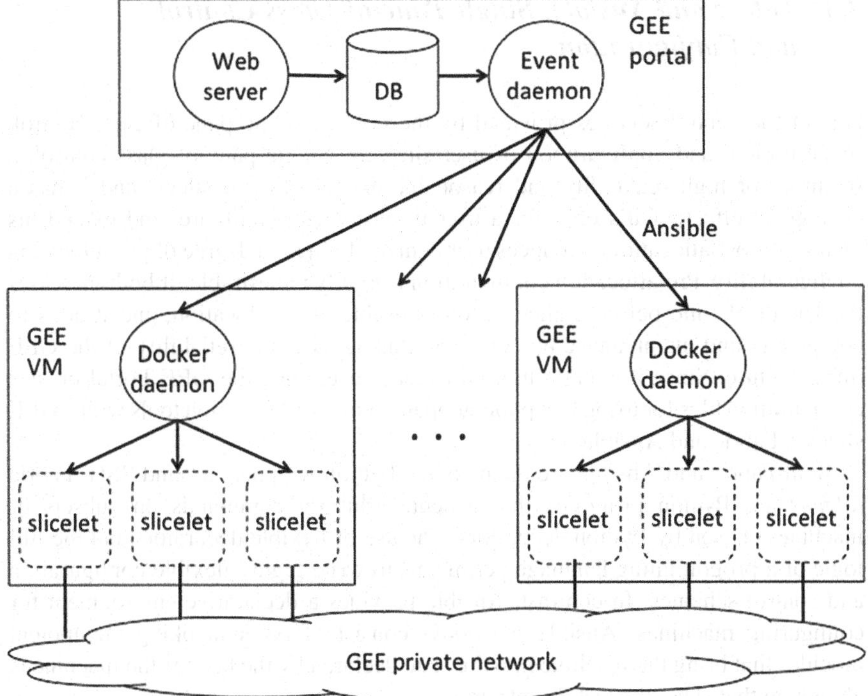

Fig. 6 The GEE's FAD architecture

on the InstaGENI racks, envisioned as a convenient way for GENI users to run experiments based on Linux that did not require modifying the kernel. To meet the "five-minute rule" requirement, GEE v1 maintained a pool of pre-allocated and pre-configured PlanetLab slices that could be handed out instantly to GEE users.

The GEE architecture described in Sect. 3 places minimal requirements on the underlying compute infrastructure, and so we were easily able to replace PlanetLab with a more modern code base. Versions 2 and later used an approach we called the FAD Architecture to create and configure GEE slicelets across the pre-allocated GENI VMs [8]. FAD stands for Fabric, Ansible, and Docker.

Figure 6 shows the current implementation of the GEE Compute Service using the FAD Architecture. The GEE Portal internally runs an Event Daemon that uses Ansible to create and destroy Docker containers for slicelets across GENI, as well as configure the SSH keys that enable the user to login to the slicelet. This process typically takes less than 30 s (the exception is when the user has specified a custom Docker image that is not already cached on the nodes and needs to be downloaded). The user then leverages Fabric to further customize his slicelet prior to running his experiment as explained below.

4.3 Fabric and Ansible: Single Pane-of-Glass Control and Configuration

One of the central services provided by the GEE is single pane-of-class control, orchestration, and configuration of user slicelets. Single pane-of-glass control is required for high scalability and reasonable extensibility of slices, and it has a synergistic effect: if it's easy for a user to customize, configure, and extend his own slice post-allocation, it reduces user demand for a high degree of pre-allocation configurability. Pre-allocation configuration is highly undesirable: it both increases the length of time before a slice becomes usable after allocation, and it adds to complexity and maintenance burden—and thus reduces the reliability of the GEE infrastructure. Since no single tool will please everyone, the GEE Portal creates configuration files for using two popular open-source orchestration tools with a GEE slicelet: Fabric and Ansible.

Both Fabric and Ansible are open-source Python wrappers around SSH. Fabric *fabfiles* are Python programs that execute arbitrary commands on subsets of machines chosen by Python decorators. The use of flexible decorators and the full logic of a programming language permits us to write highly flexible configuration and control schemes. In contrast, Ansible provides a declarative environment for configuring machines. Ansible *playbooks* consist of tasks invoking idempotent modules that bring the machine into the state declared by the task; if the machine is already in that state, the task is not run.

Fabric and Ansible provide the GEE with easy tools to offer additional services to GEE experimenters without either unnecessarily cluttering the image nor requiring extensive customization at or before slice allocation. For example, enabling the user to install and activate the messaging service (see below) merely required adding a little additional Fabric code to the fabfile that the user downloads from the GEE Portal. The GEE included support for Fabric from the beginning; Ansible support was added in v2.

4.4 The GEE File System

The GENI Experiment Engine File System (GEE FS) is designed to be an easy-to-use file system provided on all GEE slices. This file system is accessible both inside and outside experiments to allow users to access stored data from inside and outside GENI experiments. The GEE FS provides a persistent environment for all GENI experiments. It has the following design goals:

- Unix-like semantics
- Convenient, reliable, distributed storage
- Accessible from any GENI experiment
- Runs on any reasonable host backend

- Exposed API
- Web interface for file browsing

A file system consists of a block storage layer, and a metadata service which groups blocks into files, implements naming and directory structures and enforces access control. The GEE FS is built using the Syndicate [22, 23] wide-area file system for the metadata service. Syndicate handles metadata in the file system as well as access control, versioning, and replication, while providing a familiar Unix-like interface. Syndicate allows us to use multiple backend services distributed around the GENI network.

The most integral component of the file system is the Metadata Server (MS), which handles all file system metadata requests. For this we need a reliable service that can handle a lot of concurrent connections, and is easily accessible. The Syndicate MS [22] is implemented as a Google App Engine application and stores its data in BigTable. By using Google App Engine, Syndicate gets efficient app scaling under various loads, as well as efficient key-value lookups in BigTable. Users and Groups are handled by the MS restricting what a given user can access locally through the file system client. Users register an account through the file system client and provide a password for authentication. The password is used to authenticate subsequent file system requests.

Apart from the MS, Syndicate has client processes and storage processes. The storage service is a Python process that runs on remote nodes, and acts as a translator between Syndicate and the storage service being used. Syndicate writes blocks of data to a storage service, and replicates the data for data durability. The GEE FS uses Swift installations running in GEE slicelets as its storage service. Swift is accessed via HTTP and also provides a Python API which allows easy integration with Syndicate. Storage services can be added and removed from Syndicate on the fly as the MS handles the actual layout of data (and its replicas), which gives GEE FS the flexibility to grow its storage capacity to meet the demands of GENI users.

The file system client exists as both a Filesystem in Userspace (FUSE) [17] and a Python module. The FUSE module allows users to mount the file system directly on any Unix-like system. The Python module allows Python processes to bypass FUSE and access the file system directly. The API allows clients to control the physical location of their files.

4.5 The GEE Message Service

The GEE Message Service, available since GEE v2, is used to route job control messages within a slicelet; this is a common feature of many Cloud systems, and as a result a number of open-source message service implementations are available. Our requirements were that the software be extremely simple, configure automatically, have a rich set of client libraries, be enabled on the server side with a simple `service start` command, and be well-documented.

We chose Beanstalk [1]. Beanstalk has libraries in a wide variety of client languages, notably including Python. It installs as a service on Fedora, with a configurable port. It has an extremely simple put/get interface and supports a wide variety of use models, including pub/sub. As with many Message Service systems, Beanstalk is configured for a single-tenant environment. Its use mode is not that a multi-tenant provider offers messaging-as-a-service, but rather that each job or service instantiate its own messaging server accessible only from its own nodes: security is assumed at the network, not the service, level. This dictated our deployment choice: rather than instantiating a GEE- or GENI-wide messaging service, the GEE provides the experimenter with an Ansible playbook to turn the service on in the slicelet if appropriate.

4.6 The GEE Reverse Proxy Service

PlanetLab has hosted many public-facing distributed services. The most notable of these are the Content Distribution Networks (CoDeeN and Coral) [30], End-System Multicast [13], and the Distributed Hash Tables [27]. Clearly, for such services to use the GEE, some method must be found to enable public-facing services at each site.

Most GENI member institutions have been unwilling or unable to devote large banks of routable IP addresses to GENI slices; thus we are not able to give a routable IP address to each GEE sliver in a slicelet. It isn't really feasible to assign each its own port: an http service that isn't on port 80 faces multiple logistical problems, from firewalls to configuration of client-side software. The solution we hit upon was to multiplex the http ports and isolate at the URL level, enabled and enforced by the GEE Reverse Proxy.

The GEE Reverse Proxy Service operates a reverse proxy in a sliver on each GEE site. HTTP requests of the form http://<hostname>/<sliceletname>/<request> are caught by the reverse proxy and sent to the HTTP server in the slicelet's sliver over the GEE private network; the returned value is sent back to the requester.

The initial version of the GEE Reverse Proxy Service was deployed in GEE v2. By default, the GEE Reverse Proxy Service is disabled for a slicelet, to prevent the slicelet's server from dealing with unanticipated requests. The experimenter selects the proxy service for his slicelet from the GEE Portal dashboard; the portal then sends an authenticated request to enable proxy service for this slicelet to the reverse proxy. This is disabled on experimenter request or when the slicelet is destroyed.

4.7 The GEE Network

The GEE Network is a private layer-2 network spanning the infrastructure on which the GENI Experiment Engine is deployed. Each GEE Sliver has a single interface on this network, with a RFC 1918 address. Intra-slicelet communication on the GENI Experiment Engine is primarily through the private network.

Since the network is allocated by the GEE rather than set up by the user as is standard in GENI, the user won't know the IP addresses of his slicelet until he acquires it. To simply link these private IP addresses to symbolic names, GEE makes it easy to create a Python file for each slicelet that defines constants, one per sliver, by symbolic location; for example, `Northwestern = '10.64.136.1'`. The programmer can then import this file into code.

GEE slicelet networks are not completely isolated from one another. One of the use cases for PlanetLab is slices providing services for other slices: e.g., PsEPR [9] provided monitoring information and Stork [10] loaded software packages for other slices efficiently. The PlanetLab mantra for services is "put it in a slice", which led to a micro-kernel architecture for a distributed system: if it didn't absolutely need to be in the PlanetLab controller, it was in a services slice. This greatly simplifies the design of PlanetLab, permits experimentation in utilities and services, and contributes to the lifespan and maintainability of the PlanetLab infrastructure.

The GENI Experiment Engine adopted the same design philosophy. The GEE File System is deployed in a slicelet, as is the GENI Reverse Proxy Service. Our original intent was to offer the messaging service in a slicelet, but the requirement for a secure multi-tenant service restricted our choices and added unnecessary complexity to what was otherwise a simple mechanism: hence our choice to add a service to the slicelet rather than offer a multi-tenant service in its own slicelet. Slicelets can contact the APIs of these services over the GEE Network, so it provides both intra- and inter-slicelet connectivity.

Implementation of the GEE Network has changed significantly over time. GEE v1 used a collection of temporary circuits from Internet2 ION with a spanning tree topology. However occasionally a circuit would go down or disappear, which was not ideal for a long-lived overlay like GEE. For version 2 we moved GEE to VMs and used GENI's GRE tunneled topology to interconnect them into a full mesh. With this scheme the most straightforward way to interconnect a slicelet's containers was at L3, which required assigning unique IP address blocks for Docker's private network at each node, and maintaining IP routing tables with entries for all the other nodes. Also the GENI tools at the time did not support incrementally modifying topologies, e.g., to grow the GEE by adding new nodes without re-installing the old ones.

Version 3 of the GEE includes a L2 topology over EGRE tunnels that we construct ourselves using Open vSwitch (OvS). Since the v3 topology is a full mesh at L2, we faced the problem of L2 broadcast packets being continually forwarded on all virtual links and bringing down the network. We solved this by adding OpenFlow rules to the OvS bridges to suppress rebroadcasting of these packets. Creating the topology ourselves gives us complete control and enables us to dynamically change the topology without issues. Eventually, we may be able to expose SDN functionality to GEE slicelets.

5 Deploying an Application on GEE

Five minutes to set up, deploy, run, and tear down "Hello, World" makes a great demo; but the real question is whether the GEE can be used to get significant applications deployed, and how long it takes to do that. So the major test for us, which drove a number of feature decisions, was deploying and running the Ignite Distributed Collaborative Visualization System.

This system is described in another chapter in this book. Here, we restrict ourselves to the relevant characteristics for deployment on the GEE. The salient characteristic of the Distributed Collaborative Visualization System, and the Pollution Visualizer application, was the distribution of three major pieces of software and data:

1. The Lively Web network application platform.
2. The Pollution visualizer data set, consisting of some 9 GB (uncompressed) spread over 600,000 files.
3. A special-purpose data server written for the application.

To accomplish the first, on slicelet allocation the experimenter requested a purpose-built Docker image containing the Lively web. The Visualization System also requires access to three external TCP ports. Once the GEE administrators had approved the request to deploy the custom Docker image and allocated the ports, the slicelet was deployed across the GEE. The database and data server were automatically started by the Ansible script in Fig. 7. The authors report that the deployment was entirely automated by this script, and took under an hour—almost all of which was spent in copying the tarball with the data.

6 Related Work

The GENI Experiment Engine is a Platform-as-a-Service (PaaS) operated on top of an Infrastructure-as-a-Service (IaaS) base. In this, it is not unique: after all, to a first approximation offering PaaS on IaaS is simply populating component VMs with a set of programming environments and platforms.

Commercial PaaS offerings focus on scalability and automatically scaling applications. For example, the Google App Engine is a heavily-used PaaS offering on Google's infrastructure; OpenShift [24] from RedHat orchestrates application deployment on the public cloud and offers PaaS on the enterprise cloud. In the GENI context, auto-scaling is not a consideration: we do not have arbitrary resources on any single rack to scale the application; for GENI applications, location matters far more than scalability. Our primary concern is communication across the wide area and network design, concerns that are not relevant for data-center oriented PaaS systems.

A key goal of Seattle [11, 12] (a.k.a. Million Node GENI) is to provide a platform for wide deployment of networked applications on end-user systems such as PCs and smartphones; to this end, Seattle leverages a constrained PaaS in the form of a safe, restricted code execution environment based on Python. Seattle has similar motivation to the GEE, but Seattle runs on crowdsourced resources rather than the GENI infrastructure, and offers a more limited runtime environment (Python rather than a Linux container). AptLab [3] is also similar in spirit to GEE: it provides a set of pre-configured "profiles" (essentially, pre-defined slices) on Emulab to get users up and running quickly.

```
- hosts: nodes
  remote_user: slice15
  vars:
    repo: https://github.com/rickmcgeer/PollutionVisualizerDataServer.git
    pvdest: /data/PollutionVisualizerDataServer
    supconf: "{{ pvdest }}/pollutionServer-supervisor.conf"
  tasks:
     - name: unpack tarball
       shell: cd /data; tar -xzf /data/db-strip.tar.gz
     - name: move to db
       shell: mv /data/db-strip /data/db
     - name: set up data server
       git: repo="{{ repo }}"
          dest="{{ pvdest }}"
          accept_hostkey=yes
     - name: install Flask's CORS module
       pip: name='flask-cors'
     - name: install supervisor
       apt: name=supervisor update_cache=yes
     - name: install supervisorctl
       easy_install: name=supervisor
     - name: copy configuration file
       shell: cp {{ supconf }} /etc/supervisor/conf.d
     - name: start supervisor
       service: name=supervisor state=restarted
     - name: start server
       supervisorctl: name='PollutionDataServer' state=started
```

Fig. 7 Ansible script to deploy and run the pollution data server

7 Conclusions

The GEE has been brought up and deployed in stages, as the various services mature. As of October 2015, the GEE Portal is available to all users that can authenticate with the GENI Portal. The GEE Compute and Network Services are mature. We demonstrated the GEE Compute Service and the Fabric-based single-pane-of-glass experiment control at GEC-19 [20]. All code necessary to bring up a new GEE from scratch is open sourced and available on GitHub [18, 19].

The GEE File System is nearly as mature. The integration between the Swift proxies and the Syndicate metadata service is complete, and has been tested on GENI and Emulab. The GEE File System browser exists in prototype form. The GEE Proxy Service and the GEE Message Service are available for use.

GEE is functional and stable because it is built on well-tested and deployed infrastructure services and off-the-shelf components. The initial base for our compute service was PlanetLab, a 24/7 infrastructure that has run continuously for more than a decade; later we refactored the service to leverage Fabric, Ansible, and Docker. For storage, we used Swift, the block store for OpenStack, and Syndicate; for messaging, we used Beanstalk; and for single-pane-of-glass control, we used Fabric and Ansible again. Networking builds on Open vSwitch.

GEE lives light on the land. Our interface to ID providers like the GENI Portal is an OpenID callback. A GEE compute node (for hosting GEE slicelets, i.e., Docker containers) requires only a VM with a public IP address and sufficient resources. Most GEE services either run inside their own GEE slicelets or can be instantiated inside the user's slicelet. Docker images exist for bringing up new GEE Portals. As a result of these design decisions, we should be able to bring up the GEE on any distributed infrastructure with key-based access to the allocated VM's, such as an OpenStack-based facility or a commercial Cloud. This gives the GEE vast growth potential. We envision a future where the GEE becomes the go-to platform for lightweight, multi-tenant edge computing.

Acknowledgements The authors thank Leigh Stoller and Rob Ricci of the University of Utah and Niky Riga and Mark Berman of the GPO for much logistical assistance in setting up and maintaining the GEE Slicelets; Niky Riga, Vic Thomas, and Sarah Edwards of the GPO for counsel and logistical assistance in setting up GEE tutorials; Marshall Brinn of the GPO and Nick Bastin of Barnstomer Softworks for productive conversations; Chip Elliott of the GPO for years of guidance; Bill Wallace, Joe Kochan and the staff at US Ignite; Patrick Scaglia, Dan Ingalls, Alan Kay, Sanjay Rajagopalan, and Carlie Pham of SAP/CDG for financial and logistical support and mentoring; Robert Krahn and Marko Roder of SAP/CDG for assistance with our Lively Web monitoring front end; and Jack Brassil of the InstaGENI project and HP Labs for invaluable help. This chapter is an extension and update of previous workshop [6] and conference [7] papers, and we thank our co-authors: Jim Chen, Yvonne Coady, Joe Mambretti, Jude Nelson, Sean McGeer, Pat O'Connell, Glenn Ricart and Stephen Tredger. Sean McGeer designed the GENI Experiment Engine logo and is responsible for the look and feel of the GEE portal, and the aesthetically-challenged authors thank him for our portal. The students of CS 462/CS 662 at the University of Victoria, Canada, used the GEE for assignments and class projects in the spring of 2015, and gave us invaluable feedback. Of particular note was Matt Hemmings, who used the GEE to deploy the Distributed Visualizer and whose feedback was very helpful. This project was partially funded by the GENI Project Office under subaward from the National Science Foundation.

References

1. About - Beanstalkd. https://kr.github.io/beanstalkd/ (2016)
2. Ansible. http://docs.ansible.com/ (2016)
3. AptLab. http://www.aptlab.net/ (2016)
4. Bavier, A.C., Bowman, M., Chun, B.N., Culler, D.E., Karlin, S., Muir, S., Peterson, L.L., Roscoe, T., Spalink, T., Wawrzoniak, M.: Operating systems support for planetary-scale network services. In: Proceedings of NSDI, pp. 253–266. USENIX (2004)
5. Bavier, A., Coady, Y., Mack, T., Matthews, C., Mambretti, J., McGeer, R., Mueller, P., Snoeren, A., Yuen, M.: GENICloud and TransCloud: towards a standard interface for cloud federates. In: Proceedings of the 2012 Workshop on Cloud Services, Federation, and the 8th Open Cirrus Summit, FederatedClouds '12, pp. 13–18. ACM, New York (2012)
6. Bavier, A., Chen, J., Mambretti, J., McGeer, R., McGeer, S., Nelson, J., O'Connell, P., Ricart, G., Tredger, S., Coady, Y.: The GENI experiment engine. In: 2014 26th International Teletraffic Congress (ITC), pp. 1–6. IEEE, New York (2014)
7. Bavier, A., Chen, J., Mambretti, J., McGeer, R., McGeer, S., Nelson, J., O'Connell, P., Ricart, G., Tredger, S., Coady, Y.: The GENI experiment engine. In: Proceedings of TRIDENTCOM (2015)
8. Bavier, A., Chen, J., Mambretti, J., McGeer, R., McGeer, S., Nelson, J., O'Connell, P., Ricart, G., Tredger, S., Coady, Y.: The GENI experiment engine. In: Proceedings of TRIDENTCOM'15 (2015)
9. Brett, P., Knauerhase, R., Bowman, M., Adams, R., Nataraj, A., Sedayao, J., Spindel, M.: A shared global event propagation system to enable next generation distributed services. In: Proceedings of WORLDS '04 (2004)
10. Cappos, J., Baker, S., Plichta, J., Nyugen, D., Hardies, J., Borgard, M., Johnston, J., Hartman, J.: Stork: package management for distributed VM environments. In: The 21st Large Installation System Administration Conference '07 (2007)
11. Cappos, J., Beschastnikh, I., Krishnamurthy, A., Anderson, T.: Seattle: a platform for educational cloud computing. In: Proceedings of the 40th ACM Technical Symposium on Computer Science Education, SIGCSE '09, pp. 111–115. ACM, New York (2009)
12. Cappos, J., Dadgar, A., Rasley, J., Samuel, J., Beschastnikh, I., Barsan, C., Krishnamurthy, A., Anderson, T.: Retaining sandbox containment despite bugs in privileged memory-safe code. In: Proceedings of the 17th ACM Conference on Computer and Communications Security, CCS '10, pp. 212–223. ACM, New York (2010)
13. Castro, M., Druschel, P., Kermarrec, A.-M., Nandi, A., Rowstron, A., Singh, A.: Splitstream: high-bandwidth multicast in a cooperative environment. In: Proceedings of SOSP '03 (2003)
14. Christensen, C.M.: The Innovator's Dilemma: When New Technologies Cause Great Firms to Fail. Harvard Business School Press, Boston (1997)
15. Docker - Build, Ship, and Run Any App, Anywhere. https://www.docker.com/ (2016)
16. Fabric Api Documentation. http://docs.fabfile.org/en/1.8/ (2016)
17. Fuse. http://fuse.sourceforge.net/ (2016)
18. GEE Node Install Scripts. https://github.com/rickmcgeer/geni-expt-engine (2016)
19. GEE Portal Code. https://github.com/rickmcgeer/geni-expt-engine (2016)
20. McGeer, R.: GEC 19 GEE demo video. https://www.youtube.com/watch?v=RDnWIqtatkA (2016)
21. MongoDB for GIANT Ideas. https://www.mongodb.org/ (2016)
22. Nelson, J., Peterson, L.: Syndicate: democratizing cloud storage and caching through service composition. In: Proceedings of the 4th Annual Symposium on Cloud Computing, SOCC '13, pp. 46:1–46:2. ACM, New York (2013)
23. Nelson, J.C., Peterson, L.L.: Syndicate: virtual cloud storage through provider composition. In: Proceedings of the 2014 ACM International Workshop on Software-defined Ecosystems, BigSystem '14, pp. 1–8. ACM, New York (2014)
24. OpenShift by RedHat. http://www.openshift.com/ (2016)

25. Peterson, L., Bavier, A., Fiuczynski, M.E., Muir, S.: Experiences building PlanetLab. In: Proceedings of the 7th USENIX Symp. on Operating Systems Design and Implementation (OSDI) (2006)
26. Recordon, D., Reed, D.: OpenID 2.0: a platform for user-centric identity management. In: Proceedings of the Second ACM Workshop on Digital Identity Management, DIM '06, pp. 11–16. ACM, New York (2006)
27. Rhea, S.C.: OpenDHT: a public DHT service. PhD thesis, University of California at Berkeley, Berkeley, CA (2005)
28. Ricci, R., Duerig, J., Stoller, L., Wong, G., Chikkulapelly, S., Seok, W.: Designing a federated testbed as a distributed system. In: Proceedings of the 8th International ICST Conference on Testbeds and Research Infrastructures for the Development of Networks and Communities (Tridentcom) (2012)
29. Sanderson, D.: Programming Google App Engine: Build and Run Scalable Web Apps on Google's Infrastructure (Animal Guide). O'Reilly Media, Sebastopol, CA (2009)
30. Wang, L., Park, K.S., Pang, R., Pai, V., Peterson, L.: Reliability and security in the CoDeeN content distribution network. In: Proceedings of the Annual Conference on USENIX Annual Technical Conference, ATEC '04, pp. 14–14. USENIX Association, Berkeley, CA (2004)

Part III
The GENI National Buildout

Once the GENI architectural elements were in place and had been deployed and tested at small scale, it was time for a full GENI deployment. There was a significant tradeoff to consider: a broad national deployment was required, both because a number of the potential experiments and services required broad distribution, and because GENI's own raison d'etre was to be a prototype of the next Internet, and this goal could only be explored with broad geographic distribution. However, one lesson of PlanetLab was that low-provisioned sites could easily be oversubscribed. Given finite resources, there was a "beefy and small" vs "thin and broad" tradeoff—more, less-powerful sites or fewer, more-powerful sites? Moreover, PlanetLab used the routable Internet for inter-site connectivity, which excluded programmable inter-site networking. GENI had to offer a private inter-site network.

There were other, more minor, tradeoffs to consider. PlanetLab had strongly encouraged participating sites to put their PlanetLab nodes outside the University firewall, to ensure that PlanetLab traffic could not conceivably pose a security risk (unlikely in any event) or trigger spurious alarms (almost certain; security appliances are designed to detect unusual traffic, and experimental traffic is always benign but often unusual). GENI nodes obviously faced the same constraints. But in addition, a number of GENI services and experiments (see the next section) anticipated the use of campus resources, which are generally inside the firewall.

The ultimate decision was to plan for a deployment of approximately 50 sites across the United States, with a "GENI Rack" as the basic unit of deployment, interconnected by a programmable layer-2 network provided by Internet-2. The rationale for the deployment choices and the design of the underlying Mesoscale network is described by Heidi Picher Dempsey in chapter "The GENI Mesoscale Network". Each rack would function as a standalone GENI site, but experiments and services were expected to construct slices across a number of sites.

Two rack designs were chosen, representing different points in the tradeoff space and following the GENI principle of preferring multiple implementations of major capabilities. The InstaGENI rack, described by Rob Ricci and Rick McGeer in chapter "The InstaGENI Project", is a minimal-resource expandable, affordable design based on a proven and deployed aggregate manager, ProtoGENI, the direct descendant of the Emulab software stack which had run the Emulab cluster for a decade. As its portmanteau implies, it was designed to be "instant-on" and deployed widely across the United States; in the event, 34 were deployed. Each, however, was relatively resource-poor, at 80 worker cores/site.

The ExoGENI rack, described by Ilya Baldin, Claris Castillo, Jeff Chase, Victor Orlikowski, Yufeng Xin, Chris Heermann, Anirban Mandal, Paul Ruth, and Jonathan Mills in chapter "ExoGENI: A Multi-Domain Infrastructure-as-a-Service Testbed", is a more resource-rich design managed by a newer aggregate manager, ORCA, which was described in the previous section. Since ORCA was newer than ProtoGENI, it took advantage of modern underlying cluster-management technologies, notably OpenStack, and its ten sites were primarily designed for the deployment of Virtual Machines and containers as the execution environments.

It should be noted that these prototype rack designs are simply that: working prototypes. A GENI rack is simply a small cluster with an Aggregate Manager compatible with the GENI AM API, so that an experimenter or developer can construct slices or attach local resources to slices from that rack. In addition to the rack designs described in chapters "ExoGENI: A Multi-Domain Infrastructure-as-a-Service Testbed" and "The InstaGENI Project", rack designs based on native OpenStack have been built. ExoGENI is a largely IBM-based rack; InstaGENI, HP-based. In addition, Cisco UCS and Dell-based racks have been built, and there's no reason in principle why one couldn't build a rack from heterogenous equipment. Just as the Internet protocol didn't prescribe switches from a specific vendor, neither does the GENI AM API.

The GENI Mesoscale Network

Heidi Picher Dempsey

Abstract GENI is a national network of computation, storage, and networking resources interconnected by a deeply programmable nationwide infrastructure. The GENI mesoscale infrastructure was not built from scratch in a green-field design, but was a truly cooperative design, integration and operations effort. The challenge confronting the design and development team was to combine existing capabilities to virtualize individual resources across resource types to create an environment that supports smoothly interoperating "slices" of the shared GENI infrastructure.

1 Introduction

GENI is a national network of computation, storage, and networking resources interconnected by a deeply programmable nationwide infrastructure. The GENI mesoscale infrastructure was not built from scratch in a green-field design, but was a truly cooperative design, integration and operations effort. The National Science Foundation inaugurated this effort by contributing support for the project to select and integrate commercial hardware with open-source software that allowed compute, storage, and network resources to be "sliced" such that multiple experimenters could use the shared resources concurrently for independent experiments, applications, and education.

At the time the GENI project began, it was already possible to slice standalone resources individually: a single server could be divided into multiple Virtual Machines (VMs) with software such as VMware or Xen, a network could be divided into multiple concurrent active connections on a single node interface with VLANs or MPLS, and there were many proprietary and open source options for sharing storage. The challenge of building the mesoscale infrastructure was combining these independent capabilities in a research infrastructure that could be used 24×7 365 days a year. Three major factors drove the requirements, design, and integration efforts for the mesoscale: (1) resources were contributed by many different organizations, each of whom chose the vendor and technology that was most efficient for

H.P. Dempsey (✉)
GENI Project Office, Raytheon BBN Technologies, 10 Moulton St. Cambridge, MA 02138, USA
e-mail: hdempsey@bbn.com

© Springer International Publishing Switzerland 2016
R. McGeer et al. (eds.), *The GENI Book*, DOI 10.1007/978-3-319-33769-2_12

themselves, and managed their infrastructure independently, so interoperability was essential; (2) experiments and applications required a compatible way to specify and slice all three types of resources, while ensuring that they all were all connected only to each other on the dataplane, so developing standard interfaces and resource descriptions for widely varied technologies was required; and (3) most GENI users did not own or control any resources, and could not be expected to know the native platform commands for provisioning them, so GENI tools needed to provide an abstraction for requesting and provisioning resources into slices regardless of who was providing the resource or what technology they were using to deliver it.

2 Early Design Activities

Early in GENI development, the community explored software standards for specifying resources and negotiating access on the GENI control plane to meet these challenges. Five original GENI integration clusters explored this space,[1] and eventually settled on a single interoperable GENI Resource Specification (RSPEC) and GENI Application Program Interfaces (APIs) for resource aggregates, the clearinghouse, and portal. GENI used credentials from InCommon or other existing Identity Providers (IdPs), so that GENI experimenters could use resources from any of the groups that implemented the GENI APIs (PlanetLab, ProtoGENI, ExoGENI, InstaGENI and OpenFlow), combining resources from any of those systems. (Each cluster also supported additional features that were not interoperable for experimenters who only operated inside one cluster, but that was not a factor in building the mesoscale, so is not described here. See individual project chapters for more information on their unique capabilities.) Participants in the mesoscale agreed on a common rough policy for usage and security that was reflected in the GENI Aggregate Provider Agreement,[2] and agreed to a cooperative operations approach involving several operations and development groups around the country.

Mesoscale network design, integration and testing proceeded in year-long "spirals," where the community explored the best way to create, maintain, and operate the resources that would eventually be accessible using the GENI APIs and Resource Specifications (RSPECs). The mesoscale network provided separate control plane and data plane connections to each GENI resource, and eventually also to the GENI clearinghouse and portal, which provided an easy-to-use graphical experimenter interface. During the first GENI Spiral design activities, participants agreed that the control plane should rely on resource owner's existing IP connectivity to reach other resource owners and experimenters. Resource owners could leverage the effort they'd already invested in making their existing production IP

[1]Emulab (ProtoGENI/InstaGENI), PlanetLab (SPP), ORCA (ExoGENI), TIED (DETER), and ORBIT (WiMAX and LTE) were the original clusters. OpenFlow was a separate early project, but not a cluster because it did not include a complete control framework for experimenters.

[2]See http://groups.geni.net/geni/attachment/wiki/ComprehensiveSecurityPgm/Aggregate%20Provider%20Agreement%20v04.pdf.

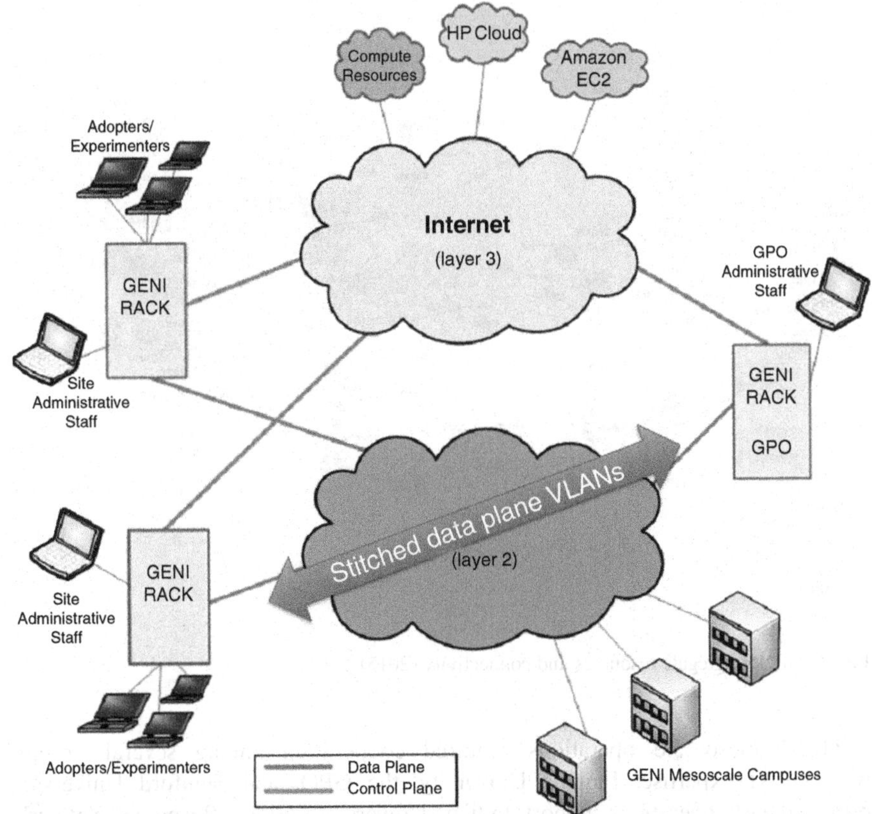

Fig. 1 GENI mesoscale functional design. GENI mesoscale racks supported various mixes of OpenFlow adopters and administrators at each campus (*left side* example racks), along with standard interfaces for remote GENI experimenters and administrators (GPO example). Racks separated control plane and data plane interfaces. Dynamic VLAN stitching was supported in later stages of the mesoscale deployment (2013)

networks resilient, and the additional control plane traffic load was negligible. Participants also agreed to base the GENI dataplane on Layer 2 Ethernet connections between GENI participants. Doing so gave experimenters the freedom to use protocols other than IP, or to design their own protocols to connect experimental compute and storage slices. Participating networks used MPLS tags or VLAN IDs to slice connections between nodes. VLANs were most widely deployed, and they became the dataplane *lingua franca* for requesting and allocating network resources to experimenters. (Some GENI experimenters also took advantage of IP connectivity on the control plane to build IP tunnels between their resources, which could support limited-bandwidth alternate connections for some GENI resources.) Figure 1 shows the basic functional design concept of the mesoscale network.

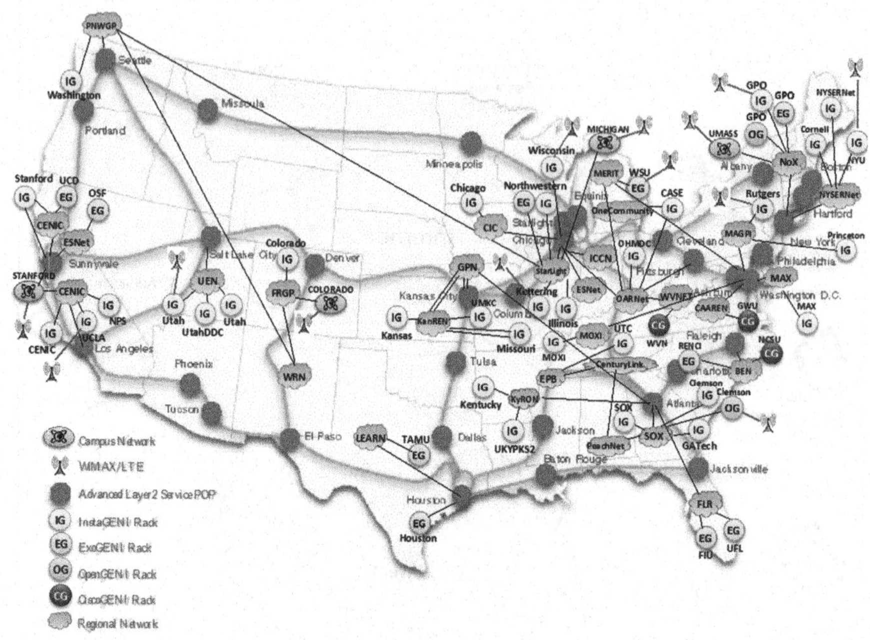

Fig. 2 GENI aggregate resources and connectivity (2015)

GENI mesoscale operations required cooperation among several groups with GENI expertise. Indiana University, the GPO, and Stanford University provided early operations support. Indiana University's Global Resources Network Operations Center, which provided network operations for many other R&E networks and campuses, provided 24 × 7 ticketing and escalation service, and experienced inputs to GENI operations designs. "GENI Meta Operations" and the resultant GENI Meta Operations Center (GMOC) grew out of Indiana's early participation. Security engineers from NCSA, SPARTA, and ISI contributed expertise to security design and security reviews for the mesoscale, as well support for incident escalation and analysis. Eventually operations groups from RENCI, the University of Utah and the University of Kentucky also became part of the shared operations activities.

During each of seven spirals, network engineers and developers who owned and developed resources collaborated with many national, regional and campus IT and network engineers and the GENI Project Office (GPO) engineers to deploy, interconnect, integrate and test connections between the GENI aggregates. These efforts began with five resource owners (eventually called aggregates) and grew to include over 90 in 2015 (see Fig. 2). These engineers also collaborated in operating and supporting the infrastructure through all spirals, developing shared tools, documentation and procedures for making sure that the mesoscale functioned continuously, kept pace with new software and systems as they became available to GENI, and responded to events or emergencies as needed.

3 Nationwide Layer-2 Dataplane Network in GENI

GENI mesoscale prototyping in the field began in Spiral 2 with initial build-outs through more than a dozen US campuses, ten regional networks and two national research backbones: Internet2 (I2) and National LambdaRail (NLR). The initial GENI mesoscale build installed GENI-enabled equipment in 14 campuses, linking them through their adjacent regional networks and the Internet2 and NLR core networks. (Internet2 and NLR were coexisting but separate nationwide Research and Engineering (R&E) networks at the time, and most campuses were members of only one core network.) The initial build-out focused on connecting OpenFlow switches in several participating campuses and in the network core with Layer 2 VLANs from several participating network providers. In parallel, the mesoscale supported a campus WiMAX buildout (described in the 4G Cellular Systems in GENI chapter) that also used the Layer 2 mesoscale VLANs to connect outdoor wireless networks to other GENI resources and to WiMAX sites at other campus.

From the beginning, GENI included participants from the academic and commercial worlds, and this combined approach was critical to GENI's success at building a shared network that was easy to expand and maintain, but still usable for innovative network research. Early mesoscale deployments in the network core included the ProtoGENI backbone nodes and Supercharged PlanetLab Processor (SPP) nodes, both of which were installed at five Internet2 Points of Presence (PoPs). Both nodes included processing, storage, and switching resources in their aggregate, all connected to the shared Dense Wavelength Division Multiplexing (DWDM) wave that Internet2 allocated to GENI. Figures 3 and 4 show the first sketch of the core mesoscale infrastructure, worked out by participants in an early GENI working group meeting. The ProtoGENI switch at each site had one 10 Gbps Ethernet interface per out-degree of the PoP connected to Internet2's DWDM equipment (see Fig. 5). In addition to connecting its own processing and storage resources to the network, the ProtoGENI backbone switch also connected multiple 1 Gbps Ethernet ports to the SPP (see Fig. 6). ProtoGENI separated control and experimenter traffic between the between nodes on the dataplane by allocating VLANs from a pre-defined reserved Internet2 range to individual flows or aggregated groups of flows (depending on traffic source and destination). Unlike the SPP's NetFPGA custom programmable switches, PG nodes used HP commercial switches capable of supporting OpenFlow, eventually allowing programmable OpenFlow experimentation between ProtoGENI and Emulab nodes.[3]

[3]Although ProtoGENI hardware switches supported OpenFlow, the nodes were originally deployed without OpenFlow configured.

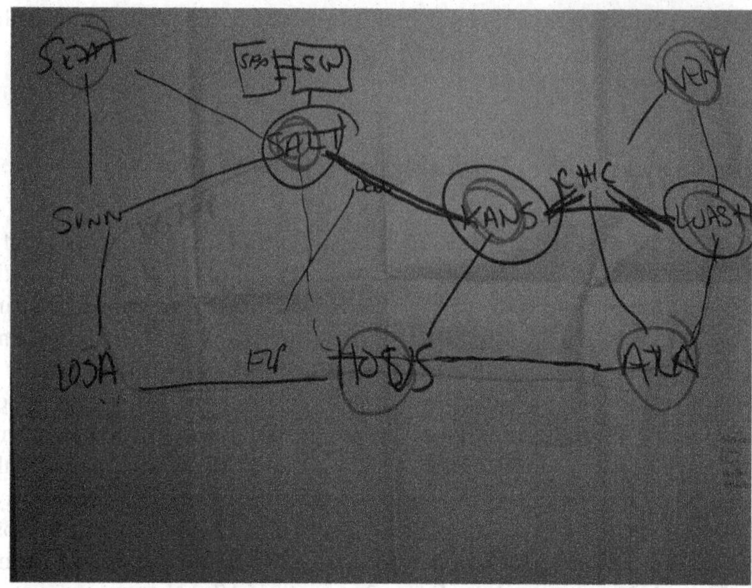

Fig. 3 First sketch of GENI Internet2 mesoscale connectivity in the Internet2 backbone, with example SPP and ProtoGENI connections (2008)

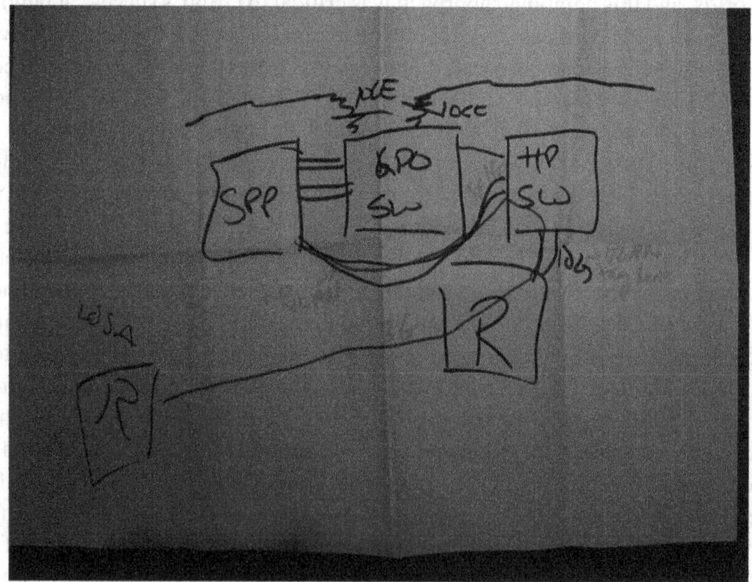

Fig. 4 Detail sketch of the Salt Lake City GENI node, showing various connect types for SPP node, ProtoGENI switches (GPO and HP) and Internet2 routers at Salt Lake City and Los Angeles

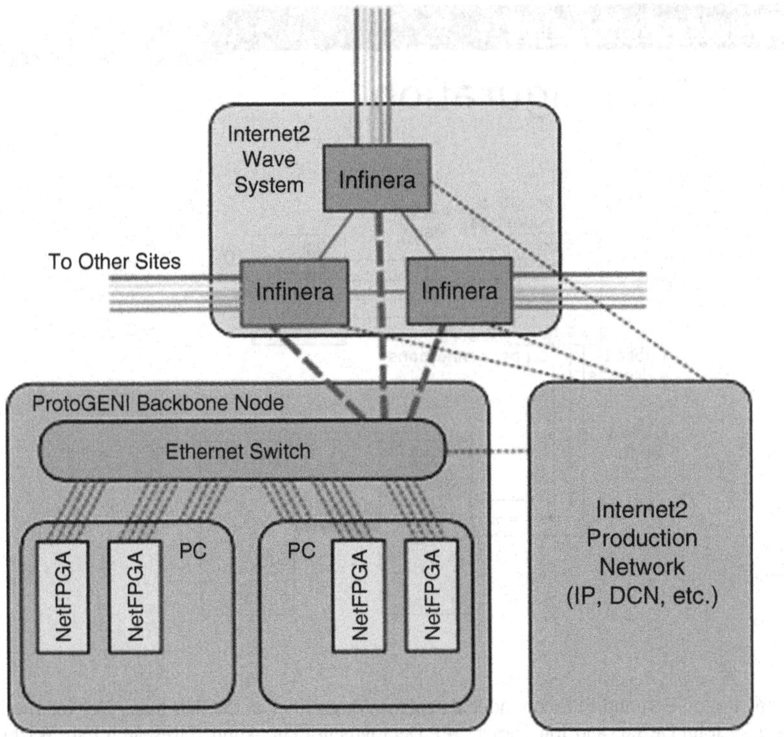

Fig. 5 ProtoGENI Backbone Node Design. Courtesy Rob Ricci, ProtoGENI wiki Backbone Node Design page (http://www.protogeni.net/ProtoGeni/wiki/BackboneNode). University of Utah Flux Research Group (2008)

The original campuses participating in the mesoscale buildout were Clemson, Columbia, UCLA, University of Colorado at Boulder, Georgia Institute of Technology, Indiana University, University of Kentucky, University of Massachusetts Amherst, NYU Polytechnic, Princeton, Rutgers, Stanford University, University of Washington, University of Wisconsin-Madison, and BBN Technologies (GPO). Stanford University collaborated with all GENI campuses involved in the earliest deployments, providing valuable lead expertise along with prototype software for slicing network resources (FlowVisor) and controlling OpenFlow access (Expedient and Opt-in Manager) based on their original Stanford deployment work [1]. OpenFlow switches from many vendors were evaluated and used during mesoscale deployments (listed in roughly chronological order of their involvement): HP, NEC, Indigo, Quanta, Juniper, Brocade, Arista (evaluated, but not deployed), Pica8, IBM, Cisco, and Dell. The Open vSwitch (OVS) software-only switch was also used extensively by GENI experimenters on various mesoscale hardware platforms.

After the initial OpenFlow deployments demonstrated that Layer 2 dataplane and GENI standard IP control plane design operated successfully to support researchers, the mesoscale expanded to include GENI racks in 2012. (For details on the

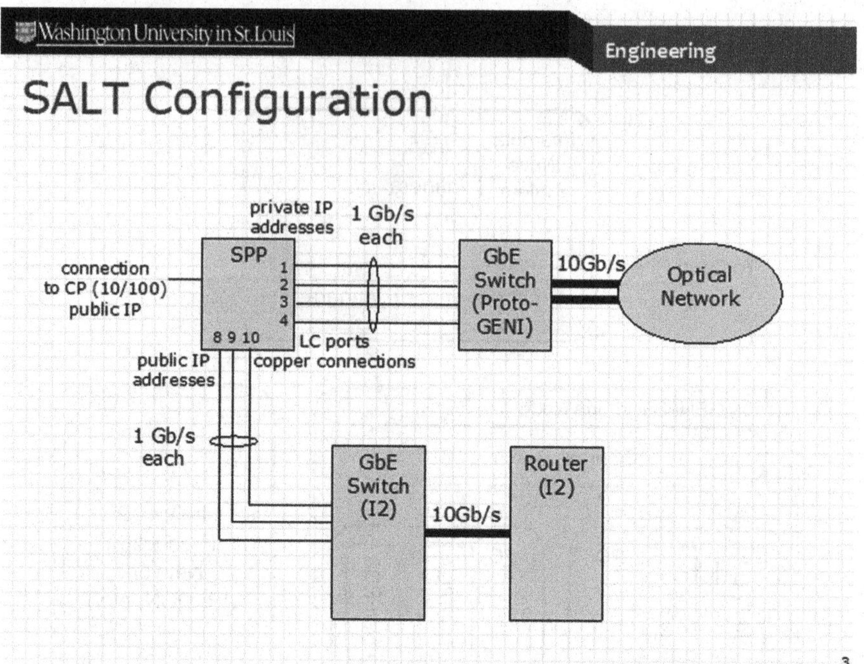

Fig. 6 ProtoGENI and SPP core node dataplane connections at the Salt Lake City Internet2 PoP. Courtesy John DeHart and Jon Turner, SPP Deployment Plan. Applied Research Lab, Washington University (2008)

design of GENI racks, see the appropriate GENI Rack chapters.) Each GENI rack included an OpenFlow-capable dataplane switch and a separate IP control plane switch. The selection of processing, data, and network resources available to experimenters in each rack varied, depending on the rack design. InstaGENI racks could support either 1 Gbps or 10 Gbps Layer 2 Ethernet interfaces. Because these racks were meant to be affordable, most were built with 1Gbps interfaces. ExoGENI racks could support 10Gbps or 40Gbps Layer 2 Ethernet interfaces (most were built with 10Gbps interfaces). Later racks built by Cisco and Dell also offered 1 Gbps or 10 Gbps interfaces. A 100 Gbps Ciena rack was under construction and partially integrated into the mesoscale at this writing. To date, 61 racks have been deployed in GENI (13 ExoGENI, 42 InstaGENI, 2 OpenGENI (Dell), 3 Cisco, 1 Ciena). (More rack documentation is available at http://groups.geni.net/geni/wiki/GENIRacksHome.)

The GPO also built three quick-turnaround "starter" racks in Cleveland, Chattanooga and Cambridge to demonstrate that virtualized experimenter resources based on a Layer 2 Ethernet VLAN dataplane, an IP control plane, and GENI resource allocation and control software would integrate successfully with metropolitan broadband networks. These racks, deployed as part of early USIgnite efforts, were based on the Eucalyptus software for creating and managing virtual machines provisioned on PowerEdge R510 Dell servers running Ubuntu 10.04 OS. The racks provided a Eucalyptus head node and two worker nodes for provisioning experimenter virtual machines along with a bare-metal application host, and a monitoring host. A Cisco 2901 router for IP control plane and commodity Internet access, and an HP ProCurve 6600 OpenFlow switch provided 1 Gbps or 10 Gbps Layer 2 dataplane connections to an upstream research network participating in GENI and the downstream local broadband infrastructure. Starter racks also included an IOGEAR KVM switch to support remote access from the GENI operations center. Starter racks were eventually decommissioned or replaced with standard GENI racks.

3.1 Internet2

The Internet2 core network infrastructure in GENI developed in three different phases. Initially Internet2 dedicated a 10 Gbps wave on their nationwide fiber infrastructure to the GENI project via a memorandum of understanding that was in place from 2008 to 2010. Internet2 configured access to the wave using an Infinera DTN packet-optical transport platform. Projects connecting to the optical infrastructure had to provide a dedicated switch on one of the DTN digital ports, or share access with an existing Internet2 digital switch or router in the production network. Some Internet2 PoPs did not have any spare digital ports, and so were not available to the GENI mesoscale.

Internet 2 migrated their core infrastructure for the mesoscale (along with some other Internet2 services) from the Infinera platform to the Ciena Core Director platform, beginning in 2010. The Ciena hardware allowed Internet2 to offer a Dynamic Circuit Network (DCN) service, with a member-programmable API and GUI called ION (Interoperable On-demand Network). Members could connect to Juniper MX-960 routers at several ION access points located in Internet2 router PoPs throughout the country (see Fig. 7). All connections from a regional network into Internet2 shared the same physical ports on the MX-960s, which meant that traffic from multiple campuses and multiple projects including GENI shared available ION bandwidth. MPLS circuits provisioned with QoS guarantees between the MX-960s routers limited the total available ION bandwidth and provided some priority service for ION traffic on those circuits. Any ION traffic queued in the router was served before any L3 traffic, as long as the offered traffic was within the parameters of the member's ION request. Internet2 supported ION connection requests of up to 2 Gbps by default, and up to 4 Gbps at one connector or 10 Gbps in the core by special arrangement. ION connections were originally expected to be in place for no more than a few hours, but Internet2 supported longer-duration connections when it became clear that most GENI experiments ran longer than a week. If an experiment offered more traffic that what was specified in its associated ION request, Internet2 could shape, drop, or deliver the excess traffic at a lower priority than other traffic, depending on what competing traffic was present

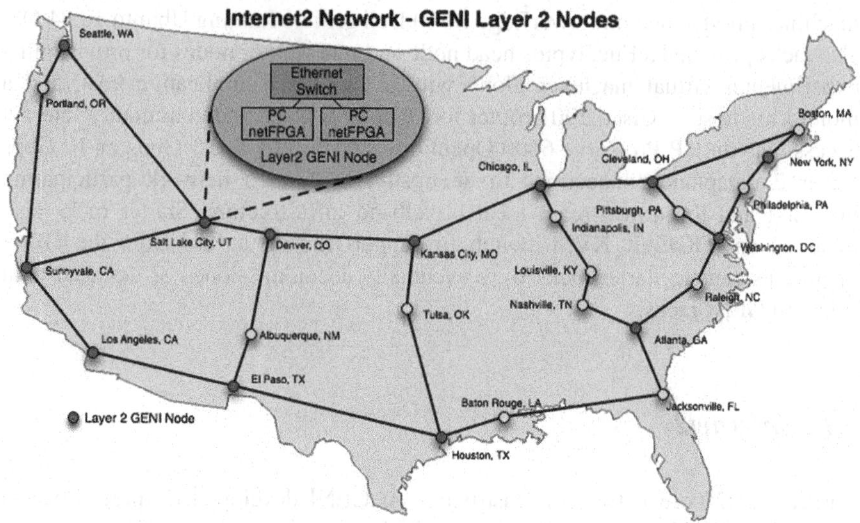

Fig. 7 Internet2 Initial Mesoscale Deployment. Courtesy Rick Summerhill, Internet2 (2008)

simultaneously on Internet2 routers supporting the experiment. This design also allowed Internet2's production IP traffic to burst into unused ION circuit capacity for efficiency, although this was a rare occurrence in practice. Figure 8 shows how the initial OpenFlow switch deployment in Internet2 combined static VLANs configured between Internet 2 PoPs with ION VLAN access to regional network members to create the Internet2 OpenFlow mesoscale core network. Note that Fig. 8 also shows statically configured VLANs that interconnected the Internet2 and NLR OpenFlow core networks in Atlanta.

Internet2 deployed NEC IP8800 OpenFlow switches with 1GigE interfaces at PoP locations in New York, Salt Lake City, Washington, DC, Atlanta, and Chicago. The Houston deployment shown in Fig. 8 was replaced with Salt Lake City due to power issues in Houston. Internet2 also deployed an OpenFlow Aggregate Manager (FOAM) in 2010. After some experimentation with FlowVisor (the software that virtualized experimenter flows on shared VLANs), Internet2 determined that it would not scale to support a nationwide production network, and began developing alternative SDN slicer called FlowSpace Firewall (FSFW). The FSFW software was eventually used in Internet2's production OpenFlow network to segregate large numbers of flows by VLANID in a network with over 35 switches from two vendors (Brocade and Juniper). Internet2 eventually released FSFW as open-source software for use on GENI and other projects, and development on FlowVisor ceased.

Originally GENI engineers used the GUI ION interface to set up persistent VLANs between GENI campuses over Internet2 instead of manually engineering static VLANs with Internet2, campuses and regional networks. I2 could provide VLAN translation in ION if needed, or match the ION VLAN ID to one statically configured in the member regional network or campus aggregate. Because

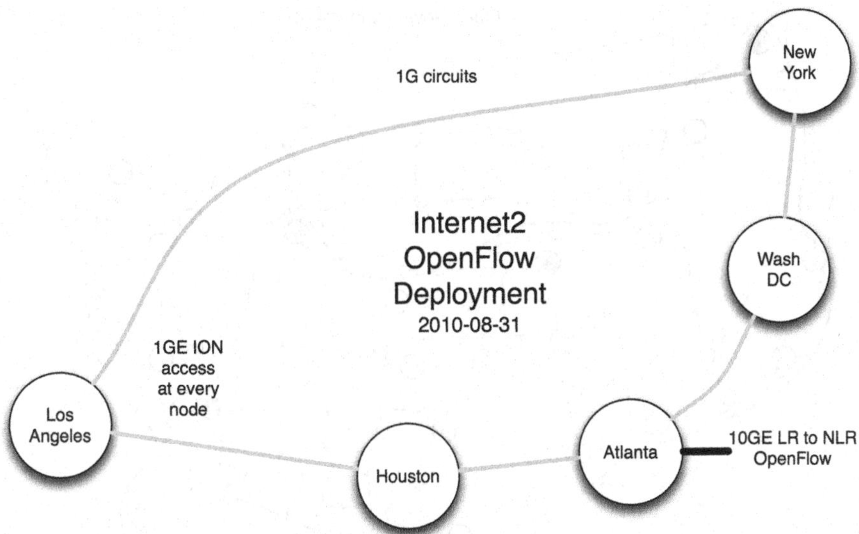

Fig. 8 Internet2 Initial OpenFlow Core Deployment. Courtesy Matt Zekauskas, Internet2 (2008)

GENI aggregates did not have a programmable interface to ION, it was still not possible for experimenters to set up their own dynamic end-to-end experimental VLANs through Internet2. When the GENI stitching project produced prototype stitching Aggregate Manager and Stitching Computation Service software, along with stitching extensions to the GENI API and RSPECs, Internet2 began running a production Stitching Aggregate Manager and SCS to support dynamic stitching for experimenters. Stitching also used the Inter-Domain Controller (IDC) and OSCARs software that Internet2 was supporting for international dynamic connections for the GLIF project.

I2 decommissioned ION in early 2015 and replaced it first with the NDDI and then with the AL2S network and OESS service, which was a production deployment of OpenFlow switches supporting all of Internet2's Layer 3 and Layer 2 services, including GENI. By 2015, GENI was also moving away from the several well-known original shared mesoscale VLANs used for OpenFlow to a more production service design based on GENI stitching, that allocated a VLAN per network sliver in each aggregate. Rather than share a single VLAN for multiple experiments with FlowVisor slicing for flows, the AL2S core network allocated a VLAN per network sliver, and relied on FlowSpace Firewall to provide traffic separation and adequate bandwidth for each of the various VLANS supported in AL2S. The OESS GUI and API software used OpenFlow to manipulate the Brocade switch flow tables in the I2 core network to create the dynamic end-to-end VLAN connections, along with out-of-band manipulation, when necessary, to ensure VLAN isolation. (Note that it is also still possible to configure static VLANs or multipoint VLANs on Internet2, both of which have been used for projects like MobilityFirst that manage

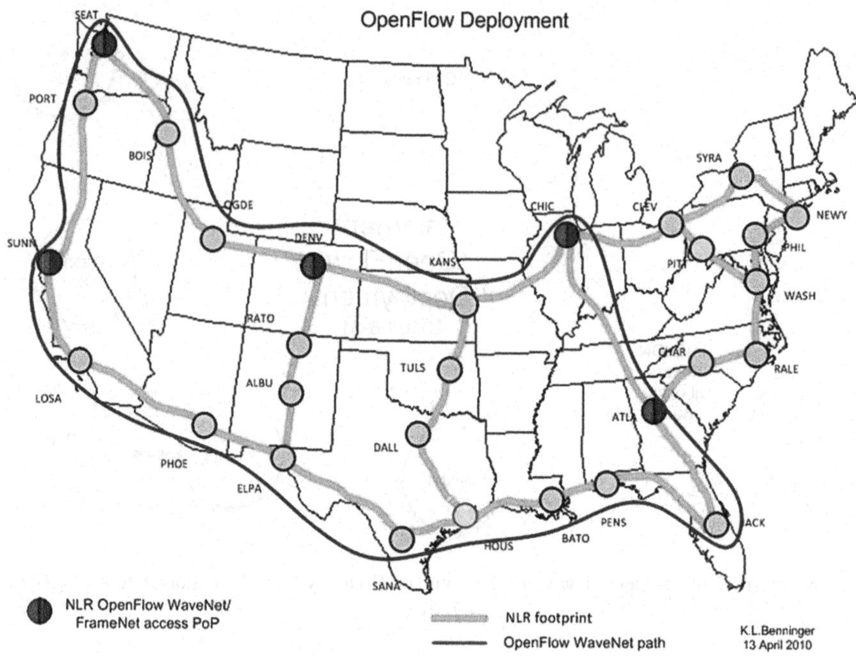

Fig. 9 Initial NLR Mesoscale Deployment. Courtesy Kathy Benninger, NLR (2010)

their own connections, or other projects that require well-known static VLAN IDs.) The flexibility of the AL2S design combined with GENI stitching also allowed Internet2 to offer higher-throughput connections for GENI, which had previously been limited to 1Gbps throughput in most of the core network.

3.2 *NLR*

NLR provided GENI access to NLR's FrameNet Layer 2 and Layer 3 services, along with their WaveNet optical services. NLR deployed and operated OpenFlow-enabled HP ProCurve 6600 switches at five NLR PoPs (Chicago, Denver, Seattle, Sunnyvale and Atlanta) interconnected at up to 10 Gbps, permitting members and non-members of NLR to connect to GENI OpenFlow services. Like Internet2, NLR allowed experimenters to run their own controllers to manage their OpenFlow traffic in NLR. NLR supported higher bandwidth connections to their core OpenFlow switches than those that were available in Internet2, making NLR more attractive to experimenters initially. Figure 9 shows the initial NLR core mesoscale infrastructure.

The NLR OpenFlow switch in Atlanta also connected to the Internet2 OpenFlow core network switch in Atlanta at 10 Gbps. This constituted the first SDN peering

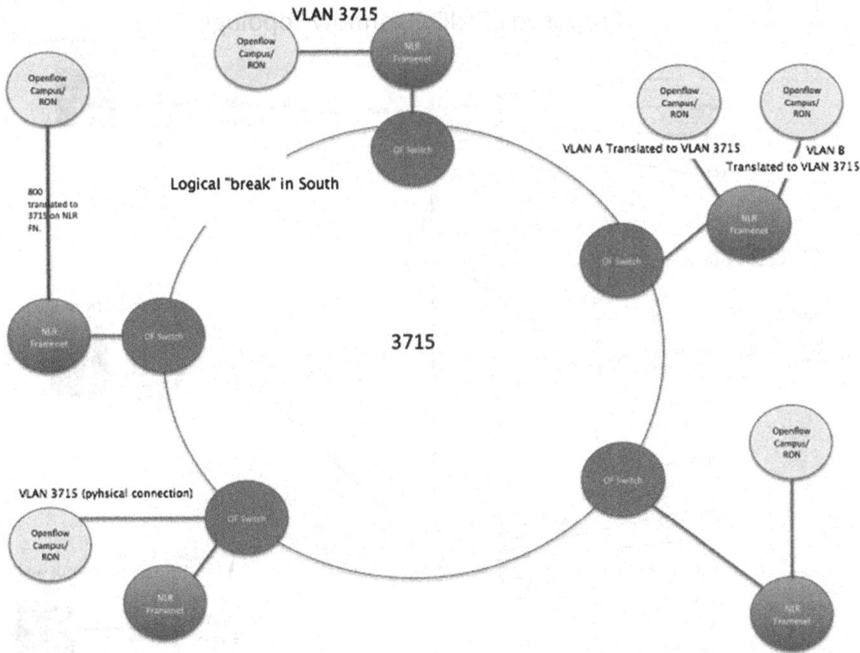

Fig. 10 Shared Mesoscale VLAN for OpenFlow Provisioned in NLR, RONs, and Campuses. Courtesy Kathy Benninger, NLR (2010)

point, later christened an SDN Exchange, between SDN networks. This exchange was particularly notable because NLR and Internet2 had no common IP peering point at the time. NLR supported connections between their OpenFlow switches with the FrameNet VLAN provisioning service. Figure 10 shows an example of how a shared OpenFlow VLAN was engineered to connect multiple mesoscale campuses to NLR.

In the second phase of the mesoscale build, NLR planned to expand their SDN deployment with more 10Gbps switches in El Paso, Houston and Kansas City, along with an extra North/West fiber connection between El Paso and Denver to increase path diversity. NLR planned to use the less-expensive Pica8 switches for their expanded deployment, maintaining interoperability between the original HP-based SDN core and the new Pica8 switches. NLR was purchased by the Chan Soon-Shiong Institute of Advanced Health in 2011, and became less able to support OpenFlow deployments, which were not part of their new owner's mission. NLR thus halted OpenFlow expansion before completing deployment. (Two switches shipped to the field.) Eventually, NLR ceased operations in March 2014, and Internet2 became the sole mesoscale core network provider for GENI.

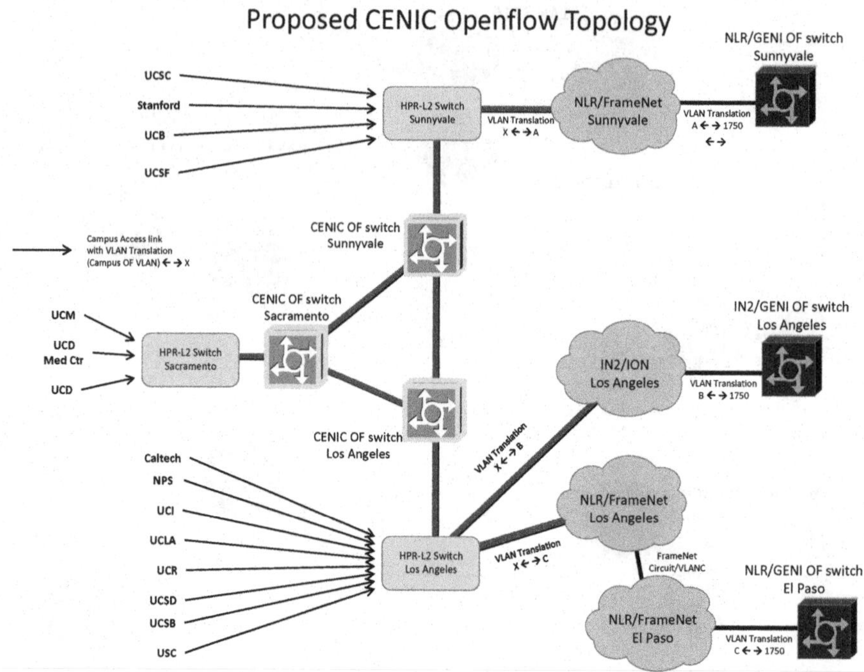

Fig. 11 CENIC Mesoscale OpenFlow Infrastructure. Courtesy Erick Sizelove, CENIC (2012)

4 Regional and International Networks in GENI

Several regional networks were early participants in GENI. In most cases, they were supporting their member campuses with regional OpenFlow deployments. The original regional networks participating in the OpenFlow mesoscale build-out were CENIC (California), KanREN plus the Great Plains Network (GPN) (Kansas, Nebraska, Missouri), MAX (Maryland, Virginia, Washington, DC), MOXI (midwest region), SoX (southern region) and UEN (Utah). Other regionals—NYSERNet, GPN, KyRON, LEARN and NCREN and MERIT—deployed their own OpenFlow infrastructure and connected to the OpenFlow mesoscale in the second phase of the OpenFlow buildout, beginning in 2012. Additional regional networks—3ROX, MAGPI, MOREnet, NOX, OneNet, OSCnet and PNWGP—supported GENI integration by helping to engineer VLANs, IP routes, or MPLS paths between GENI campuses and core R&E networks, and by supporting dynamic connection services to Internet2. (NLR did not offer dynamic connection services during the mesoscale build.) Regional network designs varied widely, depending on the existing infrastructure. Figure 11 shows the initial CENIC OpenFlow mesoscale design, which provided multiple Brocade OpenFlow switches with 10Gbps connectivity between them for experimenters to use.

The Quilt, which is a consortium of regional networks with a particular emphasis on networking for research and education, was also active in early phases of GENI mesoscale work. The Quilt sponsored a GENI workshop in 2010 to encourage more regionals to build GENI infrastructure for OpenFlow, and that resulted in several regional networks applying to start projects with GENI. The Quilt also organized an RFP effort for its members to help them get consistent and favorable quotes from OpenFlow switch vendors. These activities helped increase the diversity and availability of regional OpenFlow infrastructure.

StarLight specialized in enabling international dataplane connections between research organizations from its base in Chicago at Northwestern University. Although by no means a regional network, StarLight often functioned the same way as a regional in GENI integration because StarLight engineered connections to campuses that had no direct connections to Internet2 or NLR. Such campuses appeared to access the GENI core through StarLight's network. At times StarLight also served the same function for locations inside the US that were not already connected to GENI core networks (e.g. Oak Ridge National Laboratory). StarLight had the capability to make optical, digital (MPLS), routed or switched connections to international or US destinations. Although not one of the original OpenFlow campuses, StarLight added several different types of OpenFlow switches to their suite of network equipment during the mesoscale work, and became one of the first two SDN Exchange projects in GENI. StarLight was also the only site besides the GPO to host both an ExoGENI and an InstaGENI rack.

International experiments were an important part of GENI. GENI engineered international connectivity by special arrangement, rather than supporting it as a standard part of the mesoscale, which covered only US infrastructure. In fact, there was at least one "special" international connection to GENI active during all phases of the mesocale build and integration work, even though international federation was not included in the GENI software architecture until 2012. The first international connections to GENI were to Korea (ETRI), Australia (NICTA) and Japan (University of Tokyo), and were all engineered differently, based on the specifics on the networks and exchanges involved. NICTA resources connected to GENI solely via Layer 3 IP, while ETRI and University of Tokyo both used static Layer 2 VLANs individually engineered with each involved network provider. European researchers involved with FIRE also participated in international federation efforts with GENI, resulting in additional Layer 2 engineering efforts with European R&E networks such as GÉANT. The Brazilian RNP network was also actively involved in OpenFlow and participated in GENI, eventually becoming part of an SDX project with StarLight. A 2012 GENI demonstration called the "slice around the world" highlighted many of these international participants, connecting them to the GENI Engineering Conference floor at NYU. In each international case, network engineering was supporting a close partnership between GENI PIs and international researchers that allowed for similar close network collaboration.

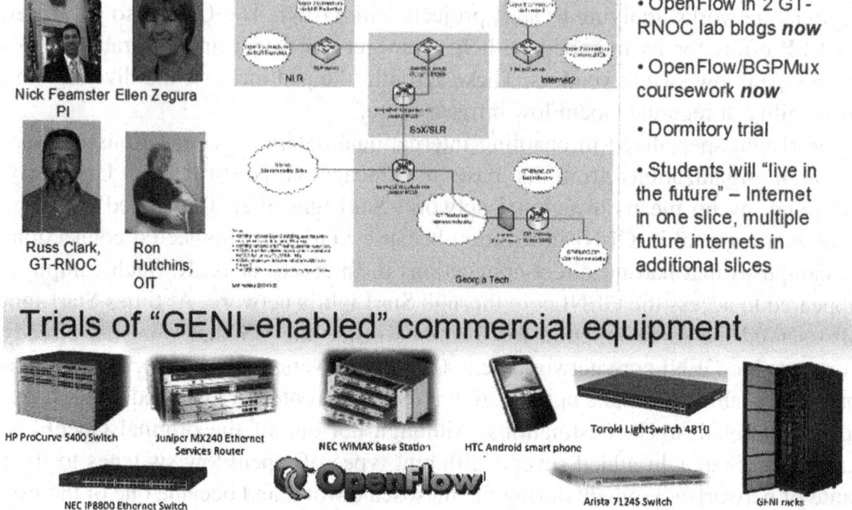

Fig. 12 Georgia Tech Mesoscale Campus Deployment (from a presentation highlighting the importance of collaboration between research and IT staff in GENI). Courtesy Chip Elliott, Heidi Dempsey, BBN Technologies (2010)

5 Campus Networks in GENI

From the earliest stages, the GENI mesoscale design and buildout included close collaborations between researchers, developers, and the network engineers and IT staff on many campuses. In particular, the OpenFlow infrastructure and GENI rack deployments required close coordination with campuses. Although the local campus network infrastructure varied widely among different universities, all GENI campuses were able to provide upstream connectivity to a research network and downstream connectivity to local compute, storage, or more network resources for experimenters to use. During the early stages of the GENI project, connectivity could be as simple as one OpenFlow switch, with an upstream connection to the local Layer 2 regional network, and a downstream connection to an Ethernet with one or two shared servers. USIgnite downstream infrastructure included entire broadband networks, providing commercial Internet and experimental dataplane service to multiple homes and businesses. The downstream connectivity associated with GENI racks could also include dormitory networks, research labs, connections to other specialized optical network testbeds, or GENI WiMAX and LTE wireless networks covering several miles and users. Figure 12 shows an example of an early GENI campus network deployment at Georgia Tech.

Network engineering for campus networks required close collaboration with the GPO and with regional and core networks supporting GENI. In fact, the campus, regional and GPO network engineers met regularly in teleconferences and at GENI Engineering Conferences to exchange information and continuously improve their integration with each other and with the rest of GENI. The practical experience of the campus engineers helped the GENI design to be sustainable, supportable, and more robust for operations than a pure research testbed. The close cooperation between researchers, developers, and IT staff was a very successful example of the "DevOps" concept that later became popular in commercial networking, especially for cloud environments.

6 VLAN Stitching

Early mesocale deployments relied on carefully engineered VLAN connections between participating networks and campuses in GENI, which was labor-intensive and not scalable to larger deployments. This solution also often required GENI experimenters to know many details about how particular VLANs interconnected in GENI. In fact, all VLAN engineering information was tracked in detailed configuration files and on the GENI wiki to allow experimenters and network engineers to reference it when changing the network or setting up new experiments. Early experimenters tended to already be very familiar with network infrastructure, because most early GENI builders were also GENI users, so understanding the low-level complexity that made GENI work was not an issue.

As GENI expanded, researchers from different domain sciences besides computer science and students without much networking experience began to use the infrastructure. Application developers who were only peripherally interested in the network also became more active, particularly as the USIgnite project solicited new application ideas. It was unreasonable to expect these new experimenters and students to understand the details of the engineered network in order to use GENI. The systems and software development efforts associated with GENI racks provided a more standardized interface for describing and accessing GENI resources than what had been available early in the mesoscale work. The standard interfaces in turn, allowed us to develop a more standardized way of connecting GENI dataplane resources dynamically, called GENI stitching.

Engineers from the MAX regional network led the earliest GENI stitching efforts, using a dynamic network approach based on earlier work from the GLIF and DRAGON projects. MAX researchers developed a stitching prototype based on the ION/DYNES GENI Aggregate Manager, and successfully demonstrated end-to-end stitching in a demo between the MAX and CRON networks.

The ExoGENI development team at RENCI and Northwestern University also developed an early type of VLAN stitching mediated by the ORCA control framework. This stitching functioned cooperatively with NLR and StarLight to connect VLANs between ExoGENI racks, and to allocate experimental traffic between the ExoGENI processing resources on different racks with a centralized service manager that ensured provided topology and bandwidth matched experimenter requests. Similarly, the InstaGENI development team at the University of Utah managed traffic allocation and delivery using a pool of VLANs in Internet2 reserved for the ProtGENI/Emulab control framework and connected to other networks using the QinQ protocol. This early InstaGENI and ExoGENI stitching could not interoperate, nor stitch Layer 2 connections to OpenFlow resources. Thus, the GENI community developed and adopted interoperable RSPEC extensions and GENI AM API features for stitching, based on early concepts from MAX, RENCI, Northwestern, the University of Utah, and Internet2. (NLR did not participate fully in this effort because it changed missions during this time, and did not field a stitching aggregate manager.) The GPO conducted stitching operations trials for an extended period, and incorporated stitching tests as part of GENI rack validation and site acceptance testing.[4]

Internet2 implemented the Open Exchange Software Suite (OE-SS) on their Advanced Layer 2 Service to support dynamic VLAN stitching with a programmable user interface (as well as via a web-based GUI). Internet2 and the GPO jointly demonstrated crossover connections passing traffic over paths that combined the GENI mesoscale and AL2S infrastructures at the Interent2 Joint Techs meeting in January 2013. Internet2 also supported stitching operations trials jointly with MAX and the GPO, culminating in a plenary demonstration at GEC17 (July 2013). CENIC, NoX, MAX. KyRON, UEN, the University of Utah, the University of Kentucky, RENCI and Stanford also participated in the initial stitching Operations Trials.

At a high level, GENI stitching provides cross-aggregate dynamic Layer2 datapaths between GENI endpoints in an experiment. The "stitcher" tool was originally part of the OMNI command line client (stitcher.py) and has now been integrated into GENI aggregates and GUI tools. A topology service, called the Stitching Computation Service (SCS), determines whether there is an available path that can satisfy the experimenter's request with GENI Layer 2 VLANs. Together, these components take a simple cross-aggregate request and compute the network hops required to establish that connectivity. This capability was demonstrated at GEC17 and continues to be refined for scalability, performance and reliability. Stitching also requires that the experimenter have GENI credentials, which are used to request GENI VLAN resources from the aggregates that are part of the stitched dataplane, so experimenters must be known to the GENI Clearinghouse. In addition, GENI aggregates participating in stitching must have delegated VLAN IDs from their available pools to GENI for exclusive use in stitching.[5]

[4]See http://groups.geni.net/geni/wiki/GeniNetworkStitchingTestPlan and http://groups.geni.net/geni/wiki/GeniNetworkStitchingTestStatus.

[5]VLAN delegation list: https://wiki.maxgigapop.net/twiki/bin/view/GENI/StaticNetworksView.

Reference

1. Kobayashi, M., Seetharaman, S., Parulkar, G., Appenzeller, G., Little, J., Van Reijendam, J., Weissmann, P., McKeown, N.: Maturing of OpenFlow and software-defined networking through deployments. Comput. Netw. **61**, 151–175 (2014)

ExoGENI: A Multi-Domain Infrastructure-as-a-Service Testbed

Ilya Baldin, Jeff Chase, Yufeng Xin, Anirban Mandal, Paul Ruth, Claris Castillo, Victor Orlikowski, Chris Heermann, and Jonathan Mills

Abstract This chapter describes ExoGENI, a multi-domain testbed infrastructure built using the ORCA control framework. ExoGENI links GENI to two advances in virtual infrastructure (IaaS) services outside of GENI: open cloud computing (OpenStack) and dynamic circuit fabrics. It orchestrates a federation of independent cloud sites and circuit providers through their native IaaS interfaces, and links them to other GENI tools and resources. ExoGENI slivers are instances of basic IaaS resources: variously sized virtual machines, bare-metal nodes, iSCSI block storage volumes, and Layer 2 network links with optional OpenFlow control.

ExoGENI offers a powerful unified hosting platform for deeply networked, multi-domain, multi-site cloud applications. ExoGENI operates its own stitching engine and Layer 2 (L2) network exchanges that work in concert to interconnect the sites with dynamic point-to-point and multi-point L2 links via multiple circuit providers. It also supports *stitchports*—named attachment points enabling direct L2 connections to resources outside the system's control.

ExoGENI is seeding a larger, evolving platform linking third-party cloud sites, transport networks, new resource types, and other infrastructure services. It facilitates real-world deployment of innovative distributed services, leading to a new vision of a future federated, more resilient, and deeply networked cyber-infrastructure. This chapter explores the unique features of ExoGENI and, in particular, how it differs from other GENI testbeds.

1 Introduction

ExoGENI is a new testbed that federates distributed resources for innovative projects in networking, operating systems (OSs), future Internet architectures, and deeply networked, data-intensive cloud computing. It supports novel applications

I. Baldin (✉) • Y. Xin • A. Mandal • P. Ruth • C. Castillo • C. Heermann
Renaissance Computing Institute (RENCI)/UNC Chapel Hill, Chapel Hill, NC, USA
e-mail: ibaldin@renci.org

J. Chase • V. Orlikowski
Duke University, Durham, NC, USA

J. Mills
NASA Center for Climate Simulation, Goddard Space Flight Center, Greenbelt, MD, USA

© Springer International Publishing Switzerland 2016
R. McGeer et al. (eds.), *The GENI Book*, DOI 10.1007/978-3-319-33769-2_13

and services, e.g., for the U.S. Ignite initiative. NSF's GENI project, the major U.S. program to develop and deploy integrated network testbeds, funds ExoGENI.

Deployment of ExoGENI began in 2012. The initial ExoGENI deployment consists of close to 20 site "racks" on host campuses around the world, linked with national research networks and other L2 circuit networks. The individual cloud sites include hardware from multiple vendors: IBM, Cisco, Dell and Ciena. ExoGENI also includes the Breakable Experimental Network (BEN), a multi-layered network testbed that connects several sites in North Carolina with a dynamic L2 circuit service. The control software for BEN and ExoGENI is built using ORCA (Open Resource Control Architecture), which is described in a separate chapter of this book [7].

ExoGENI is based on an extended Infrastructure-as-a-Service (IaaS) cloud model with orchestrated provisioning across sites. Each ExoGENI site is a private IaaS cloud using a standard cloud stack to manage a pool of servers. The sites federate by delegating to common coordinator services a portion of their resource capacity and various functions for identity management, authorization, and resource management. This structure enables a network of private clouds to operate as a hybrid community cloud. Thus, ExoGENI is an example of a *multi-domain* or *federated* cloud system, which some have called an *intercloud*.

ExoGENI combines this structure with a high degree of control over networking functions. These include VLAN-based and OpenFlow networking within each site, multi-homed cloud servers that can act as virtual routers, site connectivity to national circuit backbone fabrics through host campus networks, and linkages to international circuits through programmable exchange points.

The key contribution of ExoGENI to US research cyberinfrastructure capabilities lies in developing and deploying a *topology embedding as-a-service* approach, which allows users to realize their virtual network topologies easily on top of a distributed multi-domain substrate. A user specifies the desired topology, providing some hints as to its properties along with the locations of key elements. The system, in turn, finds available resources, produces a homeomorphic embedding of the request graph into the substrate, and coordinates the allocation of resources across multiple providers.

This capability is enhanced by addition of other advanced features, such as:

- **On-ramps to advanced network fabrics.** ExoGENI succeeds in using campus clouds to bridge from campus networks to national transport network fabrics, overcoming a key limitation identified by the National Science Foundation (NSF) in its vision of a Cyberinfrastructure Framework for 21st Century Science and Engineering (CF21). ExoGENI cloud sites can act as *virtual colocation centers* that offer on-demand cloud services adjacent to fabric access points. Sites at fabric intersection points can also act as *virtual network exchanges* to bridge "air gaps" stemming from lack of direct connectivity or from incompatible circuit interfaces among fabrics.
- **Cloud peering and data mobility.** ExoGENI can facilitate the peering and sharing of private clouds. It offers a means to bring data and computation together by migrating datasets to compute sites or placing computation close to data at rest.

- **Support for application-driven topology management.** By providing a toolkit capable of communicating with the platform APIs, ExoGENI allows applications to manage their own resources and their topologies *autonomously*. This concept enables development of a new class of *resource-aware* networked cloud applications that can track their resource needs and use available resources to move and place data and computation efficiently, improving time-to-discovery for domain scientists.

This chapter gives an overview of the ExoGENI testbed, its control software, deployment, and uses. In the following sections, we discuss the deployment of ExoGENI (Sect. 2); details of its software infrastructure operation (Sect. 3); methods and tools for its administration and use (Sects. 4 and 5); and, finally, its integration with the larger GENI ecosystem (Sect. 6).

2 Overview: A Testbed of Federated IaaS Providers

2.1 Operational Principles

ExoGENI supports virtual infrastructure resources, instances of "fundamental computing resources, such as processing, storage, and networks", in accordance with the definition of Infrastructure-as-a-Service [22] by the National Institute of Standards and Technology (NIST). Testbed users may instantiate and program multiple slices, each containing a private virtual topology consisting of virtual machines (VMs), physical (bare-metal) nodes, block storage (iSCSI) volumes, programmable-switch datapaths, and virtual network links at various layers (L0, L1, and L2). Deployment is based on an evolving set of technologies including point-to-point Ethernet circuits, OpenFlow-enabled hybrid Ethernet switches, and standard cloud-computing software—OpenStack and xCAT [10].

The "Exo" (outside) prefix reflected our view early in the project of how GENI should evolve and what capabilities would be needed to deliver on the promise of GENI to "explore networks of the future at scale". GENI was evolving alongside cloud technologies and open network control systems whose functions and goals overlap with GENI. Even at that time, the rate of investment in developing and deploying these systems was quite a bit more than an order of magnitude larger than the GENI effort.

One purpose of ExoGENI was to define reliable methods to leverage these technologies and substrates in the GENI project. At the same time, GENI control software offers new ways to combine and to extend such systems as a unified deployment platform for advances in network science and engineering. ExoGENI establishes paths by which GENI control software can leverage IaaS advances while successfully addressing important orchestration challenges for networked cloud computing.

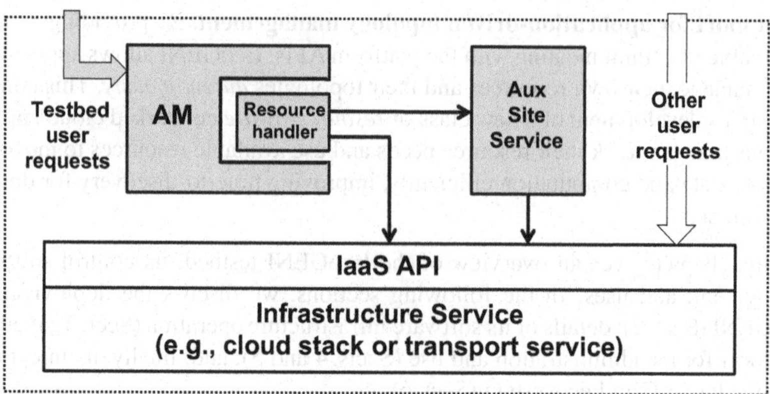

Fig. 1 Structure of a resource provider or *aggregate*. Each provider runs a native infrastructure service (IaaS) of its choice, which may serve requests from local users through a native application programming interface (API). To join a federation, the aggregate is fronted with a generic Aggregate Manager (AM) service. The AM validates user requests against local policy and serves them by invoking the native IaaS API through resource-specific plugin *handlers*. A handler may incorporate other auxiliary services for some functions, e.g., image loading, OpenFlow authorization, or network proxying

The "Exo-" prefix in "ExoGENI" captures four related principles that are illustrated in Fig. 1:

E1 **Decouple infrastructure control from orchestration.** Each provider domain (*aggregate*) runs a generic front-end service (an *Aggregate Manager* or AM) that exports the testbed APIs. The AM cooperates with other services in the federation, invoking a back-end infrastructure service to manage the resources in the domain.

E2 **Use off-the-shelf software and IaaS services for infrastructure control.** Standard IaaS software and services offer a ready back-end solution to instantiate and release virtual resources in cloud sites, circuit services, and other virtual infrastructure services. The generic AM interfaces to these standard APIs use plugin *handler* modules.

E3 **Leverage shared third-party substrates through their native IaaS interfaces.** This compatibility with standard back-end infrastructure control services facilitates inclusion of independent resource providers in federations. The provider deploys an off-the-shelf IaaS service and "wraps" it with an AM to link it into the testbed federation.

E4 **Enable substrate owners to contribute resources on their own terms.** Participating providers are autonomous: they may approve or deny any request according to their policies. Providers allocate virtual resources that have specified Quality of Service (QoS) properties for defined intervals. The callers choose the resources to request, and specify how to expose them to applications. Resource allotments are visible to both parties and are controlled by the providers.

These principles presume explicit resource control, in which all parties can determine and quantify the resources they receive or provide, and are empowered to control or influence their resource exchanges according to local policies. This idea echoes the principles of the *exokernel* extensible operating system [18] developed at the Massachusetts Institute of Technology (MIT) in the 1990s. The name of ExoGENI also pays homage to that project [17].

Based on these principles, ExoGENI provides a framework to incorporate "outside" resources and infrastructure services into a federation and to orchestrate their operation. For example, providers may deploy new cloud sites using open-source cloud stacks, such as Eucalyptus [24] or OpenStack [26], which support IaaS cloud APIs. Similarly, for transport-network circuit services, common APIs such as OSCARS [12] and the Network Service Interface (NSI) [25] are emerging. Providers may deploy these and other systems independently. Once a system is deployed, we can install a front-end orchestration service (AM) to link it into the federation without interfering with its other functions and users. The AM may be operated by the provider itself, by a delegate, or by an authorized client.

2.2 *Hardware and Topology*

Each ExoGENI cloud site includes a packaged rack with a small cloud-server cluster and an integrated OpenFlow network. Most of the existing ExoGENI sites are based on server racks preassembled by IBM—our partner in the initial NSF-funded deployment. ExoGENI now also incorporates a number of racks built by Dell, Cisco and Ciena and contributed to the ExoGENI effort by their respective campus owners.

Figure 2 depicts the rack components and connections. The nodes in the 2012 IBM racks are M4 IBM X-series servers. The *worker nodes* are the server substrate for dynamic provisioning of nodes for slices. A single *management node* (head node) runs the control servers for the site, including the OpenStack head and ORCA servers. The rack also includes an iSCSI storage unit for compute OS images and instrumentation data. Crucially, it also serves as a resource that is sliverable by users. Software developed jointly by RENCI and the University of Amsterdam allows for flexible slivering of iSCSI volumes. This action is built on varying combinations of open and proprietary APIs, supporting LVM, ZFS, Gluster, and several vendor appliances. Some racks contain commercial storage appliances, like the IBM DS-3512, and others provide storage from specially configured worker nodes upgraded with large amounts of tiered storage (SSD and/or rotating disks) managed by the Linux-based open-source software stack.

All components are connected to a *management switch*, which has a L3 connection to the campus network and, from there, to the public Internet. This switch is used for intra-site access to the iSCSI storage, for remote management by the testbed operator (RENCI) through a VPN appliance (not shown), and for slice connectivity to the public Internet. Each worker has multiple 1Gbps ports for these uses.

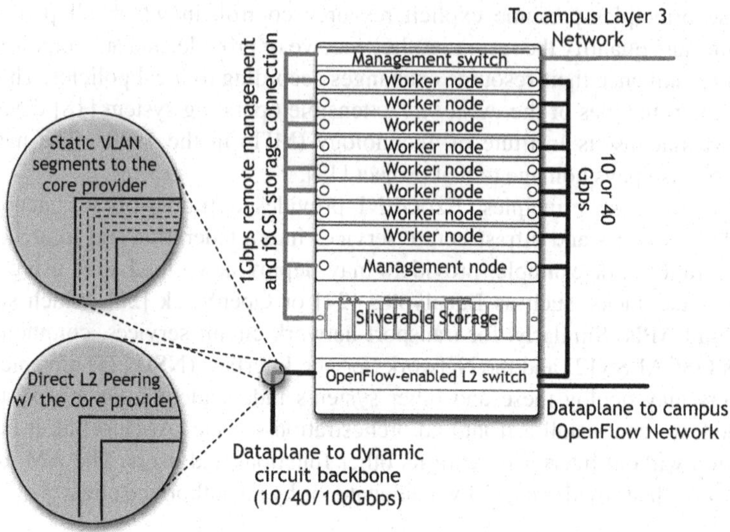

Fig. 2 Structure of an ExoGENI site rack for the initial deployment. Each rack has low-bandwidth Internet Protocol (IP) connectivity for management and a high-bandwidth hybrid OpenFlow switch for the slice dataplanes. The site ORCA server controls L2 dataplane connections among local nodes and external circuits

A separate *dataplane* switch carries experiment traffic on the slice dataplanes. It is the termination point for L2 links to external circuit providers and to VLANs or subnets on the host campus network, if permitted by the host campus. The IBM racks have an IBM/BNT G8264R 10Gbps/40Gbps OpenFlow-enabled hybrid L2 switch with VLAN support. Cisco racks use a combination of Unified Computing Systems (UCS-B) Fabric Interconnect and Nexus-series switches. Dell racks use PowerConnect-series switches or Ciena 8700 switches. Each worker node typically has two 10Gbps links to the dataplane switch, although Ciena rack servers have 40Gbps interfaces.

Slice owners can access their nodes over public IP through a management interface. As in other GENI systems, the slice owner specifies a set of named accounts and associated public keys, and the control software configures these accounts on the nodes. Nodes typically have public IP addresses when allowed them by the host campus. For cloud site racks, public IP access is managed by OpenStack and is proxied through its head node.

The racks are interconnected by regional and global transit providers. To leverage network fabrics operated by third parties (e.g., Internet2 and ESnet), ExoGENI deploys AMs with plug-in handlers that invoke the native APIs of the network providers to establish dynamic L2 connections, as described below.

The rollout of ExoGENI coincided with the push by network providers to expose more of their networks' capabilities through OpenFlow. Thus, ExoGENI control software continues to evolve to make use of those capabilities.

2.3 ORCA ExoGENI Control Software

The control software for ExoGENI was developed and refined in an ongoing collaboration between RENCI and Duke University to create a GENI testbed "island" around the Breakable Experimental Network [3] (BEN), a multi-layer optical testbed built by the State of North Carolina and managed by RENCI. Some early results of the project have been reported in previous publications [4, 5, 19, 34].

An initial goal of the project (2008) was to build a native multi-layered circuit service for BEN, based on the Open Resource Control Architecture (ORCA [7]). From these initial steps, the project grew into a more comprehensive effort to support a federated IaaS system linking BEN and other infrastructure systems under orchestrated control within a common authorization framework. The project developed a set of plugin modules for ORCA, which were later used in ExoGENI. Table 1 gives an overview of the various plugins, which are described in more detail below.

Building the native BEN circuit service presented an interesting test case for the ORCA control framework. We built policy plugins to plan requested paths through the BEN network by issuing queries on *semantic resource models*, which are logic-based declarative descriptions of the network expressed in an extended variant of the Network Description Language (NDL) [9, 13, 14]. The paths are multi-layered, i.e., in order to create a desired L2 path, the underlying L1 (Dense Wavelength Division Multiplexer, or DWDM, wavelengths) and L0 (fiber) paths must be established first. Alternatively, the system must utilize the available bandwidth efficiently, cross-connecting resources by overlaying new L2 connections onto existing L1 paths whenever capacity is available. All of this was achieved by developing a BEN plugin for ORCA to manage this logic. We also built handler plugins that manage paths over the BEN management network by forming and issuing commands to a variety of network devices, such as Cisco and Juniper L2 switches, Infinera DWDM equipment, Polatis fiber switches, based on path descriptions generated by the queries.

Later, we implemented new ORCA plugins to interface with external circuit APIs used in the various national circuit fabrics around the globe. These include National LambdaRail's Sherpa FrameNet service (now defunct), Internet2 Open Exchange Software Suite (OESS), ESnet On-demand Secure Circuits and Advance Reservation System (OSCARS), and GÉANT NSI. These plugins allow for the abstraction of provisioning operations and for stitching among providers, making the topology-embedding generic and leaving the details of provisioning actions to the plugins.

Table 1 ORCA plugins developed to support ExoGENI

Plugin name	Description	In use
Eucalyptus	Supports calling EC2 API in Eucalyptus to instantiate virtual machines. Interfaces with custom code that we introduced into Eucalyptus to support arbitrary networking configurations. Includes an AM control policy and a handler capable of interfacing to EC2 tools	N
OpenStack	Supports calling Nova API to instantiate virtual machines. Passes additional information to guests using user data. Nova has been modified to allow asynchronous updates to user data	Y
OpenStack quantum	Supports generating the necessary networking configuration for virtual machines once they are created by Nova. Interfaces to our custom implementation of a Quantum plugin to OpenStack	Y
Network drivers	Mainly handler plugins that include implementations of abstract operations, like *create VLAN*, *add port to VLAN*, and their opposites for a variety of vendors and switches: Cisco 6509 and 3400, Juniper EX2400 and QFX3500, IBM G8264, Infinera DWDM, and Polatis fiber switches. The handlers operate over a variety of interfaces (CLI, XML, Netconf) to program the switch behavior	Y
OpenFlow	Includes an AM control for issuing VLAN tags to avoid conflicts between slices. Allows stitching to external VLANs. Supports interfacing with FlowVisor to create per-VLAN slices in the rack switches and optionally attaching a controller which emulates learning switch behavior	N
Hybrid VLAN/ OpenFlow plugin	Includes an updated AM control for issuing VLAN tags and determining whether the best mode for the switch to instantiate a particular connection in a slice is VLAN or OpenFlow. Includes the functionality of OpenFlow plugin. An additional interface to the switch supports native VLAN provisioning (e.g., in G8264 above)	Y
iSCSI storage	Includes an AM control that can provision iSCSI LUNs on demand. Supports IBM DS3512 appliances as well as server-based solutions using Linux iSCSI software stack	Y
NLR Sherpa	Includes an AM control and a handler for now-defunct NLR Sherpa circuit service	N
OSCARS/ OESS/NSI	Includes an AM control that passes the correct parameters for end-to-end circuit creation in OSCARS, OESS, or NSI domains. The handler plugin distinguishes among domain types and invokes the proper driver tasks to drive the domain-specific APIs	Y
BEN	A complex control that implements multi-layered connection provisioning inside BEN and calls on a custom handler interfacing to Polatis fiber switches, Infinera DWDM gear, and the available Layer-2 switches to create on-demand connections in BEN at multiple layers	Y
VLAN SDX	A complex AM control that allows for creation of multi-point connections in slices using a series of switches connected to each other to support complex stitching and multipoint arrangements of VLANs from multiple providers. Can use any of the switches supported by network drivers above	Y
Topology-embedding controller	A plugin that interfaces to the experimenters, implementing GENI and ORCA-native outward-facing APIs. Internally processes slice topology descriptions. Uses ORCA APIs to communicate with brokers and AM controls to provision the slice. There is a single controller code that works both for intra-rack and inter-rack topologies	Y

We also extended other handlers to drive cloud APIs. With the emergence of Eucalyptus [24], we focused our development on integration with standard EC2-compatible cloud stacks, replacing our older cloud software, Cluster-on-Demand [8]. ExoGENI rack sites currently use these plugins to drive the OpenStack cloud service, which exposes a similar interface. Due to better modularity of OpenStack software, however, ExoGENI enhancement also included a new *Quantum* plugin for OpenStack that enables dynamic attachment of L2 networks to VM instances. In addition, we developed an ORCA handler plugin to unify these provisioning tasks under a common abstraction. We developed a similar plugin for the xCAT open-source provisioning system to accommodate bare-metal provisioning.

For work with OpenFlow, we produced a set of embedding extensions and plugins that communicate with FlowVisor, using its API to create OpenFlow slices. This capability was used in two ways. In situations in which a user wants explicit control over forwarding, ORCA allows the user to specify an external OpenFlow controller, which then attaches to the FlowVisor slice. To emulate traditional VLANs, ORCA dynamically creates VLAN-based flowspace slices and attaches them to an internal controller that emulates the behavior of a traditional VLAN with learning switches.

Many of these ORCA plugins process logic-based semantic resource descriptions (NDL-OWL models). Early in the project we augmented ORCA with a common library (NDL-Commons, ~13,000 lines of code) with software to process NDL-OWL models, and another larger library to handle various model query tasks including topology embedding. This library simplifies the development of ORCA plugins that use the native semantics models to describe resources. For example, the plugins to manage OpenStack VMs, which include the control policy for OpenStack AM as well as the handlers to perform operations on OpenStack and Quantum, take ~300 lines of code (LOC) for the control policy; this is because OpenStack offers its own placement and scheduling of VMs in worker nodes, relieving ORCA of the need to perform these tasks. The handlers responsible for interfacing to OpenStack and to Quantum total ~6000 LOC in Ant, bash, and Python to manage the creation of virtual machines and their network interface configurations. In contrast, BEN, which lacks its own IaaS system, requires direct management of the various network elements in its topology and is significantly more complex. It includes a ~500 LOC control policy and a handler requiring a combined ~8000 LOC in Ant and Java.

The structure of software deployed on an individual rack is layered as shown in Fig. 3. In the bottom layer, there are IaaS software stacks of various providers, exposing their native APIs to ORCA plugins. Users and their tools interact with the infrastructure *via* interfaces exposed by ORCA (marked as *GENI API* and *ORCA API*). Similarly, monitoring infrastructure is built around the open-source Nagios system, with custom plugins for various pieces of infrastructure that are specific to ExoGENI, e.g., to monitor the liveness of individual ORCA actors.

This modular architecture has had significant benefits for ExoGENI and its users. It reduces the time needed to bring new technologies into GENI, as ORCA developers are not concerned with the inner workings of an IaaS system, but rather with its provisioning APIs. The amount of code that must be written to support a new

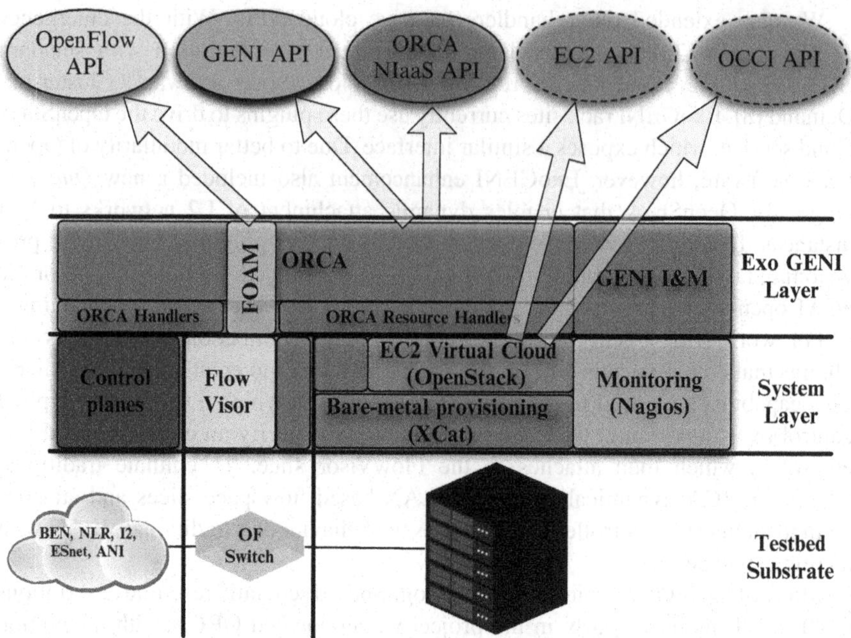

Fig. 3 Conceptual view of the ExoGENI software stack deployed in a rack. An IaaS layer consists of standard off-the-shelf software and services for server clouds, network transport services, and OpenFlow networking, which control the testbed substrate. Testbed users and tools access the testbed resources through GENI APIs and an alternative API (labeled ORCA NIaaS API in the figure) based on semantic resource models. The ORCA resource-leasing system tracks resource allocation and orchestrates calls to *handler* plugins, which invoke the APIs for services in the IaaS layer. The monitoring system has a similar structure

technology is reduced, as is its testing time. The design leverages the significant efforts of other teams and vendors in developing IaaS solutions, which are then federated under ORCA.

Users and their tools invoke GENI or ORCA APIs to instantiate and program virtual resources from participating providers. A *slice* is a set of virtualized resources under common user control. A slice may serve as a container or execution context to host an application or a network service. An ExoGENI slice may contain a network topology with programmable nodes, storage, and links woven together into a virtual distributed environment [31]. The links in the topology comprise the slice *dataplane*. Software running within a slice may manage its dataplane as a private packet network. To do so, it may use either IP or alternative protocol suites, as specified by the slice owner. A slice may span multiple sites and link to further GENI resources or to other external resources as permitted by peering and interconnection agreements.

2.4 Deployed Software Ecosystem

Requests from users enter an ORCA control system through an ORCA server called a *Slice Manager* (SM), which invokes the AMs to obtain the requested resources. Each SM runs *controller* plugins that encapsulate application-specific or slice-specific logic for managing a slice. In essence, the SM is an extensible user tool running as a recoverable server. In general, an SM runs on behalf of slice owners with no special trust from other services. However, the ExoGENI SMs are trusted: the ExoGENI AMs accept requests only from SMs endorsed by the testbed root. The ExoGENI SMs run a controller plugin that offers standard GENI APIs to GENI users and tools. The controller validates the GENI authorizations for the slice and for the user identity in each request, translates between GENI's RSpec-XML resource descriptions and the logic-based semantic resource models used in ORCA, and encapsulates the topology embedding logic.

Within ExoGENI, SM controllers interact with other ORCA coordinators called *brokers*. Brokers collect and share information about ExoGENI aggregates, including their advertised resources, services, and links to circuit providers. Each ORCA broker is trusted by some set of aggregates to advertise and to offer shares of their resources. A broker may also coordinate resource allocation across its aggregates, guiding or limiting the flow of requests to them. It does this based on, e.g., scheduling policies and capacity constraints, as described in previous work on SHARP resource-peering [11, 16]. Each request for resources is approved by a broker before the SM passes it to an AM.

Two distinct SM configurations in ExoGENI are shown in Fig. 4. Each ExoGENI rack site runs a SM called a *rack SM*. Each rack SM exposes its rack site as a distinct GENI aggregate. Requests to a rack SM can operate only on the local rack; each rack SM uses a local broker with a share of local site resources.

In addition, a global ExoGENI SM called an *ExoSM* can access all aggregates within the ExoGENI testbed. There may be any number of ExoSMs, but the initial deployment has a single ExoSM. The ExoSM exposes ExoGENI as a single GENI aggregate. The ExoSM offers a unified view of the testbed, supporting virtual topology-mapping and circuit-path planning across the ExoGENI aggregates, including the circuit providers and network exchanges. User requests to ExoSM may request a complete slice topology spanning multiple sites and circuit providers. The ExoSM coordinates construction and stitching of end-to-end slices.

A controller plans and sequences the resource requests and stitching actions to the AMs, based on declarative semantic models that advertise the resources and capabilities of the participating aggregates [4, 5, 34]. It obtains these domain models from a common testbed-wide broker (*ExoBroker*) that is accepted by all ExoGENI aggregates and receives all of their advertisements. Notably, both types of SMs use *the same* controller plugin: the topology embedding logic embeds user requests in whatever resources are available to it—either within one rack or across the entire ExoGENI testbed. The SMs differ only in the resources advertised to them in their declarative models.

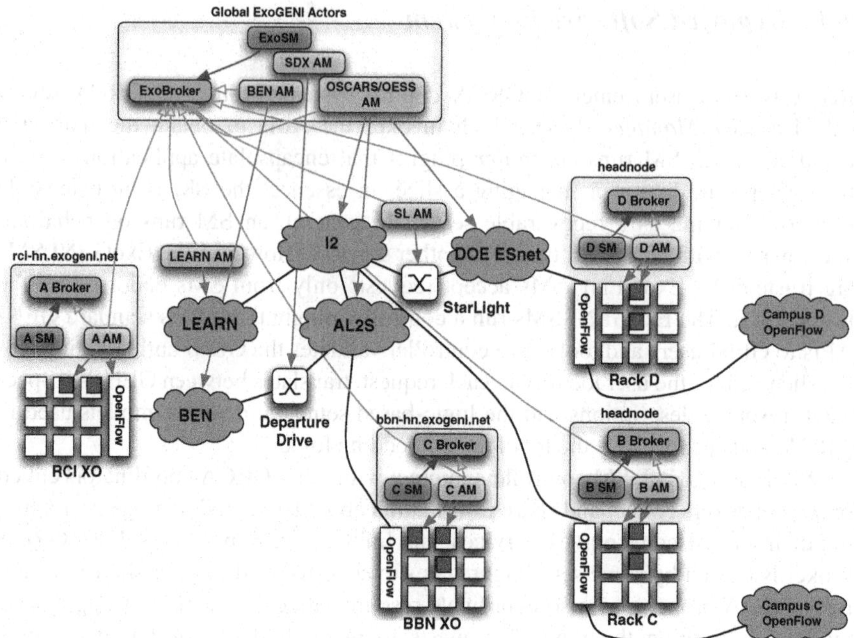

Fig. 4 ExoGENI software deployment on several ExoGENI (XO) sites connected to campus OF (OpenFlow) networks. In addition to ORCA actors responsible for carrying out the provisioning actions on the available resources, each rack runs its own ORCA controller/SM to support the GENI model of operation. The controller views the rack as a separate individually programmable substrate. In addition, we deployed the ExoSM controller, which has access to a fraction of resources from each rack as well as access to inter-rack circuit providers. The ExoSM performs complex topology-embedding tasks across the entire testbed

Thus, a user can request resources from an individual rack, using its local controller, and then use GENI tools to attempt to create a slice that spans multiple racks. Alternatively, the user can submit the full inter-rack request to an ExoSM actor. The latter has a global view of resources across the testbed. It can embed the user's request into multiple racks and transit providers. The ExoSM then attempts to fulfill the request by binding it to the available resources and stitching connections together into a linked topology. The ExoSM actor provides critical features that many ExoGENI experimenters find useful: (I) load-balancing their requests across the system, thus avoiding manual "hunting" for available resources; and (II) advanced stitching among multiple racks, which supports point-to-point and multi-point SDX L2 connections. Supporting these features and adding new capabilities to ExoSM remains a significant focus of ongoing ExoGENI/ORCA development efforts.

Over time, we added to ORCA many components and services that help manage the large, distributed infrastructure of ExoGENI in a scalable manner. They go beyond developing plugins for ORCA, adding functionality that is required in a world-wide deployment of ExoGENI to be managed by a small team of operators and developers. Shown in Fig. 5, those include:

- **ImageProxy**: Support for compute-instance image retrieval and dissemination across racks, as described in Sect. 3.1.
- **Distributed registry services** for ORCA actors and other elements of the infrastructure, based on CouchDB, as described in Sect. 4.1.
- **Scalable monitoring and notifications** using XMPP pub sub (XEP-0060), as described in Sect. 4.2.
- **Administration tools** for site and federation operators, as described in Sect. 4.3.
- **ORCA user tools** for creating slices using its native APIs and resource description mechanisms, as described in Sect. 5.1.
- **Support for interoperability with GENI**: translation between GENI RSpecs and ORCA native NDL-OWL resource descriptions, as described in Sect. 5.2.

Figure 5 shows the communications among these various components using a variety of protocols, including SOAP, XMLRPC, and REST, depending on the availability of protocol implementations and the needs of the particular use scenario within ExoGENI. This ecosystem largely relies on off-the-shelf open-source software, sometimes with custom plugins to serve ExoGENI needs. This reduces our development time and allows us to leverage the best solutions that the open-source community has to offer.

3 ExoGENI Services

Each ExoGENI rack site is reconfigurable at the physical layer, providing a flexible substrate for repeatable experiments. Most testbed users employ a virtual machine (VM) service; however, we also support bare-metal provisioning systems. In addition, most racks offer sliverable storage in the form of dynamically configurable iSCSI volumes and, of course, VLANs that can interconnect slice components together within- or between racks, including optional OpenFlow controls. VMs are offered using OpenStack, while bare-metal provisioning is done using xCAT [10], which is developed and maintained by IBM. iSCSI is offered using a variety of interfaces supported by storage appliances. Networking configuration is done directly on the switches of various brands, while OpenFlow support is accomplished by interfacing with FlowVisor. All of these separate IaaS solutions operate in a single rack, separated on operator-determined resource boundaries such as pre-assigned worker nodes and VLAN ranges as well as LUN tags and sizes. Their APIs are hidden underneath ORCA and the users are never exposed to the differences, or indeed know which particular solution is being used to provide them with the requested resources. Depending on the rack implementation, similar services can be provided in different ways in different racks without intruding on the user's awareness.

In this section, we discuss in more detail each type of resource, and how ExoGENI accomplishes the provisioning and orchestration needed to create user slices.

3.1 Compute Resources

Many of the ways in which ExoGENI provides compute services to its users are similar to those used by other testbeds. In this section, we concentrate on a few aspects of our approach that, in our opinion, are original and create value-adding services for the experimenters. These aspects include compute-image management, instance network-interface configuration, and post-boot application configuration as well as -scripting.

Compute instances are provisioned for users using a combination of OpenStack and xCAT, depending on whether the type is VM or bare-metal. A key orchestration challenge for multi-domain networked clouds is uniform management of images to

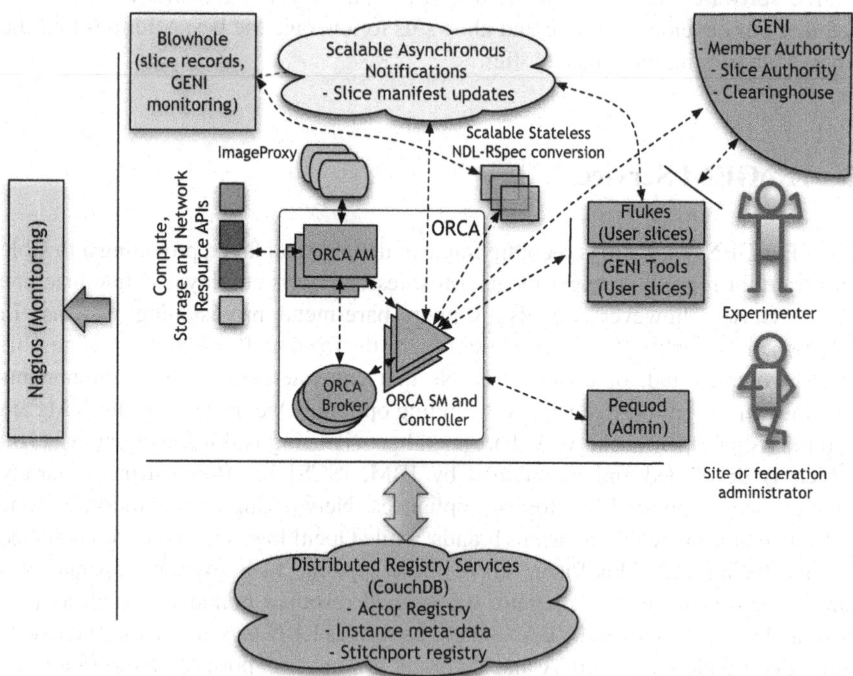

Fig. 5 The structure of ExoGENI software components and tools, in addition to the core ORCA components, includes a number of ancillary software components that act as registries, notification services, and management tools. Critically, the majority of them leverage common libraries and off-the-shelf open-source software, with extensions for ExoGENI

program the node instances. With standard IaaS cloud stacks, following the Amazon Elastic Compute Cloud (EC2) model, each cloud site requires a local user to pre-register each image with the site's cloud service, which then generates an image token that is local to that site. Networked clouds need a way to manage images across the member sites of a federation.

To increase flexibility, ExoGENI practices the BYOI (Bring Your Own Image) approach. Compute-instance images are merely cached inside each rack's object store, leaving their permanent storage to the users. Each rack AM includes a cloud-handler plugin to invoke EC2/OpenStack APIs and an *ImageProxy* server to obtain node images named by a URL in the user request. An *image* file specifies a canned operating system and application stack selected by the user in the form of a kernel file, an optional ramdisk file, and a filesystem. All components of the image are named independently by public URLs, thus allowing for mixing and matching of kernels, filesystems, and ram disks, possibly created by different users. ImageProxy is a stand-alone caching server that enables the cloud site to import images on demand from the network. In order to avoid unauthorized tainting, the image descriptor file specifies the SHA1 hash of each file, in addition to its URL.

In our approach, the creator of an image registers it at some shared image depository and names it by a URL. A request to instantiate a VM names the image descriptor file by a URL and a content hash for validation. The ORCA AM's cloud-handler plugin passes the image URL and hash to the local *ImageProxy* server. The ImageProxy fetches and caches any image components in the local object store if they are not already cached, and registers the image with the local cloud service if it is not already registered. It then returns a local token that the AM cloud handler may use to name the image to the local cloud service when it requests a VM instance. If a particular image is cached, ImageProxy merely returns the identifiers of the image components in the object store, skipping the download and registration steps. In principle, ImageProxy can work with any image server that supports image fetch by URL, e.g., the various Virtual Appliance Marketplaces operating on the web, Dropbox, Google Drive, and other cloud-storage services. It also supports BitTorrent URLs to allow scalable content swarming of images across many cloud sites.

To stitch various elements of slices together using network links, the compute instances must be informed of the specific VLANs picked to represent specific links, such that the virtual topology mimics the requested one. OpenStack Quantum presents a convenient API through which to inform virtual machines of their networking configuration, for which we implemented a custom Quantum plugin that is invoked by the AM handler code. Based on the configuration created by the controller from the user request, it creates the necessary Ethernet interface configuration for the VM. It then attaches the tap interfaces to tagged interfaces of the worker node in which the VM is being created. Notably, the interface configuration is generated *after* the VM starts to boot, such that the network interfaces appear in the VM as it is booting. This approach clears the way for us to have fully dynamic slices in which nodes and links can come and go independently of each other, rather than assuming a static slice topology.

Layer-3 (L3) configuration, i.e., IP-addressing and routing, is done guest-side (from inside the compute instance) by a set of guest extensions that we call NEuca-Quantum. The name is a combination of the legacy name "NEuca" for Networked Eucalyptus, originally developed for the Eucalyptus cloud platform, and the name "Quantum". The extensions comprise a set of scripts and a running service that are installed in every ExoGENI image. They perform necessary guest-side configurations based on metadata available to them *via* a REST interface. The metadata is presented to the service in the form of an INI file with a number of sections related to global, network, and storage configurations along with post-boot scripts. This information helps to configure application behavior inside the compute instance. Our plans include supporting fully dynamic behavior for slices, including network interfaces that can come and go as links are created and destroyed in a slice. Thus, we designed this guest-side service to expect the metadata to change over time. It periodically compares the metadata to what it has seen before and reconfigures the guest if there are any changes. This allows the user to update the behavior of the slice, attaching or removing links and storage, in the knowledge that the configuration of the instance OS will follow such user-specified changes in topology or behavior.

Post-boot scripts run on a compute instance at startup. They are used to install and configure software that is specific to each compute instance. Specifying post-boot scripts for each compute instance in an experiment enables other users to repeat the experiment using the same configuration. It is extremely desirable that scientists and experimenters should not have to rewrite these post-boot scripts when they scale their experiments to a larger number of resources, or when they make changes to topologies. Hence, ExoGENI offers *templated* post-boot scripts that allow scientists to scale their experiments from small test runs to large production runs with minimal effort.

A scalable post-boot script can configure the compute instance correctly regardless of the size of the experiment. This is achieved by allowing various elements of the slice named by the user (nodes, links, interfaces) to be referenced directly in the post-boot script templates. ORCA controller then takes care of substituting the templated variables with their slice-specific values. ORCA thus relieves users of the need to remember assigned values such as interface IP addresses. It also adds values that user scripts may need to know but that they cannot control, such as compute instance interface MAC addresses.

A common use of a scalable post-boot script is to create entries in `/etc/hosts` for each virtual machine so that machine names will resolve to IP addresses. High-performance and high-throughput distributed systems often require name resolution for proper functioning (e.g., HTCondor [32] and Hadoop [2, 19]). A slice typically consists of a *group* of worker nodes. A group may vary in size; however, all nodes are configured identically. Other nodes may be present in the topology as well, e.g., a head node. Since a slice has its own private network with user-assigned IP address space, it is desirable to enable the user software to learn this assignment easily. The remainder of this section is an example of a scalable post-boot script that configures `/etc/hosts`.

This example includes a head-node server (named Server) and a varying number of clients (named Client1, Client2, etc.) in a group, and a broadcast network (named VLAN0) connecting them. The clients might use the following post-boot script to configure /etc/hosts and to start the client software that connects to the server using its resolvable name. Note that this post-boot script has been simplified for formatting. It is written in *bash*, although any other scripting language executable in the guest VM would also work. Template variables have $ before them and lines beginning with # are commands for the Velocity templating engine, which is Turing-complete, allowing for rich capabilities for code auto-generation.

```
#!/bin/bash

echo $Server.IP("VLAN0") $Server.Name() >> /etc/hosts
#set ( $max = $Client.size() - 1 )
#foreach ( $i in [0..$max] )
 echo $Client.get($i).IP("VLAN0") `echo $Client.get
        ($i).Name()  | \ sed 's/\//-/g'` >> /etc/
         hosts
#end

/path/to/my/software/client-start $Server.Name()
```

In this example, respective entries for the server and for each client are added to the /etc/hosts file. The templated post-boot script uses variables to reference properties of resources in the request. For example, the IP address on the server assigned to the interface on VLAN0 can be referred to as $Server.IP("VLAN0"). The user submits the script with $Server.IP("VLAN0"), and the template engine replaces it with the actual IP address of the interface that is configured on the virtual machine towards the link named *VLAN0* in the slice request. In addition, node groupings of various sizes can be referenced using template language loops and conditionals, which expand to reference any number of compute nodes.

The example uses template variables to reference hostnames and IP addresses of a group of clients that varies in size. For the server and for each client, an entry is added to /etc/hosts, allowing each node's name to resolve to the IP address that is assigned to its interface on VLAN0. This templated script does not contain the actual names and IP addresses of the nodes, allowing a slice to be recreated with different IP addresses assigned to the compute nodes without modifying the script. After template expansion, the post-boot script adds an entry to /etc/hosts for the server and for each virtual machine in the group. The two output scripts displayed in Fig. 6 show the template expansions in cases in which the group has two and four virtual machines, respectively. The post-boot script makes the experiment scalable because it can be repeated with any number of compute nodes by simply modifying the number of nodes in the group, without affecting the template specified by the user.

The templating mechanism allows access to the following variables available to ORCA during topology-embedding: slice name and GUID, link and node names,

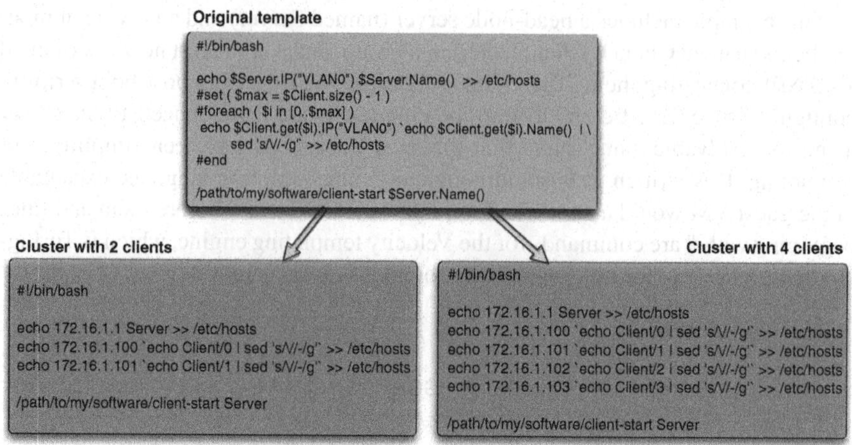

Fig. 6 Template expansion of a post-boot script with variable size compute node groups

interface MAC and IP addresses, and the size of a node group. The list of template variables is growing with user requirements.

3.2 Storage Resources

Many ExoGENI racks contain a 6-TB or larger iSCSI network-storage device that can be slivered by users; i.e., upon request, portions of this storage can be allocated dynamically to iSCSI targets that are accessible to compute nodes in the slices. In addition, the ExoGENI request can instruct the system to attach the iSCSI target to a particular compute node, to format the target block device with a specified filesystem, and to mount the filesystem at a specified point within a compute node.

Storage requirements of data-intensive applications often greatly exceed the available capacities of individual compute nodes, each of which typically is limited to a few tens of gigabytes. Such applications normally use dedicated network storage facilities available to dedicated compute resources. Typical multi-tenant cloud environments provide abstracted block devices that are virtualized by the hypervisor. Virtual machines' method of access to remote block storage through these virtual block devices is similar to that of access to their root block devices. Although this abstraction simplifies access to virtual block storage from the guest, it adds computational overhead and latency to I/O. Further, it limits the network storage to that which is managed by the local cloud system. It does not permit attaching multiple compute instances to the same block device without an intervening compute node.

In contrast, in ExoGENI, the racks contain network storage devices (iSCSI) that are sliverable and that can be partitioned into virtual iSCSI targets. The latter can be allocated to individual slices and accessed over dynamic L2 circuits in a manner that is *independent of compute resources*. These iSCSI targets are provisioned using mechanisms similar to those applied to compute and network resources. When the user requests a virtual storage device from ExoGENI, the storage target is created dynamically. The iSCSI target is stitched directly to the provisioned L2 network, opening it to access by other computational resources within the slice from the same or other racks. Storage stitching allows remote access over high-bandwidth low-latency networks. It also allows remote storage to use low-overhead network-virtualization technologies (e.g., single-root I/O virtualization, or SR-IOV).

Importantly, this approach allows infrastructure providers to expose different types of storage, depending on their capabilities. Broadly, we separate them into two models: a 'shared channel' model, in which the pool of storage is available to all tenants over a single shared VLAN, and a 'dedicated channel', in which individual storage targets can be reached by the individually bandwidth-provisioned channels. The provider's choice of the model to support is determined largely by the capability of the deployed storage technology. Appliances that are capable of tagging traffic to or from a particular target, e.g., to a specific channel or VLAN, can support the 'dedicated channel' model. Others must rely on the 'shared channel' model.

This choice has implications for performance and functionality. The 'dedicated channel' model isolates tenants' individual I/O-performance levels more effectively than the 'shared channel' model. Further, stitching of storage targets into slices is more robust when the 'dedicated VLAN' model is chosen, as it closely resembles the stitching of compute nodes.

To support guest-side configuration, the NEuca-Quantum tools described above were augmented to attach, to format, and to mount ExoGENI's iSCSI storage based on metadata parameters passed to the tools. When a virtual storage device is included in a request, the NEuca-Quantum tools can be instructed to configure the iSCSI initiator, to attach to the appropriate target, to format a filesystem, and to mount it within the machine. Alternatively, the tools can ignore remote storage, leaving the configuration to the user.

3.3 Rack-Local VLANs and OpenFlow

To connect edge resources such as storage and compute together, ExoGENI can allocate VLANs dynamically inside the rack, provision them using one of several available mechanisms, and stitch them into the edge resources. (See the chapter about ORCA for details of the stitching logic.) The capability of note that separates ExoGENI from other testbeds is its explicit support for hybrid switching, in which the users get to pick the type of VLAN that they wish to use in their slice. The worker nodes of the rack are dual-homed into the rack switches, with each switch logically separated into the "legacy VLAN" and the "OpenFlow" parts by ports. Due to limitations of switch implementations, the OpenFlow part is "uplinked" to the "VLAN" part, which is then uplinked to the outside world. When a request is submitted, the control software determines what type of connection the user is requesting and provisions the network appropriately. This approach provides the benefits of supporting reliable legacy VLAN functionality and experimental OpenFlow features in one platform, making it a user-selectable option, rather than an architectural or a deployment choice.

When the user requests an OpenFlow-type connection and passes information about its external controller, the control software allocates a VLAN tag. This tag is common to creating OpenFlow and VLAN connections. The control software proceeds to plumb the connection from the VM interface to the appropriate worker-node interface and, then, into the OpenFlow part of the switch. A separate handler interfaces with FlowVisor to create a slice matching the issued VLAN tag and points it at the user's OpenFlow controller. The user then can manage explicitly the way in which packets are forwarded inside this connection, which can be point-to-point or multipoint.

On the other hand, if the user does not request OpenFlow explicitly, the system assumes that the user prefers a legacy VLAN connection. It similarly allocates the VLAN tag. This time, however, it plumbs the VM interfaces to a different worker-node interface that is directed to the "legacy VLAN" part of the rack switch. Then, a separate handler uses one of the available interfaces, such as CLI (command-line interface) or Netconf, to provision the VLAN in the rack switch on appropriate ports.

3.4 Transit Network Providers

To stitch slices from slivers belonging to different providers or racks, ExoGENI offers its own inter-domain stitching engine. The engine can find and select paths, as well as allocate and provision, based on the model of inter-domain topology that ExoGENI maintains. This logic is encapsulated in the combination of the coordinator service and the various ExoGENI AMs interfacing to the transit network providers that provide L2 connectivity among the various racks. Figure 7 demonstrates the logical topology of ExoGENI with its various transit providers.

Notably, the transit providers fall into two separate categories. Some provide dynamic bandwidth-on-demand *via* a form of API and some do not. The former case includes the core providers, such as Internet2 AL2S (Advanced Layer-2 Service) and ION services, ESnet OSCARS, and the GÉANT NSI. The latter case includes the majority of the regional providers or connectors that support the connectivity between the racks and the core providers. These providers typically provide pools of static VLANs provisioned with resources from the rack switch, the interface to the core provider, and other points. Each of these VLANs is identified by a single tag, with ExoGENI software managing the allocation of tags to specific requested connections. In some cases, the providers require that the tags be remapped internally within their infrastructures, such that VLAN is seen as a tag X at the rack, while it appears at the interface to the core provider as tag Y.

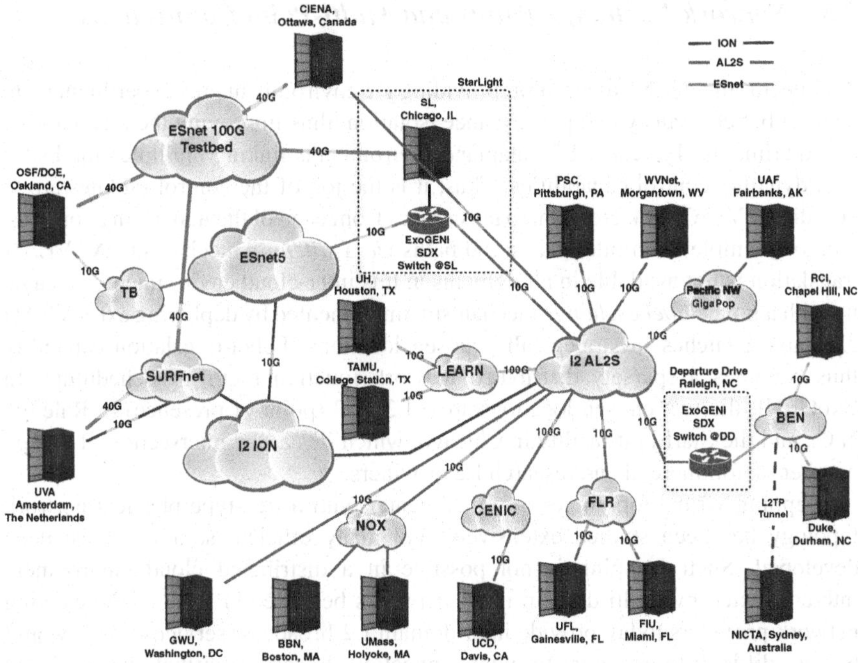

Fig. 7 Logical topology of ExoGENI racks as of mid-2014. The racks are interconnected by a variety of regional and global providers (Internet2, ESnet) with either static or programmable paths. ORCA native stitching manages the connectivity across ExoGENI racks, supporting point-to-point and multi-point connections, using stitching switches in Raleigh and Chicago

Some providers support Q-in-Q (provider bridging). ExoGENI accommodates all of these options using its coordination logic and topology descriptions. In addition to point-to-point connections, it also supports multi-point inter-rack connections, as described in detail in the following section.

ExoGENI's pathfinding and stitching guarantee path continuity, supporting dynamic tag translation when necessary to create end-to-end L2 connections among rack sites. Authorized slices also may link to other VLANs entering the dataplane switch from the campus network or backbone, such as the GENI Meso-Scale OpenFlow Waves on Internet2. Further, for connections to resources lying outside of ExoGENI control, we support the notion of a *"stitchport"*, a named peering L2 point with the outside world, described in Sect. 3.6. The topology-embedding logic of ExoGENI can create slices that include a connection from a slice element, like a compute node, to a known stitchport.

3.5 Network Exchange Points and Multi-Point Connections

Architecturally, GENI focused on providing L2 environments for experimenters to support better security and performance isolation, thus improving the repeatability of experiments. Typically, L2 dataplane environments linking multiple cloud sites lack dynamic control capabilities. Thus, it is the job of the control middleware to coordinate VLAN link creation, to tag choices from each of the underlying domains, and to assemble them into end-to-end paths *via* a *stitching* mechanism. As VLAN translation is not available in all domains in this inter-cloud environment, we came up with a novel *label exchange* mechanism, implemented by deploying extra VLAN exchange switches in strategically chosen locations. Label-translation capability thus becomes a sparsely distributed network function requiring scheduling. In ExoGENI, the two chosen locations are a L3 PoP (point of presence) in Raleigh, N.C., and the StarLight facility in Chicago, which lies at the intersection of a large number of commercial and research L2 providers.

Mapping virtual topologies to a data center with a tree-type physical network topology has been studied extensively, and many efficient solutions have been developed. Such mapping is not possible in a distributed cloud environment interconnected by multi-domain mesh networks because (1) none of the existing network providers today provide inter-domain L2 broadcast services; (2) it would be very difficult to use point-to-point connections to serve a virtual cluster placed in N cloud sites, either by setting up a full-mesh virtual topology among the N sites or by setting up extra routing nodes for a non-mesh virtual topology, as $O(N^2)$ connections are required; (3) dynamic circuit service and VLAN translation are only available in a few advanced providers among some of the cloud sites. As a result, an advanced stitching mechanism, by itself, is not enough to achieve the maximum capacity.

We designed our *VLAN exchange* mechanism and deployed it in strategically chosen locations to serve two purposes. It supports dynamic VLAN circuit service among cloud sites that can only preconfigure static VLAN circuits to other places, i.e., a *L2 exchange service*. Critically, the same architectural approach allows us to set up dynamic large-scale virtual L2 broadcast domains among *multiple VLAN segments*, which we call a *L2 broadcast service*. Figure 8 displays just such a *VLAN exchange*, consisting of two high-capacity Ethernet switches controlled by the provisioning software as part of the distributed cloud-provisioning software system. In our multi-domain multi-cloud network, the *exchange* is modeled as a separate domain, as it is owned not by transit network providers but by a separate entity offering the exchange service.

Our architecture uses switches in pairs due to a typical limitation of Ethernet switches, which are unable to translate more than one label to- and from the same tag on the same port. In Fig. 8, these labels are shown as V_2 and V_3. This is typical in our environment, in which the exchange sits next to a large transit provider, such that connections from multiple sites can coalesce simultaneously on a single port connecting the exchange to this provider. The addition of the second

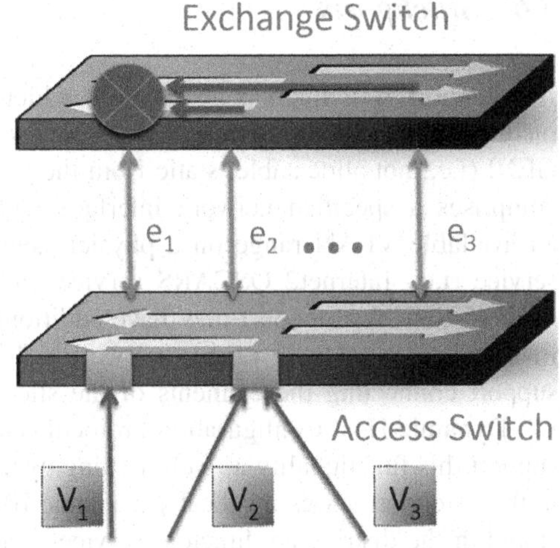

Fig. 8 Inter-domain VLAN exchange comprised of two switches that support remapping from one tag to another and creation of multi-point broadcast connections

Exchange Switch, connected to the Access Switch by several cables in parallel, allows each connection arriving at the Exchange Switch to be allocated a separate port on that switch, thus overcoming the tag-translation limitation. The number of cables connecting the two switches effectively limits the maximum cardinality of the multipoint connection. The Access switch provides tag translation for simple point-to-point connections. For multi-point broadcast requests, it forwards each incoming connection to a separate port on the Exchange Switch, which then internally creates a common tag to- and from which the incoming tags are translated.

The advantages of this solution are several. (1) Instead of $O(M^2)$, only $O(M)$ static connections are needed for M static sites in a federation. Consequently, instead of negotiating with preconfiguring static connections to every other site, a static transit provider only needs to do so with the *exchange*. (2) The basic stitching mechanism in the exchange is *VLAN or other label translation*, i.e., each site can preconfigure connections to the exchange domain with different VLAN ranges. The domain can pick up two arbitrary VLAN tags and connect them together *via* VLAN translation. This mechanism relieves operators of the obligation to enforce the stringent VLAN continuity requirement and to deal with the complexity of creating multi-domain VLAN connections. It also reduces the dependency on other advanced carrier Ethernet capabilities (e.g., Q-in-Q) in order to set up end-to-end L2 connections. (3) The additional benefit of the enabled *L2 broadcast service* makes it a relatively simple process to create a virtual L2 broadcast domain for a virtual cluster embedded in multiple cloud sites. Each participating cloud site creates a connection to the *exchange* domain, and the exchange switch translates the incoming VLANs into one common VLAN. (4) L2 network isolation is ensured between different users' virtual systems and bandwidth guarantees. (5) Every cloud site still can implement, independently, its own virtual networking solution (provisioning V_i), as long as it can interface with the exchange service *via* the negotiated set of services, e.g., VLANs.

3.6 Stitchports

A stitchport is a flexible abstraction, which gives ExoGENI the capability of including in a slice a number of resources outside the direct control of ExoGENI (i.e., not allocatable; static from the perspective of ExoGENI). A stitchport comprises a specified network interface within ExoGENI topology along with an available VLAN range on a physical switch or a URN of a network transit service (i.e., Internet2 OSCARS service), advertised as a special resource type in ExoGENI. A stitchport may have additional meta-information associated with it, such as available IP ranges, as well as local DNS and router information to support connecting the elements of the slice to the external static elements *via* L3. Dynamic Host Configuration Protocol (DHCP) can be used by the campus to support this functionality as well, issuing dynamic elements of the slice connected to the stitchport leases for locally available IP addresses. This meta-information is stored in the distributed directory services deployed in ExoGENI, as described in Sect. 4.1.

Since its deployment, the stitchport capability has found many uses in the interoperation of ExoGENI with external resources. The primary use case is so-called *cloud bursting*, which allows additional cloud computational resources (from ExoGENI) to be used on demand, for instance, when the user's campus HPC share is insufficient, or when certain data-analytic tasks need to be done, but the user has no dedicated computational and network resources. Another use case is referred to as an *on-ramp*, and is used to ferry data between campus resources and elements of user slices. This mechanism was also used for doing *hardware-in-the-loop* experiments with SmartGrid equipment emulating the behavior of a large number of PMUs. The information from the PMU emulator was delivered into the slice *via* a stitchport, then processed and stored by elements of the slice [6].

4 ExoGENI Administration

This section concentrates on the operational aspects of ExoGENI as they relate to scalable configuration as well as to the monitoring and administration of this widely distributed infrastructure. Considering the relatively small size of the ExoGENI operations team (two system administrators and a network administrator), which is responsible for maintaining, managing, and upgrading a large number of widely distributed sites, we paid special attention to procedures and tools that could be brought in to help with these tasks.

4.1 Scalable Configuration

ExoGENI comprises many layers, combining off-the-shelf and custom software with diverse hardware. Yet, it must present a relatively uniform view of its capabilities to users and, further, must remain secure from a variety of abuses. It is difficult to maintain a consistent configuration of the various elements across sites while accounting for differences among racks in hardware and software deployments. To achieve this goal, we took a two-pronged approach. On the one hand, ExoGENI reduces the amount of configuration that administrators must perform and, on the other hand, it automates configuration processes across racks.

We decided on Puppet [28] as the main supportive tool and mechanism to use in automating the configuration process. From our perspective, Puppet offered a number of advantages. It is modular and easily customizable, allowing us to reuse many existing configuration modules and to write our own, when necessary, to support configuration of custom pieces of ExoGENI. It is templated, therefore allowing us to maintain consistency while introducing customizations into each rack's template as necessary. Our initial deployment of Puppet was fully hierarchical. One Puppet master, on the testbed control node, was able to trigger changes across all racks at the same time. In the end, while being convenient in many respects, this approach did not work well in situations in which control over racks was shared between ExoGENI operators and campus owners, as is the case in a federated system. Thus, the Puppet configuration on which we settled was deployed in each rack's head node based on templates that are stored and updated using a version-control system from a central repository. This arrangement nicely balances consistency of behavior with the needs of individual substrate owners.

At this writing, each rack is set up initially using Puppet by installing it on the partially installed head node along with its templates. The latter then trigger the installation of all the software on the head node and worker nodes, the adjustment of file permissions, and the automatic generation of many of the configuration files. This significantly reduces the time it takes to set up a rack and guarantees that the setup is consistent across the racks. In normal operation, Puppet is used to trigger maintenances by managing user whitelists, deploying common changes to configuration files, and updating the software. The modularity, extensibility, and open-source nature of Puppet allow us to continue customizing it to take on an ever-larger portion of administration tasks in ExoGENI.

The second part of our strategy in reducing operator load and making configuration more scalable involved deploying a CouchDB-based [1] distributed *registry service* (Fig. 4) with which various parts of ExoGENI can communicate. The first use case for this infrastructure was the establishment of security associations among ORCA actors in various racks. ORCA depends on establishing these relationships using self-signed certificates and private keys generated by each actor in the testbed. There are dozens of these actors: four per rack, not counting those from transit providers, exchanges, and other global actors. Many of these actors need to communicate with each other. Manually propagating certificates to establish trust among them is not feasible for the operators, since the process is quite vulnerable to errors. Instead, we rely on a CouchDB-based registry to store actor certificates. Distribution of certificates to other actors is done using a simple GUI for the operators.

Upon start-up, each actor automatically registers its GUID, certificate, and other meta-data with the registry service. The operator can approve the actor as "trusted" through a CouchDB custom plugin interface. Other actors can request information about approved actors, including their certificates, thus automatically creating the necessary trust relationships and removing the need for the operator to do it. Additionally, this registry service is used to store metadata about stitchports. We anticipate that other uses will be found in future, wherever structurally simple metadata needs to be stored in a resilient fashion and to be made accessible across the testbed.

We reviewed a number of NoSQL (eventually) consistent database solutions. We decided on CouchDB because of its relative maturity and its ability to support master-master replication, which is important in a federated environment such as ExoGENI. Its distributed nature makes the infrastructure resilient to failures of individual instances, while the ability to support master-master replication means that individual clients can choose to contact any of the replicas with their updates (first available) and the changes eventually will propagate to other masters. Currently, ExoGENI operates several instances of CouchDB, deployed on geographically distributed head nodes in the testbed.

4.2 Monitoring

Another important aspect of ExoGENI operation is monitoring the hardware and software infrastructure. There are two primary tasks that the monitoring infrastructure performs. One is the liveness checks of components in order to send alarms to operators when a particular component is no longer reachable. The other is the monitoring of mappings between virtual infrastructure created by users and the physical infrastructure of the testbed. This latter task supports multiple goals. From the point of view of operators, it is critical to know which user slivers correspond to which physical components to support debugging and troubleshooting. It also fulfills LLR requirements in case unauthorized activities like network scans or storage and distribution of illegal content are traced back to user slivers. From the point of view of users, this information is part of the overall picture that they may need to analyze the performance of the slices in their experiments.

ExoGENI relies on two primary solutions. Nagios CheckMK [21] is used for liveness monitoring and collection of performance information from the substrate. At the same time, a combination of XMPP pubsub and a custom tool called Blowhole helps to collect, distribute, and store information about the mappings between virtual and physical resources (see Fig. 4).

As is the case with Puppet, CheckMK is deployed independently in each rack to monitor rack components. An additional instance of it monitors the state of other components of ExoGENI, like ORCA actors operating on transit providers, CouchDB instances, LDAP, and other logically centralized testbed functions. Also as with Puppet, we rely on a mix of standard and custom modules to monitor the various parts of the infrastructure. Modularity and the open-source nature of CheckMK make this an easily extensible solution that grew scalably with the testbed.

Timely information about the mapping between slivers and physical components can only come from the control framework software performing the topology-embedding. In our case, this information is held by the controller, either in the rack SM or in the ExoSM. The ExoGENI slice manifest holds the definitive information.

Importantly, since slices in ExoGENI change over time, the manifest can change over time as well. Another critical consideration is that the number of consumers of this information can be quite large, with various tools acting on behalf of different users and of various components in the monitoring framework. Thus, we needed a solution that would scale both with the number of racks and with the number of users, even as it reduced the communication load on the system by proactively propagating any changes in the manifests to those entities that were interested in them.

This brought us to a design based on a publish-subscribe paradigm, which has been implemented in a number of protocols. One of the most common of these is XMPP (XEP-0060) [23], a subset of the XMPP protocol that is used in Jabber Instant Messaging. Another candidate was AMQP. At the time, however, XMPP implementations were more robust, so we selected the OpenFire [15] open-sourced XMPP server with robust support for XEP-0060. Using the XMPP protocol with pubsub provided us with the necessary features. XMPP servers can be federated to create resilient distributed infrastructure to use for publishing and updating manifests. Their support of pubsub means that clients can express interest in specific manifests for particular slices and can be notified whenever a controller publishes an updated manifest. This approach also creates an architectural separation between controllers (the originators of the manifests) and their consumers. Since both are relying on a well defined protocol, both types of entities can evolve independently.

We then developed Blowhole, an XMPP client that could operate on the manifests generated by controllers and perform various actions on them via a system of plugins. The currently supported actions include saving the manifest into a history database, converting it to RSpec, publishing it to the GMOC (GENI Meta-Operations Center), and parsing it out to provide detailed mapping information for GENI monitoring. A number of Blowhole instances are deployed in the testbed; a central instance collects all manifests from all coordinator controllers to preserve historical information about testbed use. Each head node also has an instance of Blowhole supplying the mapping information to GENI for monitoring purposes. This information is correlated with the data collected by CheckMK to create a complete picture of the health of the testbed. We anticipate using this capability in user tools to notify them of changes in their slices and to avoid polling.

4.3 Administration

A vital requirement to operate the testbed is support for the management of the states of ORCA actors, both by administrators at campuses hosting the ExoGENI racks and by our team. Our chosen solution needed to operate in a federated environment featuring shared responsibilities for rack management while supporting separate management of varying subsets of aggregates, ranging from the entire ExoGENI to a single rack, according to need.

Like many other projects, ORCA started with a portal-based solution, in which a JSP-based portal was attached to each container hosting ORCA actors. This solution rapidly became non-scalable, since completing even simple tasks involving multiple actors required the operator to be logged into several portals simultaneously. As a result, we replaced the portal with an ORCA management Web-services-based API, which then permitted us to design a command-line tool called Pequod (named after the ship in "Moby Dick"). Pequod could communicate with any number of containers and actors as well as perform complex actions without having the operator think about how those are deployed. Our approach does not preclude the construction of a portal; however, we found that operators prefer command-line tools to portals.

The configuration file of Pequod accepts specification of any number of ORCA containers to which the configuring operator holds authorizing credentials. This allows several operators to share responsibilities for the same rack at a granularity of a single container. Once logged in, the operator can inspect the state of ORCA actors in the containers, see active slices and slivers, extend and close slivers on demand, and perform other management tasks. The operator uses actor names to refer to them without the need to specify in which container they reside. Importantly, Pequod is scriptable using its own language, making it possible to automate many of the frequently performed operations and simplifying the jobs of the operators.

5 ExoGENI User Tools

5.1 Flukes

When designing tools for users, we were guided by two main concerns. We wanted to expose the full capabilities of ExoGENI, and to make it easy for experimenters using ExoGENI to run large multi-domain experiments at a variety of scales. To fulfill those requirements, we developed Flukes (see Fig. 9), a Java Webstart GUI-based tool that allows the experimenter to draw the desired slice graphically, to request it from the system, and to interact with the manifest. To improve the usability of Flukes, we applied automatic graph-layout algorithms to manifests so that they could be displayed more intuitively and more easily. Flukes operates using NDL-OWL resource descriptions (discussed in more detail in Sect. 5.2) and an ORCA native API, as shown in Fig. 3.

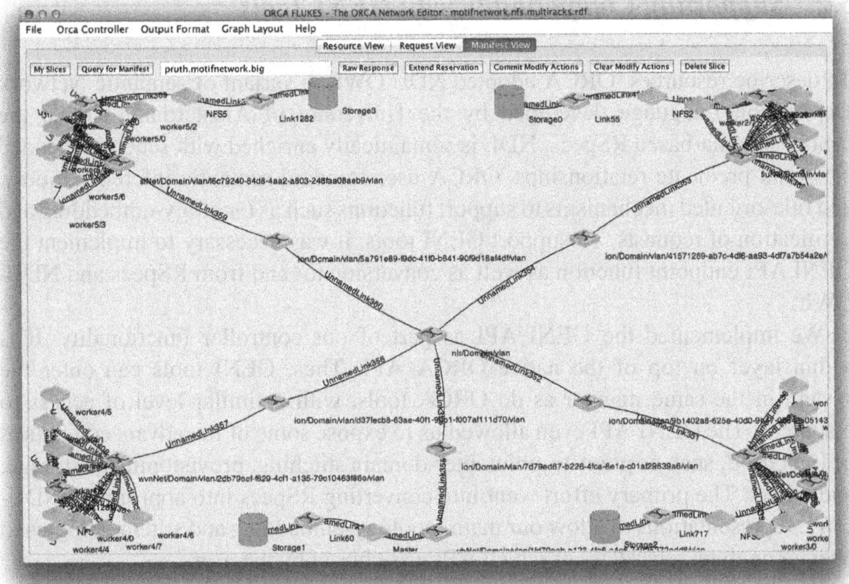

Fig. 9 Flukes, the ExoGENI graphical user tool, is designed to exercise the most advanced features that ExoGENI offers to support scalable experiments. It uses ORCA APIs and resource descriptions. Flukes supports automatic graph layouts to help users to construct and to display complex slices. The slice in the figure includes a five-way broadcast connection across multiple sites, each of which instantiates a large number of compute instances (*yellow*) and an iSCSI storage volume (*green*). *Grey* crossconnect icons indicate transit network domains on each path (Color figure online)

The approach to user tools is one of the deep differences between ExoGENI and other GENI testbeds. In GENI, the portion of logic responsible for stitching and embedding resides with the user tools. In contrast, in ExoGENI, the tools are very simple. Flukes is merely a converter from a graphical representation of a slice on the screen to NDL-OWL and back. The main logic is contained in the controller. The controller acts on behalf of the user; however, it is integrated more closely with the rest of the control framework. It has access to resource-scheduling decisions via the broker, a component not present in GENI architecture. Thus, while the controller performs all of the complex functions of topology-embedding and stitching, it can do so reliably because it can request resources from the broker directly, rather than performing two-phase commit- or request/release cycles, as is done, e.g., in GENI stitching. To support diverse user needs, multiple controller plugins with different logics can be present, and they can interact with end-user tools in different ways using the APIs and resource descriptions of their choice.

5.2 Supporting Compatibility with GENI Tools

To describe resources, ORCA adopted NDL-OWL, a variant of semantic Network Description Language developed by the University of Amsterdam. Unlike the syntax-schema-based RSpecs, NDL is semantically enriched with schema-encoded class and predicate relationships. ORCA uses standard semantic inference, query, and rule-oriented mechanisms to support functions such as topology-embedding and verification of requests. To support GENI tools, it was necessary to implement the GENI API endpoint function as well as conversion to- and from RSpecs and NDL-OWL.

We implemented the GENI API as part of our controller functionality. It is a thin layer on top of the native ORCA API. Thus, GENI tools can enter the system in the same manner as do ORCA tools, with a similar level of access to resources. The GENI API even allowed us to expose some of the advanced features of ExoGENI, such as point-to-point inter-domain stitching, provisioning of storage, and others. The primary effort went into converting RSpecs into appropriate NDL-OWL representations to allow our main topology-embedding and stitching logics to operate on them regardless of which API was chosen by the user.

We developed an on-line facility that translates GENI RSpec documents into NDL-OWL and back (see Fig. 5). Three types of documents had to be converted: slice requests, slice manifests, and aggregate advertisements. It is worth noting that the full two-way conversion was not required in all cases. We did need to convert RSpec requests into NDL-OWL requests as well as to convert NDL-OWL manifests and advertisements into RSpec. We insisted that translation be stateless, i.e., that one document go in and one document come out without any state stored in the system or acquired from elsewhere. Acquiring extra information might have required the conversion process to hold the same authorization level as the user. Such a requirement would have prevented the decoupling of the conversion from the rest of the logic, resulting in a less scalable system.

The "progressive annotation" approach used in RSpecs for conversions from requests to manifests was mostly compatible with this vision. However, we encountered difficulties with some user tools that wanted to use their private RSpec schema extensions. These extensions would not normally be interpreted by the control framework, but simply would emerge as part of the manifest unchanged, attached to the same topology elements. The XML schema approach practiced in RSpecs explicitly supports this behavior, essentially turning the RSpec into a channel of communication that the user tool can use to pass information to itself or other user tools from a slice request into its manifest. On the other hand, the conversion of the request to NDL-OWL relies on an established information model; it omits anything that it does not recognize as an extension important for the control framework. The reverse conversion of NDL-OWL manifest to RSpec then loses the information presented by the tool in the request.

We addressed this problem with proper modeling. The proposed use cases from the experimenters for this facility amounted to using RSpec request as a property graph that could be annotated arbitrarily on edges and links, with the ability to add dependencies between elements of the request. The annotation needed to re-emerge in the manifest in the same way to be useful to the tool. We defined a *coloring* extension to our NDL-OWL information model that encodes annotations to nodes and links, allowing the establishment of "colored" dependencies between them. The color is represented by an arbitrary label and denotes a namespace of the application. The application then can add property lists or XML documents as annotations to elements of the RSpec graph. These annotations would be converted into a NDL-OWL model, pass through the control framework as expected, and re-emerge as RSpec manifests after conversion.

In the following example, request excerpt node geni1 is annotated under color "gemini" with an XML blob that is understood by the application, but not by the control framework:

```
<node client_id="geni1" component_manager_id="urn:
   publicid:IDN+rcivmsite+authority+cm">
 <sliver_type name="xo.medium">
   <disk_image
        name="http://geni-images.renci.org/images/
           standard/debian/deb6-neuca-v1.0.9.xml"
        version="e1972b5a5b30fa1adbd42f2df1effbd40084fb3e" />
 </sliver_type>
 <interface client_id="geni1:0">
    <ip address="172.16.22.1" netmask="255.255.255.0" />
 </interface>
   <color:resource_color color="gemini">
     <color:xmlblob xmlns:gemini="http://geni.net/
         resources/rspec/ext/gemini/1">
      <color:blob>
      <gemini:node type="mp_node" >
        <gemini:services>
          <gemini:active install="yes" enable="yes"/>
          <gemini:passive install="yes" enable="yes"/>
        </gemini:services>
      </gemini:node>
     </color:blob>
     </color:xmlblob>
    </color:resource_color>
</node>
```

Relying on the stateless nature of the translation, we were able to deploy a number of RSpec-to-NDL-OWL translators across the testbed to support highly scalable demands from the various GENI tools. ORCA agents communicating with the tools requested translation as necessary from these translators, with the available translators picked randomly from the preconfigured available list. This approach proved both scalable and easy to manage, as converters could be upgraded easily, separately from the rest of ORCA. It allows gradual introduction of new features from ORCA into GENI by adding new translation capabilities into the translators simply by updating their code.

6 ExoGENI and the GENI Federation

6.1 Relationship to Other GENI Testbeds

ExoGENI is significant in part because it offers our first opportunity to evaluate the federated IaaS model in a production testbed. The ExoGENI principles represent a departure in the GENI effort, whose current standards evolved from testbeds that were established and accepted by the research community at the start of the GENI program in 2007: PlanetLab [27], Emulab [33], and ORBIT [29]. Each testbed developed its own control software to manage substrates that are dedicated permanently to that testbed and remain under the direct control of its central testbed authority.

The ExoGENI testbed is the first GENI-funded substrate whose control software preserves provider autonomy by departing from that model. Instead, it uses standard virtual infrastructure services that may be deployed and administered independently and/or shared with other applications. Unlike other GENI testbeds, ExoGENI does not mandate GENI as its only access interface; native OpenStack interface is still available on each ExoGENI installation, and providers can choose the exact portion of their capacity that they wish to delegate to GENI uses.

We intend that ExoGENI serve as a nucleus for a larger, evolving federation encouraging participation from independent cloud sites, transport networks, and testbed providers, beyond the core GENI-funded substrate. An important goal of the project is to provide a foundation for intuitive and sustainable growth of a networked intercloud through a flexible federation model that allows private cloud sites and other services to interconnect and to share resources on their own terms. The ExoGENI model offers potential to grow the power of the testbed as infrastructure providers join the effort and as their capabilities continue to advance. In time, these advances may lead not just to real deployment of innovative distributed services but also to new visions of a Future Internet.

This goal requires an architecture that supports and encourages federation of sites and providers. It must expose their raw IaaS capabilities, including QoS, to testbed users through common APIs. It requires a structure that differs from those of the GENI predecessors, whose primary uses have been to evaluate new ideas under controlled conditions (for Emulab and ORBIT) and to measure the public Internet as it currently exists (for PlanetLab).

ExoGENI may be viewed as a group of resource providers (aggregates) within a larger GENI federation. ExoGENI itself is an instance of the GENI architecture and supports GENI APIs as well as additional native ORCA interfaces and capabilities that are not yet available through standard GENI APIs. This section outlines the integration of ExoGENI with GENI and concomitant extension of the GENI federation model.

6.2 Aggregates

GENI aggregates implement standard GENI APIs for user tools to request resources. ExoGENI AMs, as described in this chapter, are orchestration servers built with the ORCA toolkit. They support internal ORCA protocols rather than the standard GENI aggregate APIs. The GENI APIs are evolving rapidly to support more advanced control and interconnection of slices as well as of rich resource representations and credential formats. Rather than implementing the GENI API directly in the AMs, the initial ExoGENI deployment proxies them through a GENI API implementation on its controllers.

This approach enables ExoGENI to present a secure and flexible interface to the GENI federation and to support standard user tooling for GENI. At the same time, ExoGENI supports end-to-end slice construction across ExoGENI aggregates, based on native ORCA capabilities. The AM operator interface also allows local policies that limit the local resources available to the testbed over time. This feature enables ExoGENI providers to hold back resources from the testbed for other uses, according to their own policies.

6.3 GENI Federation: Coordinators

GENI aggregates delegate certain powers and trust to *coordinator* services. The coordinators help aggregates to cooperate and to function as a unified testbed. For example, GENI currently defines coordinators to endorse and to monitor participating aggregates; to authorize and to monitor use of the testbed for approved projects; and to manage user identities along with their associations with projects. The GENI coordinators are grouped together under the umbrella of a GENI *Clearinghouse*, but they act as a group of distinct services endorsed by a common GENI root authority.

The GENI federation architecture allows participating aggregates to choose for themselves whether to accept any given coordinator as well as what degree of trust to place in it. These choices are driven by federation governance structure. For reasons of safety and simplicity, GENI opted for a hierarchical governance structure for its initial trial deployment. To join the GENI federation, an aggregate must enter into certain agreements, including compliance with various policies and the export of monitoring data to GENI coordinators.

The aggregates in the initial ExoGENI deployment enter into mandated agreements with GENI. They must accept and trust all GENI coordinators. Specifically, ExoGENI controllers trust GENI-endorsed coordinators to certify users, issue keypairs to users, authorize projects, approve creation of new slices for projects, authorize users to operate on approved slices, and endorse other aggregates in GENI. The GENI coordinators, in turn, delegate some identity management functions to identity systems operated by other GENI testbeds and participating institutions (Shibboleth/inCommon).

6.4 Integration with GENI

For each slice, GENI users and their tools choose whether to access the ExoGENI testbed as a single aggregate through the ExoSM or as a collection of distinct site aggregates through the site SMs. External GENI tools can interact with ExoGENI site aggregates based on the current GENI architecture and API, in which aggregates are loosely coupled except for common authorization of users and slices. Each ExoGENI slice may also link to other GENI resources to the extent that the standard GENI tools and APIs support that interconnection.

At the same time, the ExoSMs and ExoBroker allow ExoGENI to offer automated cross-aggregate topology-embedding, stitching, and allocation of resources within the ExoGENI testbed. These capabilities are currently unique within GENI. They are based on coordinator services, APIs, resource representations, and tools that are not part of a GENI standard. In particular, GENI defines no coordinators for resource management, so cooperation among GENI aggregates is based on direct interaction among AMs or exchanges through untrusted user tools. GENI is developing new extensions that would offer capabilities similar to those of ExoGENI, such as automated configuration of cross-aggregate virtual networks.

ExoGENI also differs from current GENI practice with respect to the usage model for OpenFlow networks. GENI views an OpenFlow datapath as a separate aggregate that allocates the right to direct network traffic flows matching specified packet header (flowspace) patterns, which are approved by an administrator. In ExoGENI, OpenFlow is an integrated capability of the ExoGENI rack aggregates, rather than a distinct aggregate in itself. ExoGENI slices may designate OpenFlow controllers to direct network traffic within the virtual network topology that makes up the dataplane of the slice. ExoGENI is VLAN-sliced: each virtual link corresponds to a unique VLAN tag at any given point in the network. The handler plugins of the ExoGENI rack AMs authorize the controllers automatically, so that the designated controllers may install flow entries in the datapath for VLANs assigned to the slice's dataplane. We believe that this approach can generalize to other OpenFlow use cases in GENI and cloud networks.

7 Conclusion

This chapter describes the design of the ExoGENI testbed, which addresses the goals of GENI by federating diverse virtual infrastructure services and providers. The approach taken in ExoGENI offers a means by which to leverage IaaS advances and infrastructure deployments occurring outside of GENI. It also harnesses GENI technologies to address key problems of interest outside of the GENI community, such as linking and peering cloud sites, deploying multi-site cloud applications, and controlling cloud network functions.

Many experimenters, including ourselves, are actively exploring the use of ExoGENI-like infrastructure for domain science. The list of projects include the aforementioned SmartGrid "hardware-in-the-loop" [6], a project called Adaptive Data-Aware Multi-domain Application Network Topologies (ADAMANT) [20], funded by NSF, to couple closely the Pegasus Workflow Management System and a class of NIaaS infrastructures, the latter represented by ExoGENI.

We and other teams are exploring how to run computational science applications on ExoGENI. For example, we have demonstrated successfully how ADCIRC [30] storm-surge modeling software can run in a GENI slice. Other teams have explored running Hadoop in a widely distributed environment across multiple ExoGENI sites, aided by its advanced topology-embedding features. The list of applications is growing every day, which we consider a very positive development and a confirmation of our vision. ExoGENI offers an architecture for federating cloud sites, linking them with advanced circuit fabrics, and deploying multi-domain virtual network topologies. The initial deployment combines off-the-shelf cloud stacks, integrated OpenFlow capability, linkages to national-footprint research networks, and exchange points with international reach.

ExoGENI and its ORCA control framework make possible the construction of elastic Ethernet/OpenFlow networks across multiple clouds and circuit fabrics. Built-to-order virtual networks are suitable for flexible packet-layer overlays that use IP or other protocols selected by the owner. IP overlays may be configured with routed connections to the public Internet through gateways and flow switches. ExoGENI can also serve a broader role as a model and platform for future deeply networked cloud services and applications.

Acknowledgements We thank NSF, IBM, and the GENI Project Office (GPO) at BBN-Raytheon for their support. Many colleagues at GPO and other GENI projects have helped work through issues relating to ExoGENI. We'd like to thank our colleagues from the EU: University of Amsterdam SNE Group and Ghent University/iMinds for their help and code contributions.

This work is supported by the US National Science Foundation through the GENI initiative and NSF awards OCI-1032873, CNS-0910653, and CNS-0720829; by IBM and NetApp; and by the State of North Carolina through RENCI.

References

1. Apache Foundation. CouchDB (2016). http://couchdb.apache.org/
2. Apache Hadoop (2016). http://hadoop.apache.org/core
3. Baldine, I.: Unique optical networking facilities and cross-layer networking. In: Proceedings of IEEE LEOS Summer Topicals Future Global Networks Workshop (2009)
4. Baldine, I., Xin, Y., Evans, D., Heermann, C., Chase, J., Marupadi, V., Yumerefendi, A.: The missing link: putting the network in networked cloud computing. In: ICVCI: International Conference on the Virtual Computing Initiative (an IBM-Sponsored Workshop) (2009)
5. Baldine, I., Xin, Y., Mandal, A., Heermann, C., Chase, J., Marupadi, V., Yumerefendi, A., Irwin, D.: Autonomic cloud network orchestration: A GENI perspective. In: 2nd International Workshop on Management of Emerging Networks and Services (IEEE MENS '10), in Conjunction with GLOBECOM'10 (2010)

6. Chakrabortty, A., Xin, Y.: Hardware-in-the-loop simulations and verifications of smart power systems over an exo-geni testbed. In: 2013 Second GENI Research and Educational Experiment Workshop (GREE), pp. 16–19 (2013)
7. Chase, J., Baldin, I.: A retrospective on ORCA: Open resource control architecture. In: GENI: Prototype of the Next Internet. Springer, New York (2016)
8. Chase, J.S., Irwin, D.E., Grit, L.E., Moore, J.D., Sprenkle, S.E.: Dynamic virtual clusters in a grid site manager. In: Proceedings of the 12th International Symposium on High Performance Distributed Computing (HPDC) (2003)
9. Dijkstra, F.: Framework for path finding in multi-layer transport networks. Ph.D. thesis, Universiteit van Amsterdam (2009)
10. Ford, E.: From Clusters To Clouds: xCAT 2 Is Out Of The Bag. Linux Magazine, Jan 2009
11. Fu, Y., Chase, J., Chun, B., Schwab, S., Vahdat, A.: SHARP: an architecture for secure resource peering. In: Proceedings of the 19th ACM Symposium on Operating System Principles (2003)
12. Guok, C., Robertson, D., Thompson, M., Lee, J., Tierney, B., Johnston, W.: Intra and interdomain circuit provisioning using the OSCARS reservation system. In: Proceedings of the 3rd International Conference on Broadband Communications, Networks and Systems (BROADNETS) (2006)
13. Ham, J.V.: A semantic model for complex computer networks. Ph.D. thesis, University of Amsterdam (2010)
14. Ham, J., Dijkstra, F., Grosso, P., Pol, R., Toonk, A., Laat, C.: A distributed topology information system for optical networks based on the semantic web. J. Opt. Switch. Netw. **5**(2–3), 85–93 (2008)
15. Ignite Realtime. OpenFire (2016). http://www.igniterealtime.org/projects/openfire/
16. Irwin, D., Chase, J.S., Grit, L., Yumerefendi, A., Becker, D., Yocum, K.G.: Sharing networked resources with brokered leases. In: Proceedings of the USENIX Technical Conference (2006)
17. Irwin, D., Chase, J., Grit, L., Yumerefendi, A.: Underware: an exokernel for the Internet? Technical report, Duke University Department of Computer Science (2007)
18. Kaashoek, M.F., Engler, D.R., Ganger, G.R., Briceno, H.M., Hunt, R., Mazieres, D., Pinckney, T., Grimm, R., Janotti, J., Mackenzie, K.: Application performance and flexibility on exokernel systems. In: Proceedings of the Sixteenth Symposium on Operating Systems Principles (SOSP) (1997)
19. Mandal, A., Xin, Y., Ruth, P., Heerman, C., Chase, J., Orlikowski, V., Yumerefendi, A.: Provisioning and evaluating multi-domain networked clouds for Hadoop-based applications. In: Proceedings of the 3rd International Conference on Cloud Computing Technologies and Science 2011 (IEEE Cloudcom '11) (2011)
20. Mandal, A., Ruth, P., Baldin, I., Xin, Y., Castillo, C., Rynge, M., Deelman, E.: Leveraging and adapting ExoGENI infrastructure for data-driven domain science workflows. In: 2014 Third GENI Research and Educational Experiment Workshop (GREE), pp. 57–60. IEEE, New York (2014)
21. Mathias Kettner. CheckMK (2016). https://mathias-kettner.de/check_mk.html
22. Mell, P., Grance, T.: The NIST definition of cloud computing. Special Publication 800-145, Recommendations of the National Institute of Standards and Technology (2011)
23. Millard, P., Saint-Andre, P., Meijer, R.: XEP-0060: Publish-Subscribe (2010). http://www.xmpp.org/extensions/xep-0060.html
24. Nurmi, D., Wolski, R., Grzegorczyk, C., Obertelli, G., Soman, S., Youseff, L., Zagorodnov, D.: The eucalyptus open-source cloud-computing system. In: Proceedings of the 9th IEEE/ACM International Symposium on Cluster Computing and the Grid (CCGRID) (2009)
25. OGF NSI WG. Network Service Interface (2012). http://redmine.ogf.org/projects/nsi-wg
26. OpenStack (2016). http://www.openstack.org
27. Peterson, L., Bavier, A., Fiuczynski, M.E., Muir, S.: Experiences building PlanetLab. In: Proceedings of the 7th Symposium on Operating Systems Design and Implementation (OSDI) (2006)
28. PuppetLabs. Puppet Configuration Management tool (2016). http://puppetlabs.com/puppet/what-is-puppet

29. Raychaudhuri, D., Seskar, I., Ott, M., Ganu, S., Ramachandran, K., Kremo, H., Siracusa, R., Liu, H., Singh, M.: Overview of the ORBIT radio grid testbed for evaluation of next-generation wireless network protocols. In: Proceedings of the IEEE Wireless Communications and Networking Conference (WCNC) (2005)
30. Ruth, P., Mandal, A.: Toward evaluating GENI for domain science applications. In: International Workshop on Computer and Networking Experimental Research using Testbeds (2014)
31. Ruth, P., Jiang, X., Xu, D., Goasguen, S.: Virtual distributed environments in a shared infrastructure. Computer **38**(5), 63–69 (2005)
32. Thain, D., Tannenbaum, T., Livny, M.: Distributed computing in practice: the condor experience. Concurr. Pract. Exp. **17**(2–4), 323–356 (2005)
33. White, B., Lepreau, J., Stoller, L., Ricci, R., Guruprasad, S., Newbold, M., Hibler, M., Barb, C., Joglekar, A.: An integrated experimental environment for distributed systems and networks. In: Proceedings of the 5th Symposium on Operating Systems Design and Implementation (OSDI), pp. 255–270 (2002)
34. Xin, Y., Baldine, I., Mandal, A., Heermann, C., Chase, J., Yumerefendi, A.: Embedding virtual topologies in networked clouds. In: 6th ACM International Conference on Future Internet Technologies (CFI) (2011)

The InstaGENI Project

Rick McGeer and Robert Ricci

Abstract In this chapter we describe *InstaGENI*, built in response to the GENI Mesoscale initiative (Berman et al., Comput Netw 61:5–23, 2014). InstaGENI was designed both as a distributed cloud, to permit experimenters to run distributed systems and networking experiments, across the wide area, and as a meta-cloud, to permit systems researchers to build experimental clouds within the underlying InstaGENI cloud. InstaGENI consists of more than 36 sites spread across the GENI infrastructure, interconnected by a nationwide, deeply-programmable layer-2 network. Each site is capable of functioning as an autonomous, standalone cloud, with builtin HaaS, IaaS, and OpenFlow (The Openflow Switch Specification. http://OpenFlowSwitch.org; McKeown et al., ACM SIGCOMM CCR 38(2):69–74, 2008) native support. Sites are also and by default linked, to offer slices across the entire GENI Mesoscale infrastructure. InstaGENI targeted and has realized its key design goals of expandability, reliability, resistance to partition, ease of maintenance upgrade, high distribution, and affordability. InstaGENI offers a highly-scalable infrastructure with OpenFlow native both between and across sites. It has demonstrated a high degree of autonomy and remote management, and has demonstrated its meta-cloud properties by hosting an IaaS and PaaS service within it, GENI PlanetLab and the GENI Experiment Engine (Bavier et al., The GENI experiment engine. In: Proceedings of Tridentcom, 2015).

1 Introduction and Motivation

The GENI [13] Mesoscale deployment was a first-in-its-kind infrastructure: small clouds (called, collectively "GENI Racks") spread across the United States, interconnected over a private, programmable layer-2 network with OpenFlow [49, 68] networking native to each cloud. The GENI Mesoscale combined the essential elements of the two principal precursors to GENI [27, 34]: the wide-area distribution

R. McGeer (✉)
Chief Scientist, US Ignite, 1150, 18th St NW, Suite 900, Washington, DC 20036, USA
e-mail: rick.mcgeer@us-ignite.org

R. Ricci
Flux Research Group, University of Utah, Salt Lake City, UT, USA
e-mail: ricci@cs.cs.utah.edu

© Springer International Publishing Switzerland 2016

R. McGeer et al. (eds.), *The GENI Book*, DOI 10.1007/978-3-319-33769-2_14

and scale of PlanetLab [1, 10, 23, 59] with Emulab's [40, 72] ability to do controlled, repeatable experimentation, and added the entirely novel feature of programmable networking through the OpenFlow protocol. The Mesoscale was to offer the ability to allocate customizable virtual machines, containers, and physical machines at any of 50 or more sites across the United States, interconnected with deeply-programmable networking on a private layer-2 substrate and with programmable networking on each rack.

In this, the Mesoscale envisioned three features not available in any commercial or operational academic cloud: geographic distribution, highly-customizable computing elements, and deeply-programmable networking [2]. The primary challenge is that each node in the network is hosted by a separate individual donor institution, which offers bandwidth and maintenance services as a donation to the community. Under these circumstances, the node and each experiment running on it must behave as a polite guest; it should not unduly consume resources, it should require hands-on maintenance very infrequently, and it should not do harmful things to the institution, or to third parties who will interpret the damage as emanating from the institution. This means every node in the network must have two distinct administrators: a central authority representing GENI, and the host institution. It further means that the node must be heavily instrumented, and either the central authority or the local authority must be able to shut it down immediately in the event of abuse or excessive resource consumption.

All of this is well known from PlanetLab [60]; GENI merely deepens the requirement, since it introduces programmable networks and heterogeneous computing environments (VMs, containers, and bare-metal machines running a variety of operating systems; PlanetLab permitted only Linux containers under VServers [66], which simplified administration significantly).

The geographic distribution and need to reduce the burden of on-site maintenance implied a requirement for highly-autonomous operation, intensive in situ measurement, and an ability to shut down slices rapidly, automatically, and selectively.

Since no facility prior to the Mesoscale had combined these features, we had no way of knowing precisely what experiments and services would be run on the Mesoscale. We drew on both the PlanetLab and Emulab experiences, but since both omitted some essential features present in the Mesoscale, it was certain that entirely new experiments would be run on this facility. As a result, flexibility became a crucial design criterion: we need to be able to rapidly customize the testbed to meet users' needs.

In InstaGENI, this implied a decision to go broad and small rather than narrow and heavy: given a choice between a lighter-weight rack that could be deployed to more sites vs a heavier-weight rack with more features, we opted for the former once we'd achieved a critical mass of functionality within the rack. There were multiple rationales for this decision. First, is far easier to add resources to an existing site than it is to bring up a new one, since the later task involves identifying and training administrative personnel at the site, arranging network connections, acceptance and site tests, physical siting, installation, and plumbing, and so on; adding resources to an existing site is usually "just" a matter of buying computers and line cards and plugging them in. Second, we didn't know the right size for a site until we had some experience. Experiments which required more resources than were present at a particular site could always simply add more sites to the experiment. The only real penalty for using other sites rather than adding resources from the same sites was latency. However, the penalty for missing a site was an area which lacked geographic coverage from GENI, and this is problematic from a coverage perspective for applications and services that require distribution. Further, GENI already had sites with many concentrated resources, notably the various Emulab-based testbeds spread around the country; experiments requiring concentrated resources could go there. Third, we wanted the community and the sites to be able to expand the racks in ways that we couldn't anticipate. Leaving space in the racks and letting people install devices with interesting properties made for a variegated and rich testbed. For example, sites have installed electrical grid monitoring devices in GENI racks, offering virtual power grid labs on GENI [21]. Room for expansion offered the community the ability to offer such services very easily.

The next dictated design decision was to use a proven software base (ProtoGENI [62]) rather than install a new, experimental software base. Building a distributed, flexible infrastructure is a large challenge; building one on a new software base would be extremely challenging, both for developers and users. Moreover, fidelity and repeatability of experiments required that the infrastructure on which they were run was stable and, to the extent that we could make it so, artifact-free. Even subtle bugs and peculiarities in the infrastructure can lead to misleading experimental results. The ProtoGENI software stack is descended from Emulab, which had operated its eponymous Utah cluster 24/7 for a decade; we knew it was stable. Finally, GENI and the Mesoscale initiative represented a large bet for the systems community and the National Science Foundation. We needed to make the GENI Mesoscale a reliable experimental facility, good for constant use by a large community, within the lifetime of the GENI project. Spending a year debugging the infrastructure was contraindicated.

While we could not anticipate all the research that would be done on the Mesoscale, research and experiments in cloud management systems was certain to be a dominant theme. This is a topic of great current interest in both the systems community and in the industry, and an area of extremely active development. Moreover, while it's relatively straightforward to do applications research on operational clouds, even at large scale, it's difficult to do cloud management research. For this reason, a key requirement was to ensure that InstaGENI would be a platform for research into the management and monitoring of both centralized and distributed clouds. Experimenters needed to be able to build and manipulate their own cloud platforms within the InstaGENI architecture. Thus, InstaGENI had to be a *meta-cloud*, the first of its kind: a cloud that permitted within it the nesting of other clouds. Indeed, it was anticipated from the first design of InstaGENI that at least one cloud would be instantiated inside the underlying ProtoGENI base code: the GENI PlanetLab infrastructure would be nested within the rack. This would serve as the prototype for our clouds-within-clouds strategy, simultaneously satisfying the need for an infrastructure that would support seamlessly long-running distributed experiments and services using lightweight, end-system resources. The architecture that we designed together with the GENI PlanetLab team remained unchanged as GENI PlanetLab evolved into the GENI Experiment Engine [11]. Indeed, the switch from GENI PlanetLab to the GENI Experiment Engine was entirely transparent to InstaGENI: the GEE team used the standard InstaGENI tools unchanged, just as any other experimenter would.

This combination of a distributed cloud and meta-cloud strategy—small clouds everywhere and embedded clouds-within-clouds—is also used by projects National Science Foundation's NSF Future Cloud concept, notably in CloudLab [63]. InstaGENI served as a validation point and proof-of-concept for this design.

A final requirement was nothing exotic: components can and do break, and must be easily replaced. Further, interest in InstaGENI racks is evident around the world, from Japan, Korea, Germany, Taiwan, and Brazil: hence the racks must be easily delivered anywhere. Further, though our original design was based entirely on HP equipment, we wanted to ensure that both expansion in the current racks and designs for future racks could incorporate equipment from other manufacturers as circumstances warranted. These considerations led to a COTS design philosophy: we would use only commodity components and build as little hardware dependence into our design as possible.

Thus, InstaGENI: a network of 35 small clouds, spread across the United States. The design requirements, discussed above, dictate the global architecture of InstaGENI: each cloud is small, expandable, built from commodity components, with a high degree of remote management and monitoring built-in. The network itself is designed to withstand partition: each rack is capable of acting autonomously. OpenFlow is native to the racks and the racks are interconnected in the control plane across both the routable Internet and the private GENI network, and the data plane across the racks is interconnected over layer 2. Then-GENI Project Director Chip Elliott's vision for the GENIRacks was to be the successor to the router in the new network architecture implied by GENI: we believe that the InstaGENI design is a good start on that.

The remainder of this chapter is organized as follows: in Sect. 2 we consider the role that GENIRacks, and specifically InstaGENI, play in the Mesoscale and in the architecture of the future Internet. In Sect. 3 we describe the architecture of InstaGENI. In Sect. 4 we describe the architecture and implementation of InstaGENI. In Sect. 5 we describe the hardware and software implementation of InstaGENI. In Sect. 6 we describe deployment considerations and concerns. In Sect. 7 we describe operating and maintaining an InstaGENI rack. In Sect. 8 we describe the current status of the InstaGENI deployment. In Sect. 9 we describe related work, particularly our cousin project ExoGENI [3], and in Sect. 10 we conclude and offer thoughts on further work.

2 InstaGENI's Place in the Universe

Testbeds and experiments are all very well; however, the implications of InstaGENI's design are much broader than experimental facilities for systems computer scientists. Though this was and remains GENI's primary mission, it was always far more than that: put simply, GENI is a prototype of the next Internet—and the GENIRacks were always envisioned as the "software routers" of that next generation of the Internet. This is a sufficiently ambitious goal, and a sufficiently deep topic, to warrant some discussion here.

We should start with the obvious: why do we need a new Internet at all? The fundamental answer is that both the fundamental underlying technology of the Internet and the use cases which informed its design point have changed radically in the generation since its architecture was finalized. In the founding era of the Internet, memory and computation were expensive relative to data transmission, and the fundamental use case was bulk, asynchronous data transfer. Today, computation and memory are cheap relative to networking, and the bulk of Internet traffic is in latency-sensitive high-bandwidth applications: video, real-time interactive simulation, high-bandwidth interactive communication, and the like.

Even the fundamental use case, bulk data transfer, has been significantly affected by the change in underlying technology. When computation and memory were expensive, moving data to computation—no matter how slow or painful—was a necessity. Now, however, cheap computation and memory are ubiquitous: it is feasible to move computation to data. And when it is feasible, it is almost always attractive. Programs are generally small relative to the data they process, and many programs reduce data. Some simple examples demonstrate the point. The CASA Nowcasting experiment [46] looks for significant severe weather events; local processing, sited at the weather radar, can find events of interest and propagate them to a cluster which can do detailed processing. Doing the reduction locally, at the weather radar, saves enormously on bandwidth and focuses the network on those events of interest.

This is a simple example, but many more in the same vein can be described; and as the Internet of Things becomes dominant, many more examples of this sort will emerge. The CASA radar is merely one example of a very large class of device: the high-bandwidth, high-capacity sensor. Choose virtually any Internet of Things use case that involves such sensors, from driverless cars to real-time crime detection. A straightforward, back-of-the-envelope calculation will demonstrate that the take from the various sensors will overwhelm the network; the IoT will require not just higher-capacity networks, but an entirely new architecture, with pervasive local computing.

Other examples include latency-sensitive computation. Real-time Interactive Simulation (RTIS) has long been a staple of computing entertainment and technical training; it is also now being used more generally in Science, Technology, Education, and Mathematics (STEM) education, educational assessment, and maintenance applications. "Gamification" is largely the deployment of RTIS for non-gaming applications. This has been spurred by the sophistication of the HTML5 platform, which has meant that the browser can now support significant, intensive 3D interactive applications.

Use of the browser as a rendering platform is preferred for a variety of reasons: ubiquity of access implies that demand on the client be minimized, and use of a standard browser platform is the best that we can do to minimize demands on the client. Further, for many use cases (educational assessment, for example) one wants to protect the application from client interference: the student shouldn't be able to cheat on the test. These requirements imply the need for cloud-based hosting of at least some RTIS applications: in general, as much as one can get away with.

However, the Achilles' Heel of cloud-based RTIS is latency; in general, the computing engine should be no more than a 50 ms round-trip from the user. Any latency more than that invites significant artifacts from a user's perspective: jitter, jumpy displays, out-of-order event sequencing, and so on. The combination of the need for cloud-based hosting of the service with the application requirement of low latency to the end-user points at the need for a pervasive cloud [20].

An excellent example of a low-latency high-bandwidth application delivered to the user through a thin-client web browser is the Ignite Distributed Collaborative Scientific Visualization System [15, 16], described in another chapter in this book [39]. A combination of a large data set (9 GB), required high data rate between visualization client and data server (100 Mb/s–1 Gb/s) and low required round trip time (<20 ms) required the use of a distributed, pervasive cloud. It is a prime example of the kinds of applications that require the InstaGENI distributed cloud.

This pervasive cloud, driven by the twin needs for in-situ data reduction and low latency between application host and application consumer, is the next generation of the network. The fundamental architecture of the current Internet is centered around moving data between fixed computation sites; the architecture of the next generation may well be centered around sending programs to be executed near data sources or users. In the argot of networking, provision of in-network layer 7 services will be the dominant use case for the network in the coming decades.

Provision of layer 4–7 services in the network is nothing new, of course: this has been the province of middleboxes and proxies almost since the inception of the Internet. What is different now is degree rather than kind: rather than being an ad-hoc appendage to the Internet, the pervasive cloud will make proxies and middleboxes the central component of the emerging new Internet architecture. In this architecture, universal, programmable middleboxes will play the role that routers played in the first generation of the Internet architecture. Fundamentally, a GENIRack is a platform for the deployment of universal, highly-programmable middleboxes; in other words, the prototype of this new central component of the emerging Internet.

3 Architecture of InstaGENI

Above all, InstaGENI is designed to meet the primary goals of the GENI project, which are directed at creating a highly customizable environment for innovative research, without restrictions and pre-conditions and with complete direct control over all resource elements. Consequently, InstaGENI is a deployment platform for GENI control frameworks, which enable researchers to discover, integrate, and experiment with GENI resources. Fundamentally, GENI is a platform for the deployment of virtual networks interconnecting virtual computational resources. "Virtual" here is used in the classic sense: details of the physical attributes and specific implementations are, to the extent possible and appropriate for the use of the resource, hidden from the programmer; further, the programmer is given the abstraction of an isolated network of isolated resources, all of which have guaranteed properties. The computational resources are generally but not always virtual machines or their close cousins, OS containers. On occasion the virtual resources are physical machines, radio nodes, specialized instruments, etc. [14, 33, 57, 67, 69]. GENI has been designed to enable novel edge components to be integrated into the experimental environment. In sum, InstaGENI is a distributed systems cluster with sliceability at the end-host, distributed systems, and network level.

3.1 The InstaGENI Software Architecture

The InstaGENI software architecture is designed to provide deeply configurable and deeply-programmable Infrastructure-as-a-Service and customizable OpenFlow networks as a service. A critical design consideration was user familiarity: an InstaGENI is essentially a small Emulab, with an embedded OpenFlow switch to permit the construction of virtual networks. Further, a collection of InstaGENI racks should behave as a distributed Emulab. While new capabilities and functions are provided within the InstaGENI rack—GENI Experiment Engine nodes hosted on InstaGENI Racks, more virtualization options, a Network Aggregate Manager, and the ability to run long-term slices within networks of virtual machines—it was critical that a user unfamiliar with InstaGENI be able to use InstaGENI just as he was used to using Emulab. Further, each rack must be independently manageable.

Management and control functions for nodes in InstaGENI racks are primarily provided by the ProtoGENI software stack. Each rack has its own installation of the control software, and is capable of operating as an independent unit.

The software architecture of InstaGENI is shown in Fig. 1. The important thing to take away from this diagram is nested and distributed control. The key element is the ProtoGENI Base Manager (or ProtoGENI Controller, or Boss Node) on the rack, which plays essentially the same role for an element of a distributed cloud that a node manager does in a cloud: it orchestrates resources on the individual rack. Nested controllers, whether they be entities such as the central GENI Portal or other

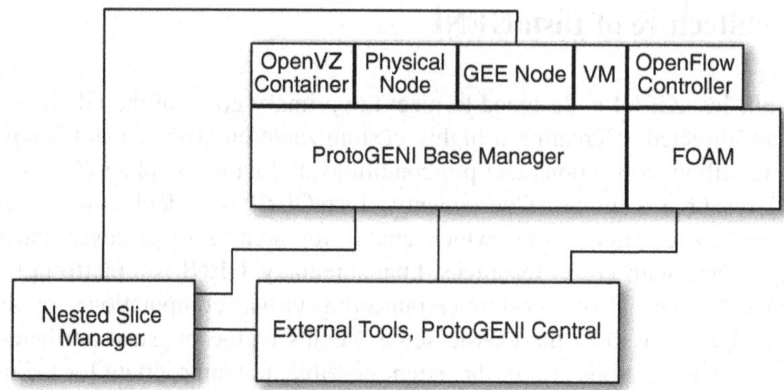

Fig. 1 The InstaGENI software architecture

controllers such as the GENI Experiment Engine, use the ProtoGENI Base Manager as an agent to manipulate resources on the individual rack: allocate and free VM's and bare metal nodes, load images, etc.

The Control Node in each rack runs Xen. This allows multiple pieces of control software to run side-by-side in different virtual machines, with potentially different operating systems and administrative control. This configuration also eases the deployment, backup, and update of the control software. At installation, there are four such virtual machines:

1. An Emulab/ProtoGENI boss node: this is a database, web, and GENI API server, and also manages boot services for the nodes
2. A local fileserver and give users shells so that they can manage and manipulate the data on the fileserver even if they do not currently have a sliver. This VM can also act as a gateway for remote logins to sites that do not have sufficient Internet Protocol (IP) addresses to give every experiment node a publicly routable address.
3. An OpenFlow Aggregate Manager (FOAM) node to control the OpenFlow resources on the in-rack switch
4. A FlowVisor controller to provide support for control-plane multi-tenancy on the OpenFlow network.

Node Control and Imaging

The experiment nodes in the InstaGENI rack are managed by the normal ProtoGENI/Emulab software stack, which provides boot services, account creation, experimental management, etc. Users have full control over physical hosts, including loading new operating system images and making changes to the kernel, in particular, to the network stack. The ProtoGENI/Emulab software uses a combination of network booting, locked down BIOS, and power cycling to ensure that nodes can be returned to the control of the facility and to a clean state, meaning that accidental or intentional changes that render a node's operating system unbootable or cut off from the network can be corrected. Nodes are scrubbed between uses; after a sliver is terminated, the node is re-loaded with a clean image for the next user.

Images for OSes popular with network researchers, including at least two Linux distributions and FreeBSD, are provided. Users may customize these images and make their own snapshots. Installation of other operating systems is possible, but involves significant expertise on the part of the experimenter and manual intervention on the part of the rack administrators. Users making images in this fashion are strongly encouraged to do so on the InstaGENI installation at the University of Utah, where the most assistance is available. One use of this capability is to boot nodes into images that support other control frameworks: e.g., to create slivers that act as GEE nodes or OpenFlow controllers.

In addition to raw hardware nodes, ProtoGENI also provides the ability to create multiple virtual machines (VMs) on the experimental nodes. ProtoGENI supports this in two forms; in the first, an experimenter can allocate a dedicated physical machine, and then slice that into any number of virtual containers. All of the containers are part of the one slice that is being run by the user. In the second form, one or more of the physical nodes are placed into shared mode, which allows multiple users to allocate containers alongside other experimenters. Typically, nodes running in shared mode exhibit better utilization. Physical nodes may be dynamically moved in and out of the shared pool at any time. InstaGENI racks typically allocate three nodes per rack as shared hosts; more nodes may be moved into this pool as required.

The slicing technology used for ProtoGENI virtual hosts is the Xen [6] virtual machine monitor. Earlier in its history, InstaGENI used OpenVZ, a Linux container technology for slicing shared hosts. OpenVZ has the advantage of being very lightweight and booting quickly [31], but we found that it was too restrictive for the types of experiments that GENI users wanted: many wanted the ability to run different Linux kernels, to move images back and forth between physical and virtual hosts, etc. Using a single kernel, as is done in OpenVZ, also proved to be less stable when exposed to the types of workloads offered by systems and network researchers.

Administration, Clearinghouse, and Local Control

The InstaGENI racks are registered as aggregate managers with the GENI clearinghouse, which provides for registration and resolution of metadata associated with users, slices, and component managers. The clearinghouse also serves as a "root of trust," exchanging root cryptographic materials (such as CA certificates) between all parties, so that they do not have to do so pairwise. This means that these entities are visible to, and usable from, existing tools that support the GENI APIs and clearinghouse; these tools include the GENI portal, ProtoGENI command line tools, Jacks (a graphical experiment design tool) [26], GENI Desktop [19], and Omni(a command line tool for reserving resources across control frameworks) [36].

Details on the GENI Desktop and the GENI Architecture are given in other chapters in this volume [17, 38]. Local administrators are given several policy knobs, which allow the administrator to make the following simple policy decisions:

- Allow all GENI users access to the rack
- Allow GENI users to access the rack, but limit how many nodes each user may allocate at a time
- Block all external users (e.g. those who do not have accounts registered on the particular rack) from using the rack
- Issue credential to specific users that allow them to bypass the policies above

Other policies of these types (e.g. user and resource restrictions) can be added as required by sites.

Each rack is given its own Certification Authority (CA) certificate; to establish trust with the rest of the GENI and ProtoGENI federations, a bundle of these certificates are available from the ProtoGENI clearinghouse. ProtoGENI federates fetch this bundle nightly, so all current member of the ProtoGENI federation will, by default, accept the InstaGENI racks as members of the federation. An InstaGENI rack can participate in any number of federations by registering at more than one clearinghouse and adding CAs from other federates to its local set. This feature has been used to prototype federations that cross international boundaries [12, 18].

Nested Control Frameworks and PlanetLab/GENI Experiment Engine Integration

The ability to nest control frameworks was a major design goal of InstaGENI. There are two major drivers for this design goal: first, to enable researchers in cloud technologies to bring up their own clouds within InstaGENI; and, second, to offer customized, simplified clouds for specific purposes, utilizing the mechanisms of the underlying meta-cloud for various services (network configuration, image load, and so on). PlanetLab was always designed as our prototype nested cloud. It, and its successor, the GENI Experiment Engine, are described more fully in another chapter [9] in this volume. Here, we cover some simple basics.

InstaGENI provides a GEE node image. Fundamentally, this is simply an Ubuntu 14.04 LTE image with the Docker container management system installed. GEE nodes use a container-based virtualization technology that provides an isolated Linux environment, rather than a standard VM, to a sliver. Containers can offer better efficiency than VMs, particularly for I/O, because a hypervisor typically introduces an extra layer in the software stack relative to a container-based OS. In the PlanetLab model, all slivers on a physical host run on an underlying shared kernel that slices cannot change. However it is possible to base the Linux environment offered to slivers on different Linux distributions.

The GEE uses Linux Containers (LXC) [48] running under Docker [25] as the core virtualization technology. LXC extends end-host networking with integrated network namespaces. Network namespaces provide each Linux container with its own view of the network. Within each container it is possible to customize many aspects of the network stack, including virtual device information such as IP and MAC address, IP forwarding rules, packet filtering rules, traffic shaping, Transmission Control Protocol (TCP) parameters, etc.

GEE nodes are managed through the GENI Experiment Engine portal and head node at http://www.gee-project.org. A full description of the GEE and its administration may be found in the GEE Chapter in this volume.

The essential elements of the GEE form the recipe for future nested clouds: form a base image which is deployed by the InstaGENI underlying service; choose which nodes to allocate in each rack to the nested cloud; use the InstaGENI service to allocate them and deploy the boss images; and write a separate, standalone controller to allocate slices on the nodes. We believe that this recipe can be followed for a large number of future nested Clouds on the InstaGENI infrastructure. Notably, we believe that it would be not only possible but easy to instantiate a distributed OpenStack-administered Cloud on the InstaGENI racks. Good examples of such OpenStack-administered distributed Clouds are the Canadian SAVI Network [5, 43, 44, 47] and the OpenCloud/XOS [61] from Stanford's ONLab, so this feature offers a potential area of expansion for both these infrastructures.

4 The InstaGENI Network

InstaGENI features two networks: a control network over the routable Internet and a private layer-2 data plane network provided over the GENI Mesoscale [24], transitioning to Internet 2's Advanced Layer 2 Service (AL2S). Experimenters have access to the raw network interfaces on nodes allocated to their slices.

A diagram of the rack network is shown in Fig. 2. Control plane connections are through a dedicated, relatively low-bandwidth conventional switch. This handles

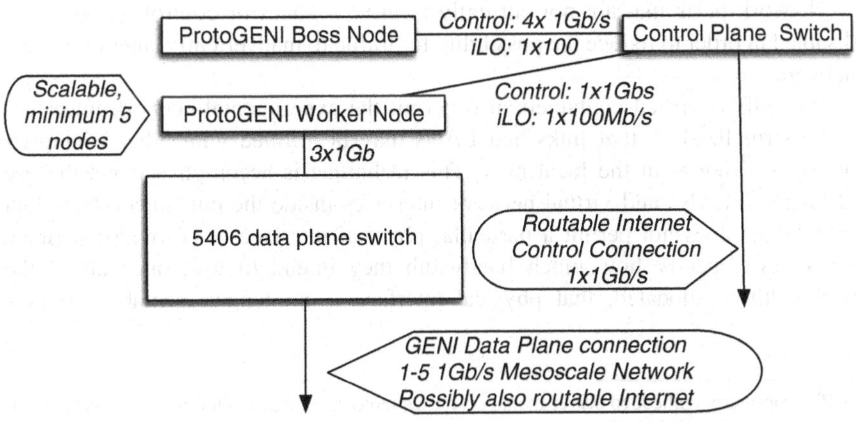

Fig. 2 The InstaGENI rack network

boss/worker control communications, Integrated Lights-Out (iLO) connections[1] and external control connections. The external control interface is over the routable Internet and is a single 1 Gb/s connection.

Data plane switching is over an OpenFlow switch. Each worker node has three 1 Gb/s connections to the data plane switch. There must be at least one 1 Gb/s connection to the GENI Mesoscale network, and with a single 20-port linecard the switch can support up to five external dataplane connections (assuming the minimum of five worker nodes in the rack, and a single 20 x 1 Gb/s linecard). The additional connections can either be to the GENI layer-2 network, or to the routable Internet, or to another network. The switch can also be configured with 10 Gb/s or above optical connections.

Virtual Local Area Networks (VLANs) are created on the rack's switch to instantiate links requested in users' Resource Specifications (RSpecs), and to provide isolation for each experiment's traffic. A small number of the available 4096 VLAN numbers are reserved for control purposes, leaving most available to experiments. Using 802.1q tagging, each physical interface has the ability, if requested, to act as many virtual interfaces, making use of many VLANs. With the exception of stitching, user traffic within racks is segregated by VLAN. The InstaGENI rack's switch is capable of providing full line-rate service to all ports simultaneously, avoiding artifacts due to interference between experiments.

OpenFlow is separately enabled or disabled for individual VLANs. VLANs requested by users default to having OpenFlow disabled. Users are able to request OpenFlow for particular VLANs; in this case, the OpenFlow controller for the VLAN is pointed to the address supplied by the user. Some shared OpenFlow VLANs are available (such as those with access to other shared resources such as Wide-Area Network (WAN) connectivity), and requests for slices of those VLANs are regulated by FOAM [7] and sliced via FlowVisor [65]. A single switch is shared for experiment traffic and control traffic, so experimenters are able to enable OpenFlow only on the VLANs that are part of their slices; OpenFlow is not enabled on VLANs used for control traffic or connections to campus or wide-area networks.

Network ports that are not currently in use for slices or control purposes are disabled in order to reduce the possibility for traffic to inadvertently enter or exit the network.

ProtoGENI virtual containers also permit the experimental network interfaces to be virtualized so that links and LANs may be formed with other containers or physical nodes in the local rack. This technique is accomplished via the use of tagged VLANs and virtual network interfaces inside the containers. Note that ProtoGENI does not permit a particular physical interface to be oversubscribed; users must specify how much bandwidth they intend to use; once all of the bandwidth is allocated, that physical interface is no longer available for new

[1]HP's proprietary embedded server management technology, similar to Dell Remote Access Controller, Oracle/Sun iLOM, Cisco Integrated Management Controller and IBM Remote Supervisor Adapter.

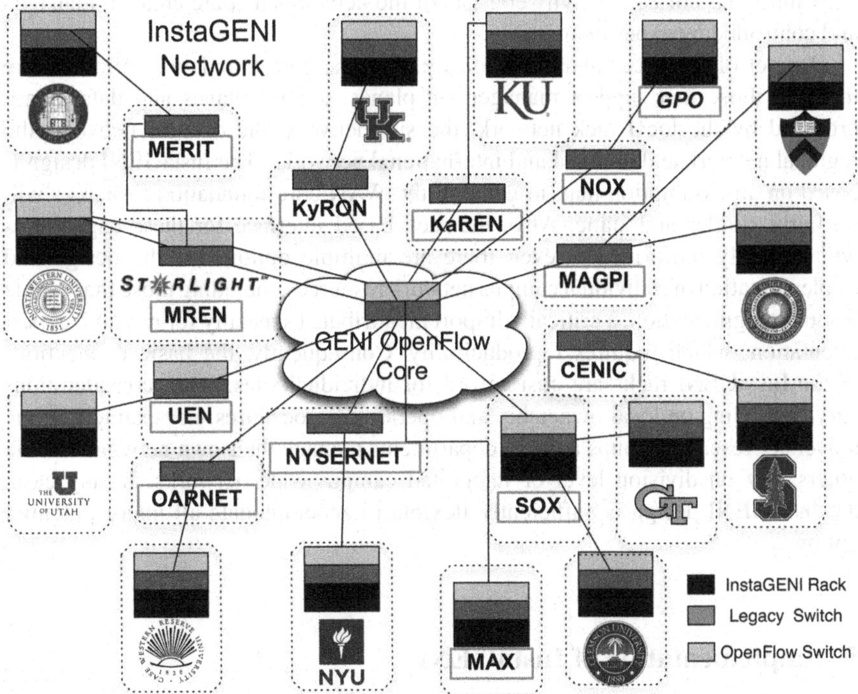

Fig. 3 The InstaGENI external network

containers. Bandwidth limits are enforced through the use of traffic shaping rules in the outer host environment. In addition to VLANs between nodes, ProtoGENI also supports Generic Routing Encapsulation (GRE) tunnels [29] that can be used to form point-to-point links between nodes residing in different locations.

The initial nationwide InstaGENI Network is shown in Fig. 3.[2] The InstaGENI Network Architecture was driven by two principal considerations: the need to offer layer-2 services across the wide area and the need to permit deep programmability and end-to-end OpenFlow capability across the entire Mesoscale.

The InstaGENI design required close consideration of three major classes of WAN connectivity. One class of WAN resources consists of those that constitute core foundation infrastructure, including those that support management planes, control planes and data planes beyond the support provided by the local rack network, which includes support provided by the local site, campus, regional, national, and international networks. A second class of WAN connectivity consists of the actual management plane, control plane, and data plane channels, which will be supported by the core infrastructure.

[2]Diagram thanks to Joe Mambretti and Jim Chen at iCAIR, primary designers of the InstaGENI network.

A third class of connectivity consists of the networks that are created, managed and controlled by experimenters.

One set of resources that constitutes part of the core foundation infrastructure includes those that support management planes, control planes and data planes provided by the local rack network, the site network, the campus network, the regional network and national and international networks. The InstaGENI design is based on an assumption that, in general, the WAN core foundation resources will be fairly similar and static. Also, the rack based interface for these capabilities will be fairly uniform. However, there are multiple options for the design and implementation of individual campus network resources, including those that enable resource segmentation, a critically important attribute especially for research experimentation, which requires reproducibility. Consequently, the basic connections to the InstaGENI racks are customized for individual sites. Also, considerations vary depending on local ownership and operations procedures. For example, some university research groups and CS departments manage their own networks, while others rely on division level or integrated campus-wide networks. In any case, the InstaGENI design is sufficiently flexible to accommodate all major potential options.

5 Implementation of InstaGENI

The InstaGENI hardware design was driven by three principal considerations. First, the goal was to support the software architecture described above; InstaGENI is fundamentally characterized by code, not boxes. Second, commodity off-the-shelf hardware was to be used, for reasons of maintenance and operations. When something broke, it had to be easy to fix or replace. Finally, a large collection of inexpensive racks is preferred to a smaller collection of more capable racks. It is relatively easy to add capacity—more worker nodes, more switch ports—to a modest rack, and somewhat more difficult to install a new rack. Therefore, our goal was to get a broad footprint of modest, but usable racks early, and make them more capable later. This strategy turned out to have unexpected benefits: lots of blank area in the rack made it possible for experimenters to install specialized hardware in individual racks, and use GENI to make that hardware available to experimenters nationwide.

The base design of the InstaGENI rack is shown in Fig. 4. It consists of five experiment nodes, one control node, an OpenFlow switch for internal routing and data plane connectivity to the Mesoscale infrastructure and thence to the Internet, and a small control plane switch/router for control plane communication with the Internet.

Figure 5 shows how the software architecture shown in Fig. 1 maps onto the physical racks. The embedded rack controller and user storage are on the control node. Each worker node can use any ProtoGENI image, including (but not limited to) XEN VMs, the GENI Experiment Engine node image, designed to host GEE

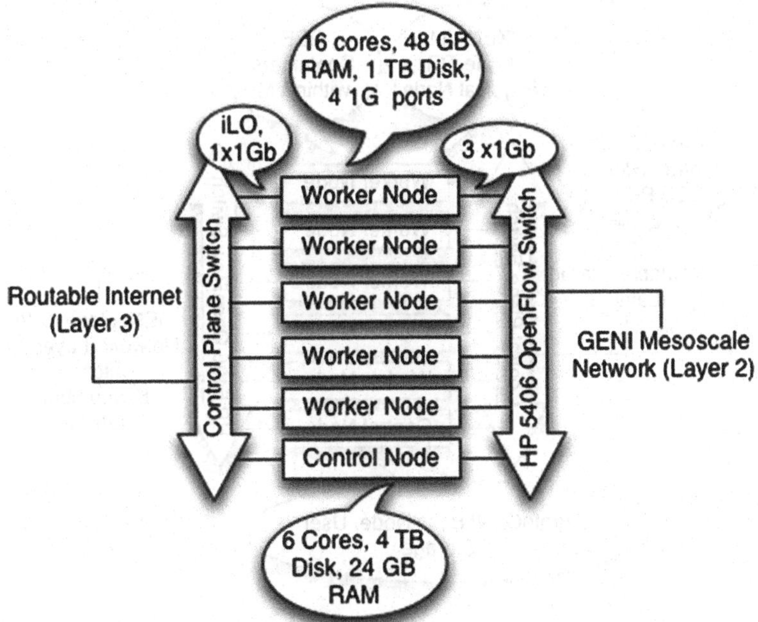

Fig. 4 Hardware diagram of the base InstaGENI rack

slivers, or physical nodes running the image of the experimenter's choice. Data plane connectivity is through GENI VLANs on the Internet-2 Advanced Layer-2 Service (AL2S) or the Mesoscale, and control connections to external embedded managers, such as the GEE, to ProtoGENI Central, and to the GENI Meta-Operations Center for logging and monitoring, are through the control connection.

The InstaGENI rack has been designed for expandability, while providing standalone functionality capable of running most ProtoGENI experiment or an exceptionally capable PlanetLab [10] site. As with all designs, the result is a compromise, yet with much potential for revision and expansion.

The base computation node is the HP ProLiant DL360e Gen8, which is used for both experiment and control nodes. The control node features a six-core, 1.9GHz processor. The experiment node has dual 2.10 GHz eight-core processors. InstaGENI therefore has 80 experimental cores/rack and six cores in the control node. Experiment nodes are configured for images and transient storage: hence disk (1 TB/node) is relatively light. Permanent user and image storage is on the control node, with features 4 TB/disk in a RAID array. Nodes in InstaGENI racks have local disk rather than a Storage-Area Network (SAN): this configuration enables isolation, when required, by allocating an entire physical node to a single slice, avoiding contention for disk or controller resources.

The experiment nodes and switch have been designed for highly flexible, rather than high-performance networking. The experiment node features four 1 Gb/s ports

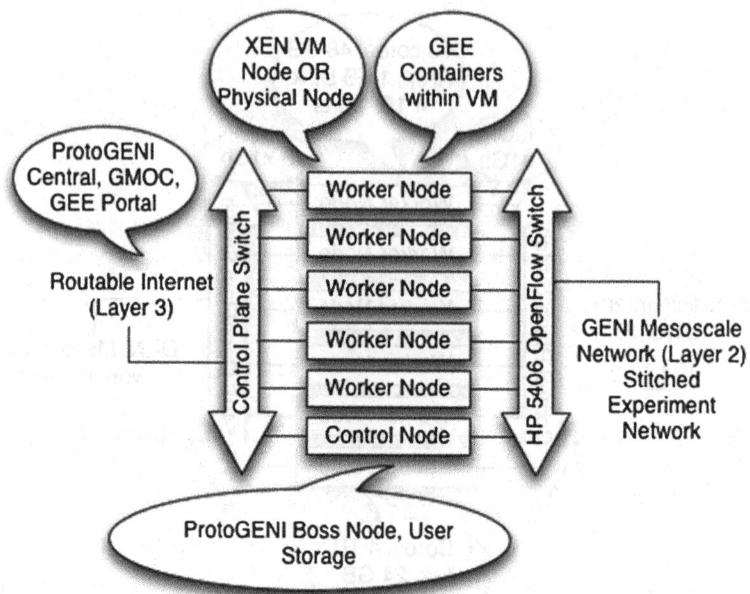

Fig. 5 Software diagram of the base InstaGENI rack

total with TCP/IP Offload Engine. The control node is configured with 12 GB of memory. The experiment node has been specified for 48 GB of memory. The nodes may be extended by the use of two Peripheral Component Interconnect (PCI) express cards.

The primary network device shipped with the InstaGENI rack is the HP ProCurve (now E-Series) 5406 switch with v2 linecards. The 5406 offers rich OpenFlow matching capabilities.

The control connection for the wide area goes through the HP 2610-24 switch. The ProCurve 2610-24 provides 24-port 10/100Base-TX connectivity, and includes two dual personality (RJ-45 10/100/1000 or Small Form-factor Pluggable (SFP)) slots for Gigabit uplink connectivity. An optional redundant external power supply also is available to provide redundancy in the event of a power supply failure. The 2610 switch also carries the six (one control Node and five experiment nodes) iLO connections.

Remote monitoring and management is an especially important InstaGENI consideration. Hence all nodes, experiment and control, ship with HP integrated Lights Out Advanced remote management, version 4 (iLO4). HP iLO is a separately-powered card with separate network connection in the server chassis, with the ability to reboot, setup the server and do power and thermal optimization, and enable embedded health monitoring.

iLO connections to both control and data nodes go through the small 2610 control switch, as does the Control Plane connection into the boss node switch from the external world. The three ProtoGENI/FOAM control connections from the boss node are wired into the HP 5406 rack switch, as are the 15 (3/node x 5 experiment nodes) experiment node data connections and 5 (1/node x 5 nodes) experiment node control connections. Finally, the data plane egress link to the wide area is hosted on the 5406 rack switch.

6 Deployment of InstaGENI

A critical design objective for the InstaGENI racks was that InstaGENI live up to its portmanteau: the InstaGENI racks had to be up and running out of the box, and instantly connected through appropriate communication services to GENI. This is the cornerstone of the InstaGENI design: the goal of InstaGENI was to have a working, at-scale GENI network up and running and ready for experiments, with each node up and on the network within a couple of weeks from hardware delivery. We felt that this was achievable: PlanetLab went from 0 to 300 nodes in its first 2 years [60].

Careful preparation of both the racks and the sites were required for this. We began in the proposal stage: each prospective site filled out an extensive survey and questionnaire before the proposal went into the GENI Project Office, which determined both physical and cyber characteristics of the sites: proposed physical location of the rack, needs regarding power supply, details of incoming connectivity including available VLANs, availability of routable IP addresses, details of boundary and firewall configuration, etc. Key personnel for both technical and administrative support were identified and briefed on the installation needs for the racks. These surveys were renewed as deployment approached.

7 Operations and Maintenance

Software installation for the ProtoGENI control nodes is accomplished through virtual machine images. The local administrator first configures the iLO on the control node (e.g. its IP address, default router, etc.) Generic control nodes images, to run inside the Xen VM, are provided by the ProtoGENI team—in particular, ProtoGENI has developed software that allows the local administrators to fill in a configuration file describing the local network environment (such as IP addresses, routers, DNS severs, etc.), and to generate from that a set of virtual machine images customized to the rack. This functionality can also be used to move an InstaGENI rack to another part of the hosting institution's network, if needed. A default Xen image running the FOAM software is also supplied.

Racks arrive at sites pre-wired. The ProtoGENI stack tests connections by enabling all switch ports, booting all nodes, and sending specially crafted packets on all interfaces. Learned MAC addresses are harvested from the switches and compared against the specified wiring list. This detects mis-wired ports and potentially failed interfaces, so that they can be corrected. The ongoing health of the network is monitored by running Emulab's "linktest" program after each slice is created; this program tests the actual configured topology against the experimenter's requests.

InstaGENI racks' control software is updated frequently and in accordance with an announced schedule to keep up to date on GENI functionality and security patches; the "frequent update" strategy has proved effective on the Utah ProtoGENI site, which rarely suffers downtime due to software updates. All updates are tested first on the InstaGENI rack at the University of Utah for a minimum of 1 week before being rolled out to other sites. All racks receive at least 1 week of warning before software updates, and updates may be postponed in the face of upcoming paper deadlines, course projects, and other high-priority events. Most updates involve no disruption of running slices; updates that do carry this risk are announced ahead of time to the GENI community and scheduled for specific (off-peak) times.

A snapshot of the control VM is taken before upgrades are undertaken, so that in the case of update problems, the control node can be returned to a working state quickly. Backwards compatibility with the two previous versions of the GENI APIs is preserved at all times to avoid the need for flag days.

Most administration of InstaGENI racks is undertaken through the Emulab/ProtoGENI web interface and via command line tools on the control node; physical access to the racks for administration is therefore not required.

All nodes in InstaGENI racks, including control nodes, include HP's iLO technology, which includes power control and console access. This allows both InstaGENI and local personnel to administer the nodes without requiring a physical presence. iLO console capabilities are used for diagnosing faulty nodes (iLO continues to function in the presence of many type of hardware failures) and during the upgrade of control software. Access to iLO on experiment nodes is accomplished through the control nodes so that public IP addresses are not required. iLO on the control node itself requires a public address; this enables remote administration and minimizes downtime in the case of software failures (and many types of hardware failures) on the control node.

Full logs of resource allocations, including information about slices and users who requested them, are available to the local administrators via a web interface. The raw data used in this interface are stored in a database on the control node, should local administrators wish to process this information in their own way. Using existing ProtoGENI APIs, the GMOC are given credentials for each rack that allow them to poll the rack for slice and sliver allocation status. InstaGENI racks use the logging service that is provided by the GENI-wide clearinghouse.

InstaGENI Racks follow the Emergency Stop procedures outlined by the GMOC in [35], and will follow newer versions as they become available.

Emergency stop of slices that are suspected of misbehavior are provided through three interfaces:

- A web interface for rack administrators for cases in which they are made aware of misbehavior
- A GENI API call for use by the owner of a slice or the leader of a project, for cases when the slice may be compromised and used for purposes not intended by the experimenters.
- A GENI API call for use by the GMOC, for cases when misbehavior is GENI-wide, is reported through GMOC channels, or occurs when local administrators are not reachable

The GMOC is given a credential for each InstaGENI rack giving them full privileges to execute emergency shutdown on any slice. The GMOC is the primary point of contact for any detected problematic behavior that occurs after hours or on weekends or holidays. Three levels of emergency stop are provided:

- Cutting off the experiment from the control plane, but not the data plane: this is appropriate for cases in which as slice is having unwanted interactions with the outside Internet, but there is believed to be state within the slice worth preserving. This particular level of emergency shutdown is for cases where the unwanted communication is on the control plane, e.g. scanning/attacking external networks.
- Powering off the affected nodes and/or shutting down the affected virtual machine
- Deletion of the slice and all associated slivers

When emergency stop is invoked on a slice, the owner of the slice is prevented from manipulating it further, and administrative action is required to complete the shutdown. This property can be used to preserve forensic evidence.

8 Experience and Status

The InstaGENI deployment is now essentially complete in the Mesoscale infrastructure, and a full map of the deployed racks can be seen in Fig. 6. The racks in general are the minimum configured with five worker nodes, though the Utah Downtown Data Center rack has over 30 worker nodes. Since the basic software stack is ProtoGENI, and since the Emulab stack from which it has descended has managed clusters up to 1000 nodes, we are confident that the basic architecture of the InstaGENI rack can scale to 1000 nodes and above.

The primary obstacle to installing InstaGENI racks turned out to be the varying types of infrastructure and policies at each site. Different sites had differing types and topologies of connectivity (both to the public Internet and to the Mesoscale), different types of firewalls, different policies regarding connectivity to outside and use of resources by users external to campus, different methods of assigning public IP addresses, etc. While these did not affect the rack itself, they did affect things such as how its external connectivity had to be configured, what domain name it

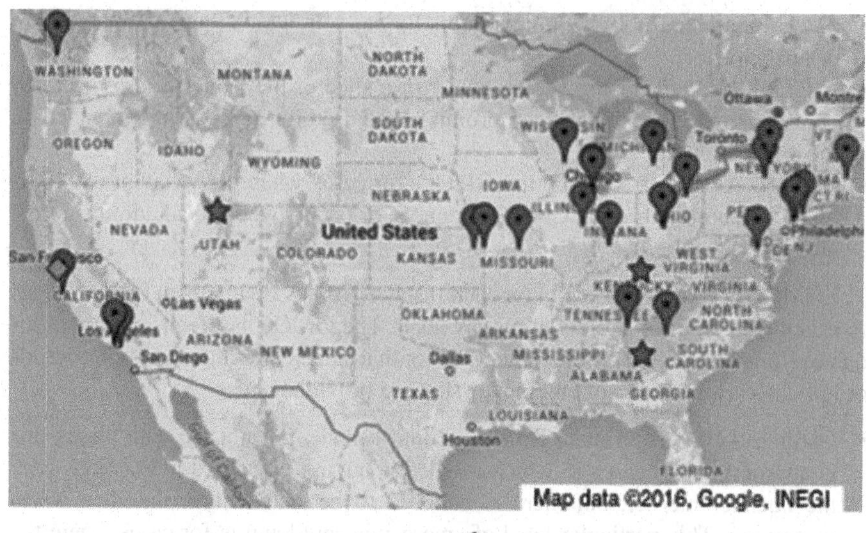

Fig. 6 The InstaGENI deployment

could be under, etc. and often involved delays while network administrators had to approve firewall bypasses, configure campus and/or regional networks, etc. Once these issues were resolved, installation of the rack software itself typically took a few days.

The InstaGENI racks are in constant use by GENI experimenters. Typical usage will have approximately 500 Xen VMs, 300 OpenVZ containers, and 30 bare-metal nodes in constant operation. This still represents a somewhat light load on the overall system; our experiments indicate that we could accommodate 2000 VMs or OpenVZ containers simultaneously with 60 bare-metal nodes in the racks themselves. Currently, about 2500 individual GENI users are creating roughly 4000 slivers monthly on the InstaGENI racks, and using them in a wide variety of experiments.

The nesting strategy has proved to be successful as well, with PlanetLab on InstaGENI and its own nested Platform-as-a-Service offering, the GENI Experiment Engine, maintaining 24/7 service at www.gee-project.net.

9 Related Work

The most prominent related work is our sister project at the GENI Project, ExoGENI, described in an adjacent chapter in this volume [4]. ExoGENI is aimed at a different design point from InstaGENI: ExoGENI supports slivers only as VM's and containers, rather than supporting the allocation of bare-metal nodes as well. In addition, ExoGENI's basic rack is somewhat richer, offering ten worker nodes rather than five, 10 Gb/s uplinks in every rack, and incorporated a storage-area network.

Each ExoGENI rack thus more resembles a conventional OpenStack-based cloud rather than the meta-cloud that forms the primary design motivation of InstaGENI. This incorporates some tradeoffs: on the one hand, InstaGENI enables some services and experiments that would be more difficult to do on ExoGENI. Conversely, ExoGENI's design permits it to easily and efficiently allocate resources for conventional cloud services and applications.

In addition to the GENI infrastructures, several other research clouds have adopted models similar to GENI and InstaGENI. BonFIRE [42] in the EU's FIRE project offers a distributed cloud with six sites, and on-request access to a substantial site at INRIA. Like InstaGENI, BonFIRE offers physical node access. Japan's V-Node [53] project under the umbrella of JGN-X, using a rack similar in many ways to the InstaGENI rack but with a different control framework. Canada's Smart Applications over Virtual Infrastructure (SAVI) [5, 28, 43, 44, 47] project operates a distributed cloud with similarities to both ExoGENI and InstaGENI. Like ExoGENI, SAVI is a VM-only infrastructure based on OpenStack. Like InstaGENI (and, we believe, like ExoGENI as well), each rack (or "site" in the SAVI terminology) can operate as a standalone cloud. SAVI is described in a subsequent chapter in this book [47]. Koren [45] is a Korean testbed primarily focused on multi-site OpenFlow experimentation, but a VM creation and orchestration capability exists at each of the six Koren sites. Ofelia [50, 56] is an EU testbed, similarly focused on multi-site OpenFlow, with VMs available at each site. The G-Lab architecture [52, 64] featured a similar distributed cloud architecture to InstaGENI, relying more heavily on the central node in Kaiserslautern than InstaGENI does: in G-Lab, boot management and resource management was done at the central node, and the local boss—HeadNode in G-Lab terminology—was mostly focused on housekeeping and low-level node management activities. NorNet [37, 55] is a PlanetLab-like infrastructure which consists of two tiers of service. NorNet Core is a twenty-site testbed, primarily of sites in Norway, each multi-homed to several network providers. NorNet Edge consists of several hundreds of smaller nodes that are connected to all mobile broadband providers in Norway. FITS [30, 51] is a joint Brazilian-European project with more than 20 sites across the two continents.

Under the Federation API all three of the FIRE, V-Node, and SAVI infrastructures should be fully interoperable with the GENI racks, creating the possibility of a instantiating a worldwide slice across these infrastructures. Indeed, full integration of GENI and SAVI has already been demonstrated this year.

10 Conclusions, Extensions and Further Work

The initial goal of the InstaGENI project was to provide a workhorse cluster design for the GENI Mesoscale project. In that, it has succeeded, as demonstrated by its usage. Installation of the racks is complete, and ongoing maintenance and troubleshooting have proven to be smooth. The software stack is stable and largely trouble-free. Nesting control frameworks has been a successful experiment, with InstaGENI PlanetLab and the GENI Experiment Engine running seamlessly under the basic ProtoGENI infrastructure.

Difficulties have largely been site-related, and specifically related to site network policies. An ongoing issue is the paucity of public IPv4 addresses at the various InstaGENI sites. Our hosts, primarily Universities, have often been reluctant to allocate IPv4 addresses. We require only the bare minimum number of addresses to give the InstaGENI maintainers, GPO staff, and GENI experimenters control plane access to the boss node on the rack. In the ideal case, each sliver could have two routable v4 addresses—one for the control plane interface, so an experimenter could directly ssh into his sliver and host public-facing services, and one for the data plane, so that localized services could be offered from each sliver. Various strategies have been employed to get around the lack of v4 addresses, primarily using application-level port-sharing and multiplexing. The GENI Experiment Engine is planning to do this with a shared http reverse proxy, to permit GEE slivers to offer public services. At the moment, the GEE offers routable ports to individual slivers on a per-request basis.

The v4 address shortage is an area that needs significant attention over the coming months and years. While the primary networking needs of GENI and other experimental testbeds can largely be met with private networks—where the private /8, combined with the VLAN address space, is more than sufficient—services offered to end-users require access to the routable Internet, since users typically don't have access to the private GENI network. There are many examples of such services: end-system multicast trees [22], wide-area stores [54, 74], virtual shared worlds [70], Content Distribution Networks [32, 71], and collaborative visualization systems [15, 16] to name five. Use of centralized servers in places where v4's are plentiful is not the answer: the whole point of putting these services on GENI, instead of, say, EC2, is to offer low-latency access to local end-users. Wholesale adoption of IPv6 would solve the problem, of course, as would the availability of more advanced network architectures such as content-centric networking [41, 58, 73]. However, it's important to note that the reason we have this problem is we're trying to offer services to people over the routable Internet, which we don't control. The problem with v6 is not that we'd have any trouble implementing or enforcing it; it's that an end-user, transiting multiple academic and commercial networks to reach his local GENI node, must be able to do so reliably. Thus, we need v6 implemented by every network, and this presents some challenges.

A second strategy is to canonize the port-sharing work that PlanetLab pioneered, using a combination of OpenFlow switching and unused header bits to run realtime NAT transparently to the external Internet.

Use and accommodation of OpenFlow, and specifically restricted forms of OpenFlow, is an area of ongoing investigation. There are a large number of use cases where developers want to direct routing at a high level, but don't need to access the full machinery of an OpenFlow controller. A restricted, high-level, easy-to-use northbound API to OpenFlow (and, more generally, the network allocation substrate) is under active investigation.

We are actively investigating both expanding the capabilities of the current GENI rack, by adding more worker nodes and by adding heterogeneous resources, in concert with related projects such as CloudLab. We are further working with partners such as US Ignite to investigate applications of GENI racks in the domain sciences, smart cities, and distributed education arenas. The fundamental purpose of the InstaGENI cluster is to permit people to create virtual machines anywhere, to reduce data, reduce latency to the end user, add application resiliency in the face of network or physical outage, or increase bandwidth to a sensor or application. In our view, the set of such applications is very large.

Acknowledgements InstaGENI is a large, complex project, and there are many people who contributed to its success. This paper is an extension of, and based heavily, on a journal paper signed by the entire InstaGENI team [8], and we would first like to express our heartfelt thanks to them. Of especial note are Joe Mambretti, Fei Yeh, and Jim Chen, who worked closely with us on network design; Narayan Krishnan, who did the original hardware design to match ProtoGENI's software specifications. It is hard to adequately describe the logistical challenges in working 35 non-standard orders through a manufacturer, ensuring their delivery to 35 separate sites, and maintaining a complex, multi-year, multi-million dollar budget. InstaGENI was an enormous project management challenge, and we were fortunate that Nicki Watts was kind enough to devote a great deal of time to this project; it literally would not have happened without her. When one of us (McGeer) moved on from HP, Jack Brassil took over as Principal Investigator on the project and completed it brilliantly. We had tremendous support from the GENI Project Office, notably Niky Riga, Heidi Dempsey, Vic Thomas, Henry Yeh, and especially Mark Berman. Leigh Stoller of the Flux research group has been extremely generous in offering operational support to InstaGENI users and experimenters. Larry Singer, then of HP Americas, offered his support for commercialization and Michaela Mezo helped enormously in that area. Shannon Champion of Matrix Integration was instrumental in making InstaGENI an HP product. Moreover, the 36+ PIs and system administrators at the InstaGENI sites have been responsive to our requests and to keep InstaGENI going. We thank all of them.

A special note is given to Chip Elliott. InstaGENI and ExoGENI were Chip's inspiration when he was GENI Project Director. It was Chip who mapped out the deployment strategy for the GENI Racks, and he worked closely with us on the initial strategic decisions that gave the project its focus. He also sharpened for us the role of these racks in the coming Internet. This project is very much his creation.

References

1. Anderson, T., Peterson, L., Shenker, S., Turner, J.: Overcoming the Internet impasse through virtualization. Computer **38**(4), 34–41 (2005)
2. Baldine, I., Xin, Y., Mandal, A., Renci, C., Chase, J., Marupadi, V., Yumerefendi, A., Irwin, D.: Networked cloud orchestration: a GENI perspective. In: 2010 IEEE GLOBECOM Workshops (GC Wkshps), pp. 573–578 (2010)
3. Baldine, I., Xin, Y., Mandal, A., Ruth, P., Heerman, C., Chase, J.: ExoGENI: a multi-domain infrastructure-as-a-service testbed. In: Testbeds and Research Infrastructure. Development of Networks and Communities, pp. 97–113. Springer, New York (2012)
4. Baldin, I., Castillo, C., Chase, J., Orlikowski, V., Xin, Y., Heermann, C., Mandal, A., Ruth, P., Mills, J.: ExoGENI: a multi-domain infrastructure-as-a-service testbed. In: GENI: Prototype of the Next Internet. Springer, New York (2016)
5. Bannazadeh, H., Leon-Garcia, A., Redmond, K., Tam, G., Khan, A., Ma, M., Dani, S., Chow, P.: Virtualized application networking infrastructure. In: Testbeds and Research Infrastructures. Development of Networks and Communities - 6th International ICST Conference, TridentCom 2010, Berlin, 18–20 May 2010, Revised Selected Papers, pp. 363–382 (2010)
6. Barham, P., Dragovic, B., Fraser, K., Hand, S., Harris, T., Ho, A., Neugebauer, R., Pratt, I., Warfield, A.: Xen and the art of virtualization. ACM SIGOPS Oper. Syst. Rev. **37**(5), 164–177 (2003)
7. Bastin, N.: Foam: an openflow aggregate manager. http://groups.geni.net/geni/wiki/OpenFlow/FOAM (2013)
8. Bastin, N., Bavier, A., Blaine, J., Chen, J., Krishnan, N., Mambretti, J., Mcgeer, R., Ricci, R., Watts, N.: The instaGENI initiative: an architecture for distributed systems and advanced programmable networks. Comput. Netw. **61**, 24–38 (2014)
9. Bavier, A., McGeer, R.: The GENI experiment engine. In: GENI: Prototype of the Next Internet. Springer, New York (2016)
10. Bavier, A.C., Bowman, M., Chun, B.N., Culler, D.E., Karlin, S., Muir, S., Peterson, L.L., Roscoe, T., Spalink, T., Wawrzoniak, M.: Operating systems support for planetary-scale network services. In: NSDI, vol. 4, pp. 19–19 (2004)
11. Bavier, A., Chen, J., Mambretti, J., McGeer, R., McGeer, S., Nelson, J., O'Connell, P., Tredger, S., Coady, Y.: The GENI experiment engine. In: Proceedings of Tridentcom (2015)
12. Berman, M., Brinn, M.: Progress and challenges in worldwide federation of future internet and distributed cloud testbeds. In: 2014 International Science and Technology Conference (Modern Networking Technologies) (MoNeTeC), pp. 1–6 (2014)
13. Berman, M., Chase, J.S., Landweber, L., Nakao, A., Ott, M., Raychaudhuri, D., Ricci, R., Seskar, I.: GENI: a federated testbed for innovative network experiments. Comput. Netw. **61**, 5–23 (2014). Special issue on Future Internet Testbeds - Part I
14. Bhanage, G., Seskar, I., Raychaudhuri, D.: A virtualization architecture for mobile WiMAX networks. SIGMOBILE Mob. Comput. Commun. Rev. **15**(4), 26–37 (2012)
15. Bhojwani, S., Hemmings, M., Ingalls, D., Krahn, R., Lary, D., Lincke, J., McGeer, R., Ricart, G., Roder, M., Coady, Y., Stege, U.: The ignite distributed collaborative scientific visualization system. In: Distributed Cloud Computing Workshop (2015)
16. Bhojwani, S., Hemmings, M., Ingalls, D., Krahn, R., Lary, D., Lincke, J., McGeer, R., Ricart, G., Roder, M., Coady, Y., Stege, U.: The ignite distributed collaborative scientific visualization system. In: Proceedings of IEEE CloudCom (2015)
17. Brinn, M.: GENI architecture foundation. In: GENI: Prototype of the Next Internet. Springer, New York (2016)
18. Brinn, M., Bastin, N., Bavier, A., Berman, M., Chase, J., Ricci, R.: Trust as the foundation of resource exchange in GENI. In: Proceedings of the 10th International Conference on Testbeds and Research Infrastructures for the Development of Networks and Communities (TRIDENTCOM) (2015)

19. Brown, D., Ascigil, O., Nasir, H., Carpenter, C., Griffioen, J., Calvert, K.: Designing a GENI experimenter tool to support the choice net internet architecture. In: 2014 IEEE 22nd International Conference on Network Protocols (ICNP), pp. 548–554 (2014)
20. Brown, D., Nasir, H., Carpenter, C., Ascigil, O., Griffioen, J., Calvert, K.: Choicenet gaming: changing the gaming experience with economics. In: Computer Games: AI, Animation, Mobile, Multimedia, Educational and Serious Games (CGAMES), 2014, pp. 1–5 (2014)
21. Chakrabortty, A., Xin, Y.: Hardware-in-the-loop simulations and verifications of smart power systems over an exo-GENI testbed. In: 2013 Second GENI Research and Educational Experiment Workshop (GREE), pp. 16–19 (2013)
22. Chu, Y.-h., Rao, S.G., Zhang, H.: A case for end system multicast (keynote address). In: ACM SIGMETRICS Performance Evaluation Review, vol. 28, pp. 1–12. ACM, New York (2000)
23. Chun, B., Culler, D., Roscoe, T., Bavier, A., Peterson, L., Wawrzoniak, M., Bowman, M.: PlanetLab: an overlay testbed for broad-coverage services. ACM SIGCOMM Comput. Commun. Rev. **33**(3), 3–12 (2003)
24. Dempsey, H.: The GENI mesoscale network. In: GENI: Prototype of the Next Internet. Springer, New York (2016)
25. Docker. https://www.docker.com/ (2015)
26. Duerig, J.: Jacks. https://www.emulab.net/protogeni/jacks-doc/ (2014)
27. Elliott, C., Falk, A.: An update on the GENI project. SIGCOMM Comput. Commun. Rev. **39**(3), 28–34 (2009)
28. Faraji, M., Kang, J., Bannazadeh, H., Leon-Garcia, A.: Identity access management for multi-tier cloud infrastructures. In: 2014 IEEE Network Operations and Management Symposium, NOMS 2014, Krakow, 5–9 May 2014, pp. 1–9 (2014)
29. Farinacci, D., Li, T., Hanks, S., Meyer, D., Traina, P.: Generic Routing Encapsulation (GRE). RFC 2784 (Proposed Standard) (2000). Updated by RFC 2890
30. FITS. Future internet testbed with security. http://www.gta.ufrj.br/fits/index.php/ (2015)
31. Fragni, C., Moreira, M.D., Mattos, D.M., Costa, L.H.M., Duarte, O.C.M.: Evaluating Xen, VMware, and OpenVZ virtualization platforms for network virtualization. Universidade Federal do Rio de Janeiro'GTA/PEE/COPPE - Rio de Janeiro (2010)
32. Freedman, M.J., Freudenthal, E., Mazieres, D.: Democratizing content publication with coral. In: NSDI, vol. 4, pp. 18–18 (2004)
33. Fund, F., Dong, C., Korakis, T., Panwar, S.: A framework for multidimensional measurements on an experimental wimax testbed. In: Korakis, T., Zink, M., Ott, M. (eds.) Testbeds and Research Infrastructure. Development of Networks and Communities. Lecture Notes of the Institute for Computer Sciences. Social Informatics and Telecommunications Engineering, vol. 44, pp. 369–371. Springer, Berlin/Heidelberg (2012)
34. GENI Planning Group. GENI design principles. Computer **39**(9), 102–105 (2006)
35. GMOC. GENI - emergency stop procedure workflow (spiral 4). http://gmoc.grnoc.iu.edu/uploads/7e/39/7e39c5ec9577a5badab80ea15419ece8/GENI-Emergency-Stop-Procedure-and-Workflow.pdf (2013)
36. GPO. The omni GENI client. http://trac.gpolab.bbn.com/gcf/wiki/OmniOverview/ (2011)
37. Gran, E.G., Dreibholz, T., Kvalbein, A.: Nornet core-a multi-homed research testbed. Comput. Netw. **61**, 75–87 (2014)
38. Griffioen, J., Fei, Z., Nasir, H., Carpenter, C., Reed, J., Wu, X., S.R.P.: The GENI desktop. In: GENI: Prototype of the Next Internet. Springer, New York (2016)
39. Hemmings, M., Lary, D., McGeer, R., Ricart, G.: The ignite distributed collaborative scientific visualization system. In: GENI: Prototype of the Next Internet. Springer, New York (2016)
40. Hermenier, F., Ricci, R.: How to build a better testbed: lessons from a decade of network experiments on Emulab. In: Proceedings of the 8th International ICST Conference on Testbeds and Research Infrastructures for the Development of Networks and Communities (Tridentcom) (2012)
41. Jacobson, V., Mosko, M., Smetters, D., Garcia-Luna-Aceves, J.: Content-centric networking. Whitepaper, Palo Alto Research Center, pp. 2–4 (2007)

42. Jofre, J., Velayos, C., Landi, G., Giertych, M., Hume, A.C., Francis, G., Oton, A.V.: Federation of the bonfire multi-cloud infrastructure with networking facilities. Comput. Netw. **61**, 184–196 (2014)
43. Kang, J., Bannazadeh, H., Leon-Garcia, A.: SAVI testbed: control and management of converged virtual ICT resources. In: 2013 IFIP/IEEE International Symposium on Integrated Network Management (IM 2013), Ghent, 27–31 May 2013, pp. 664–667 (2013)
44. Kang, J., Lin, T., Bannazadeh, H., Leon-Garcia, A.: Software-defined infrastructure and the SAVI testbed. In: Testbeds and Research Infrastructure: Development of Networks and Communities - 9th International ICST Conference, TridentCom 2014, Guangzhou, 5–7 May 2014, Revised Selected Papers, pp. 3–13 (2014)
45. Koren Future Network Testbed. http://www.koren.kr/koren/eng/ (2015)
46. Krishnappa, D., Lyons, E., Irwin, D., Zink, M.: Network capabilities of cloud services for a real time scientific application. In: 2012 IEEE 37th Conference on Local Computer Networks (LCN), pp. 487–495 (2012)
47. Leon-Garcia, A., Bannazadeh, H.: Savi testbed for applications on software-defined infrastructure. In: GENI: Prototype of the Next Internet. Springer, New York (2016)
48. Linux Containers. https://linuxcontainers.org/lxc/downloads/ (2015)
49. McKeown, N., Anderson, T., Balakrishnan, H., Parulkar, G., Peterson, L., Rexford, J., Shenker, S., Turner, J.: Openflow: enabling innovation in campus networks. ACM SIGCOMM CCR **38**(2), 69–74 (2008)
50. Melazzi, N.B., Detti, A., Mazza, G., Morabito, G., Salsano, S., Veltri, L.: An openflow-based testbed for information centric networking. In: Future Network & Mobile Summit (FutureNetw), 2012, pp. 1–9. IEEE, New York (2012)
51. Moraes, I.M., Mattos, D.M., Ferraz, L.H.G., Campista, M.E.M., Rubinstein, M.G., Costa, L.H.M., de Amorim, M.D., Velloso, P.B., Duarte, O.C.M., Pujolle, G.: Fits: a flexible virtual network testbed architecture. Comput. Netw. **63**, 221–237 (2014)
52. Mueller, P.: Europe's mission in next-generation networking with special emphasis on the German-lab project. In: GENI: Prototype of the Next Internet. Springer, New York (2016)
53. Nakao, A.: Vnode: a deeply programmable network testbed through network virtualization. http://www.ieice.org/~nv/05-nv20120302-nakao.pdf (2012)
54. Nelson, J., Peterson, L.: Syndicate: democratizing cloud storage and caching through service composition. In: Proceedings of the 4th annual Symposium on Cloud Computing, p. 46. ACM, New York (2013)
55. NorNet. A real-world, large-scale multi-homing testbed. https://www.nntb.no/ (2015)
56. Ofelia. Openflow in Europe linking infrastructure and applications. http://www.fp7-ofelia.eu/ (2015)
57. Ozcelik, I., Brooks, R.R.: Security experimentation using operational systems. In: Proceedings of the Seventh Annual Workshop on Cyber Security and Information Intelligence Research, CSIIRW '11, pp. 79:1–79:1. ACM, New York (2011)
58. Perino, D., Varvello, M.: A reality check for content centric networking. In: Proceedings of the ACM SIGCOMM Workshop on Information-Centric Networking, pp. 44–49. ACM, New York (2011)
59. Peterson, L., Anderson, T., Culler, D., Roscoe, T.: A blueprint for introducing disruptive technology into the Internet. In: Proceedings of HotNets-I, Princeton, NJ (2002)
60. Peterson, L., Bavier, A., Fiuczynski, M.E., Muir, S.: Experiences building planetlab. In: Proceedings of the 7th Symposium on Operating Systems Design and Implementation, pp. 351–366. USENIX Association (2006)
61. Peterson, L.L., Baker, S., Leenheer, M.D., Bavier, A.C., Bhatia, S., Wawrzoniak, M., Nelson, J.C., Hartman, J.H.: XOS: an extensible cloud operating system. In: Proceedings of the 2nd International Workshop on Software-Defined Ecosystems, BigSystem 2015, Portland, OR, 16 June 2015, pp. 23–30 (2015)
62. Ricci, R., Duerig, J., Stoller, L., Wong, G., Chikkulapelly, S., Seok, W.: Designing a federated testbed as a distributed system. In: Proceedings of the 8th International ICST Conference on Testbeds and Research Infrastructures for the Development of Networks and Communities (Tridentcom) (2012)

63. Ricci, R., Eide, E., The CloudLab Team.: Introducing CloudLab: scientific infrastructure for advancing cloud architectures and applications. USENIX ;login: **39**(6), 36–38 (2014)
64. Schwerdel, D., Reuther, B., Zinner, T., Müller, P., Tran-Gia, P.: Future Internet research and experimentation: the G-lab approach. Comput. Netw. **61**, 102–117 (2014)
65. Sherwood, R., Gibb, G., Yap, K.-K., Appenzeller, G., Casado, M., McKeown, N., Parulkar, G.: Can the production network be the testbed? In: Operating Systems Design and Implementation (OSDI) (2010)
66. Soltesz, S., Pötzl, H., Fiuczynski, M.E., Bavier, A., Peterson, L.: Container-based operating system virtualization: a scalable, high-performance alternative to hypervisors. SIGOPS Oper. Syst. Rev. **41**(3), 275–287 (2007)
67. Soroush, H., Banerjee, N., Corner, M., Levine, B., Lynn, B.: A retrospective look at the UMass dome mobile testbed. ACM SIGMOBILE Mobile Comput. Commun. Rev. **15**(4), 2–15 (2012).
68. The Openflow Switch Specification. http://OpenFlowSwitch.org (2009)
69. Vercher, J., Hernandez-Munoz, J., Gomez, L., Sepulveda, A.: An experimental platform for large-scale research facing FI-IoT scenarios. In: Future Network Mobile Summit (FutureNetw), pp. 1–8 (2011)
70. Virtual Worlds Framework. https://virtual.wf/ (2015)
71. Wang, L., Park, K., Pang, R., Pai, V.S., Peterson, L.L.: Reliability and security in the codeen content distribution network. In: USENIX Annual Technical Conference, General Track, pp. 171–184 (2004)
72. White, B., Lepreau, J., Stoller, L., Ricci, R., Guruprasad, S., Newbold, M., Hibler, M., Barb, C., Joglekar, A.: An integrated experimental environment for distributed systems and networks. In: Proceedings of the Fifth Symposium on Operating Systems Design and Implementation, pp. 255–270. USENIX Association, Boston (2002)
73. Yuan, H., Song, T., Crowley, P.: Scalable NDN forwarding: Concepts, issues and principles. In: 2012 21st International Conference on Computer Communications and Networks (ICCCN), pp. 1–9. IEEE, New York (2012)
74. Zurawski, J., Swany, M., Beck, M., Ding, Y.: Logistical multicast for data distribution. In: IEEE International Symposium on Cluster Computing and the Grid, 2005. CCGrid 2005, vol. 1, pp. 434–441. IEEE, New York (2005)

Part IV
GENI Experiments and Applications

The most interesting aspect of GENI is, of course, the experiments and applications that are deployed on GENI. After all, these were the purpose of the entire exercise. It is an axiom for startup companies that, after the development team, the most important element for success is an initial lighthouse customer who will not only provide a proof point for demand for the company's product, but also help the company shape it. Similarly, the success of an infrastructure such as GENI is heavily boosted by early-adopter developers and experimenters, whose feedback and participation shapes the infrastructure. Each of GENI's precursors had had early-design customers. PlanetLab was heavily shaped by the Distributed Hash Tables and Content Distribution Networks, particularly CoDeeN; DeterLab had an initial core constituency of cybersecurity experimenters and a long list of exploits and malware to offer; Emulab worked closely with investigators from a number of DARPA projects, including the Control Plane program to radically improve the performance of TCP; and ORBIT, similarly, was strongly influenced by the Mobile Ad-Hoc Network (MANET) projects that dominated wireless network research in the 2000's.

GENI has to date been used by more than 7500 experimenters, and has been (as mentioned in chapter "Programmable, Controllable Networks") the premier testbed for Software-Defined Networking, with well over 50 SDN sites installed across the United States. In this section, we have selected a few of the GENI experiments and applications to give the reader a sense of the breadth of the GENI experiments and services landscape.

We begin with an experimental focus. GENI is, at bottom, a meta-infrastructure; experimenters and developers construct their own, application-specific infrastructures from the GENI substrate. It has long been a goal of the GENI project to create a distributed laboratory that not only supports conducting computer science experiments but also facilitates and encourages good scientific discipline in their design, execution, and documentation. One of the major focuses of the GENI Project Office over the past few years has been to walk experimenters through the tasks of selecting and allocating this infrastructure, deploying an application or experiment over it, and orchestrating the action of the application across the wide area. In chapter "The Experimenter View of GENI", Niky Riga, Sarah Edwards, and Vicraj Thomas of the GPO describe this process both in the abstract and through an extended example application, the GENI Cinema.

Constructing, deploying, and orchestrating an experiment or application is obviously a substantial task, and a number of tools and services have been developed just for that—much as, in an earlier age, tools and services were developed to deploy software across the wide area on PlanetLab. In chapter "The GENI Desktop", James Griffioen, Zongming Fei, Hussamuddin Nasir, Charles Carpenter, Jeremy Reed, Xiongqi Wu and Sergio Rivera P. describe one of these tools, the GENI Desktop—a Web Portal-based front end on a GENI Experiment or Application.

Repeatability is an essential part of experimental science, both for an individual researcher who requires consistency across trials in his or her own work and for the scientific community seeking to validate and expand on published research. Unfortunately, repeatability often receives short shrift in published computer science research. In chapter "Walk through the GENI Experiment Cycle", Thierry Rakotoarivelo, Guillaume Jourjon, Olivier Mehani, Max Ott, and Michael Zink take us on "A Walk Through the GENI Experiment Cycle," using the LabWiki toolkit. LabWiki is a suite of experimenter support tools developed for GENI and other testbed environments with a strong emphasis on experiment repeatability. They argue that proper tool support can make a major difference in repeatability and make a researcher's life more pleasant at the same time.

One of GENI's most important contributions is in Computer Science and Engineering education. GENI offers a unique educational resource, and affords graduate and undergraduate students across the United States the ability to conduct experiments on wide-area systems and networks that would have been impossible without this resource. Moreover, GENI offers a unique platform for Massive Open Online Courses, for both Computer Science and for other students, including the K-12 arena. It was this unique capability that attracted the developers of The Mars Game to the GENI platform. In chapter "GENI in the Classroom", Vicraj Thomas, Niky Riga, Sarah Edwards, Fraida Fund, Thanasis Korakis describe the uses of GENI in the Classroom.

As mentioned in the introduction, interactive big-data applications are the last obstacle to Eric Schmidt's long-anticipated (and ongoing) "hollowing-out" of the personal computer—the migration of desktop applications to the Cloud. GENI's ability to offer high-bandwidth, low-latency connections to anywhere permits these applications to move to the Cloud. This does more than simply migrate the existing application to the Cloud and thus make it available on a broader range of devices—it permits the development of collaborative, interactive big data applications—an entirely new application class. In chapter "The Ignite Distributed Collaborative Scientific Visualization System", Matt Hemmings, David Lary, Rick McGeer, and Glenn Ricart describe both the first example of this class and a new application platform, The Ignite Distributed Collaborative Scientific Visualization System.

The use of GENI in the classroom, The Mars Game, and the Ignite Visualization System, along with a large body of smart-campus applications developed under the GENI program, are indications that a new network is the key to a broad range of network applications which can offer dramatic improvements to health care, public safety, education, clean energy, transportation, and manufacturing. This new, smart network powering smart cities is the heart of the US Ignite vision, described in chapter "US Ignite and Smarter GENI Cities" by Glenn Ricart, US Ignite Chief Technologist, and Rick McGeer.

The Experimenter's View of GENI

Niky Riga, Sarah Edwards, and Vicraj Thomas

Abstract GENI is a *federated* infrastructure that provides GENI experimenters with access to multiple different testbeds, enabling networking and distributed systems research. Although GENI resources are owned and operated by different organizations from a users perspective GENI appears as a unified virtual laboratory. An experimenter can instantiate custom Layer 2 topologies that include a variety of compute and network elements. This ability is achieved through the use of tools, as well as common APIs and shared authentication and authorization procedures across the federation.

GENI is a *federated* infrastructure that provides GENI experimenters with access to multiple different testbeds, enabling networking and distributed systems research. Although GENI resources are owned and operated by different organizations from a user's perspective GENI appears as a unified virtual laboratory. An experimenter can instantiate custom Layer 2 topologies that include a variety of compute and network elements. This is achieved by tools (see [5]) with the use of common APIs and shared authentication and authorization procedures across the federation.

In more detail, a GENI user can obtain compute resources from locations around the United States,[1] connect them using Layer 2 networks and can program every aspect of their topology including how traffic is routed within their network. It is important to note that all networking in GENI (wireless and wired) is done at Layer 2, allowing experimenters to run non-IP protocols.

All reservations in GENI are limited in time. When a reservation expires, resources are returned to the pool of available resources. The federated design of

[1]Through common APIs and policy agreements, GENI users can actually access resources from around the globe.

N. Riga (✉) • S. Edwards
GENI Project Office, Raytheon BBN Technologies, 10 Moulton St. Cambridge, MA 02138, USA

V. Thomas
GENI Project Office, Raytheon BBN Technologies, 5775, Wayzata Blvd., Suite 630, St. Louis Park, MN 55416, USA
e-mail: vthomas@bbn.com

© Springer International Publishing Switzerland 2016

R. McGeer et al. (eds.), *The GENI Book*, DOI 10.1007/978-3-319-33769-2_15

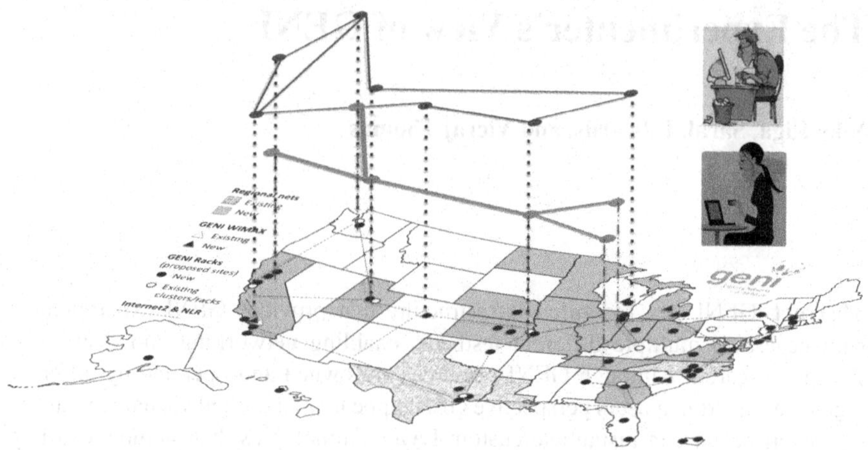

Fig. 1 Multiple GENI experiments run concurrently in isolated slices of the infrastructure

GENI makes it feasible to scale to a testbed that is much larger than one typically found in a laboratory. It provides the geographic diversity often needed in network research and the resource variety (from VMs to bare metal machines, from switches to WiMAX and LTE base stations) to make new configurations possible and to spur innovation.

Two of the major design principles in GENI that affect the interactions of users with the testbed are:

1. *Sliceability.* Each experiment is instantiated within a separate *slice* of the testbed, see Fig. 1. All slices are isolated from each other, i.e. the traffic of one experiment is not accessible or visible to an experiment running in another slice. This enables experiments that might be incompatible with each other to run concurrently on the same physical resources. For example one experimenter might be exploring the performance of a TCP variant that runs on top of IP, while a second experimenter might be investigating the feasibility of a new non-IP internet architecture in another slice. It is worth noting that several new internet architectures have been deployed and evaluated on GENI [1, 7, 15, 16, 26, 28], running concurrently in different slices. Although GENI does not provide any hard performance isolation guarantees, its architecture and resource slicing[2] provides best effort performance isolation between experiments. Sliceability not only enables different experimenters to use the testbed concurrently but also enables one user to run multiple experiments at the same time.

[2]Network slicing is done by VLAN with the appropriate bandwidth limits and there is no over-provisioning of network capacity. Some resource providers may over-provision compute resources by allocating more virtual machines on a server than available cores or memory. GENI also provides a limited number of bare metal machines that users can reserve in their experiments.

2. *Deep Programmability.* A user is allowed to program the behavior of as many elements in a slice as possible. This includes hosts at the edge of the topology (the user can choose an operating system and have full root access on the hosts to further customize them, including modifying the kernel as needed), as well as switching and computing elements in the core of the network. GENI has deployed programmable switches—mainly OpenFlow [19]—in the edge and core networks, as well as computation and storage in centrally located network exchange points, e.g. within a regional network. From the user's point of view, the slice includes a topology that can be programmed at different layers of the networking stack and allows for functionality (e.g. caching) to be placed in the middle of the network.

Accessing the GENI Testbed GENI is free for use in research and education. For users from many academic institutions accessing GENI is as simple as logging in any other service offered by their university. As described in [5], GENI has outsourced, when possible, the authentication of the users to their home institutions. This design choice not only makes user management much simpler for the federation, but also simplifies the user experience by allowing them to use their institution's credentials to login and use GENI. The single sign-on mechanism used in GENI is very similar to the prevalent practice in the web today of using well known identity providers such as Google or Facebook to access third party services. The technology used for single sign-on in GENI is Shibboleth [21]. For institutions that do not support this technology the GENI Project Office runs a Shibboleth Identity Provider to manage and authenticate users from these institutions.

1 Useful GENI Concepts

Before we delve into more details about how a user accesses GENI and instantiates experiments, we will go over some basic concepts and terminology.

1.1 GENI Resources and Resource Aggregates

Resource in GENI is used to describe elements that can be reserved by users and used in their experiments. Examples of resources include virtual machines (VMs), bare metal machines, storage, VLANs, OpenFlow datapaths, flowspace in OpenFlow-enabled devices, NetFPGAs, switches, sensors, monitoring cards and cameras. Resources can be contained within one physical device (e.g. VM) or distributed across a set of devices (e.g. VLAN), depending on the nature of the resource.

The following is a list of major GENI resource types. The elements in the list are not necessarily mutually exclusive i.e. a resource may belong to more than one type.

- *Compute resources*: Compute resources in GENI can be Virtual Machines, Linux containers or bare metal hosts. Depending on the requirements of an experiment the user can choose the resources that are most appropriate. For example, if performance is critical to the experiment, the user can reserve bare metal hosts. On the other hand if geographic diversity is more important then containers might be the right choice since they are much more widely available.
- *Wireless resources*: GENI Wireless sites have one or more WiMAX or LTE antennae that provide 4G coverage. Resources are sliced at Layer 2 by VLANs. For more information see [29]. Also each site has 2 or more wireless devices, usually referred to as *yellow nodes* that support regular *nix operating systems and can be reserved as bare metal nodes to run remote wireless experiments using WiMAX or WiFi interfaces. For users local to the sites there are 3G Android phones available for mobile experiments.
- *Storage resources*: Some sites also provide external storage that can be reserved and attached to an experiment, providing extra storage space when needed.
- *Network resources*: In addition to the wireless resources described above, GENI provides a variety of wired network resources that can be used to (1) connect resources from different locations in Layer 2 topologies and (2) allow the user to control forwarding within the network. Many of the network providers in GENI, including Internet2 [13] that provides the GENI backbone network, have deployed OpenFlow [19] switches and allow users to control the forwarding of their traffic as it traverses the network.
- *Unique resources*: The architecture GENI allows participants to connect unique resources into GENI and provide access to them to remote users. For example some of the compute nodes also have NetFPGAs or NPUs, that a user can program. This capability is not limited to networking resources and sites can connect diverse devices such as advanced microscopes and weather radars.

Resources are made available to experimenters by *Aggregate Managers*. An Aggregate Manager (AM) is a service that manages a collection (an aggregate) of physical devices in order to provide users with the requested resources. For example an AM can manage one or more VM servers providing VMs to users, it can manage a set of switches and provision links across them or manage a unique resource like a microscope. The AM is responsible for provisioning the resources, slicing shared devices ensuring isolation, providing remote access to the users when appropriate, reclaiming the resources for expired reservations, and enforcing local and global usage policies. An AM can manage any number of different devices from computation servers to switches to storage. The set of devices to be managed is an implementation and policy decision by the owners of the devices and the operator of the AM.

All AMs implement a GENI standard API called the AM API. This API is used by experimenter tools to learn about and reserve resources at the AM. See Sect. 1.2 for details.

Following are examples of some existing Aggregates that will clarify their role in the GENI ecosystem:

GENI Racks. These are the most widely deployed GENI resources. A GENI Rack consists of compute, storage and networking devices, all controlled by one or more Aggregate Managers. Details on the design and deployment of GENI Racks can be found in [3, 8, 18].

Network Providers. Several network providers that provide connectivity between GENI sites have deployed GENI-compliant Aggregate Managers for users to obtain and configure networking resources using the GENI AM API. Some characteristic examples are:

- *Internet2* and *MAX* that allow GENI users to dynamically configure Layer 2 circuits across their network.
- *SOX (Southern Crossroads), StarLight, CENIC, and MOXI regional networks* that allow GENI users to dynamically reserve flowspace on their OpenFlow switches and control the specified traffic with a user-defined OpenFlow controller.

Federated testbeds. Several existing testbeds have modified their resource managers to support the GENI AM API and thus allow users to reserve resources using the same tools they would use to resources GENI resources. Notable examples include PlanetLab [24] and Emulab [36].

1.2 GENI RSpecs and the GENI AM API

To allow interoperability among different Aggregate Managers(AMs) and a variety of tools, GENI has specified a standardized API called the GENI Aggregate Manager API (GENI AM API) [5]. The GENI AM API specifies the interactions between an AM and tools (which are a proxy for users). It defines methods to manage resource reservations (create one, expand the duration of the reservation, delete it), get status of reserved resources and discover resources offered by AMs.

Resources in GENI are described using a standardized language called *Resource Specification (RSpecs)* [27]. RSpecs are XML documents following an agreed upon schema to represent resources. The schema supports aggregate or resource specific extensions. As shown in Fig. 2, there are three different types of RSpecs:

1. **Advertisement RSpec.** Used by an AM to describe the resources it makes available to users.
2. **Request RSpec.** The document a user (usually a tool) uses to describe the resources to be reserved and how they should be configured including network topology.

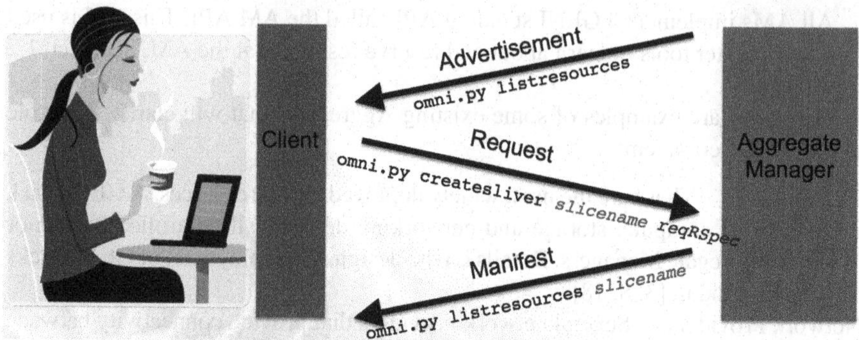

Fig. 2 There are three different types of RSpecs in GENI

3. **Manifest RSpec.** Describes the resources a user has already reserved at an AM.

Figure 2 shows the API calls that use RSpecs to communicate between the tools and the AMs. Since (1) all experimenter tools and AMs use the same API and RSpec schema, and (2) all information about resources is stored at the AMs, an experimenter may use different tools at different times for the same experiment. For example, the experimenter may reserve resources using tool A, check on their status using tool B and release them using tool C.

1.3 Slice

As mentioned earlier, one the major design pillars of GENI is *sliceability*, the ability to share the same infrastructure among multiple users and the ability to concurrently run multiple experiments. To achieve this GENI adopted, and expanded on the concept of a slice from PlanetLab [24] and the SFA architecture [23].

A GENI slice is:

- *A container for resources used in an experiment.* Users add GENI resources (compute resources, network links, etc.) to slices and run experiments that use these resources.
- *A unit of access control.* The experimenter that creates a slice can determine who has access to the slice i.e. are members of the slice.
- *The unit of isolation for experiments.* Resources assigned to the slice are dedicated to that slice and protected from access or interference from other slices, up to the capabilities of the specific virtualization technology used to slice each specific resource.

All slices in GENI are ephemeral, i.e. they have an expiration time. It is worth noting that although the resource reservations within a slice can not outlive the slice, the expiration times can be different, i.e. a slice can (and usually does) outlive the resource reservations.

A slice can contain resources from any number of federated aggregates. Although slice is an abstraction, it is the concept that allows an experiment to span multiple administrative domains. Before starting an experiment, the user has to register a slice with a trusted Slice Authority.[3] Using this registered slice, the user can request resources from individual aggregates. In some sense, the aggregates trust and grant resources to slices.

From an operator's perspective, slices are the primary abstraction for accounting and accountability—resources are acquired and consumed by slices, and experiment behavior can be traced back to a slice.

1.4 GENI Projects

GENI is a shared, federated infrastructure that is used by experimenters around the globe at no cost. However, when running experiments in GENI, a user accesses and instruments real physical devices that are located within administrative domains usually not owned by the user or his institution. To address the accountability issues that arise in such a federated environment GENI expands on Emulab's [36] concept of a *Project*.

Projects organize research in GENI. Projects contain both people and their experiments. A project is led by a single responsible individual, known as the *Project Lead*. At the time of this writing, only academic faculty, senior technical staff and GENI Rack administrators can be Project Leads. The Project Lead takes responsibility for all experiments running within his projects and agrees to respond appropriately if a problem is discovered.

Any user who meets the requirements to be a Project Lead can request to be one. Leads can create GENI projects without the need for further approvals. Although only Project Leads can create projects, a lead can choose to have other individuals administer a specific project by making them project admins. Project admins have the same privileges within a project as the Project Lead, but they can not create new projects.

A project may or may not have an expiration time depending on the purpose of the project. A project that will be used by the students of a class is typically set to expire after the end of the semester. Research projects on the other hand tend not to have an expiration date.

[3]GENI's architecture supports multiple Slice Authorities. For example GENI currently has three Slice Authorities that can register slices used in the federation: The GENI Slice Authority operated by the GPO and the PlanetLab and Emulab Slice Authorities.

2 The GENI Experimenter Workflow

The GENI Experimenter workflow is a structured approach to running experiments on GENI. It serves as a framework for experimenters to be systematic with their experimentation. For GENI tool developers it serves as a framework for describing the steps of the workflow supported by their tools and their interfaces with other tools.

The GENI experimenter workflow is illustrated in Fig. 3. The major phases of running any experiment—Design/setup, Execute and Finish—are the three large areas of the figure, The ovals represent the steps in the workflow. The white boxes are experiment artifacts produced or consumed at each step.

Even though the workflow is depicted as a linear set of steps, in reality there will be loops in the workflow with the experimenter going back to an earlier step or skipping some steps altogether.

2.1 Design and Setup Experiment

Experiment Design In this step, experimenters lay out a detailed plan in advance of running their experiments. When experimenters come to GENI to run an

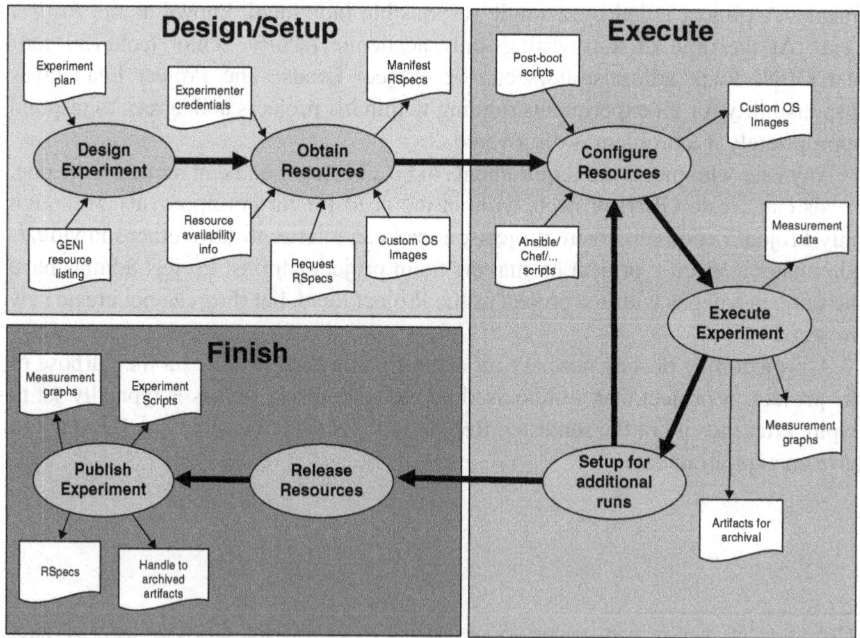

Fig. 3 Major steps in the GENI workflow and artifacts associated with each step

experiment, they typically have certain objectives in mind, the kind of experiment that will achieve the objectives and measurements needed to test if the objectives have been met.

Developing a detailed plan for an experiment on GENI includes deciding on:

1. The types, numbers and locations of resources needed. For example, experimenters must choose between virtual machines and raw-PCs for computation; configuration of these resources including memory, disk space, network interfaces and operating system; the location of the resources; the experiment topology and types of links; and the need for specialized resources such as hardware switches. Experimenters may consult the GENI web pages to find aggregates that have the resources they need.
2. How these resources will be programmed. They may choose to: (1) log into each resources separately and configure it, (2) write scripts that are automatically installed and executed when the resources come up, or (3) use custom OS images that have the needed software installed and possibly configured. Section 5 has a discussion on why experimenters should use scripts and custom images to make experiments repeatable and reproducible.
3. How the necessary network traffic will be generated and what will be measured. Options for network traffic generation include standard networking tools such as iperf and ping; more sophisticated traffic generation tools such as Tmix [35]; and the use of real user traffic (also know as opt-in user traffic). Experimenters may collect their own measurements or use GENI Instrumentation and Measurement tools such as LabWiki and GENI Desktop (Sect. 4.4).

Obtain Resources The next step in the experiment lifecycle is obtaining the resources needed for the experiments. Experimenters specify the resources they need including compute resources, network topology, operating systems to be loaded, bandwidth of network links, etc. by creating a Request RSpec. RSpecs are typically created using a tool such as Jacks Sect. 4.1. Figure 4 shows Jacks being used to create a Request RSpec for two virtual machines from two different aggregates.

Experimenters need to pick the aggregates where they want to reserve resources. This is typically a subset of the aggregates identified in the Experiment Design step of the workflow. They may pick aggregates based on factors such as availability and load. The GENI Meta-Operations Center (GMOC) maintains a calendar that shows scheduled and unscheduled maintenance events. The GENI monitoring service has information on resource utilization at each aggregate.

Experimenters submit their RSpecs to the selected aggregates using tools such as the GENI Portal and Omni. If their reservation is successful, they get a *Manifest RSpec*, an XML document with information needed to use the resources. For example, for VMs the manifest includes login information, OS installed, and MAC and IP addresses of the interfaces.

Since GENI is a shared testbed, resources have expiration times on them i.e. these resources are released when they expire. Experimenters can extend the expiration

a

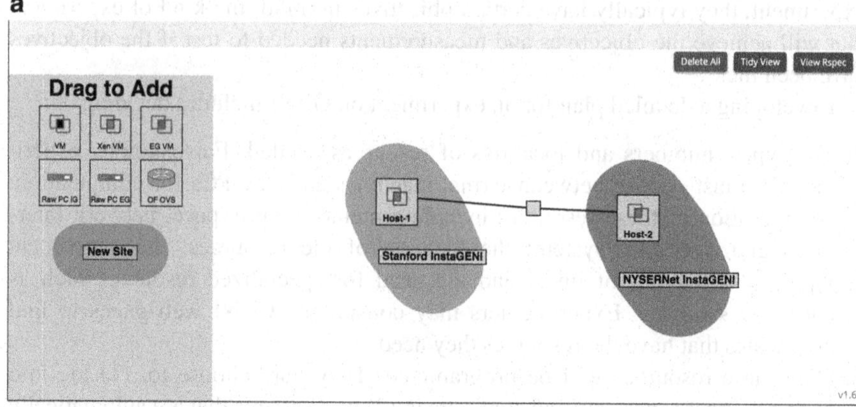

b

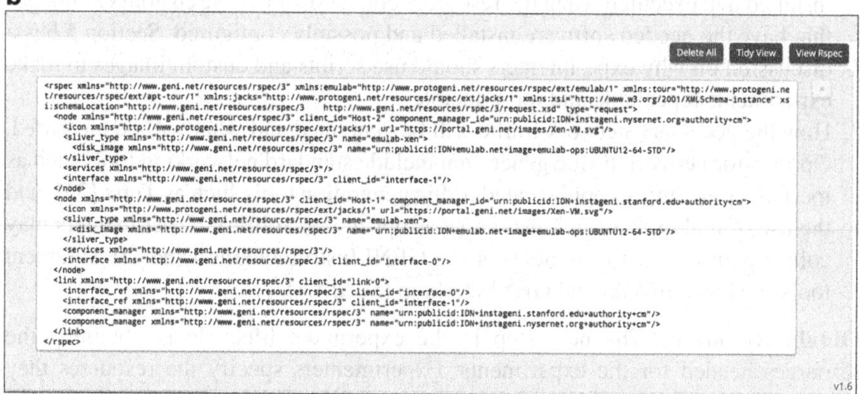


```
<rspec xmlns="http://www.geni.net/resources/rspec/3" xmlns:emulab="http://www.protogeni.net/resources/rspec/ext/emulab/1" xmlns:tour="http://www.protogeni.ne
t/resources/rspec/ext/apt-tour/1" xmlns:jacks="http://www.protogeni.net/resources/rspec/ext/jacks/1" xmlns:xsi="http://www.w3.org/2001/XMLSchema-instance" xs
i:schemaLocation="http://www.geni.net/resources/rspec/3    http://www.geni.net/resources/rspec/3/request.xsd" type="request">
  <node xmlns="http://www.geni.net/resources/rspec/3" client_id="Host-2" component_manager_id="urn:publicid:IDN+instageni.nysernet.org+authority+cm">
    <icon xmlns="http://www.protogeni.net/resources/rspec/ext/jacks/1" url="https://portal.geni.net/images/Xen-VM.svg"/>
    <sliver_type xmlns="http://www.geni.net/resources/rspec/3" name="emulab-xen">
      <disk_image xmlns="http://www.geni.net/resources/rspec/3" name="urn:publicid:IDN+emulab.net+image+emulab-ops:UBUNTU12-64-STD"/>
    </sliver_type>
    <services xmlns="http://www.geni.net/resources/rspec/3"/>
    <interface xmlns="http://www.geni.net/resources/rspec/3" client_id="interface-1"/>
  </node>
  <node xmlns="http://www.geni.net/resources/rspec/3" client_id="Host-1" component_manager_id="urn:publicid:IDN+instageni.stanford.edu+authority+cm">
    <icon xmlns="http://www.protogeni.net/resources/rspec/ext/jacks/1" url="https://portal.geni.net/images/Xen-VM.svg"/>
    <sliver_type xmlns="http://www.geni.net/resources/rspec/3" name="emulab-xen">
      <disk_image xmlns="http://www.geni.net/resources/rspec/3" name="urn:publicid:IDN+emulab.net+image+emulab-ops:UBUNTU12-64-STD"/>
    </sliver_type>
    <services xmlns="http://www.geni.net/resources/rspec/3"/>
    <interface xmlns="http://www.geni.net/resources/rspec/3" client_id="interface-0"/>
  </node>
  <link xmlns="http://www.geni.net/resources/rspec/3" client_id="link-0">
    <interface_ref xmlns="http://www.geni.net/resources/rspec/3" client_id="interface-0"/>
    <interface_ref xmlns="http://www.geni.net/resources/rspec/3" client_id="interface-1"/>
    <component_manager xmlns="http://www.geni.net/resources/rspec/3" name="urn:publicid:IDN+instageni.stanford.edu+authority+cm"/>
    <component_manager xmlns="http://www.geni.net/resources/rspec/3" name="urn:publicid:IDN+instageni.nysernet.org+authority+cm"/>
  </link>
</rspec>
```


Fig. 4 Tools such as Jacks are commonly used to created request RSpecs. (**a**) Experiment topology drawn using Jacks. (**b**) Corresponding request RSpec generated by Jacks

time using their GENI resource reservation tools. Policies on default expiration times and the maximum duration by which a reservation can be extended at one time are set by the aggregate owners.

2.2 *Execute Experiment*

Configure Resources After resources have been obtained, experimenters configure them for their experiment. They may do this by installing software, modifying configuration files, changing settings on network interfaces, etc. Experimenters may automate the configuration of their resources by:

1. The use of install and execute scripts (also called post-boot scripts). These scripts, specified in the request RSpec, are installed and executed on the resources when they are setup by the aggregate.
2. Using system administration tools such as Ansible [2] and Chef [6].

After the resources are configured, experimenters may choose to create *custom images*, which are snapshots of operating systems they have configured. For future experiments they can specify these custom images as the operating system to be loaded on their resources. The operating system to be loaded is specified in the Request RSpec.

Execute Experiment Execution can be triggered manually by logging into each resource and starting up the experiment. Experiments can also be started up automatically using execute scripts.

GENI experiment orchestration tools such as LabWiki [17] and OMF/OML [22, 25] allow experimenters to describe and instrument an experiment, execute it and collect its results.

GENI tools for instrumenting experiments and collecting measurements are the GENI Desktop and LabWiki (Sect. 4.4). These tools allow experimenters to specify measurements to be collected, view graphs of these measurements and save the measurements.

Experiments may archive measurements and other experiment related artifacts such as RSpecs and scripts using a GENI-provided iRODS service Sect. 4.6.1. Items archived on this service survive the releasing of resources used in the experiment, the expiration of slice or project.

Setup for Additional Runs Experimenters may run the same experiment multiple times with the same or with different inputs, or resource configurations. The changes they make to the experiment before each run may be based on measurements gathered during an earlier execution.

2.3 *Finish Experiment*

Release Resource Since GENI is a shared testbed, experimenters are expected to release their use as soon as they are done. Experimenters can use any of the resource reservation tools to release resources. If resources are not explicitly released by the experimenter, they will automatically be released when they expire.

Publish Results The final step of the workflow is the publication of the results of the experiment. Experiment reprodicibility is a tenet of scientific research and GENI provides mechanisms for researchers to make the experiments reproducible. The artifacts required to reproduce the experiment may be archived on the GENI iRODS service and made accessible by other researchers. The RSpecs used for the experiment may be uploaded and shared on the GENI Portal. Any custom image used in the experiment can also be made public and available for others to use.

Section 5.1 describes how experimenters can make their experiments reproducible by others.

3 Case Study: GENI Cinema, Implementing an Advanced Service on GENI

This section is based on the GENI Cinema Architecture document [14] written by Ryan Izard, Parmesh Ramanathan and KC Wang.

GENI Cinema is a persistent live video streaming service that capitalizes on the advanced capabilities of GENI. It allows any user (organization or individual) with access to the GENI network to publish a live video stream through this service. The users can also search, select and watch video streams. Being a geographically-distributed testbed, the GENI infrastructure provides an ideal platform to implement a content delivery network for efficient broadcasting of video content to customers at the edges. Combined with Software Defined Networking (SDN), this allows both network and compute resources to be conserved while users from different areas choose between the available video "channels" hosted by GENI Cinema.

In this section we describe the deployment story of GENI Cinema from an idea to a prototype service and highlight some of the design choices made. GENI Cinema was developed by teams of researchers at the University of Wisconsin (Principal Investigator Parmesh Ramanathan) and at Clemson University (Principal Investigator Kuang-Ching Wang).

3.1 Designing GENI Cinema

Designing such a complicated service is an iterative process, where the design is constantly being improved as the service is developed and deployed.

GENI Cinema consists of two main subsystems: one addressing end-to-end video/audio stream handling and the other addressing optimal forwarding in the network. Both subsystems heavily leverage GENI SDN capabilities. Each subsystem was developed separately by each of the universities on the project. Each group ran single site experiments and updated its design to optimize for performance. Once both systems were fully developed they were integrated into one system. Figure 5 shows the combined architecture that consists of the many functional blocks that comprise the GENI Cinema service. There are ingress and egress gateways for receiving and sending video streams, ingress and egress VideoLAN Client (VLC) servers for hosting video streams on the backend and providing them on the frontend, a global OpenFlow controller, hardware and software OpenFlow switches for controlling the flow of video streams internal to the GENI Cinema service, and a web server for customer interaction.

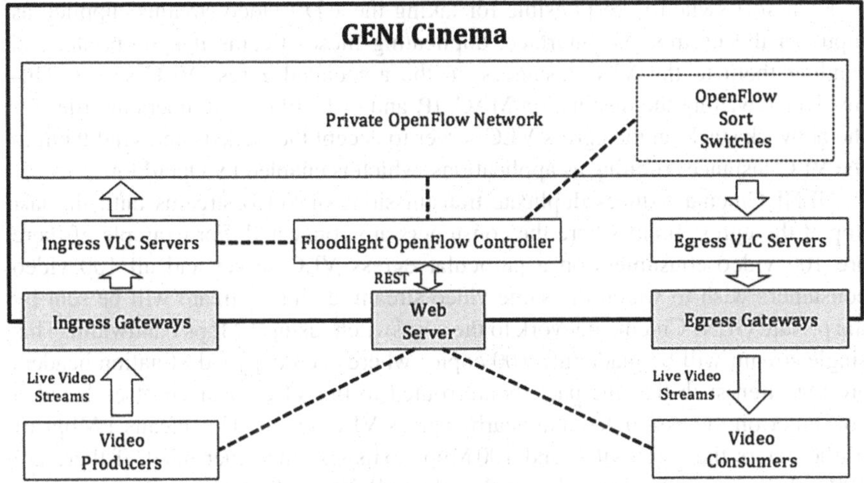

Fig. 5 Logical components of the GENI Cinema architecture

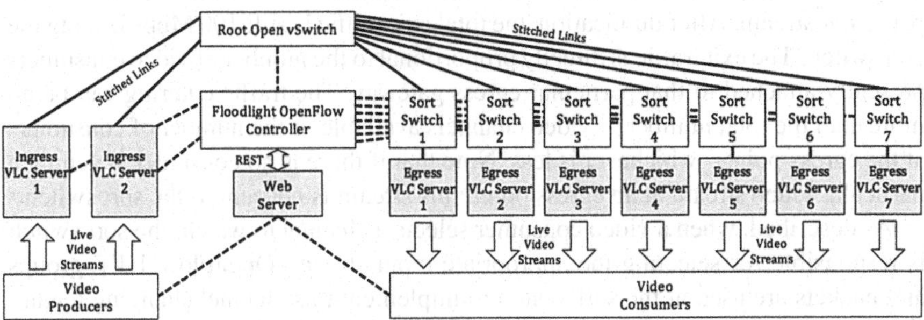

Fig. 6 The GENI Cinema SDN architecture. Note that each stitched link also contains two physical OpenFlow switches under the control of a floodlight controller—one at each end of the link

3.2 *Use of Software Defined Networking*

The GENI Cinema implementation heavily relies on the software defined capabilities of the GENI network and in particular on the deployment of OpenFlow switches.

All video traffic output from the ingress VLC servers into the private GENI Cinema network is unicast UDP in order to allow fast video stream switching without regard for connection state, sequence, or source as TCP would impose. Each UDP video stream is directed through the network toward all egress VLC servers where there is at least one video consumer wishing to watch that particular stream. Prior to each egress VLC server is an OpenFlow switch called a "sort switch", as depicted in Fig. 6.

Each sort switch is responsible for taking the UDP video streams supplied as input on the northbound interface, duplicating these streams if appropriate, and sending them to the VLC instances on the associated egress VLC server. This involves rewriting the destination MAC, IP, and/or UDP port numbers in order for the network stack on the egress VLC server to accept the packets and send them to the VLC instances running as applications, which is enabled by OpenFlow.

GENI Cinema reduces duplicate transmissions of video streams until the last hop at the egress point where the consumers are connected. For example, if there are 100 video consumers on a particular egress VLC server and all 100 video consumers wish to watch the same video stream, a single stream will be sent by the private GENI Cinema network to the sort switch, using 1 Mbps bandwidth. This single stream will be made into 100 copies where each copy's destination headers are rewritten such that the packets are routed to the VLC instance of each video consumer on the associated and nearby egress VLC server. This means 1 Mbps of traffic enters the sort switch and 100 Mbps exits. On the other hand, if there are 100 video consumers that collectively select all 20 of 20 available channels, then each channel's stream enters the sort switch for a total of 20 Mbps. The sort switch will make copies of each stream and rewrite the destination headers of each stream to send it to the VLC instance of the video consumer that wishes to watch that particular stream. After duplication, the total exit traffic is still 100 Mbps leaving the sort switch. The exit traffic is directly proportional to the number of video consumers presently attached to that particular egress gateway. The traffic entering can be no more than the total number of video channels available or the number of consumers at the egress point—whichever is less. Note that if there is no consumer watching a particular video stream at an egress point, this stream is not sent to the sort switch.

As described, when a video consumer selects a channel to watch, the sort switch is responsible for selecting the appropriate input stream. OpenFlow 1.1+ groups and buckets are used at the sort switch to implement this channel changing feature (Fig. 7). Every video is classified as an OpenFlow group, and every video consumer has a single OpenFlow bucket. An OpenFlow bucket is a list of actions, which in this case each action list rewrites the destination MAC, IP, and/or UDP port in the headers of the packets. If a video consumer switches video channels, its bucket is removed from the previous video group of the channel it was viewing, and the bucket is simply added to the group of the new video channel. In this way, only one connection and video stream per consumer is ever present at a given time within the private GENI Cinema network, and no connection is set up or torn down upon a channel change. This optimizes the bandwidth usage, as well as reduces the load on the server resources during frequent channel changes.

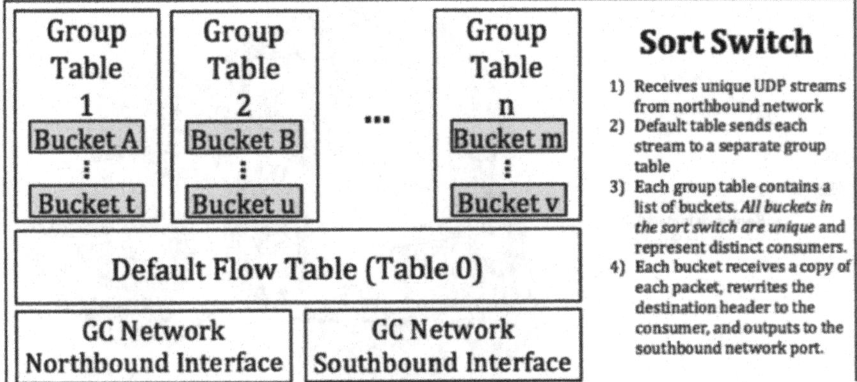
Group Table 1
Bucket A
Bucket t
Group Table 2
Bucket B
Bucket u
...
Group Table n
Bucket m
Bucket v
Default Flow Table (Table 0)
GC Network Northbound Interface
GC Network Southbound Interface
Sort Switch
1) Receives unique UDP streams from northbound network
2) Default table sends each stream to a separate group table
3) Each group table contains a list of buckets. All buckets in the sort switch are unique and represent distinct consumers.
4) Each bucket receives a copy of each packet, rewrites the destination header to the consumer, and outputs to the southbound network port.

Fig. 7 The use of OpenFlow groups in the sort switch

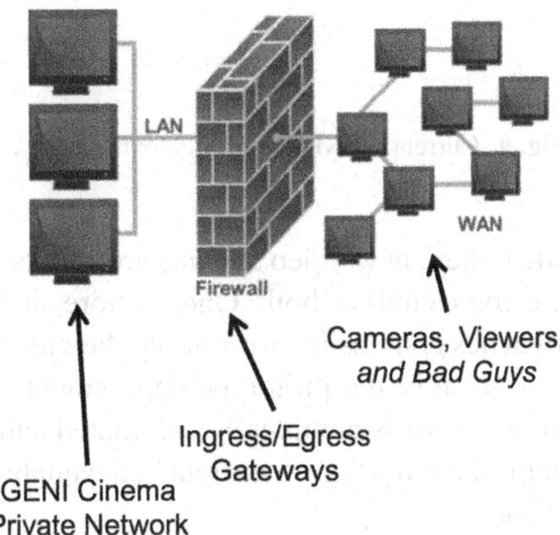
LAN
Firewall
WAN
Cameras, Viewers, and Bad Guys
Ingress/Egress Gateways
GENI Cinema Private Network

Fig. 8 The egress/ingress gateways can also serve as firewalls to the GENI Cinema private network

3.3 Deploying GENI Cinema

GENI Cinema started with a couple of single site deployments, one at Clemson University and the other at the University of Wisconsin. Originally the team broadcast local classes while debugging and enhancing the system.

While running in a single site, the team also developed the ingress and egress gateways that not only bypass any local issues due to NATing but can also secure the GENI Cinema system by acting as a firewall (Fig. 8).

Once GENI Cinema was stable, the deployment expanded to multiple sites. The first multi-site deployment was to connect the two prototype systems, the one at Clemson and the one at Wisconsin. After the two-site system was operational the team started working on a multi-site deployment. The first step was to enhance the system architecture to clearly identify which systems needed to be deployed on each site, how they are connected and how they interact with the rest of the system, i.e. they designed a distributed version of their system. Each new site can be an ingress

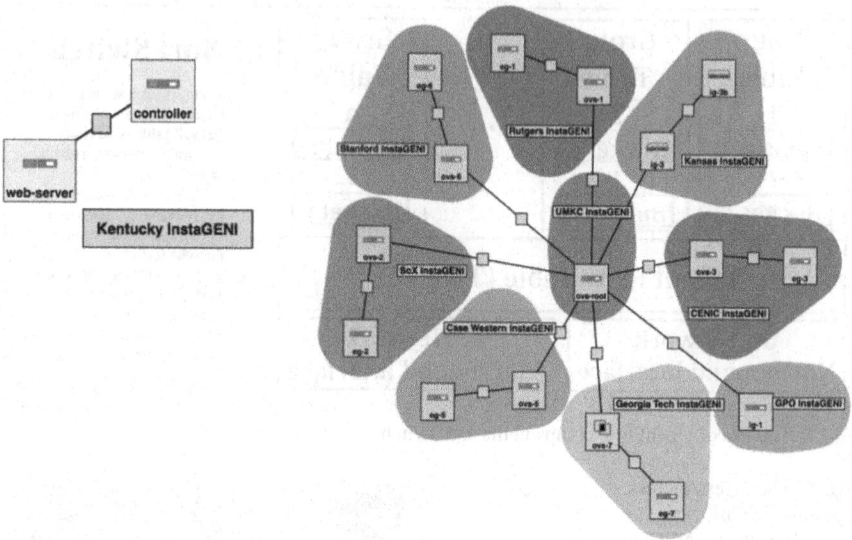

Fig. 9 Current GENI Cinema deployment

site (where new video streams are connected), an egress site (where new consumers are connected) or both. One or more sites are chosen to run central programmable switches responsible for routing the video streams from producers to consumers.

Currently the prototype deployment spans nine sites (Fig. 9). The deployment of new sites is completely automated and they can add new sites on demand. This helps them manage occasional unavailability of sites due to failures or maintenance events.

3.4 Connecting Users to GENI Cinema

GENI Cinema is open to users without GENI accounts. Connecting users to the GENI Cinema network is challenging, since the deployment lives within GENI. While the deployment was within the Clemson and Wisconsin Universities, the labs of the researchers were connected to the GENI deployment through their local GENI Rack. Classrooms in GENI-enabled campuses can be connected in a similar way, by expanding connectivity to the GENI network through the local GENI Rack. However, users should be able to access GENI Cinema from anywhere. To achieve this goal, users (producers or consumers) connect to the GENI Cinema service through the egress and ingress gateways using the public facing interface of these gateways. To avoid overloading the public interface, the users are load balanced across multiple gateways.

The workflow for video publishers and consumers is as follows:

- A producer wishing to publish a video on GENI Cinema makes a request on the GENI Cinema web service. The request is relayed to the OpenFlow controller, which responds with an ingress gateway IP address and transport port number. The producer connects and sends the video stream to the assigned address and port. The incoming video stream is relayed to an ingress VLC server where the live stream is hosted.
- When a consumer requests a video stream on the GENI Cinema web service, the request is relayed to the OpenFlow controller, which responds with an appropriate egress gateway IP address and port number where the consumer can connect and watch the video. The video selected is routed, duplicated and rewritten within the private network of GENI Cinema from the ingress VLC server on which it is being hosted to the private interface of the egress VLC server where the customer is connected. A VLC instance on this egress VLC server outputs the video on the public interface and relays it to the video consumer.

4 Experimenter Tools

The experimenters' main interface to GENI are the *experimenter tools* that serve to support the experimenter workflow (Sect. 2). Some tools support the experiment design/setup parts of the workflow by helping create Request RSpecs. Other tools support experiment execution by helping with installing and configuring software, orchestration (the automation and scheduling of the steps in the experiment), and instrumentation and measurement (the taking of and collection of data related to or produced by the experiment). Finally, other tools support the archiving and sharing of experiment results.

4.1 RSpec Creation Tools

Jacks, *Flack* and *jFed* are all graphical user interface (GUIs) tools for creating and editing Request RSpecs. They are used to define topologies and set properties of nodes and the links that connect them. Node properties include node name, OS and scripts to be installed and executed at boot time. Link properties include link type (VLAN or GRE tunnel) and IP addresses of end-points and others.

Flack and jFed can also be used as resource reservation tools (Sect. 4.2); they can be used to submit RSpecs to specified aggregates using the GENI AM API. While Jacks does not do resource reservations, the RSpecs it generates can be exported for use with other tools.

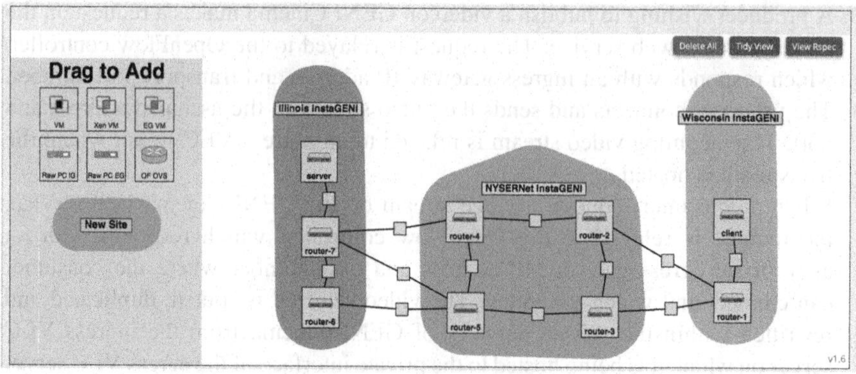

Fig. 10 Jacks GUI showing a topology spanning three aggregates

4.1.1 Jacks and Flack

Jacks and Flack are created by the Flux Research Group at the University of Utah. Both are browser-based: Jacks is written in HTML5 and Flack in Flash. Flack is no longer maintained.

Jacks is primarily an RSpec creation and viewing tool and is usually embedded in another tool such as the GENI Portal or GENI Desktop. A unique feature of Jacks is its constraint system that prevents experimenters from creating invalid RSpecs. For example, it will warn experimenters if they try to create a Layer 2 link between sites that do not support it or load an OS image in an incompatible compute resource (Fig. 10).

4.1.2 jFed

The jFed tool [12] is created by the iMinds Research Institute in Belgium. It is a Java application that runs on the experimenter's workstation. It can be used to create and view RSpecs, make resource reservations, launch ssh clients to log into nodes and do some experiment orchestration (Fig. 11).

4.1.3 geni-lib

geni-lib [4] is a python library from Barnstormer Softworks. It provides an object oriented scripting interface to both the AM API and GENI RSpecs. The purpose of geni-lib is to allow developers to build custom GENI tools. This is particularly helpful for advanced GENI experimenters. An example is the *scaleup* tool distributed with geni-lib which allows experimenters to write small topologies using standard node types (e.g. a topology might consist of multiple client, server,

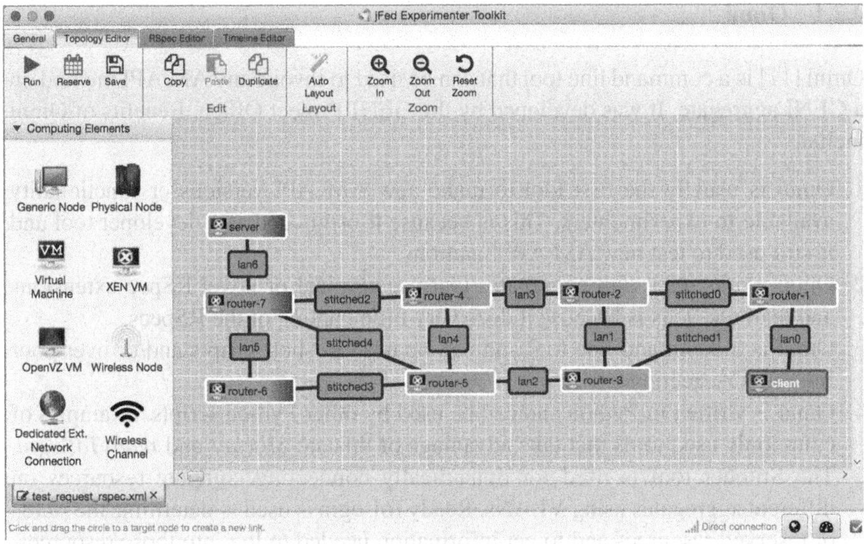

Fig. 11 jFed GUI showing a topology spanning three aggregates

```
>>> ad = IGAM.MAX.listresources(context)
>>> for node in ad.nodes:
...   if node.available and IGUtil.shared_xen(node):
...     print node.component_id
...
urn:publicid:IDN+instageni.maxgigapop.net+node+pc3
urn:publicid:IDN+instageni.maxgigapop.net+node+pc1
urn:publicid:IDN+instageni.maxgigapop.net+node+pc2
```

Fig. 12 Using *geni-lib* to list all available Xen servers at an InstaGENI rack

and router nodes) which can then be easily scaled up to a larger number of nodes in a wide range of topologies (e.g. ring, grid, random) (Fig. 12).

4.2 Resource Reservation Tools

The *Omni*, *geni-lib*, the *GENI Portal*, *jFed*, and *Flack* allow experimenters to communicate with resource providers (i.e. aggregates) using the GENI AM API (Sect. 1.2).

These tools allow experimenters to determine the resources advertised by an aggregate (i.e. to request an advertisement RSpec), to reserve resources in a slice (i.e. to submit a request RSpec), and to determine the resources reserved at an aggregate in a particular slice (i.e. to retrieve a manifest RSpec).

4.2.1 Omni

Omni [11] is a command line tool that can be used to invoke any AM API method on a GENI aggregate. It was developed by the GENI Project Office. Benefits of Omni include:

1. Omni is usually the first tool to make new AM API versions or functionality available to experimenters. This is because it originated as a developer tool and is still used to test new AM API functions.
2. Omni works well with aggregates that use atypical or novel RSpec extensions and features. This is because it does very little parsing of the RSpecs.
3. Omni is a command line tool and can be used in shell scripts and/or over poor Internet Connections.
4. Omni is written in Python and can be used by other Python scripts. Examples of commonly used tools that take advantage of this are *Stitcher* and *readyToLogin*. The Stitcher tool is used for dynamically connecting compute resources on different aggregates using VLANs. ReadyToLogin is used to determine the status of reserved resources and to get information needed to log into those resources. Additionally, tools such as the GENI Portal and GENI Desktop use Omni behind the scenes to make AM API calls.

The downside to Omni is that much of the burden of manipulating RSpecs (generating Request RSpecs, parsing Advertisement and Manifest RSpecs) falls on experimenters. Of course, experimenters can use other tools for RSpec manipulation and use Omni for resource management.

4.2.2 The GENI Portal

The GENI Portal [10] is probably the most widely used of GENI experimenter tools because it is the only tool for account and project management. It is a web-based tool that requires no software installation on the experimenters' computers, it supports much of the experimenter workflow and it serves as an identify provider for other tools and services. The GENI Portal was developed by the GENI Project Office.

The GENI Portal can be used for account management (requesting accounts, requesting Project Lead status), project and slice management (creating projects and slices, adding and removing users from projects and slices), resource management (reserving and deleting resources, extending resource reservations) and sharing of RSpecs. The GENI Portal embeds the Jacks tool for creating and viewing RSpecs.

The GENI Portal also serves as an OpenID identity provider for tools, services and testbeds hosted by other organizations. Experimenters log into the Portal and then click from the Portal to access these tools and services without having to separately log into those tools. Some examples of tools and services that are accessible from the portal inlcude JFed, GENI Desktop, the Canadiatn SAVI testned, GENI wireless, CloudLab.

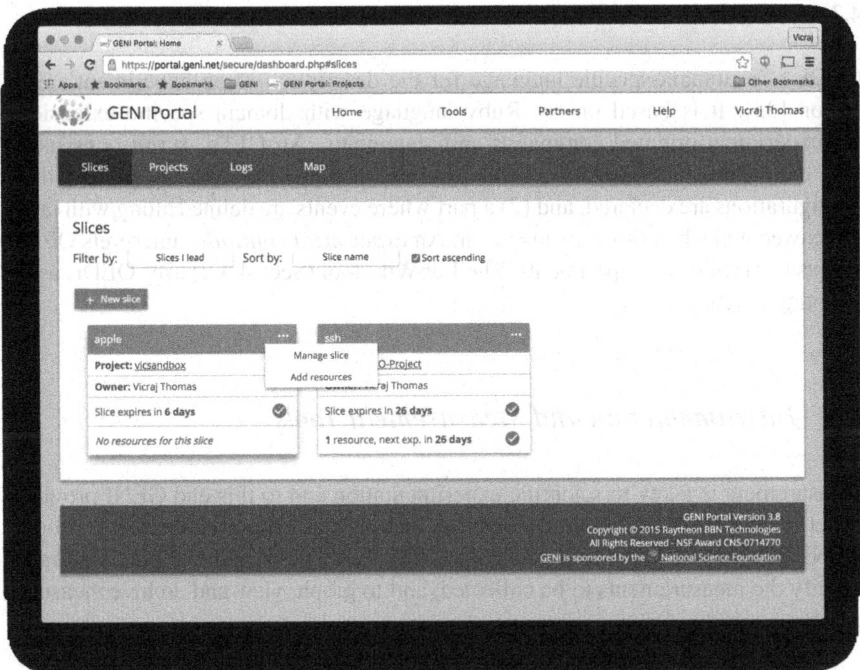

Fig. 13 Slice dashboard view of the GENI portal

Figure 13 shows the slice dashboard view for a user. In this figure the user has filtered the slices he has access, to only view the ones he leads. He can manage a slice or add resources to the slice by clicking on the dots by the name of the slice.

4.3 Experiment Orchestration and Scripting Tools

Experiment orchestration allows experimenters to automate or script their experiment procedure: start/stop data collection, start/stop traffic, schedule network events, etc. As such, orchestration is critical to the repeatability of experiments by allowing an experimenter to do multiple runs of the same procedure and to vary parameters as necessary.

While trivial procedures can be orchestrated with simple scripts (for example install scripts), GENI supports more complicated procedures using *OEDL* which is language to script and instrument data collection.

4.3.1 OEDL

OEDL is a domain-specific language for the description of an experiment's execution [33]. It is based on the Ruby language with domain specific extensions for experiment-oriented commands and statements. An OEDL script consists of two main parts: (1) A part where resources used in the experiment and their configurations are declared, and (2) a part where events are defined along with tasks to be executed when those events occur. An *experiment controller* interprets OEDL scripts to orchestrate experiments. The LabWiki tool (Sect. 4.4.2) uses OEDL as its scripting language.

4.4 Instrumentation and Measurement Tools

Measurement is a key to scientific experimentation and to this end GENI provides experimenters with a couple of Instrumentation and Measurement (I&M) tools: GENI Desktop/GEMINI and LabWiki/GIMI. Both tools allow experimenters to specify the measurements to be collected, and to graph, view and archive measurements.

4.4.1 GENI Desktop

GENI Desktop [34] is a web-based experimenter tool that, like the GENI Portal, can be used to create projects and slices, create Request RSpecs using the embedded Jacks tool, and manage resources. It was developed by the University of Kentucky.

A key feature of the GENI Desktop is the ability to *instrument* a slice to collect and view live measurements. It includes a number of pre-defined measurements such as CPU load on the hosts and number of packets sent/received on a network interface. Experimenters may also provide scripts to collect and view their own custom measurements. To select pre-defined measurements, the experimenter simply clicks on a host or link in the "Topology View" of the GENI Desktop and then selects the measurements of interest. Figure 14 shows the Topology View of an experiment and a graph of traffic on one of the interfaces attached to the link in the experiment.

4.4.2 LabWiki

LabWiki [17] is a web-based tool to design, describe and run GENI experiments. It was developed by NICTA, Australia's Information Communications Technology Research Centre of Excellence and the University of Massachusetts at Amherst. LabWiki is designed to help experimenters develop experiments that are repeatable and reproducible. LabWiki includes a panel where experimenters write experiment scripts using the OEDL scripting language (Sect. 4.3.1), a second panel for running and viewing graphs, and a third panel for recording notes and saving experiment

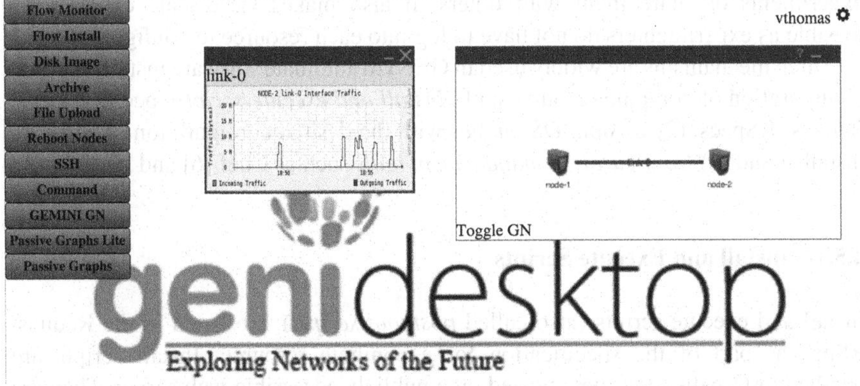

Fig. 14 GENI desktop showing graph of traffic on a network link

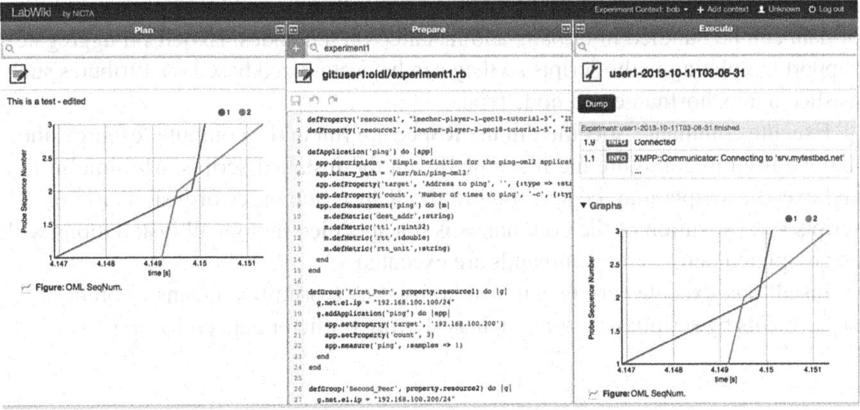

Fig. 15 Scripting and running an experiment using the LabWiki tool

results including graphs. Experiment scripts can be shared with other LabWiki users wishing to reproduce or extend the experiment.

Figure 15 shows the three panels of the LabWiki tool: The panel labeled "Prepare" is used to write or load experiment scripts, the "Execute" panel is where experimenters drop scripts to be executed and view graphs of experiment data and the "Plan" panel is used for notes, observations and saving graphs from the Execute panel.

4.5 Software Installation and Resource Configuration

Almost any experiment in GENI involves installing software or configuring compute resources. Automating this process helps experimenters easily and quickly repeat experiments or share them with others. It also makes large-scale experiments feasible as experimenters do not have to log into each resource to configure it.

Three mechanisms are widely used in GENI to automate software installation and configuration of compute resources: (1) *Install and execute scripts* specified in the Request RSpecs, (2) *Custom OS images* with the desired configuration or software installed and (3) *Configuration management tools* such as Chef [6] and Ansible [2].

4.5.1 Install and Execute Scripts

Install and execute scripts (also called *postboot scripts*), are listed in the Request RSpec as part of the specification for a compute resource. Install scripts are bundled in tarballs (.tgz files) posted on a publicly accessible web server. They are downloaded and installed on compute nodes by the Aggregate Manager when the resources are provisioned. Install scrips are typically executable scripts but any type of data can be bundled in tarballs and installed on the nodes. ExoGENI aggregates support templating in the scripts so they can be customized based on attributes such as slice name, hostname and node type.

Execute commands specified in the RSpec are run in the compute resource; they may be used to configure the resource or run the installed scripts. Multiple install and execute scripts may be specified in the RSpec; the order of installation of the scripts and execution of the commands is not specified though all installations will be completed before any commands are executed.

Install and execute scripts can also be used as a primitive means to orchestrate experiments by scripting actions such as starting traffic or data collection.

4.5.2 Custom Images

Custom images are bootable operating system images with the configurations or software needed for an experiment. Experimenters may create their own custom image by starting with a standard OS image, configuring it as needed, and taking a "snapshot" of the image. They can then specify this snapshotted image in their Request RSpecs as the operating system to be loaded when their compute resources are provisioned.

Custom images are particularly useful if configuring and installing software on a compute resource takes a long time since the experimenter has to do this just once on an instance of the operating system and then snapshot it. They are also useful if it is important that a certain version of the operating system be used for the experiment as standard images provided by the aggregates tend to keep up with newer releases of the operating system.

4.5.3 Configuration Management Tools

Industry standard configuration management tools such as Ansible and Chef are a user-friendly way of installing and configuring software on the nodes in an experiment.

Configuration management tools ensure an experiment is in a known configuration regardless of it's original state. The experimenter usually writes a *playbook* or *recipe* that describes the desired state of the node. When the playbook is run the tool uses the playbook to bring the resource to the desired known configuration. The commands in the playbooks are *idempotent* which means that the commands can be run repeatedly without concern for the initial state and no harm will result from the repeated invocations. These playbooks are usually easier to write than shell scripts or install scripts, because the experimenter is only required to describe the final intended state (e.g. Apache is installed, file.txt is present) and not how to get the node into that state (e.g. install Apache) or error handling (e.g. if Apache is not installed, then install Apache).

Configuration management tools make it easy to reproduce experiment configurations and therefore make it easy to do multiple runs with the same setup or with systematic variations such as changing parameters and scaling topologies.

4.6 Archiving

4.6.1 The GENI iRODS Service

GENI provides experimenters a long-term archival service for experiment related data such as measurements. This is the main GENI-provided storage that outlives resource reservations, slices and projects.

The GENI archival service is based on iRODS [32], an open source data management system. iRODS enables data discovery using a meta-data catalog. IRODS meta-data may be attached to files, users, groups, collections (iRODS equivalent of sub-directories), and resources (e.g., a hard drive).

The GENI iRODS service is hosted by RENCI, a research institute in North Carolina. GENI experimenters get iRODS accounts through the GENI Portal. GENI tools such as the GENI Desktop and the GENI Wireless experimentation tools can be configured to use this iRODS account to archive the measurements they collect.

5 Experiment Repeatability and Reproducibility

GENI makes it relatively easy for experimenters to recreate their setup and rerun their experiments. This is important because it encourages experimenters to collect statistics on the *repeatability* [31] of their experiments by recreating and rerunning their experiments multiple times. As a side-effect, they are less likely to hold on to resources between runs of their experiments, an important consideration in a shared testbed.

Reproducibility, an important principle of the scientific method, is the ability to run experiments created by others and verify their results [30]. GENI supports reprodicibility by: (1) providing tools and mechanisms that make it easy to recreate experiment setups, (2) defining a workflow that produces and consumes formally-defined artifacts such as experiment scripts and resource specifications, and (3) making it easy to share these artifacts for others to reproduce experiments.

5.1 Making Experiments Repeatable and Reproducible

5.1.1 Reducing Variability Across Runs of an Experiment

Picking Resources A measure of experiment repeatability is the variability in measurements across runs. Since GENI is a shared testbed this variability cannot be eliminated. However, experimenters can minimize this variability by picking non-shared resources such as bare machines and by picking the same set of aggregates for different runs of multi-aggregate experiments to minimize latency related variability.

They can also minimize variability by being specific in the Request RSpec about the characteristics of the resources being requested. For example, experimenters can specify the number of cores and memory assigned to compute resources, the locations of these resources down to the physical computer at the aggregate providing these resources and versions of operating systems installed.

Scripting Experiments To ensure resources are programmed and configured identically for every run, experimenters can use one of the techniques for software installation and resource configuration described in Sect. 4.5. In addition, experiment scripting and orchestration using tools such as OEDL and LabWiki (Sect. 4.3) can be used to reduce variability across runs of an experiment.

5.1.2 Sharing Experiment Artifacts for Reprodicibility

GENI supports experiment reprodicibility by making it easy to share artifacts such as RSpecs, custom images, experiment scripts and measurements. RSpecs and install scripts are plain files easily shared on web pages or websites such as GitHub designed for sharing programs and scripts. In addition, experimenters can choose to upload and make their RSpecs public on the GENI Portal. Experimenters can reserve resources from the Portal using RSpecs they or others have uploaded. They can also choose to make their custom images public for others to use. Likewise, the LabWiki tool allows scripts to be shared among experimenters.

Experiment related data including measurements can be archived on the GENI iRODS service (Sect. 4.6.1); experimenters can make these archives public or share them with specific people.

6 Scaling Up Experiments

GENI supports experimentation at scale by providing resources at about 50 geographical locations (as of 2015) connected with Layer 2 VLANs.

In addition, GENI makes it easy to repeatedly bring up similar topologies of different sizes. This supports best practices from software engineering and system administration. GENI experimenters can *start small* with a modest topology consisting of a trivial number of nodes which are representative of the larger topology. Then experimenters can *change one thing at a time* to bring up a sequence of larger topologies with more geographical diversity. Edwards et al. [9] provides advice for novice experimenters when dealing with these issues as well as illustrating this approach with a use case.

GENI tooling supports scaling experiments in a variety of ways. First, the use of the software installation and configuration techniques described in Sect. 4.5) makes it easy to set up and run large experiments without having to manually configure each resource.

Second, carefully crafted install scripts or configuration management playbooks, often make it possible to completely specify the configuration of a given *node type* (i.e. to use the same script to configure all nodes with the same purpose, OS image, software, and configuration). These node types can then be mixed and matched in different combinations to create topologies of different configurations and different sizes. GENI supports this with the following tooling:

- The *scaleup* tool distributed with geni-lib (Sect. 4.1.3) lets experimenters describe node types and one of several standard topologies (grid, ring, full mesh) or a custom topology using a file in INI format. The output of scaleup is a Request RSpec that can be used with any of the resource reservation tools (Sect. 4.2).
- In addition, the GENI Portal, jFed, and Flack all support a "copy and paste" feature in their graphical user interfaces so a given node type can be replicated to easily create large experiments that have a large number of a few node types.

Third, once an experiment has been tested in a single Aggregate it can be easily modified to run as a multi-Aggregate experiment. Tools such as Jacks and jFed allow a single Aggregate Request RSpec to be imported and then for different resources to be assigned to different Aggregates. This new RSpec can then be used to reserve resources and run a multi-Aggregate version of the original experiment.

7 Collaboration

GENI supports *collaborative* experimentation by allowing researchers from different institutions to operate on the same experiment and providing them the ability to add collaborators over the life of a project. This is important for large project teams such as the NSF Future Internet Architecture projects [20] and for long-running experiments.

7.1 Mechanisms for Collaboration

Research in GENI is organized into *projects*. A project contains both people and their experiments. A project may have many experimenters as its members and an experimenter may be a member of many projects. Every project has a *Project Lead* who can add or remove members. The project lead can designate one or more Admins who manage project membership as well.

Project members create slices in the context of a project; there can be many slices in a project. The person who created a slice and the Project Lead can choose to add other project members to the slice. Slice members can add and remove resources in the slice and run experiments using resources in the slice. Accounts for slice members are automatically set up on compute resources when the resources are instantiated.

This organization of research enables collaborative experimentation. A researcher can create a project and add collaborators as project members. When a new collaborator joins the team, she can be added to the project and to any slices to which she would need access.

Figure 16 shows a professor who is a project lead and has created separate projects for research and classroom use. For the class project, the professor has given his teaching assistant Admin privileges and has given the project an expiration which means the students will not be able to use the project after that date.

Figure 17 shows two slices created in the same project by the same person. The Project Lead is added to each slice by default. One of the slices contains an additional member. The two slices contain different resources and all members will have accounts to login to the resources when the resources are reserved.

References

1. Anand, A., Dogar, F., Han, D., Li, B., Lim, H., Machado, M., Wu, W., Akella, A., Andersen, D.G., Byers, J.W., Seshan, S., Steenkiste, P.: XIA: an architecture for an evolvable and trustworthy internet. In: Proceedings of the 10th ACM Workshop on Hot Topics in Networks, HotNets-X, pp. 2:1–2:6. ACM, New York (2011)
2. Ansible Inc. Ansible. http://www.ansible.com (2016). Accessed Jan 2016

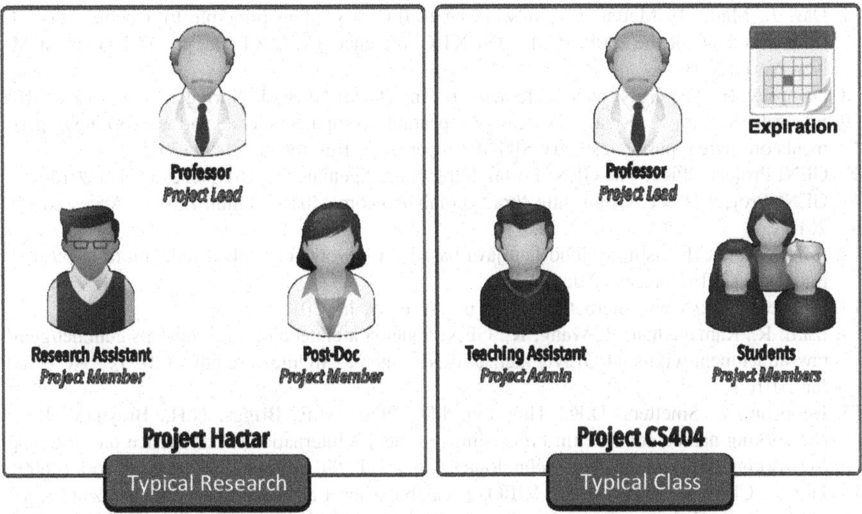

Fig. 16 A professor with separate projects for research and classroom use

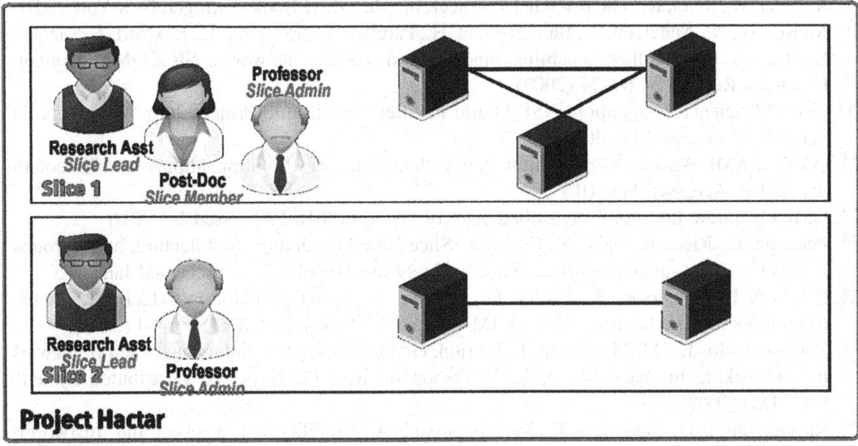

Fig. 17 Different slices can have different resources. The Project Lead is added to the slice by default as a Slice Admin. The slice creator (a.k.a. Slice Lead) can add additional people to the slice as desired. When resources are reserved, accounts will be created for all current slice members

3. Baldin, I., Castillo, C., Chase, J., Orlikowski, V., Xin, Y., Heermann, C., Mandal, A., Ruth, P., Mills, J.: ExoGENI: a multi-domain infrastructure-as-a-service testbed. In: The GENI Book. Springer, New York (2016)
4. Barnstormer Softworks. Welcome to geni-lib documentation! http://geni-lib.readthedocs.org/en/latest/ (2016). Accessed Jan 2016
5. Brinn, M.: GENI architecture foundation. In: The GENI Book. Springer, New York (2016)
6. Chef Software Inc. Chef. https://www.chef.io (2016). Accessed Jan 2016

7. Day, J., Matta, I., Mattar, K.: Networking is IPC: a guiding principle to a better internet. In: Proceedings of the 2008 ACM CoNEXT Conference, CoNEXT '08, pp. 67:1–67:6. ACM, New York (2008)
8. Dempsey, H.: The GENI mesoscale network. In: The GENI Book. Springer, New York (2016)
9. Edwards, S., Liu, X., Riga, N.: Creating repeatable computer science and networking experiments on shared, public testbeds. SIGOPS Oper. Syst. Rev. **49**(1), 90–99 (2015)
10. GENI Project Office. The GENI Portal. https://portal.geni.net (2016). Accessed Jan 2016
11. GENI Project Office. Omni. http://trac.gpolab.bbn.com/gcf/wiki/Omni (2016). Accessed Jan 2016
12. iMinds Research Institute. jFed is a java-based framework for testbed federation. http://jfed.iminds.be (2016). Accessed Jan 2016
13. Internet2. http://www.internet2.edu (2016). Accessed Jan 2016
14. Izard, R., Ramanathan, P., Wang, K.: GENI Cinema architecture. http://groups.geni.net/geni/raw-attachment/wiki/sol4/GENICinema/GENI-Cinema-Architecture.pdf (2016). Accessed Jan 2016
15. Jacobson, V., Smetters, D.K., Thornton, J.D., Plass, M.F., Briggs, N.H., Braynard, R.L.: Networking named content. In: Proceedings of the 5th International Conference on Emerging Networking Experiments and Technologies, CoNEXT '09, pp. 1–12. ACM, New York (2009)
16. Jain, S., Chen, Y., Zhang, Z.-L.: VIRO: a scalable, robust and namespace independent virtual Id routing for future networks. In: 2011 Proceedings IEEE INFOCOM, pp. 2381–2389 (2011)
17. Jourjon, G., Rakotoarivelo, T., Dwertmann, C., Ott, M.: Labwiki: an executable paper platform for experiment-based research. Proc. Comput. Sci. **4**, 697–706 (2011)
18. McGeer, R., Ricci, R.: The instaGENI project. In: The GENI Book. Springer, New York (2016)
19. McKeown, N., Anderson, T., Balakrishnan, H., Parulkar, G., Peterson, L., Rexford, J., Shenker, S., Turner, J.: Openflow: enabling innovation in campus networks. SIGCOMM Comput. Commun. Rev. **38**(2), 69–74 (2008)
20. National Science Foundation. NSF Future Internet Architecture Project. http://www.nets-fia.net (2016). Accessed Jan 2016
21. OASIS SAML Working Group. Shibboleth Federated Identity Solution. http://www.shibboleth.net (2016). Accessed Jan 2016
22. OMF Overview. http://omf.mytestbed.net/projects/omf (2016). Accessed Jan 2016
23. Peterson, L., Ricci, R., Falk, A., Chase, J.: Slice-Based Federation Architecture. http://groups.geni.net/geni/raw-attachment/wiki/SliceFedArch/SFA2.0.pdf (2016). Accessed Jan 2016
24. Peterson, L., Anderson, T., Culler, D., Roscoe, T.: A blueprint for introducing disruptive technology into the Internet. SIGCOMM Comput. Commun. Rev. **33**(1), 59–64 (2003)
25. Rakotoarivelo, T., Ott, M., Seskar, I., Jourjon, G.: OMF: a control and management framework for networking testbeds. In: SOSP Workshop on Real Overlays and Distributed Systems (ROADS) (2009)
26. Raychaudhuri, D., Nagaraja, K., Venkataramani, A.: MobilityFirst: a robust and trustworthy mobility-centric architecture for the future internet. SIGMOBILE Mob. Comput. Commun. Rev. **16**(3), 2–13 (2012)
27. Resource Specification Documents. http://groups.geni.net/geni/wiki/GENIExperimenter/RSpecs (2016). Accessed Jan 2016
28. Rouskas, G., Baldine, I., Calvert, K., Dutta, R., Griffioen, J., Nagurney, A., Wolf, T.: Choicenet: network innovation through choice. In: 2013 17th International Conference on Optical Network Design and Modeling (ONDM), pp. 1–6 (2013)
29. Seskar, I., Raychaudhuri, D., Gosain, A.: 4G cellular systems in GENI. In: The GENI Book. Springer, New York (2016)
30. Stodden, V.C.: The scientific method in practice: Reproducibility in the computational sciences. Technical Report 4773-10, MIT Sloan School of Management (2010)
31. Taylor, B.N., Kuyatt, C.E.: Guidelines for Evaluating and Expressing the Uncertainty of NIST Measurement Results, Chapter D.1.1.2 Repeatability (of results of measurements). Number Technical Note 1297. National Institute of Standards and Technology (1994)
32. The iRODS Consortium. iRODS. http://irods.org (2016). Accessed Jan 2016

33. The OMF Experiment Description Language (OEDL). https://mytestbed.net/projects/omf6/wiki/OEDLOMF6, Accessed Jan 2016
34. University of Kentucky. The GENI Desktop. http://genidesktop.netlab.uky.edu (2016). Accessed Jan 2016
35. Weigle, M.C., Adurthi, P., Hernández-Campos, F., Jeffay, K., Smith, F.D.: Tmix: a tool for generating realistic TCP application workloads in NS-2. ACM SIGCOMM Comput. Commun. Rev. **36**(3), 67–76 (2006)
36. White, B., Lepreau, J., Stoller, L., Ricci, R., Guruprasad, S., Newbold, M., Hibler, M., Barb, C., Joglekar, A.: An integrated experimental environment for distributed systems and networks. SIGOPS Oper. Syst. Rev. **36**(SI), 255–270 (2002)

The GENI Desktop

James Griffioen, Zongming Fei, Hussamuddin Nasir, Charles Carpenter, Jeremy Reed, Xiongqi Wu, and Sergio Rivera P.

Abstract The GENI Desktop supports users through the entire lifecycle of an experiment, including creating and setting up an experiment, running and interacting with the experiment, monitoring the experiment and collecting performance data, archiving the results and tearing down the experiment. It provides a single simple web-based graphical interface to access these functions. In addition, it also provides a command line interface for expert users to write scripts to control the whole process of their experiments. This chapter describes the design goals and features of the GENI Desktop. It also demonstrates usage examples showing how the GENI Desktop can help users with their experiments.

1 Running Experiments in GENI

The primary goal of the *Global Environment for Network Innovations (GENI)* [14] is to provide an infrastructure that enables research and development of new network architectures, protocols, and services at scale. Over the past few years the GENI network has developed into a large-scale network testbed infrastructure offering a wide variety of network resources that geographically span the United States and also connect to similar testbeds in other countries. While the size and reach of GENI is impressive, one can argue that the real contribution of GENI is the ability for users (experimenters) to program and control the network, or more specifically their own experimental *slice* [6] of the network, from the ground up. This capability gives users unprecedented control over the network, allowing them to redefine almost every aspect of the network, its protocols, and its services. However, redefining every aspect of a network is a massive undertaking. Even if existing protocols and services are leveraged and used as a starting point, experimenting with and testing the user's network involves running (i.e., operating) a potentially large-scale network infrastructure—something that, historically, has been done by network providers, not individual users/researchers. In other words, enabling experimenters to allocate resources and connect them together to form an experimental network

J. Griffioen (✉) • Z. Fei • H. Nasir • C. Carpenter • J. Reed • X. Wu • S. Rivera P.
Laboratory for Advanced Networking, University of Kentucky, Lexington, KY 40506, USA
e-mail: griff@netlab.uky.edu; fei@netlab.uky.edu; nasir@netlab.uky.edu; xwu@netlab.uky.edu; jeremy@netlab.uky.edu; charles@netlab.uky.edu; sergio@netlab.uky.edu

© Springer International Publishing Switzerland 2016

R. McGeer et al. (eds.), *The GENI Book*, DOI 10.1007/978-3-319-33769-2_16

(i.e., slice) is only the first step of many needed to carry out a network experiment on GENI. Moreover, the APIs used to allocate resources and create slices are anything but user-friendly—as anyone who has had to work with the GENI AM API and RSPECs can attest.

1.1 The Need for Higher-Level Tools and Services

To simplify the task of running experiments in GENI, higher-level tools and services are needed by users. In fact, the GENI architecture assumes that the GENI network consists of not only allocatable resources, but also higher-level tools and services that build on and enhance the underlying GENI API calls supported by the GENI aggregate managers (AMs). As a result, a variety of tools have been developed to help experimenters use GENI. However, the lifecycle of an experiment typically involves several steps or phases, with each step requiring tools and services designed specifically for that step in the experiment. Ideally, there would be a single tool or service that experimenters could use to perform an experiment from beginning to end.

A key goal of the *GENI Desktop (GDT)* is to provide a "one stop shop" where experimenters can go to carry out their experiment from beginning to end. In addition, the GENI Desktop fills in gaps in the experiment lifecycle where no other tools exist, offering its own tools and allowing users to create their own tool and include it in the GENI Desktop as an add-on module.

Take, for example, the experiment lifecycle illustrated in Fig. 1. An experiment's lifecycle typically involves several phases that may repeat as needed. For example, after running some initial tests and analyzing the results, the user may decide to modify the network topology and run the tests again.

During the early days of GENI, much of the focus was on the control frameworks and the ability to obtain resources. Not surprisingly, several tools quickly emerged to make these steps (i.e., steps 1–3 in Fig. 1) easier for users. Example tools included graphical tools such as Jacks [9] (previously Flack [12]) and command line tools such as OMNI [10] that helped experimenters discover the set of GENI resources that are available and reserve them. Tools such as the GENI Portal [5] provided additional services such as helping users obtain the appropriate credentials needed to access and manage slices.

In contrast, the set of tools designed to help users execute their experiment, monitor and measure its performance, and then analyze and/or save the results (i.e., all other steps in Fig. 1) were at best primitive and at worst non-existent. The GENI Desktop addresses these missing pieces of the experiment lifecycle by providing the services, software, and infrastructure needed to run experiments, instrument and measure experiments, view and analyze performance, and archive results for future analysis, comparison, and repeatability.

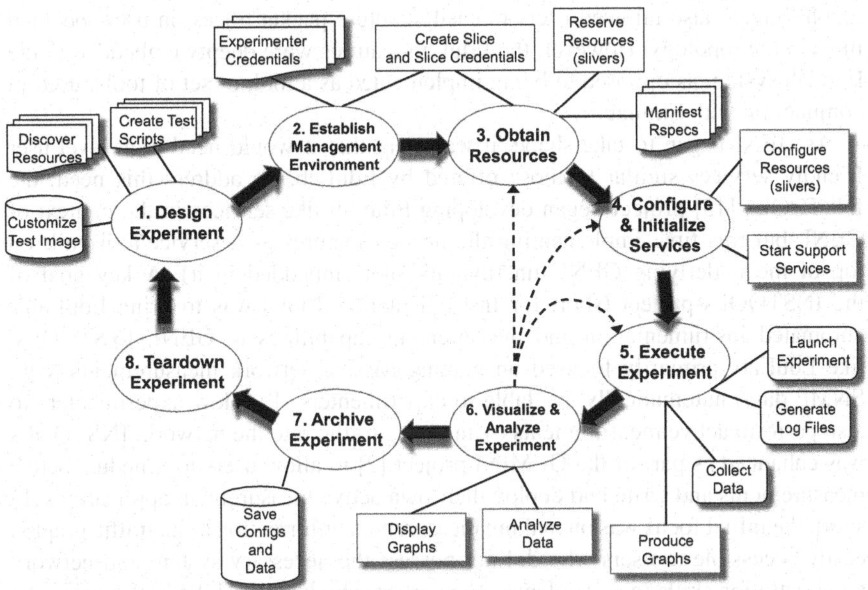

Fig. 1 Example lifecycle of an experiment (a variation of the lifecycles found in [3, 4])

1.2 *History of the GENI Desktop*

The origins of the GENI Desktop can be traced back to early user experiences running experiments on the Emulab [11] testbed. The initial Emulab system (much like the initial GENI network) enabled users to obtain resources over which they could have complete control (e.g., root access). The target user community was network and operating system researchers—users who typically were highly skilled in system administration tools and techniques not to mention network administration. In other words, the initial user community had the skills needed to "get an experiment running" and then "see what was going on". Things like monitoring the CPU or memory usage/loads, configuring routing tables, reading packet counters, collecting packet traces, adding advanced kernel modules, etc.

However, not all users possessed these skills. As Emulab began to be used as an instructional tool in operating system and networking classes, a new set of users emerged—namely students—who did not possess the system and network administration skills needed to effectively execute and monitor experiments. To make it easier for students to measure and view an Emulab experiment, Emulab was modified and enhanced, resulting in EduLab [16]—the first instantiation of the GENI Desktop. While EduLab simplified the management environment (step 2 in Fig. 1), its key contribution was the ability to automatically install, configure, and run an *instrumentation and measurement system* on behalf of the user, measuring, recording, and displaying (via a web interface) network traffic graphs, load average graphs, etc. It also allowed users to easily capture packet traces on user-specified links in the topology. However, the EduLab features were deeply embedded in the Emulab system as opposed to being implemented as a tool, or set of tools, used in conjunction with Emulab.

As GENI began to take shape, it was clear that it would need high-level user friendly services similar to those offered by EduLab. To address this need, the INSTOOLs [15] project began developing EduLab-like services in the context of GENI, but this time implementing the needed features as a service/tool built on top of the underlying GENI functionality (not embedded in it). A key goal of the INSTOOLs project (short for Instrumentation Tools) was to bring EduLab's automated instrumentation and measurement capabilities to GENI. INSTOOLs, like EduLab, primarily focused on making passive network measurements (e.g., SNMP data) automatically available to experimenters. To allow experimenters to also perform active measurements by injecting traffic into the network, INSTOOLs was enhanced as part of the GEMINI project [2] to allow users to schedule active measurements and write and deploy their own active measurement applications. In short, the initial focus was on making measurement information (e.g., traffic graphs) easily accessible to users who did not possess the necessary system and network administration skills (e.g., students). Even users who had the skills began using the tools because of their ease-of-use and simplicity.

The next step was to extend the user-friendly instrumentation and measurement interface to all other tools in all phases of an experiment's lifecycle. To meet this need, the concept of a GENI Desktop that provided a one-stop user interface for all tools was born. The goal was to offer users a single, easy-to-use, interface that ties together existing tools (e.g., Jacks, OMNI, the GENI Portal, and the INSTOOLs/GEMINI measurement tools) and adds new functionality to simplify the task of running and interacting with experiments—e.g., functionality such as loading experiment software, launch tests, logging into and directly interacting with specific resources or downloading and archiving results.

2 GENI Desktop Design Goals

The overarching goal of the GENI Desktop is to *provide an environment in which experimenters can control and interact with every phase of an experiment*, from the initial design of an experiment to its termination and teardown. In other words, *the GENI Desktop should be a "one stop shop" for experimenters*, assisting in the design of their experiment, establishing the management environment, obtaining resources, configuring and initializing services, executing the experiment, analyzing and visualizing the experiment, archiving the experiment, and eventually tearing down the experiment. Several high-level design goals follow from this objective:

Leverage and incorporate existing tools and services: Many excellent tools already exist and are highly specialized. Do not reinvent the wheel, but rather combine them into a single tool/work environment.

Develop a unified, consistent, and simple interface: In addition to being a one stop shop for all their needs, the interface to the GENI Desktop should be easy to understand and use, making it easy for experimenters—both novice users and expert users—to launch, run, and measure their experiment.

Allow access from anywhere: Experimenter's local computational environments vary widely including differences in hardware, operating systems, installed software, etc. The GENI Desktop should be usable from the largest number of experimenter environments possible. This implies placing as few requirements on the user's environment as possible.

Support extensibility: Not only should the GENI Desktop be able to support existing tools and services, but it should be easily extensible to be able to incorporate new tools and services that arise in the future.

Ensure security and accountability: In many cases, making it easy for experimenters to use GENI means doing tasks on their behalf (e.g., defining RSPECs and setting up the topology, or loading software onto all nodes in the experiment). To ensure that security is maintained and usage is properly accounted for, it is important that the GENI Desktop support and use GENI's *speaks-for* authentication and authorization mechanism when interacting with users, allowing the GENI Desktop to operate on behalf of its users.

The next section describes how these design goals were achieved in the GENI Desktop.

3 GENI Desktop

The GENI Desktop is designed to make it easy for experimenters to carry out every phase of their GENI experiment. The typical user will access the GENI Desktop using a web browser which means that it is accessible from a wide range of compute devices (e.g., desktops, laptops, and tablets) running any of the popular operating systems. Although web browser compatibility issues are a challenge for any web site, most common browsers can be used, possibly after minor configuration changes or add-on enhancements. In addition to the web interface, the GENI Desktop supports a command line interface (CLI) that users can use to script their experiments. Using the CLI, users can write programs (scripts) that set up experiments, install software, launch tests, collect data, and generate graphs that can be later viewed.

Regardless of the interface used to access the GENI Desktop, the task of running an experiment is greatly simplified because the GENI Desktop performs many (often complex) operations on the user's behalf. It does this using the GENI "speaks-for" capabilities to authenticate to GENI, indicating that it is a tool that has been given permission to act on the user's behalf. This enables GENI to associate resource usage with the user, but yet allows the GENI Desktop to be the entity that actually uses the resource.

Using the web interface, users are presented with the ability to start a new experiment (i.e., create a new slice) or work with an existing slice. The GENI Desktop leverages existing tools and services such as OMNI, Jacks, and the GENI Portal to design the experimental topology, create slices, and allocate/free resources. It leverages services previously developed as part of the INSTOOLs and GEMINI projects to support instrumentation and measurement, and it also provides new services not offered by other tools (e.g., file upload/download services, login services, verification services, archival services, etc.)

Perhaps the most obvious contribution of the GENI Desktop is the GUI it provides for working with experiments. First, the GENI Desktop gets its name from the fact that it provides a "desktop" windowing system look and feel within the context of a web browser, allowing the user to open up multiple windows used for displaying traffic graphs, visualizing the topology, logging into nodes, and generally managing all the components of the experiment. Second, the GENI Desktop provides a unified abstraction for working with the resources used in an experiment. The abstraction is based on the well-known file browser model in which users select files and then apply operations to those files (e.g., copying the files or deleting the files). In the GDT, users select resources and apply operations (tools and services) to those resources. For example, a user might select a set of nodes in the network and ask to see a graph of the TCP traffic going through those nodes. Or the user might select a set of nodes and ask to be logged into those nodes (i.e., ssh). The unified interface enables new tools and services to be directly incorporated into the interface simply by adding the tool or service as yet another operation that can be applied to a resource. This extensibility allows the GENI Desktop to adapt and evolve as new tools and services become available.

3.1 An Example Workflow

Before describing the various features and capabilities offered by the GENI Desktop, it is useful to consider an example workflow that an experimenter might use to carry out an experiment using the GENI Desktop.

Initially a user will use a browser to visit the GENI Desktop web page. The GDT welcome page asks the user to login—using, for example, their InCommon login ID and password. Having successfully logged in, the user is presented with the option to create a new slice or work with a previously created slice. Assuming the user is starting a new experiment, the user will create a new slice in which to run the experiment. Creating and setting up a slice can be accomplished by either defining the slice topology and resources graphically—dragging and dropping resources onto a canvas and drawing the network links between resources—or by selecting a predefined topology from a list (of RSPECs). The GENI Desktop will then automatically create and setup the slice on the user's behalf.

Having created a slice, the user can use the GENI Desktop to view and interact with the resources in the slice. A user often begins by selecting and loading experiment-specific software onto certain nodes in the slice. The user then runs the experiment by logging into nodes and starting up the software used to drive the experiment.

Once the experiment is running, the user can use the GENI Desktop to view graphs of the network traffic flowing through nodes or across links. The user may also change or alter the behavior of the running experiment by dynamically sending (shell) commands to certain nodes in the experiment.

When the experiment finishes, users may save (i.e., archive) the measurement data collected during the experiment (e.g., traffic graphs) for future analysis. Having completed the experiment, a user will then tear down the slice and release the resources so they can be used by other experimenters.

The following sections provide a more detailed look at the set of features supported by the GENI Desktop and then give some example usage scenarios that demonstrate how one might use these features in the various steps of an experiment's workflow.

3.2 Designing and Creating an Experimental Network (Slice)

The first step in the lifecycle of a GENI experiment is designing the network (slice) that will be used in the experiment. Designing the slice requires an understanding of the resources that can be used to create experimental networks. The GENI Desktop leverages the Jacks tool (embedded in the GENI Desktop web pages) to help users discover the types and location of resources that experimenters can use in their slices. Using the Jacks RSPEC editor, users can select the types of resources and the location of the resources (i.e., the aggregate) that they want to include in their experimental topology. After designing the network with Jacks, the resulting RSPEC is then returned to the GENI Desktop where it can be used to instantiate a slice.

The next step involves establishing a management environment for creating the experimental network. The GENI Desktop leverages existing GENI member and slice authority services to manage user identities and create slices—abstract "network containers" to which a user can add resources (slivers) needed by his/her experiment. As mentioned earlier, the GENI Desktop uses GENI speaks-for credentials to communicate with slice authority services to create and register the slice on behalf of the user. The resulting slice credentials can then be used by the GENI Desktop to assign resources to a slice.

To create an experiment network, an experimenter must add resources to a slice. The GENI Desktop offers three methods for adding resources to slices. The first method involves the user providing the GENI Desktop with an RSPEC file describing the resources and network topology to be used in the experiment. Users can upload the RSPEC file from their local machine, specify a URL where the

RSPEC file can be found, or paste the RSPEC into a web page on the GENI Desktop. The GENI Desktop will then invoke the set of OMNI tools to communicate with GENI aggregates (using the GENI speaks-for credential) and allocate resources for the user's slice as specified in the RSPEC provided by the user.

The second method makes it possible to reuse the setup of a previous experiment. The GENI Desktop allows users to save an experiment's network setup (i.e., RSPEC) either as a private network specification—if it is intended to be used exclusively by the user—or as a public specification that is made available to other users. In this model, a user simply needs to select a past RSPEC and use it to allocate resources for a slice.

The third method involves creating the experimental network (i.e., RSPEC) from scratch. To assist with the creation of new RSPECs, the GENI Desktop integrates the Jacks tool. Using the Jacks tool, users can define the set of resources they want to include in their slice. The resulting RSPEC is then allocated (by the GENI Desktop) on the user's behalf.

After creating a slice and allocating resource to the slice, experimenters need to verify that the slice was instantiated correctly and is operating correctly. To assist with this task in the lifecycle, the GENI Desktop supports a slice verification and testing service. Based on the manifest describing a slice, the verification service analyzes the topology and performs tests to determine if the interfaces of all nodes in the experiment are operating correctly. A variety of tests can be performed on the slice. The simplest test checks to see if each interface is up and whether it is reachable using a ping test. The test results are collected by the GENI Desktop verification service and are then presented on a web page showing the status of all the interfaces of all the nodes in an experiment. More advanced tests evaluate the QoS of the network to determine if the links are operating with the speed and performance specified in the RSPEC describing the network. The most advanced tests include user-defined tests in which users write their own verification scripts to test for things of importance to their experiment.

3.3 *Creating Superslices*

In addition to helping users design their experiments and create slices with resources, the GENI Desktop also allows users to combine network experiments (slices) together to create *superslices*—slices connected together to form large multi-slice network topologies.

Creating large network topologies can be challenging because the failure of a single component during the "obtain resources" phase typically requires freeing all resources (even ones that came up correctly) and starting the process all over. Moreover, trying again is no guarantee that the subsequent attempt(s) will succeed. This problem could be fixed if the underlying control framework supported dynamic addition (and deletion) of resources from a slice, but a variety of subtle implementation issues have prevented this type of feature from being widely adopted and supported.

In addition to enabling larger topologies, superslices have several other salient features and benefits. Because the superslice consists of independent slices, it is possible for certain parts of the topology (a subset of the slices) to fail without bringing down the entire topology. Superslices also allow slices from different users/experimenters to be combined into a single shared experiment. This enables independently operated slices to join or leave a shared superslice at any time, much like an internet consists of independently operated ASes that are able to join or leave at any time.

The GENI Desktop supports the concept of superslices by allowing experimenters to take individual slices and combine them into a single large superslice. In particular, the GENI Desktop provides a GUI for users to display (in a GENI Desktop web page) the topology of multiple slices and pick any pair of nodes from different slices to establish a new network link connecting the two slices together. At present the newly created link connecting slices together is a GRE tunnel which can be done without the knowledge of the underlying control frameworks. In other words, the superslice abstraction is something supported solely by the GENI Desktop—the underlying control frameworks know nothing about the fact that multiple slices have been combined together.

3.4 Running and Interacting with an Experiment

After creating the experimental network to be tested and evaluated, the GENI Desktop provides several features to help users run and interact with their experiment.

In order to interact with and control resources in a slice, users need a convenient way to identify resources in the slice. As anyone who has had to work with RSPECs will tell you, remembering and working with the identifiers in an RSPEC is anything but convenient or easy. To make it easy for users to view and control their experiment, the GENI Desktop provides three distinct "views" of the slice topology that experimenters use to think about, or picture, their slice. All views allow users to visually point at and select the resources they want to operate on (e.g., see traffic graphs or login to).

The *geographic view* shows the geographic locations of nodes on a map and visually indicates the locality or wide area features of the experimental network as well as the physical location of nodes. The *logical view* shows the logical connections between nodes without regard for their physical location. The logical view is useful when the information of importance is the connectivity between nodes. Lastly, the *list view* shows a list of all resources (i.e., nodes and links) in the topology. It supports searching the list by name and type of resource and is particularly useful for large networks consisting of many resources. Each view appears in a separate window, and the three views can be display simultaneously if desired. All three views are kept consistent and synchronized. Selecting a node in one view causes it to be selected in all views. Users can select the most appropriate view, as some operations may be performed more easily in one view than another.

Given the ability to "view" the network, users can easily upload experimental software, issue commands to start experiments, login and control particular nodes, see traffic graphs, capture packet traces, and perform various other operations supported by underlying tools and modules in the GENI Desktop. As described earlier, the GENI Desktop user interface is designed around a common unifying abstraction similar to the abstraction used in file browsers. Using any one of the "views" described above, users can select a set of resources and apply an operation to run or control their experiment. For example, a user may select several nodes in the logical view (say by dragging the mouse over the nodes much like one would select a set of files in a file browser) and apply an operation. Alternatively, a user may search for a node with a particular name in the list view and then apply an operation to that specific node. All interactions with tools and services are defined in this way, and give users easy control over their running experiments.

Several operations supported by the GENI Desktop are specifically designed to help users run their experiments. For example, the quick information window operation in the GENI Desktop pops up a window with information about the node currently selected in the view, including information about the node's name and IP address and port needed to login to the node. Once opened, information about other resources in the view can be found by simply mousing over the resource. For example, when the mouse moves over a link, the IP addresses of the end points of the link are displayed. Being able to quickly obtain information about the resources is particularly useful while running an experiment.

Another useful operation when running experiments is the ability to upload files to experimental nodes. Without the GENI Desktop, this can be a cumbersome process because users may have to upload the same set of files to many nodes in the experiment. The GENI Desktop implements a feature to allow users to select a file and select a set of nodes from the GUI. The user can then choose the destination directory on the experimental nodes, and the GENI Desktop will upload the selected file to the specified directory on the selected set of nodes.

Similarly, users may want to run their program on multiple experimental nodes. It can take a long time for users to login to each node, go to the correct directory and run their program. To address this common need, the GENI Desktop supports a "run" operation which allows the user to select a set of nodes and issue a command to be run on those nods. The GENI Desktop will run the program on all the machines selected and display the results without requiring the user to login to these nodes.

To work on a specific node, the user may want to login to it. This normally involves using an ssh-based tool with a correctly configured private ssh key. The GENI Desktop provides an easy-to-use interface for logging in to an experimental node. The user simply selects the node(s) and clicks on "ssh" operation. An ssh tab in the browser will be created for each of the selected machines, and the user will find themselves logged in without any additional effort.

3.5 *Monitoring an Experiment*

When an experiment is created with the GENI Desktop, it will be automatically instrumented with measurement software and tools for collecting data about the experiment. If an experiment is created by other tools, such as the GENI portal and Jacks, it will be instrumented the first time the GENI Desktop is used for viewing the experiment. The instrumentation process involves installing measurement software and tools at experimental nodes and configuring and initializing them to start collecting the data for users. In addition, an additional node, called the *global node* will be created for the experiment to store the collected data and process them for presentation to the user.

The default measurement software collects standard TCP/IP network performance data such as packet and byte counts of TCP and IP traffic on each interface. It also collects data about CPU load and memory usage. The data collection is configurable by the user to specify what information is to be collected at which node. In addition, the GENI Desktop can also set up NetFlow services for capturing data on a per-flow basis. The flows to be monitored are preconfigured so that the user can simply go to the web interface and see the most common types of flows. Similarly, users can use the GENI Desktop's web interface to configure what flows should be captured.

Like the INSTOOLs project, the basic monitoring capabilities of the GENI Desktop use passive measurements. However, the GENI Desktop also supports active measurement by allowing users to schedule tasks for measuring the latency and bandwidth of the network by using tools such as ping, iperf and pathchar.

The measurement data collected by the GENI Desktop is presented to the user through the web interface. The data is displayed and automatically refreshed every 5 s to give the impression of a "live" view of the running system. The GENI Desktop can display a variety of graphs, including IP traffic, TCP traffic, UDP traffic, ICMP traffic and total traffic of an experimental node or a link in the topology. The units displayed can be either byte counts or packet counts. CPU and memory usage graph can also be displayed for nodes, showing load averages over time. Users can configure which graphs are to be displayed and for which nodes or links. The user can also specify an alert condition by setting a minimum and/or a maximum value for any measure of interest in the graph. If the measure goes out of the range, an alert will be presented to the user for special attention. The user can also configure the way these graphs are displayed, such as the size of the graphs, the number of graphs per row, the scale of time unit, and the start time (offset) of the graphs.

Netflow graphs are displayed using a netflow operation. The user can select the nodes of interest and then specify the traffic of interest based on the protocol number such as protocol 169 and protocol 255, or based on the protocol name such as TCP, UDP, GRE, IPinIP, IMAP, DNS, Gnutella, and Kazza. The user can also configure what traffic should be collected and what graphs should be displayed. Similar to other graphs, the user can select the way these graphs are displayed.

The GENI Desktop also provides the ability to access X window software running on experimental nodes via the GENI Desktop's web interface. It allows users to leverage existing network monitoring tools, such as Wireshark and EtherApe, in order to observe the behavior of experimental nodes. These tools are helpful to collect packet traces, node statistics, and to visualize link traffic. However, they need X window support. To support such access, the GENI Desktop adds the ability to dynamically load X-window software onto the experimental nodes and then provides indirect access through the web browser and the VNC protocol. The GENI Desktop currently has two VNC templates that are preconfigured to run xterm and wireshark respectively on the nodes in the slice via VNC.

3.6 Tearing Down an Experiment and Archiving the Results

GENI slice and sliver resources for an experiment expire after a certain period of time for the benefit of resource sharing. The GENI Desktop provides an automatic renew function for the slices so that users do not need to do that manually (continually). Users can also explicitly renew the lifetime of their resources via the GENI Desktop. When an experiment is no longer needed, users can tear down the experiment by deleting resources explicitly through the GENI Desktop, instead of holding the resources longer than necessary.

Data collected during experimentation may be needed beyond the lifetime of the experiment. The archival service of the GENI Desktop leverages the iRODS storage service to store and later retrieve measurement data collected by the GENI Desktop. It generates the necessary metadata to describe the archived content so that the user can associate the archived data with the particular resource in the experiment that generated the data. Data can be retrieved from the archival sites independent of the GENI Desktop as long as the user can provide the credentials to the archival servers. Archived data can be downloaded to a user's machine for postmortem processing and use in documentation and publications.

To make it easier for users to quickly access, view, and make sense of archived measurement data, the GENI Desktop also supports an advanced archival service that not only archives the measurement data, but also archives the software and context needed to display the data. Because the data and the environment needed to view the data are both saved away, users can be assured that they will be able to access an archive and view the saved data using the same tools available at the time the data was collected.

To support this advanced archival service, the GENI Desktop implements an archival server that not only captures the measurement data stored on the global node (where measurement data is collected), but it also captures the state of the drupal system used to display the data, including all web server (Apache) and database (mysql) files. GENI Desktop users can request that an archive be made, which is then sent to the archive server. When a user visits the archive web page on the archive server, they can select from any of the archived snapshots. The archive server will dynamically launch a Xen VM, set up the apache, mysql, and Drupal state needed to view the measurement data, install the archived measurement data, create login credentials for the user, and share the credentials with the GENI Desktop so the user is automatically logged into the archive VM. The result is that the user is presented with the same look-and-feel as if they had gone to the global node at the time the snapshot was taken.

3.7 GENI Desktop Command Line Interface

The GENI Desktop's web interface greatly simplifies the task of instrumenting and monitoring a user's experiment (slice) for most users. However, expert users often find it easier to control their experiments through scripts and programs. In other words, they prefer a programmatic way to leverage the GENI Desktop functionality.

To address this need, the GENI Desktop provides a *command line interface (CLI)* that can be used to programmatically upload files, run commands, download measurement graphs, etc.—functions previously only possible via the GENI Desktop web interface. In particular, an application called the *gdcli* program running on Linux (or other Unix-based systems), Mac, and Window is available and can be used to invoke operations on the GENI Desktop. For example, the gdcli program can be used to:

- Upload files to a select set of nodes
- Run a command on a select set of nodes
- Download a traffic measurement graph (as PNG or CSV) from a select set of nodes
- Download a normal file from a select set of nodes
- Get a list of slices
- Check the status of a slice
- Get the topology of a slice
- Validate the setup of a slice
- List the nodes in a slice
- List the links in a slice

The gdcli program can be called from any scripting language (e.g., python, perl, sh (bash), .BAT files, etc.). As a result, users are able to write programs in their favorite scripting language that makes calls to the GENI Desktop to upload/download files, download measurement graphs, run commands, etc. Because the gdcli program is a python script, the only requirement to run it on a user's local machine is a python interpreter. Moreover, users that want to issue calls directly to the server can integrate the python functions found in the gdcli program into their own python scripts.

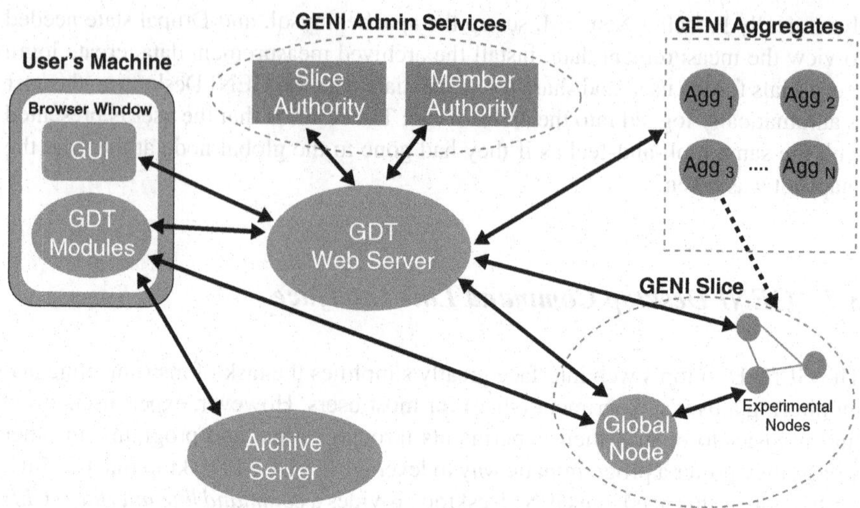

Fig. 2 GENI Desktop components

3.8 Components of the GENI Desktop System

Figure 2 illustrates the components of the GENI Desktop (shown in light blue) and their relationship to other GENI components/services (shown in orange). At the heart of the GENI Desktop is the *GDT web server*. The GUI that users see in their browser (i.e., the GENI desktop windowing system) is hosted by the GDT web server. As users work with their slices, the browser communicates with the GDT web server. The GDT web server in turn communicates with various control framework components such as member authorities, slice authorities, and aggregate managers (using the GENI speaks-for capabilities) to perform operations on the user's behalf.

To achieve scalable and efficient collection of measurement data from experimental nodes, the GENI Desktop automatically adds extra nodes to the user's slice called *global nodes (GNs)*. One GN is added to each aggregate in the slice allowing data collection to be localized and confined within an aggregate. Experimental nodes are instrumented to send their measurement information to their local GN. The GDT web server retrieves information from the appropriate GNs when asked to display measurement information (e.g., traffic graphs). The GDT web server is also capable of executing commands on experimental nodes on the user's behalf.

Extensibility in the GENI Desktop is supported through the use of *GDT modules*. GDT modules are loaded from the GDT web server into the user's browser where they execute and interact with the other windows in the GDT window system. A basic set of modules is loaded into the user's browser by default. The basic set includes modules that enable features such as viewing traffic graphs, ssh access to nodes, file upload capabilities, running commands on sets of nodes, viewing information stored on the GN, archiving measurement data, rebooting nodes, and creating a disk image. Modules may communicate with the GN, or in the case of the archive module, with an archive service that stores measurement data for later analysis.

In addition, the set of modules supported by the GENI Desktop can be easily extended and enhanced, both by the GDT operators (e.g., to extend the default set of modules) and by general users (via a "module maker" feature). Additional modules that have been developed include modules to automatically generate network traffic, support netflow data collection, modify routing paths, and insert rules in an OpenFlow controller.

4 Using the GENI Desktop

In the following, we describe and illustrate examples of how users might perform all the phases in the lifecycle of an experiment (i.e., create, run, measure/monitor, archive/analyze, and teardown an experiment) using the GENI Desktop. We first describe how to perform these tasks in the context of the GDT GUI and then in the context of the GDT CLI. For more detailed examples of how to perform an experiment using the GDT, interested readers can find additional detailed GDT tutorials on the GENI web pages (http://www.geni.net).

4.1 The GENI Desktop GUI

When the GENI Desktop page is opened, it first asks users to allow it to act on their behalf to reserve resources and have access to these resources, as shown in Fig. 3. For most university users, this will be their InCommon credentials which are their user name and password for their university account. By authorizing the GENI Desktop, the user signs the "speaks-for" certificate used by the GENI Desktop to act on their behalf.

After login, users can create a new slice by providing the slice name as shown in Fig. 4a. The slice is the container to which resources can be reserved and added for the experiment. Figure 4b shows several ways in which a user can specify the RSPEC for an experiment. For example, a user can use the Jacks tool to create a new topology from scratch (*Create New*). Alternatively, a user can choose a topology from a previous experiment (*Choose Existing*), or upload the topology from a file (*Upload File*), or get the topology from a URL (*Get Via URL*). In addition, a user can just write/paste the content of an RSPEC file into a text window (*Paste Text*).

If the user chooses to create the experiment topology from scratch, the GENI Desktop has an integrated RSPEC editing tool, Jacks, which allows users to drag and drop the nodes and links for the experiment, as shown in Fig. 5. There are a variety of types of nodes that can be requested, including raw PCs, VMs, and OVS

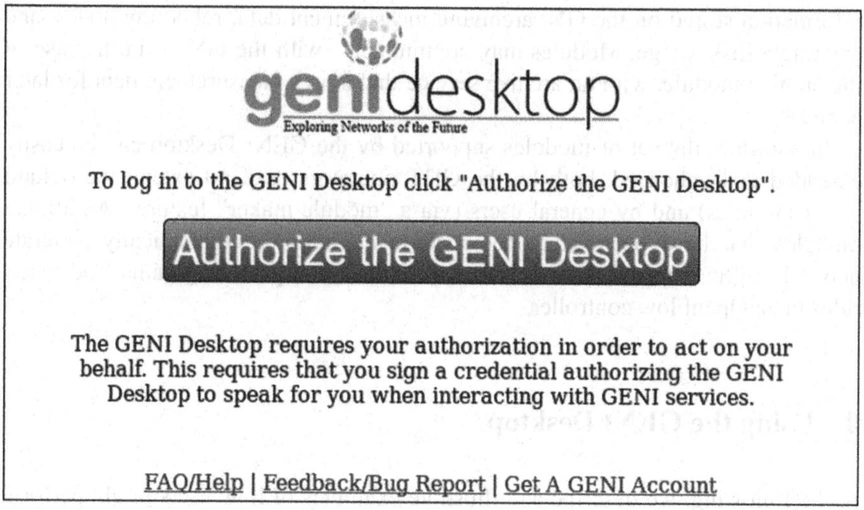

To log in to the GENI Desktop click "Authorize the GENI Desktop".

Authorize the GENI Desktop

The GENI Desktop requires your authorization in order to act on your behalf. This requires that you sign a credential authorizing the GENI Desktop to speak for you when interacting with GENI services.

FAQ/Help | Feedback/Bug Report | Get A GENI Account

Fig. 3 Authorize the GENI Desktop

a

genidesktop

Slice Name Name

Project UKGENI

Create New Slice

b

genidesktop

Your slice "mydemo" does not have any resources. You must add resources before viewing with the GENI Desktop.
Choose an RSPEC which will define the resources to be allocated to your slice.

Create New

Choose Existing

Upload File

Get Via URL

Paste Text

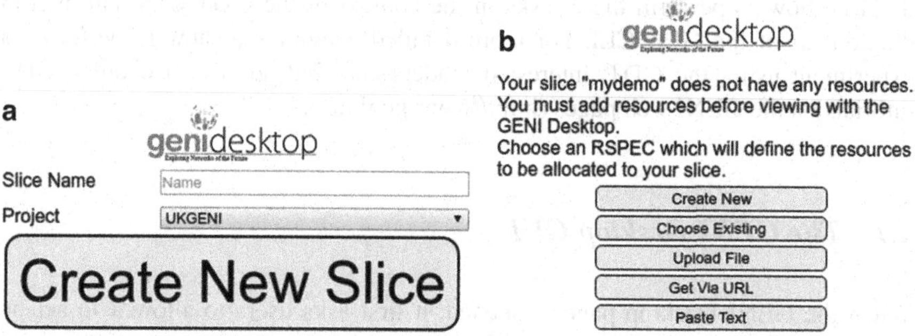

Fig. 4 Create a slice and add resources. (**a**) Create a new slice. (**b**) Add resources to a slice

nodes. In this example, the user creates a topology with two nodes connected by a link. The GENI Desktop will request resources from the aggregates specified by the user.

After reserving the resources, the GENI Desktop will then automatically instrument the slice to collect measurement data for the user. Figure 6 shows the topology of the slice on the GENI Desktop. The small information window in the upper right corner shows the IP addresses of the link highlighted in the topology window. The upper left side shows the list of modules (possible operations) that can be used in this slice. The user can select a subset of nodes from the topology and choose an operation to be performed on the selected nodes.

The first operation often performed is to upload files (programs or data files) to a selected set of nodes. In Fig. 7 we show that a selected file (curl_uploader.tgz) from the local machine will be first uploaded to the GENI Desktop, and then distributed to the selected nodes (node-0 and node-1).

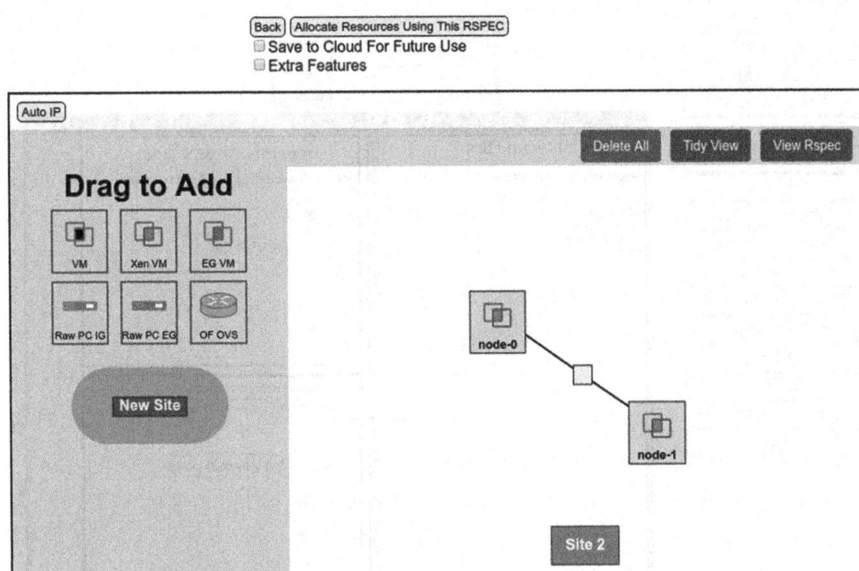

Fig. 5 Create a topology from scratch

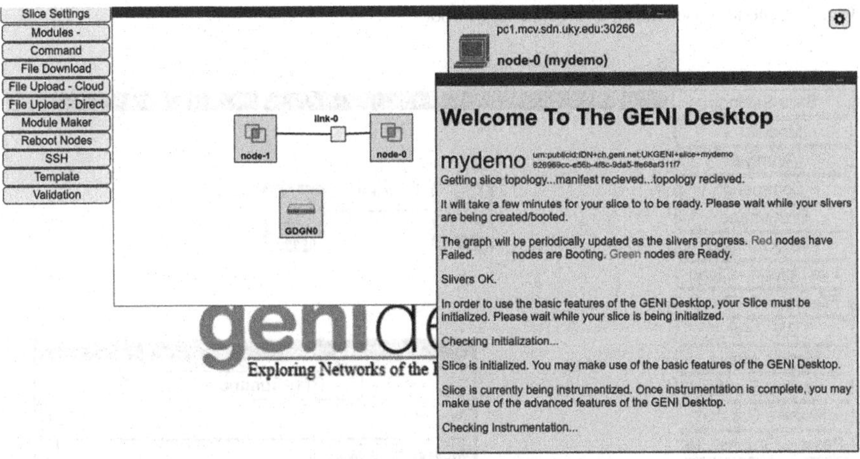

Fig. 6 View the topology and link information from the GENI Desktop

The "Command" module of the GENI Desktop allows users to run Unix commands or user programs on a selected set of nodes. Figure 8 shows that the "ls" command will be run on the nodes selected (node-0 and node-1). The results of running this command are shown in Fig. 9.

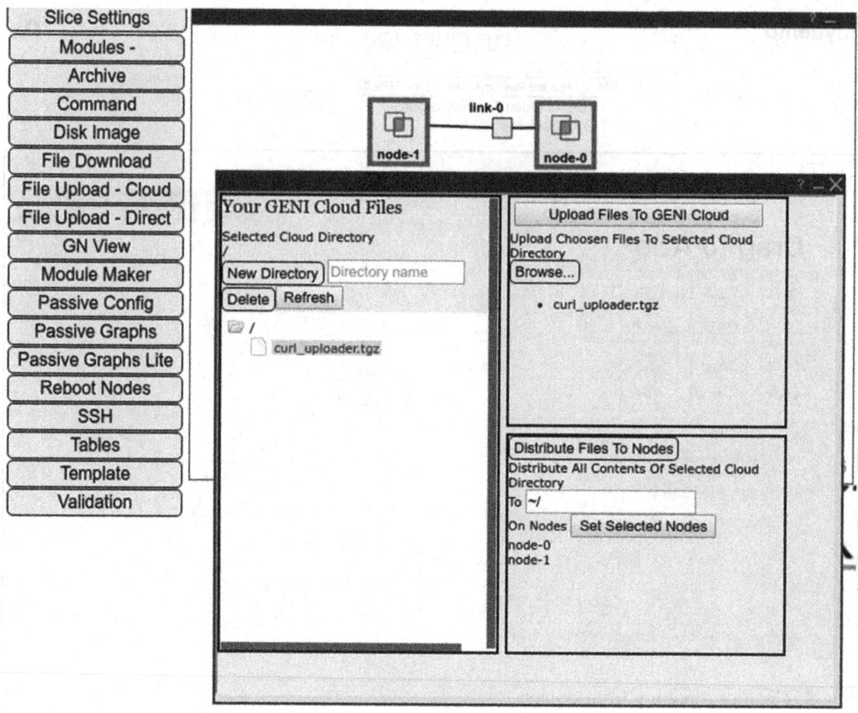

Fig. 7 Upload a user file to a selected set of nodes

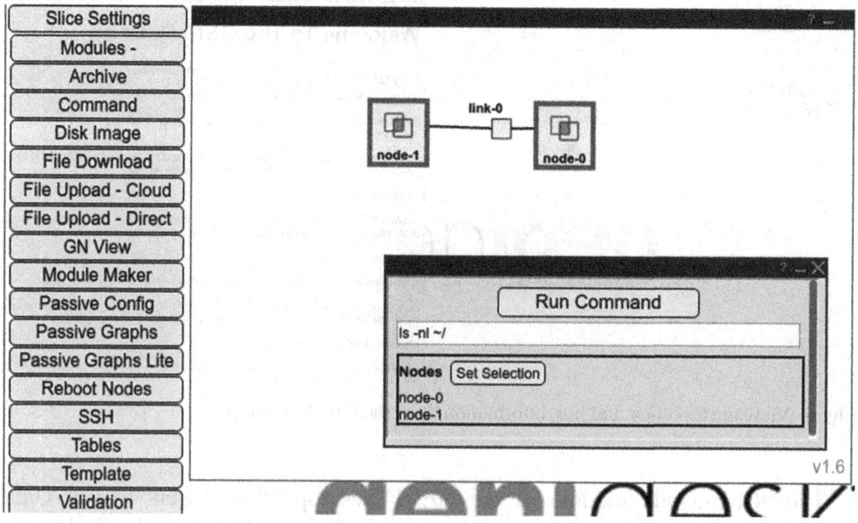

Fig. 8 Run a user program on a selected set of nodes

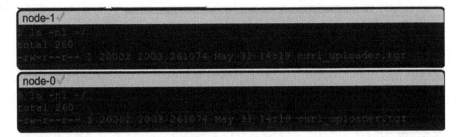

Fig. 9 Results of the Run command

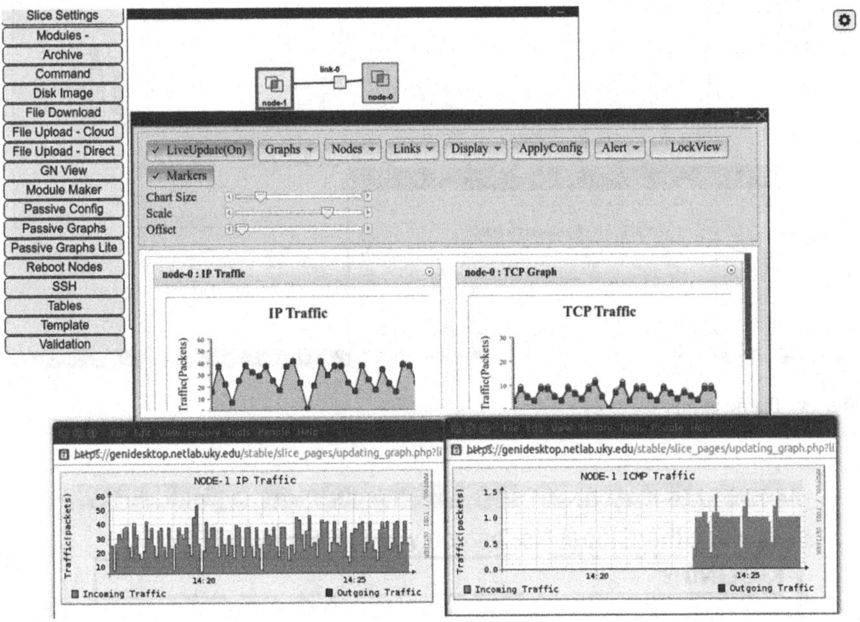

Fig. 10 Traffic at node-0

Traffic counts are automatically captured during the experiment by the GENI Desktop. Figure 10 shows the IP traffic and TCP traffic over time observed at `node-0`. The bottom two windows show the IP traffic and TCP traffic at node-1. Figure 11 shows the interface traffic observed at `link-0`. The graphs show two obvious traffic peaks that occur during the experiment. The way in which these graphs are displayed and which nodes/links should be displayed can be configured by using the drop-down menus at the top of the window. This provides great flexibility for users to observe measurements of interest to them.

The results of the experiment can be stored for long-term use in the archive servers. Figure 12 shows the menu for archiving the measurement results to an iRODS server. Users can select what data to send and which iRODS server to store

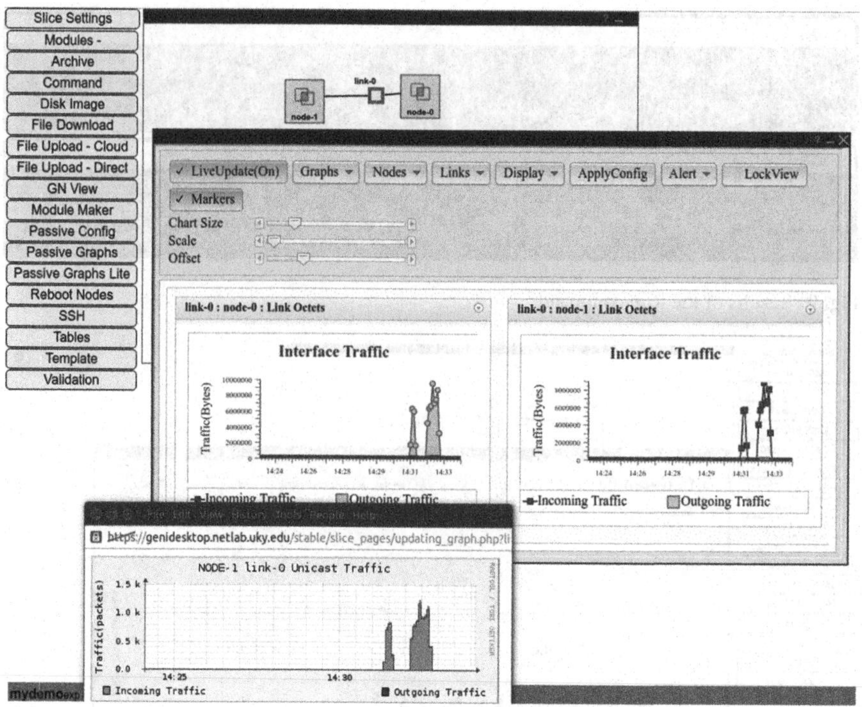

Fig. 11 Traffic at link-0

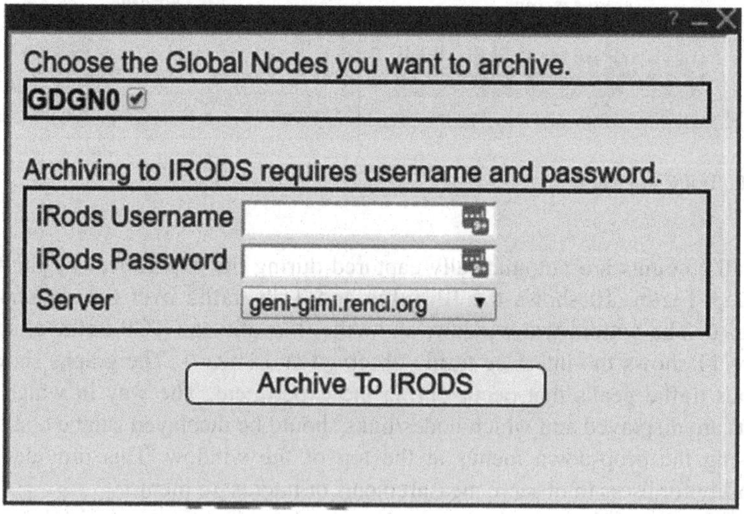

Choose the Global Nodes you want to archive.

GDGN0 ✔

Archiving to IRODS requires username and password.

iRods Username

iRods Password

Server geni-gimi.renci.org

Archive To IRODS

Fig. 12 The archive service

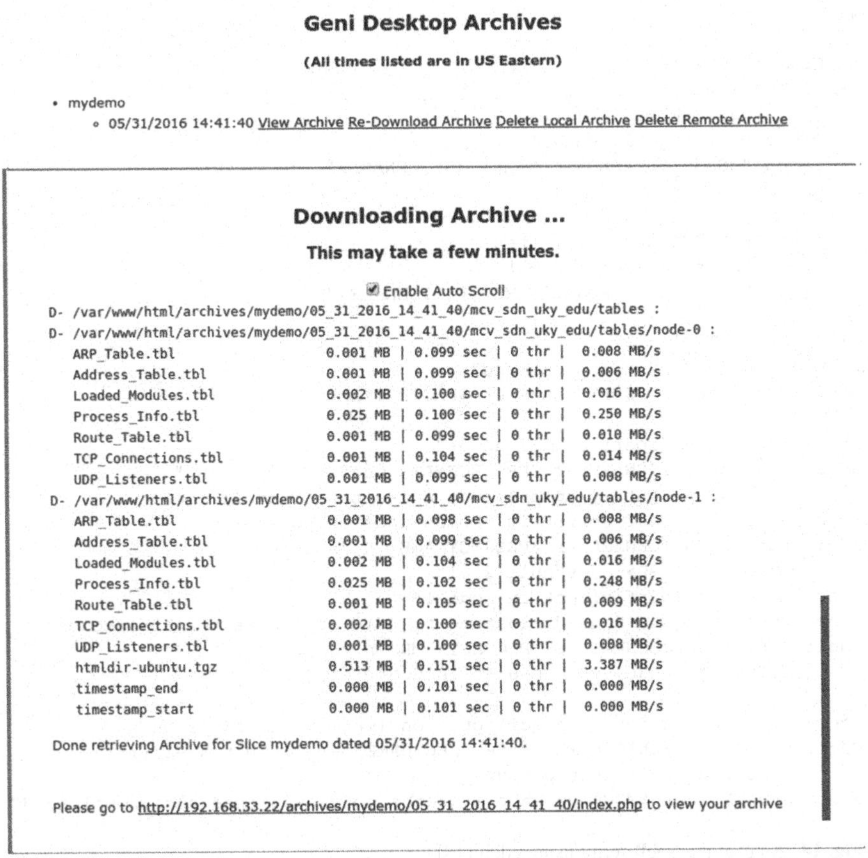

Geni Desktop Archives

(All times listed are in US Eastern)

- mydemo
  - 05/31/2016 14:41:40 View Archive Re-Download Archive Delete Local Archive Delete Remote Archive

Downloading Archive ...

This may take a few minutes.

☑ Enable Auto Scroll

```
D- /var/www/html/archives/mydemo/05_31_2016_14_41_40/mcv_sdn_uky_edu/tables :
D- /var/www/html/archives/mydemo/05_31_2016_14_41_40/mcv_sdn_uky_edu/tables/node-0 :
    ARP_Table.tbl                  0.001 MB | 0.099 sec | 0 thr |  0.008 MB/s
    Address_Table.tbl              0.001 MB | 0.099 sec | 0 thr |  0.006 MB/s
    Loaded_Modules.tbl             0.002 MB | 0.100 sec | 0 thr |  0.016 MB/s
    Process_Info.tbl               0.025 MB | 0.100 sec | 0 thr |  0.250 MB/s
    Route_Table.tbl                0.001 MB | 0.099 sec | 0 thr |  0.010 MB/s
    TCP_Connections.tbl            0.001 MB | 0.104 sec | 0 thr |  0.014 MB/s
    UDP_Listeners.tbl              0.001 MB | 0.099 sec | 0 thr |  0.008 MB/s
D- /var/www/html/archives/mydemo/05_31_2016_14_41_40/mcv_sdn_uky_edu/tables/node-1 :
    ARP_Table.tbl                  0.001 MB | 0.098 sec | 0 thr |  0.008 MB/s
    Address_Table.tbl              0.001 MB | 0.099 sec | 0 thr |  0.006 MB/s
    Loaded_Modules.tbl             0.002 MB | 0.104 sec | 0 thr |  0.016 MB/s
    Process_Info.tbl               0.025 MB | 0.102 sec | 0 thr |  0.248 MB/s
    Route_Table.tbl                0.001 MB | 0.105 sec | 0 thr |  0.009 MB/s
    TCP_Connections.tbl            0.002 MB | 0.100 sec | 0 thr |  0.016 MB/s
    UDP_Listeners.tbl              0.001 MB | 0.100 sec | 0 thr |  0.008 MB/s
    htmldir-ubuntu.tgz             0.513 MB | 0.151 sec | 0 thr |  3.387 MB/s
    timestamp_end                  0.000 MB | 0.101 sec | 0 thr |  0.000 MB/s
    timestamp_start                0.000 MB | 0.101 sec | 0 thr |  0.000 MB/s
```

Done retrieving Archive for Slice mydemo dated 05/31/2016 14:41:40.

Please go to http://192.168.33.22/archives/mydemo/05_31_2016_14_41_40/index.php to view your archive

Fig. 13 Sample archives of a user

the data on. Figure 13 shows sample archives stored at the GENI Desktop archive site. They are identified by the slice name and time the archive was stored. Multiple archives can be stored for the same slice at different times.

4.2 The GENI Desktop CLI

Many of the functions provided by the GENI Desktop graphical interface are available through the command line interface of the GENI Desktop. Because slice creation and resource allocation are provided by OMNI, the GENI Desktop CLI did not re-implement these functions. They can already be invoked from user scripts. Figure 14 shows a shell script invoking the gdcli library to list the nodes in an experiment, run a series of commands on selected nodes and download the performance results from selected nodes.

```
#!/bin/bash

SLICENAME=$2
PROJECTNAME=$1
if [ -z "$SLICENAME" ] ; then
    echo "Missing Slicename"
    echo "USAGE $0 <PROJECTNAME> <SLICENAME>"
    exit 1;
fi
if [ -z "$PROJECTNAME" ] ; then
    echo "Missing Projectname"
    echo "USAGE $0 <PROJECTNAME> <SLICENAME>"
    exit 1;
fi
echo "Running List nodes command"
gdcli listnodes -s $SLICENAME -r $PROJECTNAME

echo "Installing iperf on all nodes"
gdcli run -s $SLICENAME -r $PROJECTNAME -n "node-0,node-1" \
    -c "sudo apt-get -y update;sudo apt-get -y install iperf"
echo "Starting iperf server in daemon mode on node-0"
gdcli run -s $SLICENAME -r $PROJECTNAME -n "node-0"
    -c "iperf -s -D> /dev/null 2>&1"
echo "Starting iperf client on node-1"
echo "Connecting to node-0 blasting traffic at 10Gb/s for 30 seconds"
gdcli run -s $SLICENAME -r $PROJECTNAME -n "node-1"
     -c "iperf -t 30 -c node-0 -b 10000M"

echo "Fetch graph for data collected on link at interface on node-0"
gdcli getpng -s $SLICENAME -r $PROJECTNAME  -g linkbytes -l link-0
     -n node-0 -o ~/Desktop/mygraphs
echo "Open Folder $HOME/Desktop/mygraphs to view graphs just downloaded"
```

Fig. 14 An example shell script using gdcli [18]

The project name and slice name are the two parameters provided to the script. The first gdcli library call is `listnodes`, which gives a list of nodes in the experiment. Next, the script calls the gdcli library to install the *iperf* software on nodes `node-0` and `node-1`. It then starts the *iperf* server on `node-0` and runs the *iperf* client on `node-1` with the specified parameters. Finally, the script downloads the *png* graph generated by the monitoring functions to the desktop under directory *mygraphs* of the local machine. The graph records the byte count for `link-0` at `node-0` over time. The whole process is automated by using the script. It can be repeated to reproduce the results and modified to perform any new tests.

4.3 Common Usage Models

The GENI Desktop can support a variety of types of applications. We describe three examples to show how to make effective use of the GENI Desktop to carry out experiments.

In the first example, a user wants to use GENI to develop and test a distributed application layer service, such as a Content Distribution Network (CDN). The GENI Desktop can be used to create a slice and set up a topology with nodes distributed to the locations the user wants. This can be done using the Jacks integrated in the GENI Desktop by specifying which aggregate each node should be allocated from. The user can upload the CDN software to all the nodes with the upload feature of the GENI Desktop. Running the experiment is as easy as selecting all the nodes and typing the CDN program the user intends to run. The performance results can be collected by the monitoring and archiving services of the GENI Desktop.

In the second example, a user may want to change the "router" functionality by testing new routing protocols. One way to do that is to use the linux routers and user-level netfilter scripts to intercept traffic passing through the router. The user can write a new protocol to process the packets. After creating the topology of the experiment, the user can write a gdcli script specifying the program with the new protocol to be uploaded to the experimental nodes and also specify how the program should be run at each node. The user may try different parameters for the new protocol and repeat the experiment as many times as required. The corresponding result for each run can be collected and stored to the local machine for later analysis.

In the final example, a user wants to write an SDN application that interacts with the northbound interface of an OpenFlow controller. The user needs to create a slice with OVS images and a node to run the OpenFlow controller, set up default routes across the slice, and possibly change routes. With the tailored Jacks in the GENI Desktop, the user can create a topology using the GD OVS node, which is a special OVS image that initializes the OVS node and points the OVS node to the controller. The user can also add a special AAG Controller to the topology [17]. With the Flow Install module provided by the GENI Desktop, the user can specify a routing path for a flow and the corresponding OpenFlow rules will be installed at the OVS nodes by the Flow Install module automatically. Using the GENI Desktop Flow Monitor module, the user can monitor the performance of any selected flow.

5 Interacting with Other Tools and Services

A key goal of the GENI Desktop is to leverage existing tools and services. To that end, the GENI Desktop incorporates and uses a variety of existing tools such as OMNI [10], Jacks [9], and iRODS [8] in several of its key services. The GENI Desktop also provides an extensible framework capable of incorporating new add-on modules designed to offer enhanced functionality. In addition, the GENI Desktop has been designed to interoperate seamlessly with several peer tools and services such as the GENI Portal [5]. In the following, we briefly highlight some of these related tools and services.

The GENI Desktop windowing systems has been designed to support inter-window messages as a fundamental feature of all services running in the GENI

Desktop. Services such as displaying traffic graphs, or logging into nodes, are all driven by window messaging system. Resources selected in one of the three GENI Desktop "views" are passed as click events to the other modules (windows) in the GENI Desktop. As a result, all modules are aware of the resources that have been selected and to which the operation should be applied.

Cytoscape [1] is a graphical tool that is being used by the GENI Desktop for network topology visualization and interaction. Once a slice is instrumented using the GENI Desktop, a user can view and interact with the topology by clicking on the nodes and links in the network view drawn by Cytoscape. It has been enhanced to support the GENI Desktop window message passing service, enabling click/select events to be passed to the GENI Desktop messaging framework for further interaction with other add-on modules.

More recently we began developing an enhanced version of the Jacks [9] RSPEC editor to incorporate the GENI Desktop window messaging events. As a result, Jacks can be added as a plug-in module to the GENI Desktop and run in the user's browser. The GENI Desktop window messaging system then sends and receives events to/from JACKS which can then modify the RSPEC being drawn in it. One such app used in the GENI Desktop is the "Auto IP", which assigns non-conflicting private IP addresses to links drawn by the user.

GENI slices and slivers created by the GENI Desktop can also be used by other GENI Tools such as the GENI Portal [5] and LabWiki [7]. This is made possible by interoperability standards. Both the GENI Desktop and the GENI Portal rely on the GENI Clearing House for user and slice information. The GENI Desktop relies on the GENI Portal (and Clearing House) for authorization and authentication. The speaks-for feature of the GENI Desktop is used for the GENI users whose accounts have been issued by the GENI Clearing House. GENI slice and sliver operations, such as creation, deletion, renewal and slice membership modification, are all reported back to the GENI Clearing House and then relayed back to the GENI Portal. Thus both tools are always in sync with regard to the user account and the slice status. OMNI [10], another python based CLI tool for GENI, can also operate on the user's slices and slivers while keeping GENI Portal and GENI Desktop tools in sync. The GENI Desktop and GENI Portal both use OMNI as a library for their respective back-ends.

LabWiki [7] is a web-based interface developed as part of the GIMI project. It provides a web-based interface that can be used to design, describe and run repeatable experiments using GENI resources. GENI slices created in the GENI Desktop can also be used in the GIMI instrumentation framework. In addition, if a GIMI Disk image that is compatible with the GENI Desktop is used for a GENI slice, it becomes inter-operable with the instrumentation framework of the GENI Desktop.

Also related is the GENI Experiment Engine [13], a distributed platform-as-a-service tool that provides an easy way for a novice user to get an experiment up and running very quickly. It assigns one of a set of pre-allocated slicelets to the user to release the user from the burden of allocating virtual machines, configuring virtual machines/networks, and writing RSPECs. It is very helpful for new users when they are getting started and learning how to use GENI. However, the number of pre-allocated slicelets is limited, and the topology/configuration of these slicelets is pre-defined and may not necessarily satisfy the user's requirement. The GENI Desktop and the GENI Experiment Engine are complementary in the sense that after learning how to use GENI with the GENI Experiment Engine, a user can transition to the GENI Desktop to develop custom experiments.

GENI also provides access to archive tools such as iRODS [8]. An iRODS account can be created at the GENI Portal. Once an iRODS account is activated, a user can gain access to the iRODS interfaces (web-based or CLI) directly or use other tools such as the GENI Desktop and LabWiki that interact with the GENI iRODS server to archive user's measurement data being collected. The GENI Desktop also hosts its own archive server that provides an interface to visualize the data archived in addition to the basic archival feature.

6 Summary

The GENI Desktop provides a "one stop shop" for experimenters to carry out their experiments from beginning to end. It leverages existing tools wherever possible and provides both a web-based interface and a command line interface to support novice and expert users alike. It automates many of the tasks required by experimenters allowing them to focus on the objectives of their experiments, relying on the GENI Desktop to handle many of the details and complexities of creating, configuring, running, and measuring experiments. Moreover, the GENI Desktop makes it easy for users to run experiments by supporting a simple well-known abstraction for applying operations to resources in a common way. This abstraction, combined with a flexible and extensible event message passing system, makes it possible to add enhanced functionality as new tools and services emerge in the future.

References

1. Cytoscape: http://www.cytoscape.org/ (2015)
2. GEMINI: A GENI measurement and instrumentation infrastructure. http://groups.geni.net/geni/wiki/GEMINI (2014)
3. GENI experiment lifecycle diagram. http://groups.geni.net/geni/wiki/ExperimentLifecycle (2014)
4. GENI experiment workflows. http://groups.geni.net/geni/wiki/GeniExperiments (2013)
5. GENI Portal: https://portal.geni.net/ (2015)
6. GENI System Overview. http://www.geni.net/docs/GENISysOvrvw092908.pdf (2008)
7. Introduction to LabWiki and OEDL. https://github.com/mytestbed/labwiki (2015)
8. iRODS: http://www.irods.org/ (2015)
9. Jacks: https://www.emulab.net/protogeni/jacks-doc/ (2015)
10. Omni: http://trac.gpolab.bbn.com/gcf/wiki/Omni (2015)
11. The Emulab: http://www.emulab.net (2015)

12. The Flack GUI: http://www.protogeni.net (2012)
13. Bavier, A., Chen, J., Mambretti, J., McGeer, R., McGeer, S., Nelson, J., O'Connell, P., Tredger, S., Ricart, G., Tredger, S., Coady, Y.: The GENI experiment engine. In: Proceedings of the 2014 26th International Teletraffic Congress (ITC), Karlskrona, September 2014, pp. 1–6 (2014)
14. Berman, M., Chase, J.S., Landweber, L., Nakao, A., Ott, M., Raychaudhuri, D., Ricci, R., Seskar, I.: GENI: a federated testbed for innovative network experiments. Comput. Netw. **61**, 5–23 (2014). Special Issue on Future Internet Testbeds—Part I
15. Griffioen, J., Fei, Z., Nasir, H., Wu, X., Reed, J., Carpenter, C.: Measuring experiments in GENI. Comput. Netw. **63**, 17–32 (2014). Special Issue on Future Internet Testbeds—Part II
16. Laverell, W.D., Fei, Z., Griffioen, J.N.: Isn't it time you had an emulab? In: ACM SIGCSE 2008 Technical Symposium on Computer Science Education, Portland, OR, March 2008
17. Rivera P., S., Fei, Z., Griffioen, J.: Providing a high level abstraction for SDN networks in GENI. In: Proceedings of the 2nd International Workshop on Computer and Networking Experimental Research Using Testbeds (CNERT 2015), Columbus, OH, June 2015
18. The GENI Desktop Tutorial, http://groups.geni.net/geni/wiki/GENIExperimenter/Tutorials/GENIDesktop (2016)

A Walk Through the GENI Experiment Cycle

Thierry Rakotoarivelo, Guillaume Jourjon, Olivier Mehani, Max Ott, and Michael Zink

Abstract The ability to repeat experiments from a research study and obtain similar results is a corner stone in experiment-based scientific discovery. This essential feature has often been overlooked by the distributed computing and networking community. There are many reasons for that, such as the complexity of provisioning, configuring, and orchestrating the resources used by experiments, their multiple external dependencies, or the difficulty to seamlessly record these dependencies. This chapter describes a methodology based on well-established principles to plan, prepare and execute reproducible experiments. We propose and describe a family of tools, the LabWiki workspace, to support an experimenter's workflow based on that methodology. This proposed workspace provides services and mechanisms for each step of an experiment-based study, while automatically capturing the necessary information to allow others to repeat, inspect, validate and modify prior experiments. Our LabWiki workspace builds on existing contributions, de-facto protocols, and model standards, which emerged from recent experimental facility initiatives. We use a real experiment as a thread to guide and illustrate the discussion throughout this chapter.

1 Introduction

One of the cornerstones of scientific discovery is validation by the community. In experimental science, this requires others to repeat the experiments and obtain similar results within acceptable statistical bounds. Traditionally, the distributed computing and networking community has been largely ignoring this. There are few publications in top-tier venues, which primarily report on the successful validation of somebody else's work, while problems with repeatability are sometimes buried

T. Rakotoarivelo • G. Jourjon • O. Mehani • M. Ott
NICTA, Australian Technology Park, Eveleigh, NSW, Australia
e-mail: thierry.rakotoarivelo@nicta.com.au; guillaume.jourjon@nicta.com.au; olivier.mehani@nicta.com.au; max.ott@nicta.com.au

M. Zink (✉)
Department of Electrical and Computer Engineering, University of Massachusetts in Amherst, Amherst, MA 01003, USA
e-mail: zink@ecs.umass.edu

© Springer International Publishing Switzerland 2016

R. McGeer et al. (eds.), *The GENI Book*, DOI 10.1007/978-3-319-33769-2_17

in vague references. There are many reasons for that. Advances in the underlying technology continuously create new opportunities to explore new ideas leaving little time to reflect on the "old". But there are also very pragmatic reasons. First of all, most experiments are conducted in complex environments with many external dependencies, such as type and speed of computers and networks, size of storage, chip sets, or operating system and driver versions. Some of them will only affect the measured "utility" of the reported phenomena, while others are essential to having a successful experiment in the first place. Unfortunately, many of these dependencies are never reported and therefore making it very difficult for others to repeat an experiment.

We argue, that our inability in Computer Science to repeat reported experiments is not only bad practice, but also hampers progress in general. It reduces our ability to expand on prior work, verify and adapt it to different contexts, compare different methods in different environments and much more. We also argue that the "paper" as the traditional publication mechanism is one of the major obstacles in improving the status quo.

We are clearly not alone, initiatives, such as the Elsevier's Executable Paper Challenge [4] have been exploring new avenues for disseminating scientific results. In addition, easy access to emerging large scale experimental facilities, funded and coordinated by programs, such as GENI in the US [1], FIRE in Europe [5], and similar activities in China, Korea, and Japan, provide the community with a common "playground" in which to conduct experiments. But only providing experimental facilities is not sufficient. The sharable resources we have available today still need to get provisioned, configured and modified before they can be used in experiments. We see those steps as the crucial pieces that are needed to perform repeatable experiments.

In the remainder of this chapter we propose and describe a family of tools to support an experimenter's workflow, while also automatically capturing most of the necessary information to allow others to repeat, inspect, validate and modify prior experiments.

More specifically, we propose to model the experimenter workflow on the Scientific Method[1] which we interpret, as shown in Fig. 1, as a repeated cycle of stating a hypothesis, designing and conducting an experiment, and finally analyzing the measurements taken during the experiments with the intent to test or disprove the hypothesis.

We observed that many of these steps follow the same internal workflow of planning, preparing, and executing. We therefore built an experimenter-facing web-based tool, called LabWiki, which supports this three-step workflow in different contexts. LabWiki, as the name implies, is modeled after the traditional laboratory book, which experimenters use for a very similar workflow and purpose. LabWiki takes this further, by not only being the recording mechanism, but also the operating platform for many activities within the experiment workflow.

[1]https://en.wikipedia.org/wiki/Scientific_method.

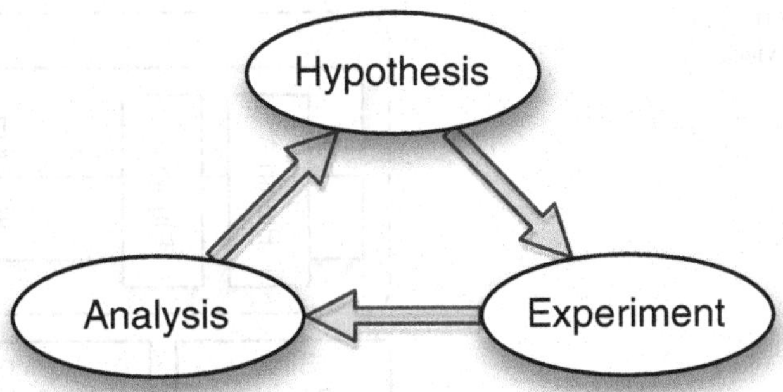

Fig. 1 Scientific method

We start the remainder of this chapter with a brief overview of LabWiki (Sect. 2) and then introduce its user interface (Sect. 3). In Sect. 4, we introduce a real experiment used to research a time synchronization approach and published in [13], which will be used as a guiding example in the remainder of this chapter. An overview on experimental facilities and testbed resources is given in Sect. 5. We will illustrate the experiment workflow by taking the reader through every step, namely the experiment design (Sect. 6), the setup of and experiment (Sects. 7 and 8), the execution of an experiment (Sect. 9), and the analysis (Sect. 10) of a previously published research result [13]. Finally, we briefly describe how the LabWiki workspace can support educators in harnessing these large facilities for lab tutorials (Sect. 11).

2 LabWiki Overview

The web-based LabWiki service strives to be the primary tool for an experimenter to plan, prepare, execute, analyse and even publish experimental-driven research. While the classical UNIX approach of "many little tools" often leads to a very rich and versatile environment it also requires great discipline on behalf of an experimenter to keep a detailed record of what combination of tools and their configurations have been used for what experimental artefact. On the other hand, a single comprehensive tool rarely works for cutting-edge research as requirements for new features often outstrip the development resources of the "mega tool" builder.

LabWiki attempts to find a sweet spot by defining a framework which a) is based on well established, unifying methodology, b) supports the tracking of artefacts, their meta data and relationships to others, c) is extensible, and d) allows for easy integration of external tasks and services. Simplistically, it can be viewed as an easily customisable glue between the many little tools and the "history keeper" on how they were all used in the pursuit of a scientific discovery.

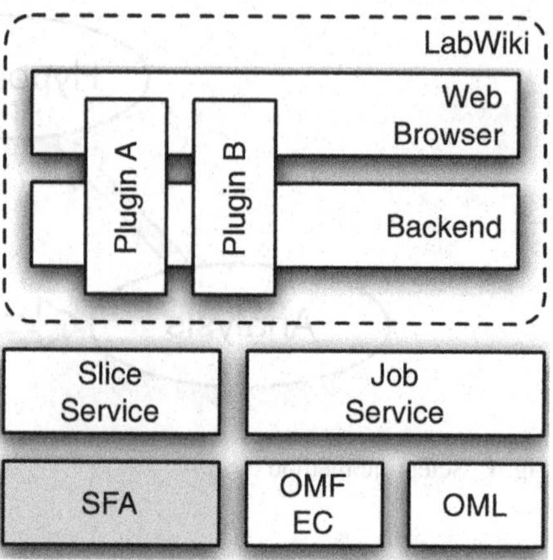

Fig. 2 LabWiki and supporting services

LabWiki is the result of the shared experience of the authors in their respective roles as tool builders, testbed operators, researchers, educators, engineers, administrators and many more. The following is a description of LabWiki's current "universe" of components, services, and capabilities.

LabWiki, as shown in Fig. 2, is sitting on top of a suite of supporting tools and services, which can be used directly by an experimenter, or more likely by other tools acting on her behalf, specifically the SliceService (Sect. 8.3) which harmonizes resource provisioning across many different testbeds; OEDL (Sect. 7), a domain-specific language for describing the orchestration of an experiment; JobService (Sect. 9) for scheduling an experiment; OMF & FRCP (Sect. 9) for executing and coordinating individual experiment runs (or trials); and OML (Sect. 7) for collecting and managing measurements during a trial.

In addition LabWiki can be easily extended through *plugins* to extend it's functionality or adapt it to a new environment. Example plugins described in this chapter are the Topology Editor (Sect. 8.2), Experiment Executer (Sect. 9.1 and right panel in Fig. 3), the Analysis Widget (Sect. 10), and the iBook Widget Creator (Sect. 11.2).

3 LabWiki User Experience

As mentioned in the Introduction, the experimenter interacts with LabWiki primarily through a web browser. After a standard login process, the user will see (Fig. 3) a browser window split into three columns, labeled "Plan", "Prepare", and "Execute". This reflects the basic workflow identified above. Each column comprises a tool & search bar, followed by a widget header, an optional widget toolbar and the widget body. The top tool & search bar allows the user to quickly locate or create resources relevant to the respective activity and choose the desired

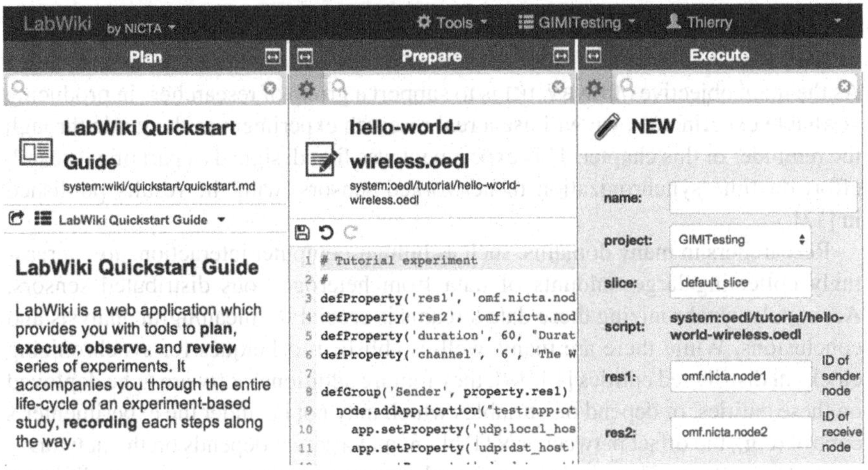

Fig. 3 Main interface of LabWiki

widget to interact with that resource. LabWiki itself is a framework with most of the functionality provided by plugins, which in turn provide one or more widgets. For instance, the *wiki* widget for the "Plan" column supports editing of rich text resources.

All widgets are stateless and only provide mechanisms for a user to interact with one or more named resources. These resources may reside on a separate service, such as the JobService (Sect. 9.2), or are file-like resources, such as a wiki entry, or an image. For these kinds of objects, LabWiki provides layered, pluggable artifact stores. Current implementations support persistence through the local filesystem, versioned and access-controlled repositories such as Git [2], and via iRODS [16]. The clean separation of stateless widgets and state-full, externally resolvable resources allows for interacting and embedding of these resources outside of LabWiki as well. For instance, plots of experiment measurements, hosted on JobService (Sect. 9.2) can be embedded into a wiki page, which then can be published from the *wiki* widget to a third-party blog service. Importantly, the link from the plot in the blog entry to the actual experiment is maintained, including access control mechanisms.

LabWiki supports multiple user accounts and uses OpenID for authentication. Resources, managed through LabWiki belong to projects and a user's membership and role in a project are the basis of LabWiki's authorization mechanism. Information about membership and respective roles are sourced from external services, such as the GENI ClearingHouse. Currently LabWiki is also facilitating the transfer of delegation and *speaks-for* credentials for the services some of the plug-ins call upon (e.g. SliceService Sect. 8.3).

4 Experiment Overview

As the main objective of LABWIKI is to support a group of researchers in producing verifiable experiments, we will use a real research experiment as the guide through the reminder of this chapter. This experiment was first designed as part of a research effort on time synchronization in networked sensors, with the results published in [13].

Researchers in many domains, such as human-computer interaction, are increasingly collecting large amounts of data from heterogeneous distributed sensors. Accurately synchronizing these data streams is crucial for meaningful analysis and conclusions. While there are many, well-established techniques for synchronizing clocks in distributed entities [14, 19], they require additional software to be deployed on these entities, or depend on variables which may not be under the experimenter's control (e.g., the offset between a NTP client and a server depends on the network's round-trip delay). The above mentioned research project proposed a different approach based on measurements of the data collection system itself and uses the obtained meta data to synchronize the original data a-posteriori.

The main experiment assumes a scenario where certain events can be measured by more than one sensor and where all sensors then forward these measurements to a common collection server. Figure 4 illustrates the resulting experiment topology. A series of events are generated by a source S and measured by two entities $E1$ and $E2$. The respective measurement samples are sent to the same collection server C. Time delays may be added at the various Dij points. $E1$, $E2$, and C add locally sourced timestamps t to all samples that they produce and receive, respectively.

5 Experimental Facilities

Major initiatives such as GENI [1] and FIRE [5] have focused on providing distributed, virtual laboratories for transformative, at-scale experiments in network science and services. Designed in response to the Internet ossification issue, these so-called *testbeds* enable a wide variety of experiments in many areas, including clean-slate networking, protocol design, distributed service offerings, social network integration, content management, and in-network service deployment. Many software tools were proposed to allow operators and experimenters to manage,

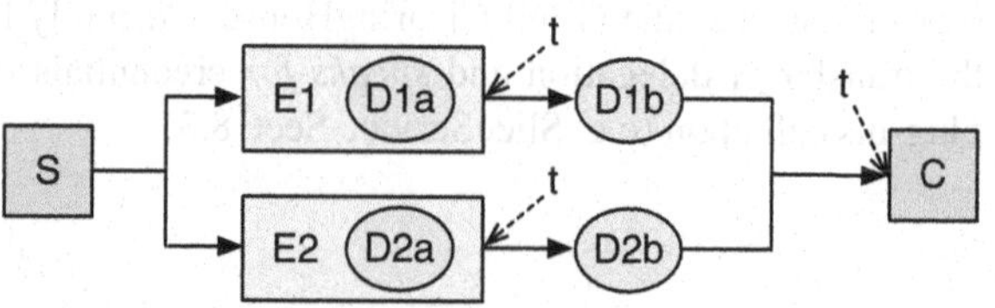

Fig. 4 Topology of the Time Calibration experiment from [13]

access and control the resources from these testbeds. The models, protocols, and APIs from some of these contributions are currently converging towards *de-facto* standards within the community.

A RSpec[2] defines a set of resources that can be used in an experiments. These resources may be requested from a variety of GENI or FIRE testbeds, such as ExoGENI, OpenFlow Mesoscale, or Fed4FIRE [20]. There are three different types of RSpecs, (a) *the Advertisement* which is sent by an Aggregate Manager (AM) to an experimenter to describe its available resources; (b) *the Request* which is sent by the experimenter to the AM to describe the resources she wants to reserve; and (c) *the Manifest* which is returned by the AM to describe which resources have been reserved by the experimenter. These RSpecs are exchanged in the previous sequence between the AM and the experimenter. The requested resources will be available to the experimenter after the successful completion of that sequence.

The Aggregate Manager APIs [1] define a common interface for software to provide, request, reserve, and provision resources over different facilities. They are based on a *slice* abstraction, which is a container for all the resources used in a project. Experimenters are associated with slices and use these APIs to interact with various entities (e.g., Clearinghouse, Aggregate Manager) in order to discover, reserve, and provision resources. These interactions are mostly performed through third-party interfaces. For example, Omni[3] is a command line tool used to specify and reserve resources from GENI facilities. It allows *stitching*, a technique to connect resources via layer 2 VLans. In contrast, Flack[4] and Jacks[5] are graphical tools, which allow experimenters to reserve resources and specify RSpecs through a visual topology editor. Finally, JFed[6] is a Java-based tool, which allows experimenters to obtain large distributed topologies using resources from both FIRE and GENI testbeds.

The Federated Resource Control Protocol (FRCP) and the OML Measurement Stream Protocol[7] (OMSP) [12, 18] are two protocols to control resources and collect data from them. They are commonly used in both GENI and FIRE facilities. FRCP defines a short set of asynchronous interactions over a *publish-and-subscribe* system, which allows experimenters to configure resources and instruct them to execute given tasks. The OMF and NEPI control tools both implement FRCP [8, 17]. OMSP defines the format and transport of measurement tuples from producers (e.g., a resource) to consumers (e.g., a storage server). It supports various types of measurements, encodings, and the use of metadata. The OML framework [12]

[2]http://geni.net/resources/rspec.

[3]http://trac.gpolab.bbn.com/gcf/wiki/Omni.

[4]http://www.protogeni.net/wiki/Flack.

[5]https://www.emulab.net/protogeni/jacks-doc/.

[6]http://jfed.iminds.be.

[7]http://oml.mytestbed.net/doc/doxygen/omsp.html.

provides an OMSP storage server and a C client library to instrument resources. Other client libraries also exist (e.g., OML4R,[8] OML4Py[9] or OML4J[10]).

6 Experiment Design

The first step of an experimental study is the design of the experiment itself. It is driven by research goals, such as testing a hypothesis, measuring performance, or demonstrating capabilities.

There have been many contributions related to experiment design since the seminal work of Fisher [3]. Examples in the area of computer science include [9, 15]. While there are many variations, a good starting point is the identification of the dependent, independent, and confounding variable sets. The *dependent* variables are measured attributes of the studied system, their analysis will provide answers to the study's questions. The *independent* variables would impact the studied system and modify its dependent variables. The third set of *confounding* variables may be unknown or uncontrollable by the experimenter and may have some effect on any of the former variables.

Given these three variable sets, the researcher then devises an experiment plan where usually the *dependent* variables are measured, the *independent* variables are controlled and varied across different repeated trial batches, and the effect of *confounding* variables are mitigated through techniques such as replication or randomization. The choice of controlled values for the *independent* variables and the number of trials and their repetition depends on the objectives of the study.

The LabWiki workspace has a set of tools to support the experiment design process. The "Plan" column on the left-hand side of its interface (Fig. 5) provides a Wiki widget that allows the experimenter to describe and record her design. This design strives to replace her pen-and-paper laboratory notebook. It currently uses the popular Markdown syntax,[11] and figures and plots from other widgets can be easily dragged-and-dropped into the write-up.

The Design of Our Example Experiment In the case of our example experiment, we identify the *dependent* variables as the arrival times of a measurement sample at different points in the system. Our *independent* variables consist of configurable clock offsets and network delays, generally referred to as *Dij* in Fig. 4. One potential *confounding* variable would be the varying delays in processing measurement samples inside the sensors.

[8]https://github.com/mytestbed/oml4r.

[9]https://github.com/mytestbed/oml4py.

[10]https://github.com/NitLab/oml4j.

[11]http://daringfireball.net/projects/markdown.

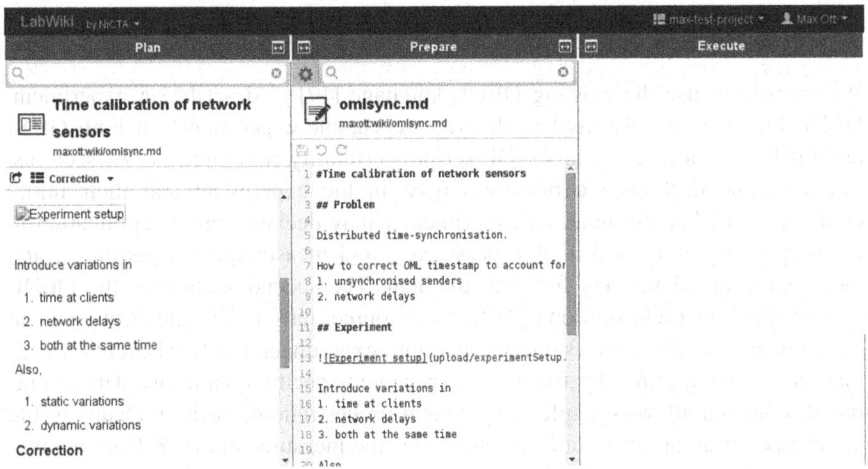

Fig. 5 The experiment design notes are shown in the "Plan" panel. They can be written down in a wiki form in the "Prepare" using Markdown source

In this particular realization of the experiment, we chose a classic ICMP ping to the network's broadcast address as the *event* generated by the source S and measured through their respective network interfaces on $E1$ and $E2$. In the first round of trials, the entities are quasi-synchronized and no delay is added on the collection network. This establishes a baseline for future results. We then planned to run a series of trials, where various known time offsets are introduced to each entity's clock and on their respective paths on the collection network. To mitigate the potential impacts of *confounding* factors, we decide to run multiple trials for each specific offset configuration and further repeat these multiple trials over different instantiations of our topology. More details about this experiment design and each series of trials are available in the original study [13].

As mentioned above, we primarily use the Wiki widget to describe the design and work plan. Dragging experiment results and other artifacts onto the wiki will allow us to keep track of progress. This will be especially important if an experiment is carried out by a team where different members are pursuing different parts of the work plan.

7 Experiment Description and Instrumentation

Once the design of the experiment is finalized and documented in LabWiki's Plan panel, the next step is to translate it into a machine-readable description.

7.1 Describing an Experiment

We propose to use the existing OEDL language [17] to describe an experiment. OEDL has been widely used to describe repeatable experiments on both GENI and FIRE facilities. A typical OEDL script is primarily composed of two sections. The first one declares the resources used in the experiment and their initial configurations. For example, an experimenter may declare a given application to be used, along with its available parameters and measurement capabilities; and the specific initial settings for both of them. The second section of the OEDL script defines the orchestration of tasks the resources have to execute throughout an experiment trial. These tasks are grouped into experimenter-defined events, which may be triggered either by timers or experiment-specific conditions. This event-based approach allows complex experiment orchestrations, such as changing the parameters of an application X seconds after the measurement of Y from another application reaches the value Z.

LABWIKI has inspired a third, optional section for a typical OEDL script. It allows an experimenter to define charts to provide quick feedback on the progress or outcome of an experiment trial. The *experiment* widget in LABWIKI uses that to display line, pie, or histogram charts in the respective column with relevant measurement data streams sourced from the JobService. We do want to note, that this is primarily to provide a graphical live feedback on an individual execution of an experiment trial, and will usually not replace a thorough data analysis over the complete result set obtained for an experiment (Sect. 10).

Listing 1 provides a shortened OEDL script for our example experiment. While the complete OEDL script[12] describes the entire experiment with all required settings as designed in Sect. 6, this shortened version only shows the experiment for the baseline trials. We will now briefly describe this script and refer the reader to the OEDL Reference document[13] for further details.

Lines 1–4 fetch and load additional OEDL scripts, similar to the *include* statement found in many programming languages. Lines 6–9 define some experiment parameters which may be modified for different trials. Lines 11–18 define a group of resources comprising of the entities $E1$ and $E2$ from Fig. 4. An ICMP packet capture application is associated to each resource in that group (line 13). The parameters and measurements to collect for this application are set using the **setProperty** and **measure** commands, respectively. Lines 20–29 define another group of resource with only the source S from Fig. 4. The ICMP ping application is associated to that resource (line 22), and configured to ping the network's broadcast address (line 23). Lines 31–39 define a third group which include all the previous resources. A time statistic reporting application is associated to all these resources (line 34). Finally, Lines 41–49 define the sequence of tasks to perform once all the resources are ready

[12]http://git.io/clock-delay-impairments.rb.

[13]http://omf6.mytestbed.net/OEDLOMF6.

```
loadOEDL('http://goo.gl/4br2MW')
loadOEDL('http://goo.gl/qg8Alo')
# From the trace-oml2 package
loadOEDL('file:///usr/share/trace-oml2/trace.rb')

defProperty('entity1','node20','1st␣entity␣ID')
defProperty('entity2','node21','2nd␣entity␣ID')
defProperty('source','node19','Event␣source␣ID')
defProperty('time',180*60,'Trial␣duration␣[s]')

defGroup('Entities',prop.entity1,prop.entity2) do |g|
  # Capture ICMP echo packets
  g.addApplication('trace') do |app|
    app.setProperty('filter', 'icmp[icmptype]=icmp-echo')
    app.setProperty('interface', 'eth1')
    app.measure('ethernet', :samples => 1)
  end
end

defGroup('Source',prop.source) do |g|
  # Broadcast ICMP echo requests every 10s
  g.addApplication('ping') do |app|
    app.setProperty('dest_addr', '10.0.0.255')
    app.setProperty('broadcast', true)
    app.setProperty('interval', 10)
    app.setProperty('quiet', true)
    app.measure('ping', :samples => 1)
  end
end

defGroup('All',prop.source,
         prop.entity1,prop.entity2) do |g|
  # Report time synchronisation every minute
  g.addApplication('ntpq') do |app|
    app.setProperty('loop-interval', 60)
    app.setProperty('quiet', true)
    app.measure('ntpq', :samples => 1)
  end
end

onEvent(:ALL_UP_AND_INSTALLED) do
  group('All').startApplications
  group('Entities').startApplications
  group('Source').startApplications
  after prop.time do
    allGroups.stopApplications
    Experiment.done
  end
end
```

Listing 1 Example of an OEDL script

and their associated applications are installed. In this simple case, all the resources first start their time reporting applications. Then the resources within the "Entities" group start their applications. Then the resource in the "Source" group does the same. After a set duration, all resources in all the groups stop their applications, and the trial is finished.

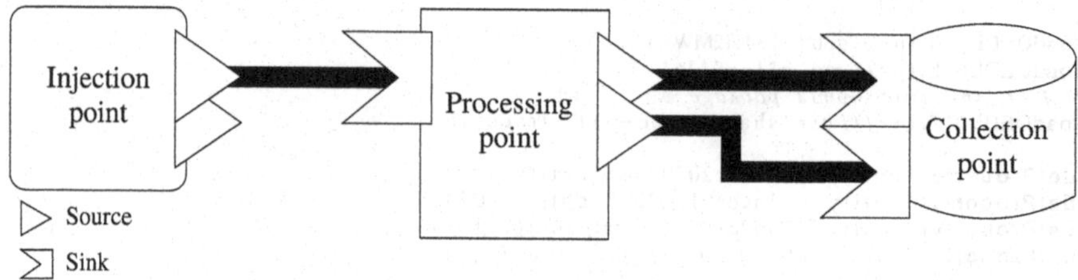

Fig. 6 Applications can be instrumented with OML to *inject* timestamped samples into various measurements streams (MS) which can be *processed* in-line (e.g., aggregated or sub-sampled) before finally being *collected* and written into a storage backend

7.2 *Instrumenting Resources*

LABWIKI leverages the OML measurement framework [12] for measurement collection and storage. OML is based on the concept of *measurement points* (MPs). The schema of an MP defines a tuple of typed metrics meaningfully linked together (e.g., sampled at the same time, or pertaining to the same group). The series of tuples generated by reporting measurements through an MP defines a *measurement stream* (MS). OML defines several entities along the reporting chain that can generate, manipulate, or consume MSs. This is illustrated in Fig. 6.

Instrumentation Process

The instrumentation of a resource consists of enabling applications to act as injection points (Fig. 6). By providing a structured way of defining MPs, OML fosters the reusability and interchangeability of instrumented applications, and simplifies the assembly of subsequent experiments. For example the "wget" and "cURL" applications report similar information about web transfers, and should therefore attempt to reuse the same MPs.

It is therefore important to first identify all the measurements that can be extracted from an application. For example, `ping` not only provides latency information, but also sequence and TTL information for each received packet from any identified host, as well as overall statistics. A rule-of-thumb is that measurements, which are calculated, measured or printed at the same time/line are good candidates to be grouped together into a single MP. For more complex cases, where samples from multiple MPs need to be linked together, OML provides a specific data type for *globally unique identifiers* (GUID). They can be used in a similar way as foreign keys in databases. For example, in the case of the `trace-oml2` application, it was decided to create one MP per protocol encapsulated in a packet (e.g., ethernet or IP). A fresh GUID is generated for each packet, which is then parsed, injecting information about each header in the relevant MP, along with the GUID.

It is also possible to report metadata about the current conditions. Such details as description, unit and precision of the fields of an MP are primordial for the later understanding of the collected measurements.[14] Other information such as the command line invocation, or application version and parameters are also worthy of inclusion as part of this metadata.

Instrumentation Libraries

The most complete OML implementation is the C `liboml2`(1,3).[15] It provides an API, which can be used to define MPs within the source code of an application, and mechanisms to process the injected samples at the source. The `oml2-scaffold`(1) tool can be used to generate most of the boilerplate instrumentation code, along with the supporting OEDL description [12, App. A]. An example application written from scratch to report network packets is `trace-oml2`,[16] used in Listing 1 (line 13).

The Ruby and Python bindings (OML4R & OML4Py) are particularly useful for writing wrappers for applications for which the source code is unavailable. Wrappers work by parsing the standard output of an application, and extracting the desired metrics to report. An example is the `ping-oml2` wrapper,[17] using OML4R[18], used in Listing 1 (line 22).

7.3 The Prepare Panel

Our LabWiki workspace has a "Prepare" panel at the center of its interface (center widget in Fig. 3), which provides a simple code editor widget. The experimenter may use this widget to edit an OEDL script, as described previously. All OEDL scripts created within this editor widget are versioned and saved within LabWiki's artefact store with group-based access control. While this widget is a convenient tool for users to edit their OEDL scripts, the may use alternate means to do so, as well. For example, they may edit their scripts in other editors and then cut & paste it into the "Prepare" panel's code editor, as illustrated in Fig. 7. Alternatively, they may directly use a git repository, and push it into LabWiki's artefact store.

[14]Base SI units should be preferred whenever possible.

[15]Manpages for OML system components can be found at http://oml.mytestbed.net/doc.

[16]http://git.mytestbed.net/?p=oml-apps.git;a=blob;f=trace/trace.c.

[17]We generally use `APPNAME-oml2` as the binary's name of OML-instrumented versions of upstream `APPNAME` utilities; the OEDL application description however only uses APPNAME for conciseness.

[18]http://git.io/oml4r-ping-oml2.

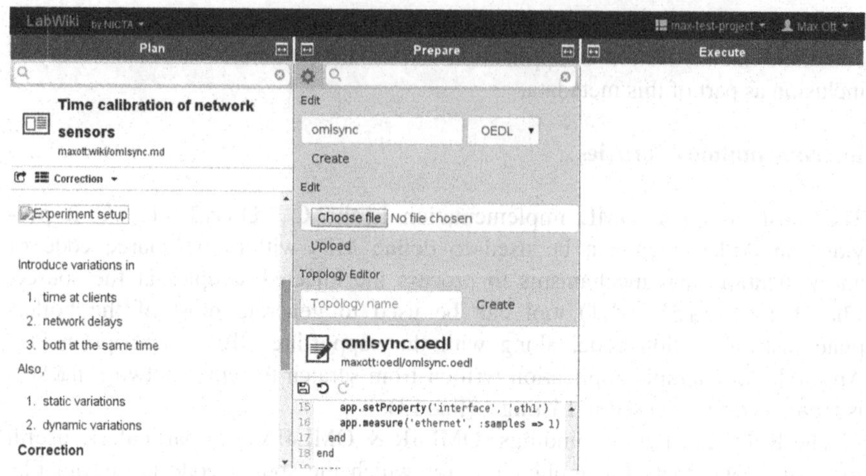

Fig. 7 The "Prepare" panel can be used to upload and edit the various files and scripts needed to describe the experiment (e.g., Markdown or OEDL)

8 Resource Selection and Provisioning

While the previous section dealt with describing the entire experiment and its resource needs, it did not consider where these resources come from. For instance, the OEDL script in Listing 1 refers to resources, such as virtual machines (nodes) and applications (trace, ping). It is the former kind of resources, which we assume will be provided by testbeds and the programmable networks, as offered by GENI, connecting them.

More specifically, an experimenter needs to first define a topology of nodes, their interconnecting networks, and their specific characteristics. For our example experiment, at least four (virtual) machines and four links are required to create the topology in Fig. 4. We note, that should the experiment be extended (e.g., by adding new entities E3 and E4), additional resources have to be reserved. Alternatively, the experimenter may reserve a larger topology and run different trials on a subset of the reserved resources.

8.1 Process Overview

Specification The very first step is the specification of resources that are required for an experiment. Section 5 described several approaches for resource description and tools to create them. The most common specification is the XML-based GENI RSpec.[19]

[19]http://groups.geni.net/geni/wiki/GENIExperimenter/RSpecs.

Reservation Once specified, a set of resources needs to be reserved. This requires a negotiation phase between the provisioning tools and the corresponding services on the testbed side. A negative outcome of this negotiation means that the requested resources can currently not be provisioned, or the requestor does not have sufficient privileges, or has exceeded her quota. For example, a VM is requested but all the physical machines' resources have been already allocated to other experimenters. In such a case, the experimenter either waits or hopes that the desired resources will become available in the near future or modifies the request for a different set of resources. A positive response means that the provisioning process will move on to the next step.

Provisioning After specific resources have been identified, they need to be provisioned before the experimenter can gain access to them. In the case of a physical machine, this may require a power up. In the case of a VM, a disk image containing the requested operating system needs to be loaded and the VM started up with the appropriate configurations. This may also include the distribution of security credentials to limit access to those resources to the requesting party. As each step may take time and in the case of large requests never fully complete successfully, proper communication between the requesting and providing services need to be maintained as even a subset of successfully provisioned services can already be used for successful experimentation.

Monitoring Most of the resources provided are virtualized in some form or depend on other services in non-obvious ways. It is therefore important for most experiments to be able to monitor their resources and potentially even the broader context in which they are provided. For instance, CPU and memory allocations to VMs may change over time, or there may be external interference in wireless testbeds. While some of these parameters can be monitored by the experimenter herself, others may need special access and therefore need to be collected by the resource provider with the results forwarded to the experimenter. For instance, the BonFIRE [7] testbed provides monitoring information on the physical server to the VM "owners" for the respective server. An experimenter can either use such infrastructure information after the completion of the experiment during the analysis phase or display this information in real time in LabWiki for actual monitoring.

8.2 Labwiki Topology Editor Plugin

LabWiki provides a topology editor plugin, which supports the experimenter in navigating the above described steps. The plugin provides two widgets, one for the "Prepare" panel to specify the topology, and one for the "Execute" panel to request the provisioning of a defined topology and its ongoing monitoring.

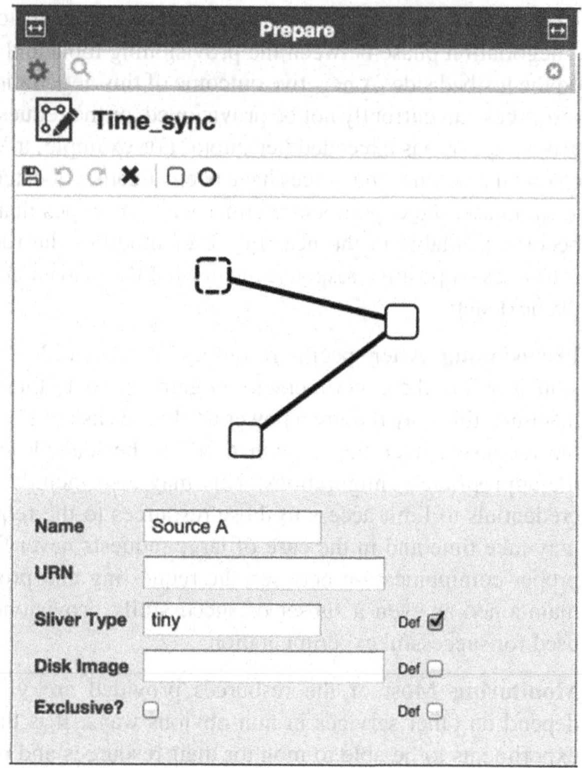

Fig. 8 LabWiki's topology widget

Figure 8 shows a screenshot of the first widget, the graphical topology editor. The widget is split vertically with the graph editor on top and the property panel for the selected resource (dotted outline) at the bottom. Interactive graphical editors are usually easy to learn, but do not scale well to large topologies. Hierarchical grouping with associated visual "collapsing" can mitigate some of these scaling issues. However, larger topologies will not be "hand crafted" but generated by tools, such as BRITE [11]. The topology editor has a text-mode, which allows the experimenter to specify a BRITE model as well as provisioning information for the nodes and links created.

The topology description can either be stored as RSpec or extended GraphJSON.[20] It is this textual representation, which the "Slice" widget is sending to the SliceService when requesting the reservation and provisioning of a specific topology. This widget uses the same graph editor (now read-only) to convey progress by animating the graph elements accordingly. Monitoring information is also overlaid to provide experimenters with quick feedback on the overall topology status.

[20]https://github.com/GraphAlchemist/GraphJSON.

8.3 *The SliceService*

The SliceService provides a REST API for requesting and provisioning of resources for a testbed federation. It is essentially a proxy service to the SFA APIs of the ClearingHouse (CH) and AggregateManager (AM). We reluctantly chose this path as the legacy decisions regarding technology (XML-RPC and client-authenticated SSL), as well as multiple versions for both API and RSpec put a considerable maintenance burden on the upstream tools. Therefore, we designed and implemented a service, which is based on current best practices for web-based services. We want to stress, that this is not a value judgement of the SFA APIs but a commonly encountered legacy problem. This decision allows us to concentrate our development as well as debugging efforts regarding testbed interactions on a single service. In fact, some design decisions for the SliceService have been heavily influenced by JFed,[21] which seems to have similar objectives.

Following the REST philosophy, SliceService defines a distinct set of resources and provides a consistent set of actions to create, modify, and delete those resources. It also takes advantage of the recently introduced delegation mechanism based on credentials. Traditionally, SFA tools were assumed to have access to the requesting user's private key. However, in this context, the user is authenticated with LabWiki which in turn requests SliceService to perform certain actions on behalf of a specific user. In addition, a specific SliceService instance may serve many different users. To maintain full transparency on who is operating on whose behalf we need to ensure that every request made by SliceService to a CH or AM contains the full delegation chain back to the user. This will allow each CH or AM to decide if it trusts the intermediate services. In turn it increases the security of both LabWiki and SliceService as user authentication can be delegated to dedicated federation services, such as the CH.

The SliceService also plays a crucial role in the security mechanism of FRCP (Sect. 5) by providing the newly created resources with proper credentials in a secure manner. However, a detailed description of the overall security mechanism is beyond the scope of this chapter.

9 Running an Experiment Trial

Once an experiment is designed, described, and all necessary resources have been allocated, the next step is to execute an instance (or trial) of that experiment. Running an experiment trial should be effortless for the experimenter, as she will need to repeat this process many times in order to gather sufficient data to derive statistically meaningful conclusions.

[21] http://jfed.iminds.be.

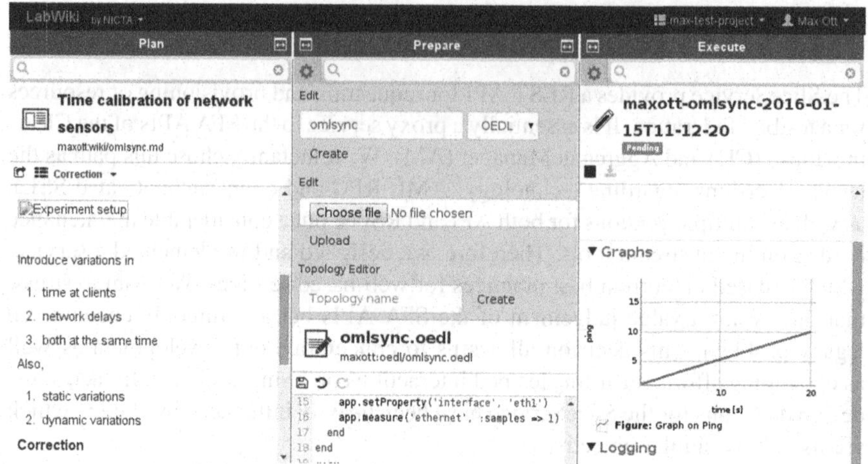

Fig. 9 The "Execute" panel shows details of the currently running experiment, including real-time graphs of the measurements being collected

9.1 The Execute Panel

Our LabWiki workspace has an "Execute" panel on the right-hand side of its interface (Fig. 3), which hosts an experiment widget. This widget allows an experimenter to configure, start, and observe individual experiment trials, as shown in Fig. 9. To initiate this process, an experimenter intuitively drags & drops the experiment's icon from the previous "Prepare" panel into the "Execute" one. This action triggers LabWiki to display a list of the experiment properties defined in the respective OEDL script which can now be configured for a specific trial. For example, the experiment design might require that 20 trials should be run with property *A* set to 1, followed by another 20 trials with *A* set to 2. Once the experimenter is satisfied with the trial's configuration, she may press the start button at the bottom of the panel, which instructs LabWiki's experiment widget to post a request for trial execution to an external JobService instance.

9.2 The JobService and Its Scheduler

Our LabWiki workspace de-couples the frontend interface used to develop and interact with experiment artifacts from the backend processes orchestrating the execution of an experiment trial. The JobService software is the backend entity in charge of supervising this execution. This decoupling enables our tools to cater for a wide range of usage scenarios, such as use of an alternative user frontend, automated trial request (e.g., software can request a given trial to be run at a periodic time), optimization of a shared pool of resources among trial requests from the same project.

The JobService provides a REST API, which allows clients such as a LabWiki instance to post trial requests (i.e. experiment OEDL scripts, property configuration). Each request is passed to an internal scheduler, which queues it and periodically decides which job to run next. This scheduler function is a plugable module of the JobService, thus a third party deploying its own JobService may define its own scheduling policies. In its simplest form, our default Scheduler is just a plain FIFO queue. However, in an education context it could be a more complex function, which could allow a lecturer to optimize the use of a pool of resources (allocated as in Sect. 8) between parallel experiment trials submitted by multiple groups of students. The JobService's REST API also allows a client to query for the execution status and other related information about its submitted trials. LabWiki uses this feature and displays the returned information in its "Execute" panel once the trial execution has started.

9.3 Orchestrating Resources

The JobService uses the existing OMF framework [17], which is available on many GENI facilities to orchestrate experiment trials. More specifically, when the JobService's Scheduler selects a given trial request for execution, it starts an OMF Experiment Controller process (EC). This EC interacts with a Resource Controller (RC) running on each of the involved resources, and have them enact the tasks required in the experiment description. This interaction is done via the FRCP protocol (Sect. 5). While the trial is being executed, the JobService constantly monitors the information from the output of the EC process and uses it as part of the status provided back to LabWiki. While our current JobService uses OMF for its underlying experiment trial execution, its design also permits the use of other alternative frameworks, such as NEPI [8].

9.4 Collecting Measurements

The applications instrumented in Sect. 7.1 inject measurement streams from measurement points as selected by the experiment description (e.g., Listing 1 line 16 or 27). In Fig. 6, the reporting chain is terminated by a collection point. The OML suite [12] provides an implementation of this element in the form of the `oml2-server(1)`. It accepts OMSP streams on a configurable TCP port, and stores the measurement tuples into SQL database backends.

In our LabWiki workspace, the EC instructs the applications on the location of the collection points to report their MSs to. One database is created per experiment, and a table is created for each MP (regardless of how many clients report into this MP). The `oml2-server` currently supports SQLite3 and PostgreSQL databases, and there are plans to extend this to semantic and NoSQL databases.

For very large deployments, the OML collection can be scaled by running multiple `oml2-servers` behind a TCP load balancer such as HAProxy.[22] Instrumented applications carry all the necessary state information in the initial connection for any server to create the tables and store the reported tuples. A PostgreSQL cluster can then be used as a backend to store data into a single centralised logical location where analysis tools can access data both in real time and retroactively.

10 Result Analysis Over Multiple Trials

As mentioned in Sect. 6, it is often necessary to execute several trials of a given experiment to gather sufficient data for a meaningful analysis. LABWIKI facilitates these replicated trials, and provides two options for the experimenter to use the produced data.

First, the "experiment" widget provides an "Export" button once the respective trial has terminated. Depending on the LABWIKI configuration, data can be exported either as a self-contained database file (SQLite3 or PostgreSQL dumps), as a compressed (e.g. ZIP) archive of comma separated (CSV) formatted files, or pushed into an existing iRODS data store. LABWIKI's plugin-based design allows other third-party export widgets to be provisioned as requested. The experimenter may then download the produced database and import it into her preferred data analysis software.

LABWIKI provides another alternative to interact with the produced data through another widget, which interfaces with a separate R server. This widget allows the experimenter to nominate an R script to be submitted to the R server, which executes the script's instructions and returns any outputs (text or graphic results) to the widget for display. Figure 10 shows a screenshot of this widget being used to analyze data from our example experiments.

In many cases, the result analysis of multiple trials will provide new insights into the subject of the study. The experimenter may then reflect on these conclusions and decide to either run further trials with an updated experiment design, description, used resources, and/or analysis as required. This step effectively closes the experiment workflow process as described in Sect. 1.

11 Store and Publish

As an experiment-based study progresses, it is essential to ensure that all generated artifacts, such as documents, data, and analysis, are permanently stored for continued and future access. Furthermore, once a study finishes, the experimenter should be able to easily share her findings with the community.

[22]http://www.haproxy.org/.

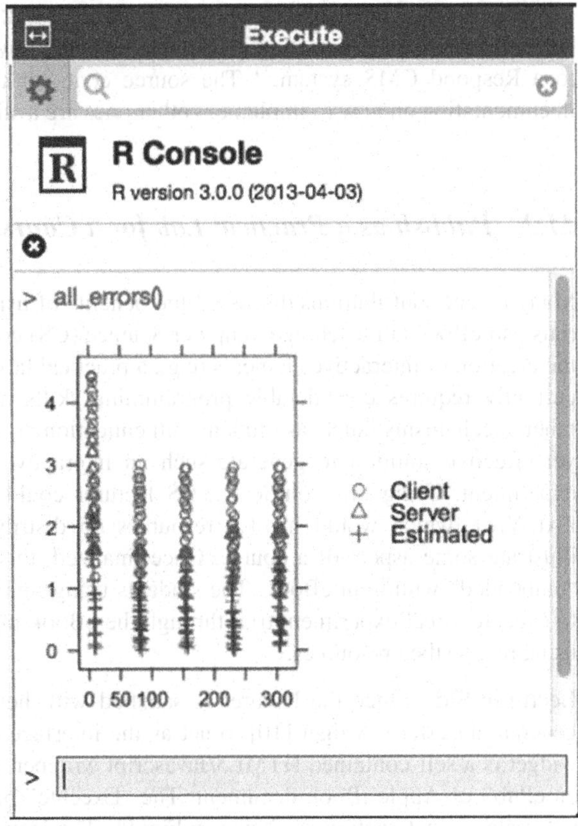

Fig. 10 LabWiki's analysis widget

11.1 Storing and Sharing with LabWiki

As mentioned previously, LabWiki supports different types of artifact stores. This allows different operators to deploy their own LabWiki instance, to select the storage solution that best suits their requirements. For example, in an evaluation or small closed institution context, a plain directory hierarchy based on the underlying file system may be enough. In contrast, for a large community spanning multiple organizations (e.g. GENI), a solution offering fine- grain access control, high service availability, and version control may be more appropriate. A list of the supported artifact stores are available in our source code repository, together with LabWiki's "Repository" API, which could be used to add interfaces to other storage alternatives.[23]

To facilitate the sharing of generated documents, LabWiki's "wiki" plugin has a pluggable export component, which allows the publication of its documents to an external Content Management System (CMS). This feature is accessed through

[23]http://mytestbed.github.io/labwiki2/.

an "Export" action displayed at the top of LABWIKI's "Plan" panel. There is currently only one export plugin available, which allows the sharing of documents to a Respond CMS system.[24] The source code for this component along with documentation on how to implement other ones are available in our repository.[25]

11.2 Publish as a Practical Lab for a Course

Many recent contributions discussed the benefits of using interactive online materials and eBooks in teaching Computer Science (CS) courses [6, 21, 22]. However, the creation of interactive elements (e.g., a practical lab) within an eBook platform currently requires considerable programming skills as well as familiarity with other mechanisms, such as student authentication. LABWIKI provides a simple yet effective solution to generate such an interactive element from an existing experiment. Using this solution, a CS lecturer could develop an experiment in LABWIKI, which would involve resources on distributed testbed facilities and illustrate some aspect of a course. Once finalized, this experiment could then be "embedded" within an eBook. The students using this eBook would then be able to execute a real experiment trial through the eBook interface, i.e., a trial actually using real testbed resources.

Lecturer Side Once the lecturer is satisfied with her experiment, she needs to generate an eBook widget [10] to act as the interface to that experiment. Such a widget is a self-contained HTML5/Javascript wrapper, which can be embedded in an ePub3 or Apple iBook document. The "Execute" panel of LABWIKI provides a "create widget" action as shown in Fig. 11. The lecturer triggers this action and provides some configuration parameters for the widget to be generated (e.g., a name, the display size of the widget in pixel, the set of allocated resources). The "create widget" process then generates a fully configured widget, which will be downloaded to the lecturer's machine. She can then include this HTML5/Javascript widget into her eBook using third-party authoring tools.

Student Side A student may then download the eBook and open the page with the forementioned widget. Once triggered inside the eBook, the widget will switch to a full-screen web container connected to the remote LABWIKI workspace. From then on, the student has access to a subset of the previously mentioned LABWIKI features, namely the experiment trial execution (Sect. 9), the result analysis (Sect. 10), and the publishing (Sect. 11.1) features. Thus, she is not allowed to modify the experiment description or the set of resources to use. Once the experiment trial finishes, she may perform any result analysis requested by the lecturer as part of the practical lab, and submit the answers as a LABWIKI generated document (Sect. 9.1).

[24] http://respondcms.com/.

[25] http://git.io/labwiki-plan.

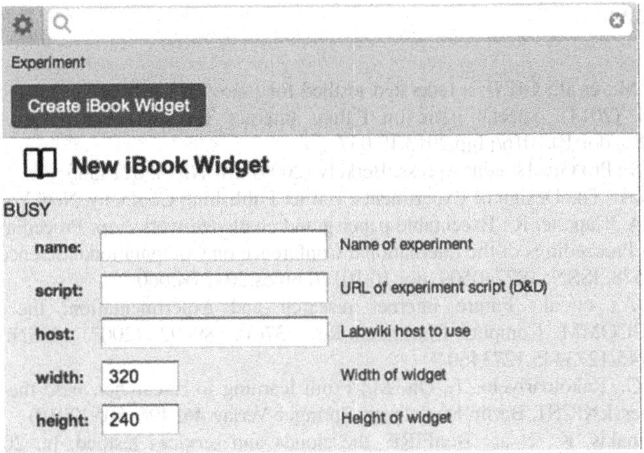

Fig. 11 Creating an eBook Widget

12 Conclusion

This chapter described a methodology based on well established principles to plan, prepare and execute experiments. We proposed and described a family of tools, the LABWIKI workspace, to support an experimenter's workflow based on that methodology. LABWIKI enables repeatable experiment-based research in Computer Networking, Distributed Systems, and to certain extends Computer Science in general. We showed how this set of tools leverages large-scale Future Internet initiatives, such as GENI and FIRE, and de-facto protocol and model standards, which emerged from these initiatives. It provides services and mechanisms for each step of an experiment-based study, while automatically capturing the necessary information to allow peers to repeat, inspect, validate and modify prior experiments. Finally, the LABWIKI workspace also provides tools for sharing all generated artifacts (e.g. documents, data, analyses) with the community. For educators, a seamless mechanism to turn an experiment in an interactive practical lab for teaching Computer Science is provided.

Acknowledgements NICTA is funded by the Australian Government through the Department of Communications and the Australian Research Council through the ICT Centre of Excellence Program. This material is based in part upon work supported by the GENI (Global Environment for Network Innovations) initiative under a National Science Foundation grant.

References

1. Berman, M., et al.: GENI: a federated testbed for innovative network experiments. Comput. Netw. **61** (2014). Special issue on Future Internet Testbeds—Part I, pp. 5–23. ISSN: 1389–1286. doi:10.1016/j.bjp.2013.12.037
2. Chacon, S.: Pro Git, 1st edn. Apress, Berkely (2009). ISBN: 1430218339
3. Fisher, R.A.: The Design of Experiments. Hafner Publishing Company, New York (1937)
4. Gabriel, A., Capone, R.: Executable paper grand challenge workshop. Procedia Comput. Sci. **4** (2011). Proceedings of the International Conference on Computational Science, ICCS 2011, pp. 577–578. ISSN: 1877–0509. doi:10.1016/j.procs.2011.04.060
5. Gavras, A., et al.: Future internet research and experimentation: the FIRE initiative. SIGCOMM Comput. Commun. Rev. **37**(3), 89–92 (2007). ISSN: 0146–4833. doi:10.1145/1273445.1273460
6. Jourjon, G., Rakotoarivelo, T., Ott, M.: From learning to researcher, ease the shift through testbed, ser. LNICST, Berlin Heidelberg: Springer-Verlag **46**, 496–505 (2010)
7. Kavoussanakis, K., et al.: BonFIRE: the clouds and services testbed. In: 2013 IEEE 5th International Conference on Cloud Computing Technology and Science (CloudCom), vol. 2, pp. 321–326 (2013). doi:10.1109/CloudCom.2013.156
8. Kim, Y.-H. et al.: Enabling iterative development and reproducible evaluation of network protocols. Comput. Netw. **63**, 238–250 (2014)
9. Krishnamurthy, B., Willinger, W., Gill, P., Arlitt, M.: A socratic method for validation of measurement based networking research. Comput. Commun. **34**, 43–53 (2011)
10. Langer, M.: iBooks Author: Publishing Your First eBook. Flying M Production, USA (2012)
11. Medina, A., Lakhina, A., Matta, I., Byers, J.: BRITE: an approach to universal topology generation. In: Proceedings of the International Workshop on Modeling, Analysis and Simulation of Computer and Telecommunications Systems, MASCOTS'01 (2001)
12. Mehani, O., Jourjon, G., Rakotoarivelo, T., Ott, M.: An instrumentation framework for the critical task of measurement collection in the future internet. Comput. Netw. **63** (2014). ISSN: 1389–1286. doi:10.1016/j.bjp.2014.01.007
13. Mehani, O., Taib, R., Itzstein, B.: Time calibration in experiments with networked sensors. In: Proceedings of IEEE of the 39th Local Computer Networks Conference (LCN) (2014). ISBN: 978-1-4799-3780-6/14
14. Mills, D., Martin, J., Burbank, J., Kasch, W.: Network time protocol version 4: protocol and algorithms specification. RFC 5905 (2010)
15. Paxson, V.: Strategies for sound internet measurement. In: The Internet Measurement Conference (IMC) (2004)
16. Rajasekar, A., et al.: iRODS primer: integrated rule-oriented data system. In: Synthesis Lectures on Information Concepts, Retrieval, and Services 2.1. Morgan and Claypool Publishers (2010). doi:10.2200/s00233ed1v01y200912icr012
17. Rakotoarivelo, T., Ott, M., Jourjon, G., Seskar, I.: OMF: a control and management framework for networking testbeds. SIGOPS Oper. Syst. Rev. **43**(4), 54–59 (2010). ISSN: 0163–5980. doi:10.1145/1713254.1713267
18. Rakotoarivelo, T., Jourjon, G., Ott, M.: Designing and orchestrating re-producible experiments on federated networking testbeds. Comput. Netw. **63** (2014). doi:http://dx.doi.org/10.1016/j.bjp.2013.12.033
19. Römer, K., Blum, P., Meier, L.: Time synchronization and calibration in wireless sensor networks. Handbook of Sensor Networks. Wiley, New York (2005). Chap. 7. doi: 10.1002/047174414x.ch7

20. Vandenberghe, W., et al.: Architecture for the heterogeneous federation of future internet experimentation facilities. Future Network and Mobile Summit. (2013). ISBN: 978-1-905824-37-3
21. Wright, A.: Tablets over textbooks? Commun. ACM **55**(3), 17–17 (2012)
22. Zhuang, Y., et al.: Taking a walk on the wild side: teaching cloud computing on distributed research testbeds. In: Proceedings of the 45th ACM Technical Symposium on Computer Science Education. SIGCSE '14. Atlanta, GA, pp. 535–540. ACM, New York (2014). ISBN: 978-1-4503-2605-6. doi:10.1145/2538862.2538931

GENI in the Classroom

Vicraj Thomas, Niky Riga, Sarah Edwards, Fraida Fund, and Thanasis Korakis

Abstract One of the great successes of GENI has been its use as a remote laboratory by instructors of networking, distributed systems and cloud computing classes. It allows instructors to provide hands-on learning experiences on a real, large-scale network. Reasons for this success include GENI's ease of use, access to resources such as programmable switches and wireless base stations that are not ordinarily available at most schools, support for collaborative experimentation and ease of recovering from mistakes. The GENI community has created and made available to instructors ready-to-use exercises based on popular networking textbooks. These exercises cover a range of topics from basic networking to advanced concepts such as software defined networking and network function virtualization. They include wired and wireless networking based exercises. GENI is also used as a platform for applications that enhance STEM education at the high-school level and as a platform for MOOC courses that use an interactive approach to teach Internet concepts to non-computer scientists.

One of the great successes of GENI has been its use as a remote laboratory by instructors of networking, distributed systems and cloud computing classes. As of the Summer of 2015, over 2500 students from 46 different institutions had used GENI for their classwork.

GENI allows instructors to provide students with hands-on learning experiences on a real, large-scale network. The value of hands-on experimentation to reinforce concepts taught in the classroom has been well understood [2]. Research has shown that lab courses significantly enhance learning [7, 11] and help students develop problem-solving and critical-thinking skills [1].

V. Thomas (✉)
GENI Project Office, Raytheon BBN Technologies, 5775, Wayzata Blvd., Suite 630 St. Louis Park, MN 55416, USA
e-mail: vthomas@bbn.com

N. Riga • S. Edwards
GENI Project Office, Raytheon BBN Technologies, Cambridge, MA, USA

F. Fund • T. Korakis
NYU School of Engineering, New York City, NY, USA

© Springer International Publishing Switzerland 2016

R. McGeer et al. (eds.), *The GENI Book*, DOI 10.1007/978-3-319-33769-2_18

The educational benefits of using research testbeds such as GENI are described in [13]. The benefits of using GENI in particular include:

- *Cost.* There is no fee to use GENI for research and education. This is of tremendous value to instructors because the construction, maintenance and operation of a laboratory for use in networking and distributed systems classes can be prohibitively expensive for most educational institutions.
- *Ease of use.* GENI experimenter tools have been designed with the philosophy of making it easy to run simple experiments and possible to run complex experiments. This emphasis on ease of use is important so students are spending time learning networking concepts and not mastering GENI tools. In the case of the GENI-based Massive Open Online Courseware (MOOC) described in Sect. 2, students with no background in Computer Science can run experiments on GENI without learning any GENI tools or programming GENI resources.
- *Accessibility.* GENI is available around the clock and accessible by students from almost anywhere with Internet connectivity. This makes GENI a convenient platform, not only for students, but also for faculty as they do not need to worry about keeping laboratories open and staffed.
- *Access to rare or expensive resources.* GENI has resources such as wireless base stations, programmable switches and Ethernet VLANs over national research networks that are expensive or not commonly available. This allows instructors to present students with opportunities for experimentation that wouldn't normally be possible.
- *Low latency and high bandwidth access to resources.* GENI resources are distributed across the United States and these resources are connected by high-speed networks. This makes possible education applications such as those described in Sect. 3.
- *Support for Collaboration.* GENI support for collaborative research [Chapter [6]], where multiple experimenters can belong to the same slice and can operate on resources in the slice, is well suited for education. Instructors and teaching assistants have access to student slices: They can run the students' experiments to grade them and log into their resources to view scripts or other configuration to help debug or grade. In addition, instructors can assign group assignments.
- *Ease of recovering from mistakes.* When students make mistakes while programming or configuring a GENI resource, they can simply delete it and start over without help from an instructor or systems administrator. This encourages exploratory learning as students can try things without worrying about causing damage or inconveniencing others.
- *Community support.* The community of GENI researchers and educators has developed ready-to-use courseware and exercises for use in classes. This is attractive to many instructors and, in fact, a large proportion of instructors using GENI rely on this courseware.

1 Instructor Resources

Resources available to instructors using GENI in their classroom include:

1. *Instructor pages on the GENI Wiki.* The GENI Wiki has a section for instructors with links to ready-to-use exercises and tips for running a class on GENI.
2. *Ready-to-use course modules.* Educators in the GENI community have developed course modules and made them available for use by others. Course modules include instructor guides, student handouts and, in some cases, videos introducing the exercise. Section 1.1 has more information on these modules.
3. *Wireless Classroom as a Service.* Two of the authors, Fraida Fund and Thanasis Korakis of the New York University School of Engineering, have developed a hosted service for instructors running wireless classes on GENI. This service, described in Sect. 1.2, includes a web portal customized for the class and login access for student to run experiments and for instructors to monitor progress of the students and testbed status.
4. *Tutorials for TAs and Instructors.* Train-the-TA is a web-based tutorial run at the start of each semester. This is a two-part tutorial run over two weekday afternoons using an online web conferencing application such as WebEx. The first part includes a presentation introducing GENI and hands-on tutorials for those with no prior GENI experience. Those with prior GENI experience usually skip the first part. The second part covers topics related to running a class on GENI: Creating a GENI Project for the class, bulk-adding students to the project and helping debug student experiments. Also covered are resources such as course modules available to instructors.
5. *Online discussion group.* The *geni-educators* Google Group [8] is an online forum for discussion related to the use of GENI in education. This forum is also used to announce new GENI-based courseware.

1.1 GENI Course Modules

The GENI-based course modules available on the instructor section of the GENI wiki may be used to teach a variety of networking concepts ranging from an introduction to TCP and IP to more advanced concepts such as software defined networking and future Internet architectures. Many of these course modules have been designed for use with the popular networking textbook by Kurose and Ross [10].

Most of the GENI course modules for the typical graduate and undergraduate networking classes have been developed by Kevin Jeffay and Jay Aikat of the University of North Carolina at Chapel Hill; Mike Zink of the University of Massachusetts, Amherst; and Sonia Fahmy and Ethan Blanton of Purdue University. Two of us, Thanasis Korakis and Fraida Fund, have developed modules for teaching networking concepts to people with no background in computer science. These

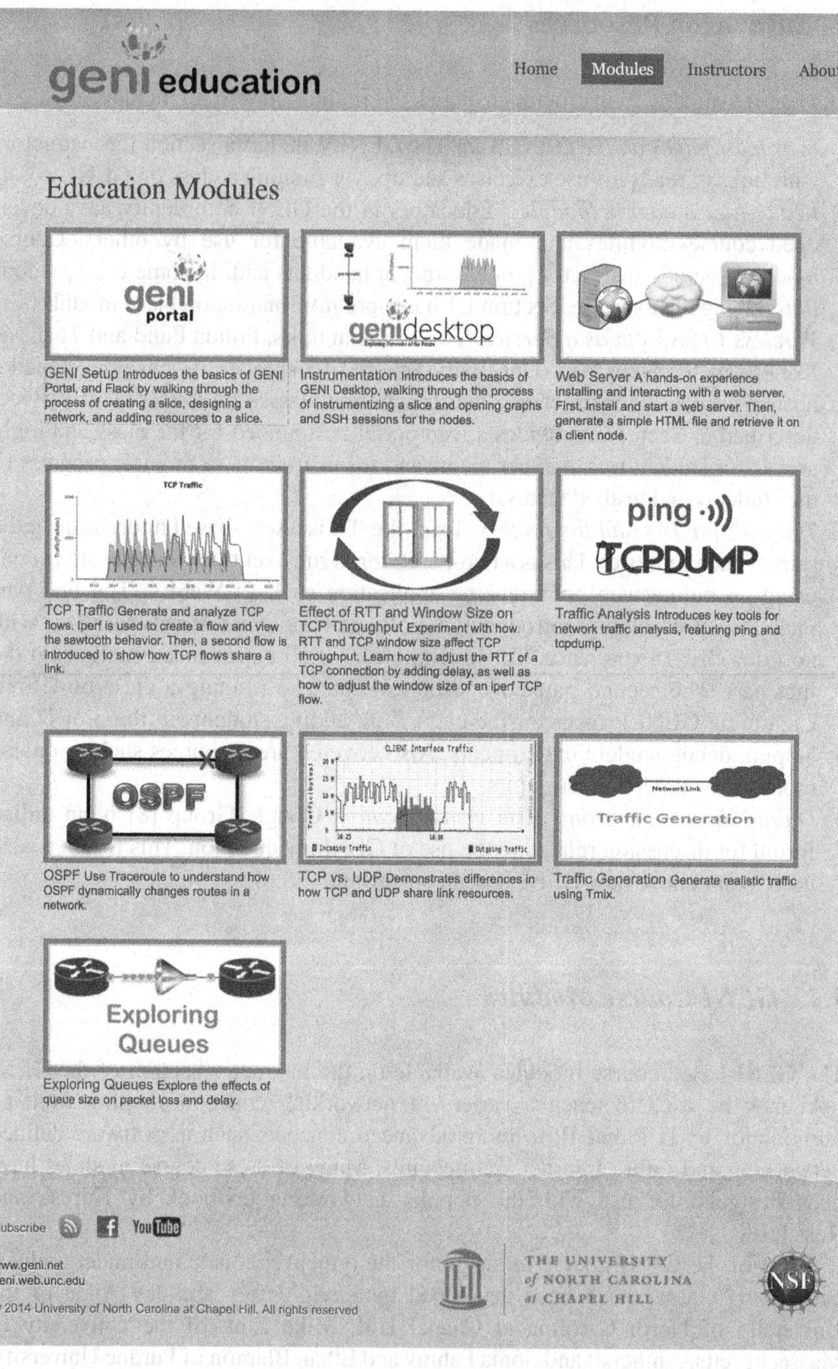

Fig. 1 GENI-based course modules developed by Jay Aikat and Kevin Jeffay of the University of North Carolina at Chapel Hill

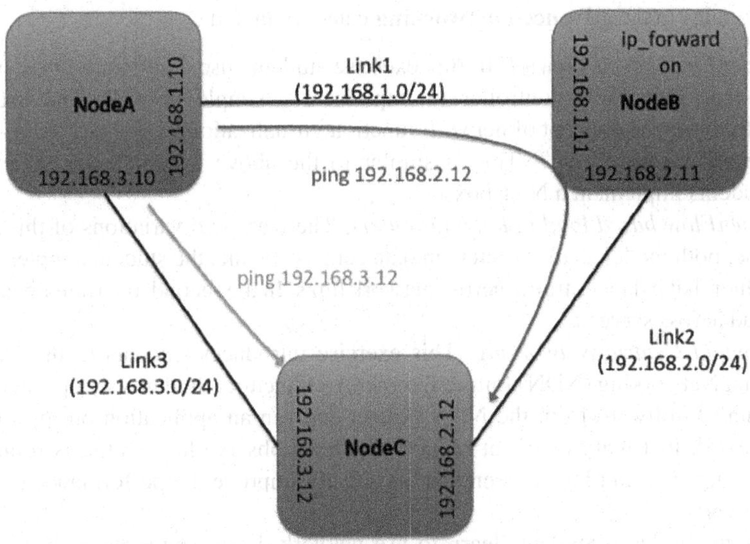

Fig. 2 This popular course module has students setting up static IP routes in a three-node network

modules, developed in MOOC format, are described in Sect. 2. Figure 1 shows course modules developed by Kevin Jeffay and Jay Aikat of the University of North Carolina at Chapel Hill.

The course modules fall into four broad categories: (1) Networking basics, (2) Advanced networking, (3) Distributed systems and (4) GENI tools. Modules in the **networking basics** category include:

1. *IP forwarding.* Students implement IP forwarding by setting up static routes in a 3-node topology. This is the most widely used of all the modules. Figure 2 shows the topology and the routes the students are asked to set up.
2. *TCP congestion control.* This exercise illustrates the TCP sawtooth behavior and how competing TCP flows share a link.
3. *Effect of RTT and window size on TCP throughput.* This module allows students to experiment with how RTT and TCP window size affect TCP throughput. They will learn how to adjust the RTT of a TCP connection by adding delay, as well as how to adjust the window size of an iperf TCP flow.
4. *FTP vs. UDT.* Students see the difference between TCP and UDP by comparing the behavior of two different file transfer applications (FTP and UDT) under varying network conditions.
5. *Learning switch.* Students implement a five-port Ethernet learning switch. The exercise uses the OVS software switch and an OpenFlow controller to implement the learning switch algorithm.

Modules in the **advanced networking** category include:

1. *OpenFlow based firewall.* In this exercise students use a software OVS switch and an OpenFlow controller to implement a simple firewall. The exercise introduces the concept of network function virtualization.
2. *OpenFlow based NAT.* This is similar to the above firewall exercise, except students implement a NAT box.
3. *OpenFlow based load balancing routers.* There are two variations of this exercise, both modeled after routers in data centers. In one, the students implement a router that balances traffic across network links. In the second, the router balances load across servers.
4. *Content Centric Networking.* This exercise introduces students to the Named Data Networking (NDN) Future Internet Architecture. Students set up a network running software from the NDN project and run an application on top of this network that makes content request. Students observe how traffic is routed in this network and how content caches greatly improve the performance of such networks.
5. *Traffic analysis.* Students learn to use network debugging tools such as TCP-Dump. Network traffic is generated using the Tmix tool

Modules that teach **distributed systems** concepts include:

1. *Web Server.* Students learn to install, configure and interact with a web server.
2. *VLC/Dynamic adaptive system.* This module how Dynamic Adaptive Streaming over HTTP (DASH) works. GENI instrumentation and measurement tools are used to visualize protocol parameters such as VLC decision bit rate (actual measured bit rate) and VLC empirical rate (instant measured throughput).

Modules that teach **GENI tools** include:

1. *Lab zero.* A simple 2-node experiment to get students started with GENI. The objective is to ensure students are able to log into GENI, understand the basics of running experiments on GENI and are able to log into their GENI resources. Most instructors have the students do this exercise in the classroom.
2. *Lab one.* This is used by instructors who want their students to gain an understanding of how GENI works and get familiar with concepts of RSpecs and the GENI Aggregate Manager API. Students also learn to use the Omni command line tool.
3. *GENI Desktop.* This module introduces the basics of the GENI Desktop, walking through the process of instrumenting a slice and opening graphs and SSH sessions for the nodes.

1.2 GENI Wireless Classroom as a Service: Testbed-Hosted Lab Exercises to Challenge Students' Assumptions About Computer Networks

For many students in Computer Science and Electrical Engineering, the first course on computer networks is also their last. While some students will complete later project work involving distributed systems or networked applications, for a subset of students this course is the last explicit instruction they will get on the subject of networks.

This first course on computer networks typically covers Internet architecture, services, and protocols. Accompanying lab assignments (using traditional in-house laboratories or on GENI) tend to focus on gaining practical experience with protocols (such as TCP, UDP, IP, HTTP, NAT, DNS, ARP, SMTP, and DHCP) and/or tools (like `dig`, `wireshark`, `ping`, and `traceroute`). However, while students may gain a good understanding of the mechanics of the protocols from this course, they often leave with a limited intuition and possibly false assumptions about their behavior of *real* networks [5, 9].

The GENI wireless lab assignments are intended to help students develop their intuition regarding the behavior of networks, by explicitly exposing them to the falsehood of commonly held assumptions in a "production" cellular network. There are two major reasons for using cellular networks as a platform for educating students about computer networks:

- They are increasingly popular as access networks.
- They are inherently subject to "interesting" complications, such as random packet loss, latency, congestion, mobility, and asymmetry.

Deploying a cellular infrastructure exclusively for use in an introductory computer networks lab is an inordinately expensive task for most instructors. Fortunately, GENI includes a set of cellular network aggregates which are available for general use in research and education. We have created a set of lab resources that are integrated with this platform, and are therefore available to educators without local lab infrastructure. Furthermore, we have made these resources available in a *Classroom as a Service* (CaaS) format, in which the material for the lab exercises, the integrated learning management elements, and the experimental infrastructure on which they run are all hosted on a GENI testbed site.

Design Considerations

Other instructors' experiences using GENI in education [3, 12, 15] and early iterations of this lab suggested a number of potential pitfalls associated with using testbeds in a lab setting. These include students' lack of basic experimental skills, the instructors' blindness into what students are doing on the testbed platform, and the difficulty of scaling exercises to work with large groups of students. The intent of the Classroom as a Service delivery model is to mitigate some of these issues.

Students' lack of experimental skills Junior/senior undergraduates or first-year M.S. students in computer science and electrical engineering are a highly heterogeneous group in terms of skills and backgrounds. In a pre-lab survey of students registered for the lab at our own institutions, close to 20 % of students said they had never used Linux before, and a similar portion said they were not comfortable writing or editing code in a scripting language. An even larger group said they could not use a non-GUI text editor. Only two students in our sample had ever used Ruby, the language that GENI experimenter tools such as OMF and LabWiki are based on.

Because our goal is to teach students about networks, *not* about testbeds, and because students do not necessarily have any experimental skills, we chose to minimize the number of testbed-specific skills required to complete the labs. We want students to expend intellectual energy reasoning about and revising mental models of network behavior, not learning how to use a highly specialized experimental platform.

Therefore, the procedure for *every* lab experiment involves the following sequence of steps:

1. Log into a remote Linux host over SSH.
2. Download a script from a given URL with `wget`.
3. Submit the script to an experiment scheduler (e.g., `omf exec /home/jdoe/orientation`)
4. Identify the unique ID assigned to the experiment in the output of the previous command.
5. Run a command (with the experiment's unique ID as the argument) to send results to a data storage site.
6. Use an online GUI integrated into the data storage system to visualize experiment results.

Experiments may require multiple iterations of steps 4–6 with different arguments. For steps 2, 3, and 5, the exact command is given to students in the lab manual, so that they may copy and paste it directly into the terminal. For step 6, explicit instructions are given for each plot that students are asked to create.

Notably, we do not ask students to reserve or provision resources, write or edit any code (whether for running experiments or visualization of results), or actually perform themselves the commands and configurations that the experiment requires.

Opacity of testbed platform to instructors With an on-site lab, instructors are able to collect information on student behavior directly from the lab infrastructure. In a testbed environment, however, the instructor of the course is not usually the same person as the manager of the lab infrastructure. The instructors are therefore limited in their ability to understand what students are doing successfully and unsuccessfully on the lab platform.

This is especially problematic in the event of an infrastructure failure, (which is inevitable, to some degree or another). Especially for students, who typically lack the experience to differentiate between an error on their part and an infrastructure failure, these can be tremendously frustrating. When lab exercises run on off-site infrastructure, instructors can not monitor and deal with these failures effectively.

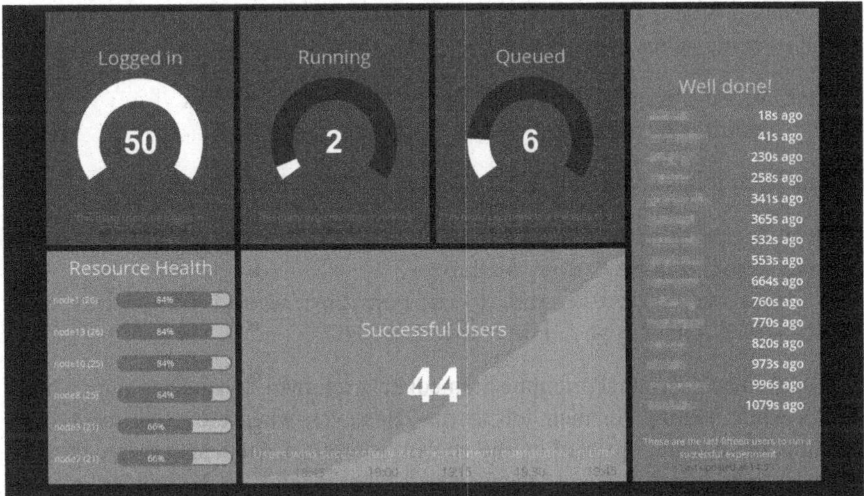

Fig. 3 During a lab session, instructors can see in real time the number of students logged in to the testbed console, the number of experiments currently running, the number of experiments submitted and waiting for available resources to run, the usernames of the students who most recently ran a successful experiment, and the "health" of the testbed resources used in the lab (in terms of the ratio of success of experiments run on each resource)

To mitigate this, we instrument the experiment controller to collect and store information about experiments that run during a group session, and provide a dashboard with which instructors and students can monitor the health of the infrastructure (Fig. 3).

Scaling to Large Class Sizes

Because of the difficulty associated with isolating wireless experiments that are co-located in space and frequency, wireless resources on GENI are usually reserved for one researcher at a time (i.e. in time quantums of 1 h or 30 min). In a classroom setting, it may be necessary for dozens of students to run an experiment involving a wireless resource within an hour or less.

To enabling sharing of testbed resources between all students in a lab section, we created a job scheduler as a transparent wrapper around OMF-based testbed utilities. The job scheduler is built on the standard `torque` batch processing system. Each testbed node is represented as a queue in the `torque` system. At the beginning of the lab session, an instructor enables some number of nodes, depending on how many concurrent experiments may be supported by the wireless channel.

When a student runs

```
omf exec /home/jdoe/script
```

this command is intercepted by a wrapper script. The wrapper checks the status of all the running queues, and assigns the experiment to the queue with the shortest length (breaking ties randomly). It does this by appending a parameter to the command, instructing the OMF experiment controller to run the experiment on the specified compute resource.

```
omf exec /home/jdoe/script -- --node nodeX |
    qsub -q nodeX
```

Finally, it prints an informational message to the student's terminal:

```
Job 5915 queued on node11:
omf exec /home/jdoe/script
3 jobs are queued and 1 is running ahead of yours.
To kill this job, run "qdel 5915"
```

The student's terminal prompt is immediately returned, so students can do other things while waiting for their job to run. However, when the experiment starts running, the output is printed on the student's terminal. Students can monitor the status of their experiment and cancel queued experiments with the standard `qstat` and related commands or using a web interface.

1.2.1 Lab Modules

To date, over 500 students have completed one or more lab exercises using the CaaS offering across six universities. They have been used most often in courses on computer networks. For a course taking a bottom-up layered approach to computer networks, we have used the following sequence:

1. Introduction and lab orientation
2. Link adaptation (PHY layer)
3. QoS of wireless networks (MAC layer)
4. IP, mobility, and handover (IP layer)
5. TCP congestion control variants (TCP layer)
6. Adaptive video streaming (application layer)

For a course taking a top-down layered approach to computer networks, we reverse the order of the last five lab assignments.

For a course on wireless systems, we have used the following sequence:

1. Introduction and lab orientation
2. Wireless signal propagation
3. Radio receiver hardware impairments
4. Link adaptation
5. QoS of wireless networks

We have also used selected lab modules in courses where students work on a design project. For example, we have had students complete the "Introduction and lab orientation" and "Adaptive video streaming" and then complete a project in which they design and test a new adaptive video delivery policy. In another course, students complete lab assignments including "Introduction and lab orientation" and "Radio receiver hardware impairments," then design a software radio pair to compete against other students' designs in a format modeled after the DARPA Spectrum Challenge.

2 GENI MOOC: Expanding Access to Lab Experiments in Computer Networks

There is a growing interest in offering some Computer Science education to non-majors, K-12 populations, under-resourced institutions, and other demographics beyond traditional Computer Science undergraduates. Computer networks are a natural choice of topic for material aimed at non-majors, given the increasing importance of the Internet in daily life.

Hands-on education in computer networks typically involves software tools such as ns-3 and Mininet, or real or virtual laboratories or testbeds such as GENI. However, mainstream GENI educational efforts remain available to a very limited population: mostly, students enrolled in higher education in the United States, whose instructors are aware of GENI, and have overcome the not-insignificant hurdles of becoming familiar with the platform and incorporating it into their course materials. This excludes a sizable group of learners potentially interested in Internet, computer networks, and related topics. Also, like the traditional computer networks laboratories and software tools, most GENI-based lab experiments require students to already have some level of technical ability and prerequisite knowledge (e.g. basic knowledge of computer networks and ability to use Linux, SSH, and other technical tools.)

The goal of the GENI MOOC project is to create learning opportunities on GENI with a very low barrier to entry, accessible to the "masses."

2.1 Design Considerations

To ensure a low barrier to access, the GENI MOOC system must meet the following criteria:

- Experiments should be open to anyone who wants to learn, without requiring academic affiliation or supervision.
- Participants should interact with the system entirely within a web browser, without needing to set up SSH keys or use special software.
- It must be possible for many users to run an experiment concurrently with minimal use of GENI resources.
- The material should be suitable for a lay audience, and cover topics of interest to a broad audience of Internet users.

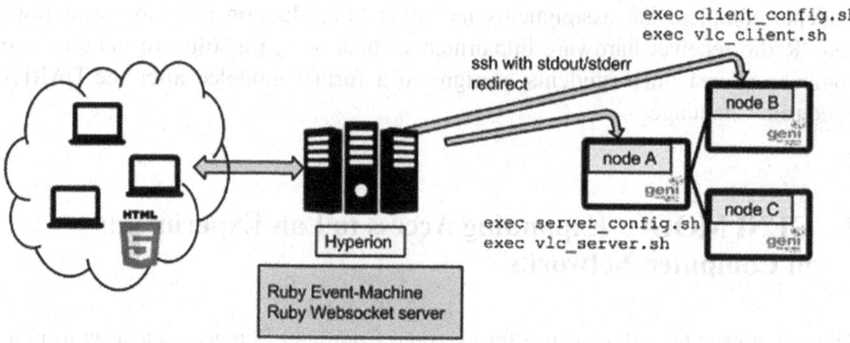

Fig. 4 GENI MOOC users interact with experiments entirely from a browser, which opens a websocket to communicate with an experiment controller on the MOOC server (`hyperion`). The experiment controller executes commands on GENI resources and returns results back to the experimenter's browser

Under normal circumstances, to access GENI resources, a GENI user must join a project supervised by a project lead, who is a principal investigator, professor or lead researcher. This is necessary to provide accountability and prevent abuse. However, this serves as a barrier to many potential learners, who may not have any academic affiliation and may not have a relationship with someone eligible to be a project lead.

To eliminate the need for supervision and also the need for specialized software, the GENI MOOC exposes a limited web-based interface to experimenters. Figure 4 shows the system architecture. A student can tune limited experiment parameters and initiate experiment execution from the browser (Fig. 5), but cannot log in to the resources directly or reserve new resources. The predefined experiment runs with the experimenter's requested configuration from a server-side experiment controller that accesses the GENI resources using the SSH credentials of the MOOC module's instructor. Experiment results are displayed to the user in tabular, graphical, or video form (Fig. 6).

The experiment controller on the MOOC server is responsible for tracking use of GENI resources and avoiding interference between concurrent experiments. The MOOC experiments are selected specifically for their ability to run with limited use of network and computing resource, so that one instance of the experiment configuration on GENI can support twenty or more concurrent experiments. During high-demand times, the MOOC experiment controller can bring up new instances of the experiment configuration on GENI to support larger numbers of concurrent experiments.

The GENI MOOC project features easy, accessible experiments that address questions an average Internet user might have.

- **Internet routing:** How does web traffic find its way from a remote site to my computer?

The various scenarios are summarized in the following table:

Scenario	Cost	Link Capacity	Other Traffic
Low Priority	$$	5	xxxx
High Priority	$$$$$$$$	10	xxx
Open Internet	$$	15	xxxxxxx

YOUR CHOICE

- (selected) Pay extra to use the fast lane
- Don't pay extra, use the slow lane

Fire Experiment

Fig. 5 In the Net Neutrality experiment shown here, the experimenter is asked to make a decision from the point of view of a content provider who, given a hypothetical scenario in which net neutrality regulations are not place, must decide whether or not to pay for priority access to customers

You selected not to pay for a priority link. Here's the outcome of this decision.

Regulation	Revenue	Expenses	Profit
No net neutrality	10901.74	113.07	10788.67
Net neutrality regime	12624.89	332.92	12291.97

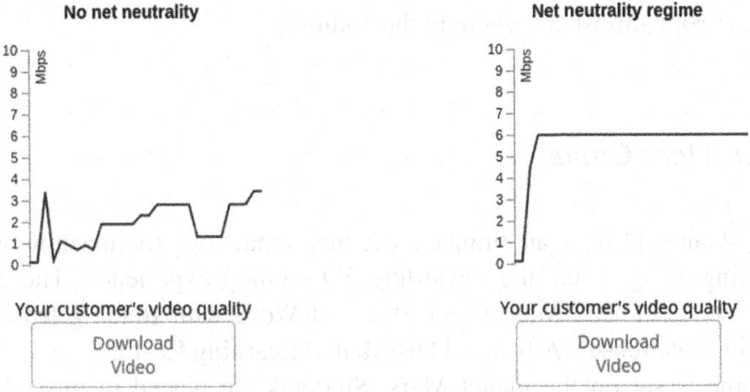

Fig. 6 Experiment results are displayed to a user in tabular, graphical, or video form. In the Net Neutrality experiment shown here, users can see relative profits of a hypothetical content provider and the video quality experienced by the content provider's customers under different net neutrality regulations. Users can also download and play back the video file from each scenario

- **Net neutrality:** What problems are net neutrality regulations trying to solve?
- **Adaptive video:** Why do I see visible quality changes when I play back Internet video?
- **Internet anonymity:** Is it possible to be *truly* anonymous on the Internet?
- **Big data with Hadoop and MapReduce:** How do companies like Amazon and Google process the massive amounts of data users generate?

The material is suitable for students who want an introduction to research topics in networking, instructors who use the experiments as in-class demonstrations or homework assignments, and anyone who wants to learn more about how the Internet works.

3 GENI in K-12 STEM Education

GENI resources have been used to enhance STEM education at the high school level. Educators use GENI compute and networking resources to provide students with learning opportunities they would not ordinarily have. These students do not directly log into GENI or run experiments on GENI. Rather, they interact with educational applications implemented on GENI and may not even be aware that these applications are running on GENI.

Two GENI-based STEM education applications are described in this section: (1) The Mars Game for teaching concepts related to coordinate geometry, and (2) The Remote Interactive Digital Cinema Microscope application for connecting students to high-resolution scientific instruments located thousands of miles away. These applications take advantage of GENI's high-speed networks and the fact that GENI resources are distributed across the US and can therefore be accessed with low-latency from almost anywhere in the country.

3.1 The Mars Game

The Mars Game [14] is an immersive educational tool for teaching math and programming using a fun and rewarding 3D gaming experience. The game was developed by Lockheed Martin and StandardsWork with funding from the U.S. Department of Defense's Advanced Distributed Learning Co-Lab.

The game is set on the planet Mars. Students are placed in this inhospitable environment and must work through a series of programming and mathematical problem-solving challenges to stay alive and save their peers. Figure 7 shows a snippet of the game where students use their knowledge of coordinate geometry to instruct a rover to perform tasks that require traversing the surface of Mars.

The Mars Game is entirely browser based with much of the compute intensive parts of the application hosted on a server. The advantages of this design are:

Fig. 7 The Mars Game requires students to use Cartesian coordinates and geometry to instruct a rover to travel to different locations and perform tasks needed to stay alive

(1) Students can play the game from almost any computer and operating system with a modern browser, and (2) School IT staff are not burdened with installing and maintaining the game software, an important consideration for most school districts.

The central server based design of the game requires low latency network connectivity between the students and the server in order to provide them with a rich interactive and immersive experience. GENI proved to be an imminently suitable platform for this because GENI Racks are available at over 50 different locations in the US and a server can be hosted on a rack close to the school using this game in their classroom. Two Mars Game servers are almost always available on GENI racks at Stanford University on the West Coast and George Washington University on the East Coast with additional servers being stood up at other locations depending on need.

Future plans for the Mars Game include a multi-player version with students from different schools playing with or against one another. Students at each school connect to a server running at the GENI rack closest to them and the servers communicate with each other over GENI's high speed backbone.

3.2 Remote Interactive Digital Cinema Microscope

The interactive, high-resolution microscopy system enables researchers at the University of Southern California (USC) to place live biological specimens under a Digital Cinema Microscope and capture ultra-high resolution (4k) movies of the microorganisms while simultaneously transmitting live, high-definition images from the microscope system to students in a STEM class in Chattanooga, Tennessee.

Fig. 8 Students at a STEM class in Chattanooga view images from and interact with a ultra-high resolution microscope at USC

Under the guidance of the USC biologists STEM students can enhance their learning further by manipulating the microscope effectively at very low latency levels. The project gives students the opportunity to see, appreciate and understand the complexity, beauty and science of the microworld that surrounds them, is critical to life on earth, and yet is otherwise invisible [4].

This project takes advantage of GENI's high speed backbone networks and GENI's ability to stitch Layer 2 VLANs over regional and national research and education networks (in this case between the GENI rack at Chattanooga and the lab at USC with the microscope. A stitched VLAN between the end-points eliminates the need for routing and eliminates latencies introduced by the routing layer.

Figure 8 shows students in Chattanooga viewing live high-definition images on a 4 K display and interacting with the microscope at USC.

Acknowledgements Portions of this work were supported by the New York State Center for Advanced Technology in Telecommunications (CATT), NSF Grant No. 1258331, and the NSF Graduate Research Fellowship Program Grant No. 1104522. The following students created course materials for GENI MOOC modules that have gone "live": Saboor Zahir, Samuel Partington, and Kimberly Devi Milner.

References

1. American Chemical Society: Importance of hands-on laboratory science (2014). http://www.acs.org/content/acs/en/policy/publicpolicies/invest/computersimulations.html. Public Policy Statement 2014–2017
2. Bruner, J.S.: The act of discovery. Harv. Educ. Rev. **31**(1), 21–32 (1961)
3. Calyam, P., Seetharam, S., Antequera, R.B.: GENI laboratory exercises development for a cloud computing course. In: Research and Educational Experiment Workshop (GREE), 2014 Third GENI, pp. 19–24 (2014)
4. Cofield, T.: USC/STEM School Chattanooga microbiology. https://vimeo.com/123082617 (2016). Accessed Jan 2016
5. Deutsch, P.: The eight fallacies of distributed computing (1994). http://goo.gl/vEKv5Z. Captured by Internet Archive: 1999-10-08
6. Edwards, S., Riga, N., Thomas, V.: The experimenter view of geni. In: GENI: Prototype of the Next Internet. Springer, Berlin (2016)
7. Freedman, M.P.: Relationship among laboratory instruction, attitude toward science, and achievement in science knowledge. J. Res. Sci. Teach. **34**(4), 343–357 (1997)
8. Google Groups. GENI Users. https://groups.google.com/forum/?hl=en#!forum/geni-educators (2016). Accessed Jan 2016
9. Kotz, D., Newport, C., Gray, R.S., Liu, J., Yuan, Y., Elliott, C.: Experimental evaluation of wireless simulation assumptions. In: Proceedings of the 7th ACM International Symposium on Modeling, Analysis and Simulation of Wireless and Mobile Systems, MSWiM '04, pp. 78–82. ACM, New York (2004)
10. Kurose, J.F., Ross, K.W.: Computer Networking: A Top-Down Approach, 5th edn. Addison-Wesley, Reading, MA (2009)
11. Magin, D.J., Churches, A.E., Reizes, J.A.: Design and experimentation in undergraduate mechanical engineering. In: Proceedings of a Conference on Teaching Engineering Designers (1986)
12. Marasevic, J., Janak, J., Schulzrinne, H., Zussman, G.: WiMAX in the classroom: designing a cellular networking hands-on lab. In: 2013 Second GENI Research and Educational Experiment Workshop (GREE), pp. 104–110. IEEE, New York (2013)
13. Riga, N., Thomas, V., Maglaris, V., Grammatikou, M., Anifantis, E.: Virtual laboratories–use of public testbeds in education. In: Proceedings of the 7th International Conference on Computer Supported Education. SciTePress, Setúbal, (2015). ISBN 978-989-758-107-6
14. The Mars Game Team. The Mars Game. http://www.themarsgame.com (2016). Accessed Jan 2016
15. Tredger, S., Zhuang, Y., Matthews, C., Short-Gershman, J., Coady, Y., McGeer, R.: Building green systems with green students: an educational experiment with GENI infrastructure. In: 2013 Second GENI Research and Educational Experiment Workshop (GREE), pp. 29–36 (2013)

The Ignite Distributed Collaborative Scientific Visualization System

Matt Hemmings, Robert Krahn, David Lary, Rick McGeer, Glenn Ricart, and Marko Röder

Abstract We describe the Ignite Distributed Collaborative Visualization System (IDCVS), a system which permits real-time interaction and visual collaboration around large data sets on thin devices for users distributed about the wide area. The IDCVS provides seamless interaction and immediate updates even under heavy load and when users are widely separated: the design goal was to fetch a 1 MB data set from a server and render it within 150 ms, for a user anywhere in the world, and reflect changes made by a user in one location to all other users within the bound given by inter-user network latency. Scientific collaboration and interaction is the initial use case for the IDCVS, since eScience is characterized by large data sets. The visualizer can be used for any application where the data can be visualized on a web page. The visualizer consists of many replicated components, distributed across the wide area, so that an instance of the visualizer is close to any user: the design goal is to place an instance of the visualizer with an 20-ms latency of any user. It is the first exemplar of a new class of application enabled by the Distributed Cloud: real-time interaction with large data sets on arbitrarily thin devices, anywhere. The IDCVS features modular design, so it functions as a specialized Platform-as-a-Service: writing a new collaborative visualization application is as simple as designing a web page and distributing a data server. The system was demonstrated successfully on a significant worldwide air pollution data set, with values on 10, 25, 50, and 100 km worldwide grids, monthly over an 18-year period. It was demonstrated on a wide variety of clients, including laptop, tablet, and smartphone. The system itself has been deployed at over 20 sites worldwide. Distribution and deployment across

M. Hemmings
Computer Sciences Department, University of Victoria, Victoria, BC, Canada
e-mail: mhemming@uvic.ca

D. Lary
Department of Physics, University of Texas at Dallas, Dallas, TX, USA
e-mail: david.lary@utdallas.edu

R. McGeer • G. Ricart
Chief Scientist, US Ignite, 1150, 18th St NW, Suite 900, Washington, DC 20036, USA
e-mail: rick.mcgeer@us-ignite.org; glenn.ricart@us-ignite.org

R. Krahn • M. Röder (✉)
Y Combinator Research, San Francisco, CA, USA
e-mail: robert.krahn@harc.ycr.org; markoroeder@cdglabs.org

© Springer International Publishing Switzerland 2016

R. McGeer et al. (eds.), *The GENI Book*, DOI 10.1007/978-3-319-33769-2_19

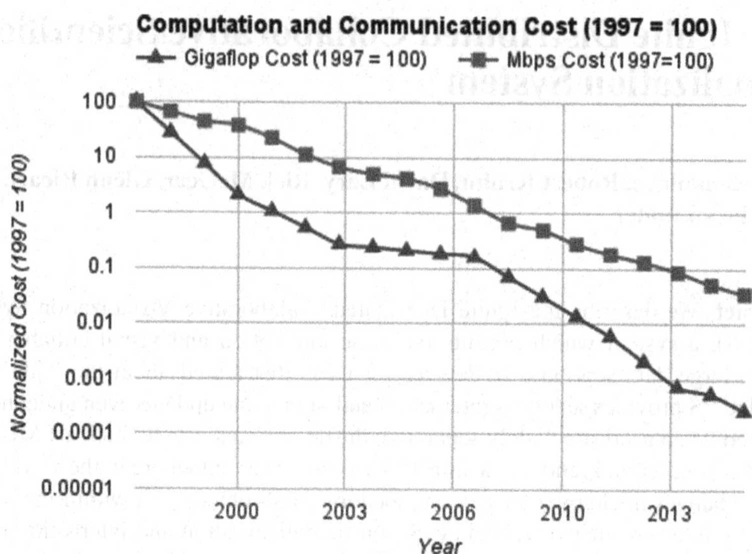

Fig. 1 Costs of communication and computing over time

the GENI Experiment Engine was accomplished in 15 min, and installing a new site is limited by ftp time for the data set.

1 Sending Programs to Data and People

The Internet was conceived around the principal application of file transfer in various guises (ftp, email, eventually Gopher and http), with some very lightweight remote computation—primarily, telnet and its various successors. This was an era where computation was expensive but communication relatively cheap. For this reason, access to remote computational resources was tightly guarded, but open-ended data transfer with few restrictions grew up as common practice.

Today the cost situation is reversed. While communication has grown cheaper, computation and storage have grown *much* cheaper. Concrete comparisons between bandwidth can be found from 1998 on. In 1998, Internet transit prices averaged \$1200/Mbps; in 2015, they are about \$0.63 per Mbps. That is impressive—a decline of a factor of 1904 in 17 years. However, computation costs decline by an order of magnitude every 4 years, so over that time computation costs have declined by a factor of over 10,000.

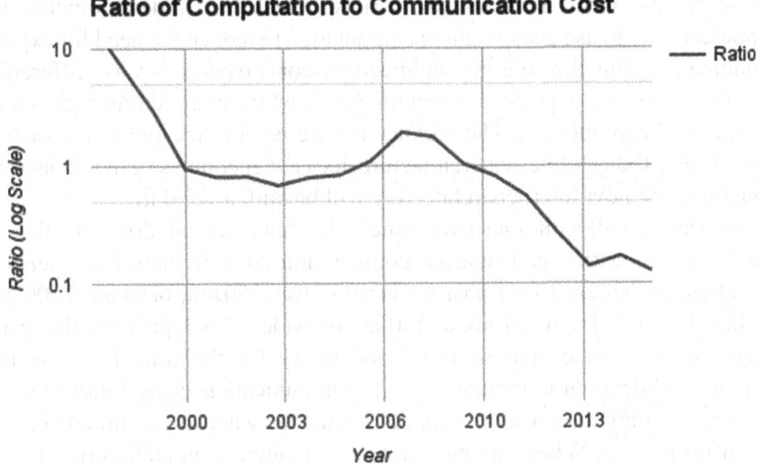

Fig. 2 Computing/communication ratio

The raw data is shown in Fig. 1.[1] We have normalized so that 1997's costs are 100 for each item ($1200 in 1998 for Mbps, $30,000 in 1997 per gigaflop). The growing gap between communications costs and computations costs is clear from the data. The vertical scale is logarithmic for legibility, which lessens the visual impact of a dramatic change in ratios. In 2015, the cost of communication was 4 % of 1997's cost of communication; but 2015's cost of computation was only 0.03 % of 1997's cost of computation.

What matters is the ratio of the cost of computation and communications, because this is the ratio between our ability to store and process data and our ability to transmit it. The observed doubling rate of communication is about 18 months: Mbps/$ doubles every 18 months or so. Just for raw computation speed, this tracks Moore's Law. But computation also declines in cost. Gigaflops/$ doubles every 11 months. Over time, this asymmetry has led to a large increase in the ratio of communications to computation costs; over the 17 year period 1998–2015, the cost of computation relative to communication has dropped by a factor of over 100. This is the fundamental driver behind the "Zettaflood", which predicts Zettabytes of monthly traffic in all categories by 2016, and traffic growth rates of 25–40 % per annum in each category [12, 14].

The ratio trends are shown in Fig. 2, using the same raw source data. The ratio of computation to communications costs has declined from a ratio of nearly 17 in 1997 to a ratio of 0.13 in 2015—a dramatic reversal in the relative costs of computation vs. communication. This has profound implications for the Internet architecture.

[1] Source of Computation Costs: http://aiimpacts.org/wikipedia-history-of-gflops-costs/. Source of Communications Costs: http://drpeering.net/white-papers/Internet-Transit-Pricing-Historical-And-Projected.php.

These trends make it clear that it makes sense to trade computation for communication—to use cheap remote computation to reduce the need for expensive communication. But the Internet architecture, conceived in a very different era, offers free access to expensive communication while maintaining tight controls on cheap local computation. The architecture we need in the future is exactly the reverse of this. Today's Internet architecture does not encompass general distributed computation; the next Internet architecture will be built around it.

These trends will continue over time. The fundamental driver behind both reduced communication and reduced computation costs is materials science and device physics: the size of a transistor in 1997 was 250 nm, or about 2500 atoms wide. In 2014, it is 14 nm, or about 140 atoms wide. This represents the distance a charge must traverse, and so is a good proxy for the time for a device to switch. But while both computation and communication costs follow the same device curve, computation is a *point* infrastructure whereas communication is a *linear* infrastructure. When one buys a new computer, computation instantly gets faster; conversely, when one buys a new communications device, communications only gets faster when everyone else similarly upgrades. Since that takes time, communication will always lag computation—until the underlying device curve flattens, at which point there will likely be a short catchup of the communication curve.

Sensors and storage follow the same device curves, of course, and are similarly point infrastructures. What this means is that over time our ability to capture, store, and process data will continue to grow exponentially over our ability to send it. And this in turn means that our need to transmit programs to data rather than data to programs will continue and increase.

GENI is a Distributed Cloud with 50 Points-of-Presence (POPs) across the United States, and, with its international federate partners, at least an equal number worldwide. The foregoing discussion suggests that use of the Distributed Cloud will make legacy network applications far more efficient and responsive, because it will permit the deployment of applications close to data and to users. But new infrastructures do more than make legacy applications better; they also enable wholly new classes of applications and services. The Ignite Collaborative Visualizer is a first example of that.

2 Collaborative Visualization Systems and Distributed Clouds

The Ignite Distributed Scientific Collaboration System grew up as an experiment in exploiting the capabilities of GENI for entirely new applications. The fundamental capability GENI offers is the ability to deploy programs close to users and data. On the data collection side, there is an obvious and important application: data reduction in situ. On the presentation and human consumption side, collaboration around scientific data is an immediate significant area of interest and impact, which exploited two new capabilities of GENI: high local area bandwidth and low latency in the local area. These are two separate effects; both are important for collaboration.

Bandwidth. *At least 99 % of bandwidth is local.* Intra-building bandwidth is at least two orders of magnitude higher than inter-building bandwidth; similarly, intracity bandwidth is several orders of magnitude higher than inter-city bandwidth. This means that high-bandwidth applications must be local. Indeed, this is common industrial practice. High-bandwidth services on the Internet today (typically large-scale software distributions and streaming video) are generally served by content distribution networks such as Akamai, LimeWire, and Coral, or homegrown service-specific networks. The effectiveness of Content Distribution Networks is why over half of all Internet traffic is expected to be carried by CDN's by 2019, and in that year 62 % of all Internet traffic will transit a CDN [13].

Latency. There are very few highly-interactive networked applications to date; collaborative games are perhaps the most obvious example. This is because high interactivity requires low latency, and physics imposes tight lower bounds on the latencies obtainable in networked applications. The speed of light in fiber is 200 km/ms, and therefore a 1 ms round-trip bounds the client-server distance above by 100 km. This is a hard physical limit, independent of queueing delays, application latency, etc. This is why gamers use servers close to them for collaborative gaming, and why "LAN Parties" are popular among gamers. Local computation is an asset for legacy applications; it is *required* for collaborative interaction.

Content distribution networks ensure high effective bandwidth between client and server, and ensure latencies low enough to maximize TCP performance. They do not ensure latency low enough for truly interactive applications [11, 22]. Highwind's Rolling Thunder network, which provides video streaming services and delivers online games for Valve software, features 70 points-of-presence (POPs) throughout Europe, Asia, and the Americas. A network capable of supporting interactive applications would require roughly that many POPs in the United States alone. Moreover, tests on a number of popular CDNs demonstrated that even a modestly-sized 12 KB object took over 370 ms to deliver within the continental US, and at least 1 s to deliver to Asia; a 1 MB object took over 2 s to deliver in North America and over 10 s to deliver to Asia [22].

Latency is a more difficult target to meet than bandwidth, because upgrading network connections increases bandwidth but does not decrease latency; colloquially, pipes can be made fat but not short. The only way to for an application meet a latency target is to be hosted on a POP sufficiently close to a user that the round-trip time is within the latency target. In [11], it was found that less than 70 % of North American hosted users were within 80 ms of a Cloud platform such as Amazon's Elastic Compute Service (EC-2), and fewer than 40 % were within 40 ms. Worse, even if the number of data centers for Amazon EC2 were quadrupled from the current five to 20, under 90 % of users would be within 80 ms and less than half of users would be within 40 ms.

Lack of POPs, and the associated high latency and low bandwidth to the end-user, has been the primary reason for the lack of deployment of truly collaborative applications across the wide area. Current collaboration systems break down into four broad classes:

1. Low-bandwidth collaborative editing applications. The primary example of this class of application is online office suites, including the Google products and Microsoft Office online. Even in these examples, the primary use case is for a single editor at any one time and multiple viewers; online simultaneous editing is supported but is not a dominant use case.
2. Presentation systems. In these applications (WebEx, Skype screen sharing, Google Hangouts, etc.) there is a primary presenter and a potentially very large audience. Latency is an issue, but not an especially significant one, because the presenter sees rapid response and members of the audience only has indirect evidence of the presenter's manipulation of the material
3. Very high end collaboration systems using dedicated, very high-performance private networks. A classic program in this area was the OptIPuter [16], which featured tiled displays ("OptIPortals") interconnected by 1–10 Gb/s fiber networks.
4. Fat-client systems which perform best if the collaborating clients are relatively close to each other in real terms. A good example of these systems are collaborative computer games, particularly rich, high-intensity games such as Counter-Strike: Global Offensive or Destiny.

Of these classes, only the third involves the transport of large scale data between client and server in real time, which is the hallmark of scientific collaboration. Collaboration around scientific data involves interactive visualizations of large data sets. These data sets are too large to be stored on a local desktop client, and complex visualization systems tax the resources of typical thin clients. Moreover, collaboration requires the transmission of information between widely-separated colleagues. For this reason, visualization systems are the primary examples of the third class of application, and are restricted to a few very well-endowed sites. While the OptIPuter achieved its objective of making a "world where distance was eliminated", they were hardly ubiquitous desktop systems, nor available everywhere. Even a low-end OptIPortal costs tens of thousands of dollars, and requires significant maintenance; similarly, a gigabit connection to a national research network is several thousand dollars a month, and commercial prices are far higher.

The objective us to permit these rich, date-intensive collaboration systems to be realized on a broad number of devices. Of course, not all of the functionality of the OptIPortal or a networked 3D immersive environment such as the Cave Automatic Virtual Environment (CAVE) [17] can be brought to the handheld device, or even a laptop; at some level, there is no substitute for Gigapixel displays. However, much of this rich interaction can be brought to a wide array of client devices today, thanks to a number of convergent technologies:

Table 1 Application classes

	Weak interaction/thin data	Rich interaction/big data
Powerful client	Anything	Desktop application
Arbitrary client	Centralized cloud	Distributed cloud

1. The emergence of the Distributed Cloud as a hosting platform, enabling the deployment of large datasets and web-based visualization systems close to the user, reducing the demand for wide-area bandwidth and permitting low-latency interaction with a server-based application.
2. Novel lightweight virtualization architectures permit the easy deployment of an application with its associated environment across this infrastructure. The Distributed Cloud will always be relatively resource-poor at each POP, and so lightweight architectures are a requirement. The development of multi-tenant container hosting environments, in particular PlanetLab and its successor, the GENI Experiment Engine, is invaluable for this.
3. The development of feature-rich ubiquitous client platforms, particularly HTML5. HTML5's impressive graphics feature set permits rich visual interaction in two and three dimensions, even on handheld devices such as phones and tablets.
4. The development of inter-web-page messaging systems, such as WebSockets, permits the active manipulation of rich media objects with very low-bandwidth inter-peer communications. Combined with the capabilities of the web browser, this opens the way to media- and graphics-based messaging systems.
5. The development of Wiki-based rich client/server in-browser application development environments such as the Lively Web [33] permits interactive application development and modification.

Of these innovations, the most significant is the Distributed Cloud. It has long been established that 150 ms is a significant threshold for an interactive application [58]. Offering an application on a ubiquitous client requires that all data be resident on a server and only that required for a client request be transmitted to the client, and one must assume that each client request will trigger a response from the server. Any nontrivial network latency between server and client will make it impossible to meet a 150 ms deadline to complete the transaction.

The Distributed Cloud thus enables an entirely new class of application: rich collaboration around large data sets using ubiquitous client devices over the wide area. Remote collaboration in the past has either required the use of fat clients (desktops with some combination of specialized hardware, massive compute, or large storage) or thin data and minor actions (collaborative editing of online documents). The distributed cloud permits rich interaction around big data from desktops, laptops, phones or tablets (Table 1). A number of authors have observed that the CDN edge rather than the Cloud is suitable for interactive, cloud-based gaming [11, 48]; as an open, distributed Cloud with ubiquitous POPs, the GENI Distributed Cloud makes this vision a reality.

3 Architecture of the IDCVS

The IDCVS is designed as a pluggable system to host any collaborative visualization, a specialized distributed platform-as-a-service. The goal is that a developer wishing to design and deploy a new visualization application merely needs to design a web page and a data server, and the IDCVS provides the scaffolding to deploy and host the application. As a result, consists of three major components: an application-specific data server; the GENI Experiment Engine; and the Lively Web. The data server is supplied by the application developer. The GENI Experiment Engine [5–7] serves as our primary deployment platform. The Lively Web [41, 57] as our deployment system and messaging service.

To develop the IDCVS, we focussed on a single driving application, the Atmospheric Quality Visualization System. We describe this application in the next section. In this section, we will describe our use of each of these components, using illustrative examples where appropriate from the driving application.

We need a server close to any client; and since clients may be anywhere, servers must be everywhere. There are three considerations which, together, completely determine the architecture of the system:

1. The need to deploy the application on a wide range of devices, including tablets and smartphones
2. The need to visualize and manipulate large datasets (in the 10 GB and beyond range)
3. The need for seamless, "desktop-like" interaction with the dataset and with remote participants.

It is important to note that a critical feature of a collaboration application is *ubiquity*: the application must be viewable on a very wide range of user devices. Expectations of the capability of the client device which rule out a broad range of devices dramatically reduce the utility of the application, which in turn drives us to relying on the server for computation and data storage.

A wide range of devices rules out resource-rich "fat" clients, and forces us to render on a web browser. The requirement of large data sets requires a significant computational resource, and since fat clients are not permitted this means a server; and the requirement for tight interaction implies low latency between this server and client. We further require tight interaction between many remote participants. This in turn means an efficient messaging architecture which rapidly transmits UI events around the network.

The considerations of client platforms is always a judgement call which depends upon the relative ubiquity of capable platforms. In the future, rich, cheap devices such as the Oculus Rift [23, 46] or SteamVR [60, 62] device will change the way

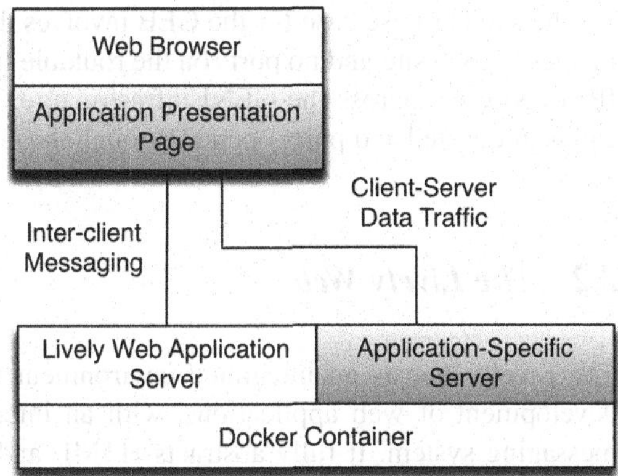

Fig. 3 Visualization system architecture

that we and others render data; however, for reasons of both power and ubiquity, we believe that a close-to-the-client client-server architecture will continue to be our deployment vehicle for the foreseeable future.

The architecture of the Visualization System is shown in Fig. 3. The shaded boxes (the Data Server, or Application-Specific Server and the Application Presentation Page) are application-specific and must be developed by the visualization application developer. The basic Lively Web server, the inter-client messaging system, and the Docker container which serves to deploy the application are the same across all Visualization System applications. The GENI Experiment Engine is our primary vehicle for deploying the Visualization System Application.

3.1 The GENI Experiment Engine

The GENI Experiment Engine [5–7] (GEE) offers deployment of applications in containers across the GENI infrastructure, leveraging the commercial ecosystem of lightweight virtualization environments. The GEE consists of a network of virtual machines, one per site across the GENI infrastructure. Each virtual machine runs the Docker [18] container system, and each application is a network of containers. This permits lightweight, rapid deployment of applications. The GEE is designed to be *embedded* (intended to be instantiated on top of underlying infrastructure-as-a-service platforms) and *cross-infrastructure* (deployed over a number of different infrastructures). This permits users to do one-click deployment of applications across multiple infrastructures.

In addition to one-click deployment of VMs across and between infrastructures, the GEE offers automated single pane-of-glass orchestration and control. In particular, the GEE creates an Ansible [1] configuration file, and an initial Python hosts file for use with Fabric [21]. This offers users both declarative and imperative wrappers around distributed computation as a service. Further details on the GEE can be found in the GEE chapter [5].

The standard use case for the GEE involves the use of one of a few pre-checked images at each site and no ports on the routable Internet, due to shortage of routable IPv4 addresses across the GENI infrastructure. However, both specialized images can be requested and ports opened through the GEE Portal interface.

3.2 *The Lively Web*

The Lively Web is an integrated environment for unified client- and server-side development of web applications, with an integrated client-server and peer-peer messaging system. It fully abstracts HTML and CSS into a graphical abstraction based on the Morphic system [42] introduced in Self [59] and deployed in Squeak [27] and Scratch [53]. As with Self, Squeak, and Scratch, Lively is a descendant of the Smalltalk [26] family of programming languages and environments.

Development of client-side applications is done by a combination of Javascript programming and manual configuration of Morphs. The developer brings up a web page (usually http://www.lively-web.org/blank.html), and then drags Morphs onto the page from a builtin repository of objects (the "Parts Bin"). Parts range in complexity from simple rectangles and ellipses to full-fledged embeddable applications, such as a Minesweeper game. Morphs are physically composed, so a developer can form complex Morphs from simpler ones.

Morphs are programmed using an object-oriented form of Javascript, with the usual forms of inheritance and self reference. Each Morph is descended from class Morph, which offers a rich library of methods to manipulate a morph's physical appearance, position, and position within the physical hierarchy of objects, the "scene graph".

A number of morphs are designed specifically for programming: there is a builtin "script editor" used to attach functions, called scripts, to morphs, and an object inspector, which is used to view an object's properties and methods. Using these scripts, an application is programmed from within the application itself, with no need for external tools such as a system console, external editor, compiler, etc.

A feature common among Smalltalk-based IDE's, but rare outside of them, is live evaluation. In any Morph that accepts text input, any chunk of text can be evaluated as a Lively expression. This notably includes the editor itself, which encourages a programming methodology of continuous test-and-evaluate. The standard compile-load-execute cycle is shortened to the minimum imaginable; program text can literally be evaluated as it is typed.

One further note of interest is that the primary point-and-click UI in Lively, and therefore in most Lively applications, is Morph-specific. A user action, typically <cmd>-click (<ctrl>-click on Windows or Linux) on a Morph brings up a halo of buttons around the Morph itself. This is a standard component of the Morphic interface, first introduced in Self, and has the effect of simplifying the global UI, since interface commands for every object no longer need to be positioned in global menus and buttons at the top of the window.

Lively integrates a Wiki [40], so deploying the client side of the application is simply a file save on the server, accomplished directly from within the page itself. Server-side programming is rarely necessary for a Lively application, but when it is required Lively comes with a pluggable node.js server which is programmed from within the browser itself. The Lively server incorporates a standard SQLite3 database, and a full API to create, access, and manipulate databases from within the Lively application itself. Using these tools, deploying a full web application, both client- and server-side, is little more complex than preparing a document using a standard desktop application.

Lively comes with a large number of pre-packaged libraries, and can incorporate any Javascript library. In particular, jQuery [31], d3.js [15], and Google Maps are all builtin. New libraries can be incorporated easily; if the library instantiates and manipulates graphical objects, this generally involves writing a small Morphic wrapper over the library itself, primarily for initialization of the object and to add a Morphic scripting overlay on the library's top-level methods.

For the Atmospheric Quality Visualization application, every library save the mapping library we chose was included in Lively. The total integration effort involved was writing a small Morphic interface over the mapping library, which only involved high-level routines to do the initial load and configuration of the map.

The Lively Messaging Service, Lively2Lively

Central to the collaborative nature of the IDCVS is messaging between the instances of the application, which permits each client to control remote instances of the application. In the next subsection, we detail the IDCVS messaging layer. It is built on Lively's builtin sophisticated client-server and peer-peer messaging system, Lively2Lively.

Lively2Lively runs identically over WebSockets and other messaging services (WebRTC, socket.io), abstracting the details of the connection. The network layer is hidden by default; users only need to specify receiver, message type, and arguments as a JSON structure. Lively2Lively also offers a centralized tracking service, for discovery of other clients attached to Lively2Lively. Participants register with a GUID to become discoverable. Participants declare their support for the receipt of messages of a specific type by specifying handlers for those messages to the tracker.

Messages are application-specific and registered by type. A message type is simply a string which indicates a handle class.

3.3 The IDCVS Messaging System

The IDCVS Messaging System is layered on top of Lively2Lively. The goals of the messaging system are to broadcast map-update events among the application participants, using low-bandwidth messages, on each UI event for any instance of the application.

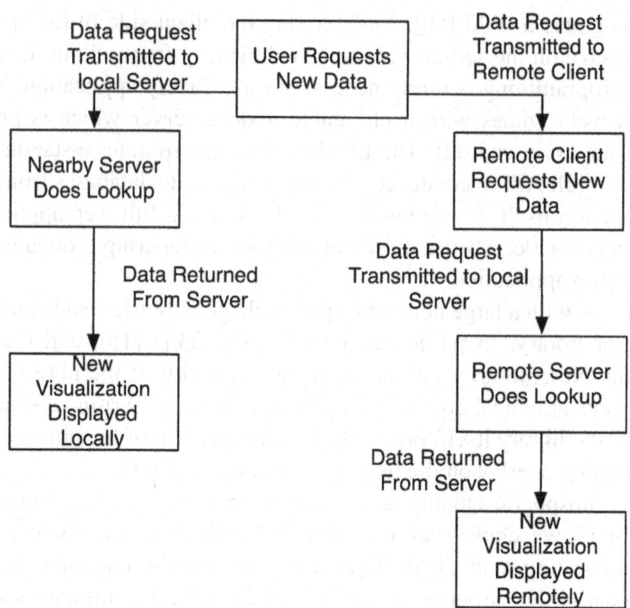

Fig. 4 Messaging and client-server dataflow

A message service is defined by a class of messages, with a given name. Each message class is associated with a specific class of Morph, and its identifier is the fully-qualified URL of the original Morph part. This suffices to ensure uniqueness of the message class. When an application instance is loaded, the relevant morph registers to receive messages from this service, and for each message registers a Javascript script on the morph as a message *handler*. The morph has now subscribed to all messages in this class. When a message is received from this class, the IDCVS messaging system calls the handler to execute the action.

The IDCVS message service broadcasts the initial registration, and the remote participant is now subscribed to the conversation. Further, participants in the conversation can now broadcast messages.

A typical workflow is used in the Atmospheric Quality Visualization System. The application page registers itself as a conversation participant in the mapping service. On each map update originated in response to a user action, the client both requests data to service the update from the local server and broadcasts an update message to other participants, with Lively2Lively handling inter-server message delivery. On receipt of an update message, the page requests data from the local server corresponding to the viewbox contained in the message, recenters the map to that viewbox, and then draws the points on receipt from the server. See Fig. 4. The remote workflow (and the perceived latency) is thus identical from a remote update and a UI event.

3.4 Deployment

The IDCVS is pre-deployed across the Lively infrastructure. The application itself is simply a Lively page. In order to deploy an application, the developer simply writes the page and uses a simple Web script to do a push to each instance of the visualizer. Lively incorporates a WebDAV server, and so this is a simple Javascript function invoked from the initial web page used to develop the application.

Performance requires that each instance of the application gets data from its local server, and therefore a copy of the application-specific data server is required at each application server. We used the orchestration and deployment services that are native to the GENI Experiment Engine to deploy the Atmospheric Quality Visualization System Data Server. This involved writing a short Ansible script to copy, and unpack the data, install necessary packaged software and deploy the application. Starting from a bare container, this process was completed autonomically in 15 min, everywhere on the GEE.

4 The Atmospheric Quality Visualization System

The Ignite Distributed Collaborative Visualization System is driven by and supports an application of significant social concern, the Atmospheric Quality Visualization System.

The Atmospheric Quality Visualization System is used to analyze two major issues: climate change and human health. Climate change is a defining issue of our generation. Currently one of the largest uncertainties in simulating climate change is the radiative forcing associated with atmospheric aerosols [28], or the quantities measured by the pollution sensor. A further compelling application is in human health. The World Health Organization [65] concluded that in 2012, seven million deaths across the world were associated with air pollution, one in eight of the total global deaths. A major component of this pollution is airborne particulate matter, which is designated by its size. $PM_{2.5}$ is particulate matter less than 2.5 μm in size; PM_{10} less than 10 μm in size. These classes of particulate matter, particularly $PM_{2.5}$, have severe consequences for human health. Approximately 50 million Americans have allergic diseases, including asthma and allergic rhinitis, both of which can be exacerbated by $PM_{2.5}$. Every day in America 44,000 people have an asthma attack, and because of asthma 36,000 kids miss school, 27,000 adults miss work, 4700 people visit the emergency room, 1200 people are admitted to the hospital, and 9 people die.

Despite its importance, ground-truth measurements of airborne particulate matter are scarce. This is because the equipment to sense it is quite expensive; even consumer-grade devices cost several hundred to a couple of thousand dollars. Industrial-grade equipment, suitable for environmental monitoring, is substantially more. Figure 5 shows current industrial-grade sensors and their deployment. As

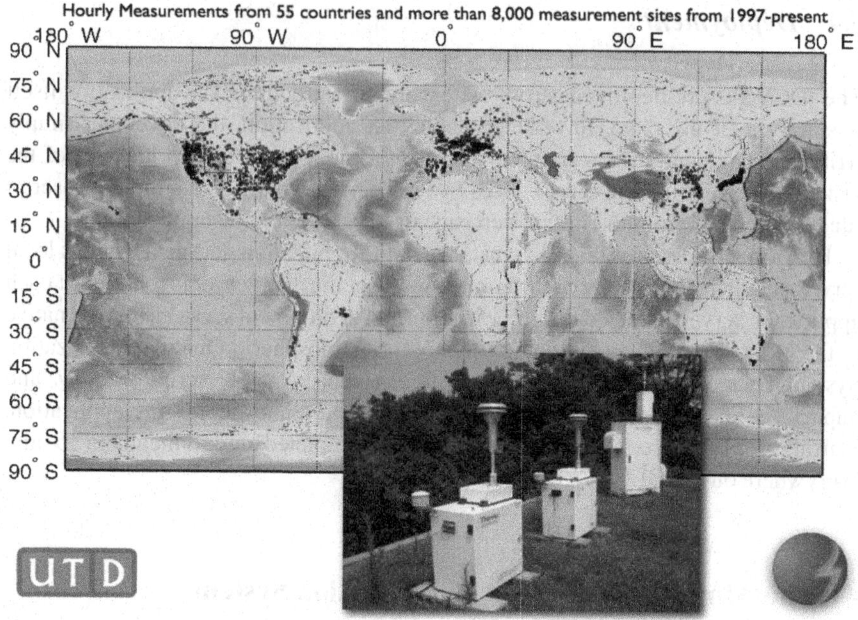

Fig. 5 $PM_{2.5}$ sensors and current deployment

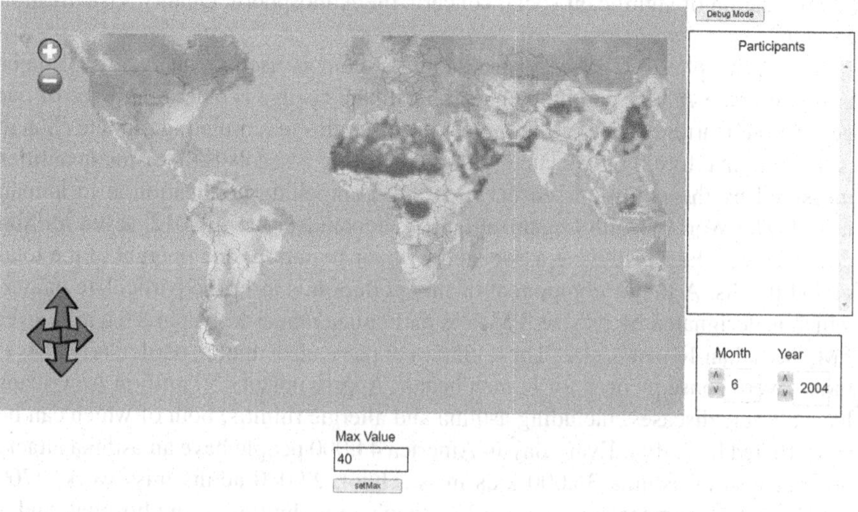

Fig. 6 $PM_{2.5}$ concentration values for June 2004

can be seen in the figure, deployment is sparse, and a comparison to the heavily-polluted regions of the world shown in Fig. 6, many of the most-polluted areas are not covered.

Fortunately, the values for areas not directly measured can be inferred from easily-available proxy measurements. By simultaneously combining around 50 massive NASA remote sensing and Earth System modeling products using multivariate, non-parametric, non-linear machine learning we have produced an accurate aerosol model on a 10 km grid, with a single value at each grid point to show the aerosol concentration within that grid cell. Details of the model derivation can be found in [36, 37], and a video presentation can be found at http://www.la-press.com/using-machine-learning-to-estimate-global-pm25-for-environmental-healt-article-a4833. The model is validated using a global sensor web of ground-based sensors and unmanned aerial vehicles. Values are calculated on a monthly basis, extending over the period 1997–present. The Atmospheric Quality Visualization System distributes the data products throughout the GENI infrastructure so it can be accessed by scientists and planners on local networks with low latency.

In the application over the IDCVS, we use this virtual sensor to produce a world-wide aerosol sensor product on a 10 km × 10 km grid, with a single scalar value at each grid point to show the aerosol concentration within that grid cell. Aerosol product concentrations are calculated on a month-by-month basis, extending over the period 1997–present. Users of the product are able to observe changes in aerosol products over time in real time, using DVR-like controls.

Data transmission and storage are well within existing capabilities throughout the distributed GENI infrastructure. The earth is an oblate spheroid 6370 km in radius; this gives a surface area of 510 million km^2, or about 5 million grid points at complete coverage. About 70 % of this is ocean, so a full data set contains about 1.5 million points. A CSV representation of a monthly data set is about 50 MB. There are 208 total months in our data set, and with coarse-resolution grids our total deployment is about 11 GB. This is well within the capabilities of servers. Efficient transmission of data between client and server is accomplished through the use of Binary JSON (BSON)-based messaging, the use of localized data placement, and intelligent prefetching. Remote and collaborative viewing are accomplished through the use of control messages from viewer to controller, which manipulate data resident on the viewer's screen and transmissions between the viewer's local server and the viewer's client.

The application (Fig. 7) displays the value in each grid square by a color, with red indicating high intensity. A unique feature of the application is that multiple users may be looking at and manipulating this visualization, even on different servers. The current users are shown in the Participant box. When a user manipulates his copy of the map, it changes for all participants, even those who have pulled the application from different servers.

Over large areas, coarse-resolution 25, 50, and 100-km grids are used, since the 10-km grid has approximately 2 million points worldwide. The total data set is 11 GB for the 19-year period.

4.1 The Data Server

The other application-specific component was the Data Server. We faced twin constraints: relatively limited disk space and a tight performance bound. Our initial experiments indicated that no existing freely-distributable database or geo information server could meet our performance requirements or fetching a 30,000-point rectangle within 20 ms without consuming hundreds of gigabytes of disk space in indexes. Moreover, we didn't have enough virtual memory to put the entire database in memory.

We therefore designed a special-purpose quad-tree server to serve the data from disk with a substantial in-memory cache, and could reliably fetch 30,000 points in 20 ms. The 100-km grid leaf cells are kept permanently in memory to provide rapid access to coarse data. Once a leaf cell at any resolution is read, it is cached; the cache is flushed when the server approaches a pre-set virtual memory limit.

5 Quantitative Analysis of the Problem Space

30,000 points represents, on a 10 km grid, a square roughly 1700 km on a side, roughly 1/4 of the continental United States. On a 25-km grid, 30,000 points represents a continent, and on a 100 km (1°) grid the world. This is a reasonable upper bound on the number of points that must be drawn in response to a user interaction.

Drawing the map is four steps: requesting data from the server, the server's search for the requested data, data return, and drawing the data on the map. We settled on polymaps.js [50] as the highest-performing mapping library, and demonstrated that

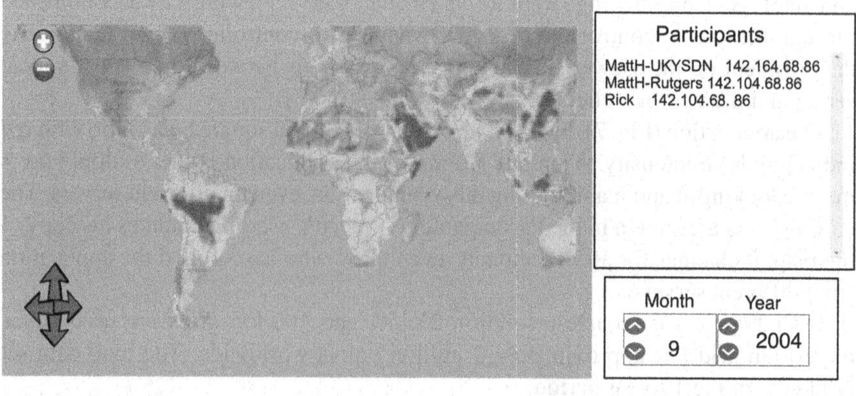

Fig. 7 Application screenshot

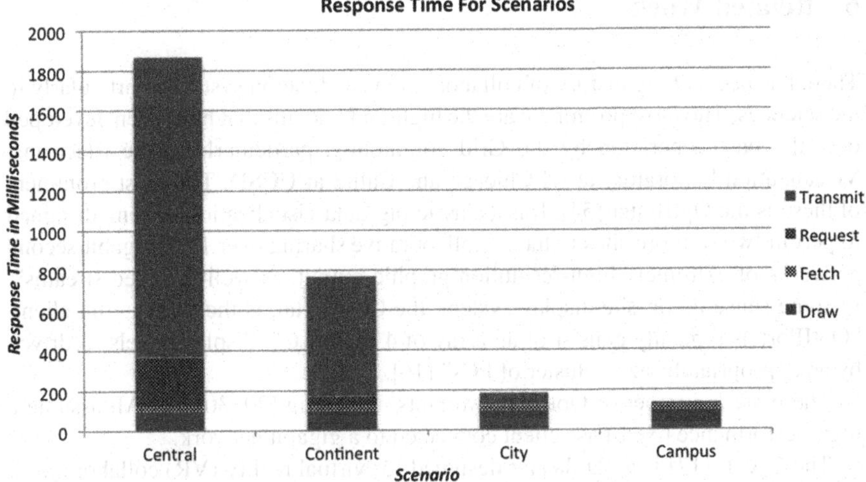

Fig. 8 Calculated response times for the scenarios

we could render about 30,000 points in 100 ms. As mentioned in the preceding section, our special-purpose quad-tree server could fetch 30,000 points in 20 ms.

The only variability is in the network. Our goal was to determine what envelope of network round-trip times and bandwidth would permit us to achieve our goal of a refresh within 150 ms. Our wire protocol had about 27 bytes/point, so 30,000 points are roughly 800 kilobytes, or 6.4 megabits. We assumed 1500 byte packets and accelerated slowstart (10-packet initial send) for our experiments, and standard AIMD window adjustment post-slowstart.

We tested four scenarios, as shown below.

Scenario	Latency	Bandwidth
Campus	1 ms	1 Gb/s
City	5 ms	1 Gb/s
Continent	50 ms	100 Mb/s
Central	250 ms	100 Mb/s

The Campus scenario assumed a server on the same campus as the client, with millisecond latency and gigabit bandwidth. City assumed a server within 100 km or so, but not resident on the same campus. Continent assumed an EC-2 like deployment, with one or two POPs per continent. Central assumes a single POP for the entire globe. Both Campus and City scenarios achieved the desired response time of 150 ms; neither Continent nor World came close (Fig. 8). We note that our results are consistent with those measured in [11] and [22].

6 Related Work

There has been a long history of collaborative visualization systems, particularly in the sciences. The most prominent are the high-end systems that have been developed over the past generation by the Grid community, particularly at the Electronic Visualization Laboratory at UI-Chicago and Calit2 at UCSD. The most prominent of these is the OptIPuter [54]. It is a classic big-data visualization system, designed to permit two or more sites to have "collaborative sharing over 1–10 gigabit/second networks of extremely high-resolution graphic output, as well as video streams". In some sense its on-site display system, the OptIPortal, is the ultimate fat client: "OptIPortals typically consist of an array of 4 to 70 LCD display panels ... driven by an appropriately sized cluster of PCs" [16].

There are a number of OptIPuter variants, including [29, 30, 51]. All assume a high-performance fixed-base client connected to a gigabit network.

The CAVE [17] is a 3D theater designed for virtual reality (VR) collaborations. Remote CAVEs can be networked through the CAVERN network [38]. Client requirements are comparable to the OptIPuter.

These high end visualization systems have been extremely popular, particularly in scientific collaborations. See, for example, the COVE, a visualization system for ocean observatories [24]. A gap analysis can be found in [25].

Computer games are a form of Real-Time Interactive Simulation and thus should be covered in this discussion. A popular collaborative game, such as Counterstrike Global Offensive or Left4Dead has a rich client, typically a PC with a high-end graphics card, with rich game software installed on the client. Game events are broadcast messages passed to all participants. Game state is resident on each client, and the only data passed between participants are messages to update the state of the game. This is similar to our messaging architecture, save that a secondary data request is not required. Game play is highly latency sensitive, leading to popular "LAN parties", where collaborators bring their PCs to play together on a single local-area network.

Thin client applications are relegated to relatively low data-rate, low-data use applications such as collaborative document editing and review, and presentation viewing. Google's office application suite (Google Docs, Google Sheets, Google Slides) are excellent examples in this space.

The PolyChrome system [2] is a generic software framework for building web-based cross-device visualizations and uses a peer-to-peer messaging framework for collaborative visualizations. In terms of the work covered here, it covers a space similar to that covered by the Lively Web and Lively2Lively. There are several fundamental distinctions:

1. PolyChrome is a specialized environment for visualization applications; the Lively Web is a general environment for Web and mobile application development.

2. PolyChrome's messaging system assumes a single server for the application, whereas we explicitly offer a network of distributed servers, offering guaranteed low client-server latency.
3. PolyChrome uses a peer-to-peer network for interclient messaging, whereas we use a client-server architecture. The peer-to-peer alternative has potentially smaller interclient latencies than the client-server architecture; the client-server architecture is more robust in the face of common inhibitions on communication, notably firewalls blocking incoming flows.
4. PolyChrome offers Operational Transform [20, 56] as a synchronization method. We currently offer no synchronization or in-order guarantees.
5. The focus of our research is on low-latency access to large data sources, whereas PolyChrome does not offer latency guarantees.

Shared Presentation Systems Shared Presentation Systems such as WebEx, Citrix GoToMeeting, or Prezi are sufficiently ubiquitous that they deserve a separate paragraph. We distinguish between *collaboration* systems and *presentation* systems. The latter feature a single presenter who manipulates the system, while the remaining participants are largely passive observers. The former, including our system and collaborative games, treat all participants symmetrically: anybody can manipulate the environment. The distinction is important because latency is only a serious concern for the presenter in a presentation system, but is a concern for all participants in a collaboration system such as the IDCVS.

To our knowledge, the Ignite system is the only interactive, thin-client, collaborative application around Big Data in the world today.

7 Demo and Future Work

7.1 Demonstration at Future Internet Summit

We demonstrated the system as a major International Demonstrator in the plenary session at the Future Internet Summit in Washington, DC, in March 2015. The system has been running and fully operational for 9 months, at seven sites across five countries on three continents. At the summit we showed this running live with participants in Washington, DC; Victoria, Canada; Tokyo, Japan; Potsdam, Germany; San Francisco, CA; and Ghent, Belgium. The local servers were located at the University of Tokyo; Stanford; the Universities of Victoria, Maryland, and Ghent; and the Hasso-Plattner Institute. In addition, we deployed the demonstrator at NICTA in Sydney, Australia, at the GPO in Cambridge, MA, and Texas A&M University. NICTA and A&M were ExoGENI sites [3]; Stanford, GPO, Ghent, and Maryland were InstaGENI sites [4], Ghent under the auspices of the Federation of Future Internet Explorations project [61]. InstaGENI and ExoGENI are the two major aggregate classes on the GENI project [8]. Tokyo was a VNode node [45]; Victoria was a site on Canada's SAVI infrastructure [32] (Fig. 9).

Participants in the demonstrator saw a mean latency of under 20 ms, and a uniform access time of just under our 150 ms target. Conversely, experiments where

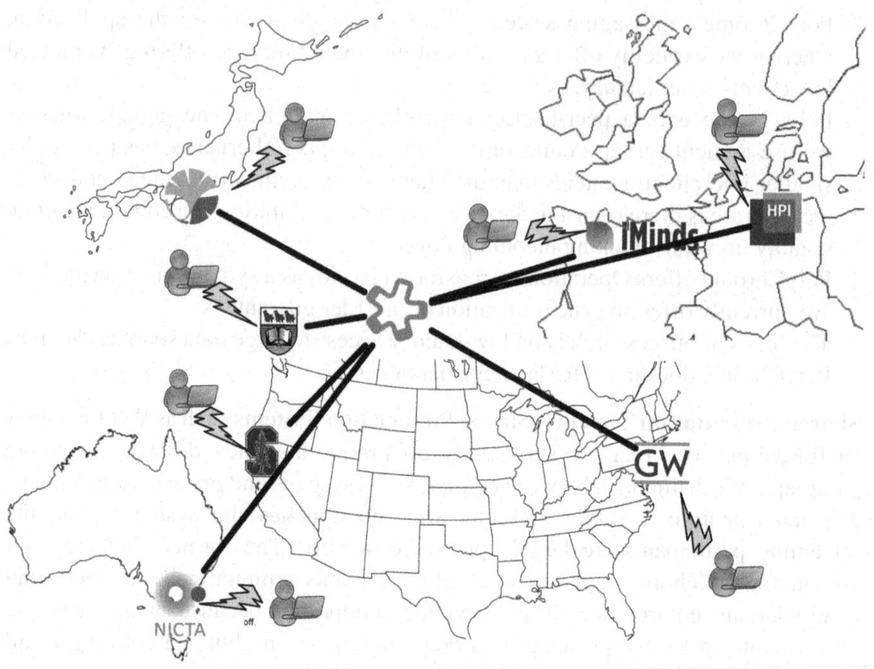

Fig. 9 Deployment and demonstration

we deliberately chose bad servers showed response times of 1000 ms or greater. Users reported that the experience was quite unsatisfactory when confronted with response times significantly above 150 ms.

The visualizer was used on eight different devices during the demonstration, running five different Operating Systems: a Nexus phone, a Nexus tablet, an iPhone, an iPad, a Chromebook, a PC, and a Mac. The OS's were Windows 7, Chrome OS, OS/X, IOS, Ubuntu, and Android, and the browsers were Chrome, Safari, and Firefox.

Preferentially, the Lively server is co-located with the geospatial database, but this is not necessary. The critical thing is that users get the data feed from a local site, over a low-latency high-bandwidth path. The application itself is relatively small and a load-once asset, and so it can be served from anywhere.

The demonstrator achieved the design goals: 150 ms response time for a user-action, complete functionality across a broad range of devices, operating systems, and browsers. This was the first time that interactive collaborative visualization of a large data set had been demonstrated in the wide area on a broad range of clients.

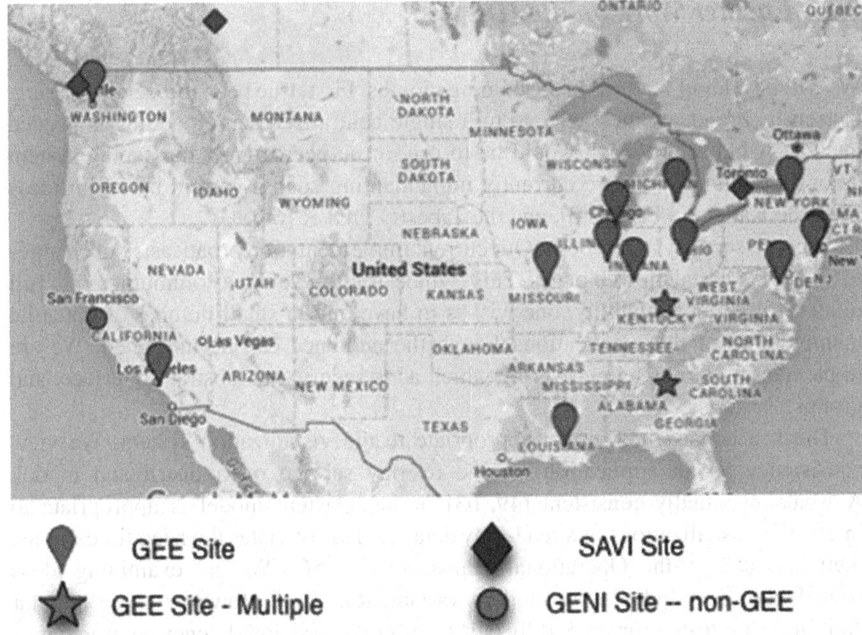

Fig. 10 Current visualizer deployment

7.2 *Current Operational Deployment*

The IDCVS is currently deployed at 20 sites across the GENI and SAVI infrastructures. Of these, 16 are GEE sites; three are SAVI [39] sites; and one is an InstaGENI [44] site not currently running the GEE. Within the 16 GEE sites we share a single VM with all other GEE applications. At the SAVI and non-GEE GENI sites, we are running within a dedicated VM. However, we use a Docker container to deploy at all sites.

The map in Fig. 10 shows the current deployment status. Due to the encapsulation offered by Docker and the wide-area orchestration tools now available, it takes us about 15 min to bring up a new site, and almost all of that time is taken by transmitting and unpacking the 2 GB tarball containing the pollution database. Over the 6 months of operational deployment, we have had outages for only two causes: hardware failure or disconnection at a site and Slice expiration or accidental deletion. In all cases, we've been able to fully restore the service within an hour of the underlying resource recovering.

7.3 Further Work on the IDCVS

We will be adding features in the coming months. First, true peer-to-peer messaging. Currently, broadcast messages go to a single centralized server, adding unneeded latency. A better approach would be to use a true peer-to-peer messaging system among the clients. We are currently implementing such a system based on ideas from the Kademlia DHT [43], and the Tapestry DOLR [64].

Second, security and groups. Our current implementation broadcasts out changes to all clients based on web pages. This is undesirable, even in a community with full mutual trust. For scalability, one wishes to have only a small number of potential manipulators of this application, though the audience can be unlimited. We are implementing a messaging overlay which adds security, an abstract interface, and groups.

Third, a consistency model appropriate to the visualization system. We view consistency as an application-specific overlay service on a distributed model. A weak, eventually-consistent [49, 63] or inconsistent model is appropriate to interactive visualizations of a read-only data set. This obviates the need for complex methods such as the Operational Transform [20, 56]. We are examining ideas from Paxos [35] and a more modern variant, Raft [47]. However, these popular consensus algorithms are generally tailored for relatively low-latency environments. An attractive alternative is a weak version of TeaTime [52, 55]. TeaTime attaches a timestamp to each event, and uses a peer-to-peer voting algorithm to establish a commit to each event and a total order on distributed events; the voting process obviates the need for a server. The voting protocol can be dispensed with if a server is used, and this was done in the shipped Croquet system. Both variants of TeaTime imply at least one round-trip time across the distributed clients. Latency can be minimized if one is prepared to accept an eventual-consistency model. The timestamp a total order on distributed events, using a server or peer-to-peer protocol to ensure a consistent total order across clients. All variants of TeaTime require at least one round-trip time across the distributed clients.

TeaTime is designed to have a global order on events largely consistent with the order of events as seen by an omniscient third-party observer. This is required when the state of the system is history-dependent. In our application, only the last event matters. So we simply attach a local timestamp to each user event, and use these timestamps to impose a total order on events at each site; ties are broken by an order on client IP address. This offers a total order; it may not be consistent with the "real" order as measured by an omniscient third-party observer, but as has long been known this is somewhat indeterminate in any event [19, 34].

The Atmospheric Quality Visualization Application served as the lighthouse application for the IDCVS, but the value of the IDCVS is in multiple visualization applications on the same platform. To this end, we and our collaborators are developing two other applications for the IDVCS platform: a genomics viewer and a visualizer for publicly-available sports data. We expect that the development of these applications will sharpen the deployment tools for the platform.

7.4 *Further Work on The Atmospheric Quality Visualization Application*

In addition, we continuously re-evaluate both the client mapping software we use and geospatial database engines. For the former, we continually seek high-performance client rendering.

The visualization system is currently used to demonstrate the capabilities of the Ignite visualization system, but there are significant public benefits possible in the future. The current virtual pollution sensor shows values on both a coarse timeline and a coarse grid, but the method can yield data on much finer time scales and data grids, enabling local specific projections of pollution values.

Cities will be able to provide real time interactive access to the abundance of airborne particulates for every census tract over the last two decades and relate this to their local policy decisions and sources of pollution, as well as the spatial and temporal (seasonal and inter-annual) changes, thus facilitating data driven policy decisions and retrospective environmental impact assessments.

A real-time online data fusion system combining the global particulate abundance as well as real-time high spatial resolution local streaming data from distributed sensors across local neighborhoods will allow the construction of real-time street level pollution views. These can then be used to provide real-time route planning tools for citizens planning outdoor exercise (walks and cycling) with minimum pollution exposure. These inexpensive wireless streaming local mobile pollution sensors ($2k) can be eventually deployed across each US Ignite city and beyond to provide a paradigm shift in neighborhood scale air quality and environmental public health applications.

The real-time online data fusion system just described can be used to provide real-time alerts to local citizens, e.g. to the over 20 million people in the USA with asthma. A timely alert to make sure asthmatics have their inhaler or avoid outdoor strenuous activity could be the difference between a normal day and a very traumatic and expensive day spent in the Emergency room. Prevention is definitely better than cure.

The visualizer can also be used to demonstrate the levels of pollution and the value of reducing pollution in our daily lives. Two dramatic stories that clearly make the case for the influence of $PM_2.5$ on our health are: A life expectancy of 27 years in the vicinity of the Aral Sea due to bioagents in the airborne particulates and a reduction in life expectancy of at least 5 years in Northern China due to the emissions from coal fired power stations. The real-time Distributed Collaborative Visualization System can be used to tell these and many other stories demonstrating the benefit of data driven decisions for local policy.

Acknowledgements The authors thank our colleagues at US Ignite, particularly Joe Kochan and William Wallace. We were assisted in deployment by the GENI Project Office, with particular thanks to Niky Riga, Mark Berman, Marshall Brinn, and Sarah Edwards. Andy Bavier set us up on the GEE, as did Rob Ricci on InstaGENI and Ilya Baldin on ExoGENI. Andi Bergen, Hausi Muller and Hadi Bannazadeh of the SAVI Project in Canada, Aki Nakao of the VNode project in

Japan, Brecht Vermuelen of Fed4Fire, Max Ott of NICTA and Robert Hirschfeld of HPI set up our international sites. Joe Mambretti iCAIR set up international networking connections. We thank our demo participants—Wim van der Meer of iMinds, Tobias Pape of HPI, and Pratama Putra of the University of Tokyo. We thank our collaborators at the Lively Web, Dan Ingalls and Jens Lincke who participated in an early version of this chapter [9, 10]. Shushil Bhojwani collaborated with us on an early version of the data server, and Ulrike Stege and Yvonne Coady supported us throughout. This work was partially supported by the GENI Project Office, by SAP and by MITACS.

References

1. Ansible api documentation: http://docs.ansible.com/ (2016)
2. Badam, S.K., Elmqvist, N.: Polychrome: a cross-device framework for collaborative web visualization. In: ACM ITS (2014)
3. Baldine, I., Xin, Y., Mandal, A., Ruth, P., Heerman, C., Chase, J.: Exogeni: a multi-domain infrastructure-as-a-service testbed. In: Testbeds and Research Infrastructure, pp. 97–113. Springer, Berlin, Heidelberg (2012)
4. Bastin, N., Bavier, A., Blaine, J., Chen, J., Krishnan, N., Mambretti, J., McGeer, R., Ricci, R., Watts, N.: The Instageni initiative: an architecture for distributed systems and advanced programmable networks. Comput. Netw. **61**(0), 24–38 (2014). Special issue on Future Internet Testbeds—Part I
5. Bavier, A., McGeer, R.: The geni experiment engine. In: The GENI Book Prototype of the Next Internet. Springer, Berlin (2016)
6. Bavier, A., Chen, J., Mambretti, J., McGeer, R., McGeer, S., Nelson, J., O'Connell, P., Ricart, G., Tredger, S., Coady, Y.: The geni experiment engine. In: Teletraffic Congress (ITC), 2014 26th International, September, pp. 1–6 (2014)
7. Bavier, A., Chen, J., Mambretti, J., McGeer, R., McGeer, S., Nelson, J., O'Connell, P., Tredger, S., Coady, Y.: The geni experiment engine. In: Tridentcom (2015)
8. Berman, M., Chase, J.S., Landweber, L., Nakao, A., Ott, M., Raychaudhuri, D., Ricci, R., Seskar, I.: Geni: a federated testbed for innovative network experiments. Comput. Netw. **61**(0), 5–23 (2014). Special issue on Future Internet Testbeds—Part I
9. Bhojwani, S., Hemmings, M., Ingalls, D., Krahn, R., Lary, D., Lincke, J., McGeer, R., Ricart, G., Roder, M., Coady, Y., Stege, U.: The ignite distributed collaborative scientific visualization system. In: Distributed Cloud Computing Workshop (2015)
10. Bhojwani, S., Hemmings, M., Ingalls, D., Krahn, R., Lary, D., Lincke, J., McGeer, R., Ricart, G., Roder, M., Coady, Y., Stege, U.: The ignite distributed collaborative scientific visualization system. In: IEEE CloudCom (2015)
11. Choy, S., Wong, B., Simon, G., Rosenberg, C.: The brewing storm in cloud gaming: a measurement study on cloud to end-user latency. In: Proceedings of the 11th Annual Workshop on Network and Systems Support for Games, p. 2. IEEE Press, Piscataway (2012)
12. Cisco: Cisco global cloud index: forecast and methodology, Cisco, Inc., San Jose, CA, 2014–2019 (2015)
13. Cisco: Cisco visual networking index: forecast and methodology, Cisco, Inc., San Jose, CA, 2014–2019 (May 2015)
14. Cisco: The zettabyte era: trends and analysis, Cisco, Inc., San Jose, CA, (May 2015)
15. D3.js: Data-driven documents. http://d3js.org/ (2015)
16. DeFanti, T.A., et al.: The optiportal, a scalable visualization, storage, and computing interface device for the optiputer. Futur. Gener. Comput. Syst. **25**(2), 114–123 (2009)
17. DeFanti, T.A., Acevedo, D., Ainsworth, R.A., Brown, M.D., Cutchin, S., Dawe, G., Doerr, K.-U., Johnson, A., Knox, C., Kooima, R., et al.: The future of the cave. Cent. Eur. J. Eng. **1**(1), 16–37 (2011)

18. Docker: https://www.docker.com/whatisdocker/. (2016)
19. Einstein, A.: On the electrodynamics of moving bodies. Ann. Phys. pp 891–921, (1905)
20. Ellis, C.A., Gibbs, S.J.: Concurrency control in groupware systems. In: Proceedings of the 1989 ACM SIGMOD International Conference on Management of Data, SIGMOD '89, pp. 399–407. ACM, New York, NY (1989)
21. Fabric api documentation: http://docs.fabfile.org/en/1.8/ (2016)
22. Dan Rayburn, Frost and Sullivan. Comparing cdn performance: Amazon cloudfront's last mile testing results. https://media.amazonwebservices.com/FS_WP_AWS_CDN_CloudFront.pdf (2016)
23. Goradia, I., Doshi, J., Kurup, L.: A review paper on http://store.steampowered.com/ universe/vr rift & project morpheus. Int. J. Curr. Eng. Technol. **4**(5) pp. 3196–3200, (2014)
24. Grochow, K., Howe, B., Stoermer, M., Barga, R., Lazowska, E.: Client + cloud: evaluating seamless architectures for visual data analytics in the ocean sciences. In: International Conference on Scientific and Statistical Database Management, SSDBM'10, pp. 114–131. Springer, Berlin, Heidelberg (2010)
25. Grochow, K., Stoermer, M., Fogarty, J., Lee, C., Howe, B., Lazowska, E.: Cove: a visual environment for multidisciplinary ocean science collaboration. In: IEEE Sixth International Conference on e-Science, ESCIENCE '10, pp. 269–276. IEEE Computer Society, Washington, DC (2010)
26. Ingalls, D.: The smalltalk-76 programming system design implementation. In: ACM Conference on Principles of Programming Languages (1978)
27. Ingalls, D., Kaehler, T., Maloney, J., Wallace, S., Kay, A.: Back to the future: the story of squeak, a practical smalltalk written in itself. In: OOPSLA, pp. 318–326. ACM Press, New York (1997)
28. IPCC: Climate Change 2013: The Physical Science Basis. Contribution of Working Group I to the Fifth Assessment Report of the Intergovernmental Panel on Climate Change. Cambridge University Press, Cambridge/New York, NY (2013)
29. Jagodic, R., Renambot, L., Johnson, A., Leigh, J., Deshpande, S.: Enabling multi-user interaction in large high-resolution distributed environments. Futur. Gener. Comput. Syst. **27**(7), 914–923 (2011)
30. Johnson, G.P., Abram, G.D., Westing, B., Navr'til, P., Gaither, K.: Displaycluster: an interactive visualization environment for tiled displays. In: 2012 IEEE International Conference on Cluster Computing (CLUSTER), pp. 239–247. IEEE, New York (2012)
31. jQuery: jquery: write less, do more. https://jquery.com/ (2015)
32. Kang, J., Lin, T., Bannazadeh, H., Leon-Garcia, A.: Software-defined infrastructure and the SAVI testbed. In: Testbeds and Research Infrastructure: Development of Networks and Communities - 9th International ICST Conference, TridentCom 2014, Guangzhou, 5–7 May 2014. Revised Selected Papers, pp. 3–13 (2014)
33. Krahn, R., Ingalls, D., Hirschfeld, R., Lincke, J., Palacz, K.: Lively Wiki a development environment for creating and sharing active web content. In: WikiSym '09: Proceedings of the 5th International Symposium on Wikis and Open Collaboration, pp. 1–10. ACM, New York, NY (2009)
34. Lamport, L.: Time, clocks, and the ordering of events in a distributed system. Commun. ACM **21**(7), 558–565 (1978)
35. Lamport, L: The part-time parliament. ACM Trans. Comput. Syst. **16**(2), 133–169 (1998)
36. Lary, D.J., Faruque, F.S., Malakar, N., Moore, A., Roscoe, B., Adams, Z.L., Eggelston, Y.: Estimating the global abundance of ground level presence of particulate matter ($pm_{2.5}$). Geospat. Health **8**(3), 611–630 (2014)
37. Lary, D.J., Lary, T., Sattler, B.: Using machine learning to estimate global $pm_{2.5}$ for environmental health studies. Environ. Health Insights **2015**(1), 41–52 (2015). doi:10.4137/EHI.S15664
38. Leigh, J., Johnson, A., DeFanti, T.A.: Cavern: a distributed architecture for supporting scalable persistence and interoperability in collaborative virtual environments. Virtual Reality: Res. Dev. Appl. **2**(2), 217–237 (1997)

39. Leon-Garcia, A., Bannazadeh, H.: Savi testbed for applications on software-defined infrastructure. In: The GENI Book Prototype of the Next Internet. Springer, Berlin (2016)
40. Leuf, B., Cunningham, W.: The Wiki Way: Collaboration and Sharing on the Internet. Addison-Wesley Professional, Reading (2001)
41. Lively: http://www.lively-web.org/ (2016)
42. Maloney, J.H., Smith, R.B.: Directness and liveness in the morphic user interface construction environment. In: Proceedings of User Interface and Software Technology (UIST 95), pp. 21–28. ACM Press, New York (1995)
43. Maymounkov, P., Mazières, D.: Kademlia: a peer-to-peer information system based on the xor metric. In: Revised Papers from the First International Workshop on Peer-to-Peer Systems, IPTPS '01, pp. 53–65. Springer, London (2001)
44. McGeer, R., Ricci, R.: The instageni project. In: The GENI Book Prototype of the Next Internet. Springer, Berlin (2016)
45. Nakao, A.: Deeply programmable network through advanced network virtualization. In: IEICE International Symposium on Network Virtualization (2012)
46. Oculus: rift: Next generation virtual reality. https://www.oculus.com/en-us/rift/ (2015)
47. Ongaro, D., Ousterhout, J.: In search of an understandable consensus algorithm. In: Proceedings of the 2014 USENIX Conference on USENIX Annual Technical Conference, USENIX ATC'14, pp. 305–320. USENIX Association, Berkeley, CA (2014)
48. Passarella, A.: A survey on content-centric technologies for the current internet: Cdn and p2p solutions. Comput. Commun. **35**(1), 1–32 (2012)
49. Petersen, K., Spreitzer, M.J., Terry, D.B., Theimer, M.M., Demers, A.J.: Flexible update propagation for weakly consistent replication. In: Proceedings of the Sixteenth ACM Symposium on Operating Systems Principles, SOSP '97, pp. 288–301. ACM, New York, NY (1997)
50. Polymaps: http://polymaps.org/ (2016)
51. Ponto, K., Doerr, K., Kuester, F.: Giga-stack: a method for visualizing giga-pixel layered imagery on massively tiled displays. Futur. Gener. Comput. Syst. **26**(5), 693–700 (2010)
52. Reed, D.P.: Designing croquet's teatime: a real-time, temporal environment for active object cooperation. In: Companion to the 20th Annual ACM SIGPLAN Conference on Object-Oriented Programming, Systems, Languages, and Applications, OOPSLA '05, p. 7. ACM, New York, NY (2005)
53. Resnick, M., Maloney, J., Monroy-Hernández, A., Rusk, N., Eastmond, E., Brennan, K., Millner, A., Rosenbaum, E., Silver, J., Silverman, B., Kafai, Y.: Scratch: programming for all. Commun. ACM **52**(11), 60–67 (2009)
54. Smarr, L.L., Chien, A.A., DeFanti, T., Leigh, J., Papadopoulos, P.M.: The optiputer. Commun. ACM **46**(11), 58–67 (2003)
55. Smith, D., Kay, A., Raab, A., Reed, D.P., et al.: Croquet-a collaboration system architecture. In: Proceedings. First Conference on Creating, Connecting and Collaborating Through Computing, 2003. C5 2003, pp. 2–9. IEEE, New York (2003)
56. Sun, C.: Undo as concurrent inverse in group editors. ACM Trans. Comput. Hum. Interact. **9**(4), 309–361 (2002)
57. Taivalsaari, A., Mikkonen, T., Ingalls, D., Palacz, K.: Web browser as an application platform: the lively kernel experience, Technical Report. Sun Microsystems, Inc., Mountain View, CA, USA (2008)
58. Tolia, N., Andersen, D.G., Satyanarayanan, M.: Quantifying interactive user experience on thin clients. IEEE Comput. **39**(3), 46–52 (2006)
59. Ungar, D., Smith, R.B.: Self: the power of simplicity. In: Conference Proceedings on Object-Oriented Programming Systems, Languages and Applications, OOPSLA '87, pp. 227–242. ACM, New York, NY (1987)
60. Valve: Steamvr. http://store.steampowered.com/universe/vr (2015)

61. Vandenberghe, W., Vermeulen, B., Demeester, P., Willner, A., Papavassiliou, S., Gavras, A., Sioutis, M., Quereilhac, A., Al-Hazmi, Y., Lobillo, F., Schreiner, F., Velayos, C., Vico-Oton, A., Androulidakis, G., Papagianni, C.A., Ntofon, O., Boniface, M.: Architecture for the heterogeneous federation of future internet experimentation facilities. In: 2013 Future Network & Mobile Summit, Lisboa, 3–5 July 2013, pp. 1–11 (2013)
62. Vlachos, A.: Advanced vr rendering. In: Game Developers Conference (2015)
63. Vogels, W.: Eventually consistent. Commun. ACM **52**(1), 40–44 (2009)
64. Zhao, B.Y., Kubiatowicz, J., Joseph, A.D.: Tapestry: an infrastructure for fault-tolerant wide-area location and routing. Technical Report, UC-Berkeley (2001)
65. 7 million premature deaths annually linked to air pollution. http://www.who.int/mediacentre/news/releases/2014/air-pollution/en/ (May 2014)

US Ignite and Smarter Communities

Glenn Ricart and Rick McGeer

What will the next generation of the Internet do? How will it change healthcare, education, public safety, clean energy, transportation, and advanced manufacturing?

These are the questions that launched US Ignite.

Computer Science research led by the National Science Foundation, DARPA, and corporate R&D labs have led to powerful new concepts. The NSF programs FIND (Future Internet Design), GENI (Global Environment for Network Innovation), and FIA (Future Internet Architectures) have led to a number of advanced networking concepts that could be transformational. US Ignite invites a wide range of application developers an opportunity to play with these new ideas to see what kinds of applications they make possible. US Ignite also aims to make trial deployments of these applications in testbed communities possessing the necessary advanced infrastructure, and to encourage more communities to deploy the necessary advanced infrastructure.

1 Genesis of the US Ignite Initiative

US Ignite began as an idea and partnership between the Office of Science and Technology Policy (OSTP) and the National Science Foundation (NSF). Prof. Jon Peha was on tour at OSTP from Carnegie-Mellon University, and, at OSTP Deputy Director Tom Kalil's urging, wrote a short white paper on the economic importance of getting beyond the "chicken and egg" problem created by the amazing abilities of the current Internet. The current Internet has become pervasive and good enough to handle a wide range of applications. App stores have tens of thousands of

G. Ricart (✉) • R. McGeer
Chief Scientist, US Ignite, 1150, 18th St NW, Suite 900, Washington, DC 20036, USA
e-mail: Glenn.Ricart@us-ignite.org; rick.mcgeer@us-ignite.org

© Springer International Publishing Switzerland 2016

R. McGeer et al. (eds.), *The GENI Book*, DOI 10.1007/978-3-319-33769-2_20

applications available. However, further improvements in the infrastructure will have to be justified by new applications. And new applications won't appear until the infrastructure will support them.

In 2009, OSTP/NSF/GENI asked Glenn Ricart to look at this problem and suggest specific advanced infrastructure and the cost of deploying it. What kind of investment in advanced infrastructure would the country need to make to create a market for next-generation applications?

The numbers were in the hundreds of millions of dollars, according to Ricart's calculations. The US Ignite idea was being worked by Dr. Nick Maynard at OSTP and Dr. Suzanne Iacono at the NSF. For about a year, there were some hopes that a portion of spectrum auctions by the FCC could be used to fund US Ignite infrastructure, but those hopes faded as a congressional consensus grew around funding public safety communications infrastructure.

Since this was a chicken-and-egg problem, Ricart pointed out that if we couldn't afford the chicken, let's focus on hatching some eggs. The NSF and corporate participants could provide funding for application development teams to begin the process. A wide range of teams would be asked to be as innovative as possible. It wouldn't be necessary to pick and choose specific new Internet infrastructure; the innovative developers would get to pick. Corporate labs would be invited to participate along with academics, open source developers, and community groups such as Code for America brigades.

Working with corporate labs and community groups would be awkward for a government-only effort. It was decided to use an existing nonprofit, One Economy, whose mission was to use technology to connect underserved communities to online information and resources, to be US Ignite's main interface to corporations, cities, and communities. One Economy's executives suggested that corporate funding would be strongest if US Ignite operated with former executives from the telecom industry. Susan Spradley, formerly of Nokia and Alcatel-Lucent, was tapped for executive director. William Wallace and Joe Kochan also had telecom industry pedigrees. Together with Ricart, the team was ready to go.

There was a significant period of sounding out potential sponsors, cities, and finding appropriate applications and services. During this time, a new US Ignite nonprofit was spun out of One Economy and established offices near Dupont Circle.

With strong assistance from OSTP and NSF, US Ignite had its official coming-out party on June 14, 2012 in the White House South Auditorium. John Holdren, the President's Science Advisor, welcomed the standing-room-only group to the White House. From the government side, speakers included Subra Suresh, director of the NSF, Julius Genachowski, the chair of the Federal Communications Commission, Larry Strickling, head of the National Telecommunications and information Agency of the Department of Commerce, and Todd Park, then CTO of Health and Human Services but later to become CTO for the U.S. Government.

There was also strong private sector support. John Donovan from AT&T spoke, and representatives were present from many of the larger industry players. US Ignite contributing industry partners included Verizon, Comcast, Juniper Networks, Cisco, Corning, Extreme Networks, JDSU, Google, HP, and Big Switch Networks.

Communities with advanced infrastructure were also critical. There were videos from Chattanooga with 170,000 homes and small businesses connected by gigabit-capable fiber, Cleveland Ohio with its "Beta Block" of gigabit fiber-connected homes conducting medical experiments with Case Western Reserve University and the Cleveland Clinic, and Clemson University working with Portland, Oregon on less-polluting smart transportation.

To help show the kinds of next-generation applications that might be possible, Case and the Cleveland Clinic gave a demonstration of simulated remote surgery which would require a low-latency and highly reliable Internet. And John Underkoffler of Oblong, Inc. talked about how hand gestures might control augmented reality in perceptually real time.

The launch was a success. The US Ignite nonprofit would end up with 16 corporate partners and 31 communities participating by its second anniversary in 2014. That year, it held an Applications Summit with 30 demonstrations of next-generation applications. The NSF allocated $10 millions of its budget to support US Ignite applications development and deployment. The NSF brought the Mozilla Foundation on board to cultivate grass-roots community applications. US Ignite forged relationships with the Open Networking Foundation and Internet2.

2 What's New about the Next Generation of the Internet

Some themes began to develop out of the first three years of US Ignite. The next generation of Internet applications would take advantage of seven strong trends:

1. Symmetric (same rates for upload and download) gigabit to the end-user.
   The simplest and easiest-to-conceive apps involved those that were bandwidth-constrained. A radiologist in Chattanooga could be sent a full-resolution 6 GB Digital Imaging and Communications in Medicine (DICOM) medical image at his home in the middle of the night in less than a minute for a stat reading that could help save a life. A disadvantaged middle-school student in Kansas City connected to Google Fiber could create and edit HD video using professional video editing software running on a server in the Kansas City Public Library. Smart phones generating uncompressed HD video would actually have the upstream bandwidth needed to allow anyone to originate high quality live video.

   Furthermore, symmetric gigabit to the end-user was going to be one of the most easily-deployed pieces of advanced infrastructure. The lower lifecycle costs of fiber were already beginning to tilt in the direction of fiber to the home. But providers effectively were limiting the bandwidth to traditional levels, not because it was expensive to provision gigabit to the home, but because the *upstream* costs of gigabit access were relatively high and because there was no perceived need at the provider level for symmetrical bandwidth.

   There is a symbiosis between symmetrical gigabit and locavore computing, discussed below. Locavore computing services handle requests within the community in which they originate, reducing the demand for upstream, or "backhaul" bandwidth. Locavore computing therefore acts to reduce the system costs of symmetrical gigabit bandwidth.

One of the biggest boosts to US Ignite came from Google. The Google Vice President for Access Services, Milo Medin, had known Glenn Ricart from the early days of the Internet when both were involved with creating the first Internet access points. Ricart ran the East coast node, and Medin the West Coast. Obviously, the two interconnection points had to be kept in sync and Ricart and Medin talked regularly. Ricart shared the US Ignite vision with Medin, but Medin was already planning to provide a local Internet infrastructure a clear step up from prevailing broadband. Medin's group at Google introduced $70/month symmetric gigabit service in Kansas City. This was the first affordable symmetrical gigabit. Based on this lead, other key US Ignite communities matched. Chattanooga went to a $70 symmetrical gigabit on its fiber, and XMission in Utah went to a $69 symmetrical gigabit available in portions of 13 cities over the UTOPIA fiber.

Big bandwidth applications and big upstream applications were now possible, opening opportunities in video and projected computing power. Dublin, Ohio became a US Ignite city and announced a municipal CAD/CAM and simulation capability available over the local fiber network to attract small advanced manufacturing companies.

It's become clear that downstream bandwidth is being driven by higher resolution screens, projections, and head-mounted displays (including direct projection into the retina). After HD, we've seen growing use of QHD (quad-HD), 4K (4-thousand pixels along the longest edge, or 8 megapixels), and 8K video (32 megapixels). While retina-level resolution was once considered to be the end of this direction, immersive experiences that surround you with (sometimes virtual) screens require retina-level stereoscopic resolution in whatever direction a user saccades his eyes or turns his head. To preserve freedom of action, everyone in the same room (physical or virtual) may have a different viewpoint and hence their own freedom to look in any direction and see retina-scale resolution with stereopsis.

Immersive VR is becoming a commercial reality thanks to the emergence of the Oculus Rift and Steam VR, and devices like Google Cardboard and Wearality, which turn cellphones into VR devices. The bandwidth requirements of stereopsis are dependent on a number of assumptions: how far away is the display surface from the eye, how many frames per second, how dense must the display be, how effective is the compression (which in turn depends on how much computing power is available for compression and decompression). We give one sample calculation here.

A stereopsis is the inner surface of a sphere, and at any time a user is looking at a window of $w \times h$ degrees on the sphere, where w and h are, respectively, the horizontal and vertical viewing angles. A standard display has $w = 120$, $h = 67.5$. The entire sphere has $w = 360$, $h = 180$, and so the total inner surface can be transmitted in $3 \times 8/3 = 8$ streams. A retina display on a handheld device typically has 1920 × 1080 pixels. Steam VR and Oculus Rift are slightly larger, with twin 1200 × 1080 displays. In the worst case, then, retina-quality surround video on a headmount display can be had for eight 2K streams, equivalent to a 16 megapixel display. This is halfway between 4K and 8K video streams. Netflix claims that 4k video can be delivered over a 15.6 Mbps/pipe and thus a projection based on this compression ratio is that retina-quality surround is feasible at 20 Mbps or so. Netflix can compress aggressively because Netflix is sending canned content. If a less-aggressive target of 99 % compression is used, we have 16 megapixel/frame × 24 bits/pixel × 30 frames/s × .01 = 100 Mbits/s. Stereopsis adds a second stream (one stream per eye) but the two streams are quite similar and therefore mutual compression is high; this should only increase bandwidth requirements modestly.

It appears that such realistic ultra-high-definition video will be the dominant use of big downstream bandwidth in the latter half of the 2010s. But an early US Ignite application called Engage3D (by Bill Brock of Chattanooga) demonstrated something that could chew up even more downstream bandwidth: realistic models of scenes that could be re-projected using 3D transformations controlled by the viewer in real-time. In Engage3D, a scene is decomposed into objects with video images painted onto the objects. All of this information on all the objects and all the surfaces is then transmitted 30 times per second in a stream to the viewers. By reducing resolution, the bandwidth was limited to fit into a 1 Gbps pipe.

One can imagine interacting realistic models (e.g., like a more realistic Second Life) exchanging information on how they are each impacting everyone else's coupled models. For example, if all the models are interacting in a given virtual geographic space, the viewpoint of each model must include the results of every other model operating in the same space.

Big upstream bandwidth is likely to be driven by self-produced video. Even a single cellphone video stream is greater than the upstream bandwidth of nearly every broadband provider in 2014. And, we expect smart houses, smart cars, smart cities, and personal medical centers to generate tons of sensor (including high resolution video sensor) upstream data. Perhaps one day everyone will real-time report their EKG via upstream wireless data to be processed against known arrhythmias for early prediction/diagnosis of cardiac events. Or, perhaps brain waves from multiple sensors or arrays of sensors will be analyzed by powerful computers to permit thought control over our environment.

For all these present and future reasons, US Ignite is suggesting a 2017 baseline of a symmetric 1 Gbps to and from a household.

2. Low Latency and Locavore Infrastructure.

   The previous section focused on bandwidth, and many people and communities do focus rather extensively on bandwidth. But low latency is destined to be the next big bandwidth.

   Latency is the time between some trigger event—waving a hand, moving the point of focus of your eyes, detecting an oncoming car, or asking to see your calendar and potentially conflicting events for next April 5th, for example, and someone seeing, hearing, touching, or feeling the results. In 2016, web pages tend to take about 1.2 s to load over a residential broadband connection. Those 1.2 s are the latency. What could we save if the latency was only half that time? If in the year 2020, say, five billion people each accessed 200 web pages (or their equivalent) per day and latency was decreased by 0.6 s, we could save 190 person-centuries per day with an economic value (at $12/h) of USD $2 billion per day. Stated another way, we could approximately increase global productivity by the yearly cost of the US Department of Defense every day.

Further, latency considerations are inhibiting the migration of applications from the desktop computer and laptop to the Cloud. In 1995, then-Sun CTO Eric Schmidt predicted that copious bandwidth would "hollow out" the PC and usher in the era of thin clients, where most applications would be accessed over, and most data would be resident in, the network. For low-data applications, where application storage can be accommodated on a thin device, and for latency-tolerant applications this is the case. However, big-data and latency-sensitive applications still, to date, must be run on PCs on the desktop. A regime of low latency to the Cloud would complete this vision of hollowing out the PC.

So, the piece of the network infrastructure perhaps most worth improving is latency. In addition to boosting productivity, new applications that appear to act in concert with people (and their music, dancing, gestures, mouse clicks, speech, vehicle driving, and their learning games) will become possible.

Musicians tell us that less than 4 hundredths of a second of latency is perceived as essentially simultaneous and allows classical ensemble playing and ensemble dancing. For sound, that corresponds to about 50 physical feet of separation on stage. In an orchestra, all of the woodwinds try to be within about 50 feet of each other; the same for brass instruments. Players cannot wait to hear the note from other players before voicing their own note; they watch the considerably faster speed of light indication from the conductor's baton. But smaller ensembles without conductors seat all players within 50 feet of one another.

The same 4 hundredths of a second is the time that light travels about 5000 miles in fiber optics. Even if one had a direct fiber optic path to a server and back, the server couldn't be located more than about 2500 miles from the source of the trigger event in order to return an action that would be perceived as simultaneous by a human user.

For example, LOLA (low latency audio/video for simultaneous musical performance) from Conservatorio di musica Giuseppe Tartini in Italy achieves a sub-4-hundreds of a second of latency by using a special industrial camera which outputs each 1/3rd of a scan line as it is being scanned. That 1/3rd of a scan line is then put into its own packet and forwarded over a specially-created network path with few obstacles. At the other end, a performer sees the result. The largest distance over which a simultaneous classical musical performance has been demonstrated with LOLA is about 750 miles.

Another theme for US Ignite applications is being highly responsive to sensors or sometimes even other applications. Today's home Internet typically loads a web page on a fast broadband connection in about 1.2 s. People who interact with web pages as their normal job (health care scheduler, telemarketer, reservation agent, etc.) may request a new web page every 10–15 s and sometimes more often. The most productive agents work in centers with high bandwidth and low-latency enterprise connections to the servers, and often the servers are located not far away in order to reduce latency further. People attempting to do such work out of their homes find themselves at a disadvantage because of the latency inherent in the Internet. That disadvantage can be up to a 10 % overall performance penalty (up to an extra 1.2 s every 12 s).

US Ignite applications often utilize several techniques to reduce latency and improve productivity and to help make the application response appear to the person to be instantaneous.

There are several techniques US Ignite applications can use to reduce latency:

(a) Use faster access pipes. A gigabit pipe to the home carries a given size packet in a hundredth of the time taken by a 10 Mbps pipe because the bits are simply clocked faster. This is often a direct gain when higher access speeds are used over fiber networks because the fiber is simply clocked at a higher rate. For cable access, the same effect is achieved by sending more bits in parallel on different cable frequencies.
(b) Use Software-Defined Networking (SDN) to choose more direct or lower-latency paths. The typical Internet connection stops at and is re-forwarded from a dozen intermediate routers. Those stops and starts aren't needed in an SDN network which uses switches that match header information to quickly forward the packet. The delays of ingest, analyze, queue, and forward are omitted. More information on SDN is in the next section below.
(c) Keep processing close. Put the servers very near the sensors and actuators and peoples. We call servers located near their users or sensors or cyberphysical systems *locavore* servers. The servers and the storage and the network architecture designed to support them are *locavore* infrastructure. We borrow the term from people who eat food grown locally when possible. Locavore infrastructure reduces latency by running applications locally when possible. In a nice side benefit, powerful locavore servers with gigabit access paths can provide a faster and smoother experience for users than their own local PC or device, particularly if that local device is a lower-cost device helping to bridge the digital divide.

In a locavore infrastructure, every element plays a role. The server is local so that it can take advantage of abundant local bandwidth. Packets don't have to travel far, saving time otherwise lost to the speed of light. Fewer intermediate Internet boxes are traversed, saving time with each box avoided. Sensitive information is present on far fewer links, reducing the attack surface. Layer 2 forwarding further reduces any attack surface and provides for application-specific economics.

It is also important to re-iterate that locavore infrastructure makes high bandwidth affordable. As was noted above, the driving cost of high bandwidth is not the cost of the local area connection, but rather the upstream cost of provisioning high-bandwidth links beyond city limits. By making computation local and servicing most requests locally, high-bandwidth links can be provisioned without incurring upstream expenses.

3. Layer 2 SDN links between the locavore infrastructure points.
Today's Internet largely operates on what's called "Layer 3," so-named for the fact that the Internet Protocol runs on layer 3 of the protocol stack. Layer 1 is the physical layer—and specifies how information is sent over fiber or wireless or copper media. Layer 2 is the logical layer which understands how the layer 1 paths interconnect with one another. Layer 3 is powerful because at layer 3, a packet can get from anyplace on the Internet to any other place.

The regular commercial Internet does long-distance routing at layer 3 almost capriciously. If more than one tier-1 provider[1] is involved, the handoff of a packet between providers will usually occur wherever it is most convenient or least expensive for the originating provider. This location may have little to do with shortest routes or end-to-end optimization.[2] The overall result is that round-trip packets often go on different routes in each direction; this is referred to as asymmetric routing.

Even when traffic isn't handed off to another tier-1 provider, there can be inefficiencies. The provider may have insufficient capacity on direct links, so the packet is sent on a long multi-hop journey reminiscent of low-cost airline trip routings. Packets may also be subject to discriminatory "traffic engineering" that favors certain types of traffic over others during congested times or over congested links. While the Internet protocols will retry until the traffic gets through, the result can be long delays and spinning icons as users wait.

Perhaps in a perfect world, there is always abundant bandwidth so that traffic engineering is unnecessary. But in real life and especially in wireless networks, there *are* hot spots that need to be handled in some way, shape, or fashion. The author prefers to let economics work its wonders. When a commodity is in short supply, the price goes up, and the additional revenue causes either an incumbent or a competitive provider to supply additional capacity to capture additional revenue.

In this respect, it's useful to be able to charge a higher price to "higher priority" traffic and give it unimpeded access while disproportionately slowing down inexpensive "salvage" traffic (like online PC backup).

But these capabilities generally don't exist in today's Internet; where they do, it's the network provider which is deciding on the traffic engineering they think will optimize the use of their network. It doesn't necessarily optimize the network for your application.

[1]A tier-1 provider has at least a national scale and agreements to exchange traffic with other tier-1 providers, often at no cost to either party.

[2]And it certainly won't optimize costs or distance for providers downstream.

To make the network respond more individually to specific flows, a new paradigm called Software Defined Networking (SDN) has become mainstream in data centers and long-haul networks and is slowly making its appearance on campuses and enterprise networks.

In SDN, instead of each router trying to make complicated decisions about how to handle packets, a controller takes a more holistic view of all the network flows, their needs, how they can be optimized, and also how the network as a whole can be optimized while satisfying the needs of the flows. From a technology standpoint, the data plane is being separated from the control plane.

While there are some mechanisms in the Internet protocols to distinguish types of traffic and deal with priorities, the introduction of an SDN Controller can centralize these functions and provide for end-to-end optimization. The controller can observe how much bandwidth each application believes it would like to use in each direction and the specific handling that traffic should receive.

For example, remote surgery traffic would do well to use guaranteed short low-latency paths and to not drop packets. We don't want to lose a scalpel movement command. Another high-priority stream is for ultra-definition public safety real-time video. It would also like to be delivered over the lowest-latency paths, but, if there are any delays, it's better to drop packets and keep queuing delays low. Web traffic is best handled with medium priority, short paths when possible but somewhat longer paths are acceptable, and it's better to queue and deliver all of the packets instead of dropping them as a contention hint. As a last example, the file backup traffic can be scavenger priority, latency is not a priority, and longer paths can be used if they have greater bandwidth or more scavenger timeslots available.

Performing end-to-end optimization in these environments suggests an SDN controller to communicate with (or at least observe) application traffic. It can do a better job if applications signal their intentions and send packets at a maximum rate which has been pre-arranged. Latency will be optimized if longer-term flows occur at an agreed-upon rate and the packets are generated and inserted into the network at time slots assigned by the controller that avoid queuing along the path. Even if an originating node is incapable of understanding the timeslot system, the first point along the network path which does understand the slot structure can re-structure the packet flow to minimize latency along the rest of the path managed by the slot controller. To accommodate stochastic traffic, certain time slots can be allocated for such traffic with the understanding that there may be queuing or other contention-management conventions employed. The value of these techniques in minimizing latency and dropped packets is enhanced under longer path lengths.

4. Virtual networks or slices.

   The phrase "The Internet" usually refers to that set of interconnected networks which can send packets end-to-end from any connected device to any other connected device based on a common IPv4 and/or IPv6 addressing structure. This architecture is both a blessing and a curse. The blessing is that new services and devices can be added by anyone at any time and when the devices and services have routable addresses, anyone can use the new services and communicate with the new devices.[3] The curse is that those with malicious intent can address and probe any and all devices addressable to them.

If the network is viewed as a large IT system, this architecture is absurd. If a file system were structured this way, then every file would be either universally accessible or only accessible by a single user. However, from the very early days of file systems, it was recognized that there needed to be a sophisticated architecture for information sharing—so that files could be viewed, edited, and executed by members of groups. This architecture continues today and is ubiquitous, and functions on an Internet-wide scale. Permissions for Google Apps files, for example, can be shared on an individualized and customized basis.

There is so much malicious probing that many devices today are located in non-routable address spaces (e.g., 10.x.x.x or 192.168.x.x per RFC1918) to protect them. Network Address Translation (NAT) is used to give them external addresses and/or port numbers to allow servers to respond to their requests. However, these devices can only request services from the rest of the Internet—they cannot provide services or participate in distributed computation. If they are sensors, they will need to have another identifying number other than their address and can only send their information to others on the same non-routable network or initiate connections to servers on the routable Internet.

For web browsing, these are not serious limitations. For an active network with distributed applications, non-routable networks create serious problems. Internet purists would point out that non-routable networks violate the end-to-end principle that states that every two endpoints on the Internet should be able to communicate with one another.

But, sometimes, the application doesn't need to communicate with the entire rest of the Internet. For security or privacy, an application may want to have a private enclave of participants. For example, access to medical records may well be best done by isolating all of the holders and requesters of the medical record in a slice or virtual network. That slice is fully routable within the slice, but, may use a private or overlapping address space which is not accessible to the ordinary Internet. The sensitive medical information has been isolated into a slice.

A slice is the network equivalent of group permissions for files or directories: it specifies an isolated, virtual private network which can only be accessed by specific users or devices. Enforcement of this isolation relies on the new capabilities of Software-Defined Networking, as outlined in this volume's chapter "Programmable, Controllable Networks". Here, the important thing is to note that this is central to the architecture of the next Internet—it will not

[3]Due to factors such as address space exhaustion and the urgency to protect customer devices that should not be used as servers from misuse, the ability to address any device is often restricted by Network Address Translation (NAT) and the use of non-routable address spaces (notably 10.x.x.x and 192.168.x.x RFC1918 spaces).

be a single globally-accessible, flat, network, but rather a large collection of virtualized, isolated application-specific networks, on a universally-accessible global substrate, with the protection and privacy that that implies.

An isolated slice may have components located in many different places on the physical Internet. A doctor at home on the weekend can use their tablet on their home network to access medical record information for their patient who has arrived at an out-of-town emergency room. Still, network virtualization provides a secure slice for the protected health information and the IP packets are not visible on the ordinary Internet.

There are several existing and new technologies for slices. Where static configuration is needed, VLANs will often suffice. VxLAN can be used where a greater range of VLAN IDs is needed. Another choice is MPLS. MPLS provides for a somewhat more dynamic and robust slice because it includes fast recovery options for failed routes.

Software-defined networking provides the most dynamic technology for slice creation and virtualization. SDN controllers or GENI software can create, extend, modify, and delete virtual networks on the fly. For example, in the doctor-at-home example given above, the invocation of the medical record access application on the doctor's home tablet may trigger the dynamic inclusion of that application into the medical record slice.

In work done by Jacobus van der Merwe at the University of Utah, the slice needn't involve the entire device or server. The slice can be targeted down to the level of a single application running on a single tablet. The slice can be mapped onto the public Internet but still provide high security for the application(s) running in the slice.

It is important to note that a slice is more than isolated network connections; it is a virtual network, including distributed VMs and endpoints. This notion of a slice: "a virtual network of virtual machines" was first introduced by PlanetLab in the early 2000s, permitting users to create a network of virtual machines spanning the globe. It was broadened and deepened by GENI, which incorporated the idea of a truly isolated virtual network connecting the virtual machines. In this, PlanetLab and GENI prototyped the Internet of the Future: not a collection of computers connected by application-agnostic pipes, but as a collection of application-specific slices, which offer holistic networking, storage, and computation services on an application-specific basis.

5. Unconstrained local gigabit interchange.
Today's Internet is optimized for one kind of traffic: web traffic from end-users to massive cloud datacenters. Requests from users are aggregated and sent over progressively bigger pipes on their way to the appropriate massive data center. There, hundreds of servers may be involved with each request, taking a few tenths of a second to compose a web page response to the user which is sent back down the aggregation network.

There are only three problems: (1) Delays are too long for highly responsive applications such as those that respond the movement of the user or instantaneous changes in her or his environment. (2) Aggregation networks are not designed for gigabits per user. (3) There are a multitude of ways that sensitive information can be captured, detoured, spoofed, or lost.

Fixing these issues in the whole Internet begins at home. Or at least it begins on campus or within a city or other relatively conscribed area where local academic and civic officials can influence Internet design and economics.

The new economics takes advantage of abundant local bandwidth at the physical layer. First, the increasingly popular use of fiber in local loops and last-mile situations provides a physical medium with very large capacity. Low-cost metro optics gives the ability to have extraordinarily abundant bandwidth to the businesses and homes with fiber connections. Limitations on the local bandwidth are put in place by carriers largely to control aggregation and long-distance costs and have nothing to do with the costs of providing abundant local bandwidth. Second, the ability to create slices with pre-defined bandwidth and other connection properties will increase the ability of physical network providers to sell incremental bandwidth even on their big pipes. That should encourage the creation of more backbone capacity to meet economic demand.

Some communities like the Research Triangle area in North Carolina and the city of Austin, Texas are already attracting competitive gigabit providers. Both of these areas have or soon will have both AT&T Gigapower and Google Fiber providing gigabit-speed local access networks. They each have major universities that also have gigabit networks, and major employers who have gigabit networks. But, because the current generation of the Internet is designed to load web pages from distant data centers (and play video), these gigabit providers are not interconnected locally.

To prevent losing end-to-end gigabit capability and to reduce the latency incurred by next-generation Internet applications, we need to provide unconstrained local gigabit interchange. This is a major purpose of the "Digital Town Square" to be discussed below.

The local bandwidth is there. We just have to interconnect the local bandwidth in a low-latency way in the local area.

But how about between cities? What can we do there?

Let's look at the economics of backbone networks again. Typically, those pipes are purchased at wholesale prices based on the largest amount of usage during a month. ISP costs for connecting to the backbone network can be reduced considerably by limiting that maximum usage. Therefore, many ISPs slow down traffic during those periods of highest usage. Managing bandwidth in this and similar ways is called "traffic engineering." Therefore, the raw bandwidth may be available, but the ISP constrains its use to control its backbone costs. In the future, critical applications may be able to pay the ISP a surcharge to carry that application's traffic on an inter-city basis without slowing even during periods of restrictive traffic engineering. US Ignite applications may need this ability. Examples include real-time healthcare and public safety applications.

6. Reliability.

   A future class of Internet applications may depend upon or require reliable time-bounded delivery of packets. This type of reliability is needed when the packets control time-dependent physical actions such a remote surgery, additive 3D printers, moving vehicles, safety shutoffs, pace makers, deep brain stimulation electrodes, etc. Of course, none of these things is being controlled by the Internet today in real-time because the reliability of delivery and guarantee of low latency simply aren't present. In fact, widely used protocols in today's Internet actually drop packets on purpose to "signal" that they should "please slow down." The end-user application waits and waits for the dropped packet until it "times out" and sends a message requesting the packet be re-sent. While this does slow things down, it results in very perceptible delays. When the response time of an application is quite a bit longer than usual, it's probable that a packet has been dropped on purpose.

   Today's fiber optic networks are so reliable that nearly every requested retransmission is simply the result of this "please slow down" mechanism. Why is it needed? Applications don't have any a priori idea of how fast they can or should send information on the Internet. So they keep sending faster and faster until a packet drop signals it should slow down. The packet send rate is cut dramatically, often by half, and then it begins once again sending faster and faster until there is a new packet drop and corresponding delay and timeout. The result is a sawtooth-like transmission rate which is the opposite of what is needed for deterministic low-latency constant-flow communication.

   There are two good defenses:

   (a) Provide so much bandwidth that there is no need to signal "please slow down." This is the primary mechanism used by high speed research and education networks like Internet2, which attempts to provision new bandwidth when the current bandwidth is more than 40 % used on average. In a US Ignite city, bandwidth over fiber is nearly free, so within a US Ignite city, it makes sense to "over-provision" to prevent packet drops. The additional cost to do so is a small percentage of the total cost.
   (b) Use an out-of-band signaling or negotiation to give each application an approved transmission rate. A centralized controller is used to make sure that the approved transmission rates will not overrun the capacity of the network equipment. If an application needs more bandwidth, the controller might re-route its packets over less-used channels. The notion that there would be explicit signaling between the application and the network is a hallmark of Software-Defined Networking.

US Ignite will encourage US Ignite cities to employ both methods. Copious bandwidth will be an easier first approach. Software-Defined Networking will provide additional benefits such as being able to provide a defined quality and quantity of service on a per-application basis and to be individually billable to each application.

7. Able to handle Big Data.
   We have entered the era of Big Data. A combination of dramatically falling storage costs and the proliferation of high-bandwidth sensors such as cheap, high-capacity cameras has brought us to the Zettabyte era: there are now over 10^{21} bytes on the world's disks, and the number doubles every 24 months. There has been a great deal of attention paid to the requirements for processing this data, but transmission and storage deserve at least as much attention. The doubling rate implies a worldwide data capture rate of approximately 10^{14} bits/s, or 100,000 Gb/s. Most of that data must remain at or near the device that captured it: ingress/egress bandwidth at most sites is orders of magnitude too low to capture this. The National Science Foundation announced in 2016 a high-performance networking initiative to spur big-data science at a number of University of California and other California campuses: a 100 Gb/s network. This network can transmit about 0.1 % of the data added to the world's disks every second. Moreover, the asymmetry between data capture and data transmission capacity will continue to grow over time. At the current doubling rate, a yottabyte will inhabit the world's disk by 2020, and the capture rate will be 10^{17} bits/s. It goes without saying that this production will outstrip the capacity of any conceivable wide area network. This implies that most computation on most data must be in-situ or near in-situ: there simply isn't the network capacity to transmit it across the network.

   A complementary factor is the dramatic falloff in bandwidth as one goes from the local to the wide area. Bandwidth is usually referred to as "speed," but this is a misleading term; a bit travels at the same rate on a 10 Mbps and 100 Gbps line. A more accurate term is capacity: the number of bits that can be carried per second on the line. One accurate analogy is to the road network. Within a city, there are many small arteries, but their aggregate capacity is very large. In contrast, intercity thoroughfares are large, multi-lane highways. Though the highway capacity is much larger than that of an individual road within a city, the aggregate capacity within a city dwarfs that of the highway. This analogy is only slightly misleading, because in fact the capacities of the arteries of a network are generally close to the capacities of the main trunk line. In sum, the network consists of a vast network of local highways interconnected by slightly larger highways.

   The falloff in available bandwidth between the local and the wide area is dramatic. Bandwidth in the wide area is often restricted and costly. Even low-end switches on a local area network will have bisection bandwidth of at least 50 gigabits per second (Gbps), and personal capacity of at least 1 Gbps and often 10 Gbps. Standard border routers, which define the bandwidth for the enterprise campus, are generally sized for 1–10 Mb/s/person. Therefore, moving Enterprise or personal IT services to the Cloud represents a bandwidth reduction between service and user of about 1000×. The same thing is true in the gap between the Metro Internet (which hooks up services within cities) and the global Internet.

Simple calculations demonstrate that almost all network traffic originating in a city must remain in that city: Chattanooga, TN, has 80,000 homes and offers 10-Gb connectivity. This gives total bandwidth of about 800 Tb/s; the egress links from any city in the US can handle only a small fraction of that. We must have a distributed Cloud, simply because our aggregate edge bandwidth far outstrips the bandwidth available in the core, and because the applications and services of the future cannot tolerate the delays of long-distance network transit. To a very good approximation, all network communication will be local.

A second trend is towards ever-smaller and more mobile devices for personal computation and consumption of media and information. These devices must be, and must continue to be, relatively low-powered, for reasons of energy consumption and portability. This implies that most storage and computing must be in the Cloud; but high latency and low bandwidth across the wide area means either a Cloud with a point of presence near the user in network terms, or non-interactive applications with only tightly-reduced data going to the device. By "near the user" we mean latencies in the range of 5 ms or so with user/Cloud bandwidth on the order of a gigabit.

In networking, time is distance. The speed of light in fiber is about 200,000 km/s, or, in units that are more meaningful in network terms, about 200 m/μs. A network transaction requires a round-trip, so the figure of merit is 100 m/μs. A tight constraint on latency implies many points of presence across the United States. Five milliseconds in fiber is 1000 km (500 km for a round-trip), or about the straight-line distance between Seattle and San Francisco; in practice, of course, assuring these latencies requires a much smaller distance. This means that to support truly interactive Big Data applications on small devices, a Cloud with at least 25 Points-of-Presence across the United States is required; in practice we expect that this is a significant underestimate.

These bandwidth and latency considerations drive not only the distributed cloud, but form the basis for the Mobility First network architecture underlying a fifth-generation wireless architecture. It is anticipated that 5G Wireless will offer gigabit connectivity to the user and latencies in the microsecond range, orders of magnitude beyond services offered today. Of course, there is no magic, and radio technology will not suddenly improve by orders of magnitude. The leap from 4G LTE to 5G is not given by improvements to the physical radio network, but rather by a new network design that moves the Internet point-of-presence very close to the mobile device. In this architecture, very small cells (picocells, anticipated to be about the size of a building) will be anchored by a server near the cell base station, and most requests will be handled by the server.

Handling such traffic locally is at the heart of the US Ignite metropolitan architecture.

3 The Magic of (the) GENI

US Ignite applications, by definition, are next-generation apps that are not feasible on 2012's typical broadband infrastructure. What are the new infrastructure features coming from GENI that enable new US Ignite applications?

1. Better privacy. We've seen time and again that despite precautions and encryption, today's networks and software are simply too susceptible to hacking attacks, and these attacks are frequent and relentless because it is too easy to monetize financial and personal information in the black market. When anything can talk to anything on the Internet, one end could be a cracker (a malicious hacker) probing for security weaknesses, or even just an untalented script kiddie looking for known weaknesses. This type of attack, as well as so-called man-in-the-middle attacks, can be cut off if applications needing privacy were allocated their own "slice" or virtual network that couldn't even be addressed by random crackers or anyone else not specifically admitted to the slice.
2. Predictable performance. The 2016 Internet is a "best-effort" network that succeeds largely because there is plenty of capacity on most networks. However, bandwidth demanded for applications continues to rise more quickly than new bandwidth can be supplied. This is particularly acute for wireless cellular networks. It is inevitable that high priority applications that require predictable performance will eventually be given an opportunity to acquire a "slice" that is given appropriate performance, perhaps because there is an incremental cost for such predictable performance. The predictability may alter such measures as packets dropped or lost, latency, jitter, bandwidth flow available, etc.
3. A financial model that gives carriers incremental revenue. Because of GENI's slicing technology, there exist individual identifiable and billable slices. No longer are carriers restricted to offering a fixed price for a fixed maximum speed or a fixed number of gigabytes. Instead, the carrier can provide some or all of the slice and bill for the services of the slice. Some slices might have critical functions and receive first-class priority for packet delivery, protected circuits and servers (guaranteed backup paths and servers), better response times, reliability guarantees, etc. Other slices may be lower cost, work well in normal conditions, but not be as resilient in cases of distress. Some slices may be opportunistic and use otherwise idle resources to provide for additional redundant copies or paths that add a degree of reliability without appreciably increasing costs (at least in "normal" times). A flexible, scalable, billable infrastructure should be very attractive to the carriers and acceptable to users since they are receiving specific value from the slices they invoke.
4. Ultra-high performance. In an experiment using Google Fiber in Kansas City, the author ran test copies of Adobe Photoshop on his high-end Intel i7 laptop and also on a quad-Xeon server with gigabit connection to the laptop. When manipulating sliders for contrast and brightness, the server-run copy of Photoshop outperformed the laptop and produced smoother and more continuous changes in the image compared to Photoshop running on the laptop itself. This is a simple

example of projecting ultra-high performance from servers to lower-cost end-user devices as long as there is a high capacity and low latency network between them. The performance achieved in such a "projection of power" situation brings enterprise-class processing and storage to small businesses and homes. Dublin, Ohio is a US Ignite city which brings advanced manufacturing design and simulation tools from a powerful datacenter to dozens of small design companies who rent the service. A gigabit fiber network provides copious bandwidth at low latency to make this work well. Another example is a class in video editing in Kansas city made possible by software and servers at the Kansas City Library (Digital Branch) enabling youth with cell phones to create professional-quality videos from their cell phone captures. The cost of video editing workstations would have been cost-prohibitive. The shared use of the Library's facilities made this high-end technology available in a low-income area.

5. Synchronized with the real world. With sufficiently low latency, slices can be the cyber portion of a cyberphysical system (CPS). A cyberphysical system operates partly in the real world and partly in a cyber world. Cyberphysical systems steer self-driving cars, guide machine tools, provide driving directions, optimize building energy use, and help find underground water leaks not yet apparent on the surface. In each case, there is a real world part, sensors or other mechanisms to turn the real world into digital information, a cyber part which analyzes the information about the real world and, with or without a human, makes decisions about what should be done. The final portion is either human action or some kind of cyber-driven actuator (e.g., to steer the car) in the real world. This continuous cyberphysical loop is a powerful construct in that it can bring the best of the cyber world into and impact the real world. A CPS is natural to implement in one or more connected slices, and often requires very low latency and high reliability … characteristics that can be engineered into GENI slices of US Ignite infrastructure.
6. Capacity and latency appropriate for the Internet of Things and the Industrial Internet. There is currently a lot of attention and expectation for the Internet of Things with estimates running as high as trillions of communicating smart things in the next decade. Each of these smart things has something to say (the "sensing" part of the CPS closed loop) and may be able to affect its world (the "actuator" part of the CPS closed loop). Even if every one of these smart things has very little to say itself, multiplying by a trillion would far overload our current networking infrastructure. The cost of networking to carry this traffic to remote datacenters would be daunting even if that approach didn't add too much latency and delay and add additional failure points to a critical CPS loop. This seems a case tailor-made for edge computing—the notion of putting intelligence at the edge of the network. Keeping sensor processing close to the sensors allows us to scale our current backbones while providing highly-responsive "edge" processing. This type of architecture with the "smart" part of the slices near the device is likely to be a significant hallmark of the Internet of Things. An industrial Internet has the same issues. How do manufacturing plants communicate internally and with each other with appropriate bandwidth,

responsiveness, and reliability? The answer would seem to be sliced edge computing with unconstrained local bandwidth and ultra-low latency.

7. Self-reliant. By keeping the processing and storage near the sensors and actuators and CPS loops, there are fewer cyber components that can fail or be disrupted. This makes a US Ignite city more self-reliant on its own resources and less dependent on distant resources. In case of a natural disaster, disrupting today's infrastructure means that there would be widespread blackouts of applications and services as winds or earthquakes took down fiber optic lines connecting towns to distant cloud data centers. The edge computing in GENI racks in US Ignite cities means that applications and services running on local compute and storage enjoy not only provide superior responsiveness, but also isolation from problems occurring elsewhere. A US Ignite city is a more self-reliant digital infrastructure.

4 The New Questions

The original questions that started US Ignite were "What will the next generation of the Internet do? How will it change healthcare, education, public safety, clean energy, transportation, and advanced manufacturing?"

But smarter cities answering these questions have a question of their own: "What would an Internet look like if it were designed to support:"

- The Internet of Things (IoT)?
- Billions of wireless devices?
- Industrial Internet?
- Low-cost endpoints? (closing the digital divide)

in addition to the applications in education, healthcare, public safety, etc.?

The portion of the answer that is clear is that "the cloud" becomes less important because it is often too distant, too unreliable, and too slow for the new applications. "The cloud" is dispersing and moving to the edge ... moving to the metropolitan areas where it's possible to achieve the sweet spot of ultra-high bandwidth, ultra-low latency, and tiny incremental costs. What would a design for such a network look like?

5 A Design for a Smart City Metropolitan Internet

For the reasons already given, we need to keep local traffic local and use high bandwidth and low latency interconnections, and have slicing capability. Therefore, our reference implementation uses a local GENI rack in each city and interconnects all the carriers (and especially the gigabit carriers) at a Digital Town Square. The purpose is to provide an exchange point for all local traffic. (In a production situation, you may want to have more than one Digital Town Square or even virtual Digital Town Squares to ensure that a single failure cannot bring down local traffic in a city.) The GENI rack may be co-located in the Digital Town Square or may be connected to the Digital Town Square facility with a very high bandwidth and low-latency connection. In a metropolitan area with an existing GENI rack at an in-town university, the diagram might look like this:

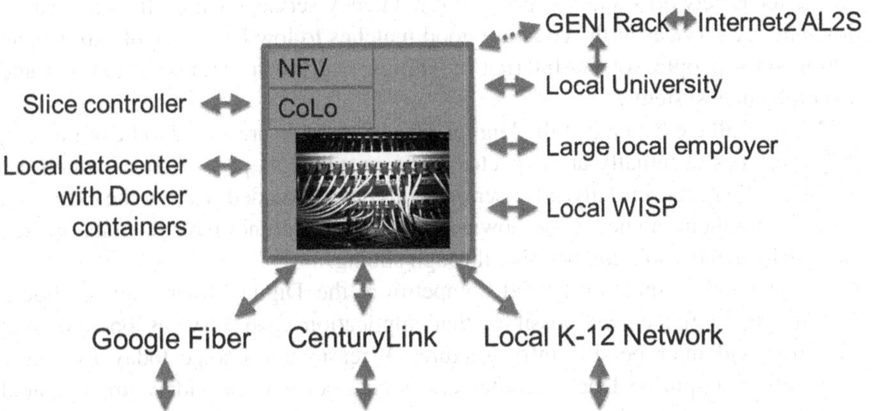

Carriers providing local access services, particularly gigabit access services, are encouraged to "drop off" their local traffic at the Digital Town Square. The advantage to the carrier is that they have less traffic to carry over their inter-city backbones. The advantage to their subscribers is lower-latency access to local services. For example, faculty, students and staff of local universities can have access similar to being on-campus. If the carrier can arrange to not throttle local services (e.g., by using SDN to shunt that traffic past bandwidth limiters or putting local traffic on a separate VLAN), the faculty, staff and students can also enjoy very high bandwidth, again, as if they were on–campus. The same can be true for large local employers which connect their enterprise LANs to the Digital Town Square. In the example above, we imagine that Google Fiber and CenturyLink connect their local access networks to the Digital Town Square along with "Large local employer" and "Local University".

The Digital Town Square exchanges local-only traffic to avoid any transit issues. In addition, ideally both layer 3 (Internet Protocol) and layer 2 (link layer or IEEE 802.2) traffic are exchanged. Layer 3 traffic helps with compatibility. Layer 2 traffic is more easily sliceable.

The Digital Town Square can be operated by the metropolitan area itself, by the local university's IT staff as a goodwill gesture to the metro, by a state or regional research and education network, or by another organization willing to provide low-cost or no-cost connections to all comers who wish to interchange local-only traffic.

The interchange of local traffic locally with better performance characteristics than can be obtained through out-of-city exchange points makes the digital town square especially valuable in middle-sized and smaller-sized cities that may not have robust traffic exchange points already present.

The Digital Town Square is not just about the network interchange. It's also a place, like GENI, for in-network services or "edge" services. These services can be provided by the GENI rack and/or other compute and storage facilities such as a local datacenter. (A local datacenter is shown in the diagram in addition to the GENI rack.) These services have access to the gigabit and higher bandwidth found in the local networks and can provide low-latency services within the city. Those low-latency services at the edge are good matches to low latency applications and services for people, for the Internet of Things, and for the Industrial Internet and cyberphysical systems.

While traffic exchange at the Digital Town Square is presumed to be at no cost, edge services eventually are expected to be provided as part of slices which are billable. Of course, initially, edge services might be provided without the extra step of isolating them in slices. The downside is that they do not have the data privacy and performance isolation possible through slicing.

To promote compatibility and competition, the Digital Town Square should encourage, to the extent possible, that applications and services operate over common and interoperable infrastructure. A reasonable choice today for many applications might be Docker containers since Docker is so widely implemented and the container can belong to a slice. Other possibilities include OpenStack and its containers (bays, pods, etc.) and orchestration (e.g., HEAT). The objective is to put each application into as broadly-interoperable an environment as possible.

6 Example Applications and Services

With these kinds of characteristics and designs, what are the new applications which become easily possible?

7 Connected Collaboration

In Cleveland, Case Western Reserve University under the direction of Marv Schwarz and Lev Gonick developed Connected Collaboration—video conferencing without the delay. We've all been in video conferences where multiple people talk over one another because the delay in the video conference is tens to hundreds of times the almost imperceptible delays in real world conference rooms. Connected Collaboration demonstrated natural video conferencing. The author found himself finishing a sentence started by another participant … something not uncommon in informal discussions among friends but not something that's very practical over today's videoconferences. A US Ignite grant powered this work.

8 Digital Cinema Microscope

Under a US Ignite grant, Dr. Richard Weinberg of the USC School of Cinematic Arts automated a microscope stage and put a 4K camera on the viewing ends. Using a low-latency gigabit connection from USC to Chattanooga, he demonstrated how middle-school kids in Chattanooga could manipulate the magnification, the position, and see very high resolution real-time images from USC. The images, in fact, had to be scaled down to high-definition (from 4K) because the gigabit bandwidth didn't permit real-time viewing of 4K images. (Although even 4K movies can be easily carried over much less than a gigabit, this is because they go through extensive and time-consuming two-pass compression techniques. The USC experiment uses readily available and lower-cost equipment which can't provide such levels of compression.) The latency must be low to allow students to control the stage in real time and to stop instantly (not a fraction of a second later) when the image of interest is under the lens. Accurate remote focusing of the microscope is similarly sensitive to low-latency so that the students can stop the focusing action as soon as they see the image become sharp on the high definition monitor in Chattanooga.

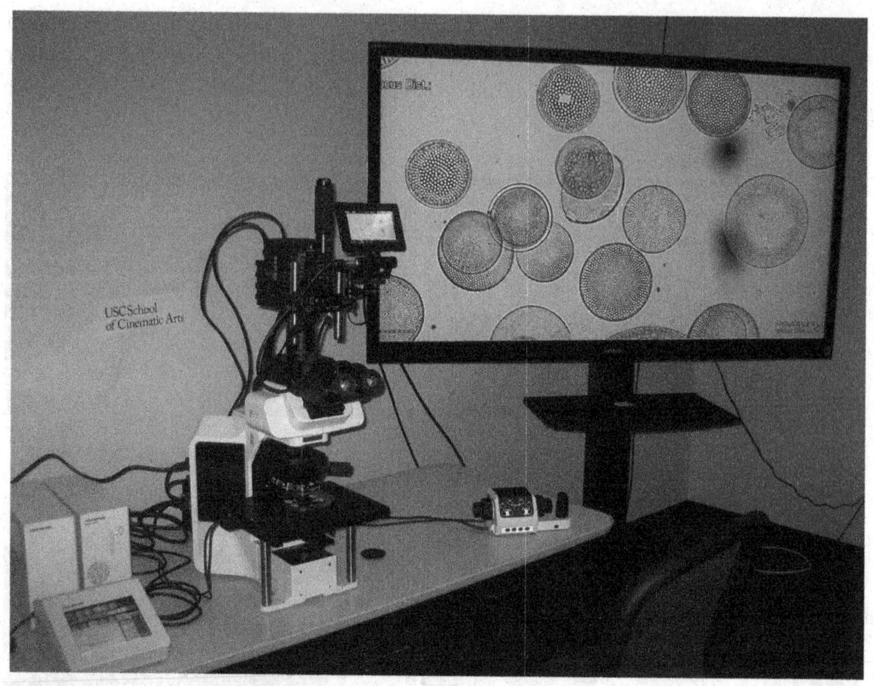

9 Virtual Reality Educational Tools in Even Low-Income Homes

The simulation of highly detailed and realistic learning environments still requires significant computational horsepower. This kind of processing is normally only done on expensive and higher-end gaming machines and workstations. In a US Ignite experiment by Lockheed-Martin and the Department of Defense, a virtual reality simulation teaching the algebraic concepts of Cartesian coordinates and slopes of lines in such coordinates has been demonstrated using gigabit connectivity between powerful servers and students in classrooms.

In 2016, there should be a new round of higher-resolution 3D virtual reality goggles that should be able to plug into gigabit networks in fiber-connected lower-income homes and provide the same kind of virtual reality education tools with the simulation being run over powerful servers at, say, the public library. (In fact, Kansas City, the first Google Fiber city, has a digital branch of its public libraries managed by David LaCrone.) The realistic projections need to be provided with ultra-low latency so that as the viewer moves her head, the projections in the goggles changes instantly. This requires both high bandwidth and ultra-low latency as can be achieved in a US Ignite city with a GENI rack.

The MARS GAME:
play, experiment, collaborate, compete, learn & more.
www.themarsgame.com

10 Reducing the Cost of 3D Printing

In a US Ignite collaboration between Purdue University, Kettering University, and Lit San Leandro, new techniques for exceptionally reliable and lower-cost remote printing were demonstrated by George Adams, Doug Comer, George Geske, and Judi Clark. A notable portion of the cost of the 3D printer is the control logic. This project put the control logic into the cloud to reduce the cost of the 3D printer. To control it from the cloud requires highly reliable communications and low latency. This project demonstrated such an arrangement. This is a good example of supporting machine-to-machine (or M2M) systems as needed for the Industrial Internet. If there are delays and glitches in communications, the additive manufacturing process cannot be stopped—material must be continuously deposited. Consequently, the process is not tolerant of dropped packets or high delays. In this work, the researchers demonstrated using multiple paths simultaneously to provide the needed degree of reliability and low latency.

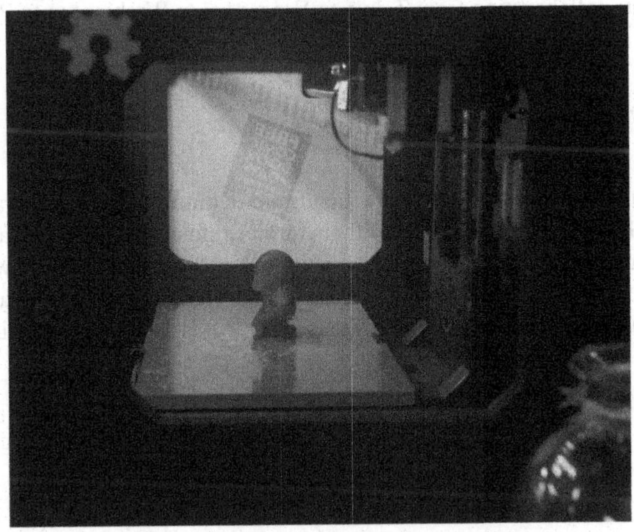

11 Collaborative Pollution Viewer

Another example of big data and low latency is the Atmospheric Quality viewer, a US Ignite project described in Chap. 19 in this volume.

It may be useful for city planners to understand how pollution in their city is affected by both factors they can change such as population and vehicle usage, and factors they cannot have such fine-grained control over such as global warming. A model of air quality developed by Dr. David Lary can help them understand those factors within their own geographic boundaries. Using his model and calibrating to existing data, Dr. Lary has an air quality model for the entire earth for each day over the past 10 years. The entire dataset is approximately 10 Gb in size. Although some compression is possible, the data is too large to fit on today's laptops and handhelds and the graphics processing needed to render it smoothly in real-time as people turn the knobs on the viewer is greater than the graphics processing present on a typical laptop computer or tablet.

However, using a GENI rack for processing and a gigabit connection to a laptop, this US Ignite application is indeed feasible. As the user moves the geography (pans the image), or changes a viewing parameter or an assumption about future temperatures or vehicle use or population, the GENI rack computes the new image and streams it to the laptop. In addition, the team has made this work collaborative across multiple locations. Each location has its own GENI rack and can process and display the results in its own metropolitan area. But each rack tells the others about user-driven parameter changes so that their racks do simultaneous and parallel computations on their racks so that the collaborators in those cities ALSO see their screens update in real time as the changes are being made. While the big data movements stay within the metropolitan area, the tiny amount of information on the parameter changes is sent significant distances to permit real-time collaboration.

At the SmartFuture 2015 event in Washington, DC, this arrangement was used to demonstrate collaboration from Tokyo to Vancouver to Washington to Postdam.

12 A Vision for Smarter Cities

Let's add up this chapter. Seven new trends are defining the next generation of the Internet. GENI research results and their further refinement by corporate labs and startups have created seven new infrastructure capabilities that create new spaces for innovative applications that touch our lives. Five signature US Ignite applications were discussed that take advantage of these new infrastructure capabilities.

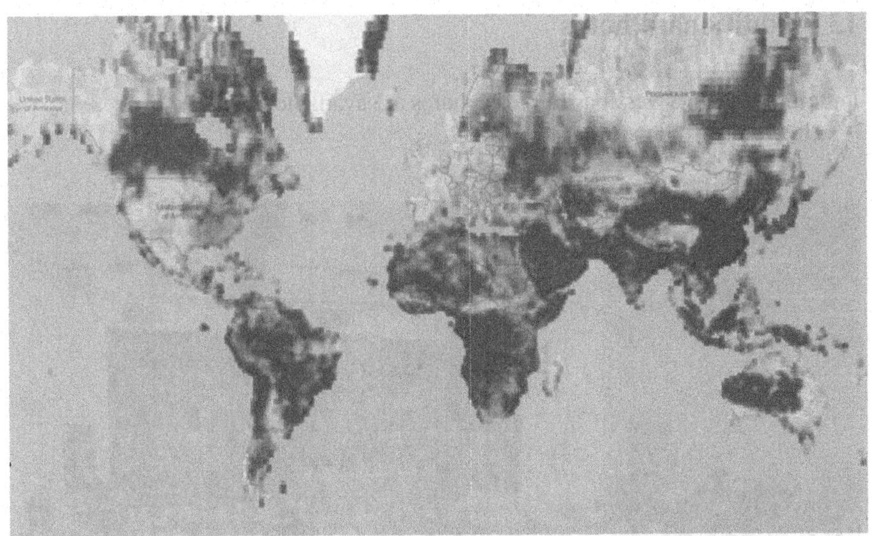

As this book goes to press, US Ignite is taking the logical next step: putting these capabilities into practice in a series of 15–20 seed cities/counties/states in a sustainable way. Digital Town Squares will be created to provide the equivalent of free trade zones but for digital information and services. In these free trade zones, there will be unconstrained local bandwidth and low latency end-to-end connections. The islands of gigabit will be interconnected. There will no longer be technical barriers to the next generation of Internet applications.

But why have disconnected free trade zones? US Ignite will also interconnect each of its seed cities with the others so that applications and services can be shared among more than two million gigabit users and with up to 94,000 fiber-connected community anchor institutions across the country. Such a marketplace should trigger substantial private capital investment. The goal is audacious: a national sustainable ecosystem of smart (city) applications.

Sustainable implies an ecosystem in which the financial incentives are aligned for rapid growth and success builds upon success. Other cities will want to reap the same advantages for themselves, and the new cities will help to create a self-perpetuating revolution in the networks that support smarter cities.

And the interconnected Digital Town Squares with locavore computing resources are a logical response to the needs of a burgeoning Internet of Things, billions of wireless devices, the Industrial Internet, and providing low-cost endpoints to help cross the digital divide.

It may very well be the next generation of the Internet.

13 Additional Photos

Letterpress-quality versions of these photos are available.
All photos by Glenn Ricart.

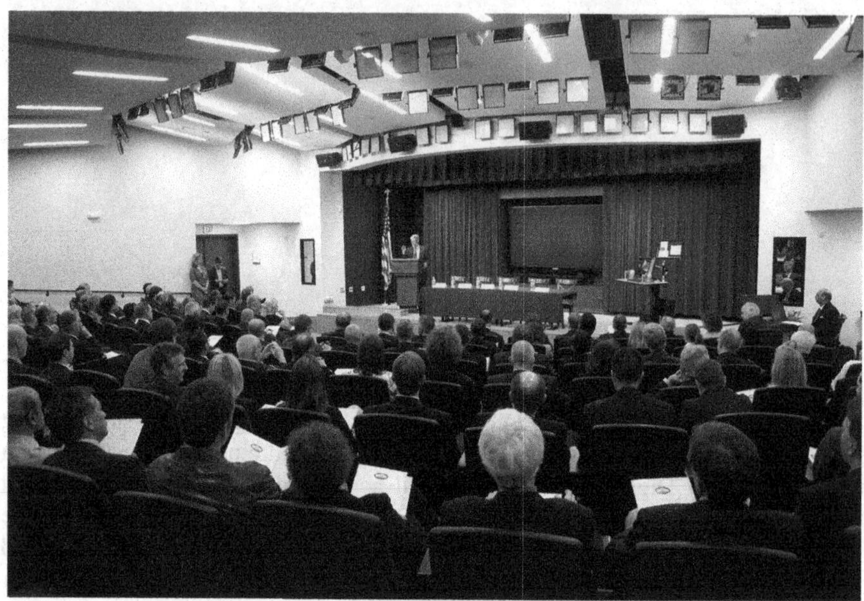

June 14, 2012—US Ignite is launched at the South Auditorium in the White House

John Holdren, the President's Science Advisor, introduces the US Ignite initiative

Tom Kalil is the sponsoring executive at OSTP for US Ignite

A capacity crowd June 12, 2012 at the White House South Auditorium is present for the launch of US Ignite

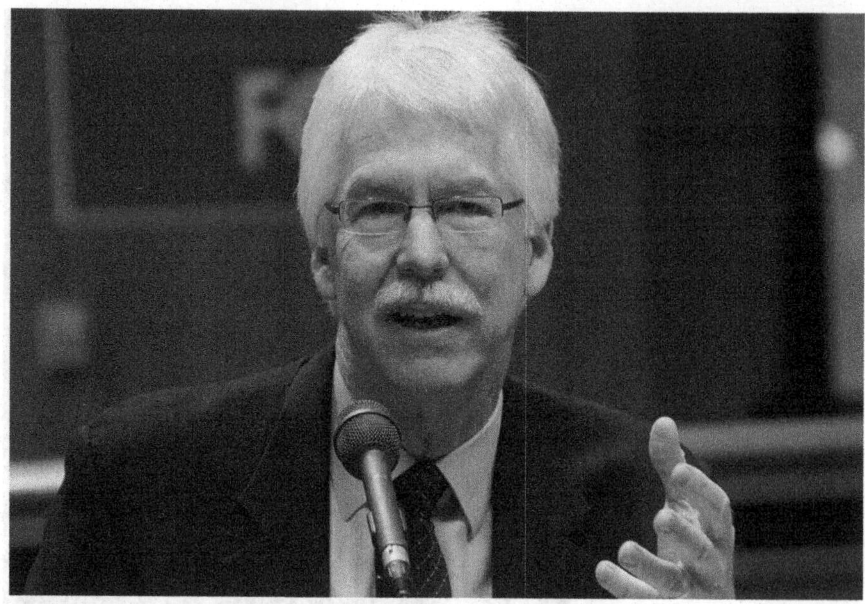

Bill Wallace, Executive Director of US Ignite, speaks at the Federal Communications Commission Gigabit Community Challenge

Glenn Ricart is founder and CTO of US Ignite [press photo given to US Ignite]

Tsege Beyene of Cisco at the US Ignite Application Summit

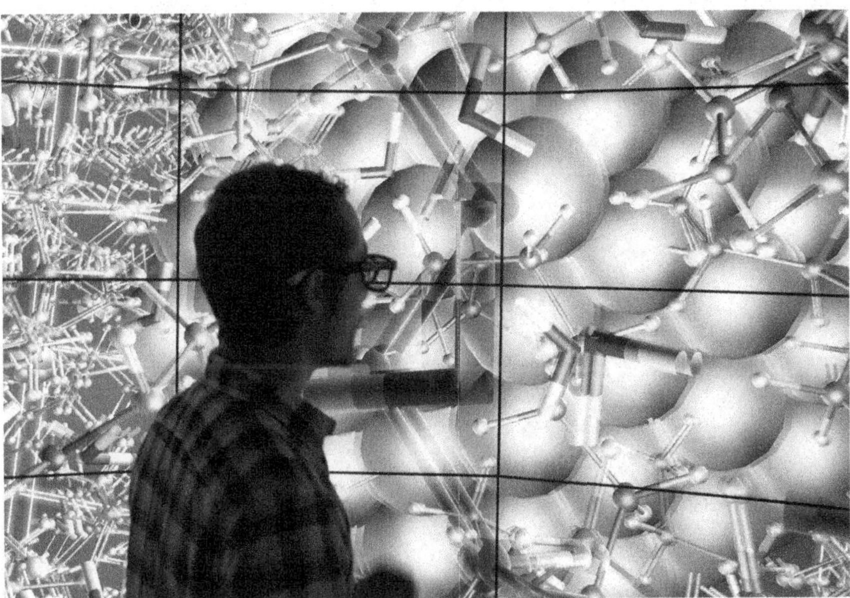

Real-time 3D visualization at Cave2 at the University of Illinois at Chicago Electronic Visualization Laboratory; the demonstration was held in conjunction with the US Ignite Application Summit

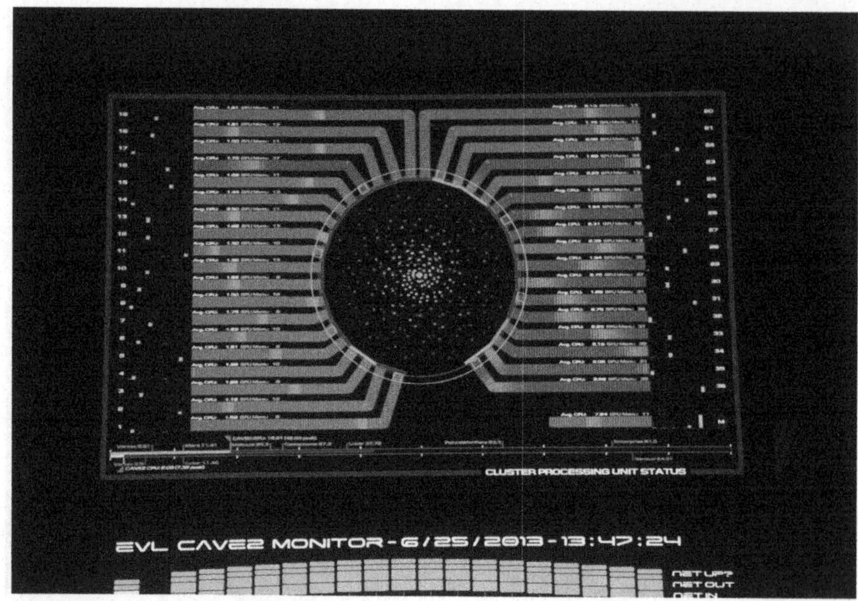

Powerful locavore graphics engines power the Cave2 facility at the University of Illinois at Chicago Electronic Visualization Laboratory

The University of North Texas displayed the use of drones for emergency Internet communications at the US Ignite Applications Summit

Jonathan Wagner's Gigabots at the US Ignite Applications Summit

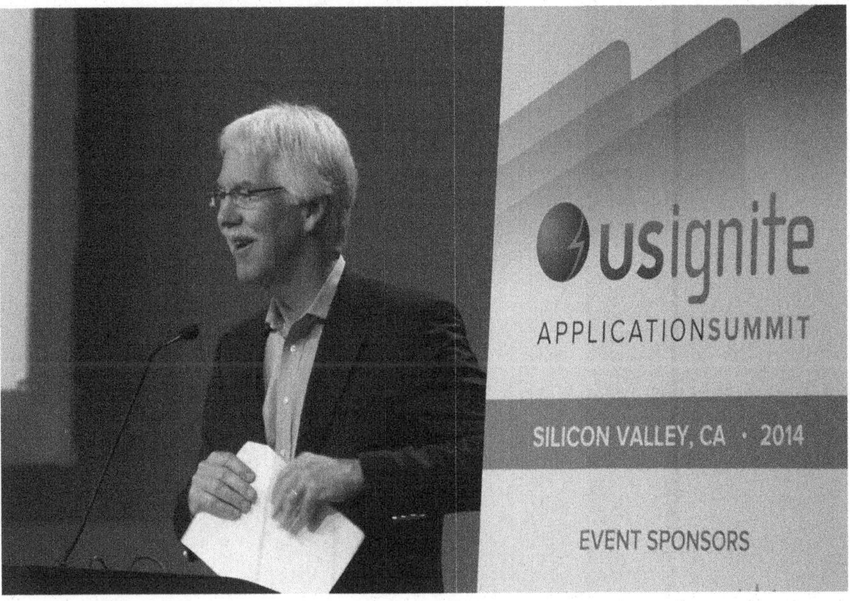

US Ignite Executive Director Bill Wallace addresses the 2014 US Ignite Application Summit

Dr. Nick Maynard of the White House Office of Science and Technology Policy was instrumental in launching US Ignte

Part V
GENI and the World

GENI is not the only infrastructure of its kind in the world. The rise of virtualization, the need for distributed computation, the dramatic drop in the cost of computation vs. bandwidth, and the emergence of software-defined networking are worldwide trends that have inspired the development of similar infrastructures on four continents. In the introduction, we discussed a number of these. In this section, several of our international colleagues describe efforts in their countries and regions that are similar to and intertwined with GENI. It is important to note that this section is intended to be representative, not a comprehensive survey. We are leaving out our colleagues in Brazil, Russia, China, Australia, South Korea, and elsewhere who are actively engaged in designing, developing, and deploying infrastructure that is influencing, and being influenced by, GENI concepts.

Europe has been a particularly active area for next-generation networking testbeds, inspired both by centralized EU programs such as the Future Internet REsearch (FIRE) program, and by national programs within the various EU countries. The first of these was the German Lab program under Prof. Paul Müller of TU-Kaiserslautern and his colleagues. Prof. Müller describes G-Lab and other European efforts in chapter "Europe's Mission in Next-Generation Networking with Special Emphasis on the German-Lab Project".

In Canada, the Smart Applications on Virtual Infrastructure (SAVI) program consists of OpenStack clusters at a number of Universities across four of Canada's major provinces, tied together with Software Defined Networking. SAVI and GENI are closely aligned in goals and objectives, and a number of researchers work collaboratively on both platforms. In 2015, federation of GENI and SAVI was accomplished, permitting GENI users to add SAVI resources to a GENI slice and SAVI users to use GENI resources in their experiments, seamlessly and easily. This will become a flagship for broader international collaboration. In chapter "SAVI Testbed for Applications on Software-Defined Infrastructure", Alberto Leon-Garcia and Hadi Bannazadeh of the University of Toronto describe SAVI.

Prof. Akihiro Nakao of the University of Tokyo was a member of the original PlanetLab team, and this experience led him to spearhead a GENI-like infrastructure in Japan. In chapter "Research and Development on Network Virtualization Technologies in Japan: VNode and FLARE Projects", Prof. Nakao and Kazuhisa Yamada describes in detail the advances in software-defined networking and distributed infrastructure enabled by the VNode and FLARE projects. They also give a perspective on the interactions between contemporaneous research efforts in the US and Japan. As with many of GENI's international collaborators, Prof. Nakao and his Japanese colleagues have been deeply engaged with GENI researchers over an extended period of time. The resulting personal interactions and interplay of technical challenges and ideas have been deeply beneficial to researchers in both countries.

The goal of our international collaborators and ourselves is to create a truly worldwide, persistent, distributed infrastructure that will bring the benefits that we have seen in GENI, our German colleagues with G-Lab, our Canadian neighbors with SAVI, our Japanese friends with V-Node and JGN-X to the entire world—to create a world-girdling distributed infrastructure that will truly be the next generation of the Internet. The person who has done the most over the past several years to make that vision a reality is Joe Mambretti, Director of the International Center for Advanced Internet Research (iCAIR) at Northwestern. Together with his colleagues at iCAIR and his many colleagues in the international advanced networking community, Mr. Mambretti is creating such a worldwide network. Joe Mambretti, Jim Chen, Fei Yeh, Jingguo Ge, Junling You, Tong Li, Cees de Laat, Paola Grosso, Te-Lung Liu, Mon-Yen Luo, Aki Nakao, Paul Müller, Ronald van der Pol, Martin Reed, Michael Stanton, and Chu-Sing Yang describe the international advanced research network in chapter "Creating a Worldwide Network for the Global Environment for Network Innovations (GENI) and Related Experimental Environments".

Europe's Mission in Next-Generation Networking with Special Emphasis on the German-Lab Project

Paul Müeller and Stefan Fischer

Abstract In this contribution we give a rough overview of the European and particularly the German approaches to next generation networking, or more specifically Future-Internet Research and Experimentation. We can identify three different classes of projects in these approaches. The first class is related to basic research that is covered by projects within Objective 1.1 (Future Networks) of Framework Program 7 (FP7) of the European Commission (EC). This can be compared to the Future-Internet Architecture (FIA) projects of the National Science Foundation (NSF) in the US. The second class of projects is related to experimentation. The FIRE (Future-Internet Research and Experimentation) projects of the EC can be considered in this context, which are more or less comparable to the GENI approach. The third class is more application-driven and covered by the Public Private Partnership (PPP) projects of the EC. This class of projects can be compared to the USIgnite program. A slightly different approach was taken by the German-Lab (G-Lab) project where basic research projects and experimentation were smoothly intertwined, and also covered application-oriented aspects like mobility, virtualization or security in its second phase. All these projects from the EU, and the G-Lab approach will be described in more detail throughout this contribution, based on typical examples.

1 Introduction

At the beginning of the twenty-first century, the first scientific discussions took place about whether current Internet architecture had ossified, and hindered innovation [1, 2] through an unintentional increase in complexity. Although current Internet architecture has successfully enabled a great deal of innovation, novel societal and

P. Müller (✉)
Integrated Communication Systems Lab., Department of Computer Science, University of Kaiserslautern, Kaiserslautern, Germany
e-mail: pmueller@informatik.uni-kl.de

S. Fischer
Institute of Telematics, University of Luebeck, Luebeck, Germany
e-mail: fischer@itm.uni-luebeck.de

© Springer International Publishing Switzerland 2016

R. McGeer et al. (eds.), *The GENI Book*, DOI 10.1007/978-3-319-33769-2_21

commercial usage is continuing to push the original Internet architecture to its limits. The first research projects were created, such as the "New Arch Project,"[1] for which former Internet pioneers published relevant articles at the ACM SIGCOMM 2003 workshop that took place in Karlsruhe, Germany.

The current Internet architecture and design, as we know it today, is nearly 50 years old and has become a global success story. It began in the late 1960s with the first ideas about packet-based communication, and became the first research network in the late 1980s, with few users, command line control and nearly no economic impact. Then in the 1990s, based on graphical front ends and web technology, the generation of real commercial services led to an explosion in traffic and increased the Internet's economic impact. Today's Internet has become the backbone of networked innovation and a hub to the globalization and circulation of services and knowledge. The Internet was originally designed for communication between a few scientific organizations but over time it has become a critical substrate from a technological, social and economic point of view. It started with four nodes, but its size and complexity have far exceeded the expectations of its inventors and architects.

As well as today's Internet architecture and design, basic Internet protocols are also more than 40 years old. In that time, driven by its ever-growing scale, new providers with new autonomous systems, hundreds of additional protocols and sub-layer extensions, have made its management more and more complex. New applications, such as VoIP, YouTube, Facebook and Twitter, with new demands arise repeatedly and drive the use of the Internet into directions which are far beyond its original intention.

Today's answers to these concerns can be seen more or less as patches [3], which treat the symptoms instead of curing the disease, and cannot last forever. This means that a radical redesign and change of paradigms will be required in the medium or long term [4] development of a Future-Internet. In recent years, quite a number of researchers and industries worldwide have started considering radical new approaches to Internet design, sometimes called "clean slate" or "revolutionary approaches" [5].

The main issues related to Future-Internet research go far beyond research dimensions and will also cover technical, economic, social and even ethical dimensions. Uncensored speech, privacy, user-generated data and information and new applications with as yet unknown network demands will have a deep impact on nearly every sector of our future society. Moreover social networking sites are attracting hundreds of millions of users worldwide, mostly young people, and have created a new means of human interaction. The ease of using "Web2.0" technology and the related availability of user-generated content generates complex challenges for the Future-Internet related to security, privacy and intellectual property rights.

From a general point of view, especially from an industry or research position, one should be aware of the positive and negative effects and impacts resulting from an evolutionary or clean-slate approach. It is important to keep a business

[1] http://www.isi.edu/newarch/.

perspective in Future-Internet research, and to involve the industry and expectations of the end-users in the research and innovation cycle, and to make sure that the regulatory and legislative requirements will be taken into account. Nevertheless, to overcome the current limitations of the Internet architecture new paradigms and ways of thinking are needed, since one cannot solve the problems with the same kinds of thinking used when creating such problems,[2] and continuing to apply patches forever is not an option. Creating sustainable business models is an important requirement but it should not hinder basic research in creating new Internet architectures. Craig Partridge gave great guidance in his editorial note about "Helping a Future-Internet architecture mature" [6].

The remainder of this paper is organized as follows: before we focus on the ICT program within the European research program we describe the different instruments for implementation. In the context of future networks we will further focus on the so-called "Challenge 1" which deals with "Pervasive and Trusted Network and Service Infrastructures". Within this challenge we concentrate on two specific objectives, Objective 1.1 "Future Networks" and Objective 1.6 "Future-Internet Research and Experimentation (FIRE)". For these objectives we describe some representative projects. The European program for Internet enabled innovation, the FI-PPP (Future-Internet Public-Private Partnership) program, will be described next, followed by the Future-Internet Assembly (FIA) as a self-organizing European body to define future directions of research. The rest of the paper is dedicated to the German G-Lab project and gives an overview of the research branch as well as the experimental facility that was established within this project.

2 The European Arena

The European Framework Programs for Research and Technological Development, also called the Framework Programs, or abbreviated to FP1 through FP7, covering a time frame from 1984 to 2013, are funding programs created by the European Union in order to support and encourage research in the European research area. The specific objectives and actions may thus vary across the different funding periods.

Framework Program projects are generally funded through a number of instruments, such as Integrated Projects (IP), Specific Targeted Research Projects (STReP) and the Network of Excellence (NoE). These instruments can be described as follows:

- **Integrated Projects (IP)**
  Medium to large-sized collaborative research projects that consist of a minimum of three partners coming from three different EU member countries and/or associated states, but can include several tens of partners. Although there is no

[2] A quote from Albert Einstein.

defined upper duration limit, such projects last typically for 3–5 years and the budget can go beyond ten million Euros, paid as a fraction of the actual costs of the participants. The main objectives of the IP's include fostering European competitiveness in industry and basic research related to the Priority Themes of the Framework Program, addressing major needs in European society. To address these needs IPs ask for a strong involvement from small or medium-sized enterprises (SMEs) to foster the translation of research results into commercially viable products or services.

- **Specific Targeted Research Projects (STReP)**
  These are small-to-medium-sized research projects that are extensively funded in the FP6 and FP7 programs. STReP projects are composed of a minimum of three partners coming from three different countries from EU member and associated states. Usually, they run for 2–3 years and typically involve between 6 and 15 partners.
- **Network of Excellence (NoE)**
  Similarly to the other instruments NoE projects also require a minimum of three EU partners from different member nations, however such projects are usually expected to involve at least different countries. An NoE is thus not strictly considered a research project, but is more an instrument to foster collaboration and exchange programs between its partners in a specified research field. Funding can cover a time period of up to 7 years.

2.1 Introducing the European FP7 ICT Approach

In the following we will focus on the last Framework Program, the FP7 ICT work program. It defines the overall priorities for the upcoming calls for proposals. The priorities are in line with the main ICT policy priorities as defined in Europe's digital agenda.[3] The work program itself is split into eight "challenges". Each challenge addresses a strategic topic of interest to European society, industry and research organizations in "future and emerging technologies". It also supports so-called horizontal actions such as international cooperation. The Future and Emerging Technologies (FET) program supplements these challenges:

- Challenge 1—Pervasive and Trusted Network and Service Infrastructures
- Challenge 2—Cognitive Systems and Robotics
- Challenge 3—Alternative Paths to Components and Systems
- Challenge 4—Technologies for Digital Content and Languages
- Challenge 5—ICT for Health, Ageing Well, Inclusion and Governance
- Challenge 6—ICT for low carbon economy
- Challenge 7—ICT for the Enterprise and Manufacturing

[3]http://ec.europa.eu/digital-agenda/.

- Challenge 8—ICT for Creativity and Learning
- Future and Emerging Technologies (FET)

The term 'Future-Internet' is therefore not about a real implementation of the Internet but is more a federated research theme. As mentioned above the current Internet architecture was developed to deal with neither the wide variety of today's application requirements and network capabilities, nor with ever-new network devices or related business models. All these shortcomings are not related to the protocols themselves, but have created an architectural problem in terms of the limitations of its flexibility, security, mobility, trust and robustness. The big challenge is to address the multiple facets of a Future-Internet, with security and energy efficiency also important societal concerns. Clean slate or evolutionary approaches, or a mix of these, should be equally considered.

From a networking research perspective, the main focus should be on rethinking the basic architecture for more flexibility in the short-term as well as from a long-term perspective. In the short-term perspective adaptivity and adaptability according to environmental conditions and user requirements should be taken into account, such that new service types can be flexible integrated and supported. Ever-new types of access networks such as wireless sensor networks or ad-hoc networks may be integrated, and the requirements of new types of media applications should be addressed. In the long-term the enhancement and exchange of functionality should be supported. Overall flexibility is one key, alongside bandwidth.

Mobility aspects and increasing end-to-end data rates must also be seen as important design drivers, as well as security and trustworthiness. At a network level, new and flexible management capabilities that have never been part of the 'best effort' design paradigm must also be taken into account. New network infrastructures need to support a new Internet architecture where services can be dynamically combined on demand, and flexibly enable the creation of opportunities for new business models.

As mentioned above, based on the Web2.0 technologies "Third party generated services" is an emerging trend moving towards user-centric services, as shown by the growing market of SOA (Service-Oriented-Architectures)-based applications. Moreover, virtualization of systems and networks in the area of cloud computing is also an important driver enabling the delivery of networked services independently from the underlying resource platform, which is an important issue (business model) for service providers. To address all these issues new architectural findings and new paradigms in software engineering are required to deal with the expected complexity in distributed, heterogeneous and dynamically composed systems.

Today's networks and service platforms become increasingly vulnerable (by patching the current Internet architecture) as current developments lead to more complex and heterogeneous networks with new cloud-based data storage and management capacities. To overcome these problems flexibility and trustworthiness are the keywords that must be addressed in a new architecture. Flexibility and trustworthiness can be seen as secure, reliable and resilient to attacks, guaranteeing quality of service and quality of experience. Moreover a new architecture should

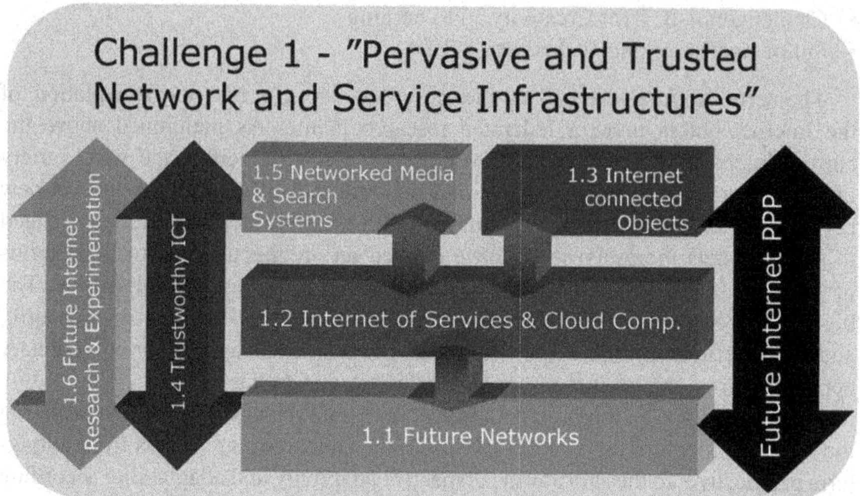

Fig. 1 Organizational chart of Challenge 1 "Pervasive and Trusted Network and Service Infrastructures" (source: http://cordis.europa.eu/fp7/ict/programme/challenge1_en.html)

ensure privacy and provide usable and trusted tools to support the end-users in its security management. All these aspects need to be considered from the outset as inherent design principles rather than being addressed as add-on feature (patches). These aspects will all be considered within Challenge 1 (see Fig. 1).

To address all the topics discussed in future network research, more than 150 projects have been started since the Seventh Framework Program (FP7) under the Objective1.1, "Future Networks", was launched. These projects are grouped into clusters of common interest, to gain synergy and critical mass. One cluster addresses the topic of Future-Internet Architectures and Network Management, the second cluster deals with Radio Access and Spectrum, and the third with Converged and Optical Networks, as depicted in Fig. 2.

2.2 *The Objective 1.1*

In the following we will focus on Challenge 1, Objective 1.1 which covers basic research in future networks and new Internet architectures. The main challenge of Objective 1.1 is to comprehensively and consistently addresses the multiple facets of a Future-Internet, addressing architectural questions as well as energy efficiency, and economic and social concerns. Clean slate or evolutionary approaches or a mix of those will be equally considered.

From a networking and distributed systems research perspective, there is a need to rethink architecture and design principles [7] ranging from the end-to-end principle to layering. From an application perspective, mobility and higher data rates also emerge as important design drivers, and so do flexibility, security

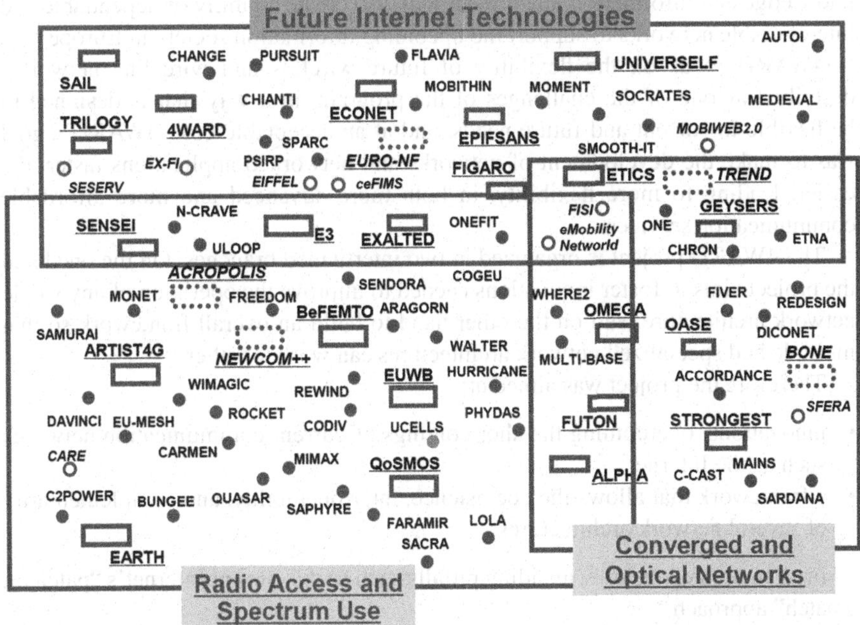

Fig. 2 The clusters of common interest (source: http://cordis.europa.eu/fp7/ict/future-networks/documents/call8/call8-project-portfolio-final.pdf)

and trustworthiness. At the lower layer of the network stack, one challenge will be dealing with the flexible and ad-hoc management capabilities from the original design of the Internet 50 years ago, which have never been taken into account.

All these challenges should be addressed by projects under this objective, ranging from new architectures, new platforms and tools for the development of yet unknown Internet applications. To encourage a close link to the current business models of providers and the software industry, a Public-Private Partnership (PPP) project for the Future-Internet was also launched. It is also expected that key technological developments in networking, wireless and wireline, digital media and service networks will be addressed. In the remainder of this section we will focus on representative projects from the different instruments.

2.2.1 4Ward[4]

The 4WARD project is a project from Call 7 intended to increase the competitiveness of the European networking industry. Providing direct and ubiquitous access to

[4]http://www.4ward-project.eu/.

knowledge and information, the project will also create a family of dependable and interoperable networks to support the upcoming information society in Europe.

4WARD focus on the flexibility of future wireless and wire line networks, described as one of the challenges of the program, in a way that is designed to be flexible to current and future needs, and at an acceptable cost. 4WARD's goal was to make the development of networks and networked applications faster and easier, leading to more flexibility, in both more advanced and more affordable communication services.

The 4WARD project is organized in two intertwined branches. On the one hand the project aims to foster innovations needed to improve the operation of any single network architecture, and on the other hand to build an overall framework so that multiple and specialized network architectures can work together.

Therefore the project was aimed at:

- innovations, overcoming the shortcomings of current communication networks such as the Internet
- a framework that allows the coexistence, inter-operability, and complementarity of several network architectures

in an integrated fashion, avoiding pitfalls such as the current Internet's "patch on a patch" approach.

The project was structured into six work packages. The first three work packages considered new network architecture models dealing with generic path topologies and network management. Another work package was based around new virtualization techniques to allow multiple networking architectures to co-exist on the same infrastructure. Another work package looked at the design and development of interoperable architectures, and finally a work package ensured that all envisaged developments took proper account of essential non-technical issues.

2.2.2 Trilogy[5]

The main focus of Trilogy as an integrated project within FP7 was to consider a new architecture for the control plane of the Internet in order to tackle the increasing complexity and technical shortages. The project thus focused on the generic control functions of the Internet, which were identified as the bottleneck of the hourglass model on the control plane. The architectural design activities focused on a radical approach to develop a Future-Internet for the next decade or more. The design was also tempered and refined by considering the need for incremental deployment.

The project was organized around three technical work packages, each dedicated to one of the three major technical areas (see Fig. 3):

[5]http://www.trilogy-project.org/.

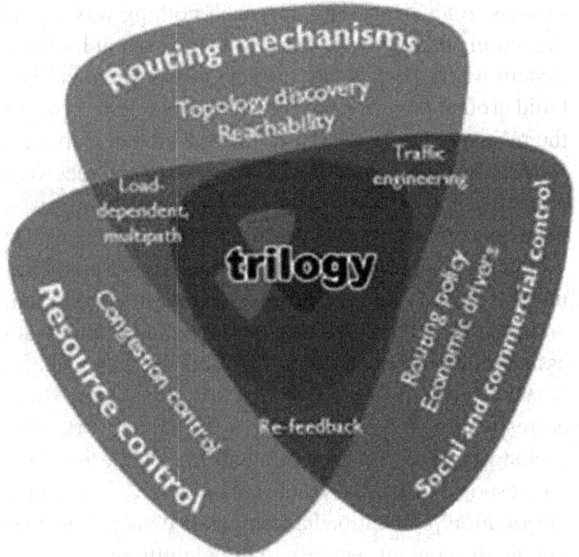

Fig. 3 The Trilogy landscape (source: http://www.trilogy-project.org/)

- Reachability mechanisms: The focus was on creating flexible and change-proof Internetworking functions that provide and manage transparent reachability in a scalable fashion. Based on these functions, communication paths between nodes can be built that address the requirements of the Future-Internet while giving different stakeholders different abilities to control reachability and transparency if needed.
- Resource control: The main focus of this work package was the development of a unified approach to resource control that is efficient, fair and incentive-compatible. Fairness is thus defined in an economic sense, and the resource control framework describes the correct incentive structure and defines how to penalize users that behave unfairly. The work package also analyzes which resources must be controlled throughout the system and creates integrated and flexible resource allocation models.
- Social and commercial control: This work package described the principles that drove the technical work in the other two technical work packages. It considered the strategic socio-economic and commercial factors related to internetworking functions to support a more flexible architecture that is designed for change.

2.2.3 Publish-Subscribe Internet Routing Paradigm (PSIRP)[6]

The main goal of the PSIRP project as a STReP was the development of a new Internet architecture which is designed from the publish/subscribe (pub/sub) point

[6]http://www.psirp.org/home.html.

of view. As a clean-slate approach nothing was considered as fixed not even IP as the routing paradigm. In contrast to the current Internet, which is a message-oriented system where packets are exchanged between end-hosts, the PSIRP architecture is build around the information where the sender is decoupled in time and space from the receiver. This is the basic idea of the pub/sub message exchange pattern where senders "publish" what they want to send and receivers "subscribe" (where a broker is responsible for managing the subscription process) to the publications that they want to receive. Based on this architectural consideration multicast and caching will be the underlying paradigms and security and mobility will be designed as an integral part into the architecture, rather than added case-by-case.

In the PSIRP architecture the information/publication is a persistent, immutable association between an identifier and the data value of the information/publication created by the publisher. If the subscriber knows the identifier it can retrieve the corresponding data value using a PSIRP network. The referential transparency and caching is based on the immutability of the information/publications and moves the responsibility for synchronization to higher layers. Moreover because each information/publication has a logical "owner" one can use self-certifying identifiers to bind the content securely to the identifiers.

The architecture of the PSIRP model focuses on four different parts, the rendezvous system (the broker), topology, routing, and forwarding. The rendezvous system decouples publishers and subscribers and matches the data hosting a certain publication where subscribers are interested in. The topology function manages the physical network topology information. Based on this each domain can configure its internal and external routes adapting to error conditions and balancing the network load. The routing function is responsible for the delivery tree for each publication and cache popular content at some points inside the domains. Finally, the information/publication is delivered to its subscribers along the delivery tree by the forwarding function.

The decision for separating the rendezvous system from forwarding system and avoid using hierarchical names in the routing tables is to keep scalability for a large number of identifiers while the payload can be routed via the shortest path. Therefore the location of data sources is irrelevant and they can easily be mobile and multihomed.

Overall the PSIRP project is focused on the development, implementation, and validation of an information-centric internetworking architecture based on the pub/sub paradigm, which can be seen as one of the recent approaches to tackling the described shortcomings of the current Internet.

2.2.4 Network of the Future (Euro-NF)[7]

In contrast to the other instruments Networks of Excellence (NoE) are created to combine and functionally integrate the research activities and capacities of different

[7] ftp://ftp.cordis.europa.eu/pub/fp7/ict/docs/future-networks/projects-euro-nf-factsheet_en.pdf.

research organizations in a given field, instead of supporting the research itself. The result can be seen as a European "virtual research center" in a given research field. To implement this idea a new mechanism called "Joint Program of Activities", which is based on the integrated and complementary use of resources from entire research units, departments, laboratories or exchange programs of students or staff, was launched. This program requires a formal commitment from the participating organizations to integrate part of their resources and their research activities into the network.

As an example of such an NoE, the Euro-Network of the Future (Euro-NF) project will be considered and explained in more detail. Euro-NF was a Network of Excellence within the FP7 framework on the Network of the Future instrument, formed of 35 institutions (from academia and industry) from 16 countries. It was a successor to the successful Euro-NGI and Euro-FGI projects carried out from 2003 to 2011.

Its main target was to coordinate the research effort on networking research of the partners. It can be seen as a think tank of excellent research organizations and a source of innovation along scientific, technological and socio-economic pathways towards the network of the future.

The project itself was organized in architectural and research domains (see Fig. 4). In the architectural domains different technical aspects of the network itself, such as convergent access, core and metro networks, beyond-IP networking, new network paradigms and services and overlays were considered, the research domains focused on the underlying technologies.

The first research activity was related to future service and network architectures, requiring the integration of the large diversity of technologies and service and networking paradigms and therefore having the capacity to manage the diversity. The future architecture has to natively integrate a global mobility of users, services, terminals and networks, an explosion of connected devices, security and dependability, inter domain networking solutions, and evolved service architectures, among other things. The second research activity was related to the quantitative issues of the networks of the future, such as new modeling, design, and dimensioning tools and contributions to the development of a network theory. In future networks, the randomness of the quality of the radio links, of the mobility of users and devices as well as of the topology of the networks, has to be considered. New quantitative methods raising complex scientific problems therefore have to be considered, in particular for the efficient processing of traffic. Finally, given the exponentially growing impact of the Internet, topics such as governance, regulation policies, privacy, trust, security, and standardization need to be considered together, and not after the design of the network. These topics were covered in the third research activity. Figure 4 summarizes the structure of the research activities.

Synergies between experts in network and service architectures on the one hand, and socio-economic experts dealing in particular with cost and business models on the other hand, are an important asset of this NoE. Overall, Euro-NF as a Network

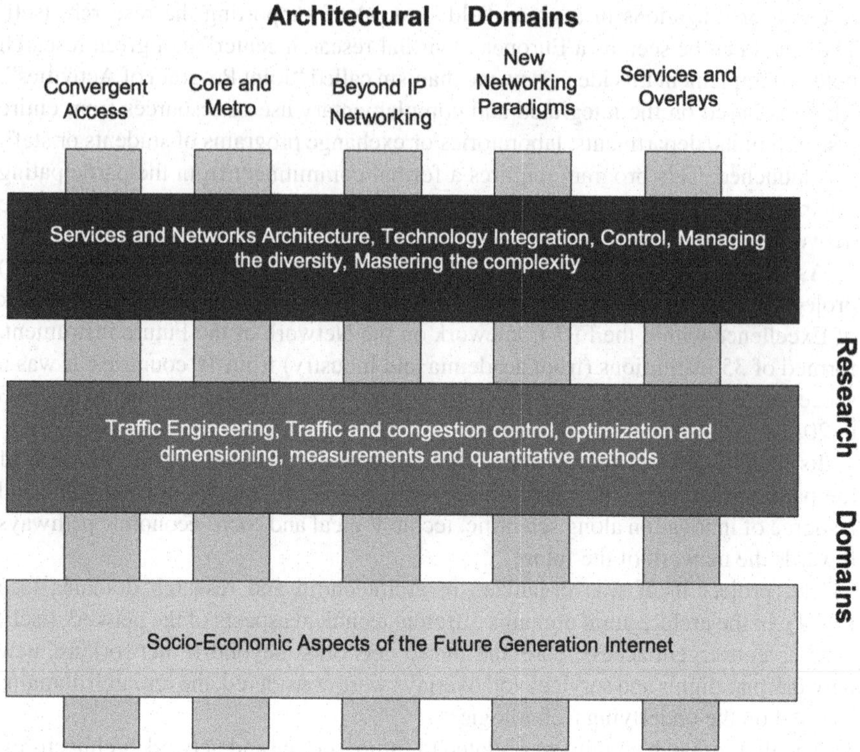

Fig. 4 The architectural and research domains of Euro-NF[8]

of Excellence could be seen as a European source of innovation and a cross-border collaboration infrastructure that is still working, especially in its idea of exchanging students and staff.

2.3 *The FIRE Projects*

Researchersaround the world have agreed on the limitations of the current Internet architecture and that it is time for research to take a fresh, long-term view, considering the Internet as a complex system [4] and offering a new paradigm for communication. This high-level approach deals with strategic and multidisciplinary research on new Internet architectures and design, including 'clean slate' or 'disruptive' ones. In order to measure, compare and validate scientific results, these new paradigms need to be tested on a large scale.

[8]Source: ftp://ftp.cordis.europa.eu/pub/fp7/ict/docs/future-networks/projects-euro-nf-factsheet_en.pdf

For this kind of research an experimental facility for validating new network algorithms, new network technologies and service architectures and paradigms is needed. It has been shown that in the past, many results have only been discovered when systems were deployed and tested in 'real-life' situations. Such experiments need to be carried out in a planned, controlled, and legal manner. An experimental facility on Future-Internet architectures, design and technologies must broadly support research on all layers from access/network to the application layer. To be credible, representative and to prove its scalability, the experimentation must be performed on a large scale that may exceed the scale of a single facility. The need for federated different experimental facilities is evident.

There is an increasing demand from academia, as well as from industry, to bridge the gap between theory and large-scale experimentation through experimentally-driven research. Consequently it is essential to set up large-scale experimental facilities, going beyond the demands of individual project testbeds. This will help with putting together different research communities (beyond computer science) in an interdisciplinary approach that may be potentially disruptive, discovering new and emerging behaviors and use patterns in an open innovation context.

The Future-Internet Research and Experimentation Initiative (FIRE) of the European Union, formulated in Challenge 1.6 as described above, addresses this need.

Its goal is to create experimental environments which will allow both innovative (evolutionary) and clean-slate (revolutionary) research for the Future-Internet. In detail, FIRE has (and still does) pushed the idea of experimentally driven research as an important tool and cornerstone of future-Internet research. It has also been very successful in creating several large-scale European experimental facilities, where federation plays an important role to create real large-scale environments and provide possibilities for experiments of various flavors. Finally, it has been very successful in creating funding opportunities for experiments to be run at the facilities.

FIRE's actions are twofold: the first dimension deals mainly with the experimental facility itself and how to support research in the Future-Internet arena at different phases of the research and development cycle, taking open coordinated federation of testbeds into account.

The second dimension of FIRE mainly supports the iterative cycles of experimentally driven research, design and large-scale experimentation of new network and service architectures for the Future-Internet from an overall system perspective (see Fig. 5).

2.3.1 Inside FIRE

Based on this two-directional approach, FIRE has, over the years, executed a number of projects. The basic financing instruments have been used as follows:

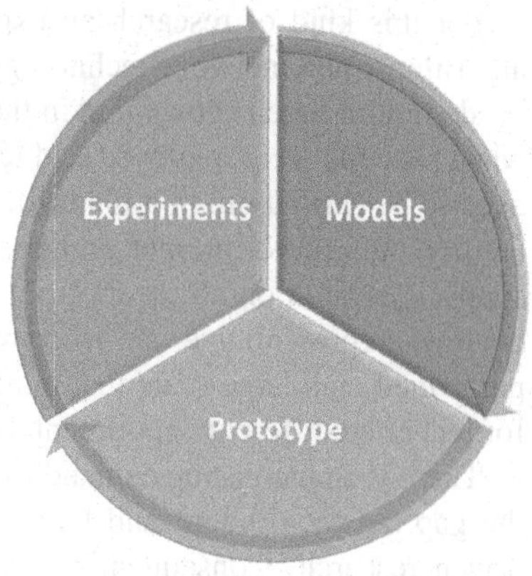

Fig. 5 Research—experimentation cycles

- The FIRE Experimental Facilities have been created mostly using the IP project instrument.
- Experimental research to be run on top of FIRE facilities has been implemented through STRePs.
- Finally, the overall coordination has been ensured through so-called Cooperation and Support Actions (CSAs).

Figure 6 presents an overview of current FIRE projects, following the three-way separation. On the left-hand side, some of the earlier FIRE projects are shown—FIRE projects have been run for 6 years now. In the following, we present brief summaries of some of the FIRE projects. A comprehensive list of FIRE projects can be found at: http://www.ict-fire.eu/home/fire-projects.html.

The **OneLab** and **PII** project were among the first wave of FIRE projects, the task of which was the establishment of initial experimental facilities. OneLab took over the task of creating and running PlanetLab Europe along with its well-known "slice architecture" as we know it from Princeton's original PlanetLab implementation. This implementation was based on a Linux 2.4-series kernel with some patches for running the vservers and the hierarchical token bucket packet scheduling. This constituted the original PlantLab OS. Because of the discontinuity of the Linux 2.4-series kernel the PlantLab core team decided to start a migration to LXC (Linux Containers) which is an implementation of container-based virtualization in the Linux kernel. Today the PlantLab OS is just a Linux node running containers without any operating system patches.

One goal of the OneLab experimental facility is to offer access to a wide range of diverse Future-Internet testbeds and research tools across Europe and beyond. It was built on the outcomes of former framework programs (especially FP 6) such as OneLab and OneLab2 (in FP7) and will continue to be developed through the OpenLab project, starting in September 2011.

OneLab thus offers three different types of federations that are realized in its experimental resources dedicated to their specific functions. First of all, the trust and resource discovery that was implemented through users can have seamless access to

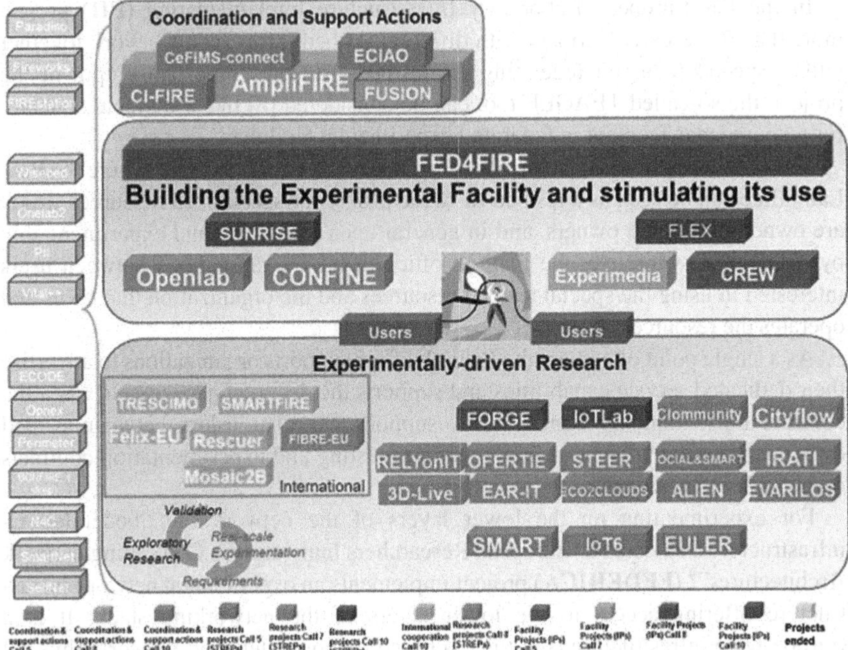

Fig. 6 Overall picture of current FIRE projects (source: http://www.ict-fire.eu)

the resources. In the second step, experiment control functionality was considered and some wireless resources and their ORBIT Management Framework (OMF) controller were added. The last step consists of adding the measurement system that is built upon the so-called Distributed Measurement Infrastructure (TopHat).

Based on the second step, users were able to request OMF resource controllers in a PlanetLab slice, and therefore OneLab allows an OMF user to run an experiment on PlanetLab. This means that an experiment can be seamlessly run on PlantLab and OMF wireless testbeds such as Rudgers/WinLab.

The resource discovery, as well as the trust infrastructure in OneLab, is based on the PlantLab idea of the slice-based federation architecture (SFA). OneLab's best known infrastructure is the PlanetLab Europe, the European branch of the global PlanetLab system. PlantLab Europe users thus have access to worldwide-distributed resources.

OneLab currently consists of more than 250 nodes, which are spread around Europe at more than 140 sites. Users thus have access to more than 1000 PlanetLab nodes worldwide, where each node runs the PlanetLab operating system. At PlanetLab itself, OneLab is also a slice-based facility where researchers can get a slice across the systems, which consist of virtual machines on any or all of the nodes.

In the Pan-European Laboratory Infrastructure Implementation (**PII**) project more than 20 research partners with different testbed infrastructures work together with a special focus on federating the testbeds. As one important output of the project, the so-called TEAGLE tool can be considered [8] like a de-facto standard for resource management in federated experimental facilities.

The project itself was developed on the concept of a PanLab office where the PanLab office can be seen as a testbed resource broker for specialized resources which are owned by testbed owners, and in general open for testing and experimentation by researchers. Moreover the PanLab office manages transactions between users interested in using the special testbed resources and the organization that owns and operates the resources.

As a single point of contact the PanLab office supports organizations to advertise their dedicated service capabilities and supports the different participants when the contract are written and signed. It also supports the relationships between testbed owners and testbed operators and negotiates testing and experimentation contracts on behalf of the customers.

For experimenting on the lower layers of the network the "Federated E-infrastructure Dedicated to European Researchers Innovating in Computing network Architectures"[9] **(FEDERICA)** project implements an experimental network infrastructure offering access to the lower layers of the networking stack. It is a Europe-wide infrastructure (see Fig. 7) based on computers and network physical resources, which are both capable of virtualization and offer a technology agnostic infrastructure. Based on dedicated user requirements FEDERICA can create virtual resources for network topologies and systems. The user can thus manage the topology, and has full control of the resources in its slice. Since the (virtual) FEDERICA infrastructure does not know anything about the specific protocols to be used, it is particularly a perfect environment for all kinds of clean-slate approaches. Security protocols and algorithms, new routing protocols and innovative distributed applications have been also experimented successfully.

The physical topology of the FEDERICA infrastructure spans Europe and offers four main points-of-presence (PoPs) at different national research networks (NRNs), for example: CESNET (Czech Republic), DFN (Germany), GARR (Italy), and PSNC (Poland), and ten non-core PoPs located in FCCN (Portugal), GRNET (Greece), HEAnet (Ireland), I2CAT (Spain), ICCS (Greece), KTH (Sweden), NIIF (Hungarnet), PSNC (Poland), RedIRIS (Spain), and SWITCH (Switzerland).[10]

The physical connectivity of the four main PoPs is based on full mesh topology runs at 1Gbps over dedicated Ethernet network circuits that are terminated on dedicated routers. The non-core PoPs are connected to core and non-core PoPs in a greatly meshed topology, also using 1 Gbps Ethernet circuits and dedicated switches, where each PoP hosts at least one powerful server. In the core PoPs the number of servers, running VMware ESXi operating system as hypervisors, is

[9]http://www.fp7-federica.eu.

[10]http://www.fp7-federica.eu/documents/FEDERICA-DSA2.3-v3-r.pdf.

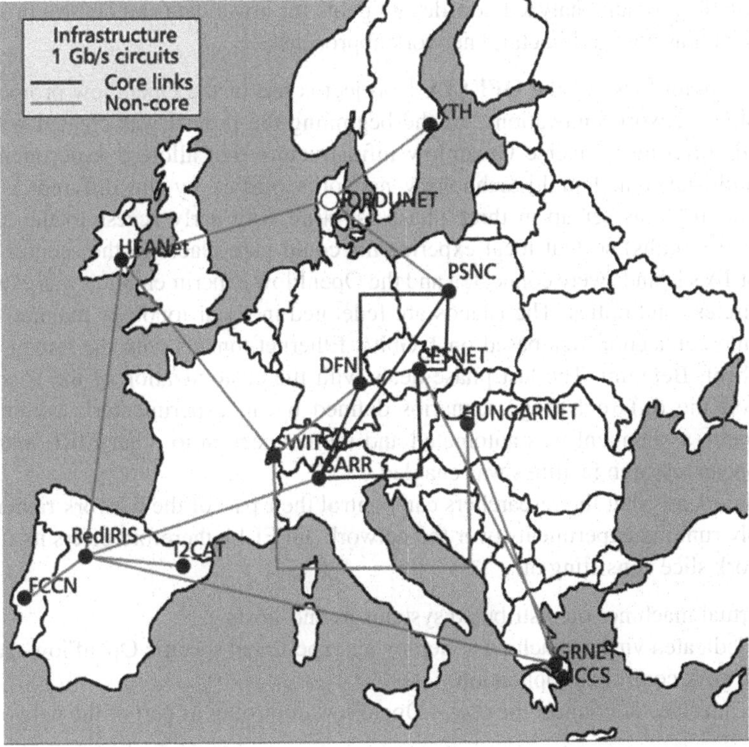

Fig. 7 FEDERICA physical topology (source: FEDERICA Project Periodic Report 09/10)

increased to at least two. The facility therefore allows disruptive testing on the lower layers of the network in a wide area environment where all components are based on virtual resources that can host any type of protocol and application.

Although the FEDERICA project ended in 2010, the substrate is still fully functional and is currently supported by the partners and partially by the "Multi-Gigabit European Research and Education Network and Associated Services" (GN3) project. For further sustainability the facility remained operational until 2012 without any upgrades.

FEDERICA served more than 20 user groups with different use cases [9] that can be divided into three areas:

- The first group of experiments was dedicated to the validation of virtual infrastructure features. The experiments were thus mainly around the general principles of virtualization capable network substrates and the measurement and configuration of reproducible virtual resources.
- The second group dealt with evaluation of multi-layer network architectures. This was based on FEDERICA's capability for federating external testbeds or virtual slices provided by other facilities.

- The third group analyzed new design goals for novel data and control protocols as well as new architectural network approaches.

The main focus of the **OFELIA**[11] project is use of the OpenFlow protocol for enabling network innovations. At the beginning the project was created with ten islands offering a diverse OpenFlow infrastructure that allowed experimentation on multi-layer and multi-technology networks offered by the different islands. The facility was set up in three phases starting with early access to the facility (after 6 months) so that local experiments could take place. In the second phase OFELIA's islands were connected and the OpenFlow experimentation was extended to wireless and optics. The islands are federated in a star topology manner where the interconnection was based on 1 Gbit/s Ethernet tunnels with the hub at IBBT in Ghent, Belgium. The last phase deals with the customization of the islands to support the different usage scenarios defined by an experimenter's community. Resource assignment was automated and the connection to other FIRE and non-European research facilities was enabled.

The idea is that the researchers can control their part of the network rather than simply running experiments over the network. OFELIA therefore offers its users a network slice consisting of:

- virtual machines on distributed systems as end-hosts
- a dedicated virtual machine to deploy an experiment specific OpenFlow-capable network controller/application
- an interface to connect the user's OpenFlow controller as part of the network.

The project period was September 2010 until September 2013 with an overall budget of about €6.3M. At the end of the project, five of the original ten islands were still providing heterogeneous network infrastructures to experiment with. Selected partners who offer different technological strength and user groups operate the different testbeds.

The OFELIA infrastructure aims at cases in which researchers need control over the complete network stack to design and analyze new network architectures and designs in terms of routing, traffic engineering etc. Under the umbrella of OFELIA the Open Calls resulted in additional projects dealing with network function virtualization which is closely related to software-defined networking (SDN) and OpenFlow, content centric networking (CCN), gaming and real-time applications, multi-domain routing and social networks.

Unlike OneLab and PII, **WISEBED** was one of only two STRePs participating in the establishment of the FIRE facilities. It was a joint effort between nine academic and research institutes across Europe and was carried out from June 2008 to May 2011. Its goal was the establishment of a facility to enable researchers to do experimental work with wireless sensor networks. The overall approach is shown in Fig. 8.

[11] www.fp7-ofelia.eu.

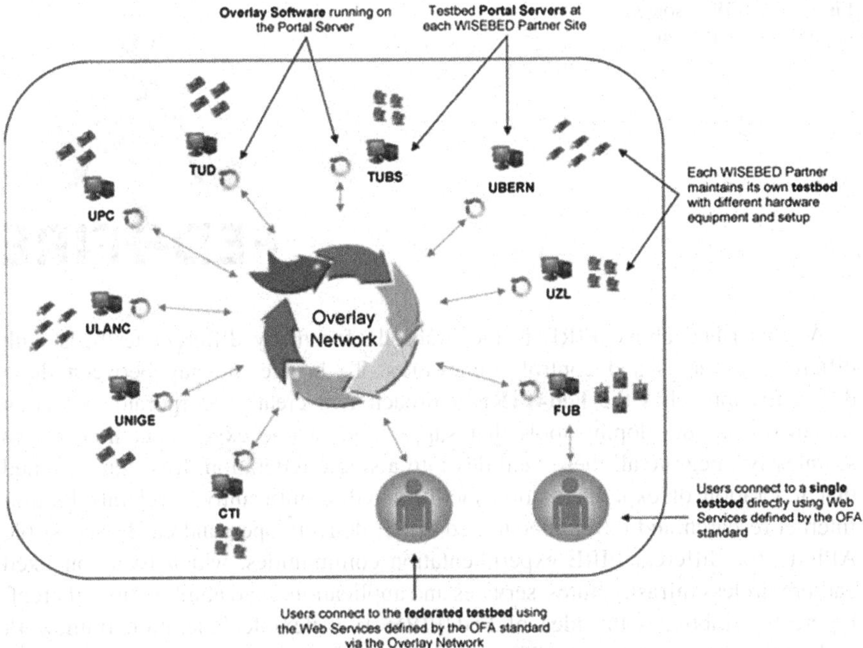

Fig. 8 The general approach of WISEBED (source: http://www.ict-fire.eu/home/fire-roadmap/fire-roadmap-v1.html)

In short, WSN testbed providers have full control over their facility, but they can also export its services to the overall WISEBED federation. This is achieved by offering the same kind of (web) service interface at all levels, be it a local testbed, a virtual testbed created from instances from several local testbeds, simulated nodes or the full federation. WISEBED developed a number of innovative federation mechanisms allowing for the very flexible setup of a desired environment. It also created a fully-fledged process to support the experimenter throughout the whole life-cycle of their experiment. Most original partners still operate their testbeds; more information is available at www.wisebed.eu.

To some extent **SmartSantander** can be seen as a successor of WISEBED. SmartSantander created an experimental facility for Internet-of-Things-related research in a real city, namely Santander in Spain. When it ended in November 2013, more than 10,000 IoT devices had been installed in Santander and its partner cities, making Santander an excellent playground for SmartCity technologies and one of the best-known SmartCity prototypes in the world. The facility is used for both basic research, for example in terms of testing new IoT protocols, and for applied research, where new applications and services are created for the city and used by its citizens. Among the best-known applications are those controlling public parking facilities, environmental monitoring and public transport.

Fig. 9 Fed4FIRE (source: http://www.fed4fire.eu/)

As described above FIRE is the umbrella for many different testbeds with different resources and control frameworks. To bridge the gap between these different approaches the **Fed4FIRE** approach will create a corporate federation framework by developing tools that support the whole experiment lifecycle to seamlessly integrate all these facilities into a single federation. It is thus assumed that federation of experimentation facilities will significantly accelerate Future-Internet research and development. Fed4FIRE delivers open and easily accessible APIs to the different FIRE experimentation communities, which focus on fixed and/or wireless infrastructures, services and applications, and combinations thereof. Figure 9 symbolizes the idea of Fed4FIRE as a flexible federation framework to bridge the gap between different and heterogeneous resources and control frameworks.

A demand-driven corporate federation framework, based on an open architecture and specifications, is an expected result of Fed4FIRE. This federation framework will be promoted internationally and it is expected that the framework will be adopted by the different experimental facilities. The framework will provide a lifecycle management of experiments including discovery mechanisms of resources, reservation mechanisms, experiment execution and control tools as well as measurement and control tools. All this should be implemented in a simple, efficient and cost effective way. Tools and services for supporting dynamic federated identity and access control, which will increase the trustworthiness of the facilities, will be created. All these services will be offered through a FIRE portal, which particularly offers brokering, user access management, and last but not least measurement capabilities.

A new instrument within the Fed4FIRE project is the idea of so-called Open Calls. Researchers from different Future-Internet research areas will be invited through such Open Calls to perform innovative experiments using the Fed4FIRE federation or integrating new resources into the federation. The expected experiences can be used to make the federation framework more mature and offer it to other experimental facilities.

Detailed information about the federated facilities, the tools adopted by the framework, and the necessary required implementation details needed from a facility when joining the federation can be found on the Fed4FIRE website, https://www.fed4fire.eu. In general Fed4FIRE is a project under the IP instrument within the Seventh Framework Program (FP7) started in October 2012 and will run for 4 years.

2.4 *Future-Internet Assembly (FIA)*[12]

The Future-Internet Assembly is the most important self-organizing European body to define future directions of research in the area of Internet technologies. The Future-Internet Assembly was launched in May 2008 with a congress in Bled (Slovenia), where all projects related to Future-Internet research signed the Bled declaration.[13] The FIA continues to expand, and encompasses, first and foremost, many projects from Challenge 1 as described above, but also quite a number of other projects[14] (and researchers).

FIA is not restricted to researchers active in a current project, but is open to all researchers engaged in Future-Internet research, also from non-European states.

There is an FIA meeting twice a year, organized by the country which in that term holds the EU presidency. A typical FIA consists of a few plenary sessions and a number of working sessions in which discussions on the state of the art in Future-Internet research, standardization and future directions take place. These sessions are framed by a large exhibition where projects from all challenges present their results. All in all, FIA is big marketplace where views are exchanged, and new connections are often made and new projects are created.

The regular work between FIA events is overseen by the FIA steering committee and organized within FIA working groups. While there are working groups for more or less "horizontal actions", such as overall road mapping, international cooperation, or the FIA book produced for every FIA event, there are also groups which deal with very technical topics, such as real-world Internet, linked open data, or trust and security. Working groups are not meant to exist forever: if they have met their objectives or no longer deliver results, they can be closed by the steering committee.

3 The Public Private Partnership (PPP) Approach

The Future-Internet Public-Private Partnership[15] (FI-PPP) is a European program for Internet-enabled innovation mostly driven by industry. The overall idea of such a PPP is to foster and accelerate product innovations of Future-Internet technologies within short time frames. It is not related to medium or long term basic research questions but it is a short term product oriented program for the European industry to boost business processes based on the Internet in the areas of smart technologies, infrastructure and industry 4.0.

---

[12] http://www.future-internet.eu/home/future-internet-assembly.html

[13] http://www.future-internet.eu/publications/bled-declaration.html.

[14] http://www.future-internet.eu/activities/fp7-projects.html or http://ec.europa.eu/information_society/activities/foi/research/fiaprojects/index_en.htm.

[15] http://www.fi-ppp.eu/.

With over a billion users worldwide, the Internet is one of history's great success stories. Its global Internetworking infrastructures and service overlays underline the backbone of the European economy and society. Yet today's Internet was designed in the 1970s within the research community with line-oriented cryptic commands far from current and future usage scenarios. The former design goals of the Internet do not match today's usage scenarios and consequently it hinders product innovation. More and more efforts are required to make today's applications secure instead of creating new ones. All this results in various challenges in a wide variety of technology, business, society and governance areas that have to be overcome if the development of the Future-Internet is to sustain the networked society of tomorrow.

To address these challenges, the European Commission has launched the industry driven Future-Internet Public-Private Partnership Program (FI-PPP). The basic idea behind this program is to foster a common vision for harmonized technology platforms on a European scale. As part of this we have to consider its implementation, as well as the integration and harmonization of the related policies, and the legal, political and regulatory frameworks of the different European countries. As described in the Digital Agenda for Europe, this can be considered as a prerequisite for achieving a European online Digital Single Market (DSM) and, overall an inclusive knowledge society.

In general the program (see Fig. 10) aims can be described as follows. Because the program is driven by industry it should first increase the effectiveness of business processes and infrastructures supporting applications in such focus areas where the European industry has its strength such as transport, health, and energy. In addition to these application areas innovative new business models in the Internet market will also be considered. Moreover it should give a push to the competitive position of the communication and media industry in Europe in the areas of telecommunication, software and services. Overall the approach of the FI-PPP is an industry-driven approach covering research and development on network and communication infrastructures, software, services, media technologies and related devices.

One main objective of the PPP is the early involvement of the users in the development/research lifecycle based on experimentation and validation of related application environments. The new platform will thus be used by a range of actors, in particular SMEs and public administrations, to validate the technologies in the context of smart applications and their ability to support user-driven innovation schemes.

4 The German-Lab Project

The German Lab (G-Lab) project aims to investigate architectural concepts and technologies for a new inter-networking architecture as an integrated approach between theoretic and experimental studies. G-Lab comprises two branches of activities: the first branch deals with research studies of future network architectures,

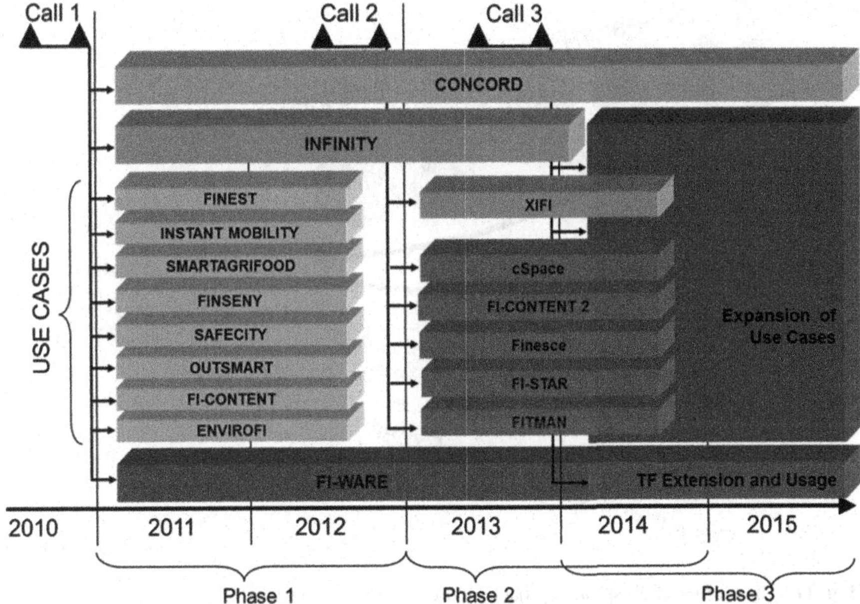

Fig. 10 Program overview of the FI-PPP (source: http://ec.europa.eu/digital-agenda/en/future-internet-public-private-partnership)

and the second branch deals with the design and setup of the experimental facility. The overall organization of the project ensures that the layout of the experimental facility really meets the demand of the researchers. This can be guaranteed because the same community controls both branches. This includes access to virtualized resources as well as exclusive access to resources if necessary.

The G-Lab project[16] started in 2008 as a distributed joint research and experimentation project for Future-Internet studies and development. This BMBF project[17] was initially distributed across six universities in Germany: Würzburg, Kaiserslautern, Berlin, München, Karlsruhe, and Darmstadt. The G-Lab project itself is divided in two parts, Future-Internet research, and the experimental facility, with the overall goal to foster experimentally driven research. The research aspects of the G-Lab project are not limited to theoretical studies and novel ideas, but also include experimental verification of the derived theoretic results. Investigation of new approaches concerning, for example, routing, addressing, monitoring and management, content distribution, and their interaction is such an intricate task that it could not be validated with only analytic research methods. Consequently, in G-Lab, theoretic research and the experimental facility are closely intertwined. This

[16]German-Lab project website: http://www.german-lab.de.

[17]Funded by the Federal Ministry of Education and Research (BMBF) 2008–2012.

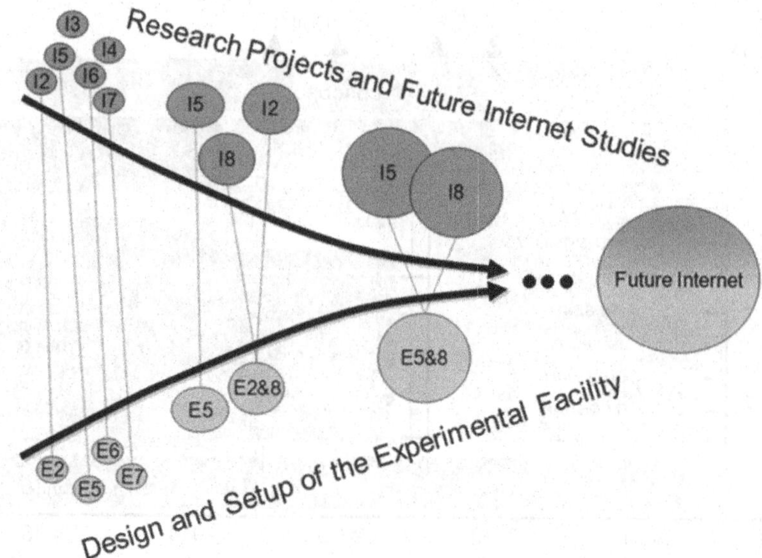

Fig. 11 The German-Lab spiral approach (source: www.german-lab.de)

means that new algorithmic approaches can be tested in the facility and the results may lead to modifications of the algorithms and models used. The resulting spiral process (see Fig. 11) can converge towards a new inter-networking architecture in the end.

Within this spiral process the experimental facility has to adapt to the actual and changing demands of the research projects within G-Lab. This process requires a certain flexibility of system and network usage. To support such flexibility research efforts are also necessary so that the experimental facility can ultimately become a research field itself. With this kind of organization the G-Lab project tries to avoid a situation where the services offered by the experimental facility do not meet the demands of the researchers.

Overall, G-Lab was structured as two phases. Phase One involves only universities where each site is responsible for different theoretical tasks, and also hosts and operates its own part of the experimental facility. This phase was structured into eight activities (ASPs) as shown in Fig. 12. ASP7 was dedicated and responsible for the setup, operation, adaption, and enhancement of the experimental facility. In its second phase[18] more academic and industrial partners extended the G-Lab project.

Most of the projects in Phase Two focused on deepening, extending, or applying the work being done in the first phase. Some partners in Phase Two enhanced the experimental facility with special equipment such as sensor networks.

[18]German-Lab Phase Two partners and projects: http://www.german-lab.de/phase-2.

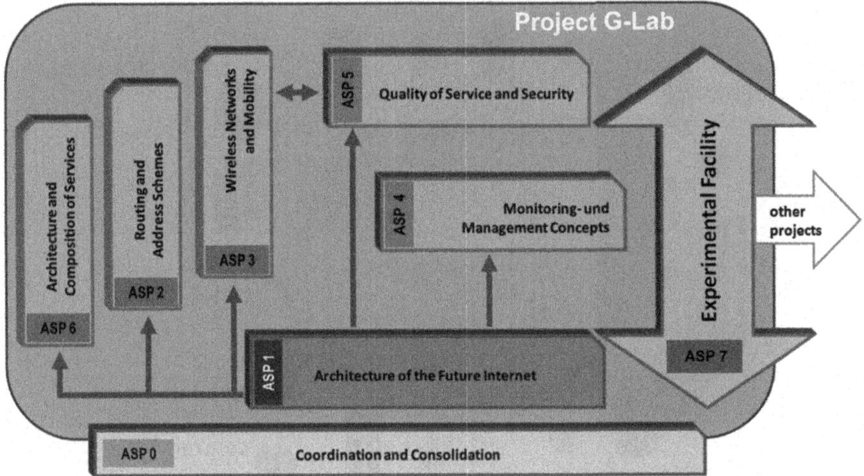

Fig. 12 German-Lab project structure (source: http://www.german-lab.de/fileadmin/Press/G-Lab_White_Paper_Phase1.pdf)

4.1 The G-Lab Experimental Facility

As well as the basic research projects covered by ASP1 to ASP6, one major outcome of the G-Lab project was the experimental facility [10].

The experimental facility operates wired and wireless network nodes with over 170 systems. All systems are fully controllable by the G-Lab partners, particularly from the head node in Kaiserslautern. The platform is distributed into clusters at six different university sites located in Germany, with Kaiserslautern as the main site, as depicted in Fig. 13. A major challenge for the experimental facility is fulfilling the changing requirements of the research projects and sometimes even integrating intermediate research results. This leads to a continuously changing infrastructure. Since the experimental facility and the research projects were launched simultaneously it was a special challenge for the experimental facility to be up and running for the first simulation requests as fast as possible. To meet this challenge ready-to-run software (such as PlantLab) was used in this phase and substituted over time by specific software developments for specific experiments.

One of the key concepts of the G-Lab experimental facility is its flexibility. While testbeds in general restrict the possible uses either by software or by policy, in G-Lab the experimental facility is open to all possible experiments, even if they impact the infrastructure of the experimental facility itself. The experimental facility in G-Lab therefore has to offer different control frameworks from which users can choose. As a consequence of this flexibility individual nodes cannot be bound to a special control framework but must be able to boot a desired control

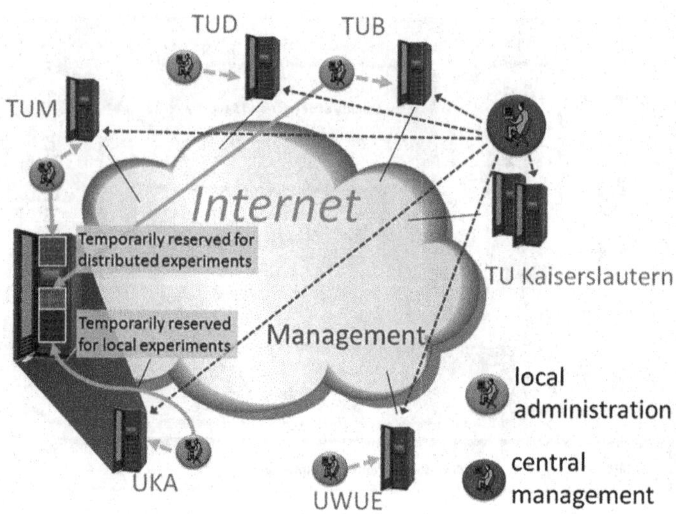

Fig. 13 Topology of the G-Lab resources

framework on demand. This ranges from the well-known PlanetLab framework (which was the standard boot image in G-Lab) to custom boot images developed by the experimenter themselves, that fit specifically to a specific experiment.

As well as the control framework setup, the hardware setup can also be changed upon request of the experimenters. Because of its large numbers of physical Ethernet ports (eight physical ports) the so-called network nodes can be flexibly reconfigured to create different network topologies, which includes manual rewiring of the network or more disk or CPU capacity. New virtualization technologies have also been introduced in order to reduce the amount of manual work, especially for creating dedicated network topologies. This work results in the development of the Topology Management Tool (ToMaTo) [11].

There is a trade-off in experimentation between access for more users and more resources or privileges for users. The control frameworks such as PlantLab offer all users access to all nodes at a high level of abstraction but enable large experiments using many nodes. This means users are restricted to a high abstraction level and do not have access to kernel modules, or raw sockets for network configuration. At the other extreme, a low abstraction level is offered based on custom boot images that offer full access to the hardware with complete control over the kernel and the network interfaces. The drawback of this solution is the limited usage of a node to one experiment at a time. But nevertheless G-Lab offers both solutions to its users and controls the fair share of general accessible nodes with policies.

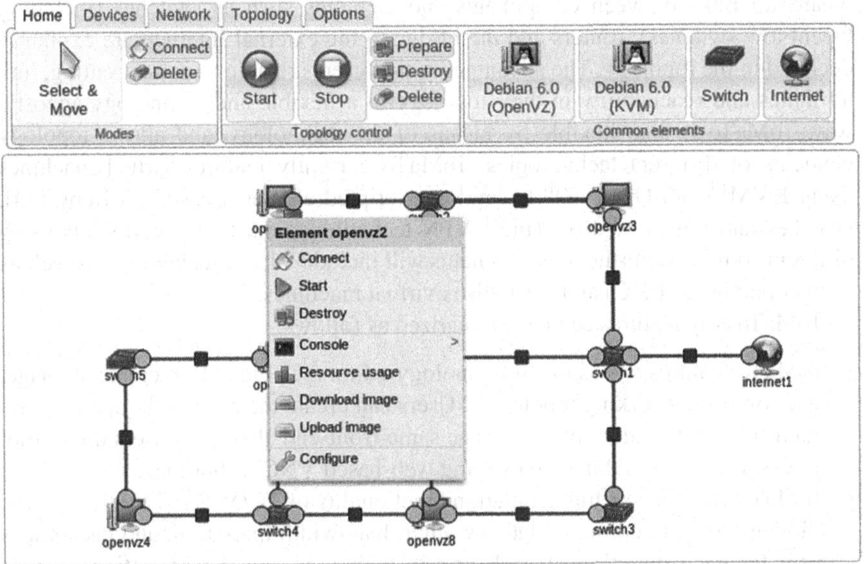

Fig. 14 ToMaTo editor

4.2 *The Topology Management Tool*

During the project the increasing number of requirements for research experiments were beyond the capabilities of the standard boot image (PlanetLab) because of their demands for lower network layer access or to define their dedicated network topology. To address these experiment-related demands a special control framework was developed, the Topology Management Tool (ToMaTo) [12, 13]. ToMaTo defines a virtual control framework that runs on virtualized nodes of the experimental facility and supplements them with additional functionalities such as network topology virtualization, packet capturing and link emulation.

The main advantage of ToMaTo is its ability to create virtual network topologies based on virtual nodes running a dedicated OS and the experimenter's software. In the following we describe the ToMaTo control framework and compare it to other frameworks. Overall the ToMaTo framework is a testbed specifically designed for experiments from the distributed systems arena. The main difference with other networking testbeds is that ToMaTo is topology-oriented, which means each experiment in ToMaTo consists of a topology of virtual components. Figure 14 shows the graphical front-end of ToMaTo and an example topology containing several components.

One of the basic concepts of The Topology Management Tool is its ability to explicitly declare the connections between the nodes and external networks. This means all links between components and elements such as switches (including OpenFlow switches), routers and the bridge to the external Internet, are explicitly declared in the topology. The advantage is that users are able to define routing, link attributes and reachability in their topology in a flexible and technology agnostic way. ToMaTo is also flexible by means of the technology, and allows topology elements of different technologies. ToMaTo currently features virtual machines using KVM[19] and OpenVZ[20] technology, scripted elements based on Repy [14], switches and hubs based on Tinc[21] VPN technology, and tunnel endpoints using different solutions. Further developments will include VLAN technology as well as Linux containers (LXC) and VirtualBox virtual machines.

ToMaTo Key features can be summarized as follows:

- ToMaTo features an easy-to-use topology-editor (see Fig. 14) to create, manage, and control networking topologies. Users can create their network topology and control it by the same interface. The same front-end also gives users direct root access to their virtual machines using web-based VNC technology.
- ToMaTo offers also a link emulation functionality of TC/NetEm [15] to its users, allowing them to configure latency, jitter, bandwidth limitation, and packet loss ratio for each direction of each link in their topology. These settings can be modified during a running experiment and become effective immediately.
- ToMaTo is able to capture and filter network packets on any link in the topology. For further analysis they can be viewed using packet analysis tools such as Wireshark[22] or CloudShark.[23]
- ToMaTo also offers live capturing of packets where copies of packets are sent directly to a running Wireshark instance where the user can control them without delay. As this functionality is independent on the operating system running inside the virtual machines, packet capturing can also capture packets that the operating system tries to hide.
- Templates. ToMaTo provides several pre-installed templates for virtual machine-based topology elements. The templates offer all common OS (Linux as well as Windows (restricted)), as well as specialized templates for network experiments such as a virtual switch (OpenVSwitch). Users are free to choose from these templates or provide their own images for their dedicated topology elements.

Similarly to the PlanetLab approach, the ToMaTo testbed will be further extended with a cooperative model in which each participating institution can provide some resources to the testbed and can use the testbed as a whole.

---

[19]KVM: Kernel-based Virtual Machine: http://www.linux.kvm.org.

[20]OpenVZ is a container-based virtualization technology, http://www.openvz.org.

[21]Tinc is a peer-to-peer virtual private network software, http://www.tinc-vpn.org.

[22]Wireshark is a packet analyzer tool, http://www.wireshark.org.

[23]Cloudshark is an online packet analyzer tool, http://www.cloudshark.org.

5 Conclusion and Outlook

In this contribution we first gave an overview of the driving forces behind the ICT part of Europe's research frameworks, especially around the network of the future. Future-Internet research is the main issue, in all its facets. The European Commission has addressed the resulting challenges with three main funding instruments: Integrating Projects (IP), Specific Targeted Research Projects (STReP) and the Network of Excellence (NoE) which were briefly described.

The main part of the paper focused on Objective 1.1 (Future Networks) within Challenge 1 (Pervasive and Trusted Network and Service Infrastructures) of the Framework Program 7. More than 150 projects have been launched since the beginning of this program, which covers topics/clusters such as Future-Internet technologies, radio access and spectrum use, and converged and optical networks. Further focusing on the cluster of Future-Internet technologies, we provided a brief glimpse of different projects that are representative of the different funding instruments.

The next part of this contribution focused on Objective 1.6 (Future-Internet Research and Experimentation), the FIRE approach. After introducing the FIRE initiative as a whole we also focused on a few concrete projects. The overall objective of FIRE is to address the increasing demand for large scale experimentally driven research through iterative cycles of research, design and experimentation. This approach needs to set up large-scale experimental facilities, going beyond specific project testbeds.

Another facet of the European ICT research arena is the so-called Future-Internet Public-Private Partnership (FI-PPP) initiative. It forms a European program for Internet-enabled innovation in order to bridge the gap between Future-Internet research and product innovation. It should boost the European market for smart technologies, and increase the effectiveness of Internet technology-based business processes. All projects under the umbrella of the FI-PPP are more or less industry driven and mainly focus on sustaining existing business models.

To subsume all these different approaches under Challenge 1, the Future-Internet Assembly (FIA) was founded in 2008. This can be seen as a loosely coupled association of the Future-Internet focused projects that try to increase European activities in Future-Internet research and development to strengthen European competitiveness in the global market. As a result of the first FIA conference held in Bled, Slovenia, the Bled Declaration[24] was signed by the major projects in this field. Today FIA is umbrella of nearly 150 projects from Challenge 1 of the ICT program of FP7.

The next wave in Europe's framework programs is Horizon 2020 that continues the work of its successful precursors from 2014 until 2020. It will bring together the main and successful ideas of research and innovation funding in a coherent form

[24]http://www.future-internet.eu/publications/bled-declaration.html.

and will integrate the main and successful parts of the former Competitiveness and Innovation Framework Program (CIP) as well as the European Institute of Innovation and Technology (EIT). This new program will make participation easier, increase scientific and economic impact, and maximize the outcome. It implements a very straightforward decision process and will explicitly encourage small and medium enterprises to take part in the program. A completely new approach in Horizon 2020 is the Open Research Data Pilot[25] that aims to organize access to, and reuse of, research data generated by the different projects.

The last part of the paper gives a short overview of the German-Lab project that was funded by the German Ministry of Research from 2008 until 2012. This project had two branches, the first was related to basic research in the area of Future-Internet, and the second to the development of an experimental facility to carry out the experimentally driven research in a spiral approach. Although the research projects finished in 2012 the experimental facility is still operating, mainly based on the Topology Management Tool (ToMaTo) which was an outcome of the research efforts within the experimental facility. Currently there is an initiative to broaden the footprint, especially of the topology management tool (ToMaTo), as a new testbed control framework worldwide. The ToMaTo consortium runs the testbed on a cooperative model. This means that academic institutions can join the testbed for free if they contribute resources. The node requirements can be seen under the tomato-lab web site.[26]

As described in this contribution, a lot of successful projects took place in the research context of next-generation networking especially in creating tools (e.g. testbeds) for evaluation [16]. But unfortunately a lot of them had been terminated just after the project stopped. It is a worldwide weakness in computer science that this community it is not able to support long-lived testbeds. There are some exceptions: Emulab is a successful example of an environment at the University of Utah, but it is done on the basis of a consecutive series of short- to medium-term grants. For others, the outlook seems much more dire: PlanetLab is getting very little new funding, V-Node in Japan has wound up, and GENI's future is uncertain. The same situation is also true in Europe as described in Sect. 2.3.1. Also the G-Lab/ToMaTo testbed lives from other project grants for replacing/extending resources and mutually offering resources.

From a perspective of creating a sustainable, digital infrastructure for the future, these developments could not come at a worse time: While the Internet of today has become the mainstay of economy and society, it approaches its technological limits. I suggest that developing new architectural paradigms to overcome these limits need to be not only a matter of theory and laboratory experiments, but must be a matter of simulation and experiments at large scale under real-world conditions. This has long been recognized by other scientific communities such as high-energy physics (Fermilab, CERN/LHC, ...) or astronomy (Palomar, Hubble, Mauna Kea,

[25] http://europa.eu/rapid/press-release_IP-13-1257_en.htm.

[26] http://tomato-lab.org/join/node_requirements/.

LIGO, ...). There, simulation has become the third pillar of scientific insight besides theory and laboratory experimentation. It is a pity to see the waste of all the financial and intellectual resources because of the missing ability to sustain research infrastructures. To overcome this challenge means facing a cultural shift as much as a technological one: we have to stop seeing funding these infrastructures as funding the groups that build them, and more as community investments in long-term community resources.

References

1. Internet Architecture Task Force. Internet Architecture for Innovation. European Commission. Information Society and Media Directorate-General (DG INFSO) (2011). doi:10.2759/54224
2. Koponen, T., Shenker, S., Balakrishnan, H., Feamster, N., Ganichev, I., Ghodsi, A., Brighten Godfrey, P., McKeown, N., Parulkar, G., Raghavan, B., Rexford, J., Arianfar, S., Kuptsov, D.: Architecting for innovation. ACM SIGCOMM Comput. Commun. Rev. **41**(3), 24–36 (2011)
3. Handley, M.: Why the internet only just works. BT Technol. J. **24**(3), 119–129 (2006)
4. Mueller, P.: Software defined networking: bridging the gap between distributed-systems and networked-systems research. 6. DFN-Forum Kommunikationstechnologien. Lecture Notes in Informatics **217** (2013)
5. Mueller, P., Reuther, B.: Future-internet architecture—a service oriented approach. IT—Inform. Technol. **50**(6), 383–389 (2008)
6. Partridge, C.: Helping a future-internet architecture mature (this article is an editorial note submitted to CCR). ACM SIGCOMM Comput. Commun. Rev. **44**(1), 50–52 (2014)
7. Mueller, P., et al.: Future-internet design principles. EC Future-Internet Architecture (FIArch) Experts Reference Group (ERG) (2012)
8. Campowski, K., Magedanz, T., Wahle, S.: Resource management in large scale experimental facilities: technical approach to federate PanLab and PlanetLab. In: 12th IEEE/IFIP Network Operations and Management Symposium (NOMS 2010), IEEE/IFIP (2010)
9. Szegedi, P., Riera, J.F., García-Espín, J.A., Hidell, M., Sjödin, P., Söderman, P., Ruffini, M., O'Mahony, D., Bianco, A., Giraudo, L., de Leon, M.P., Power, G., Cervelló-Pastor, C., López, V., Naegele-Jackson, S.: Enabling future-internet research: the FEDERICA case. IEEE Commun. Mag. **49**(7), 54–61 (2011)
10. Schwerdel, D., Reuther, B., Zinner, T., Müller, P., Tran-Gia, P.: Future-internet research and experimentation: the G-lab approach. Comput. Netw. **61**, 102–117 (2014). doi:10.1016/j.bjp.2013.12.023
11. Schwerdel, D., Cappos, J., Mueller, P.: ToMaTo a virtual research environment for large scale distributed systems research. PIK—Praxis der Informationsverarbeitung und Kommunikation (2014)
12. Schwerdel, D., Günther, D., Henjes, R., Reuther, B., Müller, P.: German-lab experimental facility. In: Proceedings of FIS 2010—Third Future-Internet Symposium, pp. 1–10 (2010)
13. Schwerdel, D., Hock, D., Günther, D., Reuther, B., Müller, P., Tran-Gia, P.: ToMaTo—a network experimentation tool. In: 7th International ICST Conference on Testbeds and Research Infrastructures for the Development of Networks and Communities (TridentCom 2011), Shanghai, China (April 2011)

14. Cappos, J., Beschastnikh, I., Krishnamurthy, A., Anderson, T.: Seattle: a platform for educational cloud computing. In: Proceedings of the 40th SIGCSE Technical Symposium on Computer Science Education, SIGCSE 2009 (2009)
15. Hemminger, S.: Network emulation with netem. In: Linux Conf Au, Citeseer, 2005, pp. 18–23 (2005)
16. Sterbenz, J.P.G., Hutchison, D., Müller, P., Elliott, C.: Special issue on future-internet testbeds—part I/II. Comput. Netw. 61/63(2014)

SAVI Testbed for Applications on Software-Defined Infrastructure

Alberto Leon-Garcia and Hadi Bannazadeh

Abstract In this chapter we introduce the Canadian project "Smart Applications on Virtual Infrastructures" that explores the design of future application platforms. First we present our original vision of a future application and content marketplace and specify requirements for application platforms. We identify multi-tier clouds that include a "Smart Edge" as essential to supporting low-latency and high-bandwidth applications. We describe a design for the Smart Edge that uses an integrated management system that virtualizes converged heterogeneous computing and networking resources and uses service orientation to provide software-defined infrastructure and platform services. Our implementation of Smart Edge clusters is presented and the deployment of these in a national testbed is described. The Janus integrated management system is introduced and we explain how it builds on OpenStack and Open Flow. We describe experiments and applications that are being conducted on the SAVI testbed. We then describe the federation of the SAVI testbed with GENI and we conclude with our plans for using the SAVI testbed as a foundation for smart city platforms.

1 Future Applications Marketplace

An application platform comprises the end-to-end software and infrastructure, from the smart phone to Internet and the computing cloud, involved in the delivery of software applications. SAVI is conducting research in two thrusts: (1) Design of application platforms that build on software-defined infrastructures; and (2) Design of frameworks for the creation of future-oriented applications. By focusing on applications that address mobility, social networking, big data, and smart industries, SAVI intends address the areas of major growth to 2020 in Information and Communications Technology.

We envision future programmable application platforms where users and providers of content, services and infrastructure interact in an open applications marketplace that is in the center of social and economic activity. This marketplace

A. Leon-Garcia • H. Bannazadeh (✉)
Electrical and Computer Engineering Department, University of Toronto, Toronto, Canada
e-mail: hadi.bannazadeh@utoronto.ca

© Springer International Publishing Switzerland 2016

R. McGeer et al. (eds.), *The GENI Book*, DOI 10.1007/978-3-319-33769-2_22

will be characterized by extremely large scale and very high churn, with new applications introduced and others retired at fast rates. Content will be produced at high rates and large volumes, and demand for content will change quickly. These attributes of the marketplace will place unprecedented demands on the supporting infrastructure for agility in resource allocation, as well as scalability, reliability, accountability and security. Cost considerations will require flexibility in the allocation of the infrastructure so that it can be re-purposed and reprogrammed to provide new capabilities. The control and management systems will need to promote efficient resource usage and high availability at low operations expense. The infrastructure will have multiple owners, so the architecture for the infrastructure must be open and allow for interconnection and federation. Crucially, the infrastructure should support rapid introduction of applications, delivery of applications with targeted levels quality, as well as rapid retirement of applications and redeployment of the supporting resources.

Sensor networks will be incorporated into or attached to application platforms to gather large volumes of information about the environment. These networks will provide the basis for monitoring and controlling the use of critical resources. Especially important will be applications that will use sensor data in smart infrastructures that control of the consumption of resources in urban regions and hence to address socioeconomic challenges in energy, air quality, water, productivity, transportation, and green house gas emissions.

The goal of SAVI is to design, deploy and use a testbed for application platforms. This testbed should demonstrate the design principles for application platform infrastructure. In addition the testbed should demonstrate futuristic applications that will be made possible by these platforms.

In 2010 we considered the following scenario as a use case for a future application platform. The local sports team is in the final game of the championship series and they are mounting a comeback. Reporters and pundits report on the game through live audio and multi-perspective video streaming feeds that are followed by the public on their super-HD tablets and smart phones. Fans in the stadium and throughout the city share their reactions and commentary using social networking apps and contribute their own audio and video streams from their personal devices to shared media repositories. Using speech and video recognition, audio and video streams are tagged in real-time with timestamps, location, and by subject. We proposed the *Kaleidoscope* app to allow viewers anywhere to create one or more customized views, e.g. from a specific vantage point in the stadium, from the perspective of key players, from the perspective of friends, etc.

As the local team wins the game, data traffic peaks. The "Smart Edge" infrastructure activates the maximum number of cells and access points to provide capacity to carry and backhaul massive data streaming content wirelessly. Local virtual circuits route data to storage in the edge and beyond. The smart edge must ensure that local security personnel continue to have their own secure and dedicated channels for the purpose of law enforcement.

As the crowd moves to local restaurants and bars to celebrate, monitoring in the smart edge and core platforms detect and predict decreasing volumes of traffic, and the power of cells and access points is dynamically and progressively turned off. Stadium traffic is reduced to fit in fewer wavelengths and power is turned down on lasers and receivers in the local backhaul. In contrast, dormant cells and access points in restaurants and bars in the area are powered up and backhaul capacity to core datacenters is increased to accommodate the demand as fans turn to Kaleidoscope on giant screens to relive and share their favorite moments.

This scenario illustrates the agility and flexibility required in future application platforms. One can readily conceive of scenarios that have similar requirements. For example, consider a similar scenario in emergency response. Suppose that a sensor system continuously monitors events in a metropolitan area feeding its data stream to an analytics system that detects the incidence of a variety of events. The detection of a catastrophic event can trigger the deployment of an emergency response system that automatically provides the applications, information support systems and communications for police, firefighters, ambulance service, hospitals, road agencies, government, and media to coordinate appropriate responses. The system is preplanned and its deployment is triggered by major events. The cost-effective deployment of this system will require the agility, flexibility, and scalability of the application platforms targeted by SAVI.

2 Applications on Multitier Clouds with a Smart Edge

Figure 1 shows the elements in the infrastructure of an application platform. Users typically access the platform through a mobile device that connects to a very-high-bandwidth, wireless access network. The application platform provides connectivity to the services that support the application of interest. Many of these services will be provided by massive core datacenters located at distant sites that are preferred because of the availability of renewable and low-cost energy. However, some services require low latency (e.g. alarms in smart grids, safety applications in transportation, monitoring in remote health) or involve large volumes of local data (e.g. video capture). These services are better provided by network and computing resources at the *edge* of the network. The edge of the network is typically on

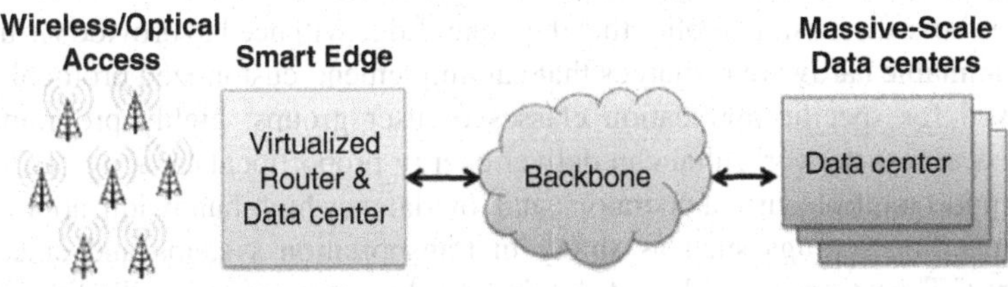

Fig. 1 Multi-tier cloud with Smart Edge

the premises of service providers, but could also be in the customer premise or elsewhere, e.g. in the cellular network infrastructure or in a stadium. For this reason, in SAVI we view the overall computing cloud as including a *Smart Edge* within a multitier cloud. Thus SAVI research investigates the role of the smart edge in multitier clouds in the delivery of applications.

In contrast to remote datacenters which give an illusion of ample resources, the network edge is a bottleneck where wireless capacity is limited, processing and computing has been inflexible, operating expenses are high, and infrastructure investment remains expensive. The cost disparity between massive datacenters and edge infrastructure is driving dramatic shifts in app usage and revenues. For example from 2010 to 2013, growth in cloud-based KakaoTalk messaging led to a drop in SMS messaging at Korea Telecom by 61 % and loss of revenues in billions of dollars. Clearly there was a need for a new architecture for the edge that can deliver radically lower operating and capital costs while supporting flexible application delivery.

We anticipate a Smart Edge, possibly replacing current telecom infrastructure, where virtualized resources, available on demand, support highly dynamic application and content requirements. Unlike datacenters, the network edge contains a broad array of networking equipment and appliances implementing a large number of protocols to handle the packet flows generated by applications. These protocols are currently provided by expensive purpose-built equipment. In addition, Internet protocols have reached an impasse where their future is limited by challenges in security, reliability, and economics. The diversity of protocols and dedicated equipment also increases the complexity and cost of managing infrastructure in the network edge. Our concept of the Smart Edge helps address these challenges by supporting a diverse and evolving mix of protocols over a common software-defined infrastructure.

In order for the future access infrastructure to support demanding content-oriented or latency-sensitive applications, in many locales it must provide ubiquitous, dense, small-cell, very high bit-rate, reliable heterogeneous wireless access possibly integrated with a pervasive optical backhaul network. Services will need to support machine-to-machine, sensor and environmental networks. For this reason, novel high-speed, energy-efficient wireless signal processing, reconfigurable networking techniques need to be explored to leverage the smart edge. SAVI is investigating adaptive mechanisms for allocating wireless network resources dynamically to demanding applications according to their QoE/QoS requirements to optimize the resource usage and power consumption.

The above discussion implies that the Smart Edge will need to provide virtualized programmable hardware resources that can implement: customized protocol stacks deployed for specific application classes or user groups; highly programmable wireless access networks that can deliver energy-proportional capacity on demand and support multiple virtual operators; and low-latency high-bandwidth applications in demanding settings such as safety in transportation systems and emergency response. The programmability of the Smart Edge infrastructure will eliminate the need for expensive purpose-built equipment. The Smart Edge will be managed using cloud-computing principles to achieve agility, flexibility and economies of scale leading to lower operations and capital costs.

The SAVI project had five research themes to achieve its objectives on application platforms:

1. **Smart Applications**. Investigate and design reusable frameworks for rapid deployment of applications enabled by virtualized application platforms.
2. **Multitier Cloud Computing**. Investigate and design adaptive resource management to provide effective, efficient and reliable support for applications across all elements of a multitier cloud while attaining economic, sustainability, and other high-level objectives.
3. **Smart Converged Edge**. Investigate and design smart-edge converged infrastructure that uses virtualization and cloud computing principles to dynamically support multiple network protocols and high-bandwidth, latency-sensitive applications.
4. **Integrated Wireless Optical Access**. Investigate and design high-bandwidth, energy-proportional adaptive virtual access networks based on dense small cells and dynamic optical backhaul.
5. **SAVI Application platform Testbed**. Build an application platform testbed that includes integrated wireless/optical access, smart converged edge, wide-area network connectivity and datacenters; Use the testbed to test future Internet architecture alternatives and smart applications; Develop and provide tools for an ecosystem of open source projects.

Before continuing, we pause to note how the SAVI plan in 2010 correctly identified what became major trends by 2016. The importance of computing in the edge in enabling applications is now widely accepted, and the evolution of cellular wireless networks and 5G are heavily focused on addressing latency, wireless capacity, and bandwidth efficiency in heterogeneous wireless access networks that include small cells [1]. The need for telecom services and networks to achieve lower cost points is now clearly evident in the virtualization of IMS and in the major effort to deploy Network Function Virtualization [2]. The need for much more flexible networking to address the needs of different services is also evident in the intense activity in deploying Software-Defined Networks [3].

3 Smart Edge Based on Software-Defined Infrastructure

The SAVI Smart Edge and testbed derive from a long-term effort investigating the design of application-oriented networks and platforms, the role of virtualization in the management of resources, the convergence of computing and networking resources, and the introduction of autonomic methods in resource management. We begin this section with an overview of the design principles underlying SAVI. We then present the design of the SAVI Smart Edge cluster.

SAVI adopts the key notions of service-oriented computing and virtualization in its approach to application platforms. Service-oriented computing uses services to support the rapid creation of interoperable large-scale distributed applications [4]. Applications are composed from services that can be accessed through networks. Given the preeminence of the Internet, web services based on open Internet standards are the most promising approach to enabling service-oriented computing. Service computing and virtualization also provide a foundation for resource management. A virtual resource reveals only the attributes that are relevant to the service or capability offered by the resource and it hides implementation details. Virtualization therefore simplifies resource management and allows operation over heterogeneous infrastructures. Virtualization is important in both computing and networking: Virtual machines play a central role in cloud computing [5] and virtual networks play a key role in future networks [6, 7].

A layered architecture for application platforms results from the combination of service-oriented computing and virtualization: (1) IaaS (Infrastructure-as-a-Service) offers virtual resources (CPU, storage, memory, bandwidth) as services; (2) PaaS (Platform-as-a-Service) offers platform services such as web, application, database, and communications servers and an associated programming model; and (3) SaaS (Software-as-a-Service) offers applications to the end user [8]. SAVI is concerned with the management of services at these different layers in the context of future infrastructure and service capabilities. Each layer offers services that are built from services from the layer below. IaaS focuses on utilization of physical resources and QoS metrics for the services it provides and is the focus of Themes 3 and 4. PaaS is concerned with platform profitability and the QoS of services it delivers to application providers and is addressed by Theme 2. Finally, SaaS, addressed in Theme 1, is focused on user experience and hence in the interplay between the services that compose an application.

The location of the Smart Edge is roughly where telecom service providers are placed, so it is natural that the design of the Smart Edge should consider the challenges of the service provider. The overarching challenge today is the need to invest huge capital expenditures to increase wireless capacity to accommodate higher traffic, while coping with slower revenue growth from competition and customer expectation for continual sustained improvement. We believe that virtualization can play a role in addressing these twin challenges.

The remote massive datacenter leverages virtualization of computing and networking resources to deliver flexibility and compelling economies of scale [5]. In contrast, the smart edge is significantly smaller in scale and much more heterogeneous in its resources. The smart edge especially when defined to include wireless and wired access networks include nonconventional computing resources, namely FPGAs, network processors, GPUs, ASICs for signal processing within purpose-built boxes as shown in Fig. 2. We believe that flexibility and economies of scale can be attained in the Smart Edge through the virtualization of converged computing, networking, and non-conventional computing resources and the introduction of a common control and management system to manage this *Software-Defined Infrastructure (SDI)*.

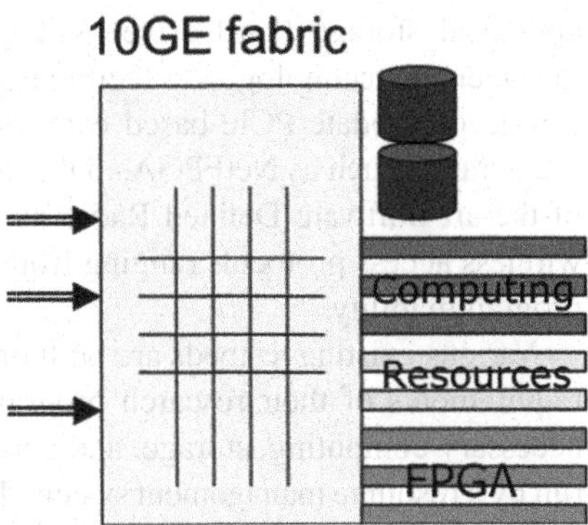

Fig. 2 SAVI Smart Edge Cluster

In SAVI we have designed a node cluster that can provide virtualized and physical computing and networking resources under SAVI's Janus SDI resource management system. The node can be adapted to serve as a SAVI core node, or with the addition of heterogenous resources, it can become a Smart Edge node.

As shown in Fig. 2, the SAVI node cluster design places the layer-3 and higher packet processing resources in virtualizable blades "on the other side" of a layer-2 fabric where they can be shared by researchers as shown in the adjacent figure. A 10GE (or later a 100GE) OpenFlow Ethernet [9] switch provides the fabric to interconnect the node resources, which include sharable resources such as programmable hardware, GPUs (graphics processing units), and Software-Defined Radio (SDR) systems. The resulting structure is a hybrid between a computing cluster (computing and storage blades + Ethernet) and a high-end router (protocol processing + fabric). The programmable hardware resources can be used for general high-performance processing including wireless, signal, packet, and application processing. We expect that GPUs will become a key edge enabler for the implementation of deep learning at the edge [10]. SAVI nodes are ideal for building virtual networks when interconnected with GE, 10GE, or 100GE WAN links. SAVI virtualization use Open Virtual Switch (OVS) [11] which provides software-based virtual switches to interconnect virtual machines, using OpenFlow control.

Multi-tier cloud computing involves the combination of a core node and a set of Smart Edges. The core node is a computing cluster with a large amount of processing resources and storage space interconnected using a 10GE networking fabric. OpenStack [12] virtualization is used to share the resources in the core node among multiple experiments and application classes.

The Smart Edges are a combination of a computing cluster and non-conventional resources that are specific to networking and applications research and experimentation. A compact compute cluster under Janus management has been chosen to provide high power and space efficiency for the Smart Edges. The cluster provides compute, networking and storage resources. The computing servers provide processing for applications that need to be close to the users such as in Software Defined Radio processing applications, video content delivery protocols and applications. These powerful servers are equipped with a high amount of memory and local storage (hard drive) and with 10GE connectivity. They also include management technologies to facilitate remote management and troubleshooting, and can accommodate PCIe-based extension cards for the (FPGA-based) specialized accelerators such as NetFPGAs, DE5-Net and GPUs used in the Smart Edge. State-of-the-art Software-Defined Radio systems are used to experiment with multiple wireless access protocols ranging from WiFi to LTE and future methods using SDI programmability.

Various existing testbeds are built on standardized nodes designed to the specific requirements of their research programs. Typically, a standard rack contains the necessary computing, storage, and networking components. These components are run by a resource management system. For example, in its recent deployments GENI has used two rack designs. The ExoGENI design emphasizes performance and is built on IBM hardware and OpenStack software. The InstaGENI rack is built on HP hardware and a combination of ProtoGENI and PlanetLab software. Our Smart Edge rack design differs from these in the choice use of the Janus SDI management, in the availability of heterogeneous resources, and in the focus on the Smart Edge. For reasons of economy and flexibility, our designs allow the racks to be built from commodity components.

4 The Janus Management System for SDI

When we use the term Software-Defined Infrastructure (SDI), by “Software-Defined” we mean providing open interfaces to: control and manage converged heterogeneous resources in different types of infrastructures for software programmability; and give an access to infrastructure resource information such as topology, usage data, etc. We have designed an SDI management system to meet these requirements [13–16].

Figure 3 shows the architecture of the Janus SDI Resource Management System (RMS). An SDI manager can control and manage resources of type A, B, and C using corresponding resource controllers A, B, or C, respectively. External entities obtain virtual resources in the converged heterogeneous resources via the SDI RMS through open Interfaces. The converged heterogeneous resources are composed of virtual resources and physical resources. Virtual resources include any resource virtualized on physical resources, such as virtual machines. Physical resources include any resource that can be abstracted or virtualized, such as computing servers, storage, network resources (routers or switches), reconfigurable hardware resources, SDR resources, sensor network resources, energy in power grids, etc.

Janus provides resource management functions for the converged heterogeneous resources to the external entities. These functions can include provisioning, configuration management, virtualization, scheduling, migration, scaling up/down, monitoring and measurement, load balancing, energy management, fault management, performance management, and security management. The external entities in Fig. 3 can be applications, users (experimenters, service developers, providers), and high-level management systems.

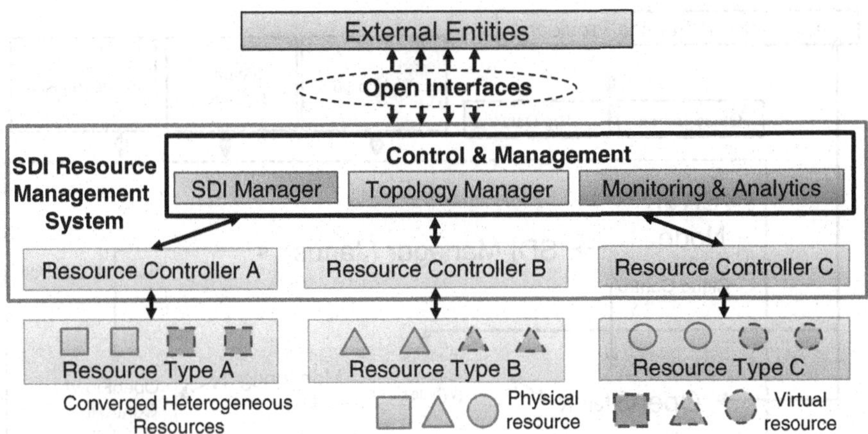

Fig. 3 SDI Resource Management System

Janus can perform coordinated resource management for converged heterogeneous resources through its SDI Manager and Whale topology manager. The SDI manager performs integrated resource management based on resource topology data provided by the Whale topology manager. Each resource controller provides high-level user descriptions and manages the resources of a given type. Whale maintains a list of the resources, their relationships, and monitoring and measurement data of each resource. Furthermore, Whale provides up-to-date resource information to the SDI manager for infrastructure-state-aware resource management. Our resource management system includes a Monitoring and Analytics subsystem, "MonArch," that provides flexible, integrated measurement and monitoring of all resources in an SDI [17]. MonArch incorporates batch and streaming analytics that can respond to queries from the operators as well as users. MonArch and Whale collaborate to provide the SDI Manager with timely, infrastructure-wide state information to assist management and control.

Figure 4 shows how the architecture in Fig. 3 was implemented in the design of Janus to manage cloud and networking resources. A SAVI node controls and manages virtual resources using OpenStack and Open-Flow controller. In the Smart Edge node, a variety of computing and networking resources are available. The SDI manager controls and manages virtual computing resources using OpenStack, e.g. through Nova for VMs, etc. The OpenFlow controller is used for controlling networking resources. The OpenFlow controller receives all events from the OpenFlow switches and creates a flow table including actions. The SDI manager performs all management functions based on data provided by the OpenStack and the OpenFlow controller, and it determines appropriate actions for computing and networking resources using management modules inside.

Janus was first implemented and deployed with the SAVI testbed in 2013 and it is continually being upgraded as OpenStack issues new releases. The RMS developed in SAVI differs from management systems used in most other testbeds. In particular,

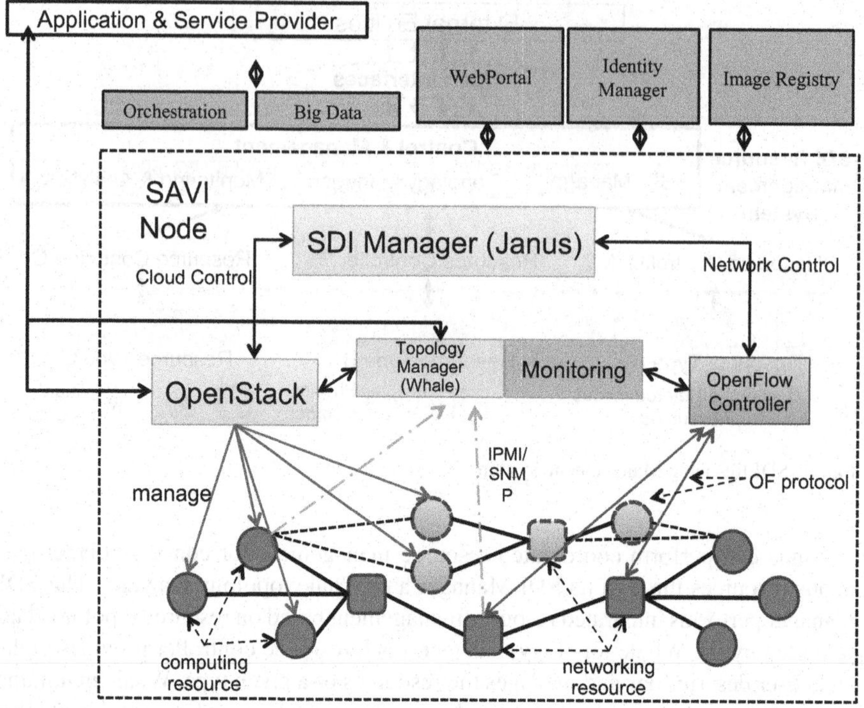

Fig. 4 Janus SDI Manager

the inclusion of Controller C allows for incorporating additional types of resources, for example, SDR resources to virtualize the wireless access networks that can be centered around the Smart Edge [18]. SAVI is also extending virtualization to sensor networks attached to the edge. To this end, SAVI has designed and deployed compact customer premise edge devices that can provide virtual resources at the user site [19].

Figure 5 shows the design for the Whale topology manager where a configuration manager monitors the state and relationship between converged heterogeneous resources through a cloud controller and a network controller, and stores the monitored data to a graph database [20]. In addition, the topology manager provides answers for the queries to the data from an SDI manager. The cloud controller (OpenStack) and network controller (Ryu) each provide physical and virtual computing or networking resource properties, as well as associated monitoring and measurement data to the configuration manager.

The configuration manager in Whale builds a model by analyzing states and relationships of monitored computing and networking resources. Whale uses a graph model because all resources and their relationships can be represented by a set of vertices and edges with flexibility and simplicity. Physical and virtual resources are represented by a vertex and their relationship is represented by an edge. Each

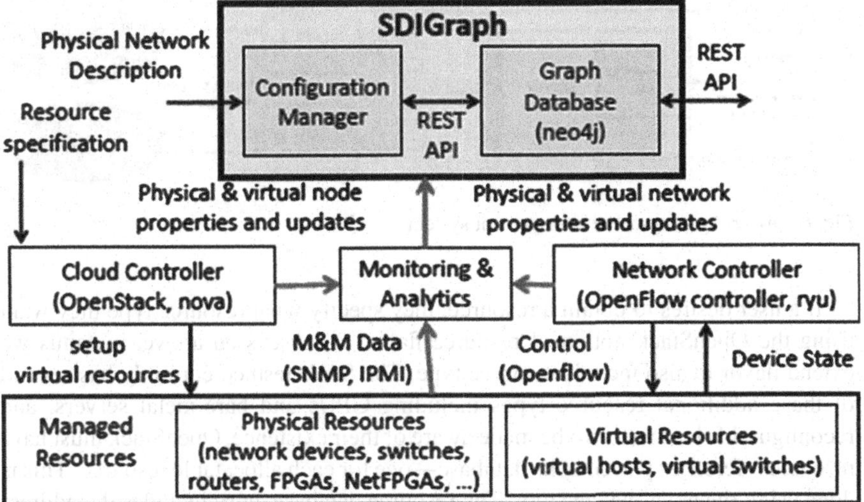

Fig. 5 Whale topology manager

vertex has its own attributes such as ID, name, and associated monitoring data. In addition, the graph model includes a set of subgraphs that represent a physical or a virtual network topology. Thus the configuration manager can store not only the state and relationship of converged heterogeneous resources, but also the physical and virtual network topology.

In the original OpenStack there is no support for virtualization of unconventional resources such as FPGA, NetFPGA or GPU. To meet our needs, we extended OpenStack Nova to support virtualization of such resources by adding their device drivers. Normally, Nova can provide a virtual resource based on a given flavor (an available hardware configuration for a server). In the original OpenStack, each flavor has a unique combination of disk space, memory capacity, and the number of virtual CPUs. We added new flavors that specify unconventional resources [21]. As a result, we can obtain both conventional and unconventional resources using the same REST APIs of Nova. Whenever a new resource needs to be introduced, we only need to add a driver and a new flavor for the resource to Nova.

A unique feature of the Janus and SAVI Smart Edge design is its ability to allocate portions of FPGA resources. Reconfigurable hardware, such as FPGAs, can allow acceleration of compute intensive tasks, provide line rate packet processing capabilities, and expand the range of experiments and applications in a testbed.

In order for OpenStack to manage different types of resources, these resources must appear homogeneous in nature. To do this, we use a Driver-Agent system shown below [21]. A driver for any resource implements required OpenStack management API methods, such as boot, reboot, start, stop and release. The driver communicates these OpenStack management commands to an Agent, which carries them out directly on the resource. In this fashion, OpenStack can manage all resources through the same interface (Fig. 6).

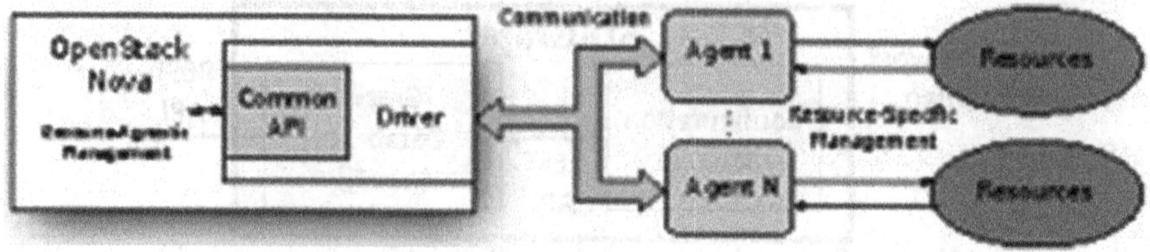

Fig. 6 Driver-Agent resource management system

If a user desires to obtain a resource, they specify what resource type they want using the OpenStack notion of resource flavor. As discussed above, in Janus we extend flavor to also include resource type. The SAVI testbed currently has several of these additional resource types including GPUs and bare-metal servers, and reconfigurable hardware. To be made aware of their existence, OpenStack must have resource references placed in its database—one for each allocatable resource. This is done using the nova-manage tool. The resource database entry includes the address of the Agent that provides the resource, a type name that can be associated with a flavor, and how many physical network interfaces the resource has.

5 The SAVI Canadian Testbed

SAVI has deployed a national application platform testbed based on the integrated management of software--defined infrastructure. The testbed consists of interconnected SAVI Smart Edge nodes of virtualized heterogeneous resources under Janus SDI management. The testbed provides slices of network and computing resources to researchers to enable them to experiment with applications, novel systems based on software-defined infrastructure, multi-tier cloud computing, and Future Internet. This type of testbed is essential for researchers to conduct experiments at scale in realistic settings. Not only is implementation and experimentation a necessary step in proving new designs, it also reduces the time to deployment and commercialization.

Figure 7 shows the currently deployed basic testbed with Smart Edge nodes at seven university sites (Victoria, Calgary, Waterloo, Toronto, York, McGill, and Carleton) and a core datacenter and the Testbed Control Center (Toronto). The SAVI Control Centre and a core datacenter are located in Toronto. A Smart Edge node with a full array of heterogeneous resources is located in Toronto. All Smart Edge nodes provide slivers of compute, storage, network, and programmable hardware resources to form virtual slices allocated to researchers as shown in Fig. 8. SAVI nodes are interconnected by high-speed links provided by partner R&E networks (CANARIE, ORION). The Testbed Control Center hosts the Testbed-wide entities such as Identity and Access Management to support experiments.

Fig. 7 SAVI testbed

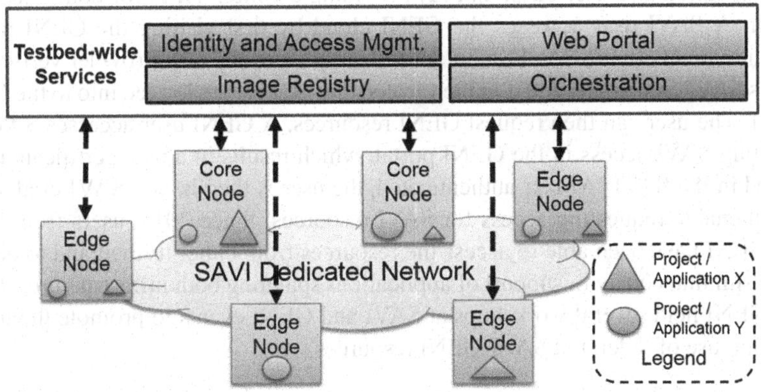

Fig. 8 SAVI offers virtual networks of converged resources

SAVI incorporates capabilities developed in the GENI and the EU FP7 FIRE initiatives:

Opting in: Enables users at multiple locations to opt in to an application being researched and trialed. This increases the user base and helps move to at-scale testing.

Clearinghouse Principle: Grants permission for slices to be created for experiments; Enables addition of resources to existing slices; Keeps track of available resources that can be added to an experiment; Can quickly tear down a slice if experimenters breach the "rules and boundaries" of experiment; Participates in federation with other testbeds.

At-Scale Capability: Large-scale tests can be set up and torn down so that proofs of scalability can be achieved and disruptive system behaviors detected and addressed. The testbed infrastructure is protected to provide a consistent test environment.

Secure Environment: Experiments are authorized using SAVI's Identity and Access Management system. Security of slices is vital since they run on infrastructure that may be shared by "operational traffic".

SAVI provides an Identity and Access Management system for a multi-tier cloud that we implemented in the SAVI basic testbed [22]. Our IAM system is built using Keystone which is OpenStack's identity manager. Each node in the SAVI testbed essentially operated autonomously, with the exception of the four testbed-wide services shown in Fig. 8: IAM, Web Portal, Image Registry, and Orchestration. The testbed-wide SAVI orchestration service enables the end-to-end deployment of resources in support of distributed applications. The Orchestrator produces a template that is used to allocate the required resources in the various nodes in the testbed. We believe that this orchestration approach is very promising in enabling end-to-end applications in future very large scale application platforms.

The SAVI and GENI teams combined efforts to establish a federation between the GENI and the SAVI testbed [23]. This work was done at two levels: (1) Federated authentication and user-base; and (2) Providing common APIs for both SAVI and GENI. A SAVI user accesses the GENI cloud by first visiting the GENI portal, which redirects to the SAVI Shibboleth Identity Provider. The Provider verifies the user's SAVI credentials and if authenticated, the user is then logged into to the GENI portal. The user can then request GENI resources. A GENI user accesses SAVI by selecting SAVI access in the GENI portal, which results in a user certificate being posted in the SAVI IAM. If authenticated, the user is then issued SAVI credentials which enable requesting access for SAVI resources. Since 2015, users from SAVI and GENI have been able to access the resources from either testbed and to deploy these simultaneously in support of applications spanning both infrastructures. SAVI and GENI held several workshops at SAVI and GENI events to promote the use of the joint use of federated SAVI-GENI resources.

6 Applications and Experiments on the SAVI Testbed

We conclude with a brief discussion of activities to proliferate skills in the use of the SAVI testbed and an overview of experiments in its use.

SAVI conducts an Annual one-day testbed workshop in conjunction with its Annual General Meeting. The purpose of the testbed workshop is to introduce SAVI researchers to the tools required to use the testbed. The workshop offers a range of exercises and tutorials that begin with basics on the use of the testbed to advanced exercises on how to use measurement and monitoring tools, big data service, and orchestration service. Topics include: Introduction to basic use of SAVI testbed; Heat orchestration and autoscaling; VINO SAVI virtual overlay network and NFV; SAVI-GENI federation; Big data on SAVI using Sahara and Spark. VINO, in particular, is a SAVI tool for deploying OpenFlow virtual networks that can be readily scaled and migrated across the testbed [24]. Another very useful tool is provided by SAVI's VNF service-chaining capability that allows users to define a service graph that determines the path that traffic must follow in traversing various VNFs [25]. The SAVI testbed workshops trained over 300 researchers in the first 4 years.

SAVI also conducts an annual Design Challenge Camp that brings together SAVI graduate students and post-doctoral fellows for a week to work in teams developing applications on the SAVI testbed. The design camp in August 2014 offered the following advanced tutorials and exercises: (1) OpenFlow: Ryu and Mininet; (2) Open VSwitch; (3) SAVI VINO; (4) Storm Stream Processing; and (5) Security Function Virtualization using Snort. The following projects where conducted by teams from McGill, Waterloo, York, Concordia, Alberta, Calgary, Carleton, and Toronto:

1. Emergency Response System using SAVI
2. Creating Multiple Virtual Wireless Access Networks on a Single Physical Network
3. Uninterrupted Migration of Low Latency Service n SAVI testbed
4. Scalable Audio/Video Live Discussion Forum using SAVI virtual infrastructure
5. Flexible and Scalable Intrusion Detection and Prevention in SAVI using SNORT
6. Kaleidoscope User Contributed Content Pub/Sub system using Twitter

The projects were selected to highlight the capabilities of the SAVI testbed and to address aspects of the use cases identified for SAVI in its research plan. Project 1 used Hadoop analytics to trigger the deployment of the Emergency Response System. Project 2 demonstrated the creation of two virtualized WIFI access networks with distinct performance characteristics using the capabilities of OpenFlow. Project 3 demonstrated the hitless migration of a video streaming application across two SAVI edge nodes. Project 4 designed a flexible and scalable conferencing system using service components that can be deployed and scaled as needed. Project 5 implemented basic Intrusion Detection System and Intrusion Protection System with SNORT exploiting the OpenFlow and service chaining in SAVI. Finally Project 6 developed a basic implementation of the Kaleidoscope app using a combination of Twitter to signal for the establishment of SAVI multicast network capabilities to support the exchange of user generated content using a publish/subscribe model.

A major project running on SAVI was the platform for Connected Vehicles and Smart Transportation [26]. The CVST platform was designed to support smart applications in transportation. The platform has three major elements: (1) A data management platform for gathering real-time data relevant to the transportation network in an urban region and disseminating these data to users in timely and secure fashion as well as to archiving systems; (2) An analytics system for addressing queries based on the archival and real-time data in the system; (3) An application platform that provides open APIs to support smart transportation applications. An interesting aspect of the CVST platform is that its data management system is built on top of an Information-Centric Networking architecture that is deployed on the SAVI testbed. Since 2014, the CVST platform has been collecting multimodal traffic information in the Greater Toronto Area (GTA), including automobile, buses, and bicycles. The platform supports a portal for real-time state information for GTA traffic from a variety of traffic sensors.

The 2015 Design Challenge Camp featured a partnership between SAVI and the EU FIWARE project. The FIWARE middleware for rapid application creation was deployed on top of SAVI and the camp participants where challenged to develop applications for smart cities, possibly using the data streams provided by the CVST platform. The following projects were presented at the camp:

1. Smart Toronto Neighborhood Ratings
2. Road Route Planner with Real-time Updates
3. Mixed Mode Bicycle-Subway Real-time Planner
4. Personal Video Recorder
5. Application Acceleration in using SAVI FPGA Orchestration
6. Collaborative Newscast Framework

7 Concluding Remarks

We have described the SAVI Research Network including its research goals and activities regarding application platforms. We presented SAVI's view on multitier clouds for application platforms, focusing on the role of the Smart Edge. We described SAVI's design for the Smart Edge and its associated Janus integrated management system for SDI. We discussed the SAVI testbed and activities to promote its use. Finally we described a broad range of experiments deploying applications on SAVI.

SAVI has successfully demonstrated the benefits of application platforms based on integrated management of software-defined infrastructure. A crucial benefit is the ability to orchestrate end-to-end applications across a multitier cloud. The federation of SAVI and GENI demonstrate that this orchestration can be done across infrastructures in different administrative domains. The Janus integrated manager has shown the power and flexibility of integrated management of converged virtualized heterogeneous resources in smoothly leveraging the advantages of virtualized computing and virtualized networking.

Our current activities build on SAVI to enhance its autonomic capabilities in terms of performance, resiliency, and especially security. The MonArch monitoring and analytics system provides the basis for developing ongoing awareness of the status of the SDI that can support these autonomics. A future direction for the SAVI architecture is to deploy it as a smart city platform. The flexibility, performance, and cost advantages of the SAVI system provide a ready-platform for deploying smart city applications. In particular, the MonArch monitoring and analytics system provides a template for the massive data gathering and analytics that will be an integral part of smart cities as the Internet-of-Things becomes pervasive in urban regions.

Acknowledgements The work is funded by the SAVI project funded under the NSERC, Canada (NETGP394424-10). We would like to thank all of the SAVI Strategic Network partners for supporting this major effort. In particular we would like to thank David Mann for his constant encouragement and wise advice.

References

1. Osseiran, A., et al.: Scenarios for 5G mobile and wireless communications: the vision of the METIS project. IEEE Commun. Mag. **52**(5), 26–35 (2014)
2. ETSI: Network Functions Virtualization: An Introduction, Benefits, Enablers, Challenges & Call for Action, SDN and OpenFlow World Congress, Darmstadt, Germany, 22 October 2012
3. Kreutz, D., Ramos, F.M.V., Veríssimo, P.E., Rothenberg, C.E., Azodolmolky, S., Uhlig, S.: Software-defined networking: a comprehensive survey. Proc. IEEE **103**(1), 14–76 (2015)
4. Papazoglou, M.P., Traverso, P., Dustdar, S., Leymann, F.: Service-oriented computing: state of the art and research challenges. IEEE Comput. **40**(11), 38–45 (2007)
5. Armbrust, M., et al.: Above the clouds: a Berkeley view of cloud computing. UC Berkeley Reliable Adaptive Distributed Systems Laboratory (10 February 2009)
6. Anderson, T., Peterson, L., Shenker, S., Turner, J.: Overcoming the internet impasse through virtualization. Computer **38**(4), 34–41 (2005)
7. Mosharaf Kabir Chowdhury, N.M., Boutaba, R.: Network virtualization: state of the art and research challenges. IEEE Commun. Mag. **47**(7), 20–26 (2009)
8. Litoiu, M., Woodside, M., Wong, J., Ng, J., Iszlai, G.: A business driven cloud optimization architecture. In: Proceedings of ACM SAC 2010, Sierre, Switzerland, 24–29 March 2010
9. McKeown, N., Anderson, T., Balakrishnan, H., Parulkar, G., Peterson, L., Rexford, J., Shenker, S., Turner, J.: Openflow: enabling innovation in campus networks. SIGCOMM Comput. Commun. Rev. **38**(2), 69–74 (2008)
10. Coates, A., et al.: Deep learning with COTS HPC systems. In: Proceedings of the 30th International Conference on Machine Learning (2013)
11. Pfaff, B., et al.: Extending Networking into the Virtualization Layer. Hotnets (2009)
12. Openstack. http://www.openstack.org
13. Kang, J.M., Bannazadeh, H., Leon-Garcia, A.: SAVI testbed: control and management of converged virtual ICT resources. In: 2013 IFIP IEEE International Symposium on Integrated Network Management (IM 2013), Ghent, Belgium, May 2014, pp. 664–667 (2014)
14. Kang, J.M., Bannazadeh, H., Rahimi, H., Lin, T., Faraji, M., Leon-Garcia, A.: Software-defined infrastructure and the future central office. In: International Conference on Communications (ICC), Budapest, Hungary, June 2013, pp. 225–229 (2013)
15. Bannazadeh, H., Leon-Garcia, A., et al.: Virtualized application networking infrastructure. In: Magedanz, T., Gavras, A., Thanh, N.H., Chase, J.S. (eds.) 6th International ICST Conference, TridentCom 2010, Berlin, Germany, 18–20 May 2010, Revised Selected Papers, Springer, pp. 363–382 (2010)
16. Park, B., Bannazadeh, H., Leon-Garcia, A.: Janus: design of a software-defined infrastructure manager and its network control architecture. IEEE NetSoft 2016, Seoul, Korea, 6–10 June 2016
17. Lin, J., et al.: Monitoring and measurement in software-defined infrastructure. IFIP IEEE Integrated Network Management (IM) 2015, Ottawa, Canada, May 2015
18. Monfared, S., et al.: Software-defined wireless access for a two-tier cloud system. IFIP IEEE Integrated Network Management (IM) 2015, Ottawa, Canada, May 2015
19. Lin, J., et al.: SAVI vCPE and internet of things. In: EAI FABULOUS 2015 Conference, Ohrid Macedonia. LNICST (2015)
20. Kang, J., Bannazadeh, H., Leon-Garcia, A.: SDIGraph: graph-based management for converged heterogeneous resources in SDI. IEEE NetSoft 2016, Seoul, Korea, 6–10 June 2016

21. Byma, S., Bannazadeh, H., Leon-Garcia, A., Steffan, J.G., Chow, P.: Virtualized reconfigurable hardware resources in the SAVI testbed. In: Magedanz, T., Gavras, A., Thanh, N.H., Chase, J.S. (eds.) 6th International ICST Conference, TridentCom 2010, Berlin, Germany, 18–20 May 2010, Revised Selected Papers, Springer, pp. 363–382 (2010)
22. Faraji, M., et al.: Identity access management for multi-tier cloud infrastructures. In: Network Operations and Management Symposium (NOMS). 2014 IEEE (2014)
23. Lin, T., et al.: SAVI testbed architecture and federation. In: EAI FABULOUS 2015 Conference, Ohrid Macedonia. LNICST (2015)
24. Bemby, S., et al.: VINO: SDN overlay to allow seamless migration across heterogeneous infrastructure. In: IFIP IEEE Integrated Network Management (IM) 2015, Ottawa, Canada, May 2015
25. Yasrebi, P., et al.: VNF service chaining on SAVI SDI. In: EAI FABULOUS 2015 Conference, Ohrid Macedonia. LNICST (2015)
26. Tizghadam, A., Leon-Garcia, A.: Application platform for smart transportation. In: EAI FABULOUS 2015 Conference, Ohrid Macedonia. LNICST (2015)

Research and Development on Network Virtualization Technologies in Japan: VNode and FLARE Projects

Akihiro Nakao and Kazuhisa Yamada

1 Introduction

The Internet currently operating on TCP/IP network protocols has its origin in ARPANET developed in the late 1960s. Although the Internet started as an experimental platform for studying inter-networking of various network protocols, after its commercialization in 1990, it has improved and been enhanced to fulfill a variety of requirements as a social infrastructure. At present, it is a critical information social infrastructure indispensable to society.

The Internetworking Principle proposed in 1974 by Cerf and Kahn has existed for more than 40 years as the core of the Internet architecture. It is simply remarkable. However, since around 2000, the Internet has been widely acknowledged to have become "ossified," i.e., it has become very hard to introduce or add new functions to the Internet; "Successful and widely adopted technologies are subject to ossification, which makes it hard to introduce new capabilities," so cited a U.S. National Research Council report [1]. In this way, the view that the Internet is inherently making evolution difficult has been acknowledged as a problem.

To address this problem, researchers around the world have therefore initiated research projects since around 2007, taking approaches such as reviewing the Internet architecture to redesign it with a clean slate. In other words, erase the blackboard and design a brand-new network as a "Clean Slate" network. In the U.S., the Global Environment for Networking (GENI) Project [2] was initiated, funded by the National Science Foundation, to construct a testbed for studies on new network architectures. Afterwards, under the coordination of BBN Technologies,

A. Nakao (✉)
The University of Tokyo, Tokyo, Japan
e-mail: nakao@nakao-lab.org

K. Yamada
NTT Network Innovation Labs, Yokosuka-shi, Japan

© Springer International Publishing Switzerland 2016

R. McGeer et al. (eds.), *The GENI Book*, DOI 10.1007/978-3-319-33769-2_23

which assumed the role of the GENI Project Office (GPO), studies on large-scale network-virtualization started across the U.S. In 2008 in Japan, the National Institute of Information and Communications Technology (NICT) assumed a similar role of the GENI Project Japan Office and started promoting R&D projects aiming at the actualization of a new generation network. NICT's actions triggered the R&D of network-virtualization technologies critical to the verification of a variety of protocols for the foundation of new generation networks.

In Japan, one of the authors of this chapter, Dr. Akihiro Nakao, a member of the GENI Community with experience working under Prof. Larry Peterson for the PlanetLab Project [3] until 2005 that had a significant impact on the initial design of GENI, served as a project leader in NICT to launch a virtualization node (VNode) project. In this project, NICT, the University of Tokyo, NTT, NEC, Hitachi, and Fujitsu collaborated to implement a platform based on VNodes from 2008 to 2010 at a faster pace than GENI. Then, in 2011, NICT, for the purpose of continuing research, contracted R&D to the six research institutes of the University of Tokyo, NTT, KDDI, NEC, Hitachi and Fujitsu; since then, NICT and the six contractors have been collaborating with a so-called "All Japan" formation, to promote the research and development and work as a counterpart in Japan of GENI.

The joint or contracted research projects conducted from 2008 to 2015 actualized a number of proposals that included innovative ideas ahead of GENI. We believe that some of these ideas had a significant impact on the GENI community. The most prominent among them is the VNode [4] architecture, which is a building component for a network-virtualization platform. In 2008, the VNode architecture design was proposed. In 2010 after the first implementation, it was introduced in a keynote presentation at the eighth GENI Engineering Conference (GEC) [5]. One year later, GENI proposed the GENI Rack architecture, which is similar to the VNode architecture.

Briefly, the VNode architecture consists of two parts mutually connected by a backplane: Redirector, a virtual network component, and Programmer, a network function implementer. The most innovative part of this architecture is that virtual network-connectivity implementation technologies and network-function implementation technologies can be scalably designed and implemented, and they are not constrained by each other. The VNode concept, originally presented in 2008, is the core of the architecture that ensures the creation of network functions through slicing a network. Some network operators have already employed this concept where Software Defined Networking (SDN), Network Function Virtualization (NFV) or their combination (orchestration) is discussed.

Another significant technology that is leading GENI even today is the technology for ensuring the programmability of the data plane. Since the start of our R&D activities, we have focused on the actualization of the next generation network. So, we have implemented a communication platform that ensures the configuration of the data plane where protocols independent of the conventional network protocols, the so-called non-IP protocols, or more precisely, protocols for layer L2 or higher, can be processed flexibly by means of programs. On the other hand, the network virtualization technologies developed in GENI project still remain at the stage where the handling of TCP/IP or Ethernet protocols are allowed, partly because they had to depend on OpenFlow [6]. Therefore, in regard to the concept of a "clean slate," although it was proposed by Stanford University and GENI, it would not be an exaggeration to say that the network virtualization technologies have been developed in Japan in advanced manner because our technologies aim at protocol modification in L2 or higher, although our R&D has been greatly influenced and encouraged by that of GENI project in the U.S.

Our concept, compared to the closest idea of Protocol Oblivious Forwarding (POF) [7] which is used in SDN today, deals with deeper programmability and ensures easier programming because POF deals with limited network functions and uses the pattern matching/action programming model. We proposed "Deeply Programmable Networking (DPN)" which has deeper programmability and ensures easier programming. [4, 11] We believe that with the application of DPN a node that can handle independent conventional TCP/IP protocols as Content Centric Networking or Information Centric Networking [8] will prove its true value due to the data-plane programmability enabled through DPN. In addition to DPN, we have an excessive number of technologies to enumerate well ahead of GENI that have made a significant impact such as the method for applying different network-virtualization technologies to fulfill different requirements for edge or core networks, the virtualization of wireless networks, the virtualization of terminals, and creating slices in an application-by-application manner.

It is worthy to note here that the projects we have promoted have had a great impact on the U.S. network-virtualization research community. Network-virtualization research conducted in Japan such as in the Architecture Journal of GENI [2] has attracted a great deal of attention and received very high evaluations.

The rest of this chapter is organized as follows. Section 2 introduces the overview and the brief history of network virtualization research world-wide. Section 3 highlights deep programmability, the focus of the research and development of network virtualization technologies in Japan. Section 4 briefly defines problems. Sections 5 and 6 introduces the overview and elaborates the detail of our research on network virtualization platform, respectively. Section 7 describes the efforts of international federation of our platform with the others, especially the ones in GENI and in Europe. Finally, Section 8 briefly concludes and addresses remarks for future.

2 Brief History of Network-Virtualization Research

In early 2000, the first-generation platforms for verifying and evaluating new network protocols were established. They were overlay systems based on IP networking, on which an experimental environment for virtual networks was implemented. A representative example is PlanetLab [3] from Princeton University. It enabled dynamic configuration into seamless slices of virtualized network-computing node resources on an overlay network. In those days, a slice was defined as a unit component with allocated resources such as computing power/storage on servers or resources existing in namespaces. Since then, a variety of network research activities have been conducted on the system involving more than 700 nodes deployed worldwide in 2000 and more than 1000 nodes currently. There was another important achievement called Super Charging PlanetLab that enhances the performance and programmability of PlanetLab nodes using network processors [9]. In Japan, CoreLab was developed by the University of Tokyo and NICT based on PlanetLab technologies. It was implemented on the JGN-X network in 2007 [10].

However, such overlay platforms have limitations in underlay network controls and are not appropriate for our research goal, i.e., the R&D of a clean slate network design. Hence, in 2008, NICT and the University of Tokyo as well as industry partners initiated an R&D project on network virtualization technologies as an industry-academia-government collaborative project.

In the project, two critical requirements are identified: *programmability* and *resource isolation*. For programmability, while conventional virtual networks implement a slice by simply multiplexing L2/L3 networks using VLAN tags or tunnels, a slice should be defined as an aggregation of independent programmable resources such as networks, computing power, storage or namespace, and such programmability ensures the handling of any type of protocol. For resource isolation, each resource used to provide such a program-execution environment should be independently allocatable.

One of the design principles in the network-virtualization node was to ensure sustainable evolvability. Following this principle, node development was conducted separately in two parts [11]: Redirector is in charge of the virtualization of network links, and Programmer is in charge of the virtualization of node functions. Redirector configures virtual networks while maintaining inter-node links and bandwidth, which is equivalent to the current SDN technologies. The Programmer performs functions, equivalent to NFV, for converting packets, filtering packets, or controlling communication paths or caching/signal processing. It can be said to integrate/enhance the functions of SDN and NFV because it can define new protocols and program their conversion/control. What has been described so far indicates that, as early as in the 2008 development phase of the joint research project, we had proposed an architecture equivalent to that of SDN/NFV. Furthermore, our architecture, by employing virtualization technologies that enable the placement of multiple slices on physical resources, ensures multiplexing of more than one network function, and can adapt to the evolution of technologies expected in the near future.

At the same time, while our development was on-going, in the U.S., the GENI project in which a total of around 30 top U.S. universities and research institutes participated, pushed forward the development of a testbed based on network-virtualization technologies for promoting R&D of a Clean Slate Network [2]. In Europe, the development of a similar testbed was promoted under the FP7 project [12]. GENI's research achievements were presented at a GEC meeting that is held three times a year. We, participating from the very beginning, gave presentations on the status of the network-virtualization technology development in Japan, as one of the achievements of the joint research on virtualization nodes that started in 2008 at

NICT. In 2011, we implemented a prototype network-virtualization node, VNode, at NICT's experimental facility in Hakusan to promote the development of the testbed based on network-virtualization technologies ahead of GENI Rack developed by the GENI project. GENI Rack has two variations. One is the InstaGENI Rack, which can be said to be an evolved type of PlanetLab, with an architecture requiring fewer resources and a large number of distributed nodes. The other is the ExoGENI Rack, which has an architecture that requires more resources but has a small number of distributed nodes and is similar in its architecture to VNode.

In 2011, for the purpose of promoting network-virtualization technology R&D, the "Network Virtualization Platform Technology Research and Development Supporting Next-Generation Network" was started as NICT contracted NTT, KDDI, the University of Tokyo, NEC, Fujitsu, and Hitachi for R&D. This research project had the following themes: (a) network-virtualization platform implementation, (b) development of a system for allocating and integrating services into the slices that are created on the platform; and (c) application development on the platform. Work on theme (c) was conducted concurrently with that on themes (a) and (b). Work on theme (a) took place from 2011 to 2014. During that period, more than 125 scheduled periodic research discussion meetings in total were held at NICT's Hakusan research site. At each of these meetings, detailed and profound discussions were held under the leadership of the University of Tokyo regarding technology and NICT regarding project management.

The status of the network-virtualization technology development was publicly presented at the Next-Generation Symposium held by NICT. In addition, for the purpose of promoting international collaboration, NICT introduced invited participants from research institutes around the world, through panel discussions at the network-virtualization symposium held by the network-virtualization technical committee established by the Institute of Electronics, Information and Communication Engineers (IEICE) [13].

With regard to international relations, participating in the abovementioned GEC every time since 2008, we opened a booth at its Live-Network Demonstration, one of their prime exhibitions, to show the validity of the functions we developed. In addition to the exhibitions, we gave presentations at the plenary sessions four times, GEC 7 in 2009, GEC 10 in 2011, GEC 13 in 2012, and GEC 15 in 2013. The participants were impressed with the degree of evolution and the advancements in our network-virtualization activities. Moreover, in 2014, at the demonstration session, our FLARE technology [14], which will be described later, was awarded "Best Demonstration." Hence, what has been described so far indicates that our activities have been attracting a very high level of interest in the GENI community, and that our technologies have been leading others and making a huge impact on researchers around the world.

On the other hand, OpenFlow, which has become the basis for actualizing SDN today, was introduced in its initial development phase. At GEC, for example, live demonstration sessions were held, and hands-on tutorials were offered to the participants to promote the technology. To catch up with the U.S. in the public relations campaign-race on network-virtualization technologies (we are a little bit behind right now), we have been putting efforts into the promotion of our technologies since FY2013 through a variety of measures, for example, offering tutorials and hands-on sessions at IEICE or Network Virtualization Study Group events.

3 Supporting Deep Programmability

OpenFlow, proposed by Stanford University, has already been acknowledged worldwide as the most popular SDN technology. Furthermore, industry has prematurely decided to employ it in the business field. It is an epoch-making architecture where the data plane and control plane that conventional network devices have both are separated and the liberated control plane is controlled by programs written in computer programming languages. However, because data plane in the OpenFlow architecture is rigidly contained in hardware, and not allowed to handle packets other than IP packets, deeper programmability especially in data plane was to be pursued.

On the other hand, being aware since as early as at the start of the project that the data plane should have such programmability as the control plane, we designed a virtualization node such that is able to execute through programs the processing of frames on L2 or higher. We refer to these frames as Any Frames. In other words, we proposed DPN (Deeply Programmable Networking) where even the data plane has an SDN feature. However, it is not easy to attain high-speed performance with a data-plane programmable node. This is the challenge for DPN. To address this problem, we employed the following configuration approach for VNode design: SlowPath programming environment that consists of Intel Architecture (IA) servers, and FastPath comprising network processors that perform network processing.

Programming of the network processors used for the VNode's FastPath is relatively difficult because the processors must execute high-speed processing in kernel mode and such difficulty in programming may be problematic. However, due to the advancements in IA processor technology and the availability of technologies for network interface controller (NIC)-offloading packet processing, upgrading VNode to increase the processing speed has become realistic. These activities however have been suspended because the study phase and the R&D contracts on the virtualization node came to a close at the end of FY2014. A joint research project was started in FY2015, and the University of Tokyo and NTT are developing their own virtualization nodes. Of course, an increase in speed to some extent should be achieved by using the Data Plane Development Kit (DPDK) [15] provided as an open source solution by Intel Corp. and other similar tools; however, such tools are applicable to a limited number of cases. An example of this is DPDK because a specific type of conventional NIC (physical layer (PHY)) must be used and DPDK cannot handle frames other than conventional Ethernet frames.

FLARE Node [14], derived from VNode, ensures simple realization of DPN. We developed it in conjunction with a research project on edge network virtualization in VNode project as described later in this chapter.

The FLARE Node is configured with SlowPath and FastPath. The network processor for FastPath employs an architecture enabling use from the user space of the Polling Mode Driver (PMD), which ensures easy programming. FLARE provides easy programmability and ensures high-scalability and high throughput by employing many-core-processor multiprocessing and recently using Intel DPDK and other technologies. Hence, in FLARE, use of such a DPN platform that can program the data plane and easily achieve expandability of the data-plane components and their API (Southbound Interface) is ensured. This means, for example, that, in addition to the byte string defined in the packet header, software can be used to control the byte string in a user specified field. The closest idea to ours mentioned above is Protocol Oblivious Forwarding (POF) [7]. However, POF has limitations. Its network function is limited, and it employs the pattern-match-and-action program model that is used in OpenFlow. On the other hand, FLARE supports deeper programmability than that provided by POF, and is much more easily programmable. Furthermore, POF or OpenFlow can be implemented on FLARE as a network function element.

While VNode uses high-speed Generic Routing Encapsulation tunneling for processing arbitrary types of frames (so called any frames or arbitrary frames), FLARE is able to process any frames without tunneling because the PHY is designed to handle any frames. This means that an edge network is not required to prepare tunnels to implement new protocols or services, and it leads to reduction of overhead in achieving high network performance. Therefore, FLARE is expected to be used in many cases, particularly in edge networks, to serve Internet of Things (IoT)/Machine-to-Machine (M2M) applications.

In summary, FLARE employs a node architecture that features deep programmability (Deeply Programmable Node Architecture). FLARE can be implemented so that many types of processors such as general purpose processors, many-core-processors or GPGPUs (general purpose graphics processor units) are integrated. As for the present version of FLARE, there are two types of implementations: (1) an integration of EZchip Tile processors, Intel general purpose servers, and GPGPUs, and (2) Intel general purpose servers by Intel DPDK. We plan to add implementations of other types in near future. In addition, in the current situation where different types of traffic from IoT, sensors, and mobile to high-precision images are exchanged, we note the usefulness of operating network functions in application-specific ways.

Regarding application communications, network devices in general, even an advanced type of SDN switch, have no way to access directly the information existing in the application layer except by conducting DPI (deep packet inspection). This is because, conventionally on the Internet, the application context is lost at the socket interface at the time when session/datagram abstraction is performed. So, on the Internet, application-specific controls are not applicable, only universal control is employed.

To enable application controls in the network layer, an architecture is required that ensures the definition of an additional field in a packet for application specific in-network control. We proposed defining a trailer that can add application information at the end of a packet instead of defining an additional header. This method, if combined with SDN-type network control, ensures packet-traffic control through application identification, which is not achievable using the conventional network equipment [16].

We implemented this method in a mobile network environment and held a demonstration at the GEC20 Demonstration Event Session. The demonstration was highly praised, and we were awarded the "Best Demo Award." [17]

4 Problem Definition

The current communication platforms are facing various problems. Advanced network virtualization is expected to be the key idea that will create a breakthrough and establish a foundation for future innovative design methodologies that could solve such problems. An information communication platform, in a broader sense, consists basically of "links" that provide network resources for transmitting data and "nodes" that provide computation/storage resources for executing programs to interpret protocols and distribute packets to nodes. Advanced network virtualization technology virtualizes whole networks based on this broad concept and contributes to network users. More specifically, advanced network virtualization technology fulfills the following five requirements: (1) Resource abstraction, (2) Resource isolation, (3) Resource elasticity, (4) Programmability, and (5) Authentication. Advance network virtualization offers two advantages. First, it enables the creation and design of new generation networks from a clean slate to inspire free-innovative thinking. Second, it enables the implementation of a new generation information communication platform that will ensure the building of a meta-architecture that can accommodate multiple virtualized networks at the same time. The following sections introduce our research and development activities on the integrated management-type network virtualization platform (Advanced Network Virtualization Platform) that will enable the actualization of such an advanced type of network virtualization.

5 Overview of Network Virtualization Platform

Figure 1 shows the overall structure of the network virtualization platform [11] that we have been studying. It consists of the following components.

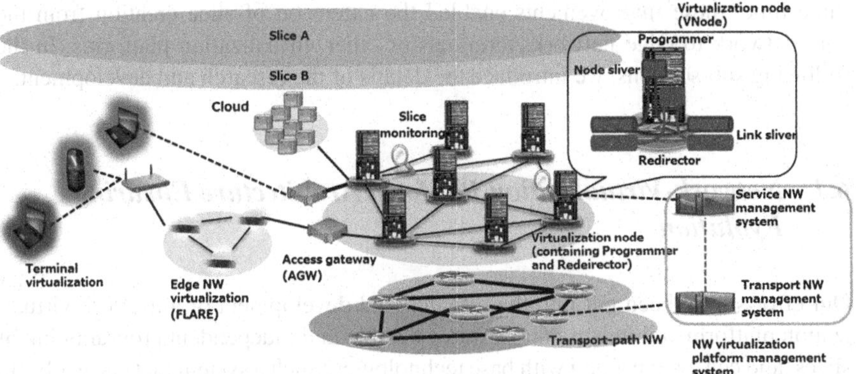

Fig. 1 Network virtualization platform: perspective

- Virtualization Node (VNode): A VNode is a component that provides a virtual network (a slice). A VNode is an integrated node unit comprising the Programmer and Redirector.
  - Programmer: The Programmer provides a virtualized node resource (Node Sliver) that actualizes network functions using software.
  - Redirector: The Redirector provides a virtualized link resource (Link Sliver) that connects the node slivers to each other.
- Access Gateway (AGW): An AGW is a component that provides a subscriber (a user) with a connection to a slice.
- Network Virtualization Platform Management System (NVMS): The NVMS manages and controls all the network virtualization platform resources and creates a slice according to a request from a slice developer (Developer).
- Edge NW virtualization (FLARE): A FLARE Node is a small form factor network virtualization node that consumes less power than the VNode. It physically comprises the following hardware components: many-core network processors for the data plane conducting packet processing and x86-architecture Intel processors for the control plane to control packet processing.

In addition, the network virtualization platform virtualizes edge networks or edge terminals to provide a slice.

6 Research and Development: Strategy, Target, and Development Items

Our research and development project actualized a network virtualization platform with higher performance and sophisticated functions by extending its functions or adding new functions from a conventional network virtualization platform. At the same time, these improvements enabled the extension of slice creation from the core network to edge networks/terminals or other virtualization platforms. In the following sub-sections, we introduce the details of the research and development.

6.1 Network-Virtualization Platform Architecture Ensuring Evolution

Our efforts have been put into the research and development of a network virtualization platform architecture that ensures flexible and independent programming of slices, and can evolve along with base technologies. Such a system that can evolve in a sustainable way must have an architecture that ensures the capability of accepting the latest developments in base technologies even if the individual technologies such as those for links, computing, or storage, evolve independently.

In regard to the VNode, its architecture is therefore divided into two components, each of which can evolve independently: Redirector, which manages link resources for defining the structure of a virtual network, and Programmer, which provides node resources and ensures "Deeply Programmable" features.

In addition, for the purpose of constructing such an architecture that, while covering the whole network from terminals to edge/core networks, makes the idea of a "slice" achievable at any place in the network, the network is configured using two types of networks individually fulfilling different requirements: a core network, which has abundant resources ensuring sufficient resource allocation and guaranteeing bandwidth with a high degree of accuracy; and an edge network, which ensures that a slice can accommodate in a scalable way different types of end terminals, and flexibly employ virtualization technologies. Hence, this architecture study was conducted under the abovementioned prerequisite conditions and an assumption that the role of an edge network would make new protocols usable end-to-end so that user terminals, including wireless mobile terminals, are easily accommodated by a virtual network and connected to the cloud network.

Figure 1 shows the overall configuration of a sliceable edge-core integrated network virtualization platform, where a VNode system is the core network. VNode will be described in detail in the following sub-sections. As for the edge node, a node named FLARE shown in Fig. 1 was developed. The FLARE Node is a smaller network virtualization node that consumes less power than the VNode. The FLARE Node architecture must have the capability to allocate sufficient computing power according to a request from a user terminal or an application, and to provide, in a programmable way, the optimum network protocol or network processing capacity.

The FLARE Node physically comprises the following hardware components: many-core network processors for the data plane conducting packet processing and x86-architecture Intel processors that are mutually connected through a PCI-Express interface for the control plane to control packet processing or to conduct complex computations. In the abovementioned configuration, slices are built on the FLARE

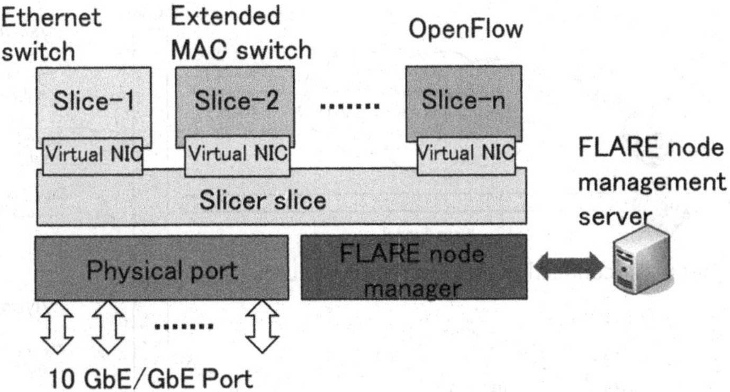

Fig. 2 FLARE node structure

Node as shown in Fig. 2. The FLARE Node Management Server, which exist outside the FLARE Node, comprise virtual machines based on Lightweight Linux Container technology and create slices by conducting virtualization processing using the abovementioned processors. As for slice creation on the many-core processors, the isolation of processes among slices is ensured by slice-by-slice allocation of a core. In addition, the individual slices are ensured to conduct independently its network processing because a slice called a "Slicer Slice," working as the controller, instructs the physical port to split a packet and distribute it to an individual slice according to the input/output port the packet uses or the tag information additionally written in the packet. Also, Fig. 2 indicates that a number of switch functions, an Ethernet Switch (allocated to slice-1 in the figure), a packet switch for switching expanded packets with extended MAC addresses of extended byte-length (allocated to slice-2), and an OpenFlow Switch (allocated to slice-n), can be allocated by program to slices.

Another new development is cooperation with the AGW. Its function is to lead traffic from an edge terminal or an edge network to the slices on the core network. Two methods were developed for leading traffic: in the first method, packets are led to the destination slice through an IPSec tunnel or via Tag-VLAN, as shown in Fig. 3a; in the second method, packets are led to a proper slice according to an information-byte added to the trailer in a packet, as shown in Fig. 3b. The trailer-byte method has an advantage in that, even if other networks exist on the route from a terminal to the destination slice, because the FLARE Node transforms the address from a trailer-byte to the VLAN ID, the packets successfully reach the proper slice in the destination core network. Furthermore, by allowing a terminal to write the information regarding the application being used by the terminal into the trailer-byte, the method enables an easy connection to the application-specific slice, while in a conventional configuration, a Deep Packet Inspection system is necessary.

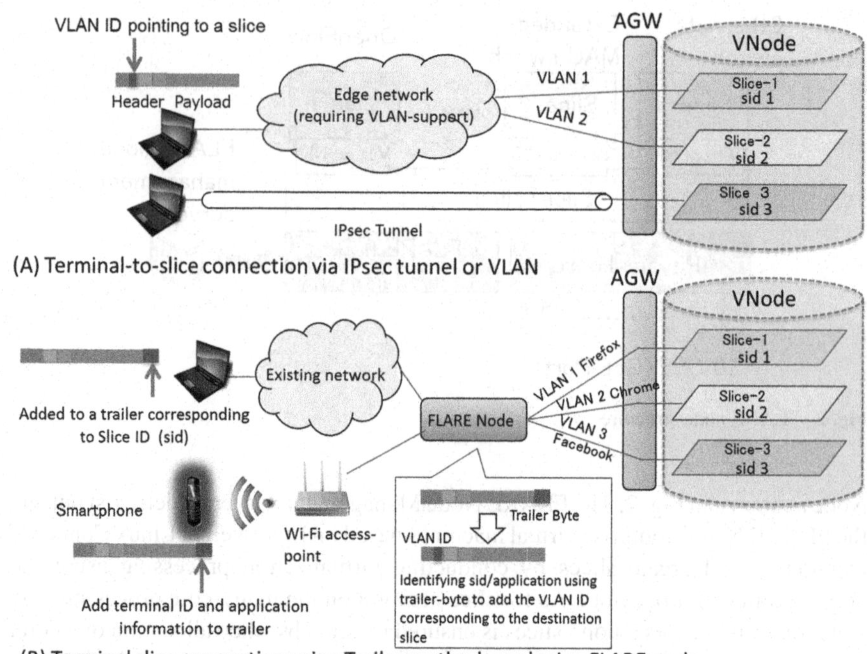

Fig. 3 Method for accommodating edge terminals in slice

6.2 Control and Management Mechanism: Abstraction and Elasticity of Resources

The purpose of an integrated management-type network virtualization platform is to enable flexible and integrated resource management through the abstraction of different types of physical network resources [18]. Our research and development achieved integrated control and management with the transport network, through the extension of such an abstract resource management scheme to the transport network. Such an integrated control and management mechanism is required for the following reasons. A network infrastructure that provides general services consists of two different networks, the "Service Network" that consists of service nodes to provide network services, and the "Transport Network" that connects and transmits packets between service nodes. The transport network transports packets independently of the service contents and is shared by multiple services. Even a network virtualization platform, because it consists of a service network comprising VNodes and a transport network comprising existing switches and routers, is required to control and manage the service network and the transport network in an integrated way. Figure 4 shows the configuration of the system implemented in our project. Its target is the actualization of the integrated control/management of a service network and a transport network. In the configuration, two components are

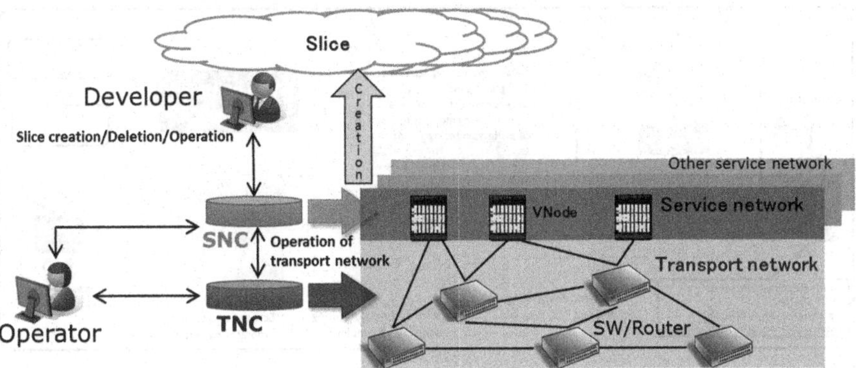

Fig. 4 Configuration of service network and transport network

integrated: the Service Network Controller (SNC), which is in charge of service-network control/management; and the Transport Network Controller (TNC), which is in charge of transport-network control/management. They work together to accomplish the following task to establish link slivers at the timing of slice creation: they allocate the required resources to the transport path or automatically create a transport path. The cooperation proceeds in the following way: the SNC, treating transport network resources as an abstract transport path, notifies the TNC of the volume of resources for a transport path required to establish a link sliver. The TNC allocates the required transport path with the prepared physical resources and provides it to the SNC.

On the other hand, in order to ensure dynamic re-configuration of the resources allocated to a slice, meaning that "resource elasticity" is secured, for the purpose of keeping the service quality at a high level or accomplishing total optimization over the network, performance monitoring of each slice is indispensable [19]. The network virtualization platform accomplishes such slice-performance monitoring in the following way to cope with the situation where individual slices have different topologies or identifiers: a NW Monitoring Manager was newly developed, which obtains the topology information and the slice information of a monitoring target slice from the SNC and TNC; determines at what points to monitor; and establishes the filtering conditions for the points. Figure 5 shows the slice monitoring scheme based on cooperation between the NVMS and the NW Monitoring Manager. The Developer who is the owner of the slice designates the slice to be performance-monitored and the points and times/dates for monitoring using the Portal screen. Then, the NVMS provides the NW Monitoring Manager with the necessary monitoring-link information and slice identifier. The NW Monitoring Manager establishes optical switches for switching monitoring points, and establishes the PRESTA10G's filter. Then, after monitoring is completed, the NW Monitoring Manager notifies the Developer of the monitoring results, through the NVMS.

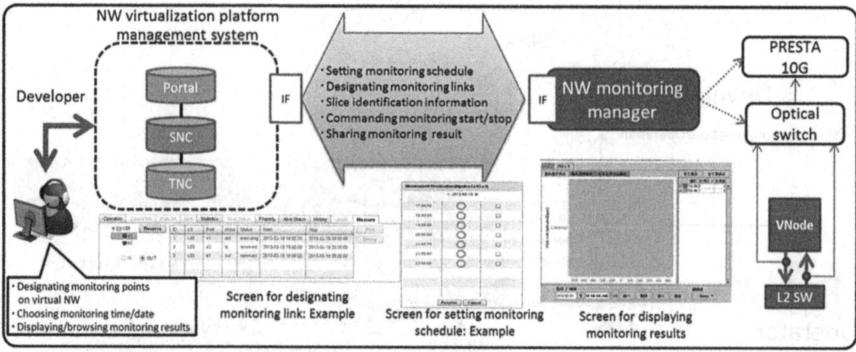

Fig. 5 Collaboration Scheme for network virtualization platform management system with NW management manager

6.3 Resource Control: Securing Resource Independence/Isolation

Avoiding resource interference between slices is critical for implementing network virtualization. For instance, a certain slice must not be allowed to monopolize bandwidth or produce unfair delays in other slices. A function for avoiding such interference is called "resource isolation." We conducted our research on avoiding resource interference, according to the following idea: "For avoiding interference among terminals (end-to-end), resource isolation must be secured entirely over the Network Virtualization Platform, and Redirector is the most important for accomplishing such resource isolation." In conventional network virtualization platforms, resource isolation had been achieved, although in a basic form [20]. We improved upon such a conventional platform in our project. We first describe below such a basic resource isolation method, and then focus on what was improved in our project.

An outline of the abovementioned basic resource isolation functions that are implemented in Redirector [20] are described hereafter.

The Redirector has a built-in L3-switch that performs such functions as a VLAN Function, IP Routing Function or QoS Function. For a QoS function, the switch has such features as Weighted Fair Queuing (WFQ) and Policing. Redirector, using these functions, achieves resource isolation. For example, the WFQ function carries out flow-output bandwidth control by putting a packet in a queue prepared for individual flows, and Redirector achieves high-performance bandwidth control for each slice using this mechanism. However, this method has been applicable only to a limited number of slices for a huge bandwidth because in this method it is relatively expensive to use high-speed memory for queuing. On the other hand, the policing function is low cost because it does not require memory. Therefore, a policing function is applicable to more than several tens of slices for resource isolation functions. However, this method degrades the performance compared to the WFQ method, so it causes higher packet loss or more severe slice jitter.

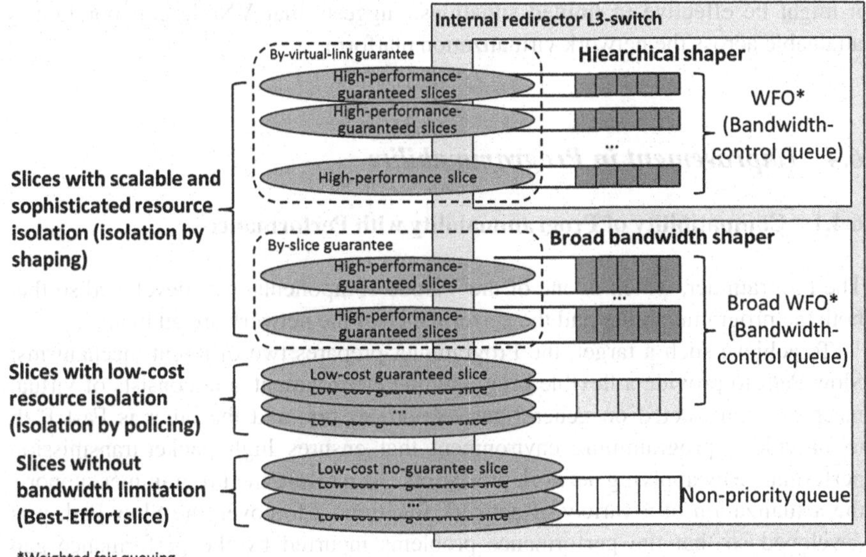

Fig. 6 Redirector resource isolation functions

There are three problems that must be addressed in regard to the basic method mentioned above. The first problem is to provide resource isolation using the WFQ function for as many slices as possible. At the same time, this problem also involves achieving more sophisticated control by improving the resource isolation function. In a conventional Redirector, the bandwidth control is carried out for all of the link slivers connected to a node sliver [20]. We improved the bandwidth control for each link sliver. We added a new hardware component called a hierarchical shaper to the internal switch of the Redirector to address this problem and achieved a scalable and sophisticated resource isolation function as shown in Fig. 6. The second problem is to suppress end-to-end interference. This was difficult because the Redirector resource isolation function works without any cooperation with the other VNode components (particularly Programmer), or other virtualization platform components such as the AGW and the transport network. To address this problem, the system was improved so that (1) the resource management of the Programmer and Redirector can collaborate with each other and (2) the NVMS has resource management capabilities for Redirector-to-AGW connection and the transport network.

The third problem is the collaboration between VNodes. As for this problem, employing the "Plug-In Architecture" described in Section "Programmability Extension by Applying Plug-in Functional Modules" has paved the way to its resolution. We successfully conducted in the network virtualization platform an experiment for measuring packet latency without interfering with the functioning of a slice applying such architecture [21]. The success of the experiment, although it might be effective in limited situations, suggests that VNode collaboration is attainable across the network virtualization platform.

6.4 *Improvement in Programmability*

6.4.1 Compatibility of Programmability with Performance

The Programmer, which is one of the VNode components, was developed so that both the programmability and the performance of the network are attainable.

To achieve such a target, the Programmer prepares two different mechanisms: Slow Path to provide a flexible programming environment, and consists of virtual machines constructed on general-purpose processors; and the other is Fast Path to provide a programming environment that ensures high packet-transmission performance by applying network processors. As a result, the Programmer supports the actualization of a variety of network functions. Moreover, the Slow Path was developed so that the performance problems incurred by the performance gap between the low I/O and high computing power were suppressed [22].

Figure 7 shows the Programmer structure.

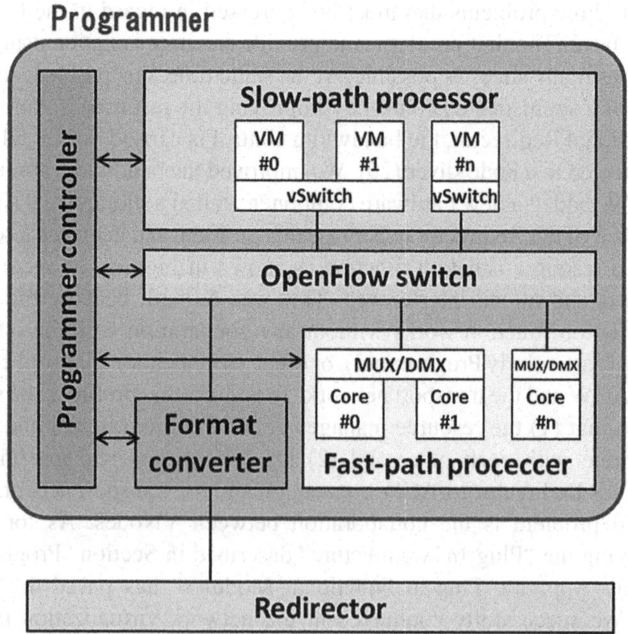

Fig. 7 Structure of programmer

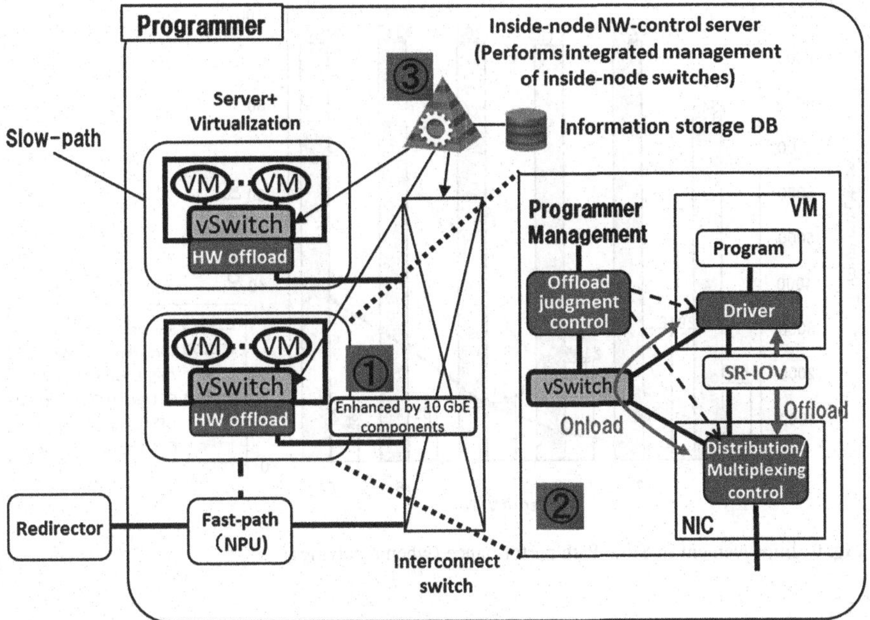

Fig. 8 Improvement in Slow-Path performance

Slow Path provides a flexible programming environment that is actualized by applying virtual machines (VM) and a high-speed network I/O environment. On the other hand, Fast Path provides a programming environment that ensures high-speed packet transmission performance by applying network processors. OpenFlow switches[23] are used in the internal connection of computing resources in the device. Therefore, we can construct a network node that can choose and combine various types of computing resources without constraint. The Format Converter converts the packet format (MAC-In-MAC format and VLAN format) between the Redirector and Programmer.

The following is a description of how the improvement in the performance of Slow Path was attained: The network I/O performance was improved, as shown in Fig. 8, through (1) the application of a switch-offload technique, and (2) by employing ten GbE compatible hardware components. Figure 9 shows the performance improvement in Slow Path achieved by (1) and (2). The plots and graphs specified by "onload" represent the results before employing (1) and (2), and those specified by "offload" represent results after employing (1) and (2). The plots designated by "CPU Load SUM" represent the sum of the CPU utilization rates of Guest OSs, and the plots designated by "CPU Load HOST" are the utilization rates of the host OS. Figure 9 indicates that, through the application of (1) and (2), the throughput performance is improved and the Host OS utilization rate is lowered.

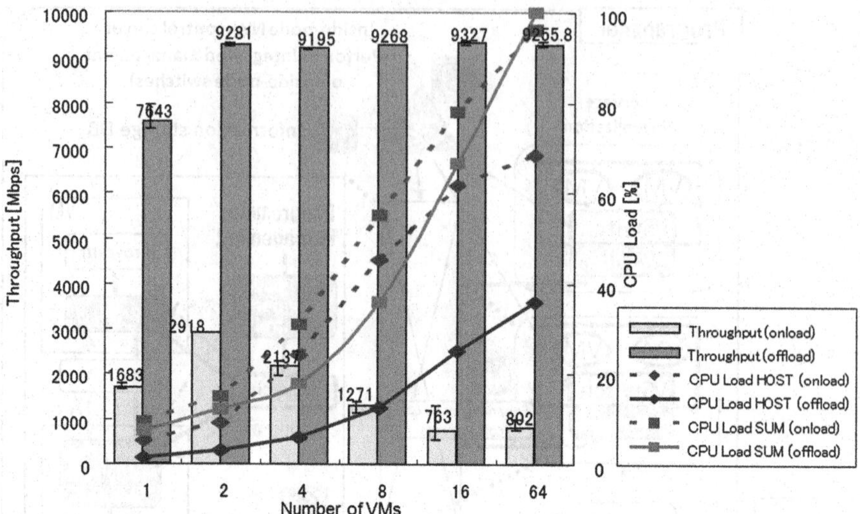

Fig. 9 Improvement in Slow-Path performance (when receiving)

6.4.2 Programmability Extension by Applying Plug-in Functional Modules

Programmability extension via plug-ins was investigated aiming at two targets. The first target is to ensure VNode evolution (adding functions) through adding programs (software) [24]; although VNode is allowed to use prepared types of node slivers and link slivers. As for node slivers, two types of Slow Path and Fast Path node slivers are prepared as mentioned earlier. As for link slivers, a single type based on Generic Encapsulation Over IP (GRE/IP) is available. We can program the Slow Path freely, but a situation will arise when extending programmability by employing other types of node slivers and link slivers to fulfill special requirements. Hence, the first target is to extend the VNode infrastructure architecture so that it can add new types of node slivers and link slivers other than those prepared into the infrastructure. The second target is to develop a method that enables simple programming of the VNode infrastructure for functional extension. VNode developers will face difficulties in programming when such functional extension is achieved through the addition of such specialized hardware components as network processors. If the second target is achieved, such programming difficulties will be removed.

The following three items were newly developed to prove the validity of the abovementioned method for removing such programming difficulty as shown in Fig. 10. The first is the programming language and development environment of the open processing platform for Fast Path [25]. The programming language is named “Phonepl” and its programming environment has been partially developed. Note that this language and programming environment are required to achieve the second

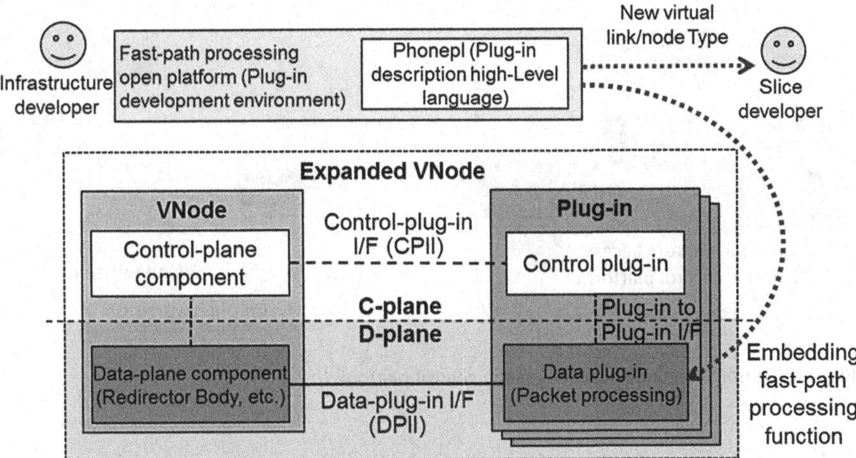

Fig. 10 Programmability expansion using plug-in functional modules: developed items for proof of concept

target. The second developed item is the mechanism for embedding the Fast-Path modules in a VNode [24]. They make it possible to incorporate the Fast-Path module or other types of Fast-Path modules. The second item enables the construction of the operation environment. The third developed item is the mechanism for embedding an offload switch/router-function in a VNode [26]. Note that the third developed item, involved in achieving the first target, enables a VNode developer to use, as a module, the functions possessed by the VNode components, particularly by the Redirector, but they are hidden to the developer and unusable from a slice.

6.4.3 Edge-Virtualization Technologies: Programming Technologies for Network Access

Edge virtualization technologies have been developed aiming at actualizing wide use of more efficient network services and providing application-oriented network functions. These are achieved by expanding the slice coverage from the core network to edge networks for the purpose of enabling the implementation of a new network function at its best position in the network structure. In this subsection, the research and development activities, for the purpose just mentioned, of a dynamic network access control platform, are introduced. The platform, handling terminals such as a mobile terminal involving more than one network with different characteristics, dynamically switches the connection to access networks with different characteristics, and collaborates with services carried out on VNode by allowing the network-side to implement such controls [27].

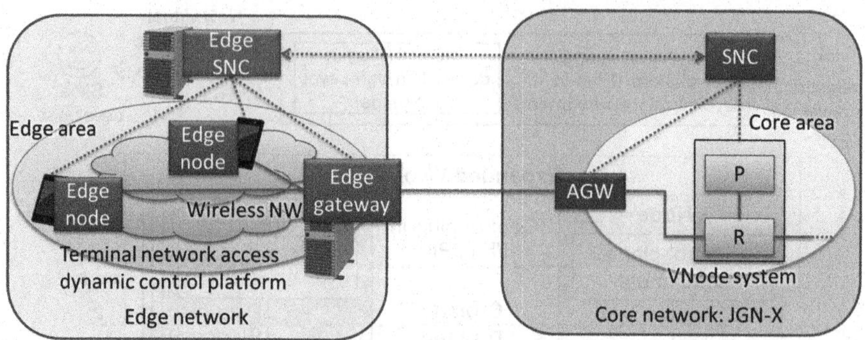

Fig. 11 Terminal network access dynamic control platform

Figure 11 shows the dynamic terminal network access control platform that was implemented in our project. The following is the outline of how the platform and other components function.

We implemented an Edge Node on an Android terminal. Our Edge Node can implement the following controls: for each application, identifying the slice that the node can communicate with according to the policy instructed by the Edge SNC; and for each slice, determining what type of mobile network to use as an access network.

The Edge Gateway is a gateway relaying the communication between an edge-network slice and a core-network slice. It connects to the AGW, a gateway device in the core area, to relay the communication between the Node Sliver on the VNode and applications on the Edge Node. The Edge SNC performs the following: creating/deleting/configuring edge-network slices by issuing an execution command of the connection to the core-network slice on VNode. Moreover, by communicating with the SNC, which is the VNode integrated management device, the Edge SNC exchanges the information necessary to establish the connection between the edge-network slice and the core-network slice.

The mechanisms mentioned above are enabled in a slice that is a combination of an edge-network slice and a core-network slice, by accommodating the applications existing in the Edge Network through an arbitrary access network.

6.4.4 Gateway Function Improvement: Authentication Capability

When using a slice on the Network Virtualization Platform, the end-to-end-connectivity covering even physical devices or networks must be ensured. Furthermore, such connections must be established using various protocols including proprietary protocols. Such connectivity is ensured by the AGW (as shown in Fig. 12) placed at the edge of a slice to ensure the following.

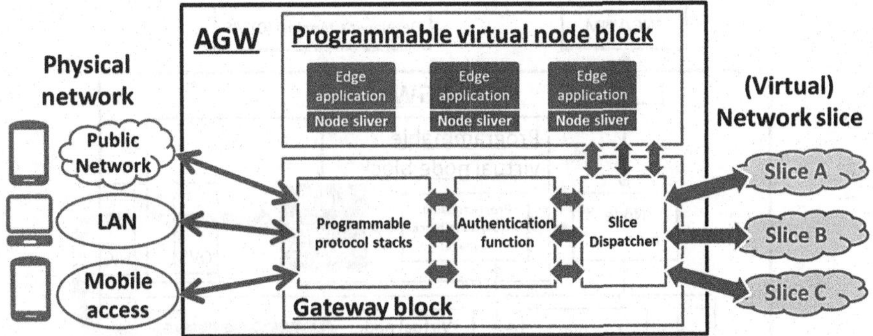

Fig. 12 Gateways: in-network positions and functions

- Connectivity: Achieved by the virtual distribution function that connects physical terminals/networks and slices
- Security: Achieved by the provision of authentication functions based on IPsec for physical terminal connection
- Programmability
  - Customization for accepting a variety of protocols is enabled by means of the programmable protocol stack in the Packet Processing Middleware (PPM) in the GW Block.
  - Arbitrary communication processing applications can work at a slice entrance (edge) because programmable virtual nodes (Node Sliver VM) are available.
- Performance: High relay processing performance using general purpose Intel architecture servers is attainable using the GW Block on the PPM.

Figure 13 shows the internal architecture and the communication path in the GW System. The following is a description of a GW: Because it is placed at the edge of a slice, a GW must be sufficiently small and inexpensive compared to a large-scale system such as a conventionally used VNode. Therefore, a GW is constructed with building blocks of which the minimum unit is a single IA chassis having all functions (conventional gateway functions plus the functions for attaining programmability), and maintaining scalability for the programmability components by allowing the addition of devices if necessary. Hence, a GW generally consists of the following major components.

- Manager: In charge of communications with SNC and VNode manager with XML-RPC control messages and configuration of the programmable virtual node block and GW block.
- Programmable Virtual Node Block: Actualizes the Node Sliver with a Virtual Machine (VM) and enables the deployment of customized programs for network or data processing applications, e.g., identification, translation, modification, and processing of the frame/packets.

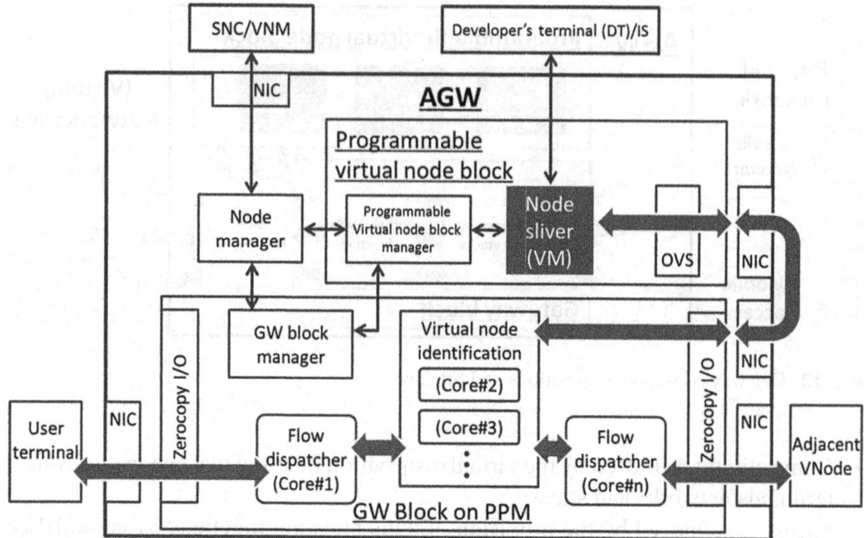

Fig. 13 Gateway architecture

Table 1 GW performance

Throughput with IPSec	Throughput with VLAN	Slow-Path node sliver throughput
1.3 Gbps[a]	4.7 Gbps[a]	1.5 Gbps[b]

[a]1372 byte frame, Intel Xeon X5690 (3.46 GHz/Core) × 2 units
[b]Using Intel Xeon X5690 (3.46 GHz/6 Cores). Allocating 2 GB of memory, two virtual CPU to VM, and four cores to vhostnet processing

- GW Block: Transmits user frames between a user terminal (UT) and VNode, transmits/format converts frames between a physical network and a slice, achieves high performance using Zero Copy I/O which is provided by the PPM Function, and parallel processes utilizing a multi-core CPU.

Table 1 gives the performance of the enhanced GW Device.

7 Deployment on Testbed and International Federation

We deployed the network virtualization platform on the JGN-X Testbed for application developers for their experiments. As shown in Fig. 14, seven VNodes, two simplified virtualization nodes, and six AGWs are deployed nationwide. As of now, the testbed is used at a rate of approximately 40 slices per month.

We conducted international federation experiments through a simplified virtualization node (NC) installed by NICT at the University of Utah, U.S. A network virtualization testbed constructed on JGN-X is connected with the GENI virtualiza-

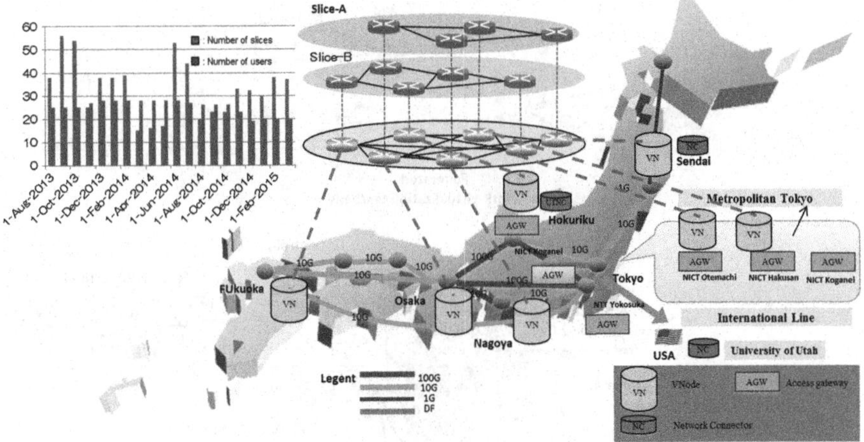

Fig. 14 Deployment of Network Virtualization Platform on JGN-X

tion testbed in North America, and the Fed4FIRE network virtualization testbed in Europe. The federation experiment proved the validity of the slice construction in a global multi-domain environment [28].

Figure 15 shows the configuration of the international federation experiment. Each network virtualization testbed is federated through the slice exchange point (SEP) installed in the NC at the University of Utah. Gaps caused by mutual differences in the testbed implementation methods are filled in at the SEP through the abstraction of the individual testbed interfaces into common application interfaces.

8 Concluding Remark and Perspectives for the Future

Since 2008, network-virtualization technologies have been developed as fundamental technologies supporting new generation networks. At the initial stage of development, these technologies were acknowledged primarily as tools to realize a clean slate experimental environment, and the method of subdividing a network into slices to build a variety of application-specific networks has been developed as a candidate of new generation network infrastructures. We believe that the network virtualization technologies we developed have had a great impact on other research activities, especially, GENI's R&D in the U.S. because of their innovativeness and uniqueness [29, 30].

Without doubt, as more and more of the things around us become connected to networks, as proposed in the concept of IoT or Internet of Everything (IoE), networks will not be able to fulfill application requirements if they continue to perform "one-size-fits-all" control on a variety of traffic types with different characteristics. Therefore, we believe that the network virtualization technologies,

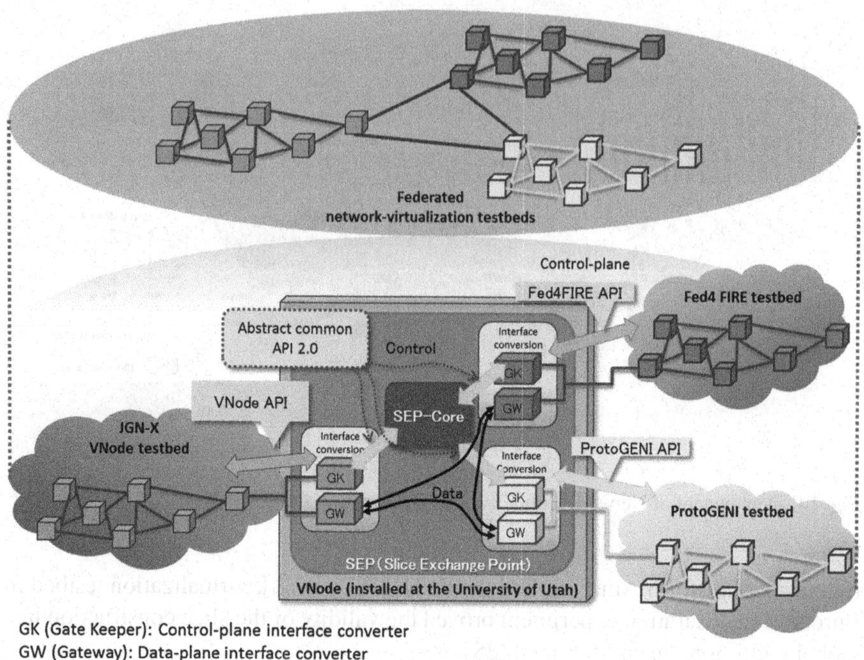

Fig. 15 Configuration of Japan-US-Europe Federation

combined with traffic-classification technologies, will support the core of new generation networks. In addition, technologies for enabling easy programming of high-speed processing are indispensable for reducing development time and cost. At present, for such programming, general servers are used. However, advanced network processor and reconfigurable ASIC technologies that may outperform the conventional processor architectures may become available as another factor contributing to the realization of network-virtualization technologies.

Considering such situations, it is reasonable to predict that the seamless consolidation of networking and computing would advance. We believe that edge networks, at least, as the idea of Mobile Edge Computing (MEC) shows, will play a crucial role and networks in near future will evolve to become "intelligent and functional pipes" where cloud computing, network functional elements at edges and user equipment collaborate naturally, instead of being "dumb pipes" connecting cloud computing and user equipment. Such network evolution will lead to a world where, as a result of seamless integration of networks and computing, computing is available any place on a network. The whole of the Internet will function as a huge distributed computer system.

We believe that our near future R&D targets should include (1) the development of use cases, a.k.a., applications of seamless integration of networks, edges, and computers, especially in the areas of IoT, IoE, M2M, and 5G mobile networks, (2) fundamental technologies to be developed to support the abovementioned use cases such as high-speed data-plane processing with many-core processors and/or general purpose processors, or high-speed SDN switches utilizing reconfigurable ASICs, and (3) advanced traffic classification and slicing technologies, e.g., according to various metrics such as application/device /context, etc. Also, considering the tradeoff between the extent of programmability and cost performance, although software based solutions will surely increase their scope because they reduce capital expense and operational expense and enable flexibility in infrastructure, we have to bear in mind that we have to apply the right technology to the right place.

On the other hand, with regard to the application to commercial networks, business issues are always challenges such as standardization, interoperability, operation and management, security measures, economic efficiency, and energy conservation. Since we believe energy conservation and interoperability, in particular, are critical issues for commercialization, we should continue our activities by participating in international standardization organizations and consortiums.

Acknowledgements The research and development achievements described in this chapter were achieved through NICT contract research entitled "Research and Development of Network Virtualization Platform Technologies Supporting New-Generation Network."

The authors express their deep appreciation to the National Institute of Information and Communications who promoted the research and development of the Next-Generation Network and gave us much support. The authors also sincerely thank the research organizations that contributed to the joint research projects, as well as GENI research community and GENI Project Office (GPO) that continuously encourage us and invite us to present our accomplishments and to grant us opportunities to discuss research directions.

References

1. Computer Science and Telecommunications Board Committee on Research Horizons in Networks. Looking Over the Fence at Networks: A Neighbor's View of Network Research. National Academy Press, Washington, D.C., (2001).
2. Berman, M., Chase, J.S., Landweber, L.H., Nakao, A., Ott, M., Raychaudhuri, D., Ricci, R., Seskar, I.: GENI: a federated testbed for innovative network experiments. Comput. Netw. **61**, 5–23 (2014)
3. Peterson, L., Roscoe, T.: The design principles of PlanetLab. SIGOPS **40**, 11–16 (2006)
4. Nakao, A.: VNode: a deeply programmable network testbed through network virtualization. In: 3rd IEICE Technical Committee on Network Virtualization (March 2012)
5. https://www.geni.net/?p=1739
6. McKeown, N., Anderson, T., Balakrishnan, H., Parulkar, G., Peterson, L., Rexford, J., Shenker, S., Turner, J.: OpenFlow: enabling innovation in campus networks. ACM SIGCOMM **38**, 69–74 (2008)
7. Song, H.: Protocol-oblivious forwarding: unleash the power of SDN through a future-proof forwarding plane. In: HotSDN'13, pp. 127–132 (August 2013)
8. Jacobson, V., Smetters, D.K., Thornton, J.D., Plass, M.F.: Networking named content. In: CoNEXT'09 (December 2009)
9. Turner, J.S., Crowley, P., DeHart, J., Freestone, A., Heller, B., Kuhns, F., Kumar, S., Lockwood, J., Lu, J., Wilson, M., Wiseman, C., Zar, D.: Supercharging planetlab: a high performance, multi-application, overlay network platform. SIGCOMM Comput. Commun. Rev. **37**(4), 85–96 (2007)

10. Nakao, A., Ozaki, R., Nishida, Y.: CoreLab: an emerging network testbed employing hosted virtual machine monitor. ACM ROADS 2008 (December 2008)
11. Nakao, A.: VNode: a deeply programmable network testbed through network virtualization. In: The 3rd Domestic Conference, IEICE Technical Committee on Network Virtualization (NV) (2 March 2012)
12. https://ec.europa.eu/research/fp7/index_en.cfm
13. http://www.ieice.org/~nv/
14. Nakao, A.: Flare: deeply programmable network node architecture. http://netseminar.stanford.edu/10_18_12.html
15. http://dpdk.org/
16. Nakao, A., Du, P., Iwai, T.: Application specific slicing for MVNO through software-defined data plane enhancing SDN. IEICE Trans. Commun. **E98-B**(11), 2111–2120 (2015)
17. http://groups.geni.net/geni/wiki/GEC20Agenda
18. Katayama, Y., Yamada, K., Shimano, K., Nakao, A.: Hierarchical resource management system on network virtualization platform for reduction of virtual network embedding calculation. APNOMS2013, P2-1 (2013)
19. Kuwabara, S., Katayama, Y., Yamada, K.: Adaptive traffic monitoring system for virtualized networks. In: Proceedings of IEICE General Conference (2013)
20. Kanada, Y., Shiraishi, K., Nakao, A.: Network-resource isolation for virtualization nodes. IEICE Trans. Commun. **E96-B**(1), 20–30 (2013). doi:10.1587/transcom.e96.b.20
21. Kanada, Y.: Extending network-virtualization platforms by using a specialized packet header and node plug-ins. In: 22nd International Conference on Software, Telecommunications and Computer Networks (SoftCom 2014) (September 2014). doi:10.1109/softcom.2014.7039092
22. Kamiya, S., et al.: A proposal of network processing node to emerging advanced network virtualization infrastructure. In: The 3rd Domestic Conference, IEICE Technical Committee on Network Virtualization (NV) (2 March 2012)
23. OpenFlow Switch Consortium, OpenFlow Switch Specification, version 1.0.0, December 2009
24. Kanada, Y.: A method for evolving networks by introducing new virtual node/link types using node plug-ins. In: 1st IEEE/IFIP International Workshop in SDN Management and Orchestration (SDNMO'14) (May 2014). doi:10.1109/noms.2014.6838417
25. Kanada, Y.: High-level portable programming language for optimized memory use of network processors. Commun. Netw. **7**(1), (2015). doi:10.4236/cn.2015.21006
26. Kanada, Y.: Providing infrastructure functions for virtual networks by applying node plug-in architecture. Workshop on Software Defined Networks for a New Generation of Applications and Services (SDN-NGAS 2014), August 2014. Procedia Comput. Sci. **34**, 661–667 (2014). doi:10.1016/j.procs.2014.07.094
27. Iihoshi, T., et al.: A slice extension architecture using smart-devices for flexible network service experiments on VNode. In: The 9th Domestic Conference, IEICE Technical Committee on Network Virtualization (NV) (11 April 2014)
28. Tarui, T., et al.: Federating heterogeneous network virtualization platforms by slice exchange point. In: IFIP/IEEE International Symposium on Integrated Network Management (IM), pp. 746–749 (2015)
29. Nakao, A.: Research and development of network virtualization technologies in Japan. J. Natl. Inst. Inform. Commun. Technol. **62**(2), 4-2 (2016). Special Issue on New-Generation Network
30. Yamada, K., Nakao, A., Kanada, Y., Saida, Y., Amemiya, K.: Integrated management for network virtualization infrastructure. J. Natl. Inst. Inform. Commun. Technol. **62**(2), 4-3 (2016). Special Issue on New-Generation Network

Creating a Worldwide Network for the Global Environment for Network Innovations (GENI) and Related Experimental Environments

Joe Mambretti, Jim Chen, Fei Yeh, Jingguo Ge, Junling You, Tong Li, Cees de Laat, Paola Grosso, Te-Lung Liu, Mon-Yen Luo, Aki Nakao, Paul Müller, Ronald van der Pol, Martin Reed, Michael Stanton, and Chu-Sing Yang

J. Mambretti (✉) • J. Chen • F. Yeh
International Center for Advanced Internet Research Northwestern University, Chicago, IL, USA
e-mail: j-mambretti@northwestern.edu; jim-chen@northwestern.edu; fyeh@northwestern.edu

J. Ge • J. You • T. Li
China Science and Technology Network, Computer Network Information Center, Chinese Academy of Sciences, Beijing, China
e-mail: gejingguo@cstnet.cn; youjunling@cstnet.cn; tongl@cstnet.cn

C. de Laat • P. Grosso
University of Amsterdam, Amsterdam, The Netherlands
e-mail: delaat@uva.nl; p.grosso@uva.nl

T.-L. Liu
National Center for High-Performance Computing, National Applied Laboratories, Hsinchu City, Taiwan
e-mail: tlliu@narlabs.org.tw

M.-Y. Luo
National Kaohsiung University of Applied Sciences, Kaohsiung, Taiwan
e-mail: myluo@kuas.edu.tw

A. Nakao
University of Tokyo, Tokyo, Japan
e-mail: nakao@nakao-lab.org

P. Müller
Integrated Communication Systems Lab., Department of Computer Science, University of Kaiserslautern, Kaiserslautern, Germany
e-mail: pmueller@informatik.uni-kl.de

R. van der Pol
SURFnet, Utrecht, The Netherlands
e-mail: Ronald.vanderPol@SURFnet.nl

M. Reed
University of Essex, Colchester, UK
e-mail: mjreed@essex.ac.uk

M. Stanton
Brazilian Research and Education Network—RNP, Rio de Janeiro, RJ, Brazil
e-mail: michael@rnp.br

C.-S. Yang
National Cheng-Kung University, Tainan City 701, Taiwan
e-mail: csyang@ee.ncku.edu.tw

© Springer International Publishing Switzerland 2016

R. McGeer et al. (eds.), *The GENI Book*, DOI 10.1007/978-3-319-33769-2_24

Abstract Many important societal activities are global in scope, and as these activities continually expand world-wide, they are increasingly based on a foundation of advanced communication services and underlying innovative network architecture, technology, and core infrastructure. To continue progress in these areas, research activities cannot be limited to campus labs and small local testbeds or even to national testbeds. Researchers must be able to explore concepts at scale—to conduct experiments on world-wide testbeds that approximate the attributes of the real world. Today, it is possible to take advantage of several macro information technology trends, especially virtualization and capabilities for programming technology resources at a highly granulated level, to design, implement and operate network research environments at a global scale. GENI is developing such an environment, as are research communities in a number of other countries. Recently, these communities have not only been investigating techniques for federating these research environments across multiple domains, but they have also been demonstration prototypes of such federations. This chapter provides an overview of key topics and experimental activities related to GENI international networking and to related projects throughout the world.

1 Introduction

It is well known that the majority of key societal activities are becoming global in scope, and as these activities expand world-wide, they require a sophisticated foundation of advanced communication services, supported by underlying innovative network architecture, technology, and core infrastructure. To continue progress in meeting these and future requirements, network research investigations cannot be limited to campus labs and small local testbeds or even to national testbeds. Researchers must be able to explore innovative concepts at a significant scale—global scale—through empirical experimentation. They must conduct experiments on world-wide testbeds that approximate the complex attributes of the real world. Today, it is possible to take advantage of several macro information technology trends, especially virtualization and capabilities for programming technology resources at a highly granulated level, to design, implement and operate network research environments across the world. In the US, the Global Environment for Network Innovations (GENI) is developing such an environment, as are research communities in a number of other countries, described in subsequent sections [1].

In the last few years, these communities have begun to federate these research environments across multiple domains, in part, to enable wide ranging exploration of innovative concepts at extremely large scales. Also, they have been demonstrating prototypes of such federations at workshops and conferences. Traditionally, network testbeds have been designed and implemented within project frameworks with limited scopes to support fairly narrowly defined research objectives over a short period of time. In contrast, GENI and related testbed environments have been planned to support experimental research across a wide range of topics, as a persistent research resource, within which many topics can be investigated at an extremely large scale—including globally. A notable aspect of these testbeds is that they not only provide a platform for innovative research, but also they incorporate architectural designs, services, and technologies that forecast the basic model of future communications infrastructure. Within these distributed environments, next generation macro trends are emerging, especially those related to the transition from limited static services and infrastructure to unlimited, highly dynamic, deeply programmable, continually evolving innovative environments. Another major transition reflected in the new models is the migration from designing networks that are controlled and managed through proprietary systems closely integrated with proprietary devices to those that are based on open architecture and open systems, for example, using approaches such as Software Defined Networking (SDN) to manage multiple generalized network resources.

2 Overview of Chapter

This chapter describes the international capabilities of GENI and related network research environments, specifically (a) required services for these types of distributed facilities, (b) basic architectural considerations, (c) existing global facilities, (d) existing international testbed facilities and examples of research experiments being conducted within those environments and (f) emerging architecture and design trends for anticipated future services, technologies, facilities, and resource expansions.

The first section provides a brief overview of the required basic services for large scale, highly distributed network science empirical research facilities. A special consideration in this chapter is one that highlights a need for ensuring flexible and programmable multi-domain L2 paths. A common networking architectural model describes seven basic layers. Of these Layer 3 (L3), is the most familiar because essentially, the Internet is based on L3 architecture. However, underlying L3 services are supporting Layer 2 (L2) and Layer 1 (L1—e.g., lightpaths within optical fiber) services and capabilities that are undergoing a rapid revolution from static to dynamic, programmable resources.

Generally, L2 and L1 paths have been implemented as static resources, implemented without change for long periods. Increasingly, L2 and L1 paths are being implemented as dynamically provisioned paths. Also, providing dynamic L2 and L1 paths across multi-domains requires special considerations because such paths transverse many difference authority, policy, management and control boundaries.

The next section describes basic architectural considerations for large scale research testbeds. Such architectural considerations include those for provisioning dynamic multi-domain L2 and L1 paths as well as hybrid networking paths comprised of services utilizing multiple network layers, e.g., L3, L2, and L1.

The next section highlights existing global facilities, with a focus on the Global Integrated Lambda Facility (GLIF) and its Open Exchanges around the world (GLIF Open Lambda Exchanges or GOLEs) as foundation resources. This distributed facility enables multiple customized production and testbed networks to be created and operated within lightpaths on terrestrial and oceanographic fiber optic cables spanning many thousands of miles. A subsequent related section describes a dynamic networking provisioning API developed by the GLIF community in partnership with a standards organization—the Open Grid Forum. This capability allows customized networks to undertake dynamic provision across paths spanning multiple domains world-wide.

The next series of sections describes existing international testbed facilities and examples of research experiments being conducted within those environments. These environments include the international GENI SDN/OpenFlow research testbed, which has been implemented by a consortium of network scientists, the Japanese led international V-Node initiative, a Virtual Research Environment for Large Scale Distributed Systems Research developed by G-Lab in Germany, an international testbed for investigating a variety of topics ranging from WAN protocol transport to Ethernet OAM, and Provider Bridging virtualization, being led by researchers in the Netherlands, a cloud/network testbed being developed in China, a large scale international tesbed for multiple research projects, such as topology management and Virtual Local Area Network (VLAN) transit, a project being led by research institutions in Taiwan, a content routing testbed in the UK, a Brazilian Future Internet testbed, and an international, advanced high performance digital media testbed.

The final sections describe emerging architecture and design trends for anticipated future services, technologies, facilities and resource expansions. Included are discussions of Software Defined Networking Exchanges (SDXs), Software Defined Infrastructure (SDI), which integrates compute resources, storage, instrumentation, sensors, and other resources, and the close integration of network research testbeds and cloud research testbeds.

3 Required Services

The majority of substantial advances in information technology have been based on innovations that have created a higher layer of abstraction than that which had existed previously. Today, many such major advances are being accomplished as a result of multiple convergent macro trends in information technology that are enabling much higher levels of abstraction and virtualization across all levels of infrastructure. Many are based on Service Oriented Architecture (SOA) and related concepts leading to—Anything-as-a-Service (XaaS), for example, Architecture-as-a-Service (AaaS), Network-as-a-Service (NaaS), Environment-as-a-Service (EaaS), Software-as-a-Service (SaaS), Infrastructure-as-a-Service (IaaS), Platform-as-a-Service (PaaS), Container-as-a-Service (CaaS), and many more. Recent work by Strijkers et al. [2] has even created a model and architecture for an "Internet Factory." Standards organizations are developing open architecture frameworks for these approaches, for example, through the Open Grid Forum's Infrastructure Services On-Demand Provisioning Research Group (ISOD-RG) and the US National Institute of Standards and Technology (NIST), which are developing XaaS open architecture standards [3, 4]. The GENI initiative leverages these trends to create highly flexible, programmable, dynamic, distributed environments. However, the goal of the GENI project is not just to leverage such innovations, but to use them to create an environment that supports services that allow experimental researchers empirically to design, develop, and test concepts that will lead to the next generation of distributed environments.

Although the attributes of next generation distributed environments are still evolving, the current macro trends in design indicate the nature of their eventual characteristics. For example, these trends will allow the creation for multiple highly differentiated networks within the same shared infrastructure, so that networks services can be precisely customized for individual requirements. These capabilities are required for many organizations and organizational partnerships that require private customized and highly specialized network services. However, it is also required by providers of large scale distributed clouds for multiple, perhaps hundreds, of individual tenants, each of which requires a private, individually managed and controlled network. Because no single centralized NOC can support many hundreds, of individual networks, new tools are being created to enable self-self networks for multi tenant networks. A number of these tools are based on "slicing" architecture, which allows for contiguous integrations of resources, including over international WAN paths, virtual and physical, to be segmented for specialized purposes. As resources are increasingly being abstracted through virtualization, new tools are also being created to discover, integrate, manage and control them, especially through new types of orchestration techniques.

Next generation distributed environments also will allow for much more dynamic network services and infrastructure environments as opposed to today's fairly static implementations, which anticipate basic resources remaining unchanged for long periods of time. They will also allow for a far more granulated control over network resources, including low level resources, such as L2 and L1 paths. As noted, much progress is being made on transitioning L2 and L1 paths from static to dynamic resources and to provide a wide range of enhancements to the capabilities of these paths within compute facilities and data centers, among such facilities and data centers, within metro regions, across nations and across the world.

In addition, they will allow for highly distributed control over network resources, including individual core elements, e.g., ports. Recent progress in virtualization and in related distributed programmable networking, give rise to opportunities for migrating from centralized network control and management to extremely distributed control and management. Management and control functions that previously were the exclusive prerogative of centralized provider NOCs now can be provided to enterprises, applications, processes at the edge of the networks, and individuals.

An especially wide range of new capabilities are being developed for flexible L2 services to meet requirements of such local and wide area deployments. Various technologies being developed include virtualized L2 services such as Virtual Extensible LAN (VXLAN), the IETF's locator/ID separation protocol (LISP), the IETF's Stateless Transport Tunneling (STT), the IETF's virtualization using generic routing encapsulation (NVCRE), the IETF's Network Virtualization Overlays initiative (NVO3), the IETF's Generic Network Virtualization Encapsulation (GENEVE), Multi Protocol Label Switching (MPLS), Virtual Private LAN Service (VPLS—Ethernet type multipoint to multipoint using IP or MPLS), Advanced VPLS (A-VPLS), Hierarchical VPLS (H-VPLS, i.e., using Ethernet bridges at edge and MPLS in the core), Pseudowire (PW—emulation over L3), PW over MPLS (PWoMPLS), PW over Generic Routing Encapsulation, a tunneling protocol (PWoGRE), PW supporting Virtual Forwarding Interfaces (VFI), Overlay Transport Virtualization (OTV), IETF Transparent Interconnection of Lots of Links (TRILL)—link state routing using a routing bridge or TRILL switch, IETF Layer Two Tunneling Protocol (L2TPv3), and others.

Also, many options are being developed for implementing virtual L2 networks that can be controlled by SDN techniques, for example, L2 VLAN Provider Bridge (802.1Q tunneling or Q-in-Q), Provider Backbone Bridge (PBB - MAC-in-MAC), MEF Access Ethernet Private Line Service (Access-EPL), MEF Ethernet Virtual Private Line (EVPL port-based point-to-point), MEF Ethernet Private LAN (EP-LAN), for port-based multipoint-to-multipoint, MEF Ethernet Private Tree (EP-Tree), MEF Ethernet Private Tree (EP-Tree), port-based rooted-multipoint, MEF Ethernet Virtual Private Tree (EVP-Tree), and MEF Ethernet Virtual Private LAN (EVPLAN).

Even though some of these capabilities are being developed for local (e.g., metro) deployments, they eventually will extend throughout the world. This attribute of extensibility world-wide is the focus of this chapter, which describes how these attributes will characterize international networking services and infrastructure at a global scale.

4 Global Environment for Network Innovations (GENI) and Related Initiatives

The GENI initiative, which was established by the National Science Foundation's Computer and Information Science and Engineering (CISE) Directorate, was formulated within the context of the policies of that organization, including those related to international partnerships [5]. Similarly, the European Union's Future Internet Research and Experimentation (FIRE) project has funded a number of major network research testbeds throughout Europe. Within and external to the FIRE program, Europe has established multiple research testbeds, including BonFIRE [6], PHOSPHORUS [7], OFELIA,[8] which is OpenFlow based [9], GEYSERS an optical integrated testbed for "GEneralized architecture for dYnamic infrastructure SERviceS" [10], FEDERICA [11], and the G-Lab testbeds [12]. G-Lab, which is presented in one of the chapters of this book, and one of the experimental areas presented in a section in this chapter, has a wide ranging agenda including research projects on a functional composition concept for a dynamic composition of functional blocks on network and service level, topology management, and investigations of federation concepts to interconnect with international Future Internet Testbeds. The European Future Internet project, FED4FiRE, is developing federation techniques for networking testbeds [13]. GENI and FIRE have been federated. In China, the Chinese Academy of Science has established network research testbeds, such as the Sea-Cloud Innovation Environment, through the China Science and Technology Network (CSTnet) to support future network research. In Taiwan, the National Center for High-Performance Computing in partnership with the Taiwan Advanced Research and Education Network (TWAREN) has established multiple network research testbeds related to Future Internet initiatives. In Japan, National Institute of Information and Communications Technology (NICT) has supported the New Generation Network initiative for many years, particularly through projects based on the JGN-X testbed. In addition, individual institutions have established a number of related projects such as the National Institute of Advanced Industrial Science and Technology's G-Lambda project [14] and the University of Tokyo's V-Node project. In Korea, the K-GENI project has been established as a persistent multi-topic research testbed [15]. In South America, the primary area of focus for these projects has been multiple Brazilian Future Internet research and development projects, including FIBRE [16].

In Canada, the Strategic Network for Smart Applications on Virtual Infrastructure (SAVI) was established as a partnership of Canadian industry, academia, research and education networks, and high performance computing centers to investigate key elements of future application platforms [17]. SAVI, which has designed and currently operates a national distributed application platform testbed for creating and delivering Future Internet applications, is described in a chapter of this book. The primary research goal of the SAVI Network is to address the design of future applications platform built on a flexible, versatile and evolvable infrastructure that can readily deploy, maintain, and retire the large-scale, possibly short-lived, distributed applications that will be typical in the future applications marketplace. GENI has been federated with SAVI, which supports multi-domain interoperability. In additional, GENI has been federated with several cloud testbeds, such as those supported by the NSFCloud program, i.e., Chameleon and Cloudlab, which are described in a later section of this chapter.

5 Basic Architectural Considerations

The GENI environment is a distributed instrument, which can be used by researchers to discover and claim resources ("slivers"), to integrate those resources within private research environments, "slices," conduct experiments using that slice, measure and analyze results, and, importantly, reproduce specific results. Note that the GENI architecture is discussed in a chapter in this book. The ability to replicate experiments is a key research requirement. A primary component of the GENI environment consists of the SDN/OpenFlow architecture, protocols and technologies [9]. The OpenFlow model is part of an instantiation of a number of broad architectural concepts.

Currently deployed digital communication services and technologies comprise the most significant advances in the history of communications. At the same time however, they are based on architectural approaches that are beginning to demonstrate major limitations that restrict future progress. For example, in today's networks, control and management functions are implemented with an assumption that the communications environment will be fairly static—that they will remain unchanged for long periods of time. Network control planes have only limited scope based on minimal state information, such as neighbor connections, reachability, and access policy. This approach cannot meet rapidly changing requirements and demands for on-going dynamic service and technology enhancements as well as for quick adjustments to network resources in response to changing conditions. Consequently, a new architectural approach is being developed to provide increased capabilities for programmable networking, especially the set of techniques termed Software Defined Networking (SDN). This model, which separates the control plane from data plane, provides for programmability and a higher level of network control abstraction and enables a more comprehensive overview of network state information, which can be used to dynamically control networks services and resources.

This comprehensive overview is made possible by an ongoing dialogue between network devices, including individual components, and controllers. Instead of having state information confined to individual devices within the network, this information is gathered by logically centralized controllers. The network devices send the controllers state information and the controllers use that information to make decisions on dynamically matching demand requirements with resources, on solving problems such as sudden congestion, and allocating resources in anticipation of demand because of network behaviors. These decisions are signaled back to the network devices for implementation, for example, by programming flow tables in network devices. Using this approach, network devices can be considered undifferentiated component resources, and the specialized capabilities can be provided by the control plane. Because these controllers are logically centralized they have a global view of the network and, consequently, they can provide for much better traffic optimization than is possible using traditional distributed approaches.

Currently, the most common implementations of these concepts are based on OpenFlow, which enables controllers to have access to a set of network primitives. The actual capabilities for programmable networking are provided by control frameworks. The general GENI architecture and its several major control frameworks (ORBIT, ProtoGENI, PlanetLab, the Open Resource Control Architecture—ORCA, and the GENI Experiment Engine - GEE) as well as an SDN implementation and also a general Aggregate Manager API developed to integrate these frameworks and its mesoscale facility implementation are described in other parts of this book.

Just as the GENI AM integrates and coordinates the control frameworks within its environment, it is possible to consider a type of international aggregate manager that functions more universally, including as part of a multi-domain federation comprised of many international network research testbeds. There are many techniques available that can be used to support such interdomain interoperability, including using the GENI control frameworks, such as ProtoGENI/InstaGENI, ORCA/ExoGENI, and combinations.

This chapter explores the implications of such federations, particularly the special considerations required for international network connections among multi-domain federations of major experimental network environments. A primary set of considerations with regard to such international federations relates to the interoperability of, and in some cases, the direct integration of, the individual control frameworks for the individual environments being connected. An important, closely related, set of considerations concerns multi-domain international connections as opposed to those dealing with international single domain networks. The majority of the core architectures currently being used for these environments, such as SDN/OpenFlow, are single domain and not multi-domain oriented. Because highly distributed multi-domain models are much more complex than single domain, the majority of the topics discussed in this chapter relate to multiple domain architecture and implementations.

To accomplish such federations, a number of elements are required, including those providing resources that can be made available to experimenters, implementing a means for advertising those resources, discovering them, claiming them, e.g., through reservations, discarding them after their use, and managing the components of the federated environment, including addressing problems. In addition, other mechanisms are required, e.g., for designing and implementing state machines, interacting with those state machines, signaling messages, interpreting those messages, sharing topologies, determining path finding, path stitching, deploying and using resource ontology and schemas, gatekeeper interfaces, federation gateways, SDN exchange points (SDXs), and more. In addition, given the objective of inter-domain resource sharing, processes must be implemented with appropriate policy to drive security mechanisms including those for authentication and authorization. Overall, mechanisms must be established for APIs, secure signaling, resource identification, advertising and discovery, trust relationship management, trust root services, federation policy enforcement, certification, monitoring and analytics, and related functions.

Although all major network research environments in various countries undertake to design and implement these basic capabilities, all approaches are somewhat different. However, because of the recent progress in virtualization of networks and in control plane capabilities, opportunities exist to develop such federations despite such differences among architectural approaches.

6 Creating a Common International Network Language and Network Programming Languages

In order to facilitate projects spanning different services, domains, and infrastructures, methods must be created that allow exchange of information about available resources and state information so that such resources can be requested and allocated for use by network services and applications. One such infrastructure information model is the Network Description Language (NDL) [18] pioneered by the University of Amsterdam that forms the basis of the Network Markup Language Workgroup in the Open Grid Forum (OGF). The NDL provides a method to describe computer networks in a meaningful way. The NDL ontology for computer networks uses the Resource Description Framework (RDF) [19]. With this ontology one can create a simple, clear, understandable description of a network. The Network Description Language (NDL) helps to reduce the complexity issues in computer networks. The goal of NDL is to allow not only network processes but also applications to have a better understanding of the network so they can more easily adapt it to their needs. NDL has been extended to include descriptions of computing and storage resources: the Infrastructure and Network Description Language (iNDL) [20].

Some research groups are interested in extending these types of languages to resources beyond networks. For example, the ORCA-BEN [21] project developed the NDL–OWL model, which uses the Web Ontology Language (OWL) instead of RDF, extends NDL to include cloud computing, in particular, software and virtual machine, substrate measurement capabilities and service procedures and protocols. This ontology models network topology, layers, utilities and technologies (PC, Ethernet, DTN, fiber switch) based on NDL. In comparison, INDL uses the latest developments in the OGF NML-WG. Standards organizations are continuing to evolve such languages so that there can be meaningful information exchanges among infrastructure related processes. Such standards are particularly important for network inter-domain provisioning.

Programmable networking requires a data modeling language. Network management protocols generally have related data modeling languages. For example, the first Internet network management tool, Simple Network Management Protocol (SNMP), utilizes Structure of Management Information (SMI), incorporating Abstract Syntax Notation One (ASN.1). When the IETF was developing the NETCONF protocol, it was observed that a data modeling language was needed to define data models manipulated by that protocol. In response, the IETF developed YANG ("Yet Another Next Generation"), a modular data modeling language for the NETCONF network configuration protocol.

The following sections describe several approaches being undertaken by research communities to create techniques for inter-domain federations. The sections immediately following this one describe some of the international foundation resources, especially those based on lightpaths implemented within world-wide optical fiber, that are being used for that research.

7 Existing International Facilities

7.1 Global Lambda Integrated Facility (GLIF)

One major global facility that is being used to support multiple distributed international environments for network research is the Global Lambda Integrated Facility (GLIF) [22]. The GLIF is a world-wide distributed facility designed, implemented and operated by an international consortium, within which participants can create many types of customized services and networks, including those that are required to support international network research environments. Unlike most communication exchange facilities today, which interconnect only at Layer 3, the GLIF was designed to enable networks to exchange traffic at all layers, including Layer 1. GLIF is based on a foundation of owned and/or leased optical fiber paths and lightpaths within optical fiber, including trans-oceanic fiber. Lightpaths are created and managed through technologies and services based on Dense Wavelength-Division Multiplexing (DWDM), which supports multiple parallel high performance, high capacity channels.

The GLIF environment is highly complementary to the GENI environment because it was designed for network programmability (Fig. 1). GLIFdomains are interconnected by the GLIF exchange facilities—Open Lambda Exchanges (GOLES), which have implemented different types of control frameworks. Current GLIF exchange points are: AMPATH (Miami), CERNLight (Geneva), CzechLight (Prague), Hong Kong Open Exchange - HKOEP (Hong Kong), KRLight (Daejoen), MAN LAN (New York), MoscowLight (Moscow), NetherLight (Amsterdam), NGIX-East (Washington, DC), NorthernLight (Copenhagen), Pacific Wave (Los Angeles), PacificWave (Seattle), PacificWave (Sunnyvale), SingLight (Singapore),

Fig. 1 Global Lambda Integrated Facility (GLIF)

SouthernLight (São Paulo), StarLight (Chicago), T-LEX (Tokyo), TaiwanLight (Taipei), and UKLight (London). A related international facility is the Global Ring Network For Advanced Applications Development (GLORIAD), which is directly interconnected with the GLIF and supports international network testbed research and other application level projects [23]. A consortium including the US, Russia, China, Korea, Canada, The Netherlands, India, Egypt, Singapore, and the Nordic countries supports GLORIAD.

8 Network Service Interface—NSI Connection Service

Because the GLIF community is comprised of multi-domains and uses multiple different control frameworks for dynamic provisioning, an initiative was established by that community with a standards development organization—the Open Grid Forum (OGF)—to create architectural standards for a generic network service interface (Network Service Interface—NSI) as an API to the multiple control frameworks within the GLIF environment, including an NSI Connection Service (currently v 2.0) [24]. When the GLIF exchange facilities around the world were implemented (GLIF Open Lambda Exchanges or GOLEs), they were established with multiple different control frameworks for resource management and control, for example, DRAC (Dynamic Resource Allocation Controller), Autobahn, Argia, OSCARS (On-Demand Secure Circuits and Advance Reservation System), G-Lambda, and many others. In other words, each open exchange point had a different control framework for reserving and establishing links through the exchange point. The NSI enables these paths to interconnect. This service was designed specifically to assist the creation of multi-domain connections across international networks and through these exchange points using a common API that would allow provisioning across multiple networks operated by many different national research and education network organizations. The NSI specifies signaling, state processes, messages, protocols, and other environmental components. A process or application at the network edge can discover and claim network resources, at this point primarily paths, within an environment comprised of heterogeneous multi-domains [24].

After several years during which the GLIF NSI participants demonstrated persistent international testbed capabilities, especially through the AutoGOLE series of demonstrations (Fig. 2), NSI implementations are being placed into production for a number of national R&E networks. This figure illustrates available VLANs implemented among multiple GLIF GOLEs that are available for use by communities participating in the AutoGOLE initiative.

Although this service is not yet being used extensively today to support network research environments, it is worth mentioning here because plans are being developed to do so, and because the NSI connection standard already has developed many of the mechanisms required to support interconnectivity among multi-domain network research environments. Also, several GOLES have started to incorporate SDN/OpenFlow capabilities to support L2 based traffic.

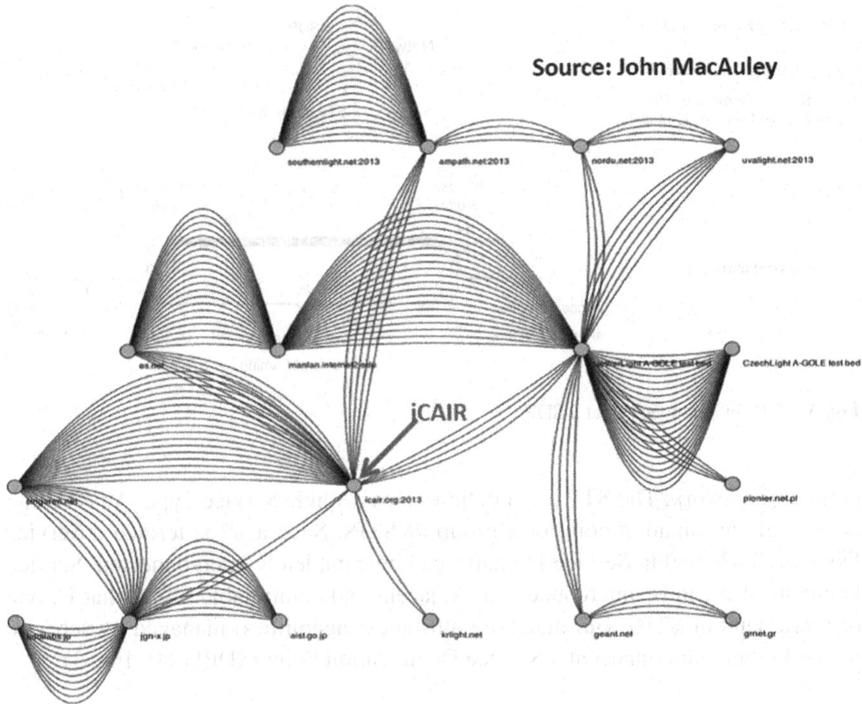

Fig. 2 AutoGOLE SC13 demonstration schematic showing addressable, programmable paths among global GLIF Open Lambda Exchanges (GOLEs)

The NSI initiative established a working group that has examined many architectural issues within the Network Services Framework (NSF) specified in the OGF GWD-R-P Network Service Framework v2.0. This framework defines an outline for a set of protocols. The NSF expects that resources and capabilities can be advertised externally through defined Network Services, and it defines a unified model for how various processes should interact with such services, for example, creating connections (Connection Service), sharing topologies (Topology Service) and performing additional services required by a federation of software agents (Discovery Service). NSI allows for implementing network paths across multiple network domains operated by disparate network providers, enabling federations. The NSI architecture specifies Service Termination Point (STP) objects, which are used by connection requests to determine connection attributes. The STP is a means to abstract resource functionality at the point where NSI services terminate from actual underlying physical resources and configurations, such as nodes or circuits. Such abstraction made possible by STP allows for use of functional options multi-domain transport termination without forcing users to deal with the complexity of the physical infrastructure and configurations at the termination points. An STP is the designation of a specific topological location that functions as the ingress/egress

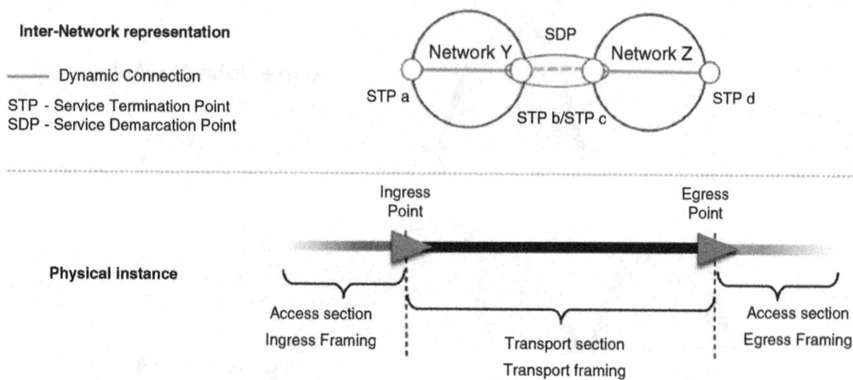

Fig. 3 STPs interconnecting at a SDP

point of a network. The STP has a definition as a single Service Type. An STP can be a single termination point or a group of STPs. Such a set is termed a Service Domain. STPs within Service Domains can be completely interconnected. Service Domains also can be interconnected. Adjacent and connectable STPs (that is, one or more pairs of STPs with matching attributes/capabilities) managed by separate networks can interconnect at a Service Demarcation Point (SDP) [24] (Fig. 3).

9 The International GENI (iGENI) and the International Advanced Network Research Facility

The International GENI (iGENI) initiative was established to create a federated international network research testbed facility. A number of iGENI participants and other international network research partners collaborated in the designed and implementation an International Advanced Network Facility based on SDN/OpenFlow as a research platform to provide network research communities with worldwide experimental resources, including addressable transport paths, that can be used to investigate many different types of topics [25]. The platform also provides options for closely integrating programmable networks with programmable clouds [26]. The figure below depicts a topology that was showcased at SC11, SC12, SC13, SC14, SC15, at multiple GENI Engineering Conferences, at other technical and research workshops and at other events (Fig. 4).

This international federated testbed was designed to enable researchers to discover, claim, integrate, use, and discard a variety of diverse resources, including core network resources, as slices across a shared infrastructure fabric. Each site has a collection of resources that are interconnected with a mix of dynamic and static L2 VLANs. Tools and methods for undertaking these tasks through orchestration frameworks are being developed. Such orchestration is one of the components

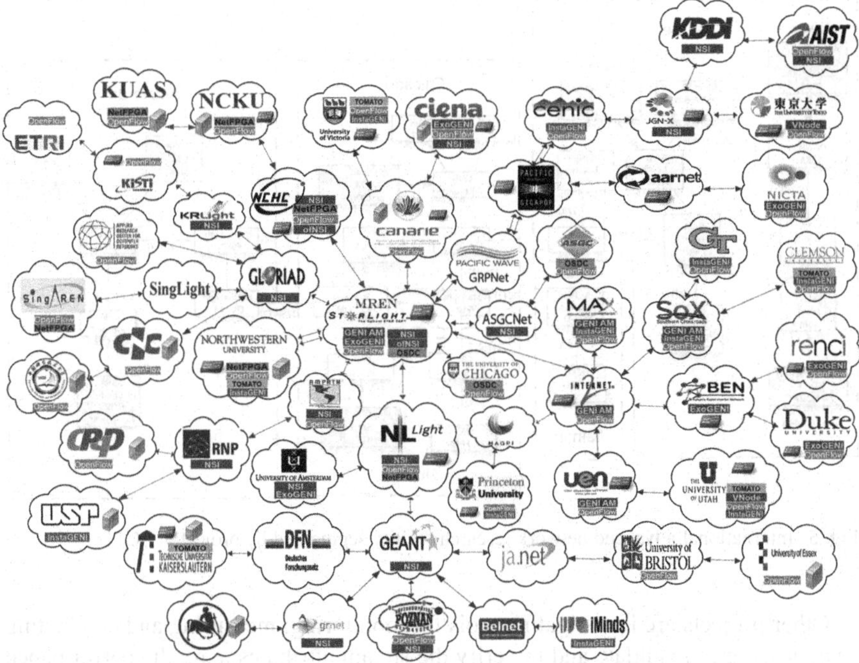

Fig. 4 International GENI (iGENI)/International OpenFlow Testbed Topology

of a hierarchical architectural stack, with edge process signaling, which could be application signaling as well as system process signaling, at the top of the stack.

This international federated testbed is being used to explore various techniques for designing and implementing orchestration processes, used to discover and claim segments of network resources, including full topologies, and options for configuring those segments, to dynamically provision paths and endpoints, and to specify the specific attributes of the services that they create. Some projects are focused on developing northbound access to the processes that control and manage the actual resources, discovery, claiming, accessing state information, configuring, provisioning, etc. Other projects are investigating southbound interfaces that provides network resource request fulfillment and state information on resources.

Several projects are investigating eastbound and westbound interfaces (E-W), which are key resources for establishing federated interoperable paths across multiple domains, for example, supporting message exchanges among controllers, including those on reachability across multiple domains, controller state monitoring and fault condition responses, and multi-domain flow coordination. E-W interoperability mechanisms for federation are discussed in more detail in a later section of this chapter. Currently, the most widely deployed E-W federation protocol at L3 is the Border Gateway Protocol (BGP), an Autonomous System (AS) peering path vector protocol used to support TCP/IP network exchanges.

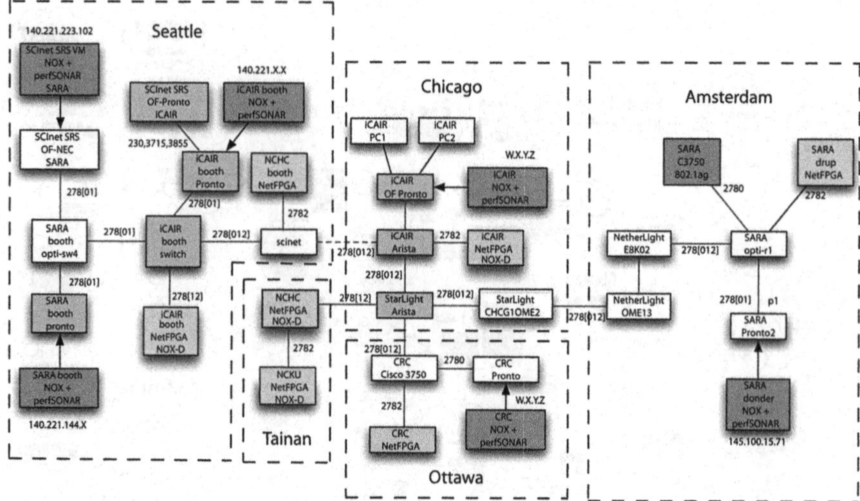

Fig. 5 International advanced network research facility segmented by project at SC11

Other projects are investigating tools for monitoring, measuring and conducting analytics, e.g., to validate and to verify the stream attributes, e.g., the performance of new types of communication services and to provide stream behavior real time information to traffic engineering and optimization processes.

Resources made available through this platform can be made available to multiple groups of researchers at the same time so that they can conduct experiments without interfering with each other. The figures below illustrate this type of resource segmentation on this platform used for SC11, SC12, and SC13 (Figs. 5, 6, and 7). Each color represents a different project undertaken by a difference research group. The figures give an indication of the distribution of resources across the various projects. Subsequent sections describe research projects that have been or that are being conducted on this platform.

10 Research Activities and Experiments Conducted Among Current International Environments

10.1 Slice Around the World Initiative

Different network research communities are developing various techniques for virtualizing distributed environments and networks, for integration, for developing control frameworks, designing network middleware and for integrating resources. Consequently, federation among these capabilities has become a major research topic.

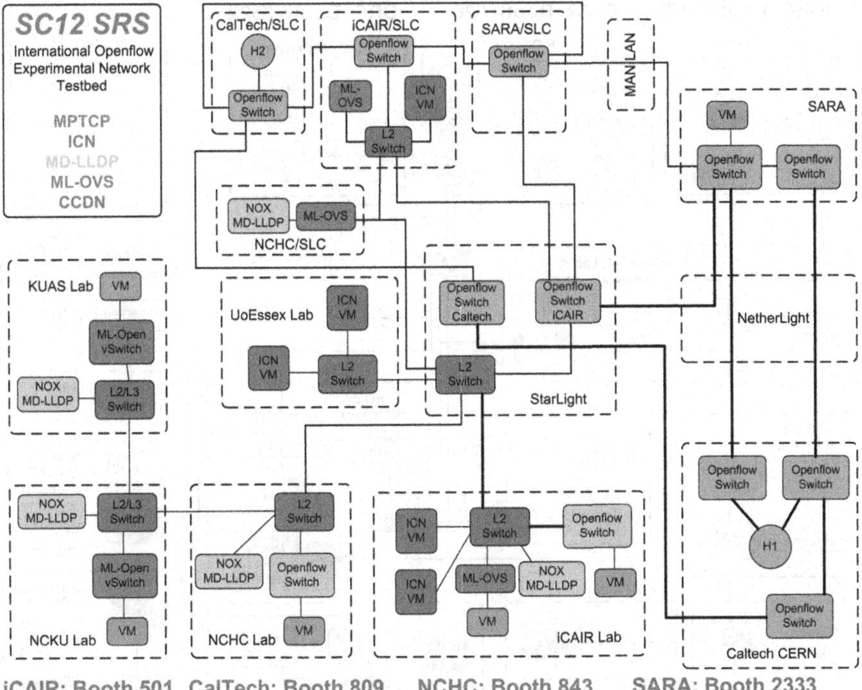

Fig. 6 International advanced network research facility segmented by project at SC12

The "Slice Around the World" initiative was established to both provide a large scale research platform and to demonstrate the powerful potential of designing and implementing world-wide environments consisting of multiple federated international computational and storage clouds closely integrated with highly programmable networks. As a basic capability, this initiative created a distributed, integrated OpenFlow environment interconnected through a customized international network. All sites have servers capable of supporting addressable VMs. Among the sites, there is a blend of static and dynamic resources. Various aspects of the design for this initiative were considered, including three primary components: (a) showcasing one or more application capabilities, for example, some aspect of federated cloud based digital media transcoding and streaming as opposed to merely showing bit-flow graphs, (b) demonstrating the capabilities of programmable networks using OpenFlow, and (c) designing a network architecture based on an international foundation infrastructure. Each of these components is further described in a subsequent section of this description. Also, participants in this initiative are developing a number of innovative architectural and basic technology concepts. Goals of the project include providing striking visuals, reflecting the potentials of a truly global environment, closely integrating programmable networking and programmable compute clouds, show capabilities not possible to

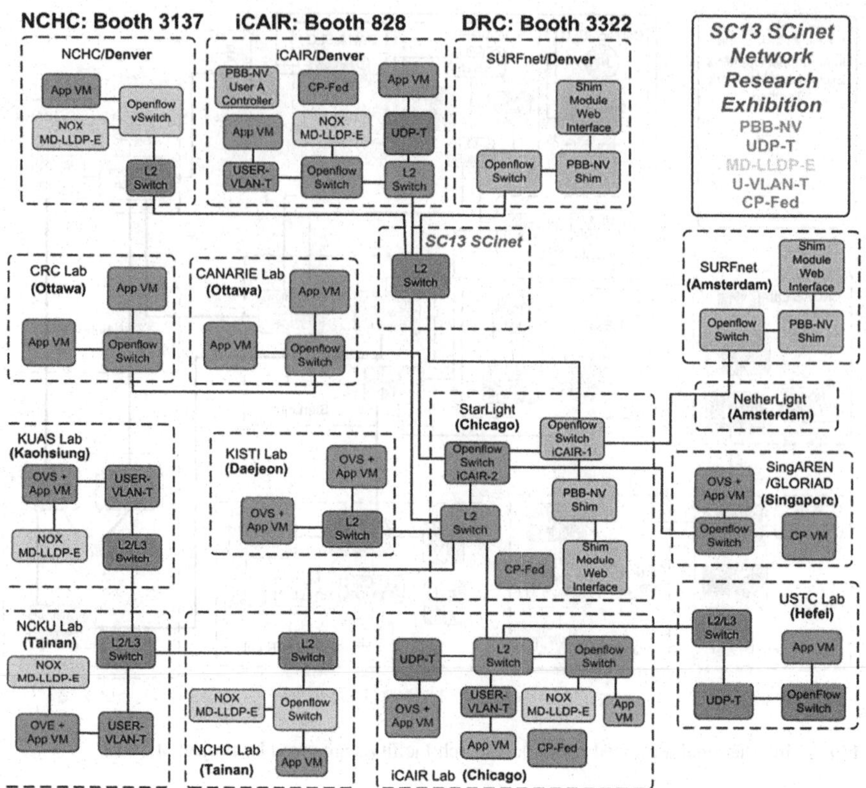

Fig. 7 International advanced network research facility segmented by project at SC13

accomplish with the general Internet or standard R&E network, highlighting the power of programmable networks, especially customization at the network edge, and showcasing a potential for resolving real current issues.

For example, for several Slice Around the World (SATW) demonstrations, Finite Difference Time Domain (FDTD) distributed simulation visualization capabilities have been demonstrated using the SATW distributed environment. FDTD is one of most commonly used computational electrodynamics modeling techniques for many research and industry simulations, such as LSI design electro verification. Under current HPC workflow techniques, researchers submit jobs, retrieve results, visualize those results and then resubmit the job with modifications, additional information, data, etc. Today this is a tedious, manual slow process, in part, because of the limitations of today's networks. These SATW demonstrations showed how by using dynamically programmable networks closely integrated with computational and storage clouds, it is possible to provide capabilities that can be used to create interactive simulation/visualization instruments to significantly improve this traditional process. Interactive real-time simulation/visualization instruments included: (a) distributed back-end MPI rendering clusters and storage, (b) a web front end to setup control parameters for rendering and to display the result, (c) customized web server to pipe rendering results to users efficiently, and (d) a program to check the rendering result and submit jobs if the results were not produced. Web interfaces were used to dynamically identify the sites around the world, where the simulation images are located, to convert the request and to send the request to the appropriate host over the private international network, and interactively visualize the simulation over the private network specifically designed for the demonstrations.

Another series of SATW demonstrations used the TransCloud international distributed testbed incorporating programmability for a range of resource infrastructure [27]. The TransCloud is a world-scale high-performance cloud testbed, incorporating a lightweight slice based federation architecture and a slice-based federation interface, with high-performance dedicated intersite networking enabling high-bandwidth data transport between physically distributed sites, the use of experimental transport protocols and guaranteed QoS among distributed clouds, and lightweight, robust isolation between components.

TransCloud supports both network researchers and researchers developing new types of efficiently managed virtualized computing aggregates, including researchers creating extensions of cloud control environments such as Eucalyptus, Tashi, OpenStack, and VICCI.

The initiative was established by a number of network research centers and labs that are participating in multiple next generation networking activities, including those developing large scale distributed experimental network research environments. Participants have included ANSP, the Applied Research Center for Computer Network at Skolkovo, Chinese Academy of Sciences/CSTNET, the Communications Research Center, Ottawa, CPqD, Duke University, ETRI (Electronics and Telecommunications Research Institute), G-Lab, TU Kaiserslautern, Hewlett Packard Research Labs, the International Center for Advanced Internet Research at Northwestern University, KISTI, KUAS/NCKU, NCHC, NICT, NICTA, Princeton University, the Renaissance Computing Institute, RNP, SURFsara, the University of Amsterdam, the University of Essex, the University of Tokyo, and the University of Utah.

11 International V-Node

Another major Slice Around the World initiative was the International V-Node project, organized by the University of Tokyo. The V-Node (Virtual Node) initiative was established to enable deeply programmable networks, especially for experimenting with arbitrary protocols, by creating extremely virtualized infrastructure, including supporting multi-domain implementations. The V-Node architecture provides for federated multi-domain control and data planes and for federation among multiple virtualization platforms. Architectural components include a Gatekeeper (GK) and a Federation Gateway (FGW), which provide for translating API messages, ensuring common data for APIs, and packet formatting. The architecture also includes a Slice Exchange Point (SEP), which supports bridge commands, control frameworks, and policies. One international V-Node SATW demonstration implemented a V-Node at the University of Utah next to a ProtoGENI node, demonstrating the integration capabilities of the V-Node federated functions. Another demonstration, based on the V-Node architecture demonstrated an innovative packet caching technique that provided for hashing of data packets to enable optimized data transport and routing, that is, converting data to hash values and determining responses based on the analytics of those values.

12 ToMaTo a Virtual Research Environment for Large Scale Distributed Systems Research

Multi-domain federations require topology management tools. The topology management tool (ToMaTo) [28, 29] has been developed as part of the German-Lab project [30], which has been funded by the German Ministry of Education and Research (BMBF) to provide a virtualized research environment for networking experiments. This tool has been implemented on a large scale international testbed. Currently, the ToMaTo testbed is run by a consortium and academic institutions can join the testbed without cost if they contribute resources. Therefore, ToMaTo continuously grows and already spans across multiple continents.

ToMaTo is a topology-oriented networking testbed designed for high resource efficiency, i.e., high parallelism where possible but high realism where needed. Topologies consist of devices (produce and consume networking data) and connectors (manipulate and forward data). Devices contain a set of networking interfaces that can be connected to connectors. Figure 8 shows a simple topology consisting of five devices (one central server and four clients) and three connectors (two switches and one Internet connector) (Fig. 8). To increase both flexibility and resource efficiency, ToMaTo offers different types of devices and connectors. Users can choose between hardware virtualization, which provides an environment nearly identical to a real computer but has a high resource usage, and container virtualization that uses fewer resources but does not suit all needs.

The default connector type is a VPN connector with a selectable forwarding policy (hub, switch or router). Public services, cloud resources or even other testbeds can be combined with ToMaTo topologies by using external network connectors. To help users with running their networking experiments, ToMaTo offers an easy-to-use, web-based front-end with an intuitive editor. Users can control their devices directly from their browser or using a VNC client of their choice. Advanced tools like link emulation and packet capturing are included in ToMaTo and can be used to run experiments (Fig. 8).

The ToMaTo software consists of three tiers as shown in Fig. 9. The hosts provide virtualization technology and a complete toolset needed for advanced features like

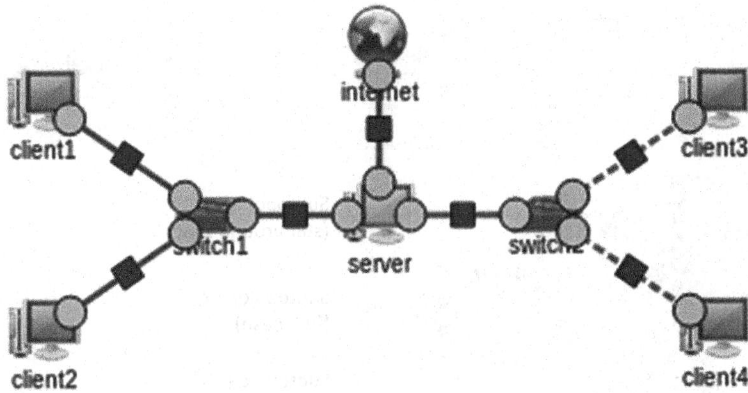

Fig. 8 Example topology

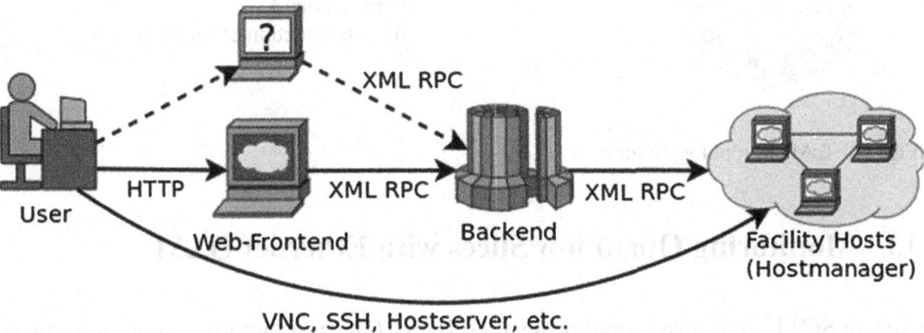

Fig. 9 ToMaTo's three tiers architecture

link emulation, packet capturing, etc. The back-end component contains all the control logic of the ToMaTo software and remotely controls the hosts. Different front-ends use the XML-RPC interface provided by the backend component. The most important front-end is the web- based user front-end that allows users to edit and manage their topologies from their browser using modern web technologies. Other front-end software includes a command line client that allows easy scripting and an adapter for the other federation frameworks.

One of the key features of ToMaTo is its graphical user interface that allows even inexperienced users to create complex network topologies by drag and drop. ToMaTo features an easy-to-use editor (Fig. 10) to create, manage, and control networking topologies. With this editor, users can configure their topology components and control them in the same intuitive interface. The editor also gives users direct access to their virtual machines using web-based VNC technology.

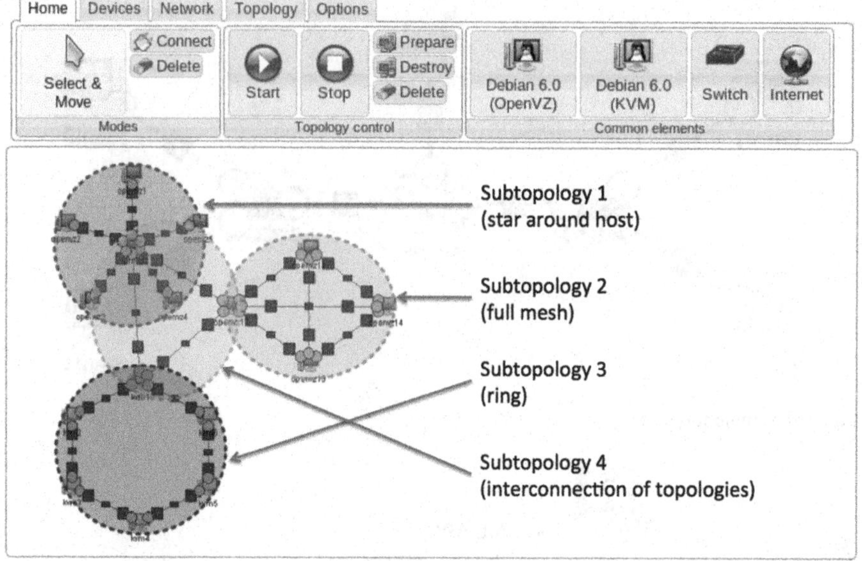

Fig. 10 ToMaTo topology editor

13 Monitoring OpenFlow Slices with Ethernet OAM

During SC11, an international project led by SARA, and including several partners (SURFnet, iCAIR and CRC) showed how the IEEE 802.1ag Ethernet OAM protocol could be used in an OpenFlow controller to monitor and troubleshoot an OpenFlow slice. The slice spanned two continents and included a couple of strategically placed OpenFlow switches in Amsterdam (NetherLight), Chicago (StarLight), Ottawa and the venue in Seattle (Fig. 5). The SDN NOX OpenFlow controller was used to manage the OpenFlow switches. An open source implementation of the 802.1ag protocol was added as a module to the NOX SDN controller. Using this implementation, the controller could send Ethernet OAM hello frames between the switches. When there was a "fiber cut," which was staged as part of the demonstration, these hello frames were lost and the OpenFlow switch at the other side of the link detected this fault and reported the link as being down. A monitoring website periodically retrieved the link status of all links from the controller and showed the status of the network in real time on a web page.

14 Multipath TCP (MPTCP)

During SC12, an international project led by SARA, and including several partners (Caltech, iCAIR, and SURFnet) demonstrated a technique for supporting large eScience data transfers over a multipath OpenFlow network [31, 32]. Data sets in eScience are increasing exponentially in size. To transfer these huge data sets, it is important to make efficient use of all available network capacity. Usually, this means using multiple paths when they are available. In this demonstration, a prototype of such a multipath network was implemented. Several emerging network technologies were integrated to achieve the goal of efficient high end-to-end throughput. Multipath TCP was used by the end hosts to distribute the traffic across multiple paths and OpenFlow was used within the network to support the wide area traffic engineering. Extensive monitoring was part of the demonstration. A website showed the actual topology (including link outages), the paths provisioned through the network and traffic statistics on all links and the end-to-end aggregate throughput.

Multipathing is usually undertaken based on flows by calculating a hash (including, e.g., Ethernet addresses, IP addresses, and TCP/UDP port numbers) of the packets. Flows with the same source and destination follow the same path. When the traffic has many different flows, the traffic will be evenly balanced over the different paths. But all the paths need to have the same bandwidth. Another disadvantage is that in large data eScience applications there are typically only a few flows and hashing will not spread the load evenly across the paths in those cases. Multipath TCP (MPTCP) tries to solve these limitations. MPTCP can handle paths of different bandwidth because there is a congestion control mechanism across the subflows. This congestion control mechanism also makes sure that traffic on a congested link is moved to a link with less congestion. Therefore, it adapts the load balancing according to the load of other traffic on the links. In this demonstration, MPTCP was used in combination with an international OpenFlow based multipath network. Data was transferred from CERN in Switzerland through the StarLight International/National Communications Exchange Facility in Chicago to the SC12 venue in Salt Lake City. OpenFlow switches were placed at CERN, NetherLight, StarLight and the SC12 booths of Caltech, iCAIR and the Dutch Research Consortium (Fig. 6). An OpenFlow application connected via a controller to the OpenFlow switches and automatically discovered the topology via LLDP. The application calculated multiple paths between the servers and the forwarding entries for the flows were pushed to the OpenFlow switches. The demonstration showed the success of using MPTCP for large scale data transport.

15 Provider Backbone Bridging Based Network Virtualization

During SC13 in Denver, SURFnet and iCAIR demonstrated how Provider Backbone Bridging (PBB) could be used as a network virtualization technology in OpenFlow networks (see Fig. 5). An important use of Software Defined Networking (SDN) is network virtualization or slicing. This technique allows multiple users to be supported by the same physical infrastructure, with each having their own virtual network or slice. FlowVisor is one of the options to achieve this result. FlowVisor is a software module that is implemented between an OpenFlow switch and OpenFlow controllers and it gives each controller a part (slice) of the flowspace. The disadvantage of this approach is that controllers do not have access to the full OpenFlow tuple space and therefore the capabilities are less than if direct access to a physical OpenFlow switch was provided. Also, at this time, FlowVisor supports OpenFlow 1.0 only and not later versions of the protocol, such as OF v.1.3.

In this demonstration, Provider Backbone Bridging (PBB) as defined in IEEE 802.1ah was used, as encapsulation technology in the OpenFlow data plane. In this way user traffic was separated and identified by the I-SID in the encapsulation header. The data part was the user's original Ethernet frame and users could create OpenFlow rules that matched on any of fields that OpenFlow 1.3 supports, except for the PBB $\sim$ I-SID because this element is used to map packets to users. This approach is a simple virtualization method that gives users access to virtual OpenFlow switches that have the same OpenFlow capabilities as physical OpenFlow switches would have. During SC13 in Denver, this PBB based network virtualization was shown on the OpenFlow enabled link between NetherLight in Amsterdam and the StarLight facility in Chicago. Pica8 3920 OpenFlow switches were used at both sides. These switches supported OpenFlow 1.3 and PBB encapsulation and decapsulation.

16 The Sea Cloud Innovation Environment

The Sea-Cloud Innovation Environment (SCIE), a national wide testbed initiative supported by the "Strategic Priority Research Program—New Information and Communication Technology" (SPRP-NICT) of the Chinese Academy of Sciences, is focused on building an open, general-purpose, federated and large-scale shared experimental facility to foster the emergence of new ICT. Recently, plans have been discussed to extend this environment to international sites using federation techniques over international network facilities such as GLIF and GLORIAD. To support the principle proposed for adaptive service-oriented experimentation, SCIE developed a wide-area testbed with hardware resources including servers, cloud services, and storage resources located geographically at five cities including Beijing, Shanghai, Shenyang, Wuxi, Hefei in China (Fig. 11).

The SCIE is connected by Smart-Flow devices which support the Open Flow 1.2 protocol. The GRE-tunnel enabled Smart-Flow device can encapsulate layer-2 protocol inside virtual point-to-point links over an Internet. Moreover, QoS feature was to Smart-Flow device to offer link efficiency mechanisms that work in conjunction with queuing and traffic shaping to improve efficiency and predictability of virtual links. To decrease the complexity of experiment device deployment, SCIE Rack was designed and prototyped. Like the ExoGENI and InstaGENI, the SCIE Rack provides integrated control and measurement software, network, computation, and storage resources in a single rack.

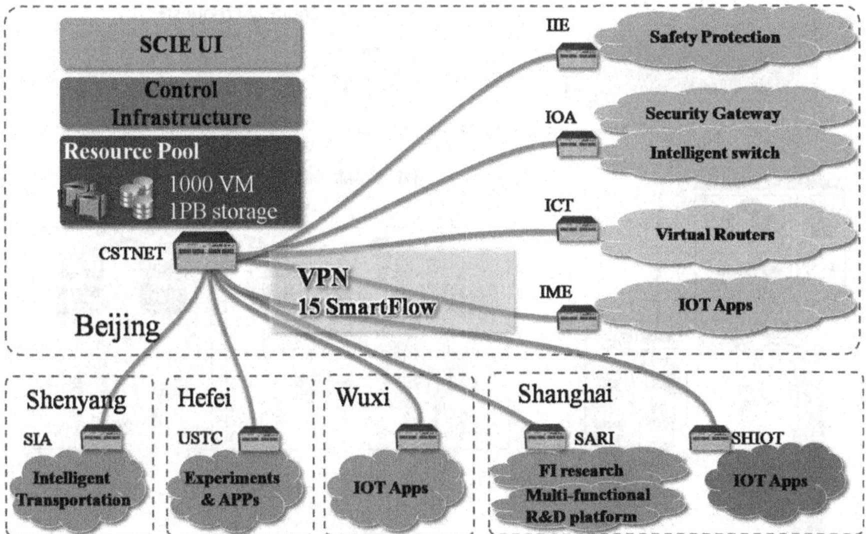

Fig. 11 The deployment of SCIE

Beyond the network and hardware layer, the SCIE offers distributed resource control, experiment measurement and an experiment service system, which was developed by the CSTNET research and development team. This system gives researchers a graphic user interface that can be used to design and use virtual network topologies consisting of virtual devices and links over the SCIE testbed. A distributed experiment traffic analysis measurement tool offers researchers a means to subscribe traffic statistics and virtual link performance data by BPF-like syntax. Moreover, to help researchers control their experiments on SCIE, Java and Python development libraries are provided for researchers who can run their experiment codes on SCIE experiment playground or their own devices. The SCIE architecture is shown as Fig. 12.

The key technologies of SCIE can be summarized as follows:

Resource virtualization: the SCIE uses both slice-based and time-based virtualization to handle sliceable and un-sliceable resources, respectively. Sliced resources can be requested and dynamically created for researchers, where the request description and visual rendering for experimentation can be achieved in SCIE (see Fig. 13).

Request description and visual rendering for an experimentation using SCIE.

High-level experiment work flow description programming language: The SCIE offers two types of libraries for high-level programming languages: Java and Python to describe the experiment work flow, measurement data subscription and experimentation visualization. With an experiment control program, the experimenters can schedule tasks running on slivers and subscribe the measurement data they need, and also, they can use all the characteristic of these languages.

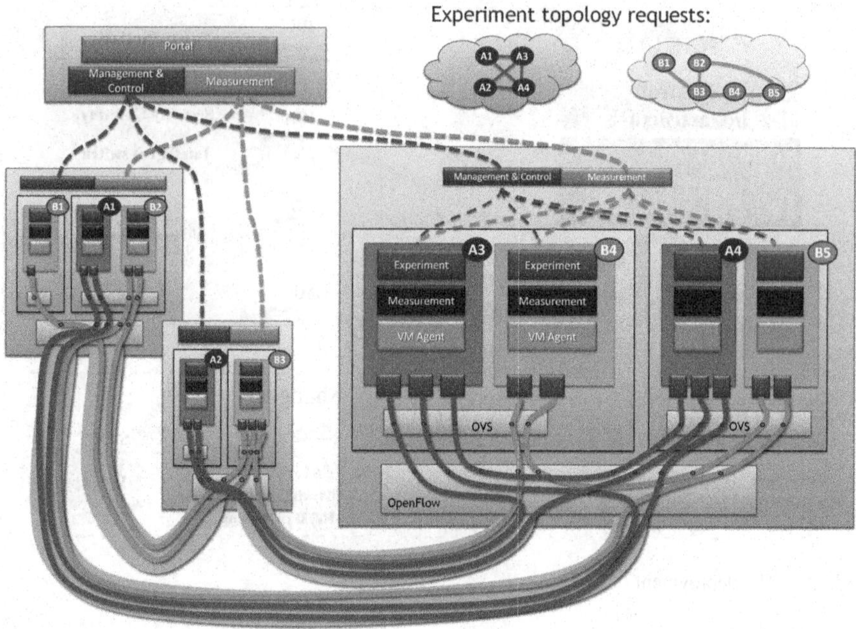

Fig. 12 SCIE architecture

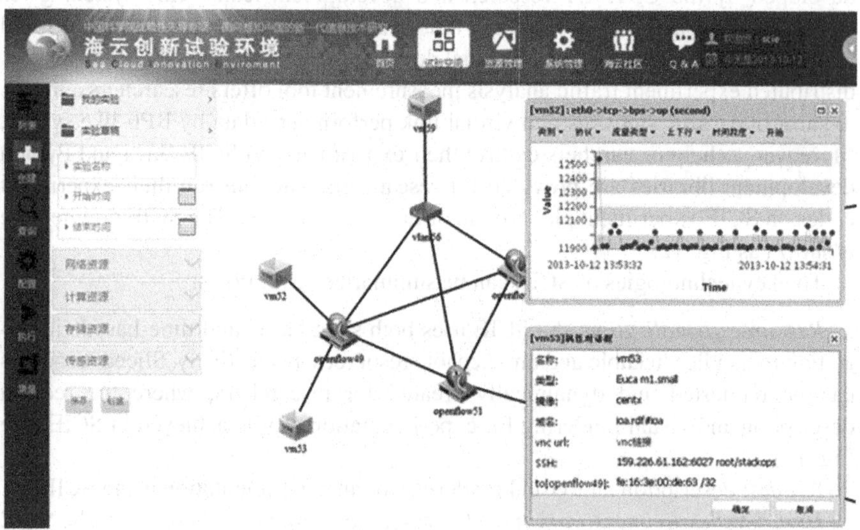

Fig. 13 Depiction of request description and visual rendering for an experimentation using SCIE

Holographic measurement: To decrease the impact on experimentation, the SCIE mainly leverages passive measurement, and still provides active measurement to design the holographic measurement system. In particular, the environment adopts several tools, e.g., sFlow and Spirent TestCenter SPT-3U testers, and uses both real-time flow handling and offline flow processing engine to provide data collection and analysis for experimentation.

17 International Multi-Domain Automatic Network Topology Discovery (MDANTD)

In almost all current OpenFlow deployments, there is a single controller managing all switches and handling network topology for routing decisions. However, for large-scale OpenFlow environments that interconnects several network domains, a single-controller may cause performance downgrade problems and policy/management issues. The National Center for High-Performance Computing is developing mechanisms to address such scenarios, by enabling each domain to deploy its own controller and to exchange topology and traffic information with other controllers [33]. Today, OpenFlow lacks an east-west interface standard (i.e. interface between controllers vs north south communications with resources within a single domain), and there are no signs that any standard body plans to work on this issue in near-term. Without such interface, when a network has problems or if the flow policy encounters some error, managers of each controller can only know what happen in their control domain, i.e., they cannot get any error information about other domains. In such situations, it is hard to investigate the problem accurately and the network manager will take more time to troubleshoot the network or flow policies.

In order to resolve the problem, NCHC has designed and implemented multi-domain automatic network topology discovery for large-scale OpenFlow/SDN environments by simply modifying Link-Layer Discovery Protocol (LLDP). LLDP, defined in IEEE 802.1AB, is a vendor-neutral link layer protocol for discovering neighbor devices. Most Openflow controllers adopt LLDP for automatic topology discovery by sending LLDP packets to switches in its domain periodically. Upon receiving the packet, OpenFlow switch will forward to its neighbor switches and then send it back to the controller. The controller will analyze the traveling path of each LLDP packet and conclude the network topology of its domain.

In Multi-Domain OpenFlow networks, when a LLDP packet travels to the controller of another domain, it will be ignored and dropped. Hence, the topology information cannot be exchanged. For example in Fig. 14a, there are two SDN Domains where Controller_1 manages Domain_1 while Controller_2 manages Domain_2. Switch A in Domain_1 has an inter-domain link with Switch B in Domain_2. LLDP packets from Domain_1 will travel from A to B and then will be sent to Controller_2. Controller_2 will ignore and drop the packet because it comes from another domain. As a result, Controller_1 only knows Controller_1's topology and Controller_2 only knows Controller_2's topology.

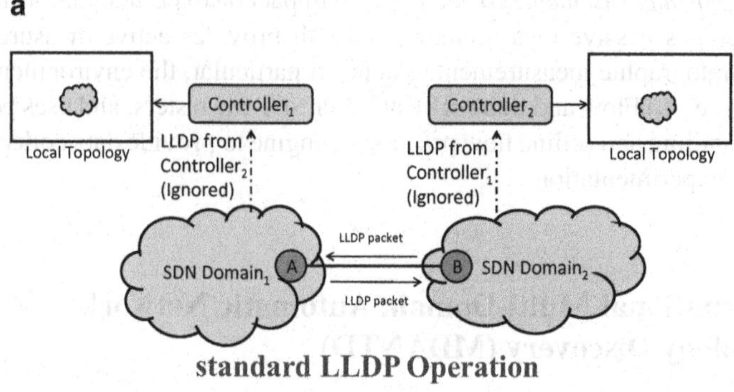

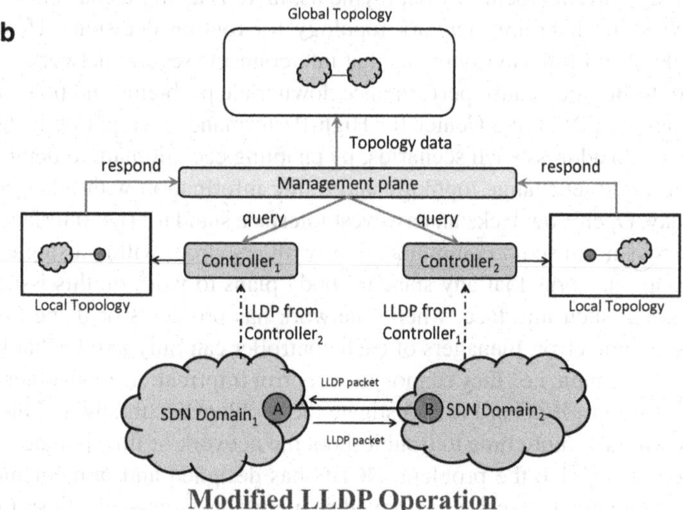

Fig. 14 Modifying LLDP for inter-domain automatic topology discovery

This new approach modifies LLDP operation as illustrated in Fig. 14b. When Controller$_2$ receives the LLDP packet from Domain$_1$, it will pick it up and analyze the traveling path. Controller$_2$ hence learns that there exists an inter-domain link from Domain$_1$'s A to Domain$_2$'s B. Finally, Controller$_2$ knows Domain$_2$'s topology with an inter-domain link to switch A. Similarly, Controller$_1$ knows Domain$_1$'s topology with an inter-domain link to switch B. A separate management plane has been designed that contacts each controller to obtain its local topology. Global topology is then understood by inter-connecting each domain's local topology and shown on a GUI console. This algorithm has been implemented in both NOX and FloodLight SDN controllers so that the management console can display multi-domain networks with a mix of NOX and FloodLight controllers.

In the latest demonstration of this technique at SC13, 7 domestic institutes in Taiwan (NARLabs/NCHC, NCKU, NCU, NTUST, NIU, NCTU, CHT-TL), StarLight/iCair, SURFnet, and JGN-X participated in a large scale international implementation. As depicted in Fig. 15a, the network was divided into a north part and a south part. Figures 15b and c show the UI display of the north part and south part separately. Each Domain is displayed in distinct color for easy identification. Circle nodes represent OpenFlow switches while square nodes represent user's end-node. NOX or FloodLight icons are placed in circle nodes so that managers can recognize which controller is adopted in this domain. In addition, a process intercepts PACKET_IN event to record the flow status and display end-to-end flow in the same user interface. In Fig. 15c, there is a ping issued from NARLabs/NCHC to JGN-X. The end-to-end path is shown in UI and the flow detail is given in the right-side panel.

In conclusion, this project developed a multi-domain automatic network topology discovery mechanism, and proved that it runs on a cross-controller platform by implementing over NOX and FloodLight. It is notable that the mechanism can be incorporated into almost any SDN controller, such as the increasingly popular OpenDayLight, and many others. For a large-scale OpenFlow/SDN environment that inter-connects several network domains, managers can easily observe network status and troubleshoot flow status with the management console. In the forthcoming year, this work will be extended to design inter-domain flow provisioning functions. These activities are key functions to implement OpenFlow/SDN network exchange centers in the near future.

18 Future Internet Virtualization Environment and VLAN Transit

With support from a project award by the National Science Council of Taiwan government, researchers at NCKU and KUAS have been designing and maintaining a network testbed that interconnects several heterogeneous datacenters across the public Internet. The current implementation and deployment has supported several research projects and applications, including cloud federation [34], international testbed interconnections [31], and network/security research experiments over production networks [35, 36]. With the system support, each sliver service could be allocated a portion of network resources that may be hosted either on the hosts within a datacenter or on a collection of physical hosts across different datacenters. As a result, a network system is required to support automated management (creation, deletion, and migration) of virtual networks, large numbers of attached physical and virtual devices, and isolated independent subnetworks for multi-tenancy. The novel contribution of this project is toward the design of a system to transform networks among a collection of independent autonomous datacenters into a manageable, flexible, multitenant transport fabric.

a

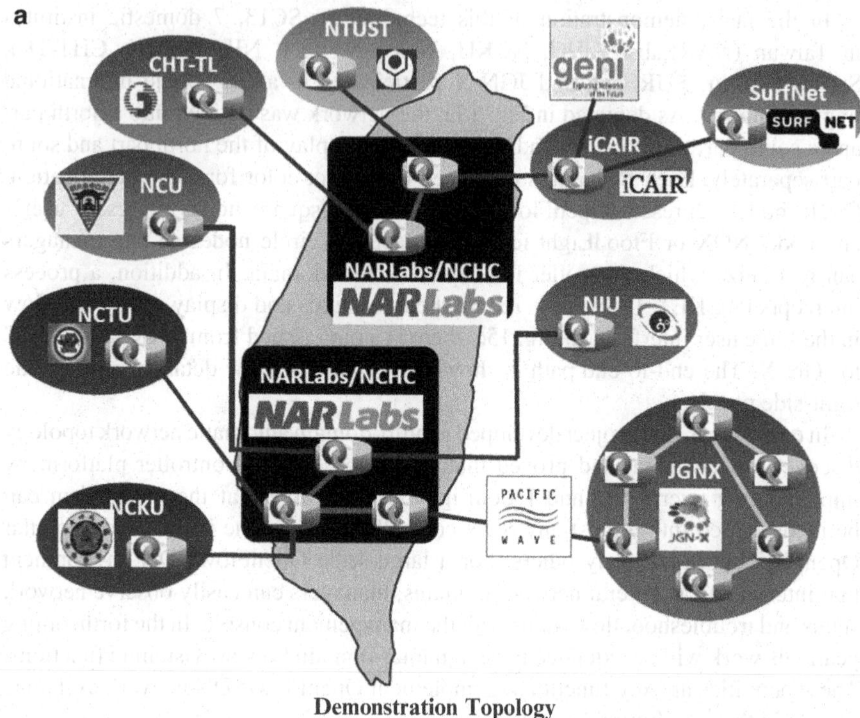

Demonstration Topology

b

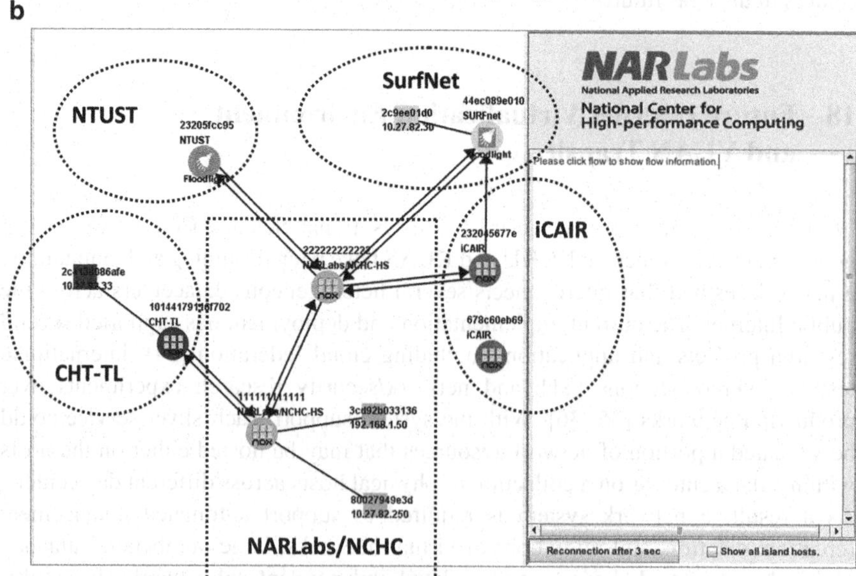

North Part Topology Shown in UI

Fig. 15 Multi-domain automatic network topology discovery in SC13 demonstration

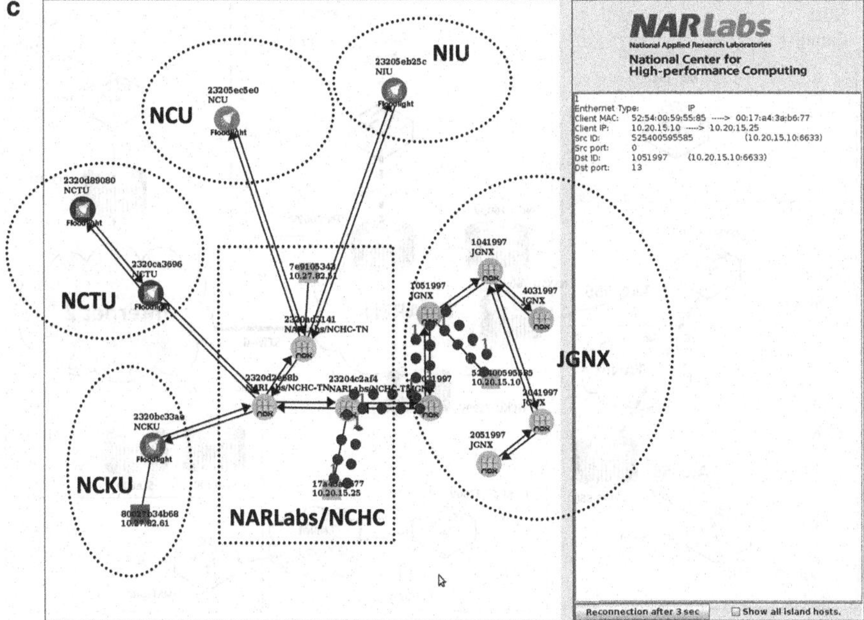

South Part Topology Shown in UI

Fig. 15 (continued)

The current deployment depicted in Fig. 16 illustrates the challenges of this project. Datacenters in this testbed are primarily interconnected with shared, public layer-3 networks (i.e., the Taiwan Academic Network (TANet) and Taiwan Advanced Research and Education Network (TWAREN)). With support from iCAIR at Northwestern University and the National Center for High-Performance Computing (NCHC), these testbeds have also been extended to connect with the iGENI project in the United States. To isolate the traffic, some of datacenters (e.g., NCKU, KUAS, NCU in Fig. 16) applied some Virtual Private LAN Service (VPLS) paths to interlink with the other systems. However, VPLS only offers a partial solution to the aforementioned problems. The first reason for this shortcoming is that the use of 12-bit VLAN IDs limits the solution's scalability. As a result, in this environment, only a few VLAN IDs are allowed to be used to interconnect all sites, which causes other serious scalability and performance problem. The system suffers from issues caused by flooding mechanisms (e.g., ARP broadcast), the huge MAC table size, and the use of Spanning Tree Protocol in a huge layer-2 network. Third, the VLAN, or VPLS, does not have an easy way to manage the configuration of paths across several network domains.

The architecture of this system can be logically viewed in three layers of a software stack, which also provides the hooks to other middleware and serves as a control plane to orchestrate all operations of virtual networks. First, they abstract the

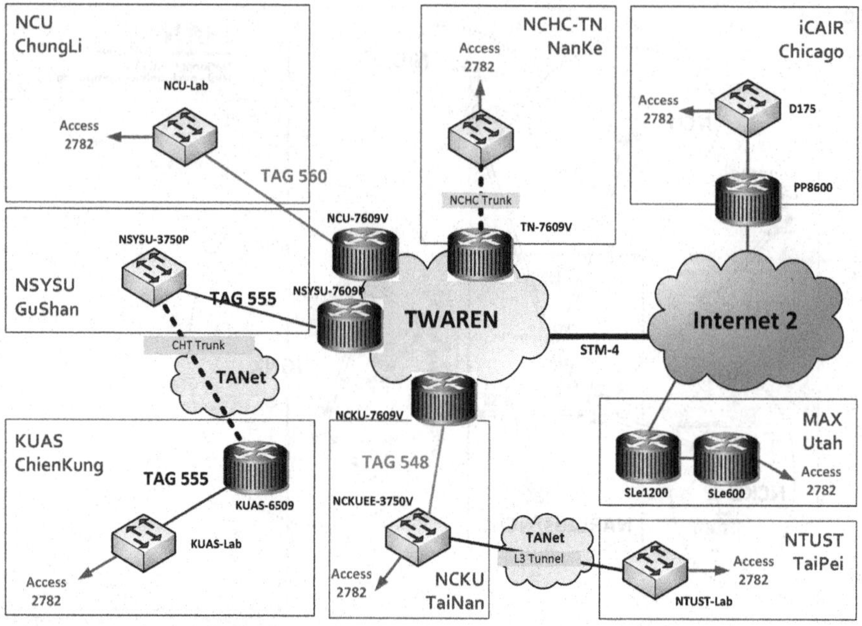

Fig. 16 Network topology of deployment

data plane of both the physical and virtual switches and synergize them coherently to enable the dynamic configuration and management of virtual networks [37]. They design a software module that resides in the kernel or hypervisor of each host in a datacenter. They also extend a POX Python based SDN controller with some added function to orchestrate the software modules to create a distributed data plane layer. The second layer is designed to deal with the interconnection issues among multiple datacenters. The major building blocks of this layer include a gateway switch that is designed based on the OpenFlow protocol. Each datacenter should at least have one gateway system located on the edge of its network to relay traffic across sites [38]. The operation of the gateway system is based on a novel mechanism termed virtualtransit [39]. Each pair of datacenters may set up one or more forwarding paths such as VPLS or MPLS. The basic idea of the virtualtransit is to abstract these paths and dynamically translated the VLAN tag to relay the traffic to the selected datacenter. A distributed control plane layer provides inter-datacenter mechanisms to create and manage virtual private networks spread across public Internet. The third layer is a logically central control framework containing the global network topology and the policy that determines the behavior of the whole system [39]. The control framework performs route computation on the virtualtransit and explicitly installs flow information to the switches along the chosen route based on OpenFlow protocol. Consolidating all flow decisions to the controller framework, a central service management system can be implemented to enforce a policy throughout the whole system. The system can also be extended further to add new functions and features by simply updating the controller. For instance, it provides an interface to allow integration with other control frameworks such as ORCA for user authentication, resource registration and discovery.

19 Interdomain ExoGENI

As noted in other chapters of this book, ExoGENI is an advanced computer and network resources virtualization project that paves the way for a new wave of virtualized applications that utilize elastic infrastructures. Basically, ExoGENI allows the IaaS paradigm to be extended to include networks integrated with the common compute and data processing cloud resources. The infrastructure information model used by the ExoGeni community for information exchange is based on the NDL, described in an earlier section of this chapter. The use of Semantic-Web technology in this approach facilitates the creation of models that can be easily connected, stacked and extended by other models. Also, as noted earlier, NDL has been extended to include descriptions of other types of resources, such as computing and storage, through the Infrastructure and Network Description Language (iNLDL). The extensibility and applicability of iNDL has been clearly demonstrated by its use as a basis for modeling efforts in three different infrastructures: the CineGrid (digital cinema production processes) infrastructure [40], the NOVI [41] federated platforms, and the GEYSERS architecture (Generalized Architecture for Dynamic Infrastructure Services) [10].

Currently, a project is being designed that will enable a distributed ExoGENI environment, with sites in the US and the Netherlands to investigate new capabilities for using SDN techniques for the international provisioning of extremely high capacity individual data streams, including streams with 40–100 Gbps capacities across thousands of miles, across nations and between continents. General networking provides support for supporting aggregations of 10s of millions of small data flows. This project is directed at creating new capabilities for supporting extremely large scale individual flows, using underlying resources described by NDL and orchestrated by SDN methods.

20 Content Routing

New large scale networking testbeds provide many opportunities for fundamentally changing traditional concepts. For example, the traditional Internet is based on physical addressing, although the Internet is primarily used for information gathering. This observation is one of the motivations behind increasing interest in alternative Internet architectures based upon content based forwarding to replace, or augment, traditional IP routing [42]. Content-based forwarding is often termed content centric networking (CCN) or information centric networking (ICN), and there is not, currently, one model that can claim to fully describe this evolving area. However, one of the central aims of most of the content-based forwarding approaches is to provide an architecture that concentrates on *what* the data is, rather than *where* it is placed. This new approach allows optimization of the network architecture to suit the delivery of the data rather than an optimization based upon end-node attachment, as is inherent in the existing IP routing strategy. Indeed it has been shown that using a content-based forwarding approach allows forms of traffic engineering that are challenging to implement in contemporary IP or IP/MPLS networks [43]. To further explore the area of ICN, the EU FP7 project PURSUIT (leading on from the EU project PSIRP) has implemented a *clean-slate* publish-subscribe ICN architecture [44]. The PURSUIT architecture has a semi-centralized mediation layer and makes use of source routing based on Bloom filters. Using this architecture, the project has demonstrated a scalable ICN solution capable of forwarding at up to 10 Gb/s using standard software based computer platforms; it promises the possibility of forwarding in custom hardware at much higher rates and with lower complexity than IP or IP/MPLS.

The PURSUIT project has created an experimental testbed that is managed at the University of Essex (UK) with partners connecting in globally (EU, US, Japan, and China). As this architecture is exploring forwarding using non-IP based mechanisms it is important that connectivity can be provided at layer-2. One of the experiments carried out as part of the project used multiple end-points and forwarding nodes hosted at StarLight (Chicago) and at the University of Essex (UK). The experiment demonstrated *information* resilience, enabled through the inherent anycast nature of the PURSUIT ICN architecture. Using this approach, a subscriber obtains content from any available publisher. If there are multiple publishers, then there is a natural resilience available as the content is fetched through the content identifier, not by the end-point. Specifically the experiment between StarLight and the University of Essex demonstrated resilience for a video transmission. It first showed traditional network resilience: if a network link becomes unavailable the architecture can route around the failed link. As a further step, all links to the video sender were cut, the architecture detected the failure and selected an alternative sender as a source of the video. A key point of this *information* resilience is that it is enabled in the network functions and requires no support from the application itself. ICN projects such as PURSUIT, which are proposing *clean-slate* Internet architectures, require flexible testbed networks that allow layer-2 connectivity. It is clear that, looking towards the future, flexible testbeds enabled through systems such as GENI will be a vital first step towards deployment of future architectures such as ICN.

21 Brazilian Future Internet Experimental Environment

FIBRE (Future Internet testbeds/experimentation between BRazil and Europe) [45] was one of five projects that were approved in response to the 2010 Brazil-EU Coordinated Call in ICT, jointly funded by CNPq (the Brazilian Council for Scientific and Technological Development) and by the Seventh Framework Programme (FP7) of the European Commission. The main objective of FIBRE was to create a common space between the EU and Brazil for Future Internet (FI) experimental research into network infrastructure and distributed applications. Prior to FIBRE, such facilities already were operated, or were being built, by partners in this project from both sides of the Atlantic Ocean. FIBRE was designed so that such a space would enable and encourage closer and more extensive BR-EU cooperation in FI research and experimentation, as well as strengthening the participation of both communities in the increasingly important global collaborations in this important area of network research and development.

The EU-side partners in FIBRE (i2CAT, NWX, NICTA, UnivBRIST, UPMC and UTH) were also participants in the EC's FIRE (Future Internet Research and Experimentation) [6] testbed projects CHANGE [46], OpenLab [47] and OFELIA [8].

An important characteristic of OFELIA was its leveraging of the OpenFlow (OF) approach from Stanford [9]. The participation of OpenLab partners allowed extensions of the OFELIA approach to new testbed environments and use cases not included in OFELIA, especially in the fields of wireless communications. In this latter area, considerable expertise in designing, building and evaluating large-scale testbed systems was brought to the project through the participation of National ICT Australia (NICTA), which has been a major contributor to the development of the OMF control framework [48]. A major objective of the Brazil-EU FIBRE project is the deployment in Brazil of FIBRE2 (Future Internet: BRazilian Experimental Environment), a wide-area network testbed to support user experimentation in the design and validation of new network architectures and applications, which interconnects experimental facilities ("islands") located at the participating institutions (CPqD, RNP, UFF, UFG, UFPA, UFPE, UFRJ, UFSCar, UNIFACS and USP), using RNP's national backbone network [49].

In such a testbed, a high degree of automated resource sharing between experimenters is required, and the testbed itself needed to be instrumented so that precise measurements and accounting of both user and facility resources could be carried out. The Control and Monitoring Framework (CMF) for the FIBRE2 testbed is based on three CMFs developed in the testbed projects OFELIA, ORBIT and ProtoGENI. In order to take best advantage of different testbed functionalities at different sites, FIBRE2 has been created as a federated testbed, which facilitates interoperation with international initiatives. Figure 17 shows diagrammatically the network topology created in the FIBRE project, interconnecting the Brazilian FIBRE2 infrastructure with that of FIBRE's European testbed sites.

22 High Performance Digital Media Network (HPDMnet)

Questions are often asked about the types of applications that will be supported by SDN. One example is provided by the High Performance Digital Media Network (HPDMnet), which is an international experimental network research testbed that has been used for over 10 years to investigate a wide range of topics, including new

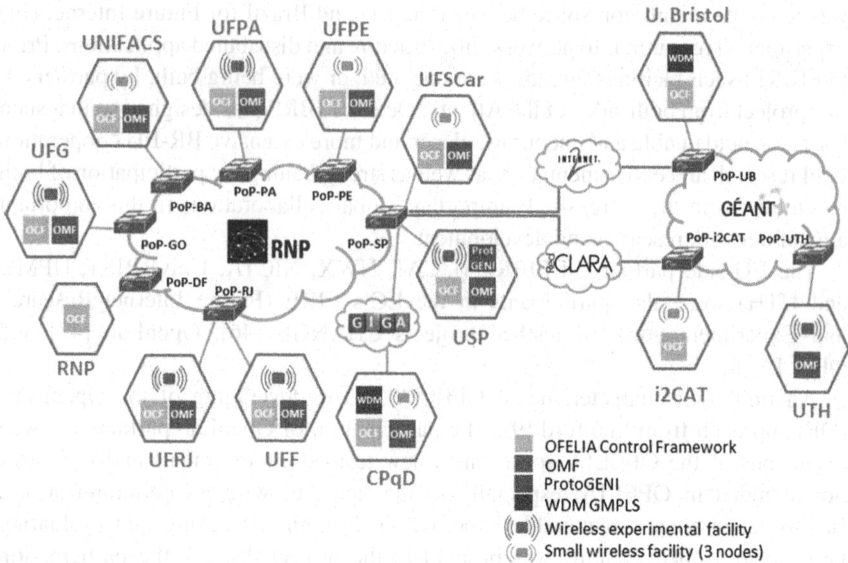

Fig. 17 The federated international testbed for the FIBRE project, including the Brazilian FIBRE2 component

global scale streaming services, architecture, techniques, and technologies for high performance, ultra high resolution digital media transport based on dynamically programmable L1/L2 paths. [50, 51] Services include those that support high volume digital media streams, required by ultra resolution HD, 4k media, and 8k media. HPDMnet is based on various mechanisms that comprise its control, management and data planes, and the majority of partners in this consortium are moving to implement SDN/OpenFlow capabilities as is shown in Fig. 2 [41–44] (Fig. **18**).

23 Anticipated Future Services and Resource Expansions

The multiple macro trends toward virtualization (e.g., everything as a service—XaaS), provides for major opportunities for continuing to design, create, and rapidly implement additional—and more innovative—international federated environments for experimental network research. Future plans include the design and implementation of a much wider set of services for such environments, and services that are more accessible, in part, through enhanced APIs with more straightforward authentication and authorization capabilities across multiple domains. The

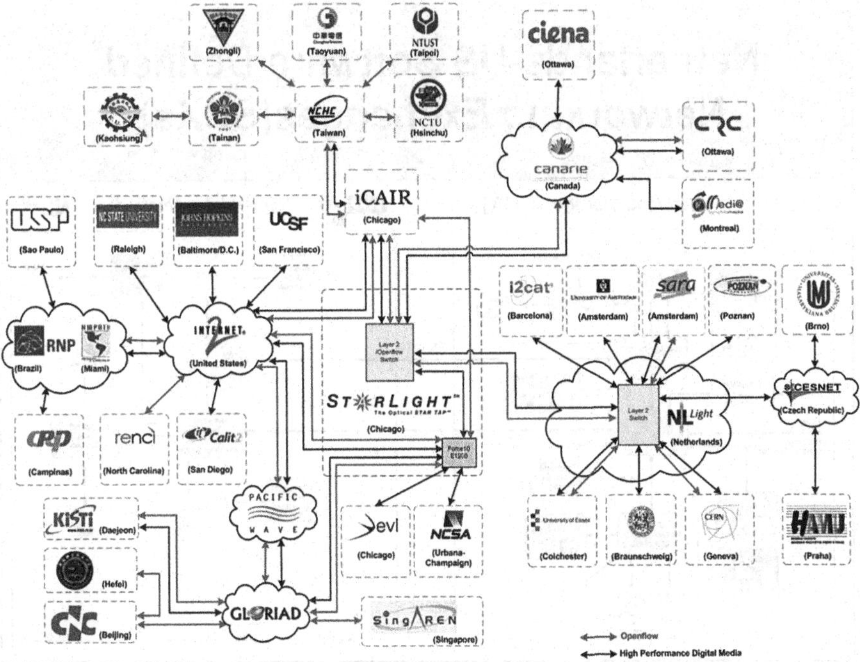

Fig. 18 The international high performance digital media network (HPDMnet)

following sections describe some of these expansion initiatives. These future testbed environments will be based on international optical networks interconnected next generation multi-service exchanges. [52].

24 Software Defined Networking Exchanges (SDXs)

As noted, Software-Defined-Networking (SDN) has fundamentally transformed networking services and infrastructure, and it will continue to do so for the foreseeable future. However, current SDN architecture and technologies are single domain oriented, and required capabilities for multi-domain SDN provisioning are fairly challenging. Consequently, the deployment of SDNs has led to multiple isolated SDN "islands." Therefore, the increasing implementations of production SDNs has highlighted the need for the design and creation of Software Defined Networking Exchanges (SDXs). Recently, several research communities have designed and implemented the world's first SDXs, including a prototype at the StarLight facility in Chicago developed by iCAIR and its research partners, initially as a GENI project, and one at the NetherLight facility in Amsterdam, developed by SURFnet. These SDXs provide multi-domain services enabled by federated controllers and

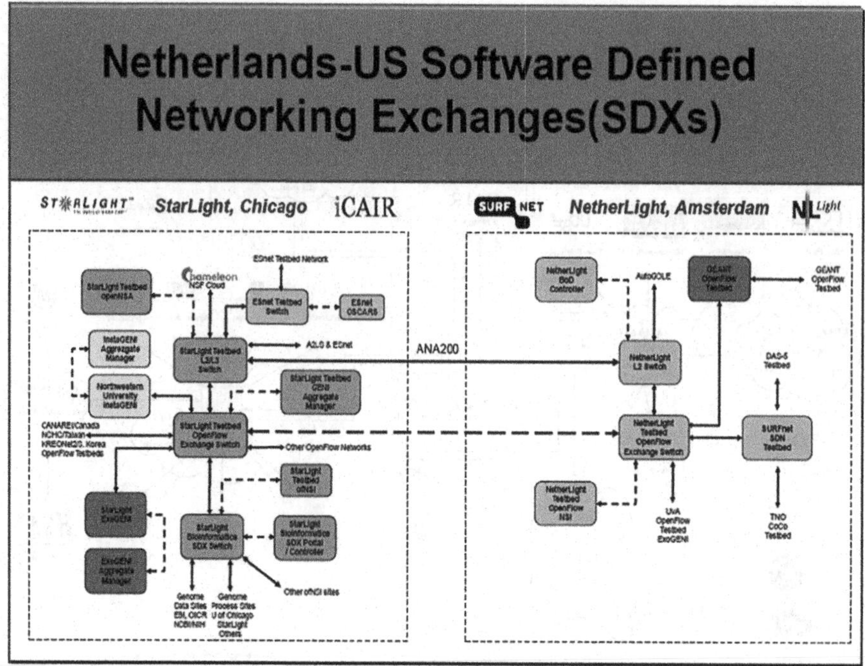

Fig. 19 International interoperability between StarLight and NetherLight SDXs

direct data plane connections. International interoperability between the StarLight and NetherLight SDXs was showcased at the TERNENA conferences in May 2014 and June 2015 (Fig. 19).

There are multiple benefits to SDXs: (a) many more options for dynamic provisioning at exchanges, including real time provisioning, (b) faster implementations of new and enhanced services, (c) enabling applications, edge processes and even individuals to directly control core exchange resources, (d) highly granulated views into individual network traffic flows through the exchanges and direct control over those flows, (e) enhanced network service and infrastructure management because of those views and (f) substantially improved options for creating customizable network services and infrastructure. The StarLight SDX, which is based on multi-domain services supported by federated controllers and high performance data planes, has been used to demonstrate the potential for creating customized SDXs for specific services and applications, including data intensive domain sciences, especially when based on programmable, segmented, high-capacity 100 Gbps paths. These capabilities, which were showcased by iCAIR and its research partners at SC14 and SC15, included a demonstration of a prototype customized computational bioinformatics SDX.

Essentially, an SDX is a type of large scale virtual switch, which can provide segmented resources for different domains, locally, nationally or internationally.

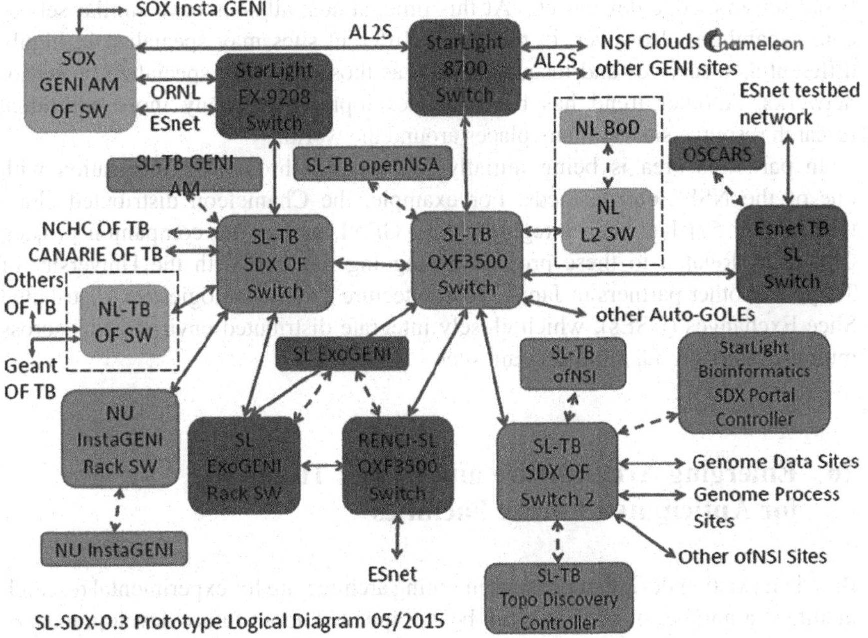

Fig. 20 Logical diagram of StarLight SDX

The substructure for the virtual switch consists of multiple other SDN/OpenFlow switches [53–55] (Fig. **20**). Recently, iCAIR and the University of Chicago, with multiple international research partners demonstrated how an SDX can be used to create a virtual exchange customized to support the complex workflows required to optimally support new techniques for precision medicine—precision medicine enabled by precision networking. iCAIR also established a project with the University of Massachusetts to explore how SDXs could be used to support new techniques for weather prediction and visualization.

25 Software Defined Infrastructure (SDI) and Cloud Testbed Integration

Other areas being investigated include SDX extensions to Software Defined Infrastructure (SDI) integrating additional resources including compute facilities, clouds, Grids, storage, instruments, mobile devices, sensors, and other resources. In general, a major trend has been seen toward the incorporation into these environments of additional types of highly programmable resources (at an extremely granulated level), e.g., compute clouds, specialized compute devices such as those based on GPUs and FPGAs, storage systems, instrumentation, wireless fabrics, RFP based sensors, edge devices etc. At this time, almost all sites have similar sets of core capabilities. However, in the future different sites may specialize in highly differentiated services and resources, such as those that may specialize on sensor networks. Another trend has been the development of many more additional research resource sites at many places around the world.

In part, this area is being initially investigated through an integration with one of the NSFCloud testbeds. For example, the Chameleon distributed cloud testbed [56, 57] has been integrated with GENI, as has the companion project, CloudLab. Related to these projects is ongoing research with the University of Tokyo and other partners in Japan on architecture and technologies for Distributed Slice Exchanges (DSEs), which closely integrate distributed environments across multiple international multi-domain sites.

26 Emerging Architecture and Design Trends for Anticipated Future Facilities

Based on existing design trends and emerging architecture for experimental research facilities, a number of premises can be extrapolated about future developments of such environments. As noted, the macro trends toward virtualization at all levels—proving every resource as a service (XaaS) is leading to an explosive growth in such services. Currently, it is possible to create "service factories," based on large scale virtualization capabilities, using as a foundation a rich array of programmable network middleware and a wide array of underlying infrastructure, which together become a platform for innovation. These platforms will continue to evolve rapidly, just as Grid platforms evolved to incorporate programmable networking architecture and technology, a progression that did much to inform today's programmable networking t [58, 59].

A key attribute of emerging and future environments is that they allow not just large organizations to create, deploy and operate networks but also they enable individuals to create their own large scale networks, customizing them to meet individualized, specialized requirements vs general requirements. The potential for these capabilities is accelerating because the underlying core infrastructure is virtualized at the same time that it is rapidly declining in cost as component technologies move to commodity.

27 Conclusions

This chapter provides an overview of the international capabilities of GENI and related network research environments, with descriptions of the services required by research communities, basic architectural approaches, existing services and facilities, and examples of current research experiments being conducted within these environments. The chapter also anticipates emerging architecture and design trends for anticipated future services for such international experimental network environments, as well as facilities, and expansions to many additional resources. Key macro trends are those that enable virtual and physical network resource to be abstracted, so that customized resource slices across international WANs can be dynamically created and implemented. Such contiguous integrations of highly distributed resources can be manipulated using new types of orchestration techniques, which are being made available not only to systems operators, but also to edge applications, processes and individuals.

Acknowledgments The authors would like to express appreciation for the support of the projects described here from many organizations, including the National Science Foundation, the GENI Program Office (GPO), the Department of Energy's Office of Energy Science, the StarLight International/National Communications Exchange Facility consortium, the Metropolitan Research and Education Network, and the Open Cloud Consortium. The International Multi-Domain Automatic Network Topology Discovery Project is sponsored by Ministry of Science and Technology, Taiwan, R.O.C. under contract number MOST 103-2221-E-492-030. The University of Amsterdam would like to thank the Dutch National Program COMMIT. This project also would like to acknowledge all participants of TWAREN SDN Testbed, including National Cheng Kung University, National Central University, National Taiwan University of Science and Technology, National Ilan University, National Chiao Tung University, National Kaohsiung University of Applied Sciences, Chunghwa Telecom Laboratories, International Center for Advanced Internet Research, JGN-X, and SURFnet, for their helping of setup, testing, and troubleshooting.

References

1. Berman, M., Chase, J., Landweber, L., Nakao, A., Ott, M., Raychaudhuri, D., Ricci, R., Seskar, I.: GENI: a federated testbed for innovative network experiments. Special issue on Future Internet Testbeds. Comput. Netw. **61**, 5–23 (2014)
2. Strijkers, R., Makkes, M.X., de Laat, C., Meijer, R.: Internet factories: creating application-specific networks on-demand. Special issue on Cloud Networking and Communications. Comput. Netw. **68**, 187–198 (2014). doi:10.1016/j.comnet.2014.01.009
3. http://www.ogf.org/gf/group_info
4. Mell, P., Grance, T.: Definition of cloud computing. Special Publication 800-145. Recommendations of the National Institute of Standards and Technology (NIST). www.nist.gov/itl/cloud
5. National Science Board. International Science and Engineering Partnerships: A Priority for U.S. Foreign Policy and Our Nation's Innovation Enterprise, National Science Foundation (2008)
6. Jofre, J., Velayos, C., Landi, G., Giertych, M., Humed, A., Francis, G., Oton, A.: Federation of the BonFIRE multi-cloud infrastructure with networking facilities. Special issue on Future Internet Testbeds. Comput. Netw. **61**, 184–196 (2014)
7. http://www.ist-phosphorus.eu/
8. Suñé, M., Bergesio, L., Woesner, H., Rothe, T., Köpsel, A., Colle, D., Puype, B., Simeonidou, D., Nejabati, R., Channegowda, M., Kind, M., Dietz, T., Autenrieth, A., Kotronis, V., Salvadori, E., Salsano, S., Körner, M., Sharma, S.: Design and implementation of the OFELIA FP7 facility: the European OpenFlow testbed. Special issue on Future Internet Testbeds. Comput. Netw. **61**, 132–150 (2014)
9. McKeown, N., et al.: OpenFlow: enabling innovation in campus networks. ACM SIGCOMM Comput. Commun. Rev. **38**(2), 69–74 (2008)

10. Belter, B., Martinez, J., Aznar, J., Riera, J., Contreras, L., Lewandowska, M., Biancani, M., Buysse, J., Develder, C., Demchenko, Y., Donadio, P., Simeonidou, D., Nejabati, R., Peng, S., Drzewiecki, L., Escalona, E., Espin, J.: The GEYSERS optical testbed: a platform for the integration, validation and demonstration of cloud-based infrastructure services. Special issue on Future Internet Testbeds. Comput. Netw. **61**, 197–216 (2014)
11. Campanella, M., Farina, F.: The FEDERICA infrastructure and experience. Special issue on Future Internet Testbeds. Comput. Netw. **61**, 176–183 (2014)
12. Schwerdel, D., Reuther, B., Zinner, T., Müller, P., Tran-Gia, P.: Future internet research and experimentation: the G-lab approach. Special issue on Future Internet Testbeds. Comput. Netw. **61**, 102–117 (2014)
13. www.fed4fire.eu
14. Thorpe, S., Battestilli, L., Karmous-Edwards, G., Hutanu, A., MacLaren, J., Mambretti, J., Moore, J., Sundar, K., Xin, Y., Takefusa, A., Hayashi, M., Hirano, A., Okamoto, S., Kudoh, T., Miyamoto, T., Tsukishima, Y., Otani, T., Nakada, H., Tanaka, H., Taniguchi, A., Sameshima, Y., Jinno, M.: G-lambda and EnLIGHTened: wrapped in middleware co-allocating compute and network resources across Japan and the US. In: Proceedings of the First International Conference on Networks for Grid Applications, Lyon, France, SIGARCH, ACM Special Interest Group on Computer Architecture, Published by Institute for Computer Sciences, Social-Informatics and Telecommunications Engineering (ICST), Brussels, Belgium (2007)
15. Kim, D., Kim, J., Wang, G., Park, J.-H., Kim, S.-H.: K-GENI testbed deployment and federated meta operations experiment over GENI and KREONET. Special issue on Future Internet Testbeds. Comput. Netw. **61**, 39–50 (2014)
16. Stanton, M.: RNP experiences and expectations in future internet research and development. In: New Network Architectures. Studies in Computational Intelligence, vol. 297, pp. 153–166. Springer, New York (2010)
17. www.savinetwork.ca
18. van der Ham, J., Stéger, J., Laki, S., Kryftis, Y., Maglaris, V., de Laat, C.: The NOVI information models. Future Gener. Comput. Syst. Available online 18 December 2013, ISSN 0167-739X. doi:10.1016/j.future.2013.12.017
19. van der Ham, J.J., Dijkstra, F., Travostino, F., Andree, H.M.A., de Laat, C.T.A.M.: Using RDF to describe networks, iGrid2005 special issue. Future Gener. Comput. Syst. **22**(8), 862–867 (2006)
20. Ghijsen, M., van der Ham, J., Grosso, P., Dumitru, C., Zhu, H., Zhao, Z., de Laat, C.: A semantic-web approach for modeling computing infrastructures. J. Comput. Electr. Eng. **39**(8), 2553–2565 (2013). doi:10.1016/j.compeleceng.2013.08.011
21. Baldine, I., Xin, Y., Mandal, A., Heermann, C., Chase, J., Marupadi, V., et al.: Networked cloud orchestration: a GENI perspective. Workshop on Management of Emerging Networks and Services (2010)
22. Global Lambda Integrated Facility (GLIF). www.glif.is
23. Global Ring Network for Advanced Applications Development (GLORIAD). www.gloriad.org
24. Roberts, G., Kudoh, T., Monga, I., Sobieski, J., MacAuley, J., Guok, C.: NSI Connection Service V2.0, Open Grid Forum, GWD-R-P, NSI-WG (2013)
25. http://groups.geni.net/geni/wiki/IGENI
26. Mambretti, J., Chen, J., Yeh, F.: International network research testbed facilities based on OpenFlow: architecture, services, technologies, and distributed infrastructure proceedings, In: 18th IEEE International Conference on Networks (ICON), pp. 234–242 (2012)
27. Bavier, A., Yuen, M., Blaine, J., McGeer, R., Young, A., Coady, Y., Matthews, C., Pearson, C., Snoeren, A., Mambretti, J.: TransCloud—Design Considerations for a High-performance Cloud Architecture Across Multiple Administrative Domains. CLOSER, pp. 120–126 (2011)
28. http://tomato-lab.org
29. Schwerdel, D., Hock, D., Günther, D., Reuther, B., Müller, P., and Tran-Gia, P.: ToMaTo—a network experimentation tool. In: 7th International ICST Conference on Testbeds and Research Infrastructures for the Development of Networks and Communities (TridentCom 2011), Shanghai, China, April 2011

30. Schwerdel, D., Reuther, B., Zinner, T., Müller, P., Tran-Gia, P.: Future internet research and experimentation: the G-lab approach. Comput. Netw. **61**, 102–117 (2014). doi:10.1016/j.bjp.2013.12.023
31. Mambretti, J., Chen, J., Yeh, F., Liu, T.-L., Luo, M.-Y., Yang, C.-S., van der Pol, R., Boele, S., Dijkstra, F., Barczyk, A., van Malensteinz, G.: Openflow Services For Science: An International Experimental Research Network Demonstrating Multi-Domain Automatic Network Topology Discovery, Direct Dynamic Path Provisioning Using Edge Signaling and Control, Integration with Multipathing Using MPTCP, 2012 SC Companion: High-Performance Computing, Networking, Storage and Analysis (SCC) (2012)
32. van der Pol, R., Boele, S., Dijkstra, F., Barczyk, A., van Malenstein, G., Chen, J.H., Mambretti, J.: Multipathing with MPTCP and OpenFlow, High Performance Computing, Networking, Storage and Analysis (SCC), 2012 SC Companion, November 2012
33. Huang, W.-Y., Hu, J.-W., Lin, S.-C., Liu, T.-L., Tsai, P.-W., Yang, C.-S., Yeh, F., Hao Chen, J., Mambretti, J.: Design and implementation of an automatic network topology discovery system for the future internet across different domains. In: Proceedings of IEEE 26th International Conference on Advanced Information Networking and Applications Workshops (AINAW'12), Singapore, March 2012
34. Luo, M.-Y., Lin, S.-W., Chen, J.-Y.: From monolithic systems to a federated e-learning cloud system. In: IEEE International Conference on Cloud Engineering, San Francisco, California, 25–28 March 2013
35. Tsai, P.-W., Cheng, P.-W., Luo, M.-Y., Liu, T.-L., Yang, C.-S.: Planning and implantation of NetFPGA platform on network emulation testbed, In: Proc. Asia Pacific Advanced Network, Network Research Workshop, Delhi, India, 22 August 2011
36. Testbed@TWISC. http://testbed.ncku.edu.tw/
37. Luo, M.-Y., Chen, J., Mambretti, J., Lin, S.-W., Tsai, P.-W., Yeh, F., Yang, C.-S.: Network virtualization implementation over global research production networks. J. Internet Technol. **14**(7), 1061–1072 (2013)
38. Luo, M.-Y., Chen, J.-Y.: Towards network virtualization management for federated cloud systems. In: IEEE 6th International Conference on Cloud Computing, Santa Clara, CA, 27 June–2 July 2013
39. Luo, M.-Y., Chen, J.-Y.: Software defined networking across distributed datacenters over cloud. In: 5th IEEE International Conference on Cloud Computing Technology and Science (IEEE CloudCom), Bristol, UK, 2–5 December 2013
40. Koning, R., Grosso, P., de Laat, C.: Using ontologies for resource description in the CineGrid exchange. Future Gener. Comput. Syst. **27**(7), 960–965 (2011)
41. NOVI—Networking Innovations Over Virtualized Infrastructures. http://www.fp7-novi.eu/
42. Ahlgren, B., Dannewitz, C., Imbrenda, C., Kutscher, D., Ohlman, B.: A survey of information-centric networking. IEEE Commun. Mag. **50**(7), 26–36 (2012)
43. Reed, M.J.: 2012 IEEE International Conference on Traffic Engineering For Information-Centric Networks, Communications (ICC), pp. 2660–2665, 10–15 June 2012
44. Trossen, D., Parisis, G.: Designing and realizing an information-centric internet. Commun. Mag. IEEE **50**(7), 60–67 (2012)
45. http://www.fibre-ict.eu
46. http://www.change-project.eu
47. http://www.ict-openlab.eu
48. Rakotoarivelo, T., Ott, M., Jourjon, G., Seskar, I.: OMF: a control and management framework for networking testbeds. ACM SIGOPS Oper. Syst. Rev. **43**(4), 54–59 (2010)
49. Abelem, A., et al.: FIT@BR—a future internet testbed in Brazil. Proc. Asia-Pacific Adv. Netw. **36**, 1–8 (2013)
50. Mambretti, J., Lemay, M., Campbell, S., Guy, H., Tam, T., Bernier, E., Ho, B., Savoie, M., de Laat, C., van der Pol, R., Chen, J., Yeh, F., Figuerola, S., Minoves, P., Simeonidou, D., Escalona, E., Amaya Gonzalez, N., Jukan, A., Bziuk, W., Kim, D., Cho, K.J., Lee, H.-L., Liu, T.L.: High performance digital media network (HPDMnet): an advanced international research initiative and global experimental testbed. Future Gener. Comput. Syst. **27**(7), 893–905 (2011)

51. Jukan, A., Mambretti, J.: Evolution of optical networking toward rich digital media services. Proc. IEEE **100**(4), 855–871 (2012)
52. Mambretti, J., Chen, J., Yeh, F.: Creating environments for innovation: designing and implementing advanced experimental network research testbeds based on the global lambda integrated facility and the starlight exchange. Special issue on Future Internet Testbeds. Comput. Netw. **61**, 118–131 (2014)
53. Feamster, N., Rexford, J., Shenkerz, S., Levin, D., Clark, R., Bailey, J.: SDX: A Software Defined Internet Exchange, White Paper, University of Maryland
54. Mambretti, J., Chen, J., Yeh, F.: Software-defined network exchanges (SDXs): enabling capabilities for distributed clouds with SDN multi-domain and multi-services techniques, accepted. In: Workshop on Future Internet Testbeds and Distributed Clouds (FIDC), Co-located with the International Teletraffic Congress in Karlskrona, Sweden, 9–11 September 2014
55. Mambretti, J., Chen, J., Yeh, F.: Software-Defined Network Exchanges (SDXs) and Infrastructure (SDI): Emerging Innovations in SDN and SDI Interdomain Multi-Layer Services and Capabilities, First International Science and Technology Conference: Modern Networking Technology: SDN and NFV—The Next Generation of Computational Infrastructure, Moscow, Russia, 28–29 October 2014. Published By IEEE, Science and Technology Conference (Modern Networking Technologies) (MoNeTeC), 2014 International
56. Mambrretti, J., Chen, J., Yeh, F.: Next generation clouds, the Chameleon cloud testbed, and software defined networking (SDN). In: Proceedings, International Conference on Cloud Computing Research and Innovation (ICCCRI 2015), 26–27 October 2015, Singapore (2015)
57. www.chameleoncloud.org
58. Travostino, F., Mambretti, J., Karmous-Edwards, G. (eds.): Grid Networks: Enabling Grids with Advanced Communication Technology. Wiley, New York (2006)
59. Doulamis, T., Mambretti, J., Tomkos, I., Varvarigou, D. (eds.): Networks for grid applications. In: Third International ICST Conference, GridNets 2009, Athens, Greece, 8–9 September 2009. Revised Selected Papers

Appendix: Additional Readings

1. Aikat, J., Hasan, S., Jeffay, K. & Smith, F. D. (2012), Discrete-Approximation of Measured Round Trip Time Distributions: A Model for Network Emulation, *in* 'First GENI Research and Educational Experiment Workshop (GREE 2012)'.
2. Alaoui, S. E., Palusa, S. & Ramamurthy, B. (2015), The Interplanetary Internet Implemented on the GENI Testbed, *in* '2015 IEEE Global Communications Conference (GLOBECOM)', IEEE, pp. 1–6. http://ieeexplore.ieee.org/stamp/stamp.jsp?tp=&arnumber=7417313&isnumber=7416057
3. Albrecht, J. & Huang, D. Y. (2010) , 'Managing distributed applications using Gush', *Proceedings of the ICST Conference on Testbeds and Research Infrastructures for the Development of Networks and Communities, Testbed Practices Session (TridentCom)* . http://dx.doi.org/10.1007/978-3-642-17851-1_31
4. Albrecht, J. R. (2009), 'Bringing big systems to small schools: distributed systems for undergraduates', *SIGCSE Bull.* **41**(1). http://dx.doi.org/10.1145/1539024.1508903
5. Albrecht, J., Tuttle, C., Braud, R., Dao, D., Topilski, N., Snoeren, A. C. & Vahdat, A. (2011), 'Distributed application configuration, management, and visualization with plush', *ACM Trans. Internet Technol.* **11**(2). http://dx.doi.org/10.1145/2049656.2049658
6. Angu, P. & Ramamurthy, B. (2011), Experiences with dynamic circuit creation in a regional network testbed, *in* '2011 IEEE Conference on Computer Communications Workshops (INFOCOM WKSHPS)', IEEE, pp. 168–173. http://dx.doi.org/10.1109/infcomw.2011.5928801
7. Antonenko, V., Smeliansky, R., Baldin, I., Izhvanov, Y. & Gugel, Y. (2014), Towards SDI-bases Infrastructure for supporting science in Russia, *in* 'Science and Technology Conference (Modern Networking Technologies) (MoNeTeC), 2014 First International', IEEE, pp. 1–7. http://dx.doi.org/10.1109/monetec.2014.6995576
8. Araji, B. & Gurkan, D. (2014), Embedding Switch Number, Port Number, and MAC Address (ESPM) within the IPv6 Address, *in* 'Research and Educational Experiment Workshop (GREE), 2014 Third GENI', IEEE, pp. 69–70. http://dx.doi.org/10.1109/gree.2014.20
9. Augé, J., Parmentelat, T., Turro, N., Avakian, S., Baron, L., Larabi, M. A., Rahman, M. Y., Friedman, T. & Fdida, S. (2014), 'Tools to foster a global federation of testbeds', *Computer Networks* **63**, 205–220. http://dx.doi.org/10.1016/j.bjp.2013.12.038
10. Babaoglu, A. C. (2014), Verification Services for the Choice-Based Internet of the Future (Doctoral dissertation), PhD thesis, North Carolina State University. http://www.lib.ncsu.edu/resolver/1840.16/9336
11. Babaoglu, A. C. & Dutta, R. (2014) , A GENI Meso-Scale Experiment of a Verification Service, *in* 'Research and Educational Experiment Workshop (GREE), 2014 Third GENI', IEEE, pp. 65–68. http://dx.doi.org/10.1109/gree.2014.13

© Springer International Publishing Switzerland 2016
R. McGeer et al. (eds.), *The GENI Book*, DOI 10.1007/978-3-319-33769-2

12. Baldine, I. (2009), Unique optical networking facilities and cross-layer networking, *in* 'Summer Topical Meeting, 2009. LEOSST '09. IEEE/LEOS', pp. 145–146. http://dx.doi.org/10.1109/LEOSST.2009.5226210
13. Baldine, I., Xin, Y., Evans, D., Heerman, C., Chase, J., Marupadi, V. & Yumerefendi, A. (2009), The missing link: Putting the network in networked cloud computing, *in* 'in ICVCI09: International Conference on the Virtual Computing Initiative'.
14. Baldine, I., Xin, Y., Mandal, A., Renci, C. H., Chase, U.-C. J., Marupadi, V., Yumerefendi, A. & Irwin, D. (2010), Networked cloud orchestration: A GENI perspective, *in* '2010 IEEE Globecom Workshops', IEEE. http://dx.doi.org/10.1109/GLOCOMW.2010.5700385
15. Baldine, I., Xin, Y., Mandal, A., Ruth, P., Yumerefendi, A. & Chase, J. (2012), 'ExoGENI: A Multi-Domain Infrastructure-as-a-Service Testbed'.
16. Bashir, S. & Ahmed, N. (2015), VirtMonE: Efficient detection of elephant flows in virtualized data centers, *in* 'Telecommunication Networks and Applications Conference (ITNAC), 2015 International', IEEE, pp. 280–285. http://dx.doi.org/10.1109/atnac.2015.7366826
17. Bastin, N., Bavier, A., Blaine, J., Chen, J., Krishnan, N., Mambretti, J., McGeer, R., Ricci, R. & Watts, N. (2014), 'The InstaGENI initiative: An architecture for distributed systems and advanced programmable networks', *Computer Networks* **61**, 24–38. http://dx.doi.org/10.1016/j.bjp.2013.12.034
18. Bavier, A., Chen, J., Mambretti, J., McGeer, R., McGeer, S., Nelson, J., O'Connell, P., Ricart, G., Tredger, S. & Coady, Y. (2014), The GENI experiment engine, *in* 'Teletraffic Congress (ITC), 2014 26th International', IEEE, pp. 1–6. http://dx.doi.org/10.1109/itc.2014.6932974
19. Bavier, A., Coady, Y., Mack, T., Matthews, C., Mambretti, J., McGeer, R., Mueller, P., Snoeren, A. & Yuen, M. (2012), GENICloud and transcloud, *in* 'Proceedings of the 2012 workshop on Cloud services, federation, and the 8th open cirrus summit', FederatedClouds '12, ACM, New York, NY, USA, pp. 13–18. http://dx.doi.org/10.1145/2378975.2378980
20. Bejerano, Y., Ferragut, J., Guo, K., Gupta, V., Gutterman, C., Nandagopal, T. & Zussman, G. (2014), Experimental Evaluation of a Scalable WiFi Multicast Scheme in the ORBIT Testbed, *in* 'Research and Educational Experiment Workshop (GREE), 2014 Third GENI', IEEE, pp. 36–42. http://dx.doi.org/10.1109/gree.2014.22
21. Berman, M. & Brinn, M. (2014), Progress and challenges in worldwide federation of future internet and distributed cloud testbeds, *in* 'Science and Technology Conference (Modern Networking Technologies) (MoNeTeC), 2014 First International', IEEE, pp. 1–6. http://dx.doi.org/10.1109/monetec.2014.6995579
22. Berman, M., Chase, J. S., Landweber, L., Nakao, A., Ott, M., Raychaudhuri, D., Ricci, R. & Seskar, I. (2014), 'GENI: A federated testbed for innovative network experiments', *Computer Networks* **61**, 5–23. http://dx.doi.org/10.1016/j.bjp.2013.12.037
23. Berman, M., Demeester, P., Lee, J. W., Nagaraja, K., Zink, M., Colle, D., Krishnappa, D. K., Raychaudhuri, D., Schulzrinne, H., Seskar, I. & Sharma, S. (2015), 'Future Internets Escape the Simulator', *Commun. ACM* **58**(6), 78–89. http://dx.doi.org/10.1145/2699392
24. Berman, M., Elliott, C. & Landweber, L. (2014), GENI: Large-Scale Distributed Infrastructure for Networking and Distributed Systems Research, *in* '2014 IEEE Fifth International Conference on Communications and Electronics (ICCE)', IEEE, pp. 156–161. http://dx.doi.org/10.1109/CCE.2014.6916696
25. Berryman, A., Calyam, P., Cecil, J., Adams, G. B. & Comer, D. (2013), Advanced Manufacturing Use Cases and Early Results in GENI Infrastructure, *in* '2013 Proceedings Second GENI Research and Educational Experiment Workshop', IEEE. http://dx.doi.org/10.1109/GREE.2013.13
26. Bhanage, G., Daya, R., Seskar, I. & Raychaudhuri, D. (2010), VNTS: A Virtual Network Traffic Shaper for Air Time Fairness in 802.16e Systems, *in* 'Communications (ICC), 2010 IEEE International Conference on', IEEE. http://dx.doi.org/10.1109/ICC.2010.5502484
27. Bhanage, G., Seskar, I., Mahindra, R. & Raychaudhuri, D. (2010), Virtual basestation: architecture for an open shared WiMAX framework, *in* 'Proceedings of the second ACM SIGCOMM workshop on Virtualized infrastructure systems and architectures', VISA '10, ACM, New York, NY, USA. http://dx.doi.org/10.1145/1851399.1851401

28. Bhanage, G., Seskar, I. & Raychaudhuri, D. (2012), 'A virtualization architecture for mobile WiMAX networks', *SIGMOBILE Mob. Comput. Commun. Rev.* **15**(4). http://dx.doi.org/10.1145/2169077.2169082
29. Bhanage, G., Seskar, I., Zhang, Y., Raychaudhuri, D. & Jain, S. (2011), Experimental Evaluation of OpenVZ from a Testbed Deployment Perspective, *in* T. Magedanz, A. Gavras, N. Thanh & J. Chase, eds, 'Testbeds and Research Infrastructures. Development of Networks and Communities', Vol. 46 of *Lecture Notes of the Institute for Computer Sciences, Social Informatics and Telecommunications Engineering*, Springer Berlin Heidelberg, pp. 103–112. http://dx.doi.org/10.1007/978-3-642-17851-1_7
30. Bhanage, G., Vete, D., Seskar, I. & Raychaudhuri, D. (2010), SplitAP: Leveraging Wireless Network Virtualization for Flexible Sharing of WLANs, *in* 'Global Telecommunications Conference (GLOBECOM 2010), 2010 IEEE', IEEE. http://dx.doi.org/10.1109/GLOCOM.2010.5684328
31. Bhat, D., Riga, N. & Zink, M. (2014), Towards seamless application delivery using software defined exchanges, *in* 'Teletraffic Congress (ITC), 2014 26th International', IEEE, pp. 1–6. http://dx.doi.org/10.1109/itc.2014.6932971
32. Bhat, D., Wang, C., Rizk, A. & Zink, M. (2015), A load balancing approach for adaptive bitrate streaming in Information Centric networks, *in* 'Multimedia & Expo Workshops (ICMEW), 2015 IEEE International Conference on', IEEE, pp. 1–6. http://dx.doi.org/10.1109/icmew.2015.7169802
33. Bhojwani, S. (2015), Interoperability in Federated Clouds (Master's thesis), Master's thesis, University of Victoria. http://hdl.handle.net/1828/6732
34. Bhojwani, S., Hemmings, M., Ingalls, D., Lincke, J., Krahn, R., Lary, D., McGeer, R., Ricart, G., Roder, M., Coady, Y. & Stege, U. (2015), 'The Ignite Distributed Collaborative Visualization System', *SIGMETRICS Perform. Eval. Rev.* **43**(3), 45–46. http://dx.doi.org/10.1145/2847220.2847234
35. Blanton, E., Chatterjee, S., Gangam, S., Kala, S., Sharma, D., Fahmy, S. & Sharma, P. (2012), Design and evaluation of the S³ monitor network measurement service on GENI, *in* '2012 Fourth International Conference on Communication Systems and Networks (COMSNETS 2012)', IEEE. http://dx.doi.org/10.1109/COMSNETS.2012.6151327
36. Brinn, M., Bastin, N., Bavier, A., Berman, M., Chase, J. & Ricci, R. (2015), Trust as the Foundation of Resource Exchange in GENI, *in* 'Proceedings of the 10th EAI International Conference on Testbeds and Research Infrastructures for the Development of Networks & Communities', ACM. http://dx.doi.org/10.4108/icst.tridentcom.2015.259683
37. Bronzino, F., Han, C., Chen, Y., Nagaraja, K., Yang, X., Seskar, I. & Raychaudhuri, D. (2014), In-Network Compute Extensions for Rate-Adaptive Content Delivery in Mobile Networks, *in* 'Network Protocols (ICNP), 2014 IEEE 22nd International Conference on', IEEE, pp. 511–517. http://dx.doi.org/10.1109/icnp.2014.81
38. Brown, D., Ascigil, O., Nasir, H., Carpenter, C., Griffioen, J. & Calvert, K. (2014), Designing a GENI Experimenter Tool to Support the Choice Net Internet Architecture, *in* 'Network Protocols (ICNP), 2014 IEEE 22nd International Conference on', IEEE, pp. 548–554. http://dx.doi.org/10.1109/icnp.2014.88
39. Brown, D., Nasir, H., Carpenter, C., Ascigil, O., Griffioen, J. & Calvert, K. (2014), ChoiceNet gaming: Changing the gaming experience with economics, *in* 'Computer Games: AI, Animation, Mobile, Multimedia, Educational and Serious Games (CGAMES), 2014', IEEE, pp. 1–5. http://dx.doi.org/10.1109/cgames.2014.6934146
40. Calyam, P., Mishra, A., Antequera, R. B., Chemodanov, D., Berryman, A., Zhu, K., Abbott, C. & Skubic, M. (2015), 'Synchronous Big Data analytics for personalized and remote physical therapy', *Pervasive and Mobile Computing* . http://dx.doi.org/10.1016/j.pmcj.2015.09.004
41. Calyam, P., Rajagopalan, S., Seetharam, S., Selvadhurai, A., Salah, K. & Ramnath, R. (2014), 'VDC-Analyst: Design and verification of virtual desktop cloud resource allocations', *Computer Networks* **68**, 110–122. http://dx.doi.org/10.1016/j.comnet.2014.02.022
42. Calyam, P., Rajagopalan, S., Selvadhurai, A., Mohan, S., Venkataraman, A., Berryman, A. & Ramnath, R. (2013) , Leveraging OpenFlow for resource placement of virtual desktop cloud

applications, *in* 'Integrated Network Management (IM 2013), 2013 IFIP/IEEE International Symposium on', pp. 311–319.

43. Calyam, P., Seetharam, S. & Antequera, R. B. (2014), GENI Laboratory Exercises Development for a Cloud Computing Course, *in* 'Research and Educational Experiment Workshop (GREE), 2014 Third GENI', IEEE, pp. 19–24. http://dx.doi.org/10.1109/gree.2014.15
44. Calyam, P., Sridharan, M., Xu, Y., Zhu, K., Berryman, A., Patali, R. & Venkataraman, A. (2011), 'Enabling performance intelligence for application adaptation in the Future Internet', *Communications and Networks, Journal of* **13**(6). http://dx.doi.org/10.1109/JCN.2011.6157475
45. Calyam, P., Venkataraman, A., Berryman, A. & Faerman, M. (2012), Experiences from Virtual Desktop CloudExperiments in GENI, *in* 'First GENI Research and Educational Experiment Workshop (GREE 2012)'.
46. Cameron, K., Brooks, R. R., Deng, J., Yu, L., Wang, K. C. & Martin, J. (2012), WiMAX: Bandwidth Contention Resolution Vulnerability to Denial of Service Attacks, *in* 'First GENI Research and Educational Experiment Workshop (GREE 2012)'.
47. Chakrabortty, A. & Xin, Y. (2013), Hardware-in-the-Loop Simulations and Verifications of Smart Power Systems Over an Exo-GENI Testbed, *in* '2013 Proceedings Second GENI Research and Educational Experiment Workshop', IEEE. http://dx.doi.org/10.1109/GREE.2013.12
48. Chen, K. & Shen, H. (2011), Global optimization of file availability through replication for efficient file sharing in MANETs, *in* 'Network Protocols (ICNP), 2011 19th IEEE International Conference on', IEEE, pp. 226–235. http://dx.doi.org/10.1109/icnp.2011.6089056
49. Chen, K. & Shen, H. (2013), Cont2: Social-Aware Content and Contact Based File Search in Delay Tolerant Networks, *in* 'Proceedings of the 2013 42Nd International Conference on Parallel Processing', ICPP '13, IEEE Computer Society, Washington, DC, USA, pp. 190–199. http://dx.doi.org/10.1109/icpp.2013.28
50. Chen, K., Shen, H. & Zhang, H. (2011), Leveraging Social Networks for P2P Content-Based File Sharing in Mobile Ad Hoc Networks, *in* '2011 IEEE Eighth International Conference on Mobile Ad-Hoc and Sensor Systems', IEEE. http://dx.doi.org/10.1109/MASS.2011.24
51. Chen, K., Xu, K., Winburn, S., Shen, H., Wang, K.-C. & Li, Z. (2012), Experimentation of a MANET Routing Algorithm on the GENI ORBIT Testbed, *in* 'First GENI Research and Educational Experiment Workshop (GREE 2012)'.
52. Chen, X., Wolf, T., Griffioen, J., Ascigil, O., Dutta, R., Rouskas, G., Bhat, S., Baldin, I. & Calvert, K. (2015), Design of a protocol to enable economic transactions for network services, *in* 'Communications (ICC), 2015 IEEE International Conference on', IEEE, pp. 5354–5359. http://dx.doi.org/10.1109/icc.2015.7249175
53. Cherukuri, R., Liu, X., Bavier, A., Sterbenz, J. P. G. & Medhi, D. (2011), Network virtualization in GpENI: Framework, implementation & integration experience, *in* '12th IFIP/IEEE International Symposium on Integrated Network Management (IM 2011) and Workshops', IEEE. http://dx.doi.org/10.1109/INM.2011.5990568
54. Chin, T., Mountrouidou, X., Li, X. & Xiong, K. (2015*a*), An SDN-supported collaborative approach for DDoS flooding detection and containment, *in* 'Military Communications Conference, MILCOM 2015 - 2015 IEEE', IEEE, pp. 659–664. http://dx.doi.org/10.1109/milcom.2015.7357519
55. Chin, T., Mountrouidou, X., Li, X. & Xiong, K. (2015*b*), Selective Packet Inspection to Detect DoS Flooding Using Software Defined Networking (SDN), *in* 'Distributed Computing Systems Workshops (ICDCSW), 2015 IEEE 35th International Conference on', IEEE, pp. 95–99. http://dx.doi.org/10.1109/icdcsw.2015.27
56. Chowdhury & Boutaba, R. (2010), 'A survey of network virtualization', *Computer Networks* **54**(5), 862–876. http://www.sciencedirect.com/science/article/pii/S1389128609003387
57. Collings, J. & Liu, J. (2014), An OpenFlow-Based Prototype of SDN-Oriented Stateful Hardware Firewalls, *in* 'Network Protocols (ICNP), 2014 IEEE 22nd International Conference on', IEEE, pp. 525–528. http://dx.doi.org/10.1109/icnp.2014.83

58. Dane, L. & Gurkan, D. (2014), GENI with a Network Processing Unit: Enriching SDN Application Experiments, *in* 'Research and Educational Experiment Workshop (GREE), 2014 Third GENI', IEEE, pp. 9–14. http://dx.doi.org/10.1109/gree.2014.27
59. Das, S., Yiakoumis, Y., Parulkar, G., McKeown, N., Singh, P., Getachew, D. & Desai, P. D. (2011), Application-aware aggregation and traffic engineering in a converged packet-circuit network, *in* 'Optical Fiber Communication Conference and Exposition (OFC/NFOEC), 2011 and the National Fiber Optic Engineers Conference', IEEE, pp. 1–3. http://ieeexplore.ieee.org/xpls/abs_all.jsp?arnumber=5875210
60. Deng, J., Brooks, R. R. & Martin, J. (2012), 'Assessing the Effect of WiMAX System Parameter Settings on MAC-level Local DoS Vulnerability', *International Journal of Performability Engineering* **8**(2).
61. Dong, M., Li, Q., Zarchy, D., Godfrey, P. B. & Schapira, M. (2015), PCC: Re-architecting Congestion Control for Consistent High Performance, *in* '12th USENIX Symposium on Networked Systems Design and Implementation (NSDI 15)', USENIX Association, Oakland, CA, pp. 395–408. https://www.usenix.org/conference/nsdi15/technical-sessions/presentation/dong
62. Duerig, J., Ricci, R., Stoller, L., Strum, M., Wong, G., Carpenter, C., Fei, Z., Griffioen, J., Nasir, H., Reed, J. & Wu, X. (2012), 'Getting started with GENI: a user tutorial', *SIGCOMM Comput. Commun. Rev.* **42**(1). http://dx.doi.org/10.1145/2096149.2096161
63. Duerig, J., Ricci, R., Stoller, L., Wong, G., Chikkulapelly, S. & Seok, W. (2012), Designing a Federated Testbed as a Distributed System.
64. Dumba, B., Sun, G., Mekky, H. & Zhang, Z.-L. (2014), Experience in Implementing & Deploying a Non-IP Routing Protocol VIRO in GENI, *in* 'Network Protocols (ICNP), 2014 IEEE 22nd International Conference on', IEEE, pp. 533–539. http://dx.doi.org/10.1109/icnp.2014.85
65. Edwards, S., Liu, X. & Riga, N. (2015), 'Creating Repeatable Computer Science and Networking Experiments on Shared, Public Testbeds', *SIGOPS Oper. Syst. Rev.* **49**(1), 90–99. http://dx.doi.org/10.1145/2723872.2723884
66. El Alaoui, S. (2015), Routing Optimization in Interplanetary Networks (Master's Thesis), Master's thesis, University of Nebraska. http://scholar.google.com/scholar_url?url=http://digitalcommons.unl.edu/cgi/viewcontent.cgi%3Farticle%3D1110%26context%3Dcomputerscidiss&hl=en&sa=X&scisig=AAGBfm3bqGZQbbqEX7SG7r5YDIw5epl3sg&nossl=1&oi=scholaralrt
67. Elliott, C. & Falk, A. (2009), 'An update on the GENI project', *SIGCOMM Comput. Commun. Rev.* **39**(3). http://dx.doi.org/10.1145/1568613.1568620
68. Elliott, S. D. (2015), Exploring the Challenges and Opportunities of Implementing Software-Defined Networking in a Research Testbed (Master's thesis), Master's thesis, North Carolina State University. http://repository.lib.ncsu.edu/ir/bitstream/1840.16/10164/1/etd.pdf
69. Erazo, M. A. & Liu, J. (2010), On enabling real-time large-scale network simulation in GENI: the PrimoGENI approach, *in* 'Proceedings of the 3rd International ICST Conference on Simulation Tools and Techniques', SIMUTools '10, ICST (Institute for Computer Sciences, Social-Informatics and Telecommunications Engineering), ICST, Brussels, Belgium, Belgium. http://dx.doi.org/10.4108/ICST.SIMUTOOLS2010.8636
70. Erazo, M. A., Rong, R. & Liu, J. (2015), 'Symbiotic Network Simulation and Emulation', *ACM Trans. Model. Comput. Simul.* **26**(1). http://dx.doi.org/10.1145/2717308
71. Esposito, F., Wang, Y., Matta, I. & Day, J. (2013), 'Dynamic Layer Instantiation as a Service', Poster and NSDI 13. https://www.usenix.org/system/files/nsdip13-paper11.pdf
72. Feamster, N., Gao, L. & Rexford, J. (2007), 'How to lease the internet in your spare time', *SIGCOMM Comput. Commun. Rev.* **37**(1), 61–64. http://doi.acm.org/10.1145/1198255.1198265
73. Feamster, N., Nayak, A., Kim, H., Clark, R., Mundada, Y., Ramachandran, A. & bin Tariq, M. (2010), Decoupling policy from configuration in campus and enterprise networks, *in* '2010 17th IEEE Workshop on Local & Metropolitan Area Networks (LANMAN)', IEEE. http://dx.doi.org/10.1109/LANMAN.2010.5507162

74. Fei, Z., Xu, Q. & Lu, H. (2014), Generating large network topologies for GENI experiments, *in* 'SOUTHEASTCON 2014, IEEE', IEEE, pp. 1–7. http://dx.doi.org/10.1109/secon.2014.6950726
75. Fei, Z., Yi, P. & Yang, J. (2014), 'A Performance Perspective on Choosing between Single Aggregate and Multiple Aggregates for GENI Experime nts', *EAI Endorsed Transactions on Industrial Networks and Intelligent Systems* **1**(1), e5+. http://dx.doi.org/10.4108/inis.1.1.e5
76. Femminella, M., Francescangeli, R., Reali, G., Lee, J. W. & Schulzrinne, H. (2011), 'An enabling platform for autonomic management of the future internet', *IEEE Network* **25**(6). http://dx.doi.org/10.1109/MNET.2011.6085639
77. Fund, F., Dong, C., Korakis, T. & Panwar, S. (2012), A Framework for Multidimensional Measurements on an Experimental WiMAX Testbed, *in* T. Korakis, M. Zink & M. Ott, eds, 'Testbeds and Research Infrastructure. Development of Networks and Communities', Vol. 44 of *Lecture Notes of the Institute for Computer Sciences, Social Informatics and Telecommunications Engineering*, Springer Berlin Heidelberg, pp. 369–371. http://dx.doi.org/10.1007/978-3-642-35576-9_32
78. Fund, F., Wang, C., Korakis, T., Zink, M. & Panwar, S. (2013), GENI WiMAX Performance: Evaluation and Comparison of Two Campus Testbeds, *in* '2013 Proceedings Second GENI Research and Educational Experiment Workshop', IEEE. http://dx.doi.org/10.1109/GREE.2013.23
79. Gangam, S., Blanton, E. & Fahmy, S. (2012), Exercises for Graduate Students using GENI, *in* 'First GENI Research and Educational Experiment Workshop (GREE 2012)'.
80. Gangam, S. & Fahmy, S. (2011), Mitigating interference in a network measurement service, *in* '2011 IEEE Nineteenth IEEE International Workshop on Quality of Service', IEEE. http://dx.doi.org/10.1109/IWQOS.2011.5931347
81. Gao, J. & Xiao, Y. (2012), ProtoGENI DoS/DDoS Security Tests and Experiments, *in* 'First GENI Research and Educational Experiment Workshop (GREE 2012)'.
82. Gember, A., Dragga, C. & Akella, A. (2012), ECOS: Practical Mobile Application Offloading for Enterprises, *in* '2nd USENIX Workshop on Hot Topics in Management of Internet, Cloud, and Enterprise Networks and Services (Hot-ICE '12)'. http://www.usenix.org/conference/hot-ice12/ecos-practical-mobile-application-of%EF%AC%82oading-enterprises
83. Ghaffarinejad, A. & Syrotiuk, V. R. (2014), Load Balancing in a Campus Network Using Software Defined Networking, *in* 'Research and Educational Experiment Workshop (GREE), 2014 Third GENI', IEEE, pp. 75–76. http://dx.doi.org/10.1109/gree.2014.9
84. Grandl, R., Han, D., Lee, S. B., Lim, H., Machado, M., Mukerjee, M. & Naylor, D. (2012), Supporting network evolution and incremental deployment with XIA, *in* 'Proceedings of the ACM SIGCOMM 2012 conference on Applications, technologies, architectures, and protocols for computer communication', SIGCOMM '12, ACM, New York, NY, USA, pp. 281–282. http://dx.doi.org/10.1145/2342356.2342410
85. Griffioen, J., Fei, Z., Nasir, H., Wu, X., Reed, J. & Carpenter, C. (2012*a*), Teaching with the Emerging GENI Network, *in* 'Proceedings of the 2012 International Conference on Frontiers in Education: Computer Science and Computer Engineering (FECS)'. http://worldcomp-proceedings.com/proc/p2012/FEC3780.pdf
86. Griffioen, J., Fei, Z., Nasir, H., Wu, X., Reed, J. & Carpenter, C. (2012*b*), The design of an instrumentation system for federated and virtualized network testbeds, *in* 'Network Operations and Management Symposium (NOMS), 2012 IEEE', IEEE. http://dx.doi.org/10.1109/NOMS.2012.6212061
87. Griffioen, J., Fei, Z., Nasir, H., Wu, X., Reed, J. & Carpenter, C. (2013), GENI-Enabled Programming Experiments for Networking Classes, *in* 'Research and Educational Experiment Workshop (GREE), 2013 Second GENI', IEEE, pp. 111–118. http://dx.doi.org/10.1109/gree.2013.30
88. Griffioen, J., Fei, Z., Nasir, H., Wu, X., Reed, J. & Carpenter, C. (2014), 'Measuring experiments in GENI', *Computer Networks* **63**, 17–32. http://dx.doi.org/10.1016/j.bjp.2013.10.016

89. Group, G. P. (2006), 'GENI Design Principles', *Computer* **39**(9), 102–105. http://dx.doi.org/10.1109/mc.2006.307
90. Guan, X., Choi, B.-Y. & Song, S. (2013), Reliability and Scalability Issues in Software Defined Network Frameworks, *in* 'Research and Educational Experiment Workshop (GREE), 2013 Second GENI', IEEE, pp. 102–103. http://dx.doi.org/10.1109/gree.2013.28
91. Gupta, A., Vanbever, L., Shahbaz, M., Donovan, S. P., Schlinker, B., Feamster, N., Rexford, J., Shenker, S., Clark, R. & Katz-Bassett, E. (2014), SDX: A Software Defined Internet Exchange, *in* 'Proceedings of the 2014 ACM Conference on SIGCOMM', SIGCOMM '14, ACM, New York, NY, USA, pp. 551–562. http://dx.doi.org/10.1145/2619239.2626300
92. Herron, J.-P. (2008), GENI Meta-Operations Center, *in* '2008 IEEE Fourth International Conference on eScience', IEEE. http://dx.doi.org/10.1109/eScience.2008.103
93. Huang, S. & Griffioen, J. (2013), Network Hypervisors: Managing the Emerging SDN Chaos, *in* 'Computer Communications and Networks (ICCCN), 2013 22nd International Conference on', IEEE, pp. 1–7. http://dx.doi.org/10.1109/icccn.2013.6614160
94. Huang, S., Griffioen, J. & Calvert, K. (2012), PVNs: Making Virtualized Network Infrastructure Usable, *in* 'ACM/IEEE Symposium on Architectures for Networking and Communications Systems (ANCS '12)'. http://dx.doi.org/10.1145/2396556.2396590
95. Huang, S., Griffioen, J. & Calvert, K. L. (2013), Fast-Tracking GENI Experiments Using HyperNets, *in* 'Research and Educational Experiment Workshop (GREE), 2013 Second GENI', IEEE, pp. 1–8. http://dx.doi.org/10.1109/gree.2013.10
96. Huang, S., Xu, H., Xin, Y., Brieger, L., Moore, R. & Rajasekar, A. (2014), A Framework for Integration of Rule-Oriented Data Management Policies with Network Policies, *in* 'Research and Educational Experiment Workshop (GREE), 2014 Third GENI', IEEE, pp. 71–72. http://dx.doi.org/10.1109/gree.2014.19
97. Javed, U., Cunha, I., Choffnes, D., Katz-Bassett, E., Anderson, T. & Krishnamurthy, A. (2013), 'PoiRoot: Investigating the Root Cause of Interdomain Path Changes', *Proceedings of the ACM SIGCOMM 2013 conference* **43**(4), 183–194. http://dx.doi.org/10.1145/2486001.2486036
98. Jin, R. & Wang, B. (2013), Malware Detection for Mobile Devices Using Software-Defined Networking, *in* 'Research and Educational Experiment Workshop (GREE), 2013 Second GENI', IEEE, pp. 81–88. http://dx.doi.org/10.1109/gree.2013.24
99. Jofre, J., Velayos, C., Landi, G., Giertych, M., Hume, A. C., Francis, G. & Vico Oton, A. (2014), 'Federation of the BonFIRE multi-cloud infrastructure with networking facilities', *Computer Networks* **61**, 184–196. http://dx.doi.org/10.1016/j.bjp.2013.11.012
100. Jourjon, G., Marquez-Barja, J. M., Rakotoarivelo, T., Mikroyannidis, A., Lampropoulos, K., Denazis, S., Tranoris, C., Pareit, D., Domingue, J., DaSilva, L. A. & Ott, M. (2015), 'FORGE Toolkit: Leveraging Distributed Systems in eLearning Platforms'. http://dx.doi.org/10.1109/tetc.2015.2511454
101. Ju, X., Zhang, H., Zeng, W., Sridharan, M., Li, J., Arora, A., Ramnath, R. & Xin, Y. (2011), LENS: resource specification for wireless sensor network experimentation infrastructures, *in* 'Proceedings of the 6th ACM international workshop on Wireless network testbeds, experimental evaluation and characterization', WiNTECH '11, ACM, New York, NY, USA. http://dx.doi.org/10.1145/2030718.2030727
102. Juluri, P. (2015), Measurement And Improvement of Quality-of-Experience For Online Video Streaming Services (Doctoral dissertation), PhD thesis, University of Missouri - Kansas City. https://mospace.umsystem.edu/xmlui/bitstream/handle/10355/46696/JuluriMeaImpQua.pdf?sequence=1&isAllowed=y
103. Juluri, P., Tamarapalli, V. & Medhi, D. (2015), SARA: Segment aware rate adaptation algorithm for dynamic adaptive streaming over HTTP, *in* 'Communication Workshop (ICCW), 2015 IEEE International Conference on', IEEE, pp. 1765–1770. http://dx.doi.org/10.1109/iccw.2015.7247436
104. Kanada, Y. & Tarui, T. (2015), Federation-less federation of ProtoGENI and VNode platforms, *in* 'Information Networking (ICOIN), 2015 International Conference on', IEEE, pp. 271–276. http://dx.doi.org/10.1109/icoin.2015.7057895

105. Kangarlou, A., Xu, D., Kozat, U. C., Padala, P., Lantz, B. & Igarashi, K. (2011), 'In-network live snapshot service for recovering virtual infrastructures', *Network, IEEE* **25**(4), 12–19. http://dx.doi.org/10.1109/mnet.2011.5958003
106. Katz-Bassett, E., Choffnes, D. R., Cunha, I., Scott, C., Anderson, T. & Krishnamurthy, A. (2011), Machiavellian Routing: Improving Internet Availability with BGP Poisoning, *in* 'Proceedings of the 10th ACM Workshop on Hot Topics in Networks', HotNets-X, ACM, New York, NY, USA. http://dx.doi.org/10.1145/2070562.2070573
107. Katz-Bassett, E., Scott, C., Choffnes, D. R., Cunha, I., Valancius, V., Feamster, N., Madhyastha, H. V., Anderson, T. & Krishnamurthy, A. (2012), 'LIFEGUARD: Practical Repair of Persistent Route Failures', *Proceedings of the ACM SIGCOMM 2012 conference* **42**(4), 395–406. http://dx.doi.org/10.1145/2377677.2377756
108. Khurshid, A., Zhou, W., Caesar, M. & Godfrey, P. B. (2012), VeriFlow: verifying network-wide invariants in real time, *in* 'Proceedings of the first workshop on Hot topics in software defined networks', HotSDN '12, ACM, New York, NY, USA, pp. 49–54. http://doi.acm.org/10.1145/2342441.2342452
109. Kim, D., Kim, J., Wang, G., Park, J.-H. & Kim, S.-H. (2014), 'K-GENI testbed deployment and federated meta operations experiment over GENI and KREONET', *Computer Networks* **61**, 39–50. http://dx.doi.org/10.1016/j.bjp.2013.11.016
110. Kim, D. Y., Mathy, L., Campanella, M., Summerhill, R., Williams, J., Shimojo, S., Kitamura, Y. & Otsuki, H. (2009), Future Internet: Challenges in Virtualization and Federation, *in* '2009 Fifth Advanced International Conference on Telecommunications', IEEE. http://dx.doi.org/10.1109/AICT.2009.8
111. Kim, H. & Lee, S. (2012), FiRST Cloud Aggregate Manager development over FiRST: Future Internet testbed, *in* 'The International Conference on Information Network 2012', IEEE. http://dx.doi.org/10.1109/ICOIN.2012.6164436
112. Kline, D. & Quan, J. (2011), Attribute description service for large-scale networks, *in* 'Proceedings of the 2nd international conference on Human centered design', HCD'11, Springer-Verlag, Berlin, Heidelberg. http://portal.acm.org/citation.cfm?id=2021672.2021735
113. Kobayashi, M., Seetharaman, S., Parulkar, G., Appenzeller, G., Little, J., van Reijendam, J., Weissmann, P. & McKeown, N. (2014), 'Maturing of OpenFlow and Software-defined Networking through deployments', *Computer Networks* **61**, 151–175. http://dx.doi.org/10.1016/j.bjp.2013.10.011
114. Krishnappa, D. K., Irwin, D., Lyons, E. & Zink, M. (2013), 'CloudCast: Cloud Computing for Short-Term Weather Forecasts', *Computing in Science & Engineering* **15**(4), 30–37. http://dx.doi.org/10.1109/mcse.2013.43
115. Krishnappa, D. K., Lyons, E., Irwin, D. & Zink, M. (2012*a*), Network capabilities of cloud services for a real time scientific application, *in* '37th Annual IEEE Conference on Local Computer Networks', IEEE, pp. 487–495. http://dx.doi.org/10.1109/lcn.2012.6423665
116. Krishnappa, D. K., Lyons, E., Irwin, D. & Zink, M. (2012*b*), Performance of GENI Cloud Testbeds for Real Time Scientific Application, *in* 'First GENI Research and Educational Experiment Workshop (GREE 2012)'.
117. Kuai, M., Hong, X. & Flores, R. R. (2014), Evaluating Interest Broadcast in Vehicular Named Data Networking, *in* 'Research and Educational Experiment Workshop (GREE), 2014 Third GENI', IEEE, pp. 77–78. http://dx.doi.org/10.1109/gree.2014.23
118. Lara, A. (2015), Using Software-Defined Networking to Improve Campus, Transport and Future Internet Architectures (Doctoral dissertation), PhD thesis, University of Nebraska - Lincoln. http://digitalcommons.unl.edu/computerscidiss/93/
119. Lara, A., Ramamurthy, B., Nagaraja, K., Krishnamoorthy, A. & Raychaudhuri, D. (2014), 'Using OpenFlow to provide cut-through switching in MobilityFirst', pp. 1–13. http://dx.doi.org/10.1007/s11107-014-0461-3
120. Lauer, G., Irwin, R., Kappler, C. & Nishioka, I. (2013), Distributed Resource Control Using Shadowed Subgraphs, *in* 'Proceedings of the Ninth ACM Conference on Emerging Networking Experiments and Technologies', CoNEXT '13, ACM, New York, NY, USA, pp. 43–48. http://dx.doi.org/10.1145/2535372.2535410

121. Lee, J. W. (2012), Towards a Common System Architecture for Dynamically Deploying Network Services in Routers and End Hosts (Doctoral dissertation), PhD thesis, Columbia University. http://academiccommons.columbia.edu/download/fedora_content/download/ac:147210/CONTENT/Lee_columbia_0054D_10773.pdf
122. Lee, J. W., Francescangeli, R., Janak, J., Srinivasan, S., Baset, S. A., Schulzrinne, H., Despotovic, Z. & Kellerer, W. (2011), NetServ: Active Networking 2.0, *in* '2011 IEEE International Conference on Communications Workshops (ICC)', IEEE. http://dx.doi.org/10.1109/iccw.2011.5963554
123. Lee, K. S., Wang, H. & Weatherspoon, H. (2013), SoNIC: Precise Realtime Software Access and Control of Wired Networks, *in* 'Proceedings of the 10th USENIX Conference on Networked Systems Design and Implementation', NSDI'13, USENIX Association, Berkeley, CA, USA, pp. 213–266. http://portal.acm.org/citation.cfm?id=2482626.2482648
124. Li, D. & Hong, X. (2011), Practical exploitation on system vulnerability of ProtoGENI, *in* 'Proceedings of the 49th Annual Southeast Regional Conference', ACM-SE '11, ACM, New York, NY, USA. http://dx.doi.org/10.1145/2016039.2016073
125. Li, D., Hong, X. & Bowman, J. (2011), 'Evaluation of Security Vulnerabilities by Using ProtoGENI as a Launchpad'. ftp://202.38.75.7/pub/%D0%C2%CE%C4%BC%FE%BC%D0%20(2)/DATA/PID1102190.PDF
126. Li, T., Van Vorst, N. & Liu, J. (2013), 'A Rate-based TCP Traffic Model to Accelerate Network Simulation', *Simulation* **89**(4), 466–480. http://dx.doi.org/10.1177/0037549712469892
127. Li, T., Van Vorst, N., Rong, R. & Liu, J. (2012), Simulation studies of OpenFlow-based in-network caching strategies, *in* 'Proceedings of the 15th Communications and Networking Simulation Symposium', CNS '12, Society for Computer Simulation International, San Diego, CA, USA. http://portal.acm.org/citation.cfm?id=2331762.2331774
128. Liu, J., Abu Obaida, M. & Dos Santos, F. (2014), Toward PrimoGENI Constellation for Distributed At-Scale Hybrid Network Test, *in* 'Research and Educational Experiment Workshop (GREE), 2014 Third GENI', IEEE, pp. 29–35. http://dx.doi.org/10.1109/gree.2014.10
129. Liu, J., O'Neil, T., Desell, T. & Carlson, R. (2012), Work-in-Progress: Empirical Verification of A Subset Sum Hypothesis in GENI Cloud, *in* 'First GENI Research and Educational Experiment Workshop (GREE 2012)'.
130. Liu, L., Peng, W.-R., Casellas, R., Tsuritani, T., Morita, I., Martinez, R., Munoz, R., Suzuki, M. & Ben Yoo, S. J. (2015), 'Dynamic OpenFlow-Based Lightpath Restoration in Elastic Optical Networks on the GENI Testbed', *Lightwave Technology, Journal of* **33**(8), 1531–1539. http://dx.doi.org/10.1109/jlt.2014.2388194
131. Liu, L., Zhu, Z., Wang, X., Song, G., Chen, C., Chen, X., Ma, S., Feng, X., Proietti, R. & Yoo, S. J. B. (2015), Field Trial of Broker-based Multi-domain Software-Defined Heterogeneous Wireline-Wireless-Optical Networks, *in* 'Optical Fiber Communication Conference', OSA. http://dx.doi.org/10.1364/ofc.2015.th3j.5
132. Liu, X. (2015), Dynamic Virtual Network Restoration with Optimal Standby Virtual Router Selection (Doctoral dissertation), PhD thesis, University of Missouri - Kansas City. https://mospace.umsystem.edu/xmlui/bitstream/handle/10355/46697/LiuDynVirNet.pdf?sequence=1&isAllowed=y
133. Liu, X., Edwards, S., Riga, N. & Medhi, D. (2015), Design of a software-defined resilient virtualized networking environment, *in* 'Design of Reliable Communication Networks (DRCN), 2015 11th International Conference on the', IEEE, pp. 111–114. http://dx.doi.org/10.1109/drcn.2015.7148999
134. Luna, N., Shetty, S., Rogers, T. & Xiong, K. (2012), Assessment of Router Vulnerabilities on PlanetLab Infrastructure for Secure Cloud Computing, *in* 'First GENI Research and Educational Experiment Workshop (GREE 2012)'.
135. Maccherani, E., Femminella, M., Lee, J. W., Francescangeli, R., Janak, J., Reali, G. & Schulzrinne, H. (2012), Extending the NetServ autonomic management capabilities using OpenFlow, *in* '2012 IEEE Network Operations and Management Symposium', IEEE. http://dx.doi.org/10.1109/NOMS.2012.6211961

136. Mahindra, R., Bhanage, G. D., Hadjichristofi, G., Seskar, I., Raychaudhuri, D. & Zhang, Y. Y. (2008), Space Versus Time Separation for Wireless Virtualization on an Indoor Grid, *in* 'Next Generation Internet Networks, 2008. NGI 2008', IEEE. http://dx.doi.org/10.1109/NGI.2008.36
137. Mahindra, R., Bhanage, G., Hadjichristofi, G., Ganu, S., Kamat, P., Seskar, I. & Raychaudhuri, D. (2008), Integration of heterogeneous networking testbeds, *in* 'Proceedings of the 4th International Conference on Testbeds and research infrastructures for the development of networks & communities', TridentCom '08, ICST (Institute for Computer Sciences, Social-Informatics and Telecommunications Engineering), ICST, Brussels, Belgium, Belgium. http://portal.acm.org/citation.cfm?id=1390609
138. Malishevskiy, A., Gurkan, D., Dane, L., Narisetty, R., Narayan, S. & Bailey, S. (2014), OpenFlow-Based Network Management with Visualization of Managed Elements, *in* 'Research and Educational Experiment Workshop (GREE), 2014 Third GENI', IEEE, pp. 73–74. http://dx.doi.org/10.1109/gree.2014.21
139. Mambretti, J., Chen, J. & Yeh, F. (2014 *a*), 'Creating environments for innovation: Designing and implementing advanced experimental network research testbeds based on the Global Lambda Integrated Facility and the StarLight Exchange', *Computer Networks* **61**, 118–131. http://dx.doi.org/10.1016/j.bjp.2013.12.024
140. Mambretti, J., Chen, J. & Yeh, F. (2014 *b*), Software-Defined Network Exchanges (SDXs) and Infrastructure (SDI): Emerging innovations in SDN and SDI interdomain multi-layer services and capabilities, *in* 'Science and Technology Conference (Modern Networking Technologies) (MoNeTeC), 2014 First International', IEEE, pp. 1–6. http://dx.doi.org/10.1109/monetec.2014.6995590
141. Mambretti, J., Chen, J. & Yeh, F. (2014 *c*), Software-Defined Network Exchanges (SDXs): Architecture, services, capabilities, and foundation technologies, *in* 'Teletraffic Congress (ITC), 2014 26th International', IEEE, pp. 1–6. http://dx.doi.org/10.1109/itc.2014.6932970
142. Mandal, A., Ruth, P., Baldin, I., Xin, Y., Castillo, C., Rynge, M. & Deelman, E. (2013), Evaluating I/O Aware Network Management for Scientific Workflows on Networked Clouds, *in* 'Proceedings of the Third International Workshop on Network-Aware Data Management', NDM '13, ACM, New York, NY, USA. http://dx.doi.org/10.1145/2534695.2534698
143. Mandal, A., Ruth, P., Baldin, I., Xin, Y., Castillo, C., Rynge, M. & Deelman, E. (2014), Leveraging and Adapting ExoGENI Infrastructure for Data-Driven Domain Science Workflows, *in* 'Research and Educational Experiment Workshop (GREE), 2014 Third GENI', IEEE, pp. 57–60. http://dx.doi.org/10.1109/gree.2014.12
144. Mandal, A., Xin, Y., Baldine, I., Ruth, P., Heerman, C., Chase, J., Orlikowski, V. & Yumerefendi, A. (2011), Provisioning and Evaluating Multi-domain Networked Clouds for Hadoop-based Applications, *in* 'Cloud Computing Technology and Science (CloudCom), 2011 IEEE Third International Conference on', pp. 690–697. http://dx.doi.org/10.1109/CloudCom.2011.107
145. Mandvekar, L., Qiao, C. & Husain, M. I. (2013), Enabling Wide Area Single System Image Experimentation on the GENI Platform, *in* 'Research and Educational Experiment Workshop (GREE), 2013 Second GENI', IEEE, pp. 97–101. http://dx.doi.org/10.1109/gree.2013.27
146. Mandvekar, L., Sathyaraja, A. & Qiao, C. (2012), Socially Aware Single System Images, *in* 'First GENI Research and Educational Experiment Workshop (GREE 2012)'.
147. Marasevic, J., Janak, J., Schulzrinne, H. & Zussman, G. (2013), WiMAX in the Classroom: Designing a Cellular Networking Hands-On Lab, *in* 'Research and Educational Experiment Workshop (GREE), 2013 Second GENI', IEEE, pp. 104–110. http://dx.doi.org/10.1109/gree.2013.29
148. Martin, V., Coulaby, A., Schaff, N., Tan, C. C. & Lin, S. (2014), Bandwidth Prediction on a WiMAX Network, *in* 'Mobile Ad Hoc and Sensor Systems (MASS), 2014 IEEE 11th International Conference on', IEEE, pp. 708–713. http://dx.doi.org/10.1109/mass.2014.75
149. Maziku, H. & Shetty, S. (2014), Network Aware VM Migration in Cloud Data Centers, *in* 'Research and Educational Experiment Workshop (GREE), 2014 Third GENI', IEEE, pp. 25–28. http://dx.doi.org/10.1109/gree.2014.18

150. Maziku, H., Shetty, S. & Rogers, T. (2012), Measurement-based IP Geolocation of Routers on Planetlab Infrastructure, *in* 'First GENI Research and Educational Experiment Workshop (GREE 2012)'.
151. McKeown, N., Anderson, T., Balakrishnan, H., Parulkar, G., Peterson, L., Rexford, J., Shenker, S. & Turner, J. (2008), 'OpenFlow: enabling innovation in campus networks', *SIGCOMM Comput. Commun. Rev.* **38**(2), 69–74. http://doi.acm.org/10.1145/1355734.1355746
152. Medhi, D., Ramamurthy, B., Scoglio, C., Rohrer, J. P., Çetinkaya, E. K., Cherukuri, R., Liu, X., Angu, P., Bavier, A., Buffington, C. & Sterbenz, J. P. G. (2014), 'The GpENI testbed: Network infrastructure, implementation experience, and experimentation', *Computer Networks* **61**, 51–74. http://dx.doi.org/10.1016/j.bjp.2013.12.027
153. Mekky, H., Jin, C. & Zhang, Z.-L. (2014), VIRO-GENI: SDN-Based Approach for a Non-IP Protocol in GENI, *in* 'Research and Educational Experiment Workshop (GREE), 2014 Third GENI', IEEE, pp. 15–18. http://dx.doi.org/10.1109/gree.2014.14
154. Mitroff, S. (2012), Lawrence Landweber Helped Build Today's Internet, Now He's Advising Its Future. http://www.wired.com/business/2012/08/lawrence-landweber/
155. Muhammad, M. & Cappos, J. (2012), Towards a Representive Testbed: Harnessing Volunteers for Networks Research, *in* 'First GENI Research and Educational Experiment Workshop (GREE 2012)'.
156. Mukherjee, S., Baid, A. & Raychaudhuri, D. (2015), Integrating Advanced Mobility Services into the Future Internet Architecture, *in* '7th International Conference on COMmunication Systems & NETworkS (COMSNETS 2015)', IEEE. http://winlab.rutgers.edu/~shreya/comsnets.pdf
157. Narisetty, R., Dane, L., Malishevskiy, A., Gurkan, D., Bailey, S., Narayan, S. & Mysore, S. (2013), OpenFlow Configuration Protocol: Implementation for the of Management Plane, *in* 'Research and Educational Experiment Workshop (GREE), 2013 Second GENI', IEEE, pp. 66–67. http://dx.doi.org/10.1109/gree.2013.21
158. Narisetty, R. & Gurkan, D. (2014), Identification of network measurement challenges in OpenFlow-based service chaining, *in* 'Local Computer Networks Workshops (LCN Workshops), 2014 IEEE 39th Conference on', IEEE, pp. 663–670. http://dx.doi.org/10.1109/lcnw.2014.6927718
159. Navaz, A., Velusam, G. & Gurkan, D. (2014), Experiments on Networking of Hadoop, *in* 'Network Protocols (ICNP), 2014 IEEE 22nd International Conference on', IEEE, pp. 544–547. http://dx.doi.org/10.1109/icnp.2014.87
160. Naylor, D., Mukerjee, M. K., Agyapong, P., Grandl, R., Kang, R., Machado, M., Brown, S., Doucette, C., Hsiao, H. C., Han, D., Kim, T. H., Lim, H., Ovon, C., Zhou, D., Lee, S. B., Lin, Y. H., Stuart, C., Barrett, D., Akella, A., Andersen, D., Byers, J., Dabbish, L., Kaminsky, M., Kiesler, S., Peha, J., Perrig, A., Seshan, S., Sirbu, M. & Steenkiste, P. (2014), 'XIA: Architecting a More Trustworthy and Evolvable Internet', *SIGCOMM Comput. Commun. Rev.* **44**(3), 50–57. http://dx.doi.org/10.1145/2656877.2656885
161. Nozaki, Y., Bakshi, P., Tuncer, H. & Shenoy, N. (2014), 'Evaluation of tiered routing protocol in floating cloud tiered internet architecture', *Computer Networks* **63**, 33–47. http://dx.doi.org/10.1016/j.bjp.2013.11.010
162. O'Neill, D., Aikat, J. & Jeffay, K. (2013), Experiment Replication Using ProtoGENI nodes, *in* '2013 Second GENI Research and Educational Experiment Workshop', IEEE, pp. 9–15. http://dx.doi.org/10.1109/gree.2013.11
163. Özçelik, I. & Brooks, R. R. (2015), 'Deceiving entropy based DoS detection', *Computers & Security* **48**, 234–245. http://dx.doi.org/10.1016/j.cose.2014.10.013
164. Ozcelik, I. & Brooks, R. R. (2011) , Security experimentation using operational systems, *in* 'Proceedings of the Seventh Annual Workshop on Cyber Security and Information Intelligence Research', CSIIRW '11, ACM, New York, NY, USA. http://dx.doi.org/10.1145/2179298.2179388
165. Ozcelik, I. & Brooks, R. R. (2012) , Performance Analysis of DDoS Detection Methods on Real Network, *in* 'First GENI Research and Educational Experiment Workshop (GREE 2012)'.

166. Ozcelik, I. & Brooks, R. R. (2013) , Operational System Testing for Designed in Security, *in* F. Sheldon, A. Giani, A. Krings & R. Abercrombie, eds, 'Proceedings of the Eighth Annual Cyber Security and Information Intelligence Research Workshop', CSIIRW '13, ACM, New York, NY, USA. http://dx.doi.org/10.1145/2459976.2460038
167. Ozcelik, I., Fu, Y. & Brooks, R. R. (2013), DoS Detection is Easier Now, *in* 'Research and Educational Experiment Workshop (GREE), 2013 Second GENI', IEEE, pp. 50–55. http://dx.doi.org/10.1109/gree.2013.18
168. Patali, R. (2011), Utility-Directed Resource Allocation in Virtual Desktop Clouds (Master's thesis), Master's thesis, The Ohio State University. https://etd.ohiolink.edu/!etd.send_file?accession=osu1306872632
169. Paul, S., Pan, J. & Jain, R. (2011) , 'Architectures for the future networks and the next generation Internet: A survey', *Computer Communications* **34**(1). http://dx.doi.org/10.1016/j.comcom.2010.08.001
170. Peter, S., Javed, U., Zhang, Q., Woos, D., Anderson, T. & Krishnamurthy, A. (2014), 'One tunnel is (often) enough', *Proceedings of the ACM SIGCOMM 2014 conference* **44**(4), 99–110. http://dx.doi.org/10.1145/2740070.2626318
171. Qin, Z., Xiong, X. & Chuah, M. (2012), Lehigh Explorer: Android Application Utilizing Content Centric Features, *in* 'First GENI Research and Educational Experiment Workshop (GREE 2012)'.
172. Qiu, C. & Shen, H. (2014), 'A Delaunay-Based Coordinate-Free Mechanism for Full Coverage in Wireless Sensor Networks', *Parallel and Distributed Systems, IEEE Transactions on* **25**(4), 828–839. http://dx.doi.org/10.1109/tpds.2013.134
173. Quan, J., Nance, K. & Hay, B. (2011), 'A Mutualistic Security Service Model: Supporting Large-Scale Virtualized Environments', *IT Professional* **13**(3). http://dx.doi.org/10.1109/MITP.2011.36
174. Rajagopalan, S. (2013), Leveraging OpenFlow for Resource Placement of Virtual Desktop Cloud Applications, Master's thesis, The Ohio State University. http://rave.ohiolink.edu/etdc/view?acc_num=osu1367456412
175. Rakotoarivelo, T., Jourjon, G., Mehani, O., Ott, M. & Zink, M. (2014), 'Repeatable Experiments with LabWiki'. http://arxiv.org/abs/1410.1681
176. Ramisetty, S., Calyam, P., Cecil, J., Akula, A. R., Antequera, R. B. & Leto, R. (2015), Ontology integration for advanced manufacturing collaboration in cloud platforms, *in* 'Integrated Network Management (IM), 2015 IFIP/IEEE International Symposium on', IEEE, pp. 504–510. http://dx.doi.org/10.1109/inm.2015.7140329
177. Randall, D. P., Diamant, E. I. & Lee, C. P. (2015), Creating Sustainable Cyberinfrastructures, *in* 'Proceedings of the 33rd Annual ACM Conference on Human Factors in Computing Systems', CHI '15, ACM, New York, NY, USA, pp. 1759–1768. http://dx.doi.org/10.1145/2702123.2702216
178. Ravi, A., Ramanathan, P. & Sivalingam, K. (2015), 'Integrated network coding and caching in information-centric networks: revisiting pervasive caching in the ICN framework', pp. 1–12. http://dx.doi.org/10.1007/s11107-015-0557-4
179. Raychaudhuri, D., Nagaraja, K. & Venkataramani, A. (2012), 'MobilityFirst: a robust and trustworthy mobility-centric architecture for the future internet', *SIGMOBILE Mob. Comput. Commun. Rev.* **16**(3), 2–13. http://dx.doi.org/10.1145/2412096.2412098
180. Ricart, G. (2014), US Ignite testbeds: Advanced testbeds enable next-generation applications, *in* 'Teletraffic Congress (ITC), 2014 26th International', IEEE, pp. 1–4. http://dx.doi.org/10.1109/itc.2014.6932975
181. Ricci, R. & Eide, E. (2014), 'Introducing Cloud Lab: Scientific Infrastructure for Advancing Cloud Architectures and Applications', *;login:* **39**(6), 36–38. http://www.usenix.org/publications/login/dec14/ricci
182. Ricci, R., Wong, G., Stoller, L. & Duerig, J. (2013), 'An Architecture For International Federation of Network Testbeds', *IEICE Transactions on Communications* . http://dx.doi.org/10.1587/transcom.E96.B.2

183. Ricci, R., Wong, G., Stoller, L., Webb, K., Duerig, J., Downie, K. & Hibler, M. (2015), 'Apt: A Platform for Repeatable Research in Computer Science', *SIGOPS Oper. Syst. Rev.* **49**(1), 100–107. http://dx.doi.org/10.1145/2723872.2723885
184. Riga, N., Thomas, V., Maglaris, V., Grammatikou, M. & Anifantis, E. (2015), Virtual Laboratories - Use of Public Testbeds in Education, *in* 'Proceedings of the 7th International Conference on Computer Supported Education', SCITEPRESS - Science and Technology Publications, pp. 516–521. http://dx.doi.org/10.5220/0005496105160521
185. Risdianto, A. C. & Kim, J. (2014), 'Prototyping Media Distribution Experiments over OF@TEIN SDN-enabled Testbed', *Proceedings of the Asia-Pacific Advanced Network* **38**(0), 12–18. http://dx.doi.org/10.7125/apan.38.2
186. Rivera, Fei, Z. & Griffioen, J. (2015), Providing a High Level Abstraction for SDN Networks in GENI, *in* 'Distributed Computing Systems Workshops (ICDCSW), 2015 IEEE 35th International Conference on', IEEE, pp. 64–71. http://dx.doi.org/10.1109/icdcsw.2015.22
187. Rohrer, J. P., Çetinkaya, E. K. & Sterbenz, J. P. G. (2011), Progress and challenges in large-scale future internet experimentation using the GpENI programmable testbed, *in* 'Proceedings of the 6th International Conference on Future Internet Technologies', CFI '11, ACM, New York, NY, USA. http://dx.doi.org/10.1145/2002396.2002409
188. Rosen, A. (2012), Network Service Delivery and Throughput Optimization via Software Defined Networking (Master's Thesis), Master's thesis, Clemson University. http://tigerprints.clemson.edu/all_theses/1332/
189. Rosen, A. & Wang, K.-C. (2012), Steroid OpenFlow Service: Seamless Network Service Delivery in Software Defined Networks, *in* 'First GENI Research and Educational Experiment Workshop (GREE 2012)'.
190. Ruth, P. & Mandal, A. (2014), Domain Science Applications on GENI: Presentation and Demo, *in* 'Network Protocols (ICNP), 2014 IEEE 22nd International Conference on', IEEE, pp. 540–543. http://dx.doi.org/10.1109/icnp.2014.86
191. Ruth, P., Mandal, A., Castillo, C., Fowler, R., Tilson, J., Baldin, I. & Xin, Y. (2015), Achieving Performance Isolation on Multi-Tenant Networked Clouds Using Advanced Block Storage Mechanisms, *in* 'Proceedings of the 6th Workshop on Scientific Cloud Computing', ScienceCloud '15, ACM, New York, NY, USA, pp. 29–32. http://dx.doi.org/10.1145/2755644.2755649
192. Schlinker, B., Zarifis, K., Cunha, I., Feamster, N. & Katz-Bassett, E. (2014), PEERING: An AS for Us, *in* 'Proceedings of the 13th ACM Workshop on Hot Topics in Networks', HotNets-XIII, ACM, New York, NY, USA. http://dx.doi.org/10.1145/2670518.2673887
193. Schwerdel, D., Reuther, B., Zinner, T., Müller, P. & Tran-Gia, P. (2014), 'Future Internet research and experimentation: The G-Lab approach', *Computer Networks* **61**, 102–117. http://dx.doi.org/10.1016/j.bjp.2013.12.023
194. Scoglio, C. M., Sydney, A., Youssef, M., Schumm, P. & Kooij, R. E. (2008), 'Elasticity and Viral Conductance: Unveiling Robustness in Complex Networks through Topological Characteristics', *CoRR* **abs/0811.3272**. http://arxiv-web3.library.cornell.edu/abs/0811.3272v3
195. Seetharam, S. (2014), Application-Driven Overlay Network as a Service for Data-Intensive Science (Master's thesis), Master's thesis, University of Missouri.
196. Seetharam, S., Calyam, P. & Beyene, T. (2014), ADON: Application-Driven Overlay Network-as-a-Service for Data-Intensive Science, Master's thesis, University of Missouri, Columbia. http://people.cs.missouri.edu/~calyamp/publications/adon-cloudnet14.pdf
197. Selvadhurai, A. (2013), Network Measurement Tool Components for Enabling Performance Intelligence within Cloud-based Applications (Master's Thesis), Master's thesis, The Ohio State University. http://rave.ohiolink.edu/etdc/view?acc_num=osu1367446588
198. Seskar, I., Nagaraja, K., Nelson, S. & Raychaudhuri, D. (2011), MobilityFirst future internet architecture project, *in* 'Proceedings of the 7th Asian Internet Engineering Conference', AINTEC '11, ACM, New York, NY, USA. http://dx.doi.org/10.1145/2089016.2089017
199. Sharma, N., Gummeson, J., Irwin, D. & Shenoy, P. (2010), Cloudy Computing: Leveraging Weather Forecasts in Energy Harvesting Sensor Systems, *in* '2010 7th Annual IEEE Communications Society Conference on Sensor, Mesh and Ad Hoc Communications and Networks (SECON)', IEEE. http://dx.doi.org/10.1109/SECON.2010.5508260

200. Shen, H. & Liu, G. (2011), Harmony: Integrated Resource and Reputation Management for Large-Scale Distributed Systems, *in* '2011 Proceedings of 20th International Conference on Computer Communications and Networks (ICCCN)', IEEE. http://dx.doi.org/10.1109/ICCCN.2011.6005739
201. Sher-DeCusatis, C. J. & DeCusatis, C. (2014), Developing a Software Defined Networking curriculum through industry partnerships, *in* 'American Society for Engineering Education (ASEE Zone 1), 2014 Zone 1 Conference of the', pp. 1–7. http://dx.doi.org/10.1109/ASEEZone1.2014.6820653
202. Shin, S., Dhondge, K. & Choi, B.-Y. (2012), Understanding the Performance of TCP and UDP-based Data Transfer Protocols using EMULAB, *in* 'First GENI Research and Educational Experiment Workshop (GREE 2012)'.
203. Singhal, M., Ramanathan, J., Calyam, P. & Skubic, M. (2014), In-the-Know: Recommendation Framework for City-Supported Hybrid Cloud Services, *in* 'Utility and Cloud Computing (UCC), 2014 IEEE/ACM 7th International Conference on', IEEE, pp. 137–145. http://dx.doi.org/10.1109/ucc.2014.22
204. Sivakumar, A., Shankaranarayanan, P. N. & Rao, S. (2012), Closer to the Cloud - A Case for Emulating Cloud Dynamics by Controlling the Environment, *in* 'First GENI Research and Educational Experiment Workshop (GREE 2012)'.
205. Soroush, H., Banerjee, N., Corner, M., Levine, B. & Lynn, B. (2012), 'A retrospective look at the UMass DOME mobile testbed', *SIGMOBILE Mob. Comput. Commun. Rev.* **15**(4). http://dx.doi.org/10.1145/2169077.2169079
206. Sridharan, M., Calyam, P., Venkataraman, A. & Berryman, A. (2011), Defragmentation of Resources in Virtual Desktop Clouds for Cost-Aware Utility-Optimal Allocation, *in* '2011 Fourth IEEE International Conference on Utility and Cloud Computing', IEEE. http://dx.doi.org/10.1109/UCC.2011.41
207. Sridharan, M., Zeng, W., Leal, W., Ju, X., Ramanath, R., Zhang, H. & Arora, A. (2010), 'From Kansei to KanseiGenie: Architecture of Federated, Programmable Wireless Sensor Fabrics', *Proceedings of the ICST Conference on Testbeds and Research Infrastructures for the Development of Networks and Communities (TridentCom)* .
208. Stabler, G., Goasguen, S., Rosen, A. & Wang, K.-C. (2012), OneCloud: Controlling the Network in an OpenFlow Cloud, *in* 'First GENI Research and Educational Experiment Workshop (GREE 2012)'.
209. Stabler, G., Rosen, A., Goasguen, S. & Wang, K.-C. (2012), Elastic IP and security groups implementation using OpenFlow, *in* 'Proceedings of the 6th international workshop on Virtualization Technologies in Distributed Computing Date', VTDC '12, ACM, New York, NY, USA, pp. 53–60. http://doi.acm.org/10.1145/2287056.2287069
210. Stavropoulos, D., Dadoukis, A., Rakotoarivelo, T., Ott, M., Korakis, T. & Tassiulas, L. (2015), Design, architecture and implementation of a resource discovery, reservation and provisioning framework for testbeds, *in* 'Modeling and Optimization in Mobile, Ad Hoc, and Wireless Networks (WiOpt), 2015 13th International Symposium on', IEEE, pp. 48–53. http://dx.doi.org/10.1109/wiopt.2015.7151032
211. Sterbenz, J. P. G., Çetinkaya, E. K., Hameed, M. A., Jabbar, A., Qian, S. & Rohrer, J. P. (2013), 'Evaluation of network resilience, survivability, and disruption tolerance: analysis, topology generation, simulation, and experimentation', *Telecommunication Systems* **52**(2), 705–736. http://dx.doi.org/10.1007/s11235-011-9573-6
212. Sterbenz, J. P. G., Egemen, Hameed, M. A., Jabbar, A. & Rohrer, J. P. (2011), Modelling and analysis of network resilience, *in* '2011 Third International Conference on Communication Systems and Networks (COMSNETS 2011)', IEEE. http://dx.doi.org/10.1109/COMSNETS.2011.5716502
213. Suñé, M., Bergesio, L., Woesner, H., Rothe, T., Köpsel, A., Colle, D., Puype, B., Simeonidou, D., Nejabati, R., Channegowda, M., Kind, M., Dietz, T., Autenrieth, A., Kotronis, V., Salvadori, E., Salsano, S., Körner, M. & Sharma, S. (2014), 'Design and implementation of the OFELIA FP7 facility: The European OpenFlow testbed', *Computer Networks* **61**, 132–150. http://dx.doi.org/10.1016/j.bjp.2013.10.015

214. Sun, P., Vanbever, L. & Rexford, J. (2015), Scalable Programmable Inbound Traffic Engineering, *in* 'Proceedings of the 1st ACM SIGCOMM Symposium on Software Defined Networking Research', SOSR'15, ACM, New York, NY, USA. http://dx.doi.org/10.1145/2774993.2775063
215. Sydney, A. (2013), The evaluation of software defined networking for communication and control of cyber physical systems (Doctoral dissertation), PhD thesis, Kansas State University. http://hdl.handle.net/2097/15577
216. Sydney, A., Nutaro, J., Scoglio, C., Gruenbacher, D. & Schulz, N. (2013), 'Simulative Comparison of Multiprotocol Label Switching and OpenFlow Network Technologies for Transmission Operations', *Smart Grid, IEEE Transactions on* **4**(2), 763–770. http://dx.doi.org/10.1109/TSG.2012.2227516
217. Sydney, A., Ochs, D. S., Scoglio, C., Gruenbacher, D. & Miller, R. (2014), 'Using GENI for experimental evaluation of Software Defined Networking in smart grids', *Computer Networks* **63**, 5–16. http://dx.doi.org/10.1016/j.bjp.2013.12.021
218. Tarui, T., Kanada, Y., Hayashi, M. & Nakao, A. (2015), Federating heterogeneous network virtualization platforms by slice exchange point, *in* 'Integrated Network Management (IM), 2015 IFIP/IEEE International Symposium on', IEEE, pp. 746–749. http://dx.doi.org/10.1109/inm.2015.7140366
219. Teerapittayanon, S., Fouli, K., Médard, M., Montpetit, M.-J., Shi, X., Seskar, I. & Gosain, A. (2012), Network Coding as a WiMAX Link Reliability Mechanism, *in* B. Bellalta, A. Vinel, M. Jonsson, J. Barcelo, R. Maslennikov, P. Chatzimisios & D. Malone, eds, 'Multiple Access Communications', Lecture Notes in Computer Science, Springer Berlin Heidelberg, pp. 1–12. http://dx.doi.org/10.1007/978-3-642-34976-8_1
220. Thomas, C., Sommers, J., Barford, P., Kim, D., Das, A., Segebre, R. & Crovella, M. (2012), A Passive Measurement System for Network Testbeds, *in* T. Korakis, M. Zink & M. Ott, eds, 'Testbeds and Research Infrastructure. Development of Networks and Communities', Vol. 44 of *Lecture Notes of the Institute for Computer Sciences, Social Informatics and Telecommunications Engineering*, Springer Berlin Heidelberg, pp. 130–145. http://dx.doi.org/10.1007/978-3-642-35576-9_14
221. Tiako, P. F. (2011), Perspectives of delegation in team-based distributed software development over the GENI infrastructure (NIER track), *in* 'Proceedings of the 33rd International Conference on Software Engineering', ICSE '11, ACM, New York, NY, USA. http://dx.doi.org/10.1145/1985793.1985905
222. Tredger, S. (2014), SageFS: The Location Aware Wide Area Distributed Filesystem (Master's thesis), Master's thesis, University of Victoria. http://dspace.library.uvic.ca/bitstream/handle/1828/5824/Tredger_Stephen_MSc_2014.pdf?sequence=3&isAllowed=y
223. Tredger, S., Zhuang, Y., Matthews, C., Short-Gershman, J., Coady, Y. & McGeer, R. (2013), Building Green Systems with Green Students: An Educational Experiment with GENI Infrastructure, *in* 'Research and Educational Experiment Workshop (GREE), 2013 Second GENI', IEEE, pp. 29–36. http://dx.doi.org/10.1109/gree.2013.15
224. Tsai, P.-W., wen Cheng, P., Yang, C.-S. & Luo, M.-Y. (2013), Supporting Extensions of VLAN-tagged traffic across OpenFlow Networks, *in* '2013 Proceedings Second GENI Research and Educational Experiment Workshop', IEEE. http://dx.doi.org/10.1109/GREE.2013.20
225. Tuncer, H., Nozaki, Y. & Shenoy, N. (2012), Virtual Mobility Domains - A Mobility Architecture for the Future Internet, *in* 'IEEE International Conference on Communications (IEE ICC 2012) Symposium on Next-Generation Networking', IEEE. ftp://lesc.det.unifi.it/pub/LenLar/proceedings/2012/ICC2012/symposia/papers/virtual_mobility_domains_-_a_mobility_architecture_for_the__.pdf
226. Turner, J. S. (2006), A proposed architecture for the GENI backbone platform, *in* 'Proceedings of the 2006 ACM/IEEE symposium on Architecture for networking and communications systems', ANCS '06, ACM, New York, NY, USA. http://dx.doi.org/10.1145/1185347.1185349

227. Turner, J. S., Crowley, P., DeHart, J., Freestone, A., Heller, B., Kuhns, F., Kumar, S., Lockwood, J., Lu, J., Wilson, M., Wiseman, C. & Zar, D. (2007), 'Supercharging planetlab: a high performance, multi-application, overlay network platform', *SIGCOMM Comput. Commun. Rev.* **37**(4). http://dx.doi.org/10.1145/1282427.1282391
228. Valancius, V. & Feamster, N. (2007) , Multiplexing BGP sessions with BGP-Mux, *in* 'Proceedings of the 2007 ACM CoNEXT conference', CoNEXT '07, ACM, New York, NY, USA. http://dx.doi.org/10.1145/1364654.1364707
229. Valancius, V., Feamster, N., Rexford, J. & Nakao, A. (2010), Wide-area route control for distributed services, *in* 'Proceedings of the 2010 USENIX conference on USENIX annual technical conference', USENIXATC'10, USENIX Association, Berkeley, CA, USA, p. 2. http://portal.acm.org/citation.cfm?id=1855842
230. Valancius, V., Kim, H. & Feamster, N. (2010), 'Transit portal: BGP connectivity as a service', *SIGCOMM Comput. Commun. Rev.* **40**(4). http://dl.acm.org/citation.cfm?id=1851265
231. Valancius, V., Ravi, B., Feamster, N. & Snoeren, A. C. (2013), Quantifying the benefits of joint content and network routing, *in* 'Proceedings of the ACM SIGMETRICS/international conference on Measurement and modeling of computer systems - SIGMETRICS '13', ACM Press, pp. 243+. http://dx.doi.org/10.1145/2465529.2465762
232. Van Vorst, N., Erazo, M. & Liu, J. (2011), PrimoGENI: Integrating Real-Time Network Simulation and Emulation in GENI, *in* 'Principles of Advanced and Distributed Simulation (PADS), 2011 IEEE Workshop on', IEEE, pp. 1–9. http://dx.doi.org/10.1109/pads.2011.5936747
233. Van Vorst, N., Erazo, M. & Liu, J. (2012), 'PrimoGENI for hybrid network simulation and emulation experiments in GENI', *Journal of Simulation* **6**(3). http://dx.doi.org/10.1057/jos.2012.5
234. Van Vorst, N., Li, T. & Liu, J. (2011), How Low Can You Go? Spherical Routing for Scalable Network Simulations, *in* 'Modeling, Analysis & Simulation of Computer and Telecommunication Systems (MASCOTS), 2011 IEEE 19th International Symposium on', IEEE. http://dx.doi.org/10.1109/MASCOTS.2011.35
235. Van Vorst, N. & Liu, J. (2012), Realizing Large-Scale Interactive Network Simulation via Model Splitting, *in* 'Principles of Advanced and Distributed Simulation (PADS), 2012 ACM/IEEE/SCS 26th Workshop on', IEEE, pp. 120–129. http://dx.doi.org/10.1109/pads.2012.35
236. Velusamy, G. (2014), OpenFlow-based Distributed and Fault-Tolerant Software Switch Architecture (Master's thesis), Master's thesis, University of Houston. http://repositories.tdl.org/uh-ir/bitstream/handle/10657/693/VELUSAMY-THESIS-2014.pdf
237. Velusamy, G., Gurkan, D., Narayan, S. & Baily, S. (2014), Fault-Tolerant OpenFlow-Based Software Switch Architecture with LINC Switches for a Reliable Network Data Exchange, *in* 'Research and Educational Experiment Workshop (GREE), 2014 Third GENI', IEEE, pp. 43–48. http://dx.doi.org/10.1109/gree.2014.17
238. Venkataraman, A. (2012), Defragmentation of Resources in Virtual Desktop clouds for Cost-aware Utility-maximal Allocation (Master's thesis), Master's thesis, The Ohio State University. https://etd.ohiolink.edu/!etd.send_file?accession=osu1339747492
239. Vulimiri, A., Michel, O., Godfrey, P. B. & Shenker, S. (2012), More is Less: Reducing Latency via Redundancy, *in* 'Proceedings of the 11th ACM Workshop on Hot Topics in Networks', HotNets-XI, ACM, New York, NY, USA, pp. 13–18. http://dx.doi.org/10.1145/2390231.2390234
240. Wallace, S. A., Muhammad, M., Mache, J. & Cappos, J. (2011), 'Hands-on Internet with Seattle and Computers from Across the Globe', *J. Comput. Sci. Coll.* **27**(1), 137–142. http://portal.acm.org/citation.cfm?id=2037151.2037181
241. Wang, H., Lee, K. S., Li, E., Lim, C. L., Tang, A. & Weatherspoon, H. (2014), Timing is Everything: Accurate, Minimum Overhead, Available Bandwidth Estimation in High-speed Wired Networks, *in* 'Proceedings of the 2014 Conference on Internet Measurement Conference', IMC '14, ACM, New York, NY, USA, pp. 407–420. http://dx.doi.org/10.1145/2663716.2663746

242. Wang, K. C., Brinn, M. & Mambretti, J. (2014), From federated software defined infrastructure to future internet architecture, *in* 'Science and Technology Conference (Modern Networking Technologies) (MoNeTeC), 2014 First International', IEEE, pp. 1–6. http://dx.doi.org/10.1109/monetec.2014.6995605
243. Wang, Q., Xu, K., Izard, R., Kribbs, B., Porter, J., Wang, K.-C., Prakash, A. & Ramanathan, P. (2014), GENI Cinema: An SDN-Assisted Scalable Live Video Streaming Service, *in* 'Network Protocols (ICNP), 2014 IEEE 22nd International Conference on', IEEE, pp. 529–532. http://dx.doi.org/10.1109/icnp.2014.84
244. Wang, Y., Akhtar, N. & Matta, I. (2014), Programming Routing Policies for Video Traffic, *in* 'Network Protocols (ICNP), 2014 IEEE 22nd International Conference on', IEEE, pp. 504–510. http://dx.doi.org/10.1109/icnp.2014.80
245. Wang, Y., Esposito, F. & Matta, I. (2013), Demonstrating RINA Using the GENI Testbed, *in* 'Research and Educational Experiment Workshop (GREE), 2013 Second GENI', IEEE, pp. 93–96. http://dx.doi.org/10.1109/gree.2013.26
246. Wang, Y., Matta, I. & Akhtar, N. (2014), Experimenting with Routing Policies Using ProtoRINA over GENI, *in* 'Research and Educational Experiment Workshop (GREE), 2014 Third GENI', IEEE, pp. 61–64. http://dx.doi.org/10.1109/gree.2014.11
247. Willner, A. & Magedanz, T. (2014), FIRMA: A Future Internet resource management architecture, *in* 'Teletraffic Congress (ITC), 2014 26th International', IEEE, pp. 1–4. http://dx.doi.org/10.1109/itc.2014.6932981
248. Wong, G., Ricci, R., Duerig, J., Stoller, L., Chikkulapelly, S. & Seok, W. (2012), Partitioning Trust in Network Testbeds, *in* 'System Science (HICSS), 2012 45th Hawaii International Conference on', IEEE. http://dx.doi.org/10.1109/HICSS.2012.466
249. Xiao, Z., Fu, B., Xiao, Y., Chen, C. L. P. & Liang, W. (2013), 'A review of GENI authentication and access control mechanisms', *International Journal of Security and Networks* **8**(1), 40+. http://dx.doi.org/10.1504/ijsn.2013.055046
250. Xin, Y., Baldin, I., Chase, J. & Ogan, K. (2014), 'Leveraging Semantic Web Technologies for Managing Resources in a Multi-Domain Infrastructure-as-a-Service Environment', *CoRR* **abs/1403.0949**. http://arxiv.org/abs/1403.0949
251. Xin, Y., Baldin, I., Heermann, C., Mandal, A. & Ruth, P. (2014*a*), Capacity of Inter-cloud Layer-2 Virtual Networking, *in* 'Proceedings of the 2014 ACM SIGCOMM Workshop on Distributed Cloud Computing', DCC '14, ACM, New York, NY, USA, pp. 31–36. http://dx.doi.org/10.1145/2627566.2627573
252. Xin, Y., Baldin, I., Heermann, C., Mandal, A. & Ruth, P. (2014*b*), Scaling up applications over distributed clouds with dynamic layer-2 exchange and broadcast service, *in* 'Teletraffic Congress (ITC), 2014 26th International', IEEE, pp. 1–6. http://dx.doi.org/10.1109/itc.2014.6932973
253. Xin, Y., Baldine, I., Mandal, A., Heermann, C., Chase, J. & Yumerefendi, A. (2011), Embedding Virtual Topologies in Networked Clouds, *in* 'Proceedings of the 6th International Conference on Future Internet Technologies', CFI '11, ACM, New York, NY, USA, pp. 26–29. http://doi.acm.org/10.1145/2002396.2002403
254. Xing, T., Huang, D., Xu, L., Chung, C.-J. & Khatkar, P. (2013), SnortFlow: A OpenFlow-Based Intrusion Prevention System in Cloud Environment, *in* 'Research and Educational Experiment Workshop (GREE), 2013 Second GENI', IEEE, pp. 89–92. http://dx.doi.org/10.1109/gree.2013.25
255. Xiong, K. & Pan, Y. (2013), Understanding ProtoGENI in Networking Courses for Research and Education, *in* 'Research and Educational Experiment Workshop (GREE), 2013 Second GENI', IEEE, pp. 119–123. http://dx.doi.org/10.1109/gree.2013.31
256. Xu, G., Amariucai, G. & Guan, Y. (2014), Delegation of Computation with Verification Outsourcing Using GENI Infrastructure, *in* 'Research and Educational Experiment Workshop (GREE), 2014 Third GENI', IEEE, pp. 49–52. http://dx.doi.org/10.1109/gree.2014.16
257. Xu, K., Izard, R., Yang, F., Wang, K.-C. & Martin, J. (2013), Cloud-Based Handoff as a Service for Heterogeneous Vehicular Networks with OpenFlow, *in* 'Research and Educational Experiment Workshop (GREE), 2013 Second GENI', IEEE, pp. 45–49. http://dx.doi.org/10.1109/gree.2013.17

258. Xu, K., Sampathkumar, S., Wang, K.-C. & Ramanathan, P. (2013), Network Coding for Efficient Broadband Data Delivery in Infrastructure-Based Vehicular Networks with OpenFlow, *in* 'Research and Educational Experiment Workshop (GREE), 2013 Second GENI', IEEE, pp. 56–60. http://dx.doi.org/10.1109/gree.2013.19
259. Xu, K., Wang, K.-C., Amin, R., Martin, J. & Izard, R. (2014), 'A Fast Cloud-based Network Selection Scheme Using Coalition Formation Games in Vehicular Networks', *IEEE Transactions on Vehicular Technology* p. 1. http://dx.doi.org/10.1109/tvt.2014.2379953
260. Yi, P. (2014), Peer-to-Peer based Trading and File Distribution for Cloud Computing (Doctoral dissertation), PhD thesis, University of Kentucky, Lexington, Kentucky. http://uknowledge.uky.edu/cs_etds/22/
261. Yi, P. & Fei, Z. (2014), Characterizing the GENI Networks, *in* 'Research and Educational Experiment Workshop (GREE), 2014 Third GENI', IEEE, pp. 53–56. http://dx.doi.org/10.1109/gree.2014.8
262. Yu, Z., Liu, X., Li, M., Liu, K. & Li, X. (2013), ExoApp: Performance Evaluation of Data-Intensive Applications on ExoGENI, *in* 'Research and Educational Experiment Workshop (GREE), 2013 Second GENI', IEEE, pp. 25–28. http://dx.doi.org/10.1109/gree.2013.14
263. Yuen, M. (2010), GENI in the Cloud (Master's Thesis), Master's thesis, University of Victoria. http://s3.amazonaws.com/marcoy_thesis/Thesis.pdf
264. Zhang, M., Kissel, E. & Swany, M. (2015), Using Phoebus data transfer accelerator in cloud environments, *in* 'Communications (ICC), 2015 IEEE International Conference on', IEEE, pp. 351–357. http://dx.doi.org/10.1109/icc.2015.7248346
265. Zhang, M., Swany, M., Yavanamanda, A. & Kissel, E. (2015), HELM: Conflict-free active measurement scheduling for shared network resource management, *in* 'Integrated Network Management (IM), 2015 IFIP/IEEE International Symposium on', IEEE, pp. 113–121. http://dx.doi.org/10.1109/inm.2015.7140283
266. Zhang, Y., Steele, A. & Blanton, M. (2013), PICCO: A General-purpose Compiler for Private Distributed Computation, *in* 'Proceedings of the 2013 ACM SIGSAC Conference on Computer & Communications Security', CCS '13, ACM, New York, NY, USA, pp. 813–826. http://dx.doi.org/10.1145/2508859.2516752
267. Zhuang, Y., Rafetseder, A. & Cappos, J. (2013), Experience with Seattle: A Community Platform for Research and Education, *in* 'Research and Educational Experiment Workshop (GREE), 2013 Second GENI', IEEE, pp. 37–44. http://dx.doi.org/10.1109/gree.2013.16
268. Zink, M. (2014), A measurement architecture for Software Defined Exchanges, *in* 'Science and Technology Conference (Modern Networking Technologies) (MoNeTeC), 2014 First International', IEEE, pp. 1–6. http://dx.doi.org/10.1109/monetec.2014.6995606

Afterword: A Fire in the Dark

Infrastructure, it is said, is not fundamentally about technology. It is a campfire in the woods on a dark night—a cheery light that spreads warmth, leavens the darkness, and most of all invites people to join the company. The invitation is not only to share the warmth and fellowship of the fire, but to contribute—to bring fuel to feed the fire, and to food to feed the band. It is our privilege to live in a community eager to contribute, one whose generosity and cheer is as warming as the fire that we have all built together.

This has been the story to date of our campfire, one which continues to grow and attract others. Some of us are fortunate enough to have watched it grow from sparks and kindling. Many others whose names do not appear in this volume have contributed to it; many more will. A very few, sadly, have left our company entirely; we take comfort in knowing that their vision was such that they saw this day, and the many days to come, with great clarity. Off in the woods we can see other fires, growing and burning festively; people regularly move between our company and theirs. Each day new people arrive at our fire, and at the others, and we are amazed and delighted by the contributions the newcomers bring.

Perhaps you will join us, too. There will always be plenty of room by our fire.

© Springer International Publishing Switzerland 2016

R. McGeer et al. (eds.), *The GENI Book*, DOI 10.1007/978-3-319-33769-2

Printed in the United States
By Bookmasters